Gauss and Jacobi Sums

CANADIAN MATHEMATICAL SOCIETY SERIES OF MONOGRAPHS AND ADVANCED TEXTS

Monographies et Études de la Société Mathématique du Canada

A complete list of titles in this series appears at the end of this volume.
Tous les titres de cette collection sont énumérés à la fin du volume.

Gauss and Jacobi Sums

Bruce C. Berndt
University of Illinois

Ronald J. Evans
University of California at San Diego

Kenneth S. Williams
Carleton University

A Wiley-Interscience Publication
JOHN WILEY & SONS, INC.
New York • Chichester • Weinheim • Brisbane • Singapore • Toronto

This book is printed on acid-free paper. ♾

Library of Congress Cataloging in Publication Data:

Berndt, Bruce C., 1939–
Gauss and Jacobi sums / Bruce C. Berndt, Ronald J. Evans, Kenneth S. Williams.
p. cm. — (Canadian Mathematical Society series of monographs and advanced texts)
"A Wiley-Interscience publication."
Includes bibliographical references and indexes.
ISBN 978-0-471-12807-6
1. Gaussian sums. 2. Jacobi sums. I. Evans, Ronald J. II. Williams, Kenneth S. III. Title. IV. Series.
QA246.8.G38B47 1997
512′.7—dc21 97-44440

To our wives (Helen, Sue, Carole)
and children

Contents

Introduction

The origins of this book lie in the work of C. F. Gauss and C. G. J. Jacobi. Gauss introduced the Gauss sum in his *Disquisitiones Arithmeticae* [2] in July, 1801, and Jacobi [1] introduced the Jacobi sum in a letter to Gauss dated February 8, 1827.

The sum introduced by Gauss in 1801 is $\sum_{n=0}^{k-1} e^{2\pi imn^2/k}$, which is now called a (quadratic) Gauss sum. This sum is not easy to evaluate, even in the special case that $m = 1$ and k is an odd positive integer. In this case, Gauss was easily able to show that this sum has the value $\pm\sqrt{k}$ or $\pm i\sqrt{k}$, according as k is of the form $4u + 1$ or $4u + 3$, respectively. Specific examples convinced Gauss that the plus sign is always correct. On August 30, 1805, Gauss wrote in his diary that he devoted some time to this problem every week for more than four years before he was able to prove his conjecture on the sign of these sums. A few years later, Gauss [3] published an evaluation of his quadratic Gauss sum for all positive integers k.

In his study of primes in arithmetic progressions, G. L. Dirichlet [4] introduced the multiplicative character χ modulo k and the sum $G(\chi) = \sum_{n=0}^{k-1} \chi(n)e^{2\pi imn/k}$; this is also called a Gauss sum, as it coincides with the quadratic Gauss sum above in the case that χ has order 2 and k is a prime not dividing m.

The sum now called a Jacobi sum, which is in essence the one Jacobi introduced in 1827, is $J(\chi, \psi) = \sum_{n \bmod p} \chi(n)\psi(1 - n)$, where χ and ψ are multiplicative characters modulo a prime p. Jacobi was already aware in 1827 that Gauss and Jacobi sums are related like gamma and beta functions, i.e., for $k = p$ and $m = 1$,

$$J(\chi, \psi) = G(\chi)G(\psi)/G(\chi\psi),$$

when $\chi\psi$ is nontrivial.

After the initial work of Gauss, Dirichlet, and Jacobi, many well-known mathematicians made contributions to the theory of Gauss and Jacobi sums, including L. Carlitz, A. Cauchy, S. Chowla, H. Davenport, G. Eisenstein, B. Gross, H. Hasse, N. M. Katz, N. Koblitz, L. Kronecker, E. E. Kummer, D. H. and E. Lehmer, L. J. Mordell, S. J. Patterson, C. L. Siegel, L. Stickelberger, and A. Weil. One major goal of this book is to offer simple proofs of many of the marvelous theorems on Gauss and Jacobi sums scattered throughout the literature. Our proofs are frequently shorter, more systematic, and/or more elementary than previous proofs. We have aimed to make this book user-friendly; in particular, it has been our wish to make classical theorems of Gauss, Jacobi, Eisenstein, and others accessible to beginning graduate students in mathematics and physics.

Our theme, Gauss and Jacobi sums, could be approached in a variety of ways. In this book, the focus is on examining basic properties of Gauss and Jacobi sums, providing systematic and explicit evaluations of these sums, and providing applications.

An amazing number of generalizations of Gauss sums have appeared in the last century: there are Gauss sums for quadratic forms over algebraic number fields, symplectic Gauss sums, Galois Gauss sums, Gauss sums for function fields, Gauss sums for general finite rings, and many others. Similarly, numerous generalizations of Jacobi sums have arisen. It would be impossible to cover the large variety of these sums in one book. We concentrate here mainly on Gauss sums over finite rings of integers (mod k) and on Gauss and Jacobi sums over general finite fields F_q, where q is a power of a prime p. References to generalizations and extensions are given in the notes following the chapters.

A considerable portion of this book is devoted to applications of Gauss and Jacobi sums in number theory. We now give five brief examples, taken from classical topics in Chapters 6, 7, 8, 9, and 10, respectively.

Let p be a prime of the form $4f+1$, so that by a theorem of Fermat, $p = a^2 + b^2$ for integers a and b with, say, $a \equiv 1 \pmod 4$. Assume that f is odd. Then $(p-3-2a)/8$ is the number of $n \pmod p$ for which $n, n+2$, and $n+4$ are all quadratic residues modulo p. We show this by evaluating a Jacobsthal sum, which in turn is determined via Jacobi sums.

Let p be a prime of the form $3f+1$, so that by a theorem of Gauss, $4p$ has the representation $4p = r^2 + 3s^2$ with integers r and s. Gauss proved that 2 is a cubic residue modulo p if and only if $2|r$, and that 3 is a cubic residue modulo p if and only if $9|s$. We employ Gauss and Jacobi sums to give simple proofs of these and similarly elegant theorems on residuacity.

Following Gauss' proofs of the law of quadratic reciprocity, Jacobi conjectured a law of cubic reciprocity and Gauss conjectured a law of biquadratic reciprocity. We give proofs of these laws based on Gauss and Jacobi sums, similiar to the first published proofs given by Eisenstein [2], [5] in 1844. Similar proofs (unpublished) were evidently communicated by Jacobi in his Königsberg lectures in 1836–1837.

Let $p = 4f+1$ be a prime, so that $p = a^2 + b^2$, with, say, $a \equiv 1 \pmod 4$. Using the evaluation of the quartic Jacobi sum, we prove Gauss' beautiful congruence

$$\binom{2f}{f} \equiv 2a \pmod p.$$

With the aid of the Gross–Koblitz formula for Gauss sums, we also prove a more general congruence for $\binom{2f}{f} \pmod{p^2}$ given by Chowla, Dwork, and Evans [1] in 1986.

Let $p = 3f+1$ be a prime with $4p = r^2 + 3s^2$, where $r \equiv 1 \pmod 3$. Jacobi sums of order 3 can be used to show that the number of solutions of the congruence

$$x^3 + y^3 + z^3 \equiv 1 \pmod p$$

with xyz not divisible by p is equal to $p^2 + 3p - 4r + 15$.

We do not wish to leave the impression that applications of Gauss and Jacobi sum evaluations are limited to number theory. In fact, as we discuss in the notes at the end of the chapters, there are numerous applications in physics (e.g., Lomont [1],

Spiridonov and Zhedanov [1], Zhedanov [1], Hannay and Berry [1], Berry and Klein [1], Leichtle, Averbukh, and Schleich [1]) as well as in diverse areas of mathematics such as graph theory and combinatorics (Yamada [1], [2]), operator theory (Farsi and Watling [1], [2]), coding theory (McEliece [1], Langevin [1]), cryptography (Koblitz [3]), combinatorial designs (Baumert [1], Jungnickel [1], [2]), analysis (Drmota and Skalba [1]), and algebra (Ishibashi [1], Sepanski [1]).

We apply our results on Gauss and Jacobi sums to study some closely related character sums, particularly the sums of Brewer, Eisenstein, and Jacobsthal. We only briefly discuss other character sums such as Kloosterman sums, which are less readily evaluable from Gauss and Jacobi sums. Moreover, as our emphasis here is on explicit determinations and their applications, we barely touch upon other important aspects of character sums such as character sum estimates.

The study of Gauss and Jacobi sums, far from being just a classical enterprise, is fruitful and thriving in contemporary research. Thus, another major goal of this book is to discuss some of the important developments in this area in modern times, and to provide references to related research.

Having offered the *raison d'être* for this book, we now summarize the contents of each of its fourteen chapters.

Chapter 1: We begin by discussing basic properties of two related types of Gauss sums defined over finite fields. The middle sections are devoted to evaluating quadratic Gauss sums defined over finite rings of integers modulo k. In the final section, we present reduction formulae for Gauss character sums modulo k and then give evaluations of these sums found by Odoni [1] in 1973 and Funakura [1] in 1992. Their proofs require some p-adic analysis, but we also give the alternative evaluations found by Mauclaire [1], [2] in 1983 which requires no knowledge of the p-adics.

Chapter 2: We present basic properties of Jacobi sums over finite fields, including a congruence relating the Jacobi sum to a binomial coefficient modulo a prime ideal. Then cyclotomic numbers and their relationships with Jacobi sums are discussed. Cyclotomic numbers are applied in this chapter and again in Chapter 7 to obtain results on residuacity, for example, a determination of the kth–power residue character of k modulo a prime $p = kf + 1$. Further applications of cyclotomic numbers to counting the number of solutions of diagonal equations over finite fields will be given in Chapter 10.

Chapter 3: We explicitly determine the values of Jacobi sums over F_p of orders 3, 4, 5, 6, 7, 8, 10, 12, 16, 20, and 24, in terms of parameters occurring in representations of p by certain quadratic forms (e.g., the parameters a and b when $p = a^2 + b^2$). These evaluations will be utilized throughout much of the book.

Chapter 4: We evaluate Gauss sums over F_p of orders 3, 4, 6, 8, and 12 in terms of the parameters defined in Chapter 3. Particular attention is given to the cubic Gauss sum (Kummer's sum) and the quartic sum.

Chapter 5: This chapter is simply devoted to applying the evaluations of Gauss sums given in Chapter 4 to determine the existence or nonexistence of nth–power residue difference sets, for some small values of n.

Chapter 6: The first section develops the elementary properties of Jacobsthal sums over F_p, including their formulations as linear combinations of Jacobi sums. Then Jacobsthal sums of orders 2, 3, 4, 5, 6, 8, 10, and 12 are explicitly evaluated. Applications are given to the distribution of quadratic residues modulo p and to the establishment of congruences for binomial coefficients modulo p.

Chapter 7: Gauss sums and cyclotomic numbers are applied to establish necessary and sufficient conditions for certain primes (especially 2 and 3) to be kth–power residues modulo a prime $p = kf + 1$, where $k = 3$, 4, 5, 8, and 16.

Chapter 8: We first prove the laws of cubic and quartic reciprocity, and the accompanying supplementary laws. Then we prove quartic, octic, and bioctic rational reciprocity laws, which have a much more recent history.

Chapter 9: Using the congruence from Chapter 2 relating Jacobi sums to binomial coefficients, we give congruences for binomial coefficients modulo p in terms of the parameters defined in Chapter 3. Then we use p-adic gamma functions and the Gross–Koblitz formula for Gauss sums to attack the more difficult problem of extending such congruences modulo p^2. Crandall, Dilcher, and Pomerance [1] have recently found a practical application for such congruences.

Chapter 10: We develop basic properties of generalized Jacobi sums over F_q, which are functions of n characters for arbitrary $n \geq 2$. There is a brief discussion of Gauss periods associated with F_p. Generalized Jacobi sums, and in a few instances Gauss periods and cyclotomic numbers, are applied to explicitly determine the number N of solutions to certain diagonal equations over finite fields, or to give upper and lower bounds for N.

Chapter 11: The Gross–Koblitz formula for Gauss sums is stated here without proof, but comparatively simple proofs are given for some of its consequences. We prove Stickelberger's congruence for Gauss sums and give the prime ideal factorizations of Gauss and Jacobi sums. We then prove two celebrated theorems of Davenport–Hasse, the product formula for Gauss sums and the theorem on lifted Gauss sums over extensions of F_q. We prove a theorem of Stickelberger evaluating pure Gauss sums of order k when -1 is a power of p (mod k) (a Gauss sum is *pure* if some positive integral power of it is real), and we give necessary and sufficient conditions in terms of p and k for all Gauss sums on F_q of orders dividing k to be pure. In the final section, a large number of the results from previous sections are applied to determine the weight distribution of certain irreducible cyclic codes.

Chapter 12: The Eisenstein sums over F_q are undeservedly neglected in the textbooks. They have many applications; for instance, in 1994, Evans [19] used Eisenstein sums to give a simple evaluation of a Soto–Andrade character sum considered by Kuang [1]. We develop basic properties of the Eisenstein sum, including its formulation as a quotient of Gauss sums, where the Gauss sum in the denominator is over F_p and that in the numerator is over F_q. Because of this formula, Eisenstein sums are sometimes called relative Gauss sums (see, e.g., Yamamoto [5]). We evaluate Eisenstein sums over

F_q for orders 3, 4, 5, 6, and 8, and in some cases for order 7. Some of these evaluations are applied to give systematic proofs of congruences of Eisenstein for binomial coefficients. We also evaluate certain Eisenstein sums of orders 12, 16, and 20 in the case that $q = p^2$; these will be applied in Chapter 13 to evaluate Brewer sums. As another application, certain Gauss periods over F_q are evaluated by expressing them as linear combinations of Eisenstein sums. There is a brief discussion of the properties of Gauss periods over F_q, which extend the Gauss periods over F_p discussed in Chapter 10.

Chapter 13: Brewer character sums, which are defined in terms of Dickson polynomials, do not occur frequently in the literature. However, they have some unexpected applications. For example, in 1986, Greene and Stanton [2] used the evaluation of a Brewer sum to prove a conjecture of Evans, Pulham, and Sheehan [1] on the value of a certain double sum of Legendre symbols. In Chapter 13, we use our knowledge of Gauss, Jacobi, Jacobsthal, and Eisenstein sums to explicitly determine several Brewer sums.

Chapter 14: In 1850, Eisenstein proved the famous Eisenstein reciprocity law for kth–power residue symbols, where k is an odd prime. In 1907–1908, Western [1], [2] extended this law to prime powers k. While Eisenstein's law is covered in textbooks, the extension to prime powers is not. In Chapter 14, we use Gauss and Jacobi sums to give a considerably simplified version of Western's result, which we call Eisenstein's reciprocity law for prime powers.

Following Chapter 14, we offer a list of open problems related to topics in this book.

For each X between 1 and 14, Chapter X is followed by a set of problems called Exercises X. Several problems offer new results, some problems are routine, and others (especially those drawn from research papers) are fairly difficult. Occasionally, details from proofs in the text are left as exercises, in which case hints are usually provided.

Following Exercises X, there are Notes on Chapter X, where we indicate the history of results established in the chapter as well as current research on related results. Thus the book contains an extensive bibliography. We have sometimes cited just the most recent paper in a string of papers on a given topic, with the idea that the interested reader can look up earlier references therein; see, for example, the reference to Burgess.

Following the bibliography, readers will find an index of notation, an author index, and a subject index.

The main prerequisites for the book are knowledge of undergraduate modern algebra (including finite fields) and basic material in elementary and algebraic number theory. The reader who wishes to review the theory of finite fields may consult the texts of Lidl and Niederreiter [1] or Ireland and Rosen [1]. A small amount of complex analysis is required in Section 1.2, and some p-adic analysis is used in Sections 1.6, 9.3, and 11.2.

This book has its genesis in course notes for graduate courses given at various times by the authors at the University of Illinois, the University of California at San Diego, and Carleton University. An introductory course on the topics examined in the first eleven chapters of this book might be drawn up from the following syllabus.

Chapter 1:	Sections 1.1–1.5.
Chapter 2:	Section 2.1.
Chapter 3:	Sections 3.1–3.2.
Chapter 4:	Sections 4.1–4.2.
Chapter 5:	Sections 5.1–5.3 through Theorem 5.3.5.
Chapter 6:	Sections 6.1–6.2 (primarily Theorem 6.1.13 through Theorem 6.2.2) and Section 6.3–6.4 through Theorem 6.4.3.
Chapter 7:	Sections 7.1–7.2.
Chapter 8:	Sections 8.1–8.3 through Corollary 8.3.4.
Chapter 9:	Sections 9.1–9.2 through Theorem 9.2.2 and Sections 9.3–9.4 through Theorem 9.4.1.
Chapter 10:	Sections 10.1 and 10.4.
Chapter 11:	Sections 11.1–11.2 through Theorem 11.2.3.

The authors are grateful to those who helped inspire in them a love for Gauss and Jacobi sums. The first author's introduction and inspiration for the subject came from the wonderful book on number theory by K. Ireland and M. I. Rosen. He also was strongly influenced by the lectures given by André Weil at the Institute for Advanced Study during the academic year 1973–74. The second author too was inspired by the book of Ireland and Rosen, as well as by the work of E. Lehmer, A. L. Whiteman, B. C. Berndt, and K. S. Williams. The third author's interest in the subject was kindled by his thesis supervisor John H. H. Chalk at the University of Toronto in the early 1960s.

The authors extend their gratitude to Sen–Shan Huang, Darrin Doud, Şaban Alaca, and Ayşe Alaca, who read the manuscript, made corrections, and offered valuable suggestions for improving the presentation of material. Special thanks go to Dunxue Liu, Şaban Alaca, and J. J. Wu, who prepared the manuscript in Latex with great efficiency and care. The staff of John Wiley & Sons Inc., especially Maria Allegra, Jessica Downey, Stephen H. Quigley, Angioline Loredo, and Angela Volan, deserve our thanks for their guidance and help. The authors are further grateful to David Aronstein, Carl Pomerance, and Samuel J. Patterson for helpful comments.

The third author would like to thank Carleton University for a Teaching Achievement Award that enabled him to work on this book during the academic year 1994-95.

The authors would appreciate being notified of misprints/mistakes. They can be reached at:

berndt@math.uiuc.edu, revans@ucsd.edu, and williams@math.carleton.ca

For the forseeable future, a list of corrections with acknowledgments will be kept on the webpage of the first author:
http://www.math.uiuc.edu/~berndt/GaussJacobiErrata

Urbana, Illinois; La Jolla, California; Ottawa, Ontario
October, 1997

Chapter 1

Gauss Sums

We shall assume that the reader has a modest knowledge of the theory of finite fields, as found for example in Ireland and Rosen [1, Chapter 7]. We refer the reader to the books by Lidl and Niederreiter [1] and Small [1] for more comprehensive treatments of finite fields. Let p denote a prime and r a positive integer. Set $q = p^r$ and denote the finite field of q elements by F_q. Throughout much of this book we are particularly interested in the case $r = 1$, and we identify F_p with the residues $0, 1, 2, \ldots, p-1 \pmod{p}$. Denote $F_q - \{0\}$ by F_q^*; this is a cyclic multiplicative group of order $q - 1$.

1.1 Elementary properties of Gauss sums over F_q

In order to define Gauss sums over F_q, we need the concept of the trace of an element of F_q.

Definition *For $\alpha \in F_q$ the trace of α from F_q to F_p, denoted by* $\mathrm{tr}(\alpha)$, *is defined by*

$$\mathrm{tr}(\alpha) = \sum_{j=0}^{r-1} \alpha^{p^j}.$$

The basic properties of the trace function are given in our first proposition.

Proposition 1.1.1 *Let $\alpha, \beta \in F_q$ and $a \in F_p$. Then*

(a) $\mathrm{tr}(\alpha) \in F_p$,

(b) $\mathrm{tr}(\alpha + \beta) = \mathrm{tr}(\alpha) + \mathrm{tr}(\beta)$,

(c) $\mathrm{tr}(a\alpha) - a\mathrm{tr}(\alpha)$,

(d) tr *maps F_q onto F_p,*

(e) $\mathrm{tr}(\alpha^p) = \mathrm{tr}(\alpha)$.

Proof (a) Employing the multinomial theorem and the fact that F_q has characteristic p, we have

$$(\alpha + \alpha^p + \cdots + \alpha^{p^{r-1}})^p = \alpha^p + \alpha^{p^2} + \cdots + \alpha^{p^r}.$$

But $\alpha^{p^r} = \alpha$, and so the equality above can be written as

$$(\mathrm{tr}(\alpha))^p = \mathrm{tr}(\alpha),$$

from which it follows that $\operatorname{tr}(\alpha) \in F_p$.

(b) Using the binomial theorem and the fact that F_q has characteristic p, we find that

$$\operatorname{tr}(\alpha+\beta) = \sum_{j=0}^{r-1} (\alpha+\beta)^{p^j} = \sum_{j=0}^{r-1} (\alpha^{p^j} + \beta^{p^j}) = \operatorname{tr}(\alpha) + \operatorname{tr}(\beta).$$

(c) As $a^p = a$, we have $a^{p^j} = a$ for every nonnegative integer j, so that

$$\operatorname{tr}(a\alpha) = \sum_{j=0}^{r-1} (a\alpha)^{p^j} = \sum_{j=0}^{r-1} a^{p^j}\alpha^{p^j} = \sum_{j=0}^{r-1} a\alpha^{p^j} = a\operatorname{tr}(\alpha).$$

(d) The polynomial $\sum_{j=0}^{r-1} x^{p^j}$ has at most p^{r-1} distinct roots in F_q. Thus the number of $\alpha \in F_q$ such that $\operatorname{tr}(\alpha) \in F_p^*$ is at least $p^r - p^{r-1} \geq 1$, so we can choose an element $\alpha \in F_q$ such that $c := \operatorname{tr}(\alpha) \in F_p^*$. Now let b be an arbitrary element of F_p. Then, by part (c),

$$\operatorname{tr}(b\alpha/c) = \frac{b}{c}\operatorname{tr}(\alpha) = \frac{b}{c}\cdot c = b.$$

(e) Since $\alpha^{p^r} = \alpha$,

$$\operatorname{tr}(\alpha^p) = \sum_{j=0}^{r-1} (\alpha^p)^{p^j} = \sum_{j=0}^{r-1} \alpha^{p^{j+1}} = \sum_{j=1}^{r} \alpha^{p^j} = \sum_{j=0}^{r-1} \alpha^{p^j} = \operatorname{tr}(\alpha).$$ ■

Definition *For $\alpha \in F_q$, define $e(\alpha) = e^{2\pi i \operatorname{tr}(\alpha)/p}$.*

Some properties of $e(\alpha)$ are given in the next proposition.

Proposition 1.1.2 *Let α, $\beta \in F_q$. Then*

(a) $e(\alpha+\beta) = e(\alpha)e(\beta)$,
(b) *there exists an element $\alpha \in F_q$ such that $e(\alpha) \neq 1$,*
(c) $\sum_{\alpha \in F_q} e(\alpha) = 0$.

Proof Parts (a) and (b) follow from parts (b) and (d) of Proposition 1.1.1, respectively. To prove (c), by part (b), we can choose $\beta \in F_q$ so that $e(\beta) \neq 1$. Then, by part (a),

$$e(\beta)\sum_{\alpha \in F_q} e(\alpha) = \sum_{\alpha \in F_q} e(\alpha+\beta) = \sum_{\gamma \in F_q} e(\gamma).$$

As $e(\beta) \neq 1$, the desired conclusion follows. ■

We assume that the reader is familiar with the elementary properties of characters, as described, for example, in Apostol [1, Chapter 6], Ayoub [1, pp. 8–13], or Ireland

and Rosen [1, pp. 88–91]. Recall that a multiplicative character χ on F_q is a map from F_q^* to the multiplicative group of complex roots of unity such that

$$\chi(\alpha\beta) = \chi(\alpha)\chi(\beta) \quad \text{for all } \alpha,\ \beta \in F_q^*.$$

The trivial character is the character χ such that $\chi(\alpha) = 1$ for all $\alpha \in F_q^*$. It is convenient to extend the domain of definition of a character χ from F_q^* to F_q by setting $\chi(0) = 1$ if χ is the trivial character and $\chi(0) = 0$ if χ is a nontrivial character. With this definition we have

$$\sum_{\alpha \in F_q} \chi(\alpha) = \begin{cases} q, & \text{if } \chi \text{ is trivial,} \\ 0, & \text{if } \chi \text{ is nontrivial.} \end{cases}$$

As the notation $\sum_{\alpha \in F_q}$ is slightly cumbersome, we often use in what follows the symbols $\sum_{\alpha}$ and $\sum_{\alpha \neq \beta}$ to indicate that the sum is over all elements in F_q and $F_q - \{\beta\}$, respectively.

We now define Gauss sums over finite fields F_q, where $q = p^r$.

Definition *Let χ be a character on F_q, and let $\beta \in F_q$. Then the Gauss sum $G_r(\beta, \chi)$ over F_q is defined by*

$$G_r(\beta, \chi) = \sum_{\alpha \in F_q} \chi(\alpha) e(\alpha\beta).$$

If $\beta = 1$, we shall write $G_r(1, \chi) = G_r(\chi)$. If $r = 1$, we shall suppress the subscript r. If χ is the trivial character, then

$$G_r(\beta, \chi) = \sum_{\alpha} e(\alpha\beta) = \begin{cases} q, & \text{if } \beta = 0, \\ 0, & \text{if } \beta \neq 0, \end{cases}$$

by Proposition 1.1.2(c).

Theorem 1.1.3 *Let χ be a nontrivial character on F_q, and let $\beta \in F_q^*$. Then*

$$G_r(\beta, \chi) = \chi(\beta^{-1}) G_r(\chi),$$

where β^{-1} denotes the inverse of β in F_q^*.

Proof Let χ^{-1} denote the inverse of χ, that is, $\chi^{-1} = \overline{\chi}$. Then

$$\begin{aligned} G_r(\beta, \chi) &= \sum_{\alpha} \chi(\alpha) e(\alpha\beta) = \chi^{-1}(\beta) \sum_{\alpha} \chi(\alpha\beta) e(\alpha\beta) \\ &= \chi^{-1}(\beta) \sum_{\gamma} \chi(\gamma) e(\gamma) = \chi(\beta^{-1}) G_r(\chi), \end{aligned}$$

since $\chi^{-1}(\beta) = \chi(\beta^{-1})$. ■

The next theorem gives four basic properties of the Gauss sum $G_r(\beta, \chi)$.

Theorem 1.1.4 *Let $\beta \in F_q^*$, and let χ be a nontrivial character on F_q. Then*
(a) $G_r(\beta,\chi)G_r(\beta,\overline{\chi}) = \chi(-1)p^r$,
(b) $\overline{G_r(\beta,\chi)} = \chi(-1)G_r(\beta,\overline{\chi})$,
(c) $|G_r(\beta,\chi)| = p^{r/2}$,
(d) $G_r(\beta,\chi^p) = G_r(\beta^p,\chi)$.

Proof By Theorem 1.1.3, as $\chi^{-1} = \overline{\chi}$,

$$\sum_{\alpha} G_r(\alpha,\chi)G_r(\alpha,\overline{\chi}) = \sum_{\alpha} \overline{\chi}(\alpha)G_r(\chi)\chi(\alpha)G_r(\overline{\chi}) = G_r(\chi)G_r(\overline{\chi})\sum_{\alpha\neq 0} 1,$$

that is,

$$\sum_{\alpha} G_r(\alpha,\chi)G_r(\alpha,\overline{\chi}) = (p^r-1)G_r(\chi)G_r(\overline{\chi}). \tag{1.1.1}$$

On the other hand,

$$G_r(\alpha,\chi)G_r(\alpha,\overline{\chi}) = \sum_{\gamma,\delta} \chi(\gamma)\overline{\chi}(\delta)e(\alpha(\gamma+\delta)).$$

Hence, summing over α, we find that, by Proposition 1.1.2(c),

$$\sum_{\alpha} G_r(\alpha,\chi)G_r(\alpha,\overline{\chi}) = \sum_{\gamma,\delta} \chi(\gamma)\overline{\chi}(\delta)\sum_{\alpha} e(\alpha(\gamma+\delta)) = p^r\sum_{\gamma} \chi(\gamma)\overline{\chi}(-\gamma),$$

that is,

$$\sum_{\alpha} G_r(\alpha,\chi)G_r(\alpha,\overline{\chi}) = p^r(p^r-1)\chi(-1). \tag{1.1.2}$$

By comparing (1.1.1) and (1.1.2), we deduce that

$$G_r(\chi)G_r(\overline{\chi}) = \chi(-1)p^r.$$

Then, appealing to Theorem 1.1.3, we obtain

$$G_r(\beta,\chi)G_r(\beta,\overline{\chi}) = \chi(\beta^{-1})\overline{\chi}(\beta^{-1})G_r(\chi)G_r(\overline{\chi}) = \chi(-1)p^r,$$

which proves (a).

To prove (b), observe that

$$\begin{aligned}\overline{G_r(\beta,\chi)} &= \overline{\sum_{\alpha} \chi(\alpha)e(\alpha\beta)} = \sum_{\alpha} \overline{\chi}(\alpha)e(-\alpha\beta)\\ &= \sum_{\alpha} \overline{\chi}(-\alpha)e(\alpha\beta) = \overline{\chi}(-1)G_r(\beta,\overline{\chi}) = \chi(-1)G_r(\beta,\overline{\chi}).\end{aligned}$$

Part (c) follows from parts (a) and (b) as

$$|G_r(\beta,\chi)|^2 = G_r(\beta,\chi)\overline{G_r(\beta,\chi)} = G_r(\beta,\chi)\chi(-1)G_r(\beta,\overline{\chi}) = \chi^2(-1)p^r = p^r,$$

so that

$$|G_r(\beta,\chi)| = p^{r/2}.$$

Finally, in view of Proposition 1.1.1(e), part (d) is proved by replacing α by α^p in the formula $G_r(\beta^p,\chi) = \sum_\alpha \chi(\alpha)e(\alpha\beta^p)$. ■

The factor $\chi(-1)$ occurring in Theorem 1.1.4 equals 1 if $p = 2$, since $-1 \equiv 1 \pmod 2$. Moreover,

(1.1.3) $$\chi(-1) = (-1)^{(q-1)/k}, \text{ if } p > 2,$$

where k is the order of the character χ on F_q. To see this, let γ be a generator of the multiplicative group F_q^*. Then,

$$\chi(-1) = \chi(\gamma^{(q-1)/2}) = e^{\pi i(q-1)j/k} = (-1)^{(q-1)j/k}$$

for some integer j coprime with k. If j is odd, then (1.1.3) follows, and if j is even, then k is odd and both sides of (1.1.3) equal 1.

We now define a second type of Gauss sum over F_q.

Definition *Let $\beta \in F_q$, and let k be a positive integer. Then the Gauss sum $g_r(\beta,k)$ over F_q is defined by*

$$g_r(\beta,k) = \sum_{\alpha\in F_q} e^{2\pi i \mathrm{tr}(\beta\alpha^k)/p}.$$

Easy calculations show that $g_r(0,k) = q$ and $g_r(\beta,1) = 0$ for $\beta \neq 0$. If $\beta = 1$, we write $g_r(1,k) = g_r(k)$, and if $r = 1$, we suppress the subscript r. The two types of Gauss sums $g_r(\beta,k)$ and $G_r(\beta,\chi)$ are connected by the relationship

(1.1.4) $$g_r(\beta,k) = \sum_{j=1}^{k-1} G_r(\beta,\chi^j),$$

where χ is a character of order k on F_q, and $\beta \in F_q^*$. To see this, note that

$$\begin{aligned} g_r(\beta,k) &= \sum_\alpha e(\beta\alpha^k) = \sum_\alpha e(\beta\alpha)\{1+\chi(\alpha)+\cdots+\chi^{k-1}(\alpha)\} \\ &= \sum_{j=0}^{k-1}\sum_\alpha e(\beta\alpha)\chi^j(\alpha) = \sum_{j=0}^{k-1} G_r(\beta,\chi^j) = \sum_{j=1}^{k-1} G_r(\beta,\chi^j), \end{aligned}$$

since for $\alpha \in F_q^*$,

$$1+\chi(\alpha)+\cdots+\chi^{k-1}(\alpha) = \begin{cases} k, & \text{if } \alpha \text{ is a } k\text{th power in } F_q^*, \\ 0, & \text{otherwise.} \end{cases}$$

We next define a quadratic Gauss sum over the ring of integers $\pmod{k}$.

Definition *Let m and k be integers with $k > 0$. Then the quadratic Gauss sum $g(m;k)$ over the ring of integers $\pmod{k}$ is defined by*

$$g(m;k) = \sum_{n=0}^{k-1} e^{2\pi imn^2/k}. \tag{1.1.5}$$

The notation for the quadratic Gauss sum in (1.1.5) should not be confused with that for the general Gauss sum appearing in (1.1.4).

For an odd prime p, let ρ denote the quadratic character $\pmod{p}$. Thus

$$G(m,\rho) = \sum_{n=0}^{p-1} \left(\frac{n}{p}\right) e^{2\pi imn/p}. \tag{1.1.6}$$

When $p \mid m$, we have $G(m,\rho) = 0$ and $g(m;p) = p$. However, when $p \nmid m$, these two Gauss sums are equal, by the following theorem.

Theorem 1.1.5 *If m is an integer not divisible by the odd prime p, then*

$$g(m;p) = \sum_{n=0}^{p-1} e^{2\pi imn^2/p} = \sum_{n=0}^{p-1} \left(\frac{n}{p}\right) e^{2\pi imn/p} = G(m,\rho).$$

Proof This is the case $r = 1,\ \beta = m,\ k = 2$ of (1.1.4). ∎

A generalization of Theorem 1.1.5 is given in Problem 24.

1.2 The reciprocity theorem for quadratic Gauss sums

The first lemma of this section is an elementary result that gives the value of the quadratic Gauss sum $g(1;p)$ up to sign, where p is an odd prime.

Lemma 1.2.1

$$g(1;p) = \sum_{n=0}^{p-1} e^{2\pi in^2/p} = \begin{cases} \pm\sqrt{p}, & \text{if } p \equiv 1 \pmod 4, \\ \pm i\sqrt{p}, & \text{if } p \equiv 3 \pmod 4. \end{cases}$$

Proof By part (b) of Theorem 1.1.4, as $\rho(n) = \left(\frac{n}{p}\right)$ is a real character,

$$\overline{G(1,\rho)} = \left(\frac{-1}{p}\right) G(1,\rho).$$

Then, by part (c) of Theorem 1.1.4, we deduce that

$$p = |G(1,\rho)|^2 = G(1,\rho)\overline{G(1,\rho)} = \left(\frac{-1}{p}\right) G(1,\rho)^2,$$

so that by Theorem 1.1.5

$$g(1;p)^2 = G(1,\rho)^2 = \left(\frac{-1}{p}\right)p = \begin{cases} p, & \text{if } p \equiv 1 \pmod 4, \\ -p, & \text{if } p \equiv 3 \pmod 4, \end{cases}$$

from which the lemma follows. ∎

In Theorem 1.2.4, we show the deeper fact that the + sign is correct in all cases of Lemma 1.2.1. This result will be seen to be a special case of the following very general theorem, which relates the two generalized Gauss sums $S(a,b,c)$ and $S(-c,-b,a)$, where

$$S(a,b,c) = \sum_{n=0}^{|c|-1} e^{\pi i(an^2+bn)/c},$$

and a, b, c are integers with $ac \neq 0$ and $ac+b$ even.

Theorem 1.2.2 (*Reciprocity theorem for generalized Gauss sums.*) *Let a, b, c be integers with $ac \neq 0$ and $ac+b$ even. Then*

(1.2.1) $$S(a,b,c) = |c/a|^{1/2} e^{\pi i\{\operatorname{sgn}(ac)-(b^2/(ac))\}/4} S(-c,-b,a).$$

Proof Let

$$f(z) = \frac{e^{\pi i(az^2+bz)/c}}{e^{2\pi i\epsilon z} - 1},$$

where $\epsilon = \operatorname{sgn}(ac)$. The function $f(z)$ is an analytic function of $z = x+iy$ except for simple poles at $z = n$, where n is an integer. Since $ac+b$ is even, a brief calculation shows that

(1.2.2) $$f(z+|c|) = e^{2\pi i|a|\epsilon z} f(z).$$

For $N > 0$ let C_N denote a positively oriented parallelogram with horizontal sides coincident with the lines $y = \pm N$ and the other two sides passing through the points $-\frac{1}{2}$ and $|c| - \frac{1}{2}$. We require that the latter sides intersect the real axis at an acute angle if $ac > 0$ and at an obtuse angle if $ac < 0$. For each integer n, the residue of $f(z)$ at the simple pole $z = n$ is

$$\operatorname{Res}_{z=n}(f(z)) = \frac{e^{\pi i(an^2+bn)/c}}{2\pi i\epsilon}.$$

By Cauchy's residue theorem, we deduce that

(1.2.3) $$\int_{C_N} f(z)dz = 2\pi i \sum_{n=0}^{|c|-1} \operatorname{Res}_{z=n}(f(z)) = \epsilon S(a,b,c).$$

Now

$$|f(z)| \le \frac{e^{-\pi(2ax+b)y/c}}{|e^{-2\pi\epsilon y}-1|}.$$

Thus, if $z = x \pm iN \in C_N$, by the positioning of the parallelogram, we easily see that the integrals over the horizontal sides of C_N tend to 0 as N tends to ∞. Let L_1 and L_2 be the infinite sides through $-\frac{1}{2}$ and $|c|-\frac{1}{2}$, respectively, that arise from the parallelogram on letting N tend to ∞, with both paths L_1 and L_2 proceeding from the lower to the upper half-plane. On letting N tend to ∞ in (1.2.3) and using (1.2.2), we find that

$$\begin{aligned}\epsilon S(a,b,c) &= -\int_{L_1} f(z)dz + \int_{L_2} f(z)dz \\ &= \int_{L_1} \{f(z+|c|)-f(z)\}dz \\ &= \int_{L_1} (e^{2\pi i|a|\epsilon z}-1)f(z)dz \\ &= \int_{L_1} e^{\pi i(az^2+bz)/c}\frac{(e^{2\pi i|a|\epsilon z}-1)}{(e^{2\pi i\epsilon z}-1)}dz \\ &= \int_{L_1} e^{\pi i(az^2+bz)/c}\sum_{n=0}^{|a|-1} e^{2\pi in\epsilon z}dz.\end{aligned}$$

We now make the change of variable $z = w-(cn\epsilon/a)$ and use Cauchy's theorem to shift the path of integration back to L_1; this yields

$$\epsilon S(a,b,c) = e^{-\pi ib^2/(4ac)}\sum_{n=0}^{|a|-1} e^{-\pi i(cn^2+b\epsilon n)/a}\int_{L_1} e^{\pi ia\{w+b/(2a)\}^2/c}dw.$$

We next make the integral independent of the parameters a, b, and c by making another change of variable $w+(b/(2a)) = (2|c/a|)^{1/2}z$. Again using Cauchy's theorem to shift the path of integration back to L_1, we obtain the following equation:

$$\begin{aligned}\epsilon S(a,b,c) &= (2|c/a|)^{1/2}e^{-\pi ib^2/(4ac)}S(-c,-b\epsilon,a)\int_{L_1} e^{2\pi i\epsilon z^2}dz \\ &= (2|c/a|)^{1/2}e^{-\pi ib^2/(4ac)}S(-c,-b,a)\int_{L_1} e^{2\pi i\epsilon z^2}dz.\end{aligned} \tag{1.2.4}$$

The last equality is trivial if $\epsilon = 1$. To establish the equality when $\epsilon = -1$, change the index of summation n by setting $m = |a|-n$ and use the fact that $ac+b$ is even. Setting $a = 2\epsilon$, $b = 0$, and $c = 1$ in (1.2.4) yields

$$\epsilon = (1+e^{-\pi i\epsilon/2})\int_{L_1} e^{2\pi i\epsilon z^2}dz,$$

or

$$\int_{L_1} e^{2\pi i\epsilon z^2}dz = \frac{\epsilon e^{\pi i\epsilon/4}}{\sqrt{2}}.$$

Using this evaluation in (1.2.4), we deduce (1.2.1). ∎

Corollary 1.2.3 *Let c be a positive integer. Then*

$$g(1;c) = \sum_{n=0}^{c-1} e^{2\pi i n^2/c} = \frac{1}{2}(1+i)(1+i^{-c})\sqrt{c}$$

$$= \begin{cases} \sqrt{c}, & \text{if } c \equiv 1 \pmod 4, \\ 0, & \text{if } c \equiv 2 \pmod 4, \\ i\sqrt{c}, & \text{if } c \equiv 3 \pmod 4, \\ (1+i)\sqrt{c}, & \text{if } c \equiv 0 \pmod 4. \end{cases}$$

Proof Set $a = 2$ and $b = 0$ in Theorem 1.2.2. ■

The following theorem resolves the ambiguous sign in Lemma 1.2.1. It will be extended to quadratic Gauss sums over general finite fields F_{p^r} in Theorem 11.5.4.

Theorem 1.2.4 *For an odd prime p,*

$$g(1;p) = \sum_{n=0}^{p-1} e^{2\pi i n^2/p} = \sum_{n=0}^{p-1} \left(\frac{n}{p}\right) e^{2\pi i n/p} = i^{(p-1)^2/4}\sqrt{p} = \begin{cases} \sqrt{p}, & \text{if } p \equiv 1 \pmod 4, \\ i\sqrt{p}, & \text{if } p \equiv 3 \pmod 4. \end{cases}$$

Proof This follows from Theorem 1.1.5 and from the case $c = p$ of Corollary 1.2.3. ■

The next lemma gives an important multiplicative property of the Gauss sum.

Lemma 1.2.5 *Let a and b be coprime positive integers, and let t be any integer. Then*

$$g(tb;a)g(ta;b) = g(t;ab).$$

Proof We have

$$g(tb;a)g(ta;b) = \sum_{m=0}^{a-1}\sum_{n=0}^{b-1} e^{2\pi i(bm+an)^2 t/ab}.$$

As $\gcd(a,b) = 1$, $bm + an$ runs through a complete residue system $\pmod{ab}$ as m and n run through a complete residue system $\pmod a$ and $\pmod b$, respectively. Hence

$$g(tb;a)g(ta;b) = \sum_{r=0}^{ab-1} e^{2\pi i r^2 t/ab} = g(t;ab).$$ ■

Gauss showed that the law of quadratic reciprocity is a consequence of the multiplicative property of the Gauss sum given in Lemma 1.2.5 and the evaluation of the Gauss sum given in Corollary 1.2.3. We now present his proof.

Theorem 1.2.6 (*Law of quadratic reciprocity.*) *Let p and p' be distinct odd primes. Then*

$$\left(\frac{p}{p'}\right)\left(\frac{p'}{p}\right) = (-1)^{(p-1)(p'-1)/4}.$$

Proof By Theorems 1.1.3 and 1.1.5 and Lemma 1.2.5,

$$g(1;pp') = g(p';p)g(p;p') = \left(\frac{p'}{p}\right) g(1;p) \left(\frac{p}{p'}\right) g(1;p').$$

Appealing to Corollary 1.2.3 with $c = pp'$, we obtain

$$i^{(pp'-1)^2/4}\sqrt{pp'} = \left(\frac{p}{p'}\right)\left(\frac{p'}{p}\right) i^{(p-1)^2/4}\sqrt{p}\, i^{(p'-1)^2/4}\sqrt{p'},$$

so that as $p^2 - 1 \equiv p'^2 - 1 \equiv 0 \pmod 8$,

$$\begin{aligned}\left(\frac{p}{p'}\right)\left(\frac{p'}{p}\right) &= i^{((pp'-1)^2-(p-1)^2-(p'-1)^2)/4} = i^{((p^2-1)(p'^2-1)-2(p-1)(p'-1))/4} \\ &= i^{-2(p-1)(p'-1)/4} = (-1)^{(p-1)(p'-1)/4}.\end{aligned}$$

■

The law of quadratic reciprocity can be proved in a much more elementary way, as follows. For a positive integer m, let R_m denote the group of $\phi(m)$ reduced residues (mod m). Let u and v be distinct odd primes, and, for brevity, write $u_2 = (u-1)/2$ and $v_2 = (v-1)/2$. Let D be the cyclic subgroup of $R_u \times R_v$ generated by $(-1,-1)$, so that D has two elements. We compute the product α of the $\phi(uv)/2$ elements of $(R_u \times R_v)/D$ in two ways. First, since

$$(R_u \times R_v)/D = \bigcup_{\substack{1 \le i \le u-1 \\ 1 \le j \le v_2}} \{(i,j)D\},$$

we have

$$\alpha = ((u-1)!^{v_2}, v_2!^{2u_2})D.$$

Since $v_2!^2 \equiv (-1)^{v_2}(v-1)! \pmod v$,

$$\alpha = ((u-1)!^{v_2}, (v-1)!^{u_2}(-1)^{u_2v_2})D. \tag{1.2.5}$$

On the other hand,

$$(R_u \times R_v)/D = \bigcup_{\substack{1 \le k < uv/2 \\ \gcd(k,uv)=1}} \{(k,k)D\}.$$

The product of the k's (mod u) is

$$(u-1)!^{v_2}v^{-u_2} \equiv (u-1)!^{v_2}\left(\frac{v}{u}\right) \pmod{u},$$

by Euler's criterion, and so by symmetry in u and v,

$$\alpha = \left((u-1)!^{v_2}\left(\frac{v}{u}\right), (v-1)!^{u_2}\left(\frac{u}{v}\right)\right)D. \tag{1.2.6}$$

Comparing (1.2.5) and (1.2.6), we see that

$$\left(\left(\frac{v}{u}\right), \left(\frac{u}{v}\right)(-1)^{u_2v_2}\right) \in D,$$

which yields the law of quadratic reciprocity

$$\left(\frac{u}{v}\right)\left(\frac{v}{u}\right) = (-1)^{(u-1)(v-1)/4}. \tag{1.2.7}$$

By the multiplicativity of the Jacobi symbol, it is easily seen that (1.2.7) holds for arbitrary odd positive integers u and v. Further, by defining

$$\left(\frac{u}{v}\right) = \left(\frac{u}{|v|}\right)$$

for negative odd integers v, we see that (1.2.7) holds for arbitrary odd integers u and v, if and only if not both are negative. This result will be used later in the book, where we simply refer to (1.2.7).

The supplementary law

$$\left(\frac{2}{v}\right) = (-1)^{(v^2-1)/8}$$

for an odd positive integer v can be deduced directly from the reciprocity law (1.2.7) as follows. Write $v = 4k + e$, where e is in $\{-1,\ 1\}$. Then $a := (v+e)/2$ is an odd positive integer, and so

$$\left(\frac{2e}{v}\right) = \left(\frac{4a}{v}\right) = \left(\frac{a}{v}\right) = \left(\frac{ev}{a}\right) = \left(\frac{-1}{a}\right).$$

Thus,

$$\left(\frac{2}{v}\right) = \left(\frac{e}{v}\right)\left(\frac{-1}{a}\right) = (-1)^k = (-1)^{(v^2-1)/8}.$$

1.3 Gauss' evaluation of a quadratic Gauss sum

In (1.1.5), we defined the quadratic Gauss sum

$$g(1;k) = \sum_{n=0}^{k-1} e^{2\pi i n^2/k}.$$

We now present Gauss' evaluation of $g(1;k)$, for an odd positive integer k. Gauss' proof makes use of the properties of certain polynomials, which are now called Gaussian polynomials. The evaluation, given in Theorem 1.3.4, is a special case of Corollary 1.2.3, but has the advantage of requiring no complex analysis in the proof. For other elementary proofs, see Section 1.4 and Problems 26–29.

Definition *For nonnegative integers m and n, define a rational function $\begin{bmatrix} n \\ m \end{bmatrix}$ of the variable x by*

$$\begin{bmatrix} n \\ m \end{bmatrix} = \frac{(1-x^n)(1-x^{n-1})\cdots(1-x^{n-(m-1)})}{(1-x)(1-x^2)\cdots(1-x^m)}, \qquad n \geq 1, m \geq 1,$$

$$\begin{bmatrix} n \\ 0 \end{bmatrix} = 1, \quad n \geq 0,$$

$$\begin{bmatrix} 0 \\ m \end{bmatrix} = 0, \quad m \geq 1.$$

Note that $\begin{bmatrix} n \\ m \end{bmatrix}$ generalizes the binomial coefficient $\binom{n}{m}$, since $\begin{bmatrix} n \\ m \end{bmatrix}$ tends to $\binom{n}{m}$ as x tends to 1. The next proposition summarizes the basic properties of $\begin{bmatrix} n \\ m \end{bmatrix}$ that we shall need; particularly important is part (e), which asserts that, in fact, $\begin{bmatrix} n \\ m \end{bmatrix}$ is a polynomial in x.

Proposition 1.3.1 *Let m and n be nonnegative integers. Then*

(a) *for $m > n \geq 0$*

$$\begin{bmatrix} n \\ m \end{bmatrix} = 0,$$

(b) *for $n \geq m \geq 0$*

$$\begin{bmatrix} n \\ m \end{bmatrix} = \begin{bmatrix} n \\ n-m \end{bmatrix},$$

(c) *for $n \geq 1$ and $m \geq 1$*

$$\begin{bmatrix} n \\ m \end{bmatrix} = \begin{bmatrix} n-1 \\ m-1 \end{bmatrix} + x^m \begin{bmatrix} n-1 \\ m \end{bmatrix},$$

(d) *for* $n \geq 1$ *and* $m \geq 1$

$$\begin{bmatrix} n \\ m \end{bmatrix} = \begin{bmatrix} n-1 \\ m \end{bmatrix} + x^{n-m} \begin{bmatrix} n-1 \\ m-1 \end{bmatrix},$$

(e) *for* $n \geq m \geq 0$, $\begin{bmatrix} n \\ m \end{bmatrix}$ *is a monic polynomial in* x *of degree* $m(n-m)$.

Proof (a) If $m > n \geq 1$, the numerator of $\begin{bmatrix} n \\ m \end{bmatrix}$ is

$$(1-x^n)\cdots(1-x^1)(1-x^0)\cdots(1-x^{-(m-n-1)}),$$

which contains the factor $1 - x^0 = 1 - 1 = 0$ so that $\begin{bmatrix} n \\ m \end{bmatrix} = 0$. If $m > n = 0$, then $\begin{bmatrix} n \\ m \end{bmatrix} = \begin{bmatrix} 0 \\ m \end{bmatrix} = 0$.

(b) For $n \geq m = 0$ clearly $\begin{bmatrix} n \\ m \end{bmatrix} = \begin{bmatrix} n \\ 0 \end{bmatrix} = 1$ and $\begin{bmatrix} n \\ n-m \end{bmatrix} = \begin{bmatrix} n \\ n \end{bmatrix} = 1$. For $n \geq m \geq 1$,

$$\begin{aligned} \begin{bmatrix} n \\ m \end{bmatrix} &= \frac{(1-x^n)\cdots(1-x^{n-m+1})}{(1-x)\cdots(1-x^m)} \\ &= \frac{(1-x^n)\cdots(1-x^{n-m+1})(1-x^{n-m})\cdots(1-x)}{(1-x)\cdots(1-x^m)(1-x^{n-m})\cdots(1-x)} \\ &= \frac{(1-x^n)\cdots(1-x^{m+1})}{(1-x)\cdots(1-x^{n-m})} \\ &= \begin{bmatrix} n \\ n-m \end{bmatrix}. \end{aligned}$$

(c) For $n \geq 2$, $m \geq 1$,

$$\begin{aligned} \begin{bmatrix} n \\ m \end{bmatrix} &= \frac{(1-x^n)}{(1-x^m)} \begin{bmatrix} n-1 \\ m-1 \end{bmatrix} \\ &= \left(1 + \frac{x^m(1-x^{n-m})}{(1-x^m)}\right) \begin{bmatrix} n-1 \\ m-1 \end{bmatrix} \\ &= \begin{bmatrix} n-1 \\ m-1 \end{bmatrix} + x^m \begin{bmatrix} n-1 \\ m \end{bmatrix}. \end{aligned}$$

For $n = 1$, $m \geq 2$,

$$\begin{bmatrix} n \\ m \end{bmatrix} = 0, \qquad \begin{bmatrix} n-1 \\ m-1 \end{bmatrix} = 0, \qquad \begin{bmatrix} n-1 \\ m \end{bmatrix} = 0,$$

so the result holds in this case, too.

For $n = m = 1$,

$$\begin{bmatrix} n \\ m \end{bmatrix} = \begin{bmatrix} 1 \\ 1 \end{bmatrix} = 1, \quad \begin{bmatrix} n-1 \\ m-1 \end{bmatrix} = \begin{bmatrix} 0 \\ 0 \end{bmatrix} = 1, \quad \begin{bmatrix} n-1 \\ m \end{bmatrix} = \begin{bmatrix} 0 \\ 1 \end{bmatrix} = 0,$$

so again the result holds.

(d) Appealing to parts (b) and (c), we have for $n \geq 1$ and $1 \leq m \leq n-1$ (so that $n - m \geq 1$)

$$\begin{aligned} \begin{bmatrix} n \\ m \end{bmatrix} = \begin{bmatrix} n \\ n-m \end{bmatrix} &= \begin{bmatrix} n-1 \\ n-m-1 \end{bmatrix} + x^{n-m} \begin{bmatrix} n-1 \\ n-m \end{bmatrix} \\ &= \begin{bmatrix} n-1 \\ m \end{bmatrix} + x^{n-m} \begin{bmatrix} n-1 \\ m-1 \end{bmatrix}. \end{aligned}$$

It remains to examine the case $1 \leq n \leq m$. If $1 \leq n < m$, then

$$\begin{bmatrix} n \\ m \end{bmatrix} = \begin{bmatrix} n-1 \\ m \end{bmatrix} = \begin{bmatrix} n-1 \\ m-1 \end{bmatrix} = 0$$

by part (a), and the result holds. If $1 \leq n = m$, then

$$\begin{bmatrix} n \\ m \end{bmatrix} = 1, \quad \begin{bmatrix} n-1 \\ m \end{bmatrix} = 0, \quad \begin{bmatrix} n-1 \\ m-1 \end{bmatrix} = 1,$$

and again the result holds.

(e) As

$$\begin{bmatrix} n \\ 0 \end{bmatrix} = \begin{bmatrix} n \\ n \end{bmatrix} = 1, \quad n \geq 0,$$

the recursion formula of (d) shows inductively that the quantities $\begin{bmatrix} n \\ m \end{bmatrix}$ are monic polynomials in x of degree $m(n-m)$. ■

The polynomial $\begin{bmatrix} n \\ m \end{bmatrix}$ is called a Gaussian polynomial. An important polynomial defined in terms of Gaussian polynomials is

$$f(x, n) = \sum_{j=0}^{n} (-1)^j \begin{bmatrix} n \\ j \end{bmatrix}.$$

When n is odd, it is clear from Proposition 1.3.1(b) that $f(x, n)$ is identically equal to 0. Thus $f(x, n)$ is interesting only when n is even, in which case $f(x, n)$ is a polynomial of degree $n^2/4$, which we determine explicitly in the next theorem.

Theorem 1.3.2 *Let n be a positive integer. Then*

$$f(x,n) = \begin{cases} 0, & \text{if } n \text{ is odd,} \\ \prod_{j=1}^{n/2}(1-x^{2j-1}), & \text{if } n \text{ is even.} \end{cases}$$

Proof We begin by showing that for any integer $n \geq 2$, $f(x,n)$ satisfies the recursion formula

$$f(x,n) = (1-x^{n-1})f(x,n-2).$$

By part (d) of Proposition 1.3.1,

$$\begin{aligned} f(x,n) &= 1 + \sum_{j=1}^{n}(-1)^j \left(\begin{bmatrix} n-1 \\ j \end{bmatrix} + x^{n-j} \begin{bmatrix} n-1 \\ j-1 \end{bmatrix} \right) \\ &= \sum_{j=0}^{n-1}(-1)^j \begin{bmatrix} n-1 \\ j \end{bmatrix} + \sum_{j=0}^{n-1}(-1)^{j+1}x^{n-j-1} \begin{bmatrix} n-1 \\ j \end{bmatrix} \\ &= \sum_{j=0}^{n-1}(-1)^j(1-x^{n-j-1}) \begin{bmatrix} n-1 \\ j \end{bmatrix}. \end{aligned}$$

Now

$$\begin{aligned} (1-x^{n-j-1}) \begin{bmatrix} n-1 \\ j \end{bmatrix} &= \frac{(1-x^{n-j-1})(1-x^{n-1})(1-x^{n-2})\cdots(1-x^{n-j})}{(1-x)(1-x^2)\cdots(1-x^j)} \\ &= (1-x^{n-1})\frac{(1-x^{n-2})\cdots(1-x^{n-j-1})}{(1-x)\cdots(1-x^j)} \\ &= (1-x^{n-1}) \begin{bmatrix} n-2 \\ j \end{bmatrix}, \end{aligned}$$

so

$$\begin{aligned} f(x,n) &= \sum_{j=0}^{n-1}(-1)^j(1-x^{n-1}) \begin{bmatrix} n-2 \\ j \end{bmatrix} \\ &= (1-x^{n-1})\sum_{j=0}^{n-2}(-1)^j \begin{bmatrix} n-2 \\ j \end{bmatrix} \\ &= (1-x^{n-1})f(x,n-2). \end{aligned}$$

By repeated applications of the recursion formula, we obtain

$$\begin{aligned} f(x,n) &= (1-x^{n-1})f(x,n-2) \\ &= (1-x^{n-1})(1-x^{n-3})f(x,n-4) \\ &= \cdots \\ &= \begin{cases} (1-x^{n-1})(1-x^{n-3})\cdots(1-x^2)f(x,1), & \text{if } n \text{ is odd,} \\ (1-x^{n-1})(1-x^{n-3})\cdots(1-x)f(x,0), & \text{if } n \text{ is even.} \end{cases} \end{aligned}$$

The assertion of the theorem now follows as $f(x,1)=0$ and $f(x,0)=1$. ■

The next step is to make the connection between the polynomial $f(x,k-1)$ and the Gauss sum $g(1;k)$.

Proposition 1.3.3 *Let k be an odd positive integer. Let $\beta = e^{2\pi i/k}$. Then*

$$g(1;k) = \beta^{(k+1)^2/4} f(\beta^{-2}, k-1).$$

Proof We first determine $\begin{bmatrix} k-1 \\ j \end{bmatrix}$ when $x=\beta^{-2}$. We have

$$\begin{aligned}\begin{bmatrix} k-1 \\ j \end{bmatrix} &= \frac{(1-x^{k-1})(1-x^{k-2})\cdots(1-x^{k-j})}{(1-x)(1-x^2)\cdots(1-x^j)} \\ &= \frac{(1-\beta^2)(1-\beta^4)\cdots(1-\beta^{2j})}{(1-\beta^{-2})(1-\beta^{-4})\cdots(1-\beta^{-2j})} \\ &= (-\beta^2)(-\beta^4)\cdots(-\beta^{2j}) \\ &= (-1)^j\beta^{2+4+\cdots+2j} \\ &= (-1)^j\beta^{j(j+1)}.\end{aligned}$$

Thus

$$\begin{aligned} f(\beta^{-2},k-1) &= \sum_{j=0}^{k-1}\beta^{j(j+1)} = \beta^{-(k+1)^2/4}\sum_{j=0}^{k-1}\beta^{(j+(k+1)/2)^2} \\ &= \beta^{-(k+1)^2/4}\sum_{j=0}^{k-1}\beta^{j^2} = \beta^{-(k+1)^2/4}g(1;k). \end{aligned}$$ ■

The final step is to use the product formula for $f(\beta^{-2},k-1)$ (given in Theorem 1.3.2) in Proposition 1.3.3 to evaluate $g(1;k)$.

Theorem 1.3.4 *Let k be an odd positive integer. Then*

$$g(1;k) = \sum_{n=0}^{k-1} e^{2\pi i n^2/k} = \begin{cases} \sqrt{k}, & \text{if } k \equiv 1 \pmod 4, \\ i\sqrt{k}, & \text{if } k \equiv 3 \pmod 4. \end{cases}$$

Proof By Theorem 1.3.2 and Proposition 1.3.3 (as $k-1$ is even),

$$\begin{aligned} g(1;k) &= \beta^{(k+1)^2/4}\prod_{j=1}^{(k-1)/2}(1-\beta^{-2(2j-1)}) \\ &= \beta^{(k+1)^2/4-(k-1)^2/4}\prod_{j=1}^{(k-1)/2}\left(\beta^{2j-1}-\beta^{-(2j-1)}\right), \end{aligned}$$

since $\sum_{j=1}^{(k-1)/2}(2j-1)=(k-1)^2/4$. Note that $\beta^{(k+1)^2/4-(k-1)^2/4}=\beta^k=1$. Replace j by $(k+1)/2-j$ in the product above to obtain

$$\begin{aligned} g(1;k) &= \prod_{j=1}^{(k-1)/2}\left(\beta^{-2j}-\beta^{2j}\right) \\ &= (-1)^{(k-1)/2}\prod_{j=1}^{(k-1)/2}\left(\beta^{2j}-\beta^{-2j}\right) \\ &= (-2i)^{(k-1)/2}\prod_{j=1}^{(k-1)/2}\sin(4\pi j/k). \end{aligned}$$

Now $\sin(4\pi j/k)>0$ if $j<k/4$ and $\sin(4\pi j/k)<0$ if $k/4<j\le(k-1)/2$. Thus

$$\operatorname{sgn}\prod_{j=1}^{(k-1)/2}\sin(4\pi j/k)=(-1)^{(k-1)/2-[k/4]}=\begin{cases} +1, & \text{if } k\equiv 1,7 \pmod 8,\\ -1, & \text{if } k\equiv 3,5 \pmod 8,\end{cases}$$

and hence

$$\begin{cases} g(1;k)>0, & \text{if } k\equiv 1 \pmod 4,\\ -ig(1;k)>0, & \text{if } k\equiv 3 \pmod 4.\end{cases} \tag{1.3.1}$$

We saw above that

$$g(1;k)=\prod_{j=1}^{(k-1)/2}\left(\beta^{2j-1}-\beta^{-(2j-1)}\right)$$

and

$$g(1;k)=(-1)^{(k-1)/2}\prod_{j=1}^{(k-1)/2}\left(\beta^{2j}-\beta^{-2j}\right),$$

so that

$$g(1;k)^2=(-1)^{(k-1)/2}\prod_{j=1}^{k-1}\left(\beta^{j}-\beta^{-j}\right).$$

Now

$$\begin{aligned} \prod_{j=1}^{k-1}\left(\beta^{j}-\beta^{-j}\right) &= \beta^{-k(k-1)/2}\prod_{j=1}^{k-1}\left(\beta^{2j}-1\right)=\prod_{j=1}^{k-1}(1-\beta^j)\left(-1-\beta^j\right) \\ &= \lim_{x\to 1}\left(\frac{x^k-1}{x-1}\right)\lim_{x\to -1}\left(\frac{x^k-1}{x-1}\right)=k, \end{aligned}$$

so that

$$g(1;k)^2=(-1)^{(k-1)/2}k. \tag{1.3.2}$$

From (1.3.1) and (1.3.2),

$$g(1;k) = \sqrt{k}, \quad \text{if } k \equiv 1 \pmod 4,$$
$$g(1;k) = i\sqrt{k}, \quad \text{if } k \equiv 3 \pmod 4. \qquad \blacksquare$$

1.4 Estermann's evaluation of a quadratic Gauss sum

In this section we reprove Theorem 1.3.4 by presenting the beautiful determination of $g(1;k)$, for an odd positive integer k, given by Estermann [1] in 1945. From (1.3.2),

$$g(1;k) = \begin{cases} \pm\sqrt{k}, & \text{if } k \equiv 1 \pmod 4, \\ \pm i\sqrt{k}, & \text{if } k \equiv 3 \pmod 4. \end{cases}$$

This formula for $g(1;k)$ can also be proved from first principles, with the use of Lemma 1.2.1 and Problem 4. The formula is equivalent to

$$\frac{1}{2}(1-i)(1+i^k)g(1;k) = \pm\sqrt{k}.$$

We prove Theorem 1.3.4 by showing that

$$\operatorname{Re}\left(\frac{1}{2}(1-i)(1+i^k)g(1;k)\right) > -\sqrt{k}.$$

First,

$$g(1;k) - 1 = \sum_{n=1}^{(k-1)/2} \left(e^{2\pi i n^2/k} + e^{2\pi i (k-n)^2/k}\right) = 2\sum_{n=1}^{(k-1)/2} e^{2\pi i n^2/k}.$$

Hence, with $f(x) = e^{2\pi i x^2/k}$,

$$\begin{aligned} \frac{1}{2}(1+i^k)(g(1;k)-1) &= (1+i^k)\sum_{n=1}^{(k-1)/2} f(n) \\ &= \sum_{n=1}^{(k-1)/2} \left(f(n) + f(\tfrac{1}{2}k - n)\right) \qquad (\text{as } f(\tfrac{1}{2}k-n) = i^k f(n)) \qquad (1.4.1) \\ &= \sum_{n=1}^{k-1} f(\tfrac{1}{2}n) = \sum_{n=1}^{k-1} e^{\pi i n^2/(2k)} = A + B, \end{aligned}$$

where

$$A = \sum_{n=1}^{[\sqrt{k}]} e^{\pi i n^2/(2k)}, \qquad B = \sum_{n=[\sqrt{k}]+1}^{k-1} e^{\pi i n^2/(2k)}.$$

Next, setting $a_n = e^{\pi i n(n+1)/(2k)}$ and $b_n = \csc(\pi n/(2k))$, we find that

$$(a_n - a_{n-1})b_n = 2ie^{\pi i n^2/(2k)},$$

and hence

$$i2B = \sum_{n=[\sqrt{k}]+1}^{k-1} (a_n - a_{n-1})b_n = \sum_{n=[\sqrt{k}]+1}^{k-1} a_n(b_n - b_{n+1}) - a_{[\sqrt{k}]}b_{[\sqrt{k}]+1} + a_{k-1}b_k.$$

Since $|a_n| = 1$ and $b_1 > b_2 > \cdots > b_k > 0$, we deduce that

$$(1.4.2) \qquad |B| \le \frac{1}{2}\left\{ \sum_{n=[\sqrt{k}]+1}^{k-1} (b_n - b_{n+1}) + b_{[\sqrt{k}]+1} + b_k \right\} = b_{[\sqrt{k}]+1} \le \frac{k}{[\sqrt{k}]+1} \le \sqrt{k}.$$

Also, as $\cos x + \sin x \ge 1$ for $0 \le x \le \pi/2$,

$$(1.4.3) \qquad \operatorname{Re}((1-i)A) = \sum_{n=1}^{[\sqrt{k}]} \left(\cos\frac{\pi n^2}{2k} + \sin\frac{\pi n^2}{2k} \right) \ge [\sqrt{k}] > \frac{1}{2}\sqrt{k}.$$

Now $\operatorname{Re}(\frac{1}{2}(1-i)(1+i^k))$ is either 0 or 1, so by (1.4.1), (1.4.3) and (1.4.2), we finally have

$$\begin{aligned} \operatorname{Re}\left(\tfrac{1}{2}(1-i)(1+i^k)g(1;k)\right) &\ge \operatorname{Re}\left((1-i)(A+B)\right) \ge \operatorname{Re}\left((1-i)A\right) - \sqrt{2}|B| \\ &> \tfrac{1}{2}\sqrt{k} - \sqrt{2}\sqrt{k} > -\sqrt{k}. \end{aligned}$$

■

1.5 Elementary determination of quadratic Gauss sums

Using the elementary evaluation

$$(1.5.1) \qquad g(1;k) = \begin{cases} \sqrt{k}, & \text{if } k \equiv 1 \pmod 4, \\ i\sqrt{k}, & \text{if } k \equiv 3 \pmod 4, \end{cases}$$

given in each of the two preceding sections, we complete here an elementary determination of

$$g(m;k) = \sum_{n=0}^{k-1} e^{2\pi i m n^2/k},$$

where m and k are coprime integers with $k > 0$. The results are presented in Theorems 1.5.1, 1.5.2, and 1.5.4. We begin with the case $k \equiv 2 \pmod 4$.

Theorem 1.5.1 *Let m and k be coprime integers with $k > 0$ and $k \equiv 2 \pmod 4$. Then $g(m,k) = 0$.*

Proof The result follows since $e^{2\pi i m n^2/k} = -e^{2\pi i m(n+k/2)^2/k}$ for $0 \le n < k/2$. ∎

We next consider the case where k is odd. Let $\left(\frac{m}{k}\right)$ denote the Jacobi symbol.

Theorem 1.5.2 *Let m and k be coprime integers with $k > 0$ and k odd. Then*

$$g(m;k) = \left(\frac{m}{k}\right) g(1;k) = \begin{cases} \left(\frac{m}{k}\right)\sqrt{k}, & \text{if } k \equiv 1 \pmod 4, \\ \left(\frac{m}{k}\right) i\sqrt{k}, & \text{if } k \equiv 3 \pmod 4. \end{cases}$$

Proof By (1.5.1),

$$g(1;k) = \pm \prod_p g(1;p), \tag{1.5.2}$$

where p runs through all the (odd) prime divisors of k, counting multiplicity. Let σ_m denote the automorphism of the cyclotomic field $Q(e^{2\pi i/k})$ defined by

$$\sigma_m(e^{2\pi i/k}) = e^{2\pi i m/k}.$$

According to Theorems 1.1.3 and 1.1.5, it follows that

$$\sigma_m(g(1;p)) = \left(\frac{m}{p}\right) g(1;p). \tag{1.5.3}$$

Now (1.5.2) and (1.5.3) show that

$$g(m;k) = \sigma_m(g(1;k)) = g(1;k) \prod_p \left(\frac{m}{p}\right) = g(1;k)\left(\frac{m}{k}\right). \qquad ∎$$

It remains to consider the case where $k \equiv 0 \pmod 4$. The following proposition deals with the special case where k is a power of 2; cf. Problem 4. Recall that for odd m,

$$\left(\frac{2}{m}\right) = \left(\frac{2}{|m|}\right) = (-1)^{(m^2-1)/8}.$$

Proposition 1.5.3 *Let m and s be integers with m odd. Then*

$$g(m;2^s) = 2g(m;2^{s-2}), \quad \textit{if } s \ge 4, \tag{1.5.4}$$

and

$$g(m;2^s) = \left(\frac{2^s}{m}\right)(1+i^m)2^{s/2}, \quad \textit{if } s \ge 2. \tag{1.5.5}$$

Proof One easily computes from the definition of $g(m;k)$ that

$$g(m;4) = 2(1+i^m) \tag{1.5.6}$$

and

$$g(m;8) = 4e^{i\pi m/4} = \left(\frac{2}{m}\right)(1+i^m)2^{3/2}. \tag{1.5.7}$$

Hence (1.5.5) holds for $s = 2, 3$. Let $s \geq 4$. To prove (1.5.4), write $b = 2^s$ and observe that

$$g(m;b) = \sum_{n=0}^{b-1} e^{2\pi imn^2/b} = \sum_{\substack{n=0 \\ n \text{ odd}}}^{b-1} e^{2\pi imn^2/b} + 2g(m;b/4).$$

The first sum on the right vanishes, since $\exp(2\pi imn^2/b)$ becomes negated when n is replaced by $n + b/4$. This proves (1.5.4). From (1.5.4),

$$g(m;2^s) = \begin{cases} 2^{(s-2)/2}g(m;4), & \text{if } s \text{ is even}, \\ 2^{(s-3)/2}g(m;8), & \text{if } s \text{ is odd}. \end{cases}$$

Substituting in the values from (1.5.6) and (1.5.7), we complete the proof of (1.5.5). ∎

Theorem 1.5.4 *Let m and k be coprime integers with $k > 0$ and $k \equiv 0 \pmod 4$. Then*

$$g(m;k) = \left(\frac{k}{m}\right)(1+i^m)\sqrt{k}.$$

Proof Write $k = bc$, where c is odd and b is a power of 2. By Theorem 1.2.6, $g(m;k) = g(mb;c)g(mc;b)$. Appealing to Theorem 1.5.2 and (1.5.5), we deduce that

$$\begin{aligned} g(m;k) &= \left(\frac{mb}{c}\right) i^{(c-1)^2/4}\sqrt{c}\left(\frac{b}{mc}\right)(1+i^{mc})\sqrt{b} \\ &= \left(\frac{m}{c}\right)\left(\frac{b}{m}\right) i^{(c-1)^2/4}(1+i^{mc})\sqrt{k}. \end{aligned} \tag{1.5.8}$$

By the law of quadratic reciprocity for the Jacobi symbol (see (1.2.7)),

$$\left(\frac{m}{c}\right) = \begin{cases} \left(\dfrac{c}{m}\right), & \text{if } c \equiv 1 \pmod 4 \text{ or } m \equiv 1 \pmod 4, \\ -\left(\dfrac{c}{m}\right), & \text{if } c \equiv m \equiv 3 \pmod 4, \end{cases} \tag{1.5.9}$$

since $c > 0$. In each of the four cases $m \equiv \pm 1 \pmod 4$, $c \equiv \pm 1 \pmod 4$, the result follows easily from (1.5.8) and (1.5.9). ■

1.6 Gauss character sums over the ring of integers (mod k)

Let k be an integer > 1. We will evaluate certain Gauss character sums $\tau_k(m, \chi)$ over the ring of integers $\pmod k$. Before defining these sums, we recall a few facts about characters. A Dirichlet character $\chi \pmod k$ is a nonzero map from the ring of integers $\pmod k$ into the complex numbers such that $\chi(ab) = \chi(a)\chi(b)$ and $\chi(a) = 0$ when $\gcd(a, k) > 1$. Let $R = (Z/kZ)^*$ be the group of reduced residues $\pmod k$. For a fixed divisor d of k with $d > 1$, let S_d be the subgroup of R defined by

$$S_d = \{a \in R : a \equiv 1 \pmod d\}.$$

If $\chi(a) = 1$ for each $a \in S_d$, then χ is definable on the quotient group R/S_d. This means that χ is well defined on cosets: $\chi(mS_d) = \chi(m)$ for each integer m coprime with k. In the case that $\chi(a) = 1$ for each $a \in S_d$, we say that χ is definable $\pmod d$.

Since reduction $\pmod d$ is a homomorphism from R onto $(Z/dZ)^*$ (see Problem 33),

$$R/S_d \approx (Z/dZ)^*.$$

Thus, if χ is definable $\pmod d$, there is a Dirichlet character $\psi \pmod d$ defined as follows:

$$\psi(a) = \begin{cases} 0, & \text{if } \gcd(a, d) > 1, \\ \chi(\overline{a}), & \text{if } \gcd(a, d) = 1, \end{cases}$$

where $\overline{a} \in R/S_d$ corresponds to $a \in (Z/dZ)^*$ under the isomorphism above.

If $k = p^r$ for a prime p and a positive integer r, then $\chi(n) = \psi(n)$ for each $n \in Z$. However, if k is not a prime power, χ and ψ do not necessarily have the same values. For example, consider the quartic character $\chi \pmod{15}$ defined by

$$\chi(a) = \begin{cases} 0, & \text{if } \gcd(a, 15) > 1, \\ i^m, & \text{if } a \equiv 2^m \pmod{15}, \\ -i^m, & \text{if } a \equiv -2^m \pmod{15}, \end{cases}$$

where $m = 0$, 1, 2, or 3. Since $\chi(1) = \chi(11) = 1$, χ is definable $\pmod 5$. The corresponding character $\psi \pmod 5$ satisfies $\psi(a) = \chi(a)$ for $a = 0$, 1, 2, and 4, but

$$0 = \chi(3) \neq \psi(3) = \chi(8) = \chi(13) = -i.$$

For a nontrivial Dirichlet character $\chi \pmod k$, the smallest d for which χ is definable $\pmod d$ is called the conductor of χ. If the conductor of this character χ is k, then χ is said to be a primitive character $\pmod{k}$. For example, if $k = p^r$ for an odd prime p and a positive integer r, then the quadratic character $\pmod{p^r}$ has conductor p, and so it is primitive if and only $r = 1$; see Problem 18.

Definition *For integers m and k with $k > 1$ and a Dirichlet character $\chi \pmod{k}$, define the Gauss sum*

$$\tau_k(m, \chi) = \sum_{n=1}^{k-1} \chi(n) e^{2\pi i m n/k}. \tag{1.6.1}$$

Note that in the case that k is a prime p and χ is nontrivial, $\tau_k(m, \chi)$ reduces to the Gauss sum $G(m, \chi)$ defined above Theorem 1.1.3. In the case that χ is the trivial character $\pmod{k}$, $\tau_k(m, \chi)$ is a Ramanujan sum (see Apostol [1, p. 160]). The sums $\tau_k(1, \chi)$ occur in the evaluation of L-functions $L(s, \chi)$ at $s = 1$ (see Koblitz [1, p. 50]).

We proceed to show that the problem of evaluating $\tau_k(m, \chi)$ can be reduced to the case where k is a prime power, χ is primitive $\pmod{k}$, and $m = 1$. Write $k = k_1 \cdots k_t$, where the k_i are pairwise coprime prime powers, and write $K_i = k/k_i$, $1 \le i \le t$. When we write K_i^{-1}, we will mean for the inverse to be taken $\pmod{k_i}$. Given integers $n_i \pmod{k_i}$, the Chinese remainder theorem provides a unique integer $n \pmod{k}$ such that $n \equiv n_i \pmod{k_i}$ for $1 \le i \le t$, namely,

$$n = \sum_{i=1}^{t} K_i K_i^{-1} n_i. \tag{1.6.2}$$

Moreover, $\gcd(n, k) = 1$ if and only if each $\gcd(n_i, k_i)$ equals 1. For each fixed i, given $n_i \pmod{k_i}$, define $b_i \pmod{k}$ by $b_i \equiv n_i \pmod{k_i}$, $b_i \equiv 1 \pmod{K_i}$. For each i, define a Dirichlet character $\chi_i \pmod{k_i}$ by $\chi_i(n_i) = \chi(b_i)$. Note that $\chi(n) = \chi_1(n_1) \cdots \chi_t(n_t)$. Thus, by (1.6.1) and (1.6.2),

$$\tau_k(m, \chi) = \prod_{i=1}^{t} \sum_{n_i=1}^{k_i-1} \chi_i(n_i) \exp(2\pi i m n_i K_i^{-1}/k_i) = \prod_{i=1}^{t} \tau_{k_i}(mK_i^{-1}, \chi_i). \tag{1.6.3}$$

Thus we may restrict our study of $\tau_k(m, \chi)$ to the case where k is a prime power p^r, $r \ge 1$.

Let $k = p^r$. If χ is trivial $\pmod{k}$, it is easily seen from (1.6.1) that

$$\tau_k(m, \chi) = \begin{cases} k - (k/p), & \text{if } p^r | m, \\ -(k/p), & \text{if } p^{r-1} \| m, \\ 0, & \text{otherwise.} \end{cases} \tag{1.6.4}$$

Thus assume that χ is nontrivial $\pmod{k}$. Let f denote the smallest power of p for which χ is a Dirichlet character $\pmod{f}$. Thus f is the conductor of χ, and χ is a primitive character (mod f). We have

$$\tau_k(m,\chi) = \sum_{n=0}^{k-1} \chi(n) e^{2\pi i m n/k} = \sum_{b=0}^{f-1} \sum_{a=0}^{(k/f)-1} \chi(af+b) e^{2\pi i a m f/k} e^{2\pi i b m/k}$$
$$= \sum_{b=0}^{f-1} \chi(b) e^{2\pi i m b/k} \sum_{a=0}^{(k/f)-1} e^{2\pi i m a/(k/f)}.$$

The inner sum vanishes unless $k \mid fm$, and so

(1.6.5) $$\tau_k(m,\chi) = \begin{cases} (k/f)\tau_f(fm/k,\chi), & \text{if } k \mid fm, \\ 0, & \text{otherwise.} \end{cases}$$

This shows that we may restrict our study to sums $\tau_k(m,\chi)$ where χ is a primitive character modulo some prime power $k = p^r$, $r \geq 1$.

By (1.6.1), $\tau_k(m,\chi) = 0$ if $k \mid m$. Assume $k \nmid m$, so we can write $m = p^u h$ with $p \nmid h$ and $0 \leq u < r$. Write $d = p^{r-u}$. Then

(1.6.6) $$\tau_k(m,\chi) = \sum_{n=0}^{k-1} \chi(n) e^{2\pi i h n/d} = \overline{\chi}(h) \sum_{n=0}^{k-1} \chi(n) e^{2\pi i n/d}$$
$$= \overline{\chi}(h) \sum_{b=0}^{d-1} e^{2\pi i b/d} \sum_{a=0}^{p^u-1} \chi(b+ad).$$

The inner sum on a equals

(1.6.7) $$\sum_{\substack{j (\mathrm{mod}\ k) \\ j \equiv b (\mathrm{mod}\ d)}} \chi(j).$$

Suppose that $p \mid m$, i.e., $d < k$. Since χ is primitive (mod k), there exists c (mod k) with $c \equiv 1 \pmod d$ such that $\chi(c) \neq 1$. Replace j by jc in (1.6.7) to see that the sum vanishes when $d < k$. It thus follows from (1.6.6) that $\tau_k(m,\chi) = 0$ when $p \mid m$. If $p \nmid m$, so that $d = k$ and $u = 0$, then (1.6.6) yields

$$\tau_k(m,\chi) = \overline{\chi}(h)\tau_k(1,\chi).$$

Thus, for $k = p^r$ and all m,

$$\tau_k(m,\chi) = \overline{\chi}(m)\tau_k(1,\chi).$$

This equality holds for arbitrary integers $k > 1$; see Problem 16.

In summary, we have reduced the problem of evaluating $\tau_k(m,\chi)$ to the study of $\tau_k(1,\chi)$, where χ is a primitive character modulo a prime power $k = p^r$, $r \geq 1$. For brevity, write

(1.6.8) $$\tau(\chi) = \tau_k(\chi) = \tau_k(1,\chi) = \sum_{n=1}^{k-1} \chi(n) e^{2\pi i n/k}.$$

For the remainder of this chapter, χ will denote a primitive character $\pmod{k}$.

While we cannot in general give an explicit determination of $\tau_k(\chi)$ in the case that k is prime, there is an evaluation of $\tau_k(\chi)$ for $k = p^r$, where $r \geq 2$ and p is an odd prime, due to Odoni [1]. Odoni's evaluation shows that for each such k, $\tau_k(\chi)$ equals $\sqrt{k}$ times a certain root of unity. We show this in Theorem 1.6.2 below.

T. Funakura [1, Section 4] extended Odoni's result to the case $p = 2$. In the case $k = 2^r$, $r \geq 2$, it turns out that $\tau_k(\chi)$ again equals $\sqrt{k}$ times a certain root of unity. This is proved in Theorem 1.6.3.

Odoni's proof (Theorem 1.6.2) and the extension to $p = 2$ (Theorem 1.6.3) require some p-adic analysis. We present another evaluation of $\tau_k(\chi)$ in Theorem 1.6.4, due to Mauclaire [1], [2], which requires no knowledge of p-adic analysis.

If k is an odd prime, it is of course still true that $\tau_k(\chi)$ has absolute value $\sqrt{k}$, but it is no longer true that $\tau_k(\chi)/\sqrt{k}$ is a root of unity, unless χ is the quadratic character, that is, the Legendre symbol $\pmod{k}$. This result was proved by Chowla [10],[11]. We will give a somewhat simpler proof in Theorem 1.6.1 below. Our proof uses the well-known fact that $\pm e^{2\pi ij/m}$ $(0 \leq j < m)$ are the only roots of unity in the cyclotomic field $Q(e^{2\pi i/m})$. For a short elementary proof of this fact, see Theorem 2.1.13.

Theorem 1.6.1 *Let p be an odd prime, and let χ be a character* $\pmod{p}$ *of order $m > 2$. Then the Gauss sum*

$$G(\chi) = \sum_{n=0}^{p-1} \chi(n) e^{2\pi in/p}$$

does not equal $\sqrt{p}$ times a root of unity.

Proof Write $\beta_n = e^{2\pi i/n}$. Choose any primitive root $t \pmod{p}$ such that $t \equiv 1 \pmod{4m^2}$. Define the automorphism σ_t of $Q(\beta_{4m^2p})$ by $\sigma_t(\beta_{4m^2p}) = \beta_{4m^2p}^t$. We have $G(\chi) \in Q(\beta_m)(\beta_p) = Q(\beta_{mp})$. By Theorem 1.1.3, $\sigma_t(G(\chi)) = \overline{\chi}(t)G(\chi)$. Thus σ_t fixes $G(\chi)^m$, so $G(\chi)^m \in Q(\beta_m)$, since $\mathrm{Gal}(Q(\beta_{mp})/Q(\beta_m))$ consists of the $p-1$ automorphisms σ_t^i restricted to $Q(\beta_{mp})$, $1 \leq i \leq p-1$. Assume for the purpose of contradiction that $\eta = G(\chi)/\sqrt{p}$ is a root of unity. Since $\eta^{2m} \in Q(\beta_m)$ and all roots of unity in $Q(\beta_m)$ are $2m$th roots of unity, we have $\eta^{4m^2} = 1$. Thus σ_t fixes η. On the other hand,

$$\sigma_t(\eta) = \frac{\sigma_t(G(\chi))}{\sigma_t(\sqrt{p})} = \frac{\overline{\chi}(t)G(\chi)}{\pm\sqrt{p}} = \pm\overline{\chi}(t)\eta.$$

Thus $\overline{\chi}(t) = \pm 1$, which contradicts the fact that χ has order > 2. ■

We now set the stage for Odoni's evaluation of the Gauss sum $\tau_k(\chi)$ (Theorem 1.6.2). A small amount of p-adic analysis will be required, as found, for example, in Koblitz [1, Chapter 4, §1].

Fix $k = p^r$ where p is an odd prime and $r \geq 2$. Let χ be a primitive character (mod k), and define the Gauss sum $\tau(\chi)$ as in (1.6.8). Write $\beta_n = e^{2\pi i/n}$. Let $R = (Z/kZ)^*$, the group of reduced residues (mod k). By induction on r,

$$(1+p)^{k/p^2} \equiv 1 + k/p \pmod{k}, \quad r \geq 2.$$

Thus $(1+p)^{k/p} \equiv 1 \pmod{k}$, so that $1+p$ is a primitive (k/p)th root of unity in R. It follows that $\chi(1+p) = \beta_{k/p}^a$ for some integer a. If $p \mid a$, then $\chi(1+k/p) = \chi(1+p)^{k/p^2} = 1$, and so, by taking bth powers, $\chi(1+bk/p) = 1$ for all positive integers b. Therefore $p \nmid a$, for otherwise χ would be definable (mod k/p), which contradicts the fact that χ is primitive (mod k). Therefore $\chi(1+p)$ is a primitive (k/p)th root of unity, so χ has order dk/p for some d dividing $(p-1)$ and $\tau(\chi) \in Q(\beta_{dk})$. Define b by $b \equiv 1 \pmod{d}$ and $ab \equiv -1 \pmod{k/p}$. Since $\gcd(b, dk) = 1$, there exists an automorphism σ_b of $Q(\beta_{dk})$ defined by $\sigma_b(\beta_{dk}) = \beta_{dk}^b$. We have $\sigma_b(\tau(\chi)) = \overline{\chi}^b(b)\tau(\chi^b)$, and so $\tau(\chi)$ is easily evaluated once $\tau(\chi^b)$ is known. Thus, without loss of generality, normalize χ by taking $a = -1$, that is, $\chi(1+p) = \beta_{k/p}^{-1}$. Define, for the p-adic logarithm,

$$\lambda = \frac{p}{\log(1+p)}\left(1 - \log\left(\frac{p}{\log(1+p)}\right)\right), \tag{1.6.9}$$

viewed as a p-adic integer. Writing $c = ph$ for some p-adic integer h, we have the p-adic Taylor expansion

$$\log(1+c) = c - \frac{1}{2}c^2 + \frac{1}{3}c^3 + O(p^4), \tag{1.6.10}$$

where the "error term" $O(p^4)$ denotes p^4 times some p-adic integer, and where the Taylor coefficients are viewed as p-adic numbers. It follows from (1.6.9) and (1.6.10) that

$$\lambda = 1 - \frac{1}{8}p^2 + \frac{1}{16}p^3 + O(p^4), \quad \text{if } p > 3. \tag{1.6.11}$$

If $p = 3$, write (1.6.10) in the form

$$\log(1+c) = c + c^2\left(h - \frac{1}{2}\right) + O(p^4)$$

(so that 3 does not occur in the denominators of the coefficients) to deduce that

$$\lambda = 1 + O(p^4), \quad \text{if } p = 3. \tag{1.6.12}$$

By taking additional terms in the expansion (1.6.10), we can, in principle, determine as many terms as we like in the p-adic expansion of λ. We could compute the Taylor expansion of λ directly from (1.6.9) to obtain

$$\lambda = 1 - \frac{1}{8}p^2 + \frac{1}{16}p^3 - \frac{23}{576}p^4 + \frac{11}{384}p^5 - \frac{2281}{103680}p^6 + \cdots, \tag{1.6.13}$$

but one cannot get an accurate error term from (1.6.13) because some Taylor coefficients may have factors of p in their denominators. For example, it is not obvious directly from (1.6.13) that $\lambda = 1 + O(p^4)$ for $p = 3$. One needs the error term $O(p^r)$ for λ in order to evaluate β_k^λ in Theorem 1.6.2.

Theorem 1.6.2 *Let $k = p^r$, where p is an odd prime and $r \geq 2$. Write $\beta_n = e^{2\pi i/n}$. Let χ be a primitive character* (mod k) *normalized such that $\chi(1+p) = \beta_{k/p}^{-1}$. Then*

$$(1.6.14)\quad \tau(\chi) = \sum_{m=0}^{k-1} \chi(m)\beta_k^m = \begin{cases} \sqrt{k}\beta_k, & \text{if } r = 2, \\ \sqrt{k}\beta_k i^{(1-p)/2}\beta_p^{(p^2-1)/8}, & \text{if } r = 3, \\ \sqrt{k}\beta_k^\lambda, & \text{for even } r > 3, \\ \sqrt{k}\beta_k^\lambda i^{(1-p)/2}, & \text{for odd } r > 3, \end{cases}$$

where λ is the p-adic integer in (1.6.9).

Odoni [1] states that the last equality in (1.6.14) holds for $r = 3$. This is true for $p > 3$ but false for $p = 3$; see (1.6.11) and (1.6.12).

Proof Since

$$R = (Z/kZ)^* = \{z(1+k/p)^w : \; 0 \leq z < k/p, \; p \nmid z, \; 0 \leq w < p\},$$

we have

$$\tau(\chi) = \sum_{0 \leq z < k/p} \chi(z)\beta_k^z \sum_{w=0}^{p-1} \chi(1 + wk/p)\beta_p^{wz}.$$

From the normalization of χ, we can conclude that

$$\chi(1 + wk/p) = \chi(1+p)^{wk/p^2} = \beta_p^{-w},$$

so that

$$\tau(\chi) = p \sum_{\substack{0 \leq z < k/p \\ z \equiv 1 (\mathrm{mod}\ p)}} \chi(z)\beta_k^z.$$

Write $z = 1 + sp, \; 0 \leq s < k/p^2$. Then

$$\tau(\chi) = p \sum_{0 \leq s < k/p^2} \chi(1 + sp)\beta_k \beta_{k/p}^s.$$

Now, $1+p$ has order k/p in R, and hence the powers of $1+p$ comprise the k/p elements of R that are congruent to 1 (mod p). In particular, each $1 + sp$ is a power of $1 + p$. That power may be given p-adically by

$$(1.6.15)\qquad 1 + sp = (1+p)^{\log(1+sp)/\log(1+p)}.$$

Formula (1.6.15) follows from a proposition in Koblitz's text [1, p. 81], and it is quite subtle. To prove our point, we remark that (1.6.15) need not hold in the case $p = 2$; see Problem 14.

From (1.6.15),

$$\tau(\chi) = p\beta_k \sum_{0 \le s < k/p^2} \beta_{k/p}^{B(s)}, \tag{1.6.16}$$

where for any p-adic integer u,

$$B(u) = u - \log(1 + up)/\log(1 + p). \tag{1.6.17}$$

Theorem 1.6.2 is now proved for $r = 2$, since for $k = p^2$ the sum in (1.6.16) equals 1.

Using (1.6.10), we see that

$$B(u) = p(u^2 - u)/2 + O(p^2)$$

(valid also for $p = 3$). Thus, for $r = 3$, (1.6.16) yields

$$\tau(\chi) = p\beta_k \sum_{s=0}^{p-1} \beta_p^{(s^2-s)/2}.$$

Completing the square in the exponent, we obtain the desired result for $r = 3$, in view of Theorem 1.5.2.

For the remainder of this proof, let $r \ge 4$. For a p-adic integer $x \ne 0$, define $\nu(x)$ by $p^{\nu(x)} \| x$ (i.e., $\nu(x)$ is the order of p in x). By (1.6.17),

$$B'(u) = 1 - \frac{p}{(1 + pu)\log(1 + p)} \equiv 0 \pmod{p}.$$

Moreover, for each integer $j \ge 2$,

$$\frac{B^{(j)}(u)}{(j-1)!} = \frac{(-p)^j}{(1 + pu)^j \log(1 + p)},$$

so $\nu(B^{(j)}(u)/(j-1)!) = j - 1$. Thus,

$$p^{j-1-\nu(j)} \;\|\; \frac{B^{(j)}(u)}{j!}, \quad j \ge 2.$$

Since $j - 1 - \nu(j) \ge 1$ for $j \ge 2$, and $j - 1 - \nu(j) \to \infty$ as $j \to \infty$, we have the Taylor expansion

$$B(u + p^t v) = B(u) + p^t v B'(u) + \frac{p^{2t} v^2}{2!} B''(u) + O(p^{3t+1}), \tag{1.6.18}$$

for any integer $t \ge 0$ and p-adic integers u and v.

Case 1: r is even. Write $r = 2t + 2$ with $t \geq 1$. In (1.6.16), set $s = u + p^t v$ with $0 \leq u, v < p^t$. Then by (1.6.18),

$$\tau(\chi) = p\beta_k \sum_{u=0}^{p^t-1} \beta_{k/p}^{B(u)} \sum_{v=0}^{p^t-1} \beta_{p^t}^{vB'(u)/p}.$$

Since $t > 0$, the inner sum vanishes unless $B'(u)/p \equiv 0 \pmod{p^t}$, i.e., unless $u \equiv \mu \pmod{p^t}$, where μ is the p-adic integer defined by

$$\mu = \frac{1}{\log(1+p)} - \frac{1}{p}.$$

Note that $B'(\mu) = 0$. We now see that

$$\tau(\chi) = p^{t+1}\beta_k^{1+pB(u)}$$

for the unique integer u satisfying $u \equiv \mu \pmod{p^t}$ with $0 \leq u < p^t$. By (1.6.18), $B(u) \equiv B(\mu) \pmod{p^{2t+1}}$, since $B'(\mu) = 0$. Thus,

$$\tau(\chi) = p^{t+1}\beta_k^{1+pB(\mu)} = \sqrt{k}\beta_k^{\lambda},$$

as desired.

Case 2: r is odd. Write $r = 2t + 1$ with $t \geq 2$. In (1.6.16), set $s = u + p^t v$ with $0 \leq u < p^t$, $0 \leq v < p^{t-1}$. Then by (1.6.18),

$$\tau(\chi) = p\beta_k \sum_{0 \leq u < p^t} \beta_{k/p}^{B(u)} \sum_{0 \leq v < p^{t-1}} \beta_{p^{t-1}}^{vB'(u)/p}.$$

Since $t > 1$, the inner sum vanishes unless $u \equiv \mu \pmod{p^{t-1}}$. Fix one such solution u by requiring $0 \leq u < p^{t-1}$; then $\{u + xp^{t-1} : 0 \leq x < p\}$ is the set of all solutions. Thus

$$\tau(\chi) = p^t\beta_k \sum_{x=0}^{p-1} \beta_{k/p}^{B(u+xp^{t-1})}.$$

Since $u \equiv \mu \pmod{p^{t-1}}$, we have

$$\{u + xp^{t-1} \pmod{p^t} : 0 \leq x < p\} = \{\mu + xp^{t-1} \pmod{p^t} : 0 \leq x < p\}.$$

Also, since $B'(\mu + xp^{t-1}) = B'(\mu + xp^{t-1}) - B'(\mu) = O(p^t)$, it follows from (1.6.18) that for any p-adic integer v,

$$B(\mu + xp^{t-1} + vp^t) \equiv B(\mu + xp^{t-1}) \pmod{p^{2t}},$$

so that

$$\beta_{k/p}^{B(\mu+xp^{t-1}+vp^t)} = \beta_{k/p}^{B(\mu+xp^{t-1})}.$$

Thus,

$$\tau(\chi) = p^t \beta_k \sum_{x=0}^{p-1} \beta_{k/p}^{B(\mu+xp^{t-1})}.$$

By (1.6.18) with $t-1$ in place of t,

$$\beta_{k/p}^{B(\mu+xp^{t-1})} = \beta_{k/p}^{B(\mu)+(p^{2t-1}x^2B''(\mu))/(2p)},$$

since $B'(\mu) = 0$. Thus

$$\tau(\chi) = p^t \beta_k^{1+pB(\mu)} \sum_{x=0}^{p-1} \beta_p^{x^2B''(\mu)/(2p)} = p^t \beta_k^{\lambda} \sum_{x=0}^{p-1} \beta_p^{x^2/2},$$

since $B''(\mu)/p \equiv 1 \pmod p$. The result now follows from Theorem 1.5.2. ∎

Next, we set the stage for the evaluation of the Gauss sum $\tau_k(\chi)$ for $p=2$ (Theorem 1.6.3). Fix $k = 2^r$ with $r \geq 2$. Let χ be a primitive character (mod k) and define the Gauss sum $\tau(\chi)$ as in (1.6.8). Write $\beta_n = e^{2\pi i/n}$. The group of $k/2$ reduced residues (mod k) consists of the $k/4$ distinct powers of 5 (mod k) (namely, the $k/4$ elements $\equiv 1 \pmod 4$) together with their negatives (the $k/4$ elements $\equiv 3 \pmod 4$). It is easily proved by induction on r that

(1.6.19) $$5^{k/8} \equiv 1 + (k/2) \pmod k, \quad \text{for } r \geq 3,$$

and

(1.6.20) $$5^{k/16} \equiv 1 - (k/4) \pmod k, \quad \text{for } r \geq 5.$$

Since the order of 5 (mod k) is $k/4$, $\chi(5) = \beta_{k/4}^a$ for some integer a. If a were even with $r \geq 3$, then, by (1.6.19), $\chi(1+k/2) = 1$, and so χ would be a character (mod $k/2$), contradicting the primitivity of χ. Thus we may assume that a is odd. Without loss of generality we may assume that $a=1$, i.e., $\chi(5) = \beta_{k/4}$; otherwise, apply an appropriate automorphism as we did in the case of odd k. Define, for the 2-adic logarithm,

(1.6.21) $$\lambda = \frac{-4}{\log 5}\left(1 - \log\left(\frac{-4}{\log 5}\right)\right),$$

viewed as a 2-adic integer. For any 2-adic integer h,

(1.6.22) $$\log(1+4h) = 4\left(h - 2h^2 + 16h^3\left(\frac{1}{3} - h\right) + \frac{256}{5}h^5 + O(2^9)\right).$$

In particular,

(1.6.23) $$\log 5 = 4\left(-1 - \frac{32}{3} + \frac{256}{5} + O(2^9)\right).$$

By (1.6.21)–(1.6.23), λ has the 2-adic expansion

(1.6.24) $$\lambda = 1 + 2^9 - 2^{10} + O(2^{11}).$$

Theorem 1.6.3 *Let* $k = 2^r$, $r \geq 2$. *Write* $\beta_n = e^{2\pi i/n}$. *Let* χ *be a primitive character* (mod k) *normalized such that* $\chi(5) = \beta_{k/4}$. *Then*

(1.6.25)
$$\tau(\chi) = \sum_{m=0}^{k-1} \chi(m)\beta_k^m = \begin{cases} i\sqrt{k}, & \text{if } r = 2, \\ \sqrt{k}\beta_8^{(1-\chi(-1))}, & \text{if } r = 3, \\ \sqrt{k}\chi(-1)\beta_{16}^{-1}, & \text{if } r = 4, \\ \sqrt{k}\beta_k^{\lambda}\beta_8, & \text{if } r \geq 5, \end{cases}$$

where λ *is the 2-adic integer in* (1.6.21).

Proof The result is easily checked for $r < 5$, so let $r \geq 5$. We have $\tau(\chi) = A + \chi(-1)B$, where

$$A = \sum_{0 \leq s < k/4} \chi(1+4s)\beta_k^{1+4s}, \quad B = \sum_{0 \leq s < k/4} \chi(1+4s)\beta_k^{-1-4s}.$$

When s is replaced by $s - (k/16)$, $1+4s \pmod k$ becomes $(1+4s)(1-(k/4)) \pmod k$. By (1.6.20), $\chi(1-(k/4)) = i$. Thus the replacement of s by $s - (k/16)$ leaves the s-summand of A unchanged while negating the s-summand of B. It follows that $B = 0$ and

$$\tau(\chi) = 4\beta_k \sum_{0 \leq s < k/16} \chi(1+4s)\beta_{k/4}^s.$$

From this, the result is easily verified for $r = 5$, and so let $r \geq 6$. From Koblitz' text [1, Proposition, p. 81],

$$1 + 4s = 5^{\log(1+4s)/\log 5}.$$

As $\chi(5) = \beta_{k/4}$,

(1.6.26)
$$\tau(\chi) = 4\beta_k \sum_{0 \leq s < k/16} \beta_{k/4}^{B(s)},$$

where for any 2-adic integer u,

(1.6.27)
$$B(u) = u + \log(1+4u)/\log 5.$$

For a 2-adic integer $x \neq 0$, let $\nu(x)$ denote the order of 2 in x. By (1.6.27),

$$B'(u) = 1 + \frac{4}{(1+4u)\log 5} \equiv 0 \pmod 4.$$

Moreover, for each integer $j \geq 2$,

$$\frac{B^{(j)}(u)}{(j-1)!} = \frac{(-1)^{j+1}4^j}{(1+4u)^j \log 5},$$

and so

$$2^{2j-2-\nu(j)} \;||\; \frac{B^{(j)}(u)}{j!}, \quad j \geq 2.$$

In particular, $2 \;||\; B''(2)/2$. Since $2j-2-\nu(j) \geq 4$ for $j \geq 3$ and $2j-2-\nu(j) \to \infty$ as $j \to \infty$, we have the Taylor expansion

$$(1.6.28) \qquad B(u+2^t v) = B(u) + 2^t v B'(u) + 2^{2t} v^2 \frac{B''(u)}{2!} + O(2^{3t+4}),$$

for any integer $t \geq 0$ and 2-adic integers u and v.

Case 1: r is odd. Write $r = 2t+3$ with $t \geq 2$. In (1.6.26), set $s = u + 2^t v$, with $0 \leq u < 2^t$ and $0 \leq v < 2^{t-1}$. Then by (1.6.28),

$$\tau(\chi) = 4\beta_k \sum_{0 \leq u < 2^t} \beta_{k/4}^{B(u)} \sum_{0 \leq v < 2^{t-1}} \beta_{2^{t-1}}^{vB'(u)/4}.$$

Since $t > 1$, the inner sum vanishes unless $B'(u)/4 \equiv 0 \pmod{2^{t-1}}$. This congruence is equivalent to $u \equiv \mu \pmod{2^{t-1}}$, where μ is the 2-adic integer defined by

$$\mu = -\frac{1}{4} - \frac{1}{\log 5}.$$

Note that $B'(\mu) = 0$. We now see that

$$\tau(\chi) = 2^{t+1}\beta_k \sum_{x=0}^{1} \beta_{k/4}^{B(u+x2^{t-1})}$$

for the unique integer u with $0 \leq u < 2^{t-1}$ and $u \equiv \mu \pmod{2^{t-1}}$. For each $x = 0, 1$, there is an integer $y \in \{0, 1\}$ and a 2-adic integer v such that $u + x2^{t-1} = \mu + y2^{t-1} + v2^t$. Since

$$B'(\mu + y2^{t-1}) = B'(\mu + y2^{t-1}) - B'(\mu) = O(2^{t+1}),$$

it follows from (1.6.28) that

$$B(\mu + y2^{t-1} + v2^t) \equiv B(\mu + y2^{t-1}) \pmod{2^{2t+1}}.$$

Thus,

$$\tau(\chi) = 2^{t+1}\beta_k \sum_{x=0}^{1} \beta_{k/4}^{B(\mu+x2^{t-1})}.$$

Then, by (1.6.28) with $v = 1$ and $t-1$ in place of t,

$$\tau(\chi) = 2^{t+1}\beta_k \beta_{k/4}^{B(\mu)} (1 + \beta_8^{B''(\mu)/2}),$$

since $B'(\mu) = 0$. It is easily checked that $B''(\mu)/2 = 2 + O(2^4)$, and so

$$\tau(\chi) = 2^{t+1}\beta_k^{1+4B(\mu)}(1+i) = \sqrt{k}\beta_k^{\lambda}\beta_8,$$

as desired.

Case 2: r is even. Write $r = 2t+4$ with $t \geq 1$. In (1.6.26), set $s = u + 2^t v$ with $0 \leq u, v < 2^t$. Then by (1.6.28),

$$\tau(\chi) = 4\beta_k \sum_{0 \leq u < 2^t} \beta_{k/4}^{B(u)} \sum_{0 \leq v < 2^t} \beta_{2^t}^{v(2^{t-1}+B'(u)/4)},$$

since $v^2 B''(u)/2 \equiv 2v \pmod 4$. Since $t > 0$, the inner sum vanishes unless $B'(u)/4 \equiv 2^{t-1} \pmod{2^t}$, i.e., unless $u \equiv \mu + 2^{t-1} \pmod{2^t}$. Thus

$$\tau(\chi) = 2^{t+2}\beta_k^{1+4B(u)}$$

for the unique u with $0 \leq u < 2^t$ and $u \equiv \mu + 2^{t-1} \pmod{2^t}$. Write $u = \mu + 2^{t-1} + v2^t$. By (1.6.28),

$$B(u) = B(\mu + 2^{t-1}) + 2^t v B'(\mu + 2^{t-1}) + 2^{2t} v^2 \frac{B''(\mu + 2^{t-1})}{2} + O(2^{3t+4}).$$

Now,

$$B'(\mu + 2^{t-1}) = B'(\mu + 2^{t-1}) - B'(\mu) = \frac{-2^{t+1}(4/\log 5)}{(1 + 4\mu + 2^{t+1})(1 + 4\mu)} \equiv 2^{t+1} \pmod{2^{t+2}}$$

and

$$v^2 B''(\mu + 2^{t-1})/2 \equiv 2v \pmod 4.$$

Therefore,

$$B(u) \equiv B(\mu + 2^{t-1}) \pmod{2^{2t+2}},$$

and so

$$\tau(\chi) = 2^{t+2}\beta_k^{1+4B(\mu+2^{t-1})}.$$

By (1.6.28) with $v = 1$ and $t-1$ in place of t,

$$B(\mu + 2^{t-1}) = B(\mu) + 2^{2t-2}B''(\mu)/2 + O(2^{3t+1}).$$

Since $B''(\mu)/2 \equiv 2 \pmod{16}$,

$$\tau(\chi) = 2^{t+2}\beta_k^{1+4B(\mu)}\beta_8 = \sqrt{k}\beta_k^{\lambda}\beta_8,$$

as desired. ■

Finally, we give Mauclaire's evaluation of the Gauss sum $\tau_k(\chi)$ (Theorem 1.6.4). Fix $k = p^r$, where p is a prime and $r \geq 2$. Let χ be a primitive character (mod k) and define the Gauss sum $\tau(\chi)$ as in (1.6.8). Write $\beta_n = e^{2\pi i/n}$.

We assumed without loss of generality in Theorems 1.6.2 and 1.6.3 that χ had a particular normalization. Similarly, we can assume here without loss of generality that

$$\chi(1+p^s) = \beta_{p^s}^{-1}, \quad \text{for even } r = 2s, \tag{1.6.29}$$

and

$$\chi(1+p^s+\frac{1}{2}p^{2s}) = \beta_{p^{s+1}}^{-1}, \quad \text{for odd } r = 2s+1 \text{ and } k \neq 8. \tag{1.6.30}$$

(If $p > 2$, then $\frac{1}{2}$ is interpreted as the integer $(p+1)/2$.) The normalization of χ in (1.6.29) and (1.6.30) is more complicated than that in Theorems 1.6.2 and 1.6.3. However, with this normalization, we give in Theorem 1.6.4 a much simpler evaluation of $\tau(\chi)$ that requires no p-adic analysis. Theorem 1.6.4 excludes the case $k = 8$; see Theorem 1.6.3 for the evaluation of $\tau(\chi)$ when $k = 8$.

Theorem 1.6.4 *Let $k = p^r \neq 8$, where p is a prime and $r \geq 2$. Write $\beta_n = e^{2\pi i/n}$. Let χ be a primitive character* (mod k), *normalized as in* (1.6.29) *and* (1.6.30). *Then*

$$\tau(\chi) = \sum_{m=0}^{k-1} \chi(m)\beta_k^m = \begin{cases} \sqrt{k}\beta_k, & \text{if } r \text{ is even}, \\ \sqrt{k}\beta_k\beta_8^{1-p}, & \text{if } p > 2 \text{ and } r \text{ is odd}, \\ \sqrt{k}\beta_k\beta_8, & \text{if } p = 2,\ r > 3, \text{ and } r \text{ is odd}. \end{cases}$$

Proof First we consider even $r = 2s$. Let w and z denote nonnegative integers. By the binomial theorem,

$$(1+p^s)^w \equiv 1 + wp^s \pmod{k},$$

so that

$$\{z(1+p^s)^w : 0 \leq z < p^s, p \nmid z, \text{ and } 0 \leq w < p^s\}$$

is a reduced residue system (mod k). Thus,

$$\tau(\chi) = \sum_{0 \leq z < p^s} \chi(z)\beta_k^z \sum_{0 \leq w < p^s} \chi^w(1+p^s)\beta_{p^s}^{zw}.$$

Then by (1.6.29),

$$\tau(\chi) = \sum_{0 \leq z < p^s} \chi(z)\beta_k^z \sum_{0 \leq w < p^s} \beta_{p^s}^{w(z-1)} = p^s\beta_k,$$

since the inner sum on w vanishes unless $z = 1$.

Now we consider odd $r = 2s + 1$ with $k \neq 8$. Since

$$\{z(1 + wp^s) : 0 \leq z < p^s, p \nmid z, \text{ and } 0 \leq w < p^{s+1}\}$$

is a reduced residue system $\pmod{k}$,

$$\tau(\chi) = \sum_{0 \leq z < p^s} \chi(z)\beta_k^z \sum_{0 \leq w < p^{s+1}} \chi(1 + wp^s)\beta_{p^{s+1}}^{zw}.$$

By the binomial theorem,

$$1 + wp^s \equiv (1 + p^s + \frac{1}{2}p^{2s})^{w(1-\frac{1}{2}wp^s)} \pmod{k}.$$

Then, by (1.6.30),

$$\chi(1 + wp^s) = \beta_{p^{s+1}}^{\frac{1}{2}w^2p^s - w},$$

and so

$$\begin{aligned}\tau(\chi) &= \sum_{0 \leq z < p^s} \chi(z)\beta_k^z \sum_{0 \leq w < p^{s+1}} \beta_{p^{s+1}}^{\frac{1}{2}w^2p^s + w(z-1)} \\ &= \sum_{0 \leq z < p^s} \chi(z)\beta_k^z \sum_{0 \leq x < p^s} \beta_{p^{s+1}}^{x(z-1)+\frac{1}{2}x^2p^s} \sum_{0 \leq y < p} \beta_p^{y(z-1)},\end{aligned}$$

where we have written $w = yp^s + x$. The inner sum on y vanishes unless $z = 1$, so that

$$\tau(\chi) = p\beta_k \sum_{0 \leq x < p^s} \beta_{p^{s+1}}^{\frac{1}{2}x^2p^s} = \begin{cases} p\beta_k \sum_{0 \leq x < p^s} \beta_p^{\frac{1}{2}x^2} = p^s\beta_k \sum_{x=0}^{p-1} \beta_p^{\frac{1}{2}x^2}, & \text{if } p > 2, \\ p\beta_k \sum_{0 \leq x < p^s} i^{x^2} = p^s\beta_k(1 + i), & \text{if } p = 2. \end{cases}$$

When $p = 2$,

$$\tau(\chi) = p^{s+\frac{1}{2}}\beta_k(1 + i)/\sqrt{2} = \sqrt{k}\beta_k\beta_8.$$

When $p > 2$, Theorem 1.5.2 gives

$$\sum_{x=0}^{p-1} \beta_p^{\frac{1}{2}x^2} = \left(\frac{2}{p}\right) i^{(p-1)^2/4}\sqrt{p} = \beta_8^{1-p}\sqrt{p},$$

so that

$$\tau(\chi) = \sqrt{k}\beta_k\beta_8^{1-p}. \quad \blacksquare$$

Exercises 1

1. If χ is a character on F_q, show that

$$\sum_{\beta} G_r(\beta, \chi) = \begin{cases} 0, & \text{if } \chi \text{ is nontrivial,} \\ q, & \text{if } \chi \text{ is trivial.} \end{cases}$$

2. Let χ and ψ be nontrivial characters on F_q, and let $\delta_1,\ \delta_2 \in F_q$. Prove that

$$G_r(\overline{\chi}) \sum_{\alpha} \chi(\alpha)\psi(\alpha - \delta_1)e(\alpha\delta_2) = G_r(\psi) \sum_{\alpha} \overline{\psi}(\alpha)\overline{\chi}(\alpha - \delta_2)e(\alpha\delta_1).$$

3. (a) Let k and n be integers with $k > 0$ and $k \not| \, n$. Prove that

$$\sum_{j=1}^{k-1} j \sin(2\pi jn/k) = -\frac{1}{2}k \cot(\pi n/k).$$

(b) Let χ be an odd character (mod p), that is, $\chi(-1) = -1$. Using part (a), prove that

$$\sum_{n} n\chi(n) = i\frac{1}{2}G(\chi) \sum_{n} \overline{\chi}(n) \cot(\pi n/p).$$

(c) For a prime $p > 2$, prove that

$$\sum_{x=1}^{p-1} \cot\left(\frac{\pi x^2}{p}\right) = -\frac{2}{\sqrt{p}} \sum_{n=1}^{p-1} n \left(\frac{n}{p}\right),$$

where $(\frac{n}{p})$ denotes the Legendre symbol.

(d) For a prime $p \equiv 3 \pmod 4$, prove that

$$\sum_{x=1}^{p-1} \cot\left(\frac{2\pi x^2}{p}\right) = -\frac{2}{\sqrt{p}} \sum_{\substack{n=1 \\ n \text{ odd}}}^{p-1} n \left(\frac{n}{p}\right)$$

and that

$$\prod_{x=1}^{(p-1)/2} \cot\left(\frac{2\pi x^2}{p}\right) = \left(\frac{2}{p}\right) (-1)^N p^{-1/2},$$

where N is the number of quadratic nonresidues (mod p) among $1, 2, \ldots, (p-1)/2$.

4. Let a and m be integers with $m > 0$. Recall from (1.1.5) that

$$g(a; m) = \sum_{n (\bmod m)} e^{2\pi i a n^2/m}.$$

(a) Let $m = m_1 m_2 \cdots m_k$, where the m_i are pairwise coprime positive integers. Let $M_i = m/m_i$ $(1 \le i \le k)$. Prove from first principles that for any integer a,

$$g(a; m) = g(aM_1; m_1) \cdots g(aM_k; m_k).$$

(b) Part (a) shows that in the study of $g(a; m)$ we may restrict ourselves to the cases when m is a prime power. Let p be an odd prime, a an integer such that $a \not\equiv 0 \pmod{p}$, and b an integer with $b \ge 2$. Prove, from first principles, the following reduction property:

$$g(a; p^b) = pg(a; p^{b-2}) = \begin{cases} p^{b/2}, & \text{if } b \text{ is even,} \\ p^{(b-1)/2} g(a; p), & \text{if } b \text{ is odd.} \end{cases}$$

(For the analogue in the case $p = 2$, see Proposition 1.5.3.)

5. Let c be a positive integer. Prove that

$$\sum_{n=0}^{c-1} e^{\pi i n^2/c} = e^{\pi i/4}\sqrt{c}, \qquad \text{if } c \text{ is even,}$$

and

$$\sum_{n=0}^{c-1} (-1)^n e^{\pi i n^2/c} = e^{\pi i(1-c)/4}\sqrt{c}, \qquad \text{if } c \text{ is odd.}$$

6. Using Lemma 1.2.5 and Theorems 1.5.2 and 1.5.4, verify the following supplement to the law of quadratic reciprocity:

$$\left(\frac{2}{p}\right) = (-1)^{(p^2-1)/8},$$

where p is an odd prime. Also deduce this supplement from Theorem 1.2.2.

7. Let m and n be positive odd integers. Prove that

(a) $$\sum_{j=0}^{m-1} (-1)^j e^{(2j+1)^2 \pi i n/(4m)} = \frac{e^{\pi i/4}}{\sqrt{mn}} \sum_{j=1}^{mn} e^{-\pi i j^2/(mn)} \sec(\pi j/m),$$

(b) $$\sum_{j=-(m-1)/2}^{(m-1)/2} (-1)^j e^{-2\pi i j^2 n/m} = \left(\frac{-1}{n}\right) i^{(mn-1)^2/4} \frac{1}{\sqrt{mn}} \sum_{j=1}^{mn} e^{2\pi i j^2/(mn)} \sec(2\pi j/m).$$

Before stating the next problem, we first recall the Poisson summation formula. Let f be a continuous function of bounded variation on $[a, b]$. Then

$$\sum_{a \le n \le b}{}' f(n) = \int_a^b f(x)dx + 2\sum_{n=1}^{\infty} \int_a^b f(x)\cos(2\pi n x)dx,$$

where the prime (′) on the summation sign indicates that if a or b is an integer, then only $\frac{1}{2}f(a)$ or $\frac{1}{2}f(b)$, respectively, is counted.

8. Use the Poisson summation formula to prove Corollary 1.2.3.

9. For Re $z > 0$, the theta function $\theta(z)$ is defined by

$$\theta(z) = \sum_{n=-\infty}^{\infty} e^{-\pi n^2 z}.$$

Let a and c be positive, coprime integers of opposite parity. Using the well-known transformation formula for $\theta(z)$, namely,

$$\theta(1/z) = \sqrt{z}\theta(z),$$

prove the following special case of Theorem 1.2.2:

$$\sqrt{a}\sum_{n=0}^{c-1} e^{\pi i a n^2/c} = e^{\pi i/4}\sqrt{c}\sum_{n=0}^{a-1} e^{-\pi i c n^2/a}.$$

Hint: Let $z = \epsilon + (ic/a)$, $\epsilon > 0$, and $n = ma + r$, $-\infty < m < \infty$, $0 \le r \le a-1$, in the definition of $\theta(z)$. Approximate the sum over m by an integral. Let ϵ tend to 0. Proceed similarly with $\theta(1/z)$.

10. Let a and c be positive, coprime integers. Prove that

(a) $$\sqrt{a}\sum_{n=0}^{c-1} e^{\pi i a n^2/c} - e^{\pi i/4}\sqrt{c}\sum_{n=0}^{a-1} e^{-\pi i c n^2/a} = \frac{1-(-1)^{ac}}{2}(\sqrt{a} - e^{\pi i/4}\sqrt{c}),$$

(b) $$\sqrt{a}\sum_{n=0}^{c-1} e^{(2n+1)^2\pi i a/(4c)} - e^{\pi i/4}\sqrt{c}\sum_{n=0}^{a-1}(-1)^n e^{-\pi i c n^2/a} = -\frac{(1-(-1)^{(c+1)a})}{2}e^{\pi i/4}\sqrt{c}.$$

11. Let n be an odd positive integer. Prove that

$$\sum_{j=1}^{n-1} \tan(\pi j^2/n) = -2 \sum_{1\le j<n/2} \operatorname{Im}(g(2j;n)).$$

12. Let M, N, and s be integers with $N > 0$ and $\gcd(M, 2N) = 1$. Write $\rho = \exp(\pi i M/N)$. Use the evaluations of Gauss sums in Section 1.5 to show that

$$\sum_{k=0}^{N-1}(-1)^k \rho^{k^2+2ks} = \begin{cases} \rho^{-(s+N/2)^2}\left(\dfrac{2N}{M}\right)(1+i^M)(N/2)^{1/2}, & \text{if } 2 \mid N, \\ \rho^{-s^2-sN}\left(\dfrac{2M}{N}\right)N^{1/2}, & \text{if } N \equiv 1 \pmod 4, \\ \rho^{-s^2-sN}\left(\dfrac{2M}{N}\right)iN^{1/2}, & \text{if } N \equiv 3 \pmod 4, \end{cases}$$

where $\left(\frac{a}{b}\right)$ denotes the Jacobi symbol.

13. Show that the Gauss sum $\tau(\chi)$ of modulus $k = p^4$ in (1.6.14) is given by

$$\tau(\chi) = \begin{cases} 9\beta_{81}, & \text{if } p = 3, \\ p^2 \beta_{p^4} \beta_{p^2}^{(p^4-1)(2-p)/16}, & \text{if } p > 3. \end{cases}$$

14. Evaluate the 2-adic number $3^{\log 5/\log 3}$, where log denotes the 2-adic logarithm.
Hint: The surprising answer is -5 (instead of 5). This shows that (1.6.15) is false for $p = s = 2$.

15. Show that the Gauss sum $\tau(\chi)$ of modulus $k = 2^r$ in (1.6.25) has the value $\sqrt{k}\beta_k^m$, where $m = 5, 9, 17, 33, 65, 641, 1793$, and 1 for $r = 5, 6, 7, 8, 9, 10, 11$, and 12, respectively.

16. In the notation of (1.6.1), show that $\tau_k(m, \chi) = \overline{\chi}(m)\tau_k(1, \chi)$ for each primitive character $\chi \pmod{k}$, where k is an arbitrary integer > 1.
Hint: Use (1.6.3) to show that $\tau_k(m, \chi) = 0$ when $\gcd(m, k) > 1$.

17. In the notation of (1.6.3), show that the character $\chi = \chi_1 \cdots \chi_t$ is primitive if and only if each character χ_i $(1 \le i \le t)$ is primitive. More generally, show that χ has conductor $f_1 \cdots f_t$, where f_i is the conductor of χ_i $(1 \le i \le t)$.

18. Let $k = p^r$, where p is an odd prime and $r \ge 1$. Show that if χ is a primitive quadratic character $\pmod{k}$, then $k = p$ and $\chi(m) = \left(\frac{m}{p}\right)$.

19. Let $k = 2^r$, where $r \ge 1$. Show that if χ is a primitive quadratic character (mod k), then either $k = 4$ and $\chi(3) = -1$ or else $k = 8$ and χ is one of the two quadratic characters (mod 8) with $\chi(5) = -1$.

20. Let χ be a primitive quadratic character $\pmod{k}$, where $k > 2$. In the notation of (1.6.3), show that

$$\prod_{i=1}^{t} \chi_i(K_i) = \begin{cases} (-1)^{(N-1)/2}, & \text{if } N \text{ is odd}, \\ (-1)^{N/2}, & \text{if } N \text{ is even}, \end{cases}$$

where N denotes the number of χ_i $(1 \le i \le t)$ with $\chi_i(-1) = -1$.
Hint: Use the law of quadratic reciprocity (1.2.7) and Problems 17–19.

21. Let χ be a primitive quadratic character $\pmod{k}$, where $k > 2$, and let $\tau(\chi)$ be the Gauss sum defined in (1.6.8). Show that

$$\tau(\chi) = \begin{cases} \sqrt{k}, & \text{if } \chi(-1) = 1, \\ i\sqrt{k}, & \text{if } \chi(-1) = -1. \end{cases}$$

Hint: Use (1.6.3), Theorem 1.2.4, and Problems 16 and 20.

22. Let $N(r,p,n)$ denote the number of solutions to the congruence

$$x_1^2 + \cdots + x_r^2 \equiv 0 \pmod{p^n}$$

in integers $x_1, \ldots, x_r \pmod{p^n}$, where p is prime and r and n are positive integers. Prove that, for $p > 2$,

$$N(r,p,1) = p^{r-1}, \text{ if } 2 \nmid r,$$

$$N(r,p,n) = p^{n(r-1)}\left\{1 + \frac{p^{(2-r)[n/2]-1}(p-1)(p^{(r-2)[n/2]}-1)}{p^{r-2}-1}\right\}, \text{ if } n > 1,\ 2 \nmid r,$$

$$N(2,p,n) = p^n + p^{n-1}(p-1)n, \text{ if } p \equiv 1 \pmod 4,$$

and for all other cases with $2 \mid r$,

$$N(r,p,n) = p^{n(r-1)}\left\{1 + \frac{p^{(1-r/2)n-1}\epsilon^n(p-1)(1-\epsilon^n p^{(-1+r/2)n})}{1-\epsilon p^{-1+r/2}}\right\},$$

where $\epsilon = \left(\frac{-1}{p}\right)^{r/2}$.

Also prove that, for $p = 2$,

$$N(r,2,n) = 2^{n(r-1)} + \left(\frac{2}{r}\right)2^{n(r-1)-(r-2)[n/2]+(r-3)/2}\frac{(2^{(r-2)[n/2]}-1)}{2^{r-2}-1}, \text{ if } 2 \nmid r,$$

$$N(r,2,n) = 2^{n(r-1)}, \text{ if } r \equiv 2 \pmod 4,$$

and

$$N(r,2,n) = 2^{n(r-1)} + (-1)^{r/4}2^{-1+r(n+1)/2}\frac{(2^{(-1+r/2)(n-1)}-1)}{2^{(-1+r/2)}-1}, \text{ if } r \equiv 0 \pmod 4.$$

Hint: Note that

$$p^n N(r,p,n) = \sum_{m=0}^{p^n-1} \sum_{x_1,\ldots,x_r=0}^{p^n-1} e^{2\pi i m(x_1^2+\cdots+x_r^2)/p^n};$$

now apply Theorems 1.5.2 and 1.5.4.

23. Let c be a positive integer, and let m be an integer coprime with c. Prove that

$$\sum_{n=0}^{c-1} \exp(2\pi i m(n+1/2)^2/c) = \begin{cases} 0, & \text{if } c \equiv 0 \pmod 4, \\ i^m \left(\frac{m}{c}\right)\sqrt{c}, & \text{if } c \equiv 1 \pmod 4, \\ (1+i^m)\left(\frac{c}{m}\right)\sqrt{c}, & \text{if } c \equiv 2 \pmod 4, \\ i^{1-m}\left(\frac{m}{c}\right)\sqrt{c}, & \text{if } c \equiv 3 \pmod 4. \end{cases}$$

Hint: If c is odd, replace $(n+1/2)^2$ by $(n+(c+1)/2)^2 - c(c+2)/4$. If $c = 2b$ with b odd, change the sum to twice a sum from $n = 0$ to $n = b-1$, and then replace n by $2n$.

24. Let m and k be positive coprime integers with k odd and squarefree. Prove that

$$g(m;k) = \sum_{n=0}^{k-1} e^{2\pi i m n^2/k} = \sum_{n=0}^{k-1} \left(\frac{n}{k}\right) e^{2\pi i m n/k},$$

which generalizes Theorem 1.1.5.

25. Let $\beta = e^{2\pi i/k}$, where k is an integer > 2. Find all quadratic subfields of $Q(\beta)$.
Hint: There are $N_k - 1$ such subfields, where N_k is a certain power of 2. Use Theorem 1.2.4.

26. Let m and k be coprime integers with k odd, $k > 1$. Prove that $|g(m;k)|^2 = k$ by making the change of variable $s \to s+t$ in the formula

$$|g(m;k)|^2 = \sum_{s=0}^{k-1}\sum_{t=0}^{k-1} \exp(2\pi i m(s^2-t^2)/k).$$

27. Let $\beta = e^{2\pi i/k}$, where k is an odd integer > 1. Let M be the $k \times k$ matrix with entry $\beta^{xy}/\sqrt{k}$ in row x, column y $(0 \le x,\ y \le k-1)$. Prove that M^2 is the permutation matrix

$$M^2 = \begin{pmatrix} 1 & 0 & 0 & 0 & \cdots & 0 \\ 0 & 0 & 0 & 0 & \cdots & 1 \\ \vdots & \vdots & \vdots & \vdots & \vdots & \vdots \\ 0 & 0 & 1 & 0 & \cdots & 0 \\ 0 & 1 & 0 & 0 & \cdots & 0 \end{pmatrix}.$$

Deduce that M^4 is the identity matrix, so that each eigenvalue of M is in $\{\pm 1, \pm i\}$. Then use the fact that $\mathrm{tr}(M^2) = 1$ to show that M^2 has eigenvalues 1 and -1 with multiplicities $(k+1)/2$ and $(k-1)/2$, respectively.

28. In the notation of Problem 27, let m_j denote the multiplicity of the eigenvalue i^j of M $(j = 0, 1, 2, 3)$. Show that

$$g(1;k)/\sqrt{k} = \mathrm{tr}(M) = (m_0 - m_2) + i(m_1 - m_3).$$

Use the facts that (see Problem 26)

$$\{|m_0 - m_2|,\ |m_1 - m_3|\} = \{0,\ 1\}$$

and $m_0 + m_2 = (k+1)/2$, $m_1 + m_3 = (k-1)/2$ (Problem 27) to conclude that for $k = 4c+1$, $m_1 = m_3 = c$ and either

$$\text{(i) } m_0 = c+1,\ m_2 = c \quad \text{or} \quad \text{(ii) } m_0 = c,\ m_2 = c+1,$$

while for $k = 4c+3$, $m_0 = m_2 = c+1$ and either

$$\text{(i) } m_1 = c+1,\ m_3 = c \quad \text{or} \quad \text{(ii) } m_1 = c,\ m_3 = c+1.$$

(In fact, (i) holds in each case; see Problem 29.)

29. Show that for each k,

$$\det M = i^{k(k-1)/2} \quad \text{or} \quad -i^{k(k-1)/2},$$

according as (i) or (ii) holds in Problem 28. Using the Vandermonde determinant formula, show that

$$\det M = k^{-k/2} \prod_{0 \le r < s \le k-1} (\beta^s - \beta^r) = k^{-k/2} AB,$$

where

$$A = \prod_{0 \le r < s \le k-1} i\beta^{(s+r)/2} = i^{k(k-1)/2}$$

and

$$B = \prod_{0 \le r < s \le k-1} 2\sin(\pi(s-r)/k) > 0.$$

Conclude that $B = k^{k/2}$ and that (i) holds in each case of Problem 28, so that

$$g(1;k) = \begin{cases} \sqrt{k}, & \text{if } k = 4c+1, \\ i\sqrt{k}, & \text{if } k = 4c+3. \end{cases}$$

30. Let m and k be coprime integers with k odd, $k > 1$. Prove that

$$\prod_{0 \le r < s \le k-1} 2\sin(\pi m(s-r)/k) = \left(\frac{m}{k}\right) k^{k/2}.$$

31. Let $q = p^r$ for a prime p. Prove that

$$\sum_{\chi} G_r(\chi) = 1 + (q-1)e^{2\pi i r/p},$$

where the sum is taken over all characters χ on F_q.

32. Let $q = p^r$ for a prime p. Let χ be a character of order k on F_q. Prove that

$$\sum_{\alpha \in F_q} e(\alpha^k) = \sum_{m=1}^{k-1} G_r(\chi^m) = 1 + k \sum_{\substack{\alpha \in F_q \\ \chi(\alpha) = 1}} e(\alpha).$$

33. Let d and k be positive integers with $d \mid k$. Prove that reduction $\pmod d$ maps $(Z/kZ)^*$ onto $(Z/dZ)^*$.

Hint: Let N be the product of those primes dividing k but not d. Given an integer c coprime with d, show that c has preimage $c + dj$, where j is chosen such that $c + dj \equiv 1 \pmod N$.

34. Let χ be a primitive Dirichlet character $\pmod k$, where $k > 1$. Prove that the Gauss sum

$$\tau(\chi) = \sum_{n=1}^{k} \chi(n)e^{2\pi i n/k}$$

has modulus $\sqrt{k}$.

Hint: Replace n by nm in the formula

$$|\tau(\chi)|^2 = \sum_{\substack{m=0 \\ \gcd(m,k)=1}}^{k-1} \sum_{n=1}^{k-1} \chi(n)\overline{\chi}(m)e^{2\pi i n/k}e^{-2\pi i m/k}$$

to obtain

$$|\tau(\chi)|^2 = \sum_{\substack{m=0 \\ \gcd(m,k)=1}}^{k-1} e^{-2\pi i m/k} \sum_{n=1}^{k-1} \chi(n)e^{2\pi i m n/k}.$$

Use Problem 16 to show that the condition $\gcd(m,k) = 1$ above can be dropped, and note that the sum over m then vanishes unless $n = 1$.

35. Let χ be a Dirichlet character $\pmod{p^r}$ of order $m > 1$, where p is prime and $r \geq 1$.

(a) If $p > 2$, prove that χ is primitive $\pmod{p^r}$ if and only if $p^{r-1} \mid m$.

(b) If $p = 2$ and $r \geq 4$, prove that χ is primitive $\pmod{2^r}$ if and only if $m = 2^{r-2}$. (For $r < 4$, see Problem 19.)

Hint: For part (a), use the fact (proved below Theorem 1.6.1) that $1+p$ is a primitive p^{r-1}th root of unity $\pmod{p^r}$. For part (b), see (1.6.19).

(c) Prove that for $p, r \geq 2$, there are $(p-1)^2 p^{r-2}$ primitive characters $\pmod{p^r}$.

36. For a positive integer k, evaluate the sums

$$\sum_{n=0}^{k-1} (-1)^n e^{2\pi i n^2/k} \quad \text{and} \quad \sum_{n=0}^{k-1} e^{\pi i n^2/k}.$$

Hint: First show that both sums equal 1 when k is odd.

Notes on Chapter 1

As we have seen from Lemma 1.2.1, it is easy to evaluate the quadratic Gauss sum $g(1;p)$ up to a determination of sign. The resolution of the ambiguous sign is considerably more difficult. Gauss conjectured the precise evaluation of $g(1;p)$ (Theorem 1.2.4) in May, 1801; see the diary of Gauss [1, Entry 118, pp. 36, 56, 77] and the translation of Gray [1, p. 124]. On August 30, 1805, Gauss wrote, "At last we have achieved a demonstration of the very elegant theorem mentioned above in May, 1801, which we had sought for more than four years with all efforts. " (Gauss [1, Entry 123, pp. 37, 57, 78], Gray [1, p. 125].) In 1811, Gauss [3] published his famous elementary evaluation of $g(1;k)$ for *all* positive integers k; the result may be found in Corollary 1.2.3. For odd k, Gauss' argument makes use of the Gaussian polynomials defined in Section 1.3, and it is this argument we have used to evaluate $g(1;k)$ for odd k (Theorem 1.3.4). In 1801, Gauss [2, art. 356] had already introduced the more general quadratic Gauss sum $g(m;k)$. Using the argument of Gauss [3] and the law of quadratic reciprocity (1.2.7), we evaluate $g(m;k)$ for all positive integers k in Section 1.5; see also the books of Nagell [2, pp. 177–180] and Davenport [3, pp. 66–68] for the case $m = 1$, the books of Hua [2, pp. 166–167] and Grosswald [1, p. 138] for the case k odd or a power of 2, and the report of H. J. S. Smith [1, p. 224]. Grosswald [1, Chapter 12] shows how the evaluation of $g(m;k)$ can be used to evaluate the singular series occurring in the circle method. (See also Barrucand [1] for the cubic case.)

In Theorem 1.3.2, we gave a product formula for a function $f(x,n)$ defined in terms of Gaussian polynomials. Theorem 1.3.2 is proved in Rademacher's book [1, pp. 85–86]. The function $f(x,n)$ is a special case of the Rogers–Szegö polynomial, which has been evaluated as a finite product in certain cases by Andrews [2] using partition–theoretic methods. Carlitz [12] gives an elegant method for evaluating $f(x, n)$, and offers a beautiful generalization of Theorem 1.3.4. Bressoud [1] proves Theorem 1.3.4 by using a variant of Theorem 1.3.2 in which $f(x, n)$ is replaced by a different function defined in terms of Gaussian polynomials. For further properties of Gaussian polynomials, see the book of Andrews [1, Chapter 3].

Since Gauss' initial evaluation of the Gauss sum (Theorem 1.3.4), many other proofs have been given. We proceed to discuss a few of them; for further references, consult the paper of Berndt and Evans [4]. In each of three papers, Dirichlet [2], [3], [4] gave proofs of Gauss' result by means of the Poisson summation formula. The details are simplest in Dirichlet [4]; see also Dirichlet's book [6]. Dirichlet's proof has been discussed by Kronecker [4] and can be found, for example, in the books of Landau [2, pp. 197–201] and Davenport [3, pp. 12–16]. See also Problem 8 of Exercises 1. Dirichlet's method has been generalized by Guinand [1] and by Berndt and Schoenfeld [1].

In 1840, Cauchy [3] gave two proofs of Gauss' result. The first of these uses the transformation formula of the theta function and is essentially the proof outlined in more general form in Problem 9 of Exercises 1. This method was generalized by Hecke; see the discussion of Theorem 1.2.2 below.

Several proofs have been given by contour integration. The first is due to Kronecker [3] in 1889 and is reproduced, for example, in the books of Landau [2, pp. 203–206] and Ayoub [1, pp. 315–317]. An especially simple proof by contour integration has been given by Mordell [1], [2]. Other proofs by contour integration will be discussed in connection with the more general Theorem 1.2.2.

The Abel–Plana summation formula was used by Genocchi [1] in 1852 to prove the theorem of Gauss. Essentially the same method was used by Weber [1] in 1903. This proof has been reproduced in Lindelöf's book [1, pp. 73–75].

The determination of the sign of Gauss' sum is generally shorter via analytic methods. However, because of the elementary nature of the statement of the result, a more elementary proof is desirable. The second method of Cauchy [3] is elementary. The proof was shortened and simplified by Kronecker [1] and Mordell [6] and can be found in Hasse's text [2, pp. 473–477]. Schur has given a beautiful linear algebraic proof, which can be found in Landau's book [2, pp. 207–212]. This proof is outlined in Problems 27–29 of Exercises 1. A short variant of Schur's proof has been given by Waterhouse [1]. The matrix $M = k^{-1/2}((e^{2\pi irs/k})), 0 \leq r, s < k$, in Problem 27 is the finite (discrete) Fourier transform. The trace of M is $g(1; k)/\sqrt{k}$. In the case where k is an odd prime, Benkö and Răduică [1] have diagonalized M with respect to an explicit basis of eigenvectors, thus giving another proof of Theorem 1.2.4. Morton [1] has given the eigenvectors of M for general k, and Tolimieri [1] has constructed an orthonormal basis of eigenvectors. Farsi and Watling [1] have given the dimensions of the eigenspaces for each of the eigenvalues $\pm 1, \pm i$ of the matrix $k^{-1/2}((e^{2\pi irsm/k}))$, where m is coprime with k. For some earlier determinations of the eigenvalues and eigenvectors of M, consult the references in Morton [1] and Tolimieri [1].

Mertens gave another elementary proof of Theorem 1.2.4, based on finite trigonometric sums; see Landau's book [2, pp. 213–218]. By approximating Gauss' sum by an integral, van der Corput [1] gave in 1921 a fairly elementary determination of the sign of Gauss' sum. Bambah and Chowla [1] simplified van der Corput's method in 1947. Using trigonometric methods, Estermann [1] determined the sign of the Gauss sum by proving inequalities such as $g(1;p) > -\sqrt{p}$ when $p \equiv 1 \pmod 4$. Estermann's proof is given in Section 1.4. See the books of S. Chowla [12, pp. 25–33] and Hua [2, pp. 165–166] for other accounts of Estermann's proof. The same basic idea was used by Landau [1] earlier to determine the sign of Gauss' sum; however, the details of the proofs are quite different. Carlitz [15] has given a proof of Theorem 1.2.4 that makes use of the determinant of the finite Fourier transform matrix as well as circulant determinants. A less elementary proof of Carlitz [21] depends on reducing the Gauss sum modulo a power of the prime ideal dividing p in the cyclotomic field $Q(e^{2\pi i/p})$. Shanks [1] has given a short elementary proof based on the identity

$$\sum_{j=1}^{2n} x^{j(j-1)/2} = \sum_{j=0}^{n-1} x^{j(2n+1)} \prod_{s=j+1}^{n} \frac{1-x^{2s}}{1-x^{2s-1}}.$$

The identity given at the end of Problem 9 is called Schaar's identity and was discovered by Schaar [1] in 1848. The proof outlined in Problem 9 was given by Kronecker [2] in 1880 and was generalized and simplified by Lerch [1] in 1903. Proofs of Schaar's identity via the use of the theta function can be found in the books of Krazer [1, pp. 184–189] and Bellman [1, pp. 38–39]. By considering arbitrary modular transformations of functions akin to theta functions, Bochner [1] discovered in 1951 a very general formula related to Gauss sums.

Theorem 1.2.2, the reciprocity theorem for generalized Gauss sums, was apparently first discovered by Genocchi [1]. The proof that we have given is a generalization of the aforementioned proof of Mordell [1], [2]. The proof of an even more general reciprocity theorem by this method has been given by Berndt [1]. Another proof of Theorem 1.2.2 by contour integration is due to Siegel [3]. This proof is given in more detail in Chandrasekharan's book [1, pp. 34–39]. In Apostol's book [1, pp. 196–200], Schaar's identity is proved by Siegel's method. Landsberg [1] has proved a generalization of Theorem 1.2.2 by contour integration. Farsi and Watling [2] have applied Theorem 1.2.2 to evaluate quadratic Gauss sums, which they needed for the computation of dimensions of certain eigenspaces.

A reciprocity theorem generalizing Schaar's identity, for quadratic Gauss sums over general algebraic number fields, is proved in Hecke's book [2, Theorem 163], via theta functions. Hecke uses this result to prove a general law of quadratic reciprocity over algebraic number fields (Hecke [2, Theorem 165]). For another proof, see Auslander, Tolimieri, and Winograd [1]. Eichler [1] has another formulation of the law of quadratic reciprocity in totally real fields, also obtained from a reciprocity law for Gauss sums. Siegel [3] gives a reciprocity law relating (multiple) Gauss sums for quadratic forms, and applies it to prove Hasse's version of the law of quadratic reciprocity, which evaluates

$(\frac{\alpha}{\beta})(\frac{\beta}{\alpha})$ for all algebraic integers α, β with $\alpha \equiv \beta \equiv 1 \pmod 2$. A reciprocity law for symplectic Gauss sums due to Stark may be found in a paper of Styer [1, Theorem 1].

The deduction of the law of quadratic reciprocity from the values of Gauss sums (Theorem 1.2.6) is due to Gauss. For a discussion of Gauss' proofs of the law of quadratic reciprocity, see the report of H. J. S. Smith [1, pp. 59–68]. Moore [1] has made a study of many proofs of the law of quadratic reciprocity. Other authors discussing this law are, for example, Brewer [1], S. Chowla [13], Cox [1], Eisenstein [4], Frame [1], Kaplan [1], Lebesgue [6], Rousseau [1], [2], and Rowe [1]. Our proof of the law of quadratic reciprocity (1.2.7) is that of Rousseau [1]. A law of quadratic reciprocity over function fields $F_q[T]$ has been given by Merrill and Walling [1]. For further references, consult the book of Lemmermeyer [3].

In (1.6.1), we defined the Gauss sums $\tau_k(m, \chi)$ over rings. Jager [1] has computed the average order of $|\tau_k(1, \chi)|$, as χ runs through the Dirichlet characters (mod k). In Section 1.6.1, we showed how to reduce the Gauss sums $\tau_k(m, \chi)$ in terms of primitive Gauss sums modulo prime powers. Such reduction is given in the books of Davenport [3, pp. 66–68], Hasse [2], and Narkiewicz [2, pp. 280–282]. See also the paper of Joris [1]. Nemchenok [1] extends the method of Joris [1] to express very general Gauss sums over finite rings of algebraic integers in terms of primitive Gauss sums of this type.

By Theorem 1.2.4, the root of unity $G(\chi)/\sqrt{p}$ is known precisely, in the case that χ is a quadratic character (mod p). Theorem 1.6.1 shows that if χ is not quadratic, then $G(\chi)/\sqrt{p}$ cannot be a root of unity. This was proved independently by S. Chowla [10], [11] and Mordell [5]. Our proof is a simplification of Chowla's. A proof is also given in Narkiewicz's book [2, p. 288]. Funakura [1] has extended the Chowla–Mordell theorem to the Gauss sums $\tau_k(1, \chi)$ over rings defined in (1.6.1). Evans [1] has generalized the Chowla–Mordell theorem and has established related results for Gauss sums over arbitrary finite fields. As an example, if χ and ψ are nontrivial characters over F_q and if the order of $\chi\overline{\psi}$ does not divide $(q-1)/(p-1)$, where $q = p^r$, then $G_r(\chi)/G_r(\psi)$ is not a root of unity; in particular, for $r = 1$, $G(\chi)/G(\psi)$ is never a root of unity if $\chi \neq \psi$. For related results over finite fields, see Evans [8] and Yokoyama [1].

In Problems 17–21, we outlined a proof that if χ is a primitive quadratic character (mod k), then

$$\tau(\chi) = \begin{cases} \sqrt{k}, & \text{if } \chi(-1) = 1, \\ i\sqrt{k}, & \text{if } \chi(-1) = -1, \end{cases}$$

where $\tau(\chi)$ is the Gauss sum defined in (1.6.8). Dirichlet [6, §116], [4] proved this for odd k using the evaluation of the quadratic Gauss sum given by Gauss [3]. Proofs are given for general k in the books of Borevich and Shafarevich [1, pp. 349–353], Hua [2, pp. 167–169], Landau [2, pp. 221–225], and Narkiewicz [2, pp. 283–287]. This evaluation of the primitive quadratic Gauss sum $\tau(\chi)$ is equivalent to the formula (Landau [2,

p. 221], Kronecker [CW, vol. 4, pp. 385, 506], Weber [2, vol. 3, §86])

$$\sum_{a=1}^{k-1} \left(\frac{d}{a}\right) e^{2\pi i n a/k} = \begin{cases} \left(\dfrac{d}{n}\right)\sqrt{k}, & \text{if } d > 0, \\ \left(\dfrac{d}{n}\right) i\sqrt{k}, & \text{if } d < 0, \end{cases}$$

where $(\frac{d}{n})$ is the Kronecker symbol for a fundamental discriminant d , $k = |d|$, and n is a positive integer. The equivalence follows readily from the fact that the primitive quadratic characters are the Kronecker symbols (see Hua [2, p. 159]). It is shown in Landau [2, p. 226] how Dirichlet essentially used the evaluation of the primitive quadratic Gauss sum $\tau(\chi)$ to deduce the L–function evaluation

$$\sum_{n=1}^{\infty} \left(\frac{d}{n}\right) \frac{1}{n} = \begin{cases} -\dfrac{1}{\sqrt{d}} \displaystyle\sum_{a=1}^{d-1} \left(\frac{d}{a}\right) \log\left(\sin\left(\frac{\pi a}{d}\right)\right), & \text{if } d > 0, \\ -\dfrac{\pi}{|d|^{3/2}} \displaystyle\sum_{a=1}^{|d|-1} \left(\frac{d}{a}\right) a, & \text{if } d < 0. \end{cases}$$

(The two sums on a above are both negative, but there are no truly elementary proofs of this fact. See the discussion in Narkiewicz's book [2, p. 472].) The L–function formula in turn yields Dirichlet's class number formula for $Q(\sqrt{d})$; see Dirichlet [4], [6, §106], Borevich and Shafarevich [1, Chapter 5, §4], or Hecke [2, §51, §52].

Theorem 1.6.2 is due to Odoni [1], and the counterpart for $p = 2$, Theorem 1.6.3, is due to Funakura [1, §4]. Theorem 1.6.4 is an elegant result of Mauclaire [1], [2].

Formulas like those in Problem 3 may be found in Lebesgue [2], [3]. For references to other proofs and generalizations, see Berndt [2]. The results in Problem 7 were discovered by S. Chowla [1], [2] and Carlitz [22]. Lerch [1] discovered the formulas of Problem 10. Halberstadt [1] related $\sum_{j=0}^{n-1} \tan(\pi j^2/n)$ and Gauss sums; see Problem 11. Zhedanov [1] used the formula in Problem 12 to find explicit weight functions for q–Hermite polynomials and q–Rogers–Szegö polynomials when q is a root of unity. See also Spiridonov and Zhedanov [1] for related applications of Gauss sum evaluations. A special case of Problem 23 was used by Farsi and Watling [2, Lemma 5.5.8] to evaluate certain quadratic sums needed for the computations of dimensions of eigenspaces. Lomont [1] has shown that nth powers of certain linear operators on complex vector spaces can be expressed in terms of quadratic Gauss sums; the sums in question can be evaluated from Problem 23 or Theorem 1.5.4 according as n is odd or even. Berry and Klein [1] give the phases of fractional Talbot images explicitly in terms of quadratic Gauss sums. The evaluations of these sums, given in Hannay and Berry [1, eq. (14)], simplify computations of Talbot images. (These evaluations can also be deduced from the Theorems of Section 1.5 and Problem 23.) In Leichtle, Averbukh, and Schleich [1, eq. (4.10), p. 5304], quadratic Gauss sums appear as weight functions in a formula for transient signals originating from multilevel quantum systems. Li and Li [1] use evaluations of quadratic Gauss sums to compute invariants of lens spaces.

The sum $f(h) = \sum_{m=0}^{h-1} e^{2\pi i m^2/k}$ is an example of a *partial* or *incomplete* quadratic Gauss sum if $h < k$. The estimate $f(h) = O(k^{1/2+\epsilon})$ is due to Hua [1], [2, p. 171]. Improved estimates have been given by D. H. Lehmer [2] and by Sullivan and Zannier [1]. Good estimates for $f(h)$ explain why, for large fixed k, the polygonal curve connecting the successive elements $f(1)/\sqrt{k}, f(2)/\sqrt{k}, \ldots, f(k)/\sqrt{k}$ approximates a spiral. For related investigations with applications to physics, see Berry and Goldberg [1], who study patterns of $\sum_{n=1}^{L} e^{i\pi\tau n^2}$ as $L \to \infty$; this sum is the classical quadratic Gauss sum when $\tau = 2/L$. A class of incomplete primitive Gauss sums stemming from (1.6.1), called "half Gauss sums," has been investigated by Berndt and Evans [3]. Burgess [1] has obtained upper estimates for partial Gauss sums of the form $\sum_{b=N+1}^{N+H} \chi(b) e^{2\pi i ab/k}$, where χ is a nontrivial Dirichlet character (mod k). Wingen [1, §4] studied partial Gauss sums of the form $\sum_{\alpha} \chi(\alpha) e(\alpha)$, where the sum is over a subgroup of F_q^*, and applied them to study the structure of integral group rings. Diamond, Gerth, and Vaaler [1] obtained conditions equivalent to the vanishing of a partial Gauss sum of the form $\sum_{m \in H} \chi(m) e^{2\pi i m s/k}$, where H is a subgroup of the group of reduced residues (mod k). Partial Gauss sums over additive subgroups of finite fields have been considered in papers of Perel'muter [1], [2]. Oskolkov [1] has obtained upper bounds for a type of partial quadratic Gauss sum. Some applications of complete and incomplete quadratic Gauss sums are given in Schroeder's book [1].

Quadratic Gauss sums over very general finite rings are evaluated in Lamprecht [1]. Gauss sums associated with prehomogeneous vector spaces are discussed in Gyoja and Kawanaka [1]. Kim [1] evaluated Gauss sums of the form

$$\sum_{g \in G} \chi(\det g) e(\operatorname{tr} g)$$

for various classical matrix groups G over F_q, where χ is a multiplicative character on F_q and e is defined above Proposition 1.1.2. See also Kondo [1]. A generalization of Gauss sums to vector spaces over finite fields has been given by Bremser [1]. Styer [1] has evaluated symplectic Gauss sums. Saito [1] has explicitly determined certain Gauss sums associated with symmetric matrices over finite fields. In several papers, Thakur has developed the theory of Gauss sums for function fields; see Thakur [1]. For representation–theoretic studies of Gauss sums connected with analogues of Barnes' identity, see papers of Li [1], Gérardin and Li [1], and Helversen–Pasotto [1]–[3]. Gauss sums for quadratic forms on finite Abelian groups are studied in connection with Witt groups in Scharlau's book [1, Chapter 5]. See also the book of Milnor and Husemoller [1, p. 127]. Siegel [1]–[4] examines Gauss sums for n–ary quadratic forms. We have already mentioned in connection with Theorem 1.2.2 that Siegel [3] proved a reciprocity law for such Gauss sums. Some evaluations of these sums are given in Siegel [1, §3], [2], [4, §5], and in Reti [1]. Composition formulas for quadratic Gauss sums are given by Siegel [2, p. 242] and Kesava Menon [1]. Two non–Abelian generalizations of Gauss sums, congruence and Galois Gauss sums, are discussed in Bushnell and Fröhlich [1] and Fröhlich [1], respectively. Dimitrakoudis [1] has obtained reduction formulas for Gauss sums associated with Grössencharacters of algebraic number fields. Lin [1] studied a double Gauss sum, which plays a role in transformation formulas for higher-order theta functions. In Sarnak's book [1, p. 12], the value of the classical quadratic Gauss sum is used to compute the multiplier for the transformation of the classical theta function.

For further references on Gauss sums, see the Notes in succeeding chapters, the paper of Berndt and Evans [4], and the books of Narkiewicz [2, pp. 319–320] and Lidl and Niederreiter [1, pp. 240–257]. Excellent references to Gauss' contributions to number theory are Dieudonné [1], Narkiewicz [1], and Neumann [1], [2], [3]. In particular, Neumann [3] recounts the important role that Gauss and Jacobi sums have played in the development of algebraic number theory in the nineteenth century.

Chapter 2

Jacobi Sums and Cyclotomic Numbers

In this chapter we present the basic properties of Jacobi sums over finite fields F_q of characteristic p. These sums are useful, for example, in the determination of the number of solutions of diagonal equations over finite fields; see Chapter 10. We also develop the basic properties of cyclotomic numbers and discuss their relationship to Jacobi sums over F_p. The cyclotomic numbers are useful, for example, in the study of residuacity (mod p); see Chapter 7.

2.1 Basic properties of Jacobi sums over F_q

We begin by defining Jacobi sums over finite fields.

Definition *Let χ and ψ denote characters on the finite field F_q, where $q = p^r$. The Jacobi sum $J_r(\chi, \psi)$ is defined by*

$$J_r(\chi,\psi) = \sum_{\alpha} \chi(\alpha)\psi(1-\alpha),$$

where α runs through the elements of F_q. The order m of the Jacobi sum $J_r(\chi,\psi)$ is the least common multiple of the orders of χ and ψ; thus a Jacobi sum of order m is an integer of the cyclotomic field $Q(\exp(2\pi i/m))$.

As $\alpha \longrightarrow 1-\alpha$ is a bijection on F_q, we see that Jacobi sums have the symmetry property

$$J_r(\chi,\psi) = J_r(\psi,\chi).$$

It is convenient to set

$$K_r(\chi) = \chi(4)J_r(\chi,\chi).$$

When $r = 1$ we drop the subscript from J_r and K_r.

Theorem 2.1.1 *Let χ and ψ be characters on F_q.*
(a) *If χ and ψ are both trivial, then $J_r(\chi,\psi) = q$.*
(b) *If exactly one of χ and ψ is trivial, then $J_r(\chi,\psi) = 0$.*
(c) *If χ is nontrivial, then $J_r(\chi,\overline{\chi}) = -\chi(-1)$.*

Proof (a) If χ and ψ are both trivial, then

$$J_r(\chi,\psi) = \sum_{\alpha\in F_q} 1 = q.$$

(b) If χ is nontrivial and ψ is trivial, then

$$J_r(\chi,\psi) = \sum_{\alpha\in F_q} \chi(\alpha) = 0.$$

(c) If χ is nontrivial, then

$$J_r(\chi,\overline{\chi}) = \sum_{\alpha\neq 0} \chi(1-\alpha)\chi(\alpha^{-1}) = \sum_{\alpha\neq 0} \chi(\alpha^{-1}-1) = \sum_{\alpha\neq 0} \chi(\alpha-1) = -\chi(-1). \quad \blacksquare$$

From now on, when $p > 2$, we denote the character on F_q of order 2 by ρ. In this book we evaluate character sums of the type $\sum_\alpha \rho(f(\alpha))$ for certain classes of polynomials $f(x) \in F_q[x]$. The next theorem yields the evaluation for all quadratic polynomials $f(x)$.

Theorem 2.1.2 *Let $p > 2$. Let $f(x) = x^2 + \beta x + \gamma \in F_q[x]$ and set $\delta = \beta^2 - 4\gamma$. Then for the quadratic character ρ on F_q,*

$$\sum_\alpha \rho(f(\alpha)) = \begin{cases} -1, & \text{if } \delta \neq 0, \\ q-1, & \text{if } \delta = 0. \end{cases}$$

Proof Since $4f(x) = (2x+\beta)^2 - \delta$,

$$\sum_\alpha \rho(f(\alpha)) = \sum_\alpha \rho(\alpha^2 - \delta).$$

Given ϵ in F_q, the number of solutions $\alpha \in F_q$ of $\alpha^2 = \epsilon$ is $1 + \rho(\epsilon)$, so that, for $\delta \neq 0$,

$$\begin{aligned}\sum_\alpha \rho(\alpha^2-\delta) &= \sum_\alpha \rho(\alpha-\delta)\{1+\rho(\alpha)\} = \sum_\alpha \rho(\alpha-\delta)\rho(\alpha) = \sum_\alpha \rho(\delta\alpha-\delta)\rho(\delta\alpha) \\ &= \rho(-1)J_r(\rho,\rho) = -1,\end{aligned}$$

by Theorem 2.1.1(c). For $\delta = 0$,

$$\sum_\alpha \rho(f(\alpha)) = \sum_\alpha \rho(\alpha^2) = \sum_{\alpha\neq 0} \rho(\alpha^2) = \sum_{\alpha\neq 0} \rho^2(\alpha) = \sum_{\alpha\neq 0} 1 = q-1. \quad \blacksquare$$

Example If p is an odd prime and a, b, c are integers with $p \nmid a$ then, from Theorem 2.1.2,

$$\sum_{x=0}^{p-1} \left(\frac{ax^2+bx+c}{p}\right) = \begin{cases} -\left(\dfrac{a}{p}\right), & \text{if } b^2 - 4ac \not\equiv 0 \pmod{p}, \\ (p-1)\left(\dfrac{a}{p}\right), & \text{if } b^2 - 4ac \equiv 0 \pmod{p}. \end{cases}$$

An important relationship between Jacobi sums and Gauss sums is given in the next theorem.

Theorem 2.1.3 *Let χ and ψ be characters on F_q. If $\chi\psi$ is nontrivial, then*

$$\text{(a)} \quad J_r(\chi, \psi) = \frac{G_r(\chi)G_r(\psi)}{G_r(\chi\psi)},$$

and if χ, ψ, and $\chi\psi$ are all nontrivial,

$$\text{(b)} \quad |J_r(\chi, \psi)| = q^{1/2}.$$

Proof (a) By the definition of $G_r(\chi)$ given in Section 1.1,

$$\begin{aligned} G_r(\chi)G_r(\psi) &= \sum_\alpha \sum_\beta \chi(\alpha)\psi(\beta)e(\alpha+\beta) = \sum_\gamma e(\gamma) \sum_{\alpha+\beta=\gamma} \chi(\alpha)\psi(\beta) \\ &= \sum_{\alpha+\beta=0} \chi(\alpha)\psi(\beta) + \sum_{\gamma\neq 0} e(\gamma) \sum_\alpha \chi(\alpha)\psi(\gamma-\alpha) \\ &= \psi(-1)\sum_\alpha \chi\psi(\alpha) + \sum_{\gamma\neq 0} e(\gamma)\sum_\alpha \chi(\gamma\alpha)\psi(\gamma-\gamma\alpha) \\ &= 0 + J_r(\chi,\psi)\sum_{\gamma\neq 0}\chi\psi(\gamma)e(\gamma) = J_r(\chi,\psi)G_r(\chi\psi). \end{aligned}$$

(b) By part (a) and Theorem 1.1.4(c),

$$|J_r(\chi, \psi)| = \frac{|G_r(\chi)||G_r(\psi)|}{|G_r(\chi\psi)|} = \frac{q^{1/2}q^{1/2}}{q^{1/2}} = q^{1/2}.$$ ∎

The following theorem is a special case of the deep Davenport–Hasse relation for Gauss sums, which is discussed in Chapter 11; see Theorem 11.3.5.

Theorem 2.1.4 *Let $p > 2$. If χ is a nontrivial character on F_q, and ρ is the quadratic character on F_q, then*

$$K_r(\chi) = J_r(\chi, \rho).$$

Proof Let β be a fixed element of F_q. Each solution $\alpha \in F_q$ of the equation $\alpha(1-\alpha) = \beta$ is a solution of the equation $(2\alpha-1)^2 = 1 - 4\beta$, and vice versa. The number of solutions of the latter equation is $1 + \rho(1-4\beta)$. Thus

$$\begin{aligned} J_r(\chi,\chi) &= \sum_\alpha \chi(\alpha(1-\alpha)) = \sum_\beta \chi(\beta)\{1+\rho(1-4\beta)\} \\ &= \overline{\chi}(4)\sum_\beta \chi(\beta)\rho(1-\beta) = \overline{\chi}(4)J_r(\chi,\rho), \end{aligned}$$

where, in the penultimate step, we replaced β by $4^{-1}\beta$. ∎

Theorem 2.1.5 *If χ and ψ are characters on F_q with χ, ψ, and $\chi\psi$ nontrivial, then*

$$J_r(\chi, \psi) = \psi(-1)J_r(\overline{\chi\psi}, \psi) = \chi(-1)J_r(\overline{\chi\psi}, \chi).$$

Proof Appealing to Theorem 2.1.3(a) and Theorem 1.1.4(a), we obtain

$$\psi(-1)J_r(\overline{\chi}\overline{\psi},\psi) = \psi(-1)\frac{G_r(\overline{\chi}\overline{\psi})G_r(\psi)}{G_r(\overline{\chi}\overline{\psi}\psi)} = \psi(-1)\frac{G_r(\overline{\chi}\overline{\psi})G_r(\psi)}{G_r(\overline{\chi})}$$
$$= \psi(-1)\frac{\chi\psi(-1)q}{G_r(\chi\psi)}\frac{G_r(\chi)}{\chi(-1)q}G_r(\psi) = \frac{G_r(\chi)G_r(\psi)}{G_r(\chi\psi)} = J_r(\chi,\psi).$$

The proof of $\chi(-1)J_r(\overline{\chi}\overline{\psi},\chi) = J_r(\chi,\psi)$ follows from the symmetry in χ, ψ. ■

Theorem 2.1.6 *Let $p > 2$. Let χ be a nontrivial character on F_q with $\chi \neq \rho$. Then*

$$\rho(-1)K_r(\overline{\chi}\rho) = K_r(\chi) = \chi(-1)J_r(\chi,\overline{\chi}\rho). \tag{2.1.1}$$

In particular, if χ has order $2j$, where $j > 1$, then

$$\rho(-1)K_r(\chi^{j-1}) = K_r(\chi) = \chi(-1)J_r(\chi,\chi^{j-1}). \tag{2.1.2}$$

Proof As χ, ρ, $\chi\rho$, and $\overline{\chi}\rho$ are nontrivial characters, we find by Theorems 2.1.4 and 2.1.5 that

$$K_r(\chi) = J_r(\chi,\rho) = \rho(-1)J_r(\overline{\chi}\rho,\rho) = \rho(-1)K_r(\overline{\chi}\rho)$$

and

$$K_r(\chi) = J_r(\chi,\rho) = \chi(-1)J_r(\overline{\chi}\rho,\chi) = \chi(-1)J_r(\chi,\overline{\chi}\rho),$$

which proves (2.1.1). The equalities in (2.1.2) follow from those in (2.1.1) as $\rho = \chi^j$ in this case. ■

We remark that with $q = p = kf + 1$, where k is even and χ has order k, the first equality in (2.1.2) becomes

$$K(\chi) = \left(\frac{-1}{p}\right)K(\chi^{(k/2)-1}).$$

If K is an algebraic number field of finite degree over Q, let O_K denote the ring of algebraic integers in K. For $\alpha, \beta, \gamma \in O_K$, the congruence $\alpha \equiv \beta \pmod{\gamma}$ means that $\alpha - \beta \in \gamma O_K$. The norm of an element $\alpha \in O_K$ is a rational integer and is denoted by $N(\alpha)$.

Theorem 2.1.7 *For $j = 1, 2$, let χ_j be a character on F_q of order $k_j > 1$, and let $\beta_j = \exp(2\pi i/k_j)$. Then*

$$J_r(\chi_1,\chi_2) \equiv -q \pmod{(1-\beta_1)(1-\beta_2)}.$$

Proof When $\alpha \neq 0, 1$, $\chi_1(\alpha)$ and $\chi_2(1-\alpha)$ are powers of β_1 and β_2, respectively. Hence

$$J_r(\chi_1, \chi_2) = \sum_{\alpha}(1-\chi_1(\alpha))(1-\chi_2(1-\alpha)) - q \equiv -q \pmod{(1-\beta_1)(1-\beta_2)}. \quad \blacksquare$$

Theorem 2.1.8 *Let $p > 2$. Let χ be a character on F_q of order $k > 1$ and define $\beta = \exp(2\pi i/k)$. Then*

$$K_r(\chi) \equiv -q \pmod{2(1-\beta)}. \tag{2.1.3}$$

Further, if $k > 2$, then

$$\text{Re } K_r(\chi) \equiv 1 + i \text{ Im } K_r(\chi) \equiv \frac{1}{2}(1+\rho(-1)) \pmod{2}. \tag{2.1.4}$$

Proof The congruence (2.1.3) follows from Theorem 2.1.4 and Theorem 2.1.7 (with $\chi_1 = \chi$, $\chi_2 = \rho$).

Next we prove (2.1.4). As $K_r(\chi) = \chi(4)J_r(\chi, \chi)$ and $K_r(\chi\rho) = \chi(4)J_r(\chi\rho, \chi\rho)$,

$$\begin{aligned} K_r(\chi) + K_r(\chi\rho) &= \sum_{\alpha} \chi(4\alpha(1-\alpha))\{1 + \rho(\alpha(1-\alpha))\} \\ &= 2 \sum_{\substack{\alpha \\ \rho(\alpha(1-\alpha))=1}} \chi(4\alpha(1-\alpha)) = 2 + 2 \sum_{\substack{\alpha \neq 2^{-1} \\ \rho(\alpha(1-\alpha))=1}} \chi(4\alpha(1-\alpha)). \end{aligned} \tag{2.1.5}$$

Since the replacement of α by $1-\alpha$ leaves $\chi(4\alpha(1-\alpha))$ unchanged, and since $\alpha \neq 1-\alpha$ when $\alpha \neq 2^{-1}$, (2.1.5) yields

$$K_r(\chi) + K_r(\chi\rho) \in 2 + 4O_K,$$

where $K = Q(\beta)$. As $k > 2$, χ is a nontrivial character with $\chi \neq \rho$, so that, by Theorem 2.1.6, we have

$$\overline{K_r(\chi)} = K_r(\overline{\chi}) = \rho(-1)K_r(\chi\rho).$$

Therefore

$$K_r(\chi) + K_r(\chi\rho) = \begin{cases} 2 \text{ Re } K_r(\chi), & \text{if } \rho(-1) = 1, \\ i2 \text{ Im } K_r(\chi), & \text{if } \rho(-1) = -1. \end{cases}$$

Thus $\text{Re } K_r(\chi) \in 1 + 2O_K$ if $\rho(-1) = 1$ and $i\text{Im } K_r(\chi) \in 1 + 2O_K$ if $\rho(-1) = -1$. Since $\text{Re } K_r(\chi) + i\text{Im } K_r(\chi) \equiv 1 \pmod{2}$ by (2.1.3), the result follows. $\blacksquare$

The next theorem gives the value of the norm of the element $1 - e^{2\pi i/k}$.

Theorem 2.1.9 *Let $\beta = \exp(2\pi i/k)$ with $k > 1$. Then*

$$N(1-\beta) = \begin{cases} t, & \textit{if } k \textit{ is a power of a prime } t, \\ 1, & \textit{otherwise}. \end{cases}$$

Proof Let $\Phi_d(x)$ denote the unique monic irreducible polynomial with integral coefficients satisfied by $\exp(2\pi i/d)$. Then

$$(x^k - 1)/(x - 1) = \prod_{\substack{d \mid k \\ d > 1}} \Phi_d(x).$$

Letting x tend to 1, we obtain

$$k = \prod_{\substack{d \mid k \\ d > 1}} \Phi_d(1).$$

By the Möbius inversion formula (see, for example, Ireland and Rosen [1, p. 20]),

$$\Phi_k(1) = \prod_{d|k} \left(k/d\right)^{\mu(d)}.$$

As $k > 1$ we can express k uniquely in the form

$$k = t_1^{m_1} \cdots t_r^{m_r},$$

where $r, m_1, \ldots, m_r$ are positive integers, and $t_1, \ldots, t_r$ are distinct primes. If $r = 1$,

$$\Phi_k(1) = \prod_{u_1=0}^{m_1} (t_1^{m_1-u_1})^{\mu(t_1^{u_1})} = \prod_{u_1=0}^{1} t_1^{(m_1-u_1)(-1)^{u_1}} = t_1.$$

If $r \geq 2$,

$$\begin{aligned} \Phi_k(1) &= \prod_{u_1,\ldots,u_r=0}^{m_1,\ldots,m_r} (t_1^{m_1-u_1} \cdots t_r^{m_r-u_r})^{\mu(t_1^{u_1} \cdots t_r^{u_r})} \\ &= \prod_{u_1,\ldots,u_r=0}^{1} (t_1^{m_1-u_1} \cdots t_r^{m_r-u_r})^{(-1)^{u_1+\cdots+u_r}}. \end{aligned}$$

The exact power of t_1 dividing $\Phi_k(1)$ is as follows:

$$\sum_{u_1,\ldots,u_r=0}^{1} (m_1 - u_1)(-1)^{u_1+\cdots+u_r} = \sum_{u_1=0}^{1} (m_1 - u_1)(-1)^{u_1} \sum_{u_2=0}^{1} (-1)^{u_2} \cdots \sum_{u_r=0}^{1} (-1)^{u_r} = 0.$$

Similarly, the exact power of each t_i $(i = 2, \ldots, r)$ dividing $\Phi_k(1)$ is 0. Hence $\Phi_k(1) = 1$ when $r \geq 2$. We have shown that

$$\Phi_k(1) = \begin{cases} t, & \text{if } k \text{ is a power of a prime } t, \\ 1, & \text{if } k \text{ has at least two distinct prime factors.} \end{cases}$$

The assertion of the theorem now follows as $N(1 - \beta) = \Phi_k(1)$. ∎

Corollary 2.1.10 *Let $\beta = \exp(2\pi i/k)$, where k is a power of a prime t. Set $K = Q(\beta)$. Then $t/(1-\beta)^{\phi(k)}$ is a unit in O_K, where ϕ denotes Euler's totient function.*

Proof By Theorem 2.1.9,

$$t = \prod_{\substack{1 \le j \le k \\ \gcd(j,k)=1}} (1-\beta^j),$$

so that

$$t/(1-\beta)^{\phi(k)} = \prod_{\substack{1 \le j \le k \\ \gcd(j,k)=1}} (1-\beta^j)/(1-\beta) .$$

Each of the $\phi(k)$ factors $(1-\beta^j)/(1-\beta)$ is in O_K, has norm 1, and is therefore a unit in O_K. This establishes the assertion that $t/(1-\beta)^{\phi(k)}$ is a unit in O_K. ■

Theorem 2.1.11 *Let χ_1 and χ_2 be characters on F_q of order $k > 2$. Put $\beta = \exp(2\pi i/k)$. Then*

$$J_r(\chi_1, \chi_2) \equiv -1 \pmod{(1-\beta)^2}.$$

Proof If k is not a prime power, then $1-\beta$ is a unit by Theorem 2.1.9, and so the result is trivial. If k is a prime power, then $(1-\beta)^2$ divides k by Corollary 2.1.10. Since k, in turn, divides $q-1$, the result follows from Theorem 2.1.7. ■

The result above shows that $J_r(\chi_1, \chi_2)$ is normalized in the sense of Leonard and Williams [2, p. 148], [4, p. 370].

In the case that k is an odd prime power > 3, the modulus $(1-\beta)^2$ in Theorem 2.1.11 may replaced by $(1-\beta)^3$. This was proved by Iwasawa [1] for prime k and Ihara [1, p. 81] for prime power k. For a generalization, see Evans [21]. For an extension to the generalized Jacobi sums defined in Section 10.1, see Gouvêa and Yui [1, Prop. 1.7] and Miki [1, Theorem 2].

For the remainder of this section, let $\beta = \exp(2\pi i/k)$ with $k > 2$. The ring O_K of integers in $K = Q(\beta)$ is $Z[\beta]$ (see, for example, Washington's book [1, p. 11]). If an ideal of O_K is generated by each of two given Jacobi sums of order k (for example, as in (3.6.2)), we would like to conclude that the two sums are equal up to a factor $\pm\mu$, where μ is a kth root of unity. In Chapter 3, we shall draw such conclusions with the help of the following two well-known results.

Theorem 2.1.12 *If all the algebraic conjugates of an algebraic integer α over Q have modulus 1, then α is a root of unity.*

Proof Suppose that α has positive degree d over Q. Consider the finite family of polynomials

$$x^d + a_{d-1}x^{d-1} + \cdots + a_1 x + a_0$$

with integral coefficients a_j such that $|a_j| \le \binom{d}{j}$, $0 \le j \le d-1$. Since the conjugates of α have modulus 1, so do the conjugates of α^m for any integer m, and thus each α^m is a zero of one of the polynomials above. Hence there are distinct integers m_1 and m_2 such that $\alpha^{m_1} = \alpha^{m_2}$. Thus $\alpha^{m_1-m_2} = 1$, and α is a root of unity as asserted. ∎

We remark that the algebraic integer $\alpha = (1 - \sqrt{2} + i\sqrt{2\sqrt{2}+1})/2$ has absolute value 1, but α is not a root of unity. The minimal polynomial of α over Q is $x^4 - 2x^3 + x^2 - 2x + 1$, which has the four zeros $(1-\sqrt{2} \pm i\sqrt{2\sqrt{2}+1})/2$, $(1+\sqrt{2} \pm \sqrt{2\sqrt{2}-1})/2$. The algebraic integer α is an example of a Salem number of degree 4; see Narkiewicz [2, pp. 76, 77].

Theorem 2.1.13 *Let $K = Q(\beta)$, where $\beta = \exp(2\pi i/k)$. The only elements of modulus 1 in O_K are $\pm\beta^j$, $0 \le j < k$. In particular, the only roots of unity in O_K are $\pm\beta^j$.*

Proof Let $u \in O_K$ satisfy $u\overline{u} = 1$. Write $u = \sum a_j\beta_j$ with $a_j \in Q$. For each σ in the Galois group $\mathrm{Gal}(Q(\beta)/Q)$,

$$\sigma(\overline{u}) = \sigma(\textstyle\sum a_j\beta^{-j}) = \sum a_j\sigma(\beta)^{-j} = \overline{\sigma(u)},$$

and so $|\sigma(u)| = 1$. Thus u is a root of unity by Theorem 2.1.12. It remains to show that the only roots of unity in $Q(\beta)$ are $\pm\beta^j$, $0 \le j < k$.

For each positive integer m, set $\beta_m = \exp(2\pi i/m)$ and $A_m = Q(\beta_m)$. For positive integers m and n, let $d = \gcd(m,n)$ and let $c = mn/d$ (the least common multiple of m and n). Since $am + bn = d$ for some integers a and b, we have $\beta_n^a\beta_m^b = \beta_c$, so that

$$A_mA_n = A_c.$$

We now prove that

$$A_m \cap A_n = A_d. \tag{2.1.6}$$

Let $H = A_m \cap A_n$. Clearly $A_n \supseteq H \supseteq A_d$. Thus

$$[A_d(\beta_m) : A_d] \ge [H(\beta_m) : H] \ge [A_n(\beta_m) : A_n]. \tag{2.1.7}$$

Now

$$[A_d(\beta_m) : A_d] = [A_m : A_d] = \phi(m)/\phi(d),$$
$$[A_n(\beta_m) : A_n] = [A_c : A_n] = \phi(c)/\phi(n).$$

It follows easily from the standard formula for $\phi(j)$ in terms of the prime factorization of j that

$$\phi(m)\phi(n) = \phi(c)\phi(d).$$

Thus equality holds throughout (2.1.7). Since

$$A_d(\beta_m) = A_m = H(\beta_m),$$

it must also follow that $A_d = H$. This completes the proof of (2.1.6).

Assume now that $\beta_m \in Q(\beta_k)$. We must show that either $m \mid k$ or else $m \mid 2k$ with k odd. By (2.1.6),

$$\beta_m \in Q(\beta_k) \cap Q(\beta_m) = Q(\beta_d),$$

where $d = \gcd(m, k)$, so that $\phi(m) \mid \phi(d)$. Since $d \mid m$ we have $\phi(d) \mid \phi(m)$. Hence $\phi(m) = \phi(d)$. It follows that either $m = d$ or else $m = 2d$ with d odd. In the first case, $m = \gcd(m, k)$, so that $m \mid k$. In the second case, $m/2 = \gcd(m, k)$ with $m \equiv 2 \pmod 4$, so that $(m/2) \mid k$ with k odd. ■

Write $G = \mathrm{Gal}(K/Q)$, where $K = Q(\beta)$ with $\beta = \exp(2\pi i/k)$, $k > 2$. If $\gcd(j, k) = 1$, define $\sigma_j \in G$ by $\sigma_j(\beta) = \beta^j$. Note that if χ is a character on F_q of order k, then $\sigma_j(J_r(\chi^m, \chi^n)) = J_r(\chi^{mj}, \chi^{nj})$.

We will need the following general facts about algebraic number fields L. If A and B are nonzero ideals of O_L, then $A \mid B$ if and only if $A \supseteq B$. The ideal A can be factored uniquely as a product of prime ideals in O_L. If E is an algebraic number field containing L, then $AO_E \cap O_L = A$. This last fact follows easily by multiplying by the fractional ideal A^{-1} in L. See Narkiewicz [2, Chapter 1, §1].

Now take p to be a prime with $k \not\equiv 0 \pmod p$. Let $r = r(p)$ be the least positive integer for which $p^r \equiv 1 \pmod k$. Then the principal ideal $p^r O_K$ is a product of $\phi(k)$ conjugate prime ideals. In other words, for any prime ideal factor P of pO_K, we have

$$p^r O_K = \prod_{\sigma \in G} \sigma(P) = \prod_{\substack{0<j<k \\ \gcd(j,k)=1}} P_j, \tag{2.1.8}$$

where $P_j = \sigma_j(P)$. There are $\phi(k)/r$ distinct prime ideals on the right side of (2.1.8), each occurring with multiplicity r; in fact, $P = \sigma_{p^i}(P)$ for $1 \le i \le r$. The residue field O_K/P has p^r elements. The reader will find proofs of these facts, for example, in Washington's book [1, p. 14].

If O_K is a principal ideal domain, then (2.1.8) yields the unique prime factorization of p in O_K. For example, when $k = 4$, (2.1.8) reduces to the well-known facts that (i) p is a prime in the ring of Gaussian integers $Z[i]$ if $p \equiv 3 \pmod 4$, and (ii) $p = \pi\bar{\pi}$, where π is a Gaussian prime, if $p \equiv 1 \pmod 4$. We remark that O_K is a principal ideal domain precisely for $1 \le k \le 22$, $24 \le k \le 28$, $k = 30$, $32 \le k \le 36$, and $k = 38, 40, 42, 44, 45, 48, 50, 54, 60, 66, 70, 84, 90$; see the paper by Masley and Montgomery [1].

Clearly, the cosets $\beta^j + P$ in O_K/P, $0 \le j < k$, are kth roots of unity in O_K/P. To show that they comprise all k of the kth roots of unity, we will show that they are distinct. Assume that they are not distinct. Then $1 - \beta^j \in P$ for some j, $0 < j < k$. Consequently, $N(1 - \beta^j) \in P$. If the order of the root of unity β^j is not a power of a prime then, by Theorem 2.1.9, $1 = N(1-\beta^j) \in P$, which is impossible. If, on the other hand, the order of β^j is a power of a prime t then, by Theorem 2.1.9, $t = N(1-\beta^j) \in P$. As $k \not\equiv 0 \pmod p$, we have $p \neq t$, but $p \in P$, so $1 \in P$, which is impossible. This proves that the cosets $\beta^j + P$ are distinct.

For $\alpha \in O_K$, $\alpha \notin P$, we claim that there is a unique power of β that is congruent to $\alpha^{(p^r-1)/k} \pmod P$. Since the nonzero elements of O_K/P form a (cyclic) group of order $p^r - 1$, $\alpha^{(p^r-1)/k} + P$ is a kth root of unity in O_K/P. Thus, there is a unique integer j, $0 \leq j < k$, such that $\alpha^{(p^r-1)/k} + P = \beta^j + P$, as asserted.

Define the function χ_P on O_K by letting $\chi_P(\alpha)$ be the (unique) power of β congruent to $\alpha^{(p^r-1)/k} \pmod P$, if $\alpha \notin P$, and by letting $\chi_P(\alpha) = 0$, if $\alpha \in P$. Observe that for $\sigma \in G$, $\chi_{\sigma(P)} = \sigma\chi_P\sigma^{-1}$. We can view χ_P as a power residue character of order k on the finite field O_K/P by defining $\chi_P(\alpha + P) = \chi_P(\alpha)$. Given a character χ on a finite field F_q with $q = p^r$, we will frequently view χ as the character χ_P on the field O_K/P by identifying a generator $\alpha + P$ of $(O_K/P)^*$ with the generator τ of F_q^* chosen such that $\chi(\tau) = \chi_P(\alpha + P)$.

The symbol $\left(\frac{\alpha}{P}\right)_k$ or $\left(\frac{\alpha}{P}\right)$ is often used to denote $\chi_P(\alpha)$. It can be extended in the same manner that the Legendre–Jacobi symbol extends the Legendre symbol. If P is a principal ideal generated by π, then $\chi_P(\alpha) = \left(\frac{\alpha}{P}\right)_k$ is written $\chi_\pi(\alpha) = \left(\frac{\alpha}{\pi}\right)_k$. See Section 14.3.

In (2.1.9) below, we give the prime ideal factorization in O_K of Jacobi sums of order dividing k, in the case $p \equiv 1 \pmod k$, i.e., in the case $r = 1$. This will be done for general r in Theorem 11.2.3. In (2.1.11), we give a congruence for Jacobi sums (mod P) for general r. It is convenient to let $L(a)$ denote the least positive integer congruent to $a \pmod k$, where a is an integer not divisible by k.

Theorem 2.1.14 *Let $K = Q(\beta)$ with $\beta = \exp(2\pi i/k)$, $k > 2$. Let p be a prime and write $q = p^r$ with $r \geq 1$ chosen minimal such that $p^r \equiv 1 \pmod k$. Let χ be the character χ_P on O_K/P of order k, for some prime ideal P in O_K dividing pO_K. Let m and n be integers with $0 < m, n < k$ and $m + n \neq k$. If $p \equiv 1 \pmod k$, i.e., $r = 1$, then*

$$J(\chi^m, \chi^n)O_K = \prod_{j\in S} P_{j^{-1}}, \tag{2.1.9}$$

where $S = \{j : 0 < j < k, \gcd(j,k) = 1$ and $L(mj) + L(nj) < k\}$ and j^{-1} denotes the inverse of $j \pmod k$; in particular,

$$K(\chi)O_K = \prod_{\substack{1 \leq j < k/2 \\ \gcd(j,k)=1}} P_{j^{-1}}. \tag{2.1.10}$$

Moreover, for general r,

$$J_r(\chi^m, \chi^n) \equiv \begin{cases} 0 \pmod P, & \text{if } m+n<k, \\ (-1)^{q+mf}\dbinom{nf}{(k-m)f} \pmod P, & \text{if } m+n>k, \end{cases} \tag{2.1.11}$$

where $f = (q-1)/k$.

Proof For any integer j such that $0 < j < k$ and $\gcd(j,k) = 1$,

$$(2.1.12)\quad J_r(\chi^{mj}, \chi^{nj}) = J_r(\chi^{L(mj)}, \chi^{L(nj)}) = \sum_{0 \neq \alpha \in O_K/P} \chi^{L(mj)}(\alpha)\chi^{L(nj)}(1-\alpha)$$
$$\equiv \sum_{0 \not\equiv \alpha (\text{mod } P)} \alpha^{fL(mj)}(1-\alpha)^{fL(nj)} \pmod P$$
$$\equiv \sum_{0 \not\equiv \alpha (\text{mod } P)} \alpha^{fL(mj)} \sum_{v=0}^{fL(nj)} \binom{fL(nj)}{v} (-\alpha)^v \pmod P$$
$$\equiv \sum_{v=0}^{fL(nj)} \binom{fL(nj)}{v} (-1)^v \sum_{0 \not\equiv \alpha (\text{mod } P)} \alpha^{fL(mj)+v} \pmod P.$$

Observe that

$$(2.1.13)\qquad \sum_{\alpha \in F_q^*} \alpha^e = \begin{cases} -1, & \text{if } e \equiv 0 \pmod{q-1}, \\ 0, & \text{if } e \not\equiv 0 \pmod{q-1}, \end{cases}$$

since α runs through the $q-1$ distinct powers of a generator of the cyclic multiplicative group F_q^*. If $j \in S$, then for $v = 0, 1, \ldots, fL(nj)$,

$$0 < fL(mj) + v \leq fL(mj) + fL(nj) < q - 1,$$

and so by (2.1.13), the rightmost inner sum in (2.1.12) vanishes for each v. Thus for $j \in S$,

$$J_r(\chi^{mj}, \chi^{nj}) \equiv 0 \pmod P,$$

or equivalently,

$$(2.1.14)\qquad J_r(\chi^m, \chi^n) \equiv 0 \pmod{P_{j^{-1}}}, \quad j \in S.$$

The first congruence in (2.1.11) follows by taking $j = 1$ in (2.1.14). The second congruence in (2.1.11) follows by taking $j = 1$ in (2.1.12), since the rightmost inner sum in (2.1.12) vanishes for each $v \neq (k-m)f$, while for $v = (k-m)f$ it is congruent to $-1 \pmod P$.

Now suppose that $r = 1$, i.e., $p \equiv 1 \pmod k$. Then all the prime ideals P_j in (2.1.8) are distinct, so that, by (2.1.14),

$$(2.1.15)\qquad J(\chi^m, \chi^n)O_K = A \prod_{j \in S} P_{j^{-1}}$$

for some ideal A of O_K. Note that for an integer $a \not\equiv 0 \pmod k$, $L(-a) = k - L(a)$. Hence, by (2.1.14) with m and n replaced by $k-m$ and $k-n$, respectively,

$$(2.1.16)\qquad J(\chi^{-m}, \chi^{-n})O_K = B \prod_{\substack{j \notin S \\ 0 < j < k \\ \gcd(j,k)=1}} P_j^{-1}$$

for some ideal B of O_K. Multiplying (2.1.15) and (2.1.16), we obtain $pO_K = ABp$, by (2.1.8). Thus $A = B = O_K$, and so (2.1.9) follows from (2.1.15). ■

For an extension of (2.1.11) to the generalized Jacobi sums in (10.1.1), see Problem 22 of Exercises 10. It is easily seen that the case $t = 2$ of that problem is equivalent to (2.1.11).

2.2 Cyclotomic numbers

In this section, let k be an integer with $k \geq 2$, and let $p = kf + 1$ be a prime. Let g be a fixed primitive root (mod p). A table of primitive roots for primes $p < 50,000$ has been given by Hardy and Williams [1]. If n is an integer not divisible by p, then the unique integer i such that

$$n \equiv g^i \pmod{p}, \quad 0 \leq i < p-1,$$

is called the index of n with respect to the primitive root g, and is denoted by $\text{ind}_g n$. Indices play a role similar to logarithms. They have the following properties:

$$\begin{aligned} \text{ind}_g m + \text{ind}_g n &\equiv \text{ind}_g mn \pmod{p-1}, \\ \text{ind}_g n^t &\equiv t\, \text{ind}_g n \pmod{p-1}, \\ \text{ind}_g 1 = 0,\ \text{ind}_g(-1) &= (p-1)/2. \end{aligned}$$

Definition *For given integers s and t, the cyclotomic number $(s,t)_k$ of order k is defined to be the number of integers n* (mod p) *for which n/g^s and $(n+1)/g^t$ are both nonzero kth-power residues* (mod p).

Observe that $(s,t)_k$ depends only on the congruence classes (mod k) to which s and t belong. Note also that $(s,t)_k$ depends upon p and g but the notation does not reflect this. The cyclotomic matrix of order k is the $k \times k$ matrix whose entry in the ith row and jth column is the cyclotomic number $(i-1, j-1)_k$. The cyclotomic number $(s,t)_k$ can also be viewed as the number of integers n satisfying

$$\text{ind}_g n \equiv s \pmod{k}, \quad \text{ind}_g(n+1) \equiv t \pmod{k}, \quad 1 \leq n \leq p-2,$$

and also as the number of pairs (u_1, u_2) of integers satisfying

$$1 + g^{ku_1+s} \equiv g^{ku_2+t} \pmod{p}, \qquad 0 \leq u_1, u_2 \leq f-1.$$

From this latter formulation it is clear that the number of solutions (u_1, u_2) of the congruence

$$1 + g^{ku_1+s} + g^{ku_2+t} \equiv 0 \pmod{p}, \qquad 0 \leq u_1, u_2 \leq f-1,$$

is

$$(s, t + \tfrac{1}{2}kf)_k = \begin{cases} (s,t)_k, & \text{if } f \equiv 0 \pmod 2, \\ (s, t + \tfrac{1}{2}k)_k, & \text{if } f \not\equiv 0 \pmod 2. \end{cases}$$

The basic properties of the cyclotomic numbers of order k are given in the next theorem.

Theorem 2.2.1 *The cyclotomic numbers of order k have the following properties:*

(2.2.1) $$(s,t)_k = (-s, t-s)_k,$$

(2.2.2) $$(s,t)_k = \begin{cases} (t,s)_k, & \textit{if } f \equiv 0 \pmod 2, \\ (t + k/2, s + k/2)_k, & \textit{if } f \not\equiv 0 \pmod 2, \end{cases}$$

(2.2.3) $$\sum_{t=0}^{k-1} (s,t)_k = \begin{cases} f-1, & \textit{if } f \equiv 0 \pmod 2 \textit{ and } s \equiv 0 \pmod k, \\ & \textit{or } f \not\equiv 0 \pmod 2 \textit{ and } s \equiv k/2 \pmod k, \\ f, & \textit{otherwise}, \end{cases}$$

(2.2.4) $$\sum_{s=0}^{k-1} (s,t)_k = \begin{cases} f-1, & \textit{if } t \equiv 0 \pmod k, \\ f, & \textit{if } t \not\equiv 0 \pmod k, \end{cases}$$

(2.2.5) $$(s,t)'_k = (Ns, Nt)_k,$$

where the prime (′) indicates that the cyclotomic number is taken with respect to the primitive root g^N *in place of* g.

Proof Consider a value of n for which ng^{-s} and $(n+1)g^{-t}$ are both (nonzero) kth-power residues $\pmod p$. Suppose that $a^k \equiv ng^{-s} \pmod p$ and $b^k \equiv (n+1)g^{-t} \pmod p$. Then $a^{-k} \equiv n^{-1}g^s \pmod p$ and $(a^{-1}b)^k \equiv (n^{-1}+1)g^{s-t} \pmod p$. Hence, (2.2.1) follows. Replacing n by $-n-1$, we obtain (2.2.2) because

(2.2.6) $$\operatorname{ind}_g(-1) = kf/2 \equiv \begin{cases} 0 \pmod k, & \text{if } f \equiv 0 \pmod 2, \\ k/2 \pmod k, & \text{if } f \not\equiv 0 \pmod 2. \end{cases}$$

The left side of (2.2.3) counts the number of values of $n \pmod p$ of the form $n = g^{s+rk} \not\equiv -1 \pmod p$, $1 \le r \le f$. The result (2.2.3) then follows from (2.2.6). Next, (2.2.4) is an easy consequence of (2.2.2) and (2.2.3). Finally, (2.2.5) follows after replacing g by g^N in the definition of $(s,t)_k$. ∎

It is interesting to observe that the properties given in Theorem 2.2.1 are sufficient to determine the cyclotomic numbers of order 2.

Theorem 2.2.2 *Let* $p = 2f+1$ *be a prime. Then*

$$4(0,0)_2 = p - 4 - (-1)^f,$$
$$4(0,1)_2 = p - (-1)^f,$$
$$4(1,0)_2 = 4(1,1)_2 = p - 2 + (-1)^f.$$

Proof When f is even, we see by (2.2.1) and (2.2.2) that the cyclotomic matrix of order 2 has the form

$$\begin{array}{c} \\ 0 \\ 1 \end{array} \begin{array}{c} \begin{array}{cc} 0 & 1 \end{array} \\ \begin{bmatrix} A & B \\ B & B \end{bmatrix} \end{array},$$

where A and B satisfy (by (2.2.3)) the linear relations

$$\begin{aligned} A + B &= f - 1, \\ 2B \quad &= f. \end{aligned}$$

Hence in this case,

$$\begin{aligned} 4(0,0)_2 &= 4A = 2f - 4 = p - 5, \\ 4(0,1)_2 &= 4(1,0)_2 = 4(1,1)_2 = 4B = 2f = p - 1, \end{aligned}$$

as asserted.

When f is odd, the cyclotomic matrix of order 2 is (by (2.2.1) and (2.2.2)) of the form

$$\begin{array}{c} \\ 0 \\ 1 \end{array} \begin{array}{c} \begin{array}{cc} 0 & 1 \end{array} \\ \begin{bmatrix} A & B \\ A & A \end{bmatrix} \end{array},$$

where A and B satisfy (by (2.2.3)) the linear relations

$$\begin{aligned} A + B &= f, \\ 2A \quad &= f - 1. \end{aligned}$$

Hence in this case,

$$\begin{aligned} 4(0,0)_2 &= 4(1,0)_2 = 4(1,1)_2 = 4A = 2f - 2 = p - 3, \\ 4(0,1)_2 &= 4B = 2f + 2 = p + 1, \end{aligned}$$

as asserted. ■

When $k > 2$ the properties (2.2.1)–(2.2.4) are not sufficient to determine the cyclotomic numbers of order k. For example, when $k = 3$, there are $k^2 = 9$ cyclotomic numbers of order 3 but (2.2.1)–(2.2.4) give only seven linear relations among them.

In the next two sections we give Gauss' evaluation of the cyclotomic numbers of order 3 and 4 and then go on to give the relationship between Jacobi sums of order k and cyclotomic numbers of order k.

2.3 Cyclotomic numbers of order 3

Let p be a prime of the form $p = 3f + 1$. Let g be a primitive root $(\bmod\ p)$. Properties (2.2.1) and (2.2.2) with $k = 3$ show that the cyclotomic matrix of order 3 has the form

$$\begin{array}{c} \\ 0 \\ 1 \\ 2 \end{array} \begin{array}{c} \begin{array}{ccc} 0 & 1 & 2 \end{array} \\ \begin{bmatrix} A & B & C \\ B & C & D \\ C & D & B \end{bmatrix} \end{array} . \tag{2.3.1}$$

According to (2.2.3), A, B, C, D satisfy the linear relations

$$\begin{cases} A + B + C = f - 1, \\ B + C + D = f. \end{cases} \tag{2.3.2}$$

Thus two more linear relations between A, B, C, and D are needed in order to determine the cyclotomic numbers of order 3. We show next how Gauss [2, §358] was able to derive two further linear relations by counting the number N of solutions of the congruence

$$1 + g^{3u_1} + g^{3u_2+1} + g^{3u_3+2} \equiv 0 \ (\bmod\ p), \quad 0 \le u_1, u_2, u_3 \le f - 1,$$

in two different ways.

First, noting that this congruence does not have a solution with $1 + g^{3u_1} \equiv 0 \ (\bmod\ p)$, we see that

$$\begin{aligned} N &= \sum_{\substack{u_1, u_2, u_3 = 0 \\ 1 + g^{3u_1} + g^{3u_2+1} + g^{3u_3+2} \equiv 0 \ (\bmod\ p)}}^{f-1} 1 \\ &= \sum_{h=0}^{2} \sum_{v=0}^{f-1} \Bigg(\sum_{\substack{u_1 = 0 \\ 1 + g^{3u_1} \equiv g^{3v+h} \ (\bmod\ p)}}^{f-1} 1 \Bigg) \Bigg(\sum_{\substack{u_2, u_3 = 0 \\ g^{3v+h} + g^{3u_2+1} + g^{3u_3+2} \equiv 0 \ (\bmod\ p)}}^{f-1} 1 \Bigg) \\ &= \sum_{h=0}^{2} \Bigg(\sum_{\substack{u_1, v = 0 \\ 1 + g^{3u_1} \equiv g^{3v+h} \ (\bmod\ p)}}^{f-1} 1 \Bigg) \Bigg(\sum_{\substack{v_2, v_3 = 0 \\ 1 + g^{3v_2+1-h} + g^{3v_3+2-h} \equiv 0 \ (\bmod\ p)}}^{f-1} 1 \Bigg) \\ &= \sum_{h=0}^{2} (0, h)_3 (1 - h, 2 - h)_3 \\ &= AD + B^2 + C^2. \end{aligned}$$

Second, writing N in the form

$$N = \sum_{\substack{u_1, u_2, u_3 = 0 \\ 1 + g^{3u_2+1} + g^{3u_1} + g^{3u_3+2} \equiv 0 \ (\bmod\ p)}}^{f-1} 1$$

$$= \sum_{h=0}^{2} \sum_{v=0}^{f-1} \Big(\sum_{\substack{u_2=0 \\ 1+g^{3u_2+1} \equiv g^{3v+h} \pmod p}}^{f-1} 1 \Big) \Big(\sum_{\substack{u_1, u_3=0 \\ g^{3v+h}+g^{3u_1}+g^{3u_3+2} \equiv 0 \pmod p}}^{f-1} 1 \Big)$$

and proceeding exactly as above, we obtain

$$N = BC + CD + BD.$$

Equating the two expressions for N, we obtain

$$AD + B^2 + C^2 = BC + CD + BD. \tag{2.3.3}$$

Now, from (2.3.2),

$$A = D - 1, \tag{2.3.4}$$

and substituting (2.3.4) into (2.3.3), we find that

$$D = B^2 + C^2 + D^2 - BC - BD - CD. \tag{2.3.5}$$

From (2.3.2) and (2.3.5),

$$\begin{aligned} 4p &= 12f + 4 = 12(B + C + D) + 4 = 36D + 12(B + C) - 24D + 4 \\ &= 36(B^2 + C^2 + D^2 - BC - BD - CD) + 12(B + C) - 24D + 4, \end{aligned}$$

that is

$$4p = (-3B - 3C + 6D - 2)^2 + 3(3B - 3C)^2.$$

This quadratic partition of $4p$ gives the two additional linear relations involving A, B, C, D that we are seeking, namely,

$$\begin{cases} -3B - 3C + 6D - 2 = r_3, \\ 3B - 3C \qquad\qquad\quad = s_3, \end{cases} \tag{2.3.6}$$

where

$$4p = r_3^2 + 3s_3^2, \quad r_3 \equiv 1 \pmod 3, \quad s_3 \equiv 0 \pmod 3). \tag{2.3.7}$$

Solving the four linear equations in (2.3.2) and (2.3.6) for A, B, C, D, we obtain

$$\begin{aligned} 9A &= p + r_3 - 8, \\ 18B &= 2p - r_3 + 3s_3 - 4, \\ 18C &= 2p - r_3 - 3s_3 - 4, \\ 9D &= p + r_3 + 1. \end{aligned}$$

It is well known (see Lemma 3.0.1) that (2.3.7) determines r_3 uniquely but determines s_3 only up to sign. Thus r_3 depends only on p and not on the choice of primitive root g. We show that s_3 is uniquely determined by proving that s_3 satisfies the congruence

$$3s_3 \equiv (g^{(p-1)/3} - g^{2(p-1)/3}) r_3 \pmod p.$$

We do this by considering the sum

$$\sum := \sum_{x=1}^{p-1} (1+x^3)^{(p-1)/3}$$

modulo p in two different ways. Set $w \equiv g^{(p-1)/3} \equiv g^f \pmod{p}$, so that

$$w^3 \equiv 1 \pmod{p},\ \ w \not\equiv 1 \pmod{p},\ \ 1+w+w^2 \equiv 0 \pmod{p},\ \ (w-w^2)^2 \equiv -3 \pmod{p}.$$

First, making use of the formulas we obtained for the cyclotomic numbers of order 3, we find that modulo p

$$\begin{aligned}
\sum &\equiv \sum_{u=0}^{3f-1} (1+g^{3u})^f \equiv 3\sum_{u=0}^{f-1} (1+g^{3u})^f \equiv 3 \sum_{\substack{u=0 \\ u \neq f/2}}^{f-1} (1+g^{3u})^f \\
&\equiv 3\sum_{h=0}^{2}\sum_{v=0}^{f-1} \sum_{\substack{u=0 \\ 1+g^{3u} \equiv g^{3v+h} (\mathrm{mod}\ p)}}^{f-1} g^{(3v+h)f} \equiv 3\sum_{h=0}^{2} g^{hf} \sum_{\substack{u,v=0 \\ 1+g^{3u} \equiv g^{3v+h} (\mathrm{mod}\ p)}}^{f-1} 1 \equiv 3\sum_{h=0}^{2} w^h (0,h)_3 \\
&\equiv \frac{1}{6}\Big((2p+2r_3-16) + w(2p-r_3+3s_3-4) + w^2(2p-r_3-3s_3-4)\Big) \\
&\equiv \frac{1}{2}(r_3 + s_3(w-w^2)) - 2.
\end{aligned}$$

Second, using (2.1.13) and the binomial theorem, we find modulo p that

$$\begin{aligned}
\sum &\equiv \sum_{x=1}^{p-1}\sum_{n=0}^{(p-1)/3} \binom{(p-1)/3}{n} x^{3n} \equiv \sum_{n=0}^{(p-1)/3} \binom{(p-1)/3}{n} \sum_{x=1}^{p-1} x^{3n} \\
&\equiv -\sum_{\substack{n=0 \\ p-1 \mid 3n}}^{(p-1)/3} \binom{(p-1)/3}{n} = -2.
\end{aligned}$$

Putting these two determinations of $\sum$ modulo p together, we see that

$$r_3 + s_3(w-w^2) \equiv 0 \pmod{p}.$$

Multiplying this congruence by $(w-w^2)$, we obtain

$$3s_3 \equiv (w-w^2)r_3 \pmod{p}$$

as asserted.

We have thus established the following theorem.

Theorem 2.3.1 *Let p be a prime of the form $p = 3f + 1$. Let g be a primitive root* $(\text{mod } p)$. *Then the cyclotomic numbers of order* 3 *are given by* (2.3.1) *and*

$$(2.3.8)\qquad \begin{cases} 9A = p + r_3 - 8, \\ 18B = 2p - r_3 + 3s_3 - 4, \\ 18C = 2p - r_3 - 3s_3 - 4, \\ 9D = p + r_3 + 1, \end{cases}$$

where the integers r_3 and s_3 are given uniquely by

$$(2.3.9)\qquad \begin{cases} 4p = r_3^2 + 3s_3^2, \quad r_3 \equiv 1 \pmod 3, \quad s_3 \equiv 0 \pmod 3, \\ 3s_3 \equiv (g^{(p-1)/3} - g^{2(p-1)/3})r_3 \pmod p. \end{cases}$$

We will encounter the partition (2.3.9) again in Theorem 3.1.3.

Example For a numerical example to illustrate Theorem 2.3.1, we take $p = 7$ and $g = 3$. Clearly, $r_3 = 1$ and $s_3 = \pm 3$. As $g^{(p-1)/3} \equiv 3^2 \equiv 2 \pmod 7$, we have $3s_3 \equiv 2 - 4 \equiv -9 \pmod 7$, so that $s_3 = -3$. Then (2.3.8) gives

$$(0,0)_3 = 0, \ (0,1)_3 = 0, \ (0,2)_3 = 1, \ (1,2)_3 = 1.$$

This is easily seen to be correct, as modulo 3 we have

$$\text{ind}_3 1 \equiv 0, \ \text{ind}_3 2 \equiv 2, \ \text{ind}_3 3 \equiv 1, \ \text{ind}_3 4 \equiv 1, \ \text{ind}_3 5 \equiv 2, \ \text{ind}_3 6 \equiv 0.$$

2.4 Cyclotomic numbers of order 4

Let p be a prime of the form $p = 4f + 1$. Let g be a primitive root $(\text{mod } p)$. Appealing to (2.2.1) and (2.2.2), we see that the cyclotomic matrix of order 4 takes the following forms according as f is even or odd:

$$(2.4.1)\qquad \begin{array}{c} f \text{ even} \\ \begin{array}{c|cccc} & 0 & 1 & 2 & 3 \\ \hline 0 & A & B & C & D \\ 1 & B & D & E & E \\ 2 & C & E & C & E \\ 3 & D & E & E & B \end{array} \end{array} \qquad \begin{array}{c} f \text{ odd} \\ \begin{array}{c|cccc} & 0 & 1 & 2 & 3 \\ \hline 0 & A & B & C & D \\ 1 & E & E & D & B \\ 2 & A & E & A & E \\ 3 & E & D & B & E \end{array} \end{array}.$$

Further, from (2.2.3), we obtain the following linear relations:

$$
\begin{array}{lcc}
 & f \text{ even} & f \text{ odd} \\
(2.4.2) & \left\{\begin{aligned} A+B+C+D &= f-1, \\ B+D+2E &= f, \\ 2C+2E &= f, \end{aligned}\right. & \left\{\begin{aligned} A+B+C+D &= f, \\ B+D+2E &= f, \\ 2A+2E &= f-1. \end{aligned}\right.
\end{array}
$$

As in the case $k = 3$, two more linear relations involving A, B, C, D, and E are needed in order to determine the cyclotomic numbers of order 4. By counting the number of solutions of the congruence

$$1 + g^{4u_1} + g^{4u_2+1} + g^{4u_3+2} \equiv 0 \pmod{p},\ 0 \le u_1, u_2, u_3 \le f-1,$$

in two different ways, as in the case $k = 3$, we obtain the following quadratic relations:

$$(2.4.3) \quad \begin{cases} AE + B^2 + CD + DE = BC + DE + CE + E^2, & \text{if } f \text{ is even}, \\ AE + BD + BC + DE = AE + E^2 + AD + BE, & \text{if } f \text{ is odd}. \end{cases}$$

It is convenient at this point to treat the two cases f even and f odd separately.

***f* even:** From (2.4.3),

$$(2.4.4) \qquad 0 = AE + B^2 + CD - BC - CE - E^2.$$

Now, from (2.4.2),

$$A = -C + 2E - 1,$$

and, using this in (2.4.4), we obtain

$$(2.4.5) \qquad 0 = (C-E)^2 + B^2 + CD - BC - C^2 - E.$$

Next, from (2.4.2),

$$C = (B + D)/2,$$

so that

$$B^2 + CD - BC - C^2 = (B - D)^2/4.$$

Thus (2.4.5) becomes

$$(2.4.6) \qquad 0 = 4(C-E)^2 + (B-D)^2 - 4E.$$

Further, from (2.4.2),

$$4E = f - 2(C - E),$$

so that (2.4.6) can be rewritten as

$$f = 4(C-E)^2 + 2(C-E) + (B-D)^2.$$

Multiplying both sides by 4 and adding 1, we obtain

$$p = 4f + 1 = (4(C - E) + 1)^2 + (2B - 2D)^2.$$

This gives the two required linear relations, namely,

$$\text{(2.4.7)} \qquad \begin{cases} 4(C - E) + 1 = -a_4, \\ 2B - 2D = b_4, \end{cases}$$

where

$$\text{(2.4.8)} \qquad p = a_4^2 + b_4^2, \quad a_4 \equiv -1 \pmod 4).$$

Solving (2.4.2) and (2.4.7) for A, B, C, D, E, we obtain

$$\begin{aligned} 16A &= p + 6a_4 - 11, \\ 16B &= p - 2a_4 + 4b_4 - 3, \\ 16C &= p - 2a_4 - 3, \\ 16D &= p - 2a_4 - 4b_4 - 3, \\ 16E &= p + 2a_4 + 1. \end{aligned}$$

Again it is well known (see Lemma 3.0.1) that (2.4.8) determines a_4 uniquely (so that a_4 does not depend on the choice of the primitive root g) but determines b_4 only up to sign. We show how to determine b_4 uniquely at the end of this section.

***f* odd:** From (2.4.3),

$$0 = E^2 + AD + BE - BD - BC - DE.$$

Now, from (2.4.2), we obtain $A + C = 2E$, so that eliminating C, we have

$$\text{(2.4.9)} \qquad 0 = E^2 + AD + AB - BD - BE - DE.$$

Further, from (2.4.2), we see that $B + D = 1 + 2A$, so that

$$\text{(2.4.10)} \qquad \begin{cases} 2B = 1 + 2A + (B - D), \\ 2D = 1 + 2A - (B - D). \end{cases}$$

Multiplying (2.4.9) by 4, and replacing $2B$ and $2D$ by means of (2.4.10), we deduce that

$$0 = 4E^2 - 4E - 1 - 8AE + 4A^2 + (B - D)^2,$$

that is

$$4E + 1 = (B - D)^2 + 4(A - E)^2.$$

Hence, from (2.4.2),

$$\begin{aligned} p &= 4f + 1 = 4(2A + 2E + 1) + 1 = 4(4E + 1) + 8(A - E) + 1 \\ &= 4(B - D)^2 + 16(A - E)^2 + 8(A - E) + 1 = (4(A - E) + 1)^2 + (2B - 2D)^2, \end{aligned}$$

so that

$$\begin{cases} 4(A - E) + 1 = a_4, \\ 2B - 2D \quad\quad = b_4, \end{cases} \tag{2.4.11}$$

where

$$p = a_4^2 + b_4^2, \quad a_4 \equiv 1 \pmod 4. \tag{2.4.12}$$

Solving (2.4.2) and (2.4.11) for A, B, C, D, and E, we obtain

$$\begin{aligned} 16A &= p + 2a_4 - 7, \\ 16B &= p + 2a_4 + 4b_4 + 1, \\ 16C &= p - 6a_4 + 1, \\ 16D &= p + 2a_4 - 4b_4 + 1, \\ 16E &= p - 2a_4 - 3. \end{aligned}$$

We conclude this section by showing that, for both f even and f odd, b_4 is determined uniquely by means of the congruence

$$b_4 \equiv g^{(p-1)/4} a_4 \pmod p.$$

Analogous to the argument at the end of Section 2.3, we do this by considering the sum

$$\sum := \sum_{x=1}^{p-1} (1 + x^4)^{(p-1)/4}.$$

Set $w \equiv g^{(p-1)/4} \pmod p$, so that $w^2 \equiv -1 \pmod p$, $w^3 \equiv -w \pmod p$ and $w^4 \equiv 1 \pmod p$.

First, we replace x $(1 \le x \le p-1)$ by g^u $(0 \le u \le p-2)$ in $\sum$. Then, proceeding exactly as in the previous section and making use of the formulas for the cyclotomic numbers A, B, C, D, E in terms of a_4 and b_4, we find that

$$\sum \equiv 2a_4 + 2b_4 w - 2 \pmod p.$$

Second, by applying the binomial theorem to $(1+x^4)^{(p-1)/4}$ in $\sum$ and then interchanging the order of summation, we obtain

$$\sum \equiv -2 \pmod p.$$

Comparing the two evaluations of $\sum$ modulo p, we deduce that

$$a_4 + b_4 w \equiv 0 \pmod p,$$

and so

$$b_4 \equiv a_4 w \pmod p$$

as asserted.

Summarizing, we have the following theorem.

Theorem 2.4.1 *Let p be a prime of the form $p = 4f + 1$. Let g be a primitive root* $\pmod p$. *Then the cyclotomic numbers of order* 4 *are given by* (2.4.1), *where if* $f \equiv 0 \pmod 2$,

$$\begin{aligned} 16A &= p + 6a_4 - 11, \\ 16B &= p - 2a_4 + 4b_4 - 3, \\ 16C &= p - 2a_4 - 3, \\ 16D &= p - 2a_4 - 4b_4 - 3, \\ 16E &= p + 2a_4 + 1, \end{aligned}$$

and if $f \equiv 1 \pmod 2$,

$$\begin{aligned} 16A &= p + 2a_4 - 7, \\ 16B &= p + 2a_4 + 4b_4 + 1, \\ 16C &= p - 6a_4 + 1, \\ 16D &= p + 2a_4 - 4b_4 + 1, \\ 16E &= p - 2a_4 - 3, \end{aligned}$$

with the integers a_4 and b_4 given uniquely by

$$(2.4.13) \qquad p = a_4^2 + b_4^2,\ a_4 \equiv -\left(\frac{2}{p}\right) \pmod 4,\ b_4 \equiv g^{(p-1)/4} a_4 \pmod p.$$

We will encounter the partition (2.4.13) again in Theorem 3.2.1.

Example For a numerical example to illustrate Theorem 2.4.1, we take $p = 17$ and $g = 3$. Clearly $a_4 = -1$ and $b_4 = \pm 4$. As $g^{(p-1)/4} \equiv 3^4 \equiv -4 \pmod{17}$, we have $b_4 \equiv (-4)(-1) \equiv 4 \pmod{17}$ so that $b_4 = 4$. Then, by Theorem 2.4.1,

$$(0,0)_4 = 0,\ (0,1)_4 = 2,\ (0,2)_4 = 1,\ (0,3)_4 = 0,\ (1,2)_4 = 1,$$

which are easily seen to be correct, as modulo 4 we have

$$\begin{aligned} &\text{ind}_3 1 \equiv 0,\ \text{ind}_3 2 \equiv 2,\ \text{ind}_3 3 \equiv 1,\ \text{ind}_3 4 \equiv 0,\ \text{ind}_3 5 \equiv 1,\ \text{ind}_3 6 \equiv 3, \\ &\text{ind}_3 7 \equiv 3,\ \text{ind}_3 8 \equiv 2,\ \text{ind}_3 9 \equiv 2,\ \text{ind}_3 10 \equiv 3,\ \text{ind}_3 11 \equiv 3,\ \text{ind}_3 12 \equiv 1, \\ &\text{ind}_3 13 \equiv 0,\ \text{ind}_3 14 \equiv 1,\ \text{ind}_3 15 \equiv 2,\ \text{ind}_3 16 \equiv 0. \end{aligned}$$

The method of Gauss used in Sections 2.3 and 2.4 to determine the cyclotomic numbers of orders 3 and 4, respectively, required transforming a quadratic relation involving cyclotomic numbers into a quadratic representation of a multiple of p with essentially a unique solution. It is far from clear how to do this for general k, although it has essentially been done for $k = 5$; see Rajwade [3] and Problems 8–14 of this chapter. A better method is therefore required in general to determine formulas for the cyclotomic numbers of order k, and that is provided by the use of Jacobi sums. This is the topic of the next section.

2.5 Relationship between Jacobi sums and cyclotomic numbers

Let k be an integer with $k \geq 2$, and let p be a prime of form $p = kf + 1$. Let g be a fixed primitive root (mod p). Let $\beta = \exp(2\pi i/k)$, and let χ be a character (mod p) of order k such that $\chi(g) = \beta$. We show in the next theorem that the cyclotomic numbers of order k can be expressed linearly in terms of Jacobi sums of order dividing k. For an extension to general finite fields F_q, see (11.6.5).

In Section 2.1, we defined the Jacobi sum $J_r(\chi, \psi)$ to be $\sum \chi(\alpha)\psi(1-\alpha)$, where the sum is over all α in F_q. It is convenient to define another Jacobi sum

$$J_r^*(\chi, \psi) = \sum_{\substack{\alpha \in F_q \\ \alpha \neq 0,1}} \chi(\alpha)\psi(1-\alpha).$$

Observe that $J_r(\chi, \psi) = J_r^*(\chi, \psi) + l$, where l is the number of trivial characters in the set $\{\chi, \psi\}$.

Theorem 2.5.1 *For* $s, t = 0, 1, \ldots, k-1$,

(2.5.1) $$k^2(s,t)_k = \sum_{u=0}^{k-1}\sum_{v=0}^{k-1}(-1)^{uf}\beta^{-su-tv}J^*(\chi^u, \chi^v)$$

Proof By (1.1.3), $\chi(-1) = (-1)^f$. Then

$$\begin{aligned}
\sum_{u=0}^{k-1}\sum_{v=0}^{k-1}(-1)^{uf}\beta^{-su-tv}J^*(\chi^u, \chi^v) &= \sum_{u=0}^{k-1}\sum_{v=0}^{k-1}\chi^u(-1)\chi^{-su-tv}(g)\sum_{n=2}^{p-1}\chi^u(n)\chi^v(1-n) \\
&= \sum_{u=0}^{k-1}\sum_{v=0}^{k-1}\chi^u(g^{-s})\chi^v(g^{-t})\sum_{n=2}^{p-1}\chi^u(-n)\chi^v(1-n) \\
&= \sum_{u=0}^{k-1}\sum_{v=0}^{k-1}\chi^u(g^{-s})\chi^v(g^{-t})\sum_{n=1}^{p-2}\chi^u(n)\chi^v(1+n) \\
&= \sum_{n=1}^{p-2}\sum_{u=0}^{k-1}\chi^u(n/g^s)\sum_{v=0}^{k-1}\chi^v((n+1)/g^t).
\end{aligned}$$

Now,

$$\sum_{u=0}^{k-1} \chi^u(n/g^s) = \begin{cases} k, & \text{if } n/g^s \text{ is a } k\text{th power (mod } p), \\ 0, & \text{otherwise,} \end{cases}$$

and

$$\sum_{v=0}^{k-1} \chi^v((n+1)/g^t) = \begin{cases} k, & \text{if } (n+1)/g^t \text{ is a } k\text{th power (mod } p), \\ 0, & \text{otherwise.} \end{cases}$$

Thus, the sum on n equals

$$\sum_{\substack{n=1 \\ n/g^s \text{ and } (n+1)/g^t \\ \text{both } k\text{th powers (mod } p)}}^{p-2} k^2 = k^2(s,t)_k. \qquad \blacksquare$$

Theorem 2.5.1 shows that if the Jacobi sums $J(\chi^u, \chi^v)$ with orders dividing k are known explicitly then by substituting their values into the right side of (2.5.1), we can obtain formulas for the cyclotomic numbers $(s,t)_k$. This method has been used to determine explicit formulas for the cyclotomic numbers of order k for all $k \le 12$ and $k = 14, 15, 16, 18, 20$, and 24. Here we just illustrate the case $k = 2$ by this method. In this case $p = 2f + 1$, $\rho(-1) = (-1)^f$, and the values of the Jacobi sums of order 2 are (by Theorem 2.1.1)

$$J^*(\rho^m, \rho^n) = \begin{cases} p-2, & \text{if } m = n = 0, \\ -1, & \text{if } m = 0,\ n = 1 \text{ or } m = 1,\ n = 0, \\ (-1)^{f+1}, & \text{if } m = n = 1. \end{cases}$$

Thus for f even (equivalently $p \equiv 1 \pmod 4$), Theorem 2.5.1 gives

$$\begin{aligned} 4(0,0)_2 &= \sum_{u=0}^{1}\sum_{v=0}^{1} J^*(\rho^u, \rho^v) = p - 5, \\ 4(0,1)_2 &= \sum_{u=0}^{1}\sum_{v=0}^{1} (-1)^v J^*(\rho^u, \rho^v) = p - 1, \\ 4(1,0)_2 &= \sum_{u=0}^{1}\sum_{v=0}^{1} (-1)^u J^*(\rho^u, \rho^v) = p - 1, \\ 4(1,1)_2 &= \sum_{u=0}^{1}\sum_{v=0}^{1} (-1)^{u+v} J^*(\rho^u, \rho^v) = p - 1; \end{aligned}$$

and for f odd (equivalently $p \equiv 3 \pmod 4$),

$$4(0,0)_2 = \sum_{u=0}^{1}\sum_{v=0}^{1} (-1)^u J^*(\rho^u, \rho^v) = p - 3,$$

$$4(0,1)_2 = \sum_{u=0}^{1}\sum_{v=0}^{1}(-1)^{u+v}J^*(\rho^u,\rho^v) = p+1,$$

$$4(1,0)_2 = \sum_{u=0}^{1}\sum_{v=0}^{1}J^*(\rho^u,\rho^v) = p-3,$$

$$4(1,1)_2 = \sum_{u=0}^{1}\sum_{v=0}^{1}(-1)^{v}J^*(\rho^u,\rho^v) = p-3,$$

which reproves Theorem 2.2.2.

In the next section we show how the cyclotomic numbers of order k can be used to determine $\text{ind}_g 2$ and $\text{ind}_g k$ modulo k.

2.6 Determination of $\text{ind}_g 2$ and $\text{ind}_g k$ (mod k)

Let k be an integer with $k \geq 2$ and let p be a prime of the form $p = kf+1$. Let g be a primitive root (mod p). Given an integer l we are interested in determining $\text{ind}_g l$ modulo k. For a general l, this determination will be done in terms of generalized cyclotomic numbers in Section 2.7. Here we examine two cases that can be handled using ordinary cyclotomic numbers of order k, namely, $l=2$ and $l=k$.

First we examine $l=2$.

Theorem 2.6.1 *For* $t = 0,1,\ldots,k-1$,

$$\text{ind}_g 2 \equiv t \pmod{k} \text{ if and only if } (0,t)_k \equiv 1 \pmod{2}.$$

Proof Let n be an integer with the property that n and $(n+1)g^{-t}$ are both nonzero kth-power residues (mod p). Then n^{-1} (mod p) also has this property. To see this, suppose that $a^k \equiv n \pmod{p}$ and $b^k \equiv (n+1)g^{-t} \pmod{p}$. Then $(a^{-1})^k \equiv n^{-1} \pmod{p}$ and $(a^{-1}b)^k \equiv (n^{-1}+1)g^{-t} \pmod{p}$. Moreover, $n \not\equiv n^{-1} \pmod{p}$ unless $n \equiv 1 \pmod{p}$, and the latter congruence occurs if and only if $2g^{-t}$ is a kth-power residue (mod p). Hence $(0,t)_k$ is odd if and only if $\text{ind}_g 2 \equiv t \pmod{k}$. ■

An immediate consequence of Theorem 2.6.1 is the following result about the parity of cyclotomic numbers.

Corollary 2.6.2 *Exactly one of the numbers* $(0,t)_k$ $(t = 0,1,\ldots,k-1)$ *is odd.*

Taking $k=2$ and $t=0$ in Theorem 2.6.1, and recalling that $4(0,0)_2 = p-4-(-1)^f$ by Theorem 2.2.2, we readily deduce the well-known fact that 2 is a quadratic residue (mod p) if and only if $p \equiv \pm 1 \pmod{8}$.

Next, taking $k=3$ and $t=0,1,2$ in Theorem 2.6.1, and making use of the formulas for the cyclotomic numbers of order 3 given in Theorem 2.3.1, we obtain the following determination of $\text{ind}_g 2$ modulo 3 in terms of congruences involving r_3 and s_3.

Theorem 2.6.3 *Let $p = 3f + 1$ be a prime. Let g be a primitive root* $\pmod p$. *Then*

$$\begin{aligned} &\text{ind}_g 2 \equiv 0 \pmod 3, && \text{if } r_3 \equiv s_3 \equiv 0 \pmod 2, \\ &\text{ind}_g 2 \equiv 1 \pmod 3, && \text{if } r_3 \equiv s_3 \equiv 1 \pmod 2 \text{ and } r_3 \equiv -s_3 \pmod 4, \\ &\text{ind}_g 2 \equiv 2 \pmod 3, && \text{if } r_3 \equiv s_3 \equiv 1 \pmod 2 \text{ and } r_3 \equiv s_3 \pmod 4, \end{aligned}$$

where the integers r_3 and s_3 are given by (2.3.9).

Proof From (2.3.9) we see that $r_3 \equiv s_3 \pmod 2$. If $r_3 \equiv s_3 \equiv 0 \pmod 2$, then by (2.3.8) we have $(0,0)_3 \equiv 9(0,0)_3 \equiv p + r_3 - 8 \equiv 1 \pmod 2$ so that, by Theorem 2.6.1, $\text{ind}_g 2 \equiv 0 \pmod 3$.

If $r_3 \equiv s_3 \equiv 1 \pmod 2$ and $r_3 \equiv -s_3 \pmod 4$, then, by (2.3.8),

$$2(0,1)_3 \equiv 2B \equiv 18B \equiv 2p - r_3 + 3s_3 - 4 \equiv 2 - r_3 - s_3 \equiv 2 \pmod 4.$$

Hence $(0,1)_3 \equiv 1 \pmod 2$ and, by Theorem 2.6.1, $\text{ind}_g 2 \equiv 1 \pmod 3$.

Finally, if $r_3 \equiv s_3 \equiv 1 \pmod 2$ and $r_3 \equiv s_3 \pmod 4$, then, by (2.3.8),

$$2(0,2)_3 \equiv 2C \equiv 18C \equiv 2p - r_3 - 3s_3 - 4 \equiv 2 - r_3 + s_3 \equiv 2 \pmod 4.$$

Hence $(0,2)_3 \equiv 1 \pmod 2$, and, by Theorem 2.6.1, $\text{ind}_g 2 \equiv 2 \pmod 3$. ■

As a special case of Theorem 2.6.3 we have the following well-known result.

Corollary 2.6.4 *Let p be a prime with $p \equiv 1 \pmod 3$. Then 2 is a cubic residue* $\pmod p$ *if and only if $p = x^2 + 27y^2$ for some integers x and y.*

Proof Recalling that $s_3 \equiv 0 \pmod 3$, we have

$$\begin{aligned} &2 \text{ is a cubic residue } \pmod p \\ &\iff \text{ind}_g 2 \equiv 0 \pmod 3 && \text{(for any primitive root } g \pmod p) \\ &\iff r_3 \equiv s_3 \equiv 0 \pmod 2 && \text{(by Theorem 2.6.3)} \\ &\iff r_3 = 2x,\ s_3 = 6y && \text{(for some integers } x \text{ and } y) \\ &\iff p = x^2 + 27y^2. \end{aligned}$$ ■

The next theorem, which determines $\text{ind}_g 2$ modulo 4, follows from Theorem 2.6.1 with $k = 4$ and the formulas for the cyclotomic numbers of order 4 (Theorem 2.4.1).

Theorem 2.6.5 *Let p be a prime of the form $p = 4f + 1$. Let g be a primitive root* $\pmod p$. *Then*

$$\text{ind}_g 2 \equiv b_4/2 \pmod 4,$$

where the integer b_4 is defined in Theorem 2.4.1.

Proof First we examine the case when f is even. In this case $p \equiv 1 \pmod 8$, $\text{ind}_g 2 \equiv 0 \pmod 2$, $a_4 \equiv -1 \pmod 4$, $b_4 \equiv 0 \pmod 4$; say, $p = 8s+1$, $a_4 = 4a-1$, $b_4 = 4b$. Then, from $p = a_4^2 + b_4^2$, we obtain $s = 2a^2 - a + 2b^2 \equiv a + 2b \pmod 4$.

If $b_4 \equiv 0 \pmod 8$, then $b \equiv 0 \pmod 2$ and $s \equiv a \pmod 4$, so that, by Theorem 2.4.1,

$$(0,0)_4 = \frac{1}{16}(p + 6a_4 - 11) = \frac{1}{2}(s+a) + a - 1 \equiv 2a - 1 \equiv 1 \pmod 2,$$

and appealing to Theorem 2.6.1 we deduce that $\text{ind}_g 2 \equiv 0 \pmod 4$.

If $b_4 \equiv 4 \pmod 8$, then $b \equiv 1 \pmod 2$ and $s \equiv a+2 \pmod 4$, so that, by Theorem 2.4.1,

$$(0,2)_4 = \frac{1}{16}(p - 2a_4 - 3) = \frac{1}{2}(s-a) \equiv 1 \pmod 2,$$

and appealing to Theorem 2.6.1 we deduce that $\text{ind}_g 2 \equiv 2 \pmod 4$.

We now turn to the case when f is odd. Here $p \equiv 5 \pmod 8$, $\text{ind}_g 2 \equiv 1 \pmod 2$, $a_4 \equiv 1 \pmod 4$, $b_4 \equiv 2 \pmod 4$; say, $p = 8s+5$, $a_4 = 4a+1$, $b_4 = 4b+2$. Then, from $p = a_4^2 + b_4^2$, we deduce that $s = 2a^2 + a + 2b^2 + 2b \equiv -a \pmod 4$.

If $b_4 \equiv 2 \pmod 8$, then $b \equiv 0 \pmod 2$, so that, by Theorem 2.4.1,

$$(0,1)_4 = \frac{1}{16}(p + 2a_4 + 4b_4 + 1) = \frac{1}{2}(s+a) + b + 1 \equiv 1 \pmod 2,$$

and, by Theorem 2.6.1, $\text{ind}_g 2 \equiv 1 \pmod 4$.

If $b_4 \equiv 6 \pmod 8$, then $b \equiv 1 \pmod 2$, so that, by Theorem 2.4.1,

$$(0,3)_4 = \frac{1}{16}(p + 2a_4 - 4b_4 + 1) = \frac{1}{2}(s+a) - b \equiv 1 \pmod 2,$$

and, by Theorem 2.6.1, $\text{ind}_g 2 \equiv 3 \pmod 4$.

Hence in every case,

$$\text{ind}_g 2 \equiv b_4/2 \pmod 4. \qquad \blacksquare$$

The following special case of Theorem 2.6.5 is another famous result.

Corollary 2.6.6 *Let p be a prime with $p \equiv 1 \pmod 8$. Then 2 is a quartic residue $\pmod p$ if and only if $p = x^2 + 64y^2$ for some integers x and y.*

Proof 2 is a quartic residue $\pmod p$

$$\begin{aligned} &\iff \text{ind}_g 2 \equiv 0 \pmod 4 && \text{(for any primitive root } g \pmod p) \\ &\iff b_4/2 \equiv 0 \pmod 4 && \text{(by Theorem 2.6.5)} \\ &\iff b_4 = 8y && \text{(for some integer } y) \\ &\iff p = a_4^2 + 64y^2. && \blacksquare \end{aligned}$$

Next we determine $\text{ind}_g k$ modulo k in terms of a congruence modulo k that involves sums of cyclotomic numbers of order k. First we prove a lemma.

Lemma 2.6.7 *Let $z \equiv g^f \pmod p$. If $v \not\equiv 0 \pmod k$, then*

$$\text{ind}_g(1 - z^v) \equiv (p-1)/2 + \sum_{j=1}^{k-1} j(j,v)_k \pmod k.$$

Proof The polynomial $x^f - z^v$ has the factorization

$$x^f - z^v \equiv (-1)^f \prod_{r=0}^{f-1} (x_r - x) \pmod p,$$

where $x_r = g^{v+rk}$. Putting $x = 1$ in the congruence above, we deduce that

$$\text{ind}_g(1 - z^v) \equiv f(p-1)/2 + \sum_{r=0}^{f-1} \text{ind}_g(x_r - 1) \pmod{p-1},$$

so that

$$\text{ind}_g(1 - z^v) \equiv (p-1)/2 + \sum_{r=0}^{f-1} \text{ind}_g(x_r - 1) \pmod k.$$

Finally

$$\sum_{r=0}^{f-1} \text{ind}_g(x_r - 1) = \sum_{r=0}^{f-1} \text{ind}_g(g^{v+rk} - 1) = \sum_{j=0}^{k-1} \sum_{s=0}^{f-1} \sum_{\substack{r=0 \\ g^{v+rk} - 1 \equiv g^{j+sk} \pmod p}}^{f-1} (j + sk)$$

$$\equiv \sum_{j=0}^{k-1} j\Bigg(\sum_{\substack{r,s=0 \\ g^{v+rk} - 1 \equiv g^{j+sk} \pmod p}}^{f-1} 1 \Bigg) = \sum_{j=1}^{k-1} j(j,v)_k \pmod k.$$ ■

Our next theorem makes use of Lemma 2.6.7 to determine $\text{ind}_g k$ modulo k.

Theorem 2.6.8

(a) $\text{ind}_g k \equiv (p-1)/2 - \sum_{j=1}^{k-1} j(j,0)_k \pmod k.$

(b) *If k is odd, then*

$$\text{ind}_g k \equiv \sum_{j=1}^{(k-1)/2} j\{(-j,0)_k - (j,0)_k\} \pmod k.$$

Proof Let $z \equiv g^f \pmod p$. Differentiating the polynomial

$$x^k - 1 \equiv \prod_{v=1}^{k} (x - z^v) \pmod p$$

and then setting $x = 1$, we see that

$$k \equiv \prod_{v=1}^{k-1} (1 - z^v) \pmod p.$$

Thus, by Lemma 2.6.7,

$$\text{ind}_g k \equiv (k-1)(p-1)/2 + \sum_{j=1}^{k-1} j \sum_{v=1}^{k-1} (j,v)_k \pmod k.$$

Now, by (2.2.3),

$$\sum_{v=1}^{k-1} (j,v)_k = f - \lambda_j - (j,0)_k,$$

where

$$\lambda_j = \begin{cases} 1, & \text{if } f \equiv 0 \pmod 2 \text{ and } j \equiv 0 \pmod k, \\ 1, & \text{if } f \not\equiv 0 \pmod 2 \text{ and } j \equiv k/2 \pmod k, \\ 0, & \text{otherwise.} \end{cases}$$

Hence

$$\text{ind}_g k \equiv (p-1)/2 + (k-1)kf/2 - \sum_{j=1}^{k-1} j\lambda_j - \sum_{j=1}^{k-1} j(j,0)_k \pmod k.$$

The congruence (a) now follows as

$$\sum_{j=1}^{k-1} j\lambda_j = \left\{ \begin{array}{ll} 0, & \text{if } f \equiv 0 \pmod 2 \\ k/2, & \text{if } f \not\equiv 0 \pmod 2 \end{array} \right\} \equiv (k-1)kf/2 \pmod k.$$

Now suppose that k is odd. Replacing j by $k - j$ in the congruence

$$\text{ind}_g k \equiv (p-1)/2 - \sum_{j=1}^{k-1} j(j,0)_k \pmod k,$$

we obtain

$$\text{ind}_g k \equiv (p-1)/2 + \sum_{j=1}^{k-1} j(-j,0)_k \pmod k.$$

Adding these congruences yields

$$2\ \text{ind}_g k \equiv \sum_{j=1}^{k-1} j\{(-j,0)_k - (j,0)_k\} \pmod k.$$

Now

$$\sum_{j=(k+1)/2}^{k-1} j\{(-j,0)_k - (j,0)_k\} \equiv \sum_{j=1}^{(k-1)/2} -j\{(j,0)_k - (-j,0)_k\}$$
$$\equiv \sum_{j=1}^{(k-1)/2} j\{(-j,0)_k - (j,0)_k\} \pmod{k},$$

and (b) follows. ■

Taking $k = 3$ in Theorem 2.6.8(b), we can obtain $\mathrm{ind}_g 3$ modulo 3, as follows.

Corollary 2.6.9 *If g is a primitive root modulo the prime $p = 3f + 1$, then*

$$\mathrm{ind}_g 3 \equiv -s_3/3 \pmod{3},$$

where s_3 is defined in (2.3.9).

Proof By Theorem 2.6.8(b) with $k = 3$,

$$\mathrm{ind}_g 3 \equiv (2,0)_3 - (1,0)_3 \pmod{3},$$

and by Theorem 2.3.1,

$$(2,0)_3 - (1,0)_3 = C - B = -s_3/3,$$

from which the asserted congruence follows. ■

A special case of Corollary 2.6.9 is the following result of Jacobi.

Corollary 2.6.10 *Let $p = 3f + 1$ be a prime. Then 3 is a cubic residue* (mod p) *if and only if $4p = x^2 + 243y^2$ for some integers x and y.*

Proof 3 is a cubic residue (mod p)

$$\Longleftrightarrow \mathrm{ind}_g 3 \equiv 0 \pmod{3} \qquad \text{(for any primitive root } g \pmod{p})$$
$$\Longleftrightarrow s_3 = 9y \qquad \text{(for some integer } y \text{ by Corollary 2.6.9)}$$
$$\Longleftrightarrow 4p = r_3^2 + 243y^2.$$
■

The next corollary is the special case $k = 5$ of Theorem 2.6.8(b). It may be applied to determine $\mathrm{ind}_g 5$ modulo 5 after we have determined the cyclotomic numbers of order 5 in terms of an appropriate quadratic partition of p. See Problem 14 of this chapter, Theorem 3.7.8, and Theorem 7.4.3.

Corollary 2.6.11 *If g is a primitive root modulo the prime $p = 5f + 1$, then*

$$\mathrm{ind}_g 5 \equiv (4,0)_5 - (1,0)_5 + 2(3,0)_5 - 2(2,0)_5 \pmod{5}.$$

2.7 Generalized cyclotomic numbers and the determination of $\text{ind}_g l \pmod{k}$

Let k be an integer with $k \geq 2$ and let $p = kf + 1$ be a prime. Let g be a primitive root modulo p. We begin by defining cyclotomic numbers that generalize those considered in the preceding sections.

Definition *Let k and l be integers with $k \geq 2$ and $l \geq 1$. Let $p = kf + 1$ be a prime. For integers $s_1, \ldots, s_l$ the generalized cyclotomic number $(s_1, \ldots, s_l)_{l,k}$ of length l and order k is defined to be the number of l-tuples $(u_1, \ldots, u_l)$ of integers satisfying*

$$g^{ku_1+s_1} + \cdots + g^{ku_l+s_l} \equiv 1 \pmod{p}, \quad 0 \leq u_1, \ldots, u_l \leq f - 1. \tag{2.7.1}$$

Clearly $(s_1, \ldots, s_l)_{l,k}$ depends only on the congruence classes (mod k) to which $s_1, \ldots, s_l$ belong.

When $l = 1$ the congruence (2.7.1) reduces to

$$g^{ku_1+s_1} \equiv 1 \pmod{p}, \quad 0 \leq u_1 \leq f - 1,$$

which has one solution if $s_1 \equiv 0 \pmod{k}$ and no solutions if $s_1 \not\equiv 0 \pmod{k}$; thus

$$(s_1)_{1,k} = \begin{cases} 1, & \text{if } s_1 \equiv 0 \pmod{k}, \\ 0, & \text{if } s_1 \not\equiv 0 \pmod{k}. \end{cases} \tag{2.7.2}$$

The next lemma determines $(s_1, s_2)_{2,k}$ in terms of ordinary cyclotomic numbers.

Lemma 2.7.1

$$(s_1, s_2)_{2,k} = (s_2 - s_1, -s_1)_k.$$

Proof When $l = 2$, the congruence (2.7.1) can be rewritten in the form

$$1 + g^{ku_1+s_1+kf/2} + g^{ku_2+s_2+kf/2} \equiv 0 \pmod{p}, \quad 0 \leq u_1, u_2 \leq f - 1,$$

which, by the remarks at the beginning of Section 2.2, has $(s_1 + \frac{1}{2}kf, s_2)_k$ solutions. Thus, appealing to (2.2.1) and (2.2.2), we have

$$(s_1, s_2)_{2,k} = (s_1 + \frac{1}{2}kf, s_2)_k = (-s_1 - \frac{1}{2}kf, s_2 - s_1 - \frac{1}{2}kf)_k = (s_2 - s_1, -s_1)_k. \quad \blacksquare$$

In the next theorem we determine $(s_1, s_2, s_3)_{3,k}$ in terms of ordinary cyclotomic numbers.

Theorem 2.7.2

$$(s_1, s_2, s_3)_{3,k} = \theta + \sum_{h=0}^{k-1} (h - s_3, -s_3)_k (s_2 - s_1, h - s_1)_k,$$

where

$$\theta = \begin{cases} f, & \text{if } s_1 \equiv s_2 + kf/2 \pmod k, \ s_3 \equiv 0 \pmod k, \\ 0, & \text{otherwise.} \end{cases}$$

Proof We have

$$(s_1, s_2, s_3)_{3,k} = \sum_{\substack{u_1, u_2, u_3 = 0 \\ g^{ku_1+s_1} + g^{ku_2+s_2} + g^{ku_3+s_3} \equiv 1 \pmod p}}^{f-1} 1 = S_1 + S_2,$$

where

$$S_1 = \sum_{\substack{u_1, u_2, u_3 = 0 \\ g^{ku_1+s_1} + g^{ku_2+s_2} \equiv 0 \pmod p \\ g^{ku_3+s_3} \equiv 1 \pmod p}}^{f-1} 1, \qquad S_2 = \sum_{h=0}^{k-1} \sum_{v=0}^{f-1} \sum_{\substack{u_1, u_2, u_3 = 0 \\ g^{ku_1+s_1} + g^{ku_2+s_2} \equiv g^{kv+h} \pmod p \\ g^{kv+h} + g^{ku_3+s_3} \equiv 1 \pmod p}}^{f-1} 1.$$

First we determine

$$S_1 = (s_3)_{1,k} \sum_{\substack{u_1, u_2 = 0 \\ g^{ku_1+s_1} + g^{ku_2+s_2} \equiv 0 \pmod p}}^{f-1} 1.$$

The congruence $g^{ku_1+s_1} + g^{ku_2+s_2} \equiv 0 \pmod p$ is equivalent to

$$g^{k(u_2-u_1)} \equiv -g^{s_1-s_2} \equiv g^{s_1-s_2+kf/2} \pmod p,$$

that is

$$k(u_2 - u_1) \equiv s_1 - s_2 + kf/2 \pmod{kf}.$$

This congruence has f or 0 solutions (u_1, u_2) according to whether k divides or does not divide $s_1 - s_2 + kf/2$. Hence, by (2.7.2), we have

$$S_1 = \begin{cases} f, & \text{if } s_1 - s_2 + kf/2 \equiv 0 \pmod k, \ s_3 \equiv 0 \pmod k, \\ 0, & \text{otherwise.} \end{cases}$$

Now we turn to the evaluation of S_2. We have

$$\begin{aligned} S_2 &= \sum_{h=0}^{k-1} \sum_{v=0}^{f-1} \sum_{\substack{u_3 = 0 \\ g^{kv+h} + g^{ku_3+s_3} \equiv 1 \pmod p}}^{f-1} \sum_{\substack{u_1, u_2 = 0 \\ g^{k(u_1-v)+(s_1-h)} + g^{k(u_2-v)+(s_2-h)} \equiv 1 \pmod p}}^{f-1} 1 \\ &= \sum_{h=0}^{k-1} \sum_{v=0}^{f-1} \sum_{\substack{u_3 = 0 \\ g^{kv+h} + g^{ku_3+s_3} \equiv 1 \pmod p}}^{f-1} \sum_{\substack{v_1, v_2 = 0 \\ g^{kv_1+(s_1-h)} + g^{kv_2+(s_2-h)} \equiv 1 \pmod p}}^{f-1} 1 \end{aligned}$$

$$= \sum_{h=0}^{k-1}(h, s_3)_{2,k}(s_1 - h, s_2 - h)_{2,k} = \sum_{h=0}^{k-1}(s_3 - h, -h)_k(s_2 - s_1, h - s_1)_k,$$

by Lemma 2.7.1. Apply (2.2.1) to the first factor to complete the proof. ∎

The next result is the special case $s_1 = s_2 = s_3$ of Theorem 2.7.2, which we shall need later (e.g., for Theorem 2.7.5).

Corollary 2.7.3

$$(s, s, s)_{3,k} = \phi + \sum_{h=0}^{k-1}(h, -s)_k(0, h)_k,$$

where

$$\phi = \begin{cases} f, & \textit{if } s \equiv 0 \pmod{k} \textit{ and } f \textit{ is even,} \\ 0, & \textit{otherwise.} \end{cases}$$

Proof If $s \equiv 0 \pmod{k}$ and f is even, then, by Theorem 2.7.2,

$$(0, 0, 0)_{3,k} = f + \sum_{h=0}^{k-1}(h, 0)_k(0, h)_k.$$

Otherwise,

$$(s, s, s)_{3,k} = \sum_{h=0}^{k-1}(h - s, -s)_k(0, h - s)_k,$$

and changing the summation variable from h to $h + s$, we deduce that

$$(s, s, s)_{3,k} = \sum_{h=0}^{k-1}(h, -s)_k(0, h)_k.$$ ∎

We are now ready to turn our attention to the problem of determining $\text{ind}_g l$ modulo k for an arbitrary integer l not divisible by k. As

$$\text{ind}_g(l_1\, l_2) \equiv \text{ind}_g l_1 + \text{ind}_g l_2 \pmod{k}$$

and

$$\text{ind}_g(-1) = kf/2 \equiv \begin{cases} 0 \pmod{k}, & \text{if } f \text{ is even,} \\ k/2 \pmod{k}, & \text{if } f \text{ is odd,} \end{cases}$$

we may suppose that l is prime. We prove the following result, which is a generalization of Theorem 2.6.1.

Theorem 2.7.4 *Let $p = kf + 1$ be a prime where $k \geq 2$, let g be a primitive root* $(\text{mod } p)$, *and let l be a prime. Then, for $s = 0, 1, \ldots, k-1$,*

$\text{ind}_g l \equiv s \pmod{k}$ *if and only if* $(-s, \ldots, -s)_{l,k} \equiv 1 \pmod{l}$.

Proof From (2.7.1) we see that $(-s,\ldots,-s)_{l,k}$ is the number of solutions of

(2.7.3) $$g^{ku_1}+\cdots+g^{ku_l}\equiv g^s \pmod{p},\ 0\le u_1,\ldots,u_l\le f-1.$$

Let N_1 denote the number of solutions of (2.7.3) with not all the u_i equal, and N_2 the number of solutions with $u_1=\cdots=u_l$. Since l is a prime, the number of distinct permutations of $\{u_1,\ldots,u_l\}$ is a multiple of l when the u_i are not all equal. Thus $N_1\equiv 0 \pmod{l}$. Further, (2.7.3) has a solution with $u_1=\cdots=u_l$ if and only if lg^{-s} is a kth-power residue (mod p). Hence

$$N_2=\begin{cases}1, & \text{if } \text{ind}_g l\equiv s \pmod{k},\\ 0, & \text{otherwise.}\end{cases}$$

Finally

$$(-s,\ldots,-s)_{l,k}=N_1+N_2\equiv\begin{cases}1 \pmod{l}, & \text{if } \text{ind}_g l\equiv s \pmod{k},\\ 0 \pmod{l}, & \text{otherwise,}\end{cases}$$

which completes the proof of Theorem 2.7.4. ■

Taking $l=2$ in Theorem 2.7.4, we obtain

$$\text{ind}_g 2\equiv s \pmod{k} \iff (-s,-s)_{2,k}\equiv 1 \pmod{2} \iff (0,s)_k\equiv 1 \pmod{2}$$

by Lemma 2.7.1, which reproves Theorem 2.6.1.

The next result determines $\text{ind}_g 3$ modulo k.

Theorem 2.7.5 *Let $p=kf+1$ be a prime with $k\ge 2$. Let g be a primitive root (mod p). Let*

$$E_t=\sum_{h=0}^{k-1}(h,t)_k(0,h)_k.$$

For $t=1,2,\ldots,k-1$ we have

$$\text{ind}_g 3\equiv t \pmod{k} \ \textit{if and only if}\ E_t\equiv 1 \pmod{3}.$$

Moreover,

$$\text{ind}_g 3\equiv 0 \pmod{k} \ \textit{if and only if}\ E_0\equiv 1+f+f(-1)^f \pmod{3}.$$

Proof By Theorem 2.7.4, $\text{ind}_g 3\equiv t \pmod{k}$ if and only if $(-t,-t,-t)_{3,k}\equiv 1 \pmod{3}$. By Corollary 2.7.3,

$$(-t,-t,-t)_{3,k}=\begin{cases}f+E_t, & \text{if } t=0 \text{ and } f \text{ is even},\\ E_t, & \text{otherwise,}\end{cases}$$

and the result follows. ■

When $k = 3$, Theorem 2.7.5 is equivalent to Corollary 2.6.9, i.e.,

$$s_3 \equiv 0 \pmod 9 \Longleftrightarrow E_0 \equiv 1 - f \pmod 3,$$
$$s_3 \equiv 3 \pmod 9 \Longleftrightarrow E_2 \equiv 1 \pmod 3,$$
$$s_3 \equiv 6 \pmod 9 \Longleftrightarrow E_1 \equiv 1 \pmod 3.$$

These equivalences are far from apparent. We prove only the first, as the other two follow similarly. We begin by applying Theorem 2.3.1 to evaluate the cyclotomic numbers $A = (0,0)_3$, $B = (0,1)_3$, $C = (0,2)_3$, and $D = (1,2)_3$ modulo 3. Since $p = 3f + 1$, it is easily checked by expanding that $4p \equiv (p^2 - 3p)^2 \pmod{27}$. Since $4p \equiv r_3^2 \pmod{27}$ and $r_3 \equiv 1 \pmod 3$ by (2.3.9), it follows that $r_3 \equiv p^2 - 3p \pmod{27}$. Thus $9A = p + r_3 - 8 \equiv 9f^2 - 9 \pmod{27}$, so that $A \equiv f^2 - 1 \pmod 3$. Similarly, $B \equiv f^2 - f - s_3/3 \pmod 3$, $C \equiv f^2 - f + s_3/3 \pmod 3$, and $D \equiv f^2 \pmod 3$. Next,

$$E_0 = A^2 + B^2 + C^2 \equiv (f^2 - 1)^2 + 2(f^2 - f)^2 + 2(s_3/3)^2 \equiv 1 - f + 2(s_3/3)^2 \pmod 3.$$

Thus $E_0 \equiv 1 - f \pmod 3$ if and only if $s_3 \equiv 0 \pmod 9$, as desired.

In the final theorem of this section, we use Theorem 2.7.5 in conjunction with the formulas for the cyclotomic numbers of order 4 derived in Theorem 2.4.1 to obtain a necessary and sufficient condition for 3 to be a fourth power modulo a prime $p \equiv 1 \pmod 4$.

Theorem 2.7.6 *Let p be a prime of the form $p = 4f + 1$. Then*

$$3 \textit{ is a fourth power} \pmod p \Longleftrightarrow \begin{cases} b_4 \equiv 0 \pmod 3, & \textit{if } f \equiv 0 \pmod 2, \\ a_4 \equiv 0 \pmod 3, & \textit{if } f \not\equiv 0 \pmod 2. \end{cases}$$

Proof Let g be a primitive root $\pmod p$. By Theorem 2.7.5 with $t = 0$ and $k = 4$ we have, making use of (2.4.1),

$$\text{ind}_g 3 \equiv 0 \pmod 4 \Longleftrightarrow \begin{cases} A^2 + B^2 + C^2 + D^2 \equiv 1 - f \pmod 3, & \text{if } f \equiv 0 \pmod 2, \\ A^2 + AC + BE + DE \equiv 1 \pmod 3, & \text{if } f \not\equiv 0 \pmod 2. \end{cases}$$

First suppose that f is even. By Theorem 2.4.1, we have $A \equiv p + 1 \pmod 3$, $B \equiv p + a_4 + b_4 \pmod 3$, $C \equiv p + a_4 \pmod 3$, and $D \equiv p + a_4 - b_4 \pmod 3$, so that

$$A^2 + B^2 + C^2 + D^2 \equiv (p+1)^2 + 2b_4^2 \pmod 3.$$

Since $p = 4f + 1$, we have $f \not\equiv 2 \pmod 3$, so that $f^2 \equiv f \pmod 3$. Hence

$$A^2 + B^2 + C^2 + D^2 \equiv 1 - f - b_4^2 \pmod 3,$$

so that $\text{ind}_g 3 \equiv 0 \pmod 4$ if and only if $b_4 \equiv 0 \pmod 3$. Finally, suppose that f is odd. By Theorem 2.4.1, $A \equiv p - 1 - a_4 \pmod 3$, $B \equiv p + 1 - a_4 + b_4 \pmod 3$, $C \equiv p + 1 \pmod 3$, $D \equiv p + 1 - a_4 - b_4 \pmod 3$, and $E \equiv p + a_4 \pmod 3$, so that

$$A^2 + AC + (B + D)E \equiv 1 - a_4^2 \pmod 3.$$

Hence $\text{ind}_g 3 \equiv 0 \pmod 4$ if and only if $a_4 \equiv 0 \pmod 3$. ■

Example (i) $p = 5$. Here $f = 1$, $|a_4| = 1$, so 3 is not a fourth power $\pmod 5$.

(ii) $p = 13$. Here $f = 3$, $|a_4| = 3$, so 3 is a fourth power $\pmod{13}$. Indeed, $3 \equiv 2^4 \pmod{13}$.

(iii) $p = 17$. Here $f = 4$, $|b_4| = 4$, so 3 is not a fourth power $\pmod{17}$.

(iv) $p = 193$. Here $f = 48$, $|b_4| = 12$, so 3 is a fourth power $\pmod{193}$. Indeed, $3 \equiv 20^4 \pmod{193}$.

Exercises 2

1. Suppose that χ is a character on F_q of order $k > 2$. Show that

$$G_r^k(\chi) = \chi(-1) q \prod_{j=1}^{k-2} J_r(\chi, \chi^j).$$

2. Let χ be a character on F_q of order $k \geq 2$, and let $\beta = \exp(2\pi i/k)$. Set $K = Q(\beta)$. Prove that

$$G_r^k(\chi) \in O_K.$$

Then use Theorem 1.1.4(d) to show that in fact $G_r^k(\chi)$ is in a subfield of K of index s, where s is the order of $p \pmod k$.

3. Let $p = 4f + 1$. Let R_p be the set of $2f$ quadratic residues $\pmod p$ and let N_p be the set of $2f$ quadratic nonresidues $\pmod p$. Fix $r \in R_p$ and $n \in N_p$. Prove that the sets $r + N_p$ and $n + R_p$ each contain f quadratic residues and f quadratic nonresidues.

Hint: Consider first $n + R_p$ and use Theorem 2.1.2.

4. Let $p = 4f + 1$. In the notation of Problem 3, prove that each of the sets $(r + R_p) \cup \{r\}$ and $(n + N_p) \cup \{n\}$ contains f quadratic residues, f quadratic nonresidues, and 0.

5. Let $p = 4f - 1$. With the notation of Problem 3, prove that each of the sets $r + N_p$ and $n + R_p$ consists of $f - 1$ quadratic residues, $f - 1$ quadratic nonresidues, and 0.

6. Let $p = 4f - 1$. Prove that each of the sets $(r + R_p) \cup \{r\}$ and $(n + N_p) \cup \{n\}$ consists of f quadratic residues and f quadratic nonresidues.

7. Prove Theorem 2.1.5 in a manner similar to the proof of Theorem 2.1.1(c).

Problems 8–14 *below deal with the cyclotomic numbers of order* 5.

8. Show that the cyclotomic matrix of order 5 is of the form

$$\begin{bmatrix} A & B & C & D & E \\ B & E & F & G & F \\ C & F & D & G & G \\ D & G & G & C & F \\ E & F & G & F & B \end{bmatrix}.$$

9. Use Theorem 2.2.1 to prove that

$$\begin{aligned} A + B + C + D + E &= f - 1, \\ B + E + 2F + G &= f, \\ C + D + F + 2G &= f. \end{aligned}$$

10. Making use of the ideas employed in Sections 2.3 and 2.4, prove that

$$\begin{aligned} AF + B^2 + CE + EG &= BC + EF + FG + G^2, \\ AG + BC + CF + D^2 &= BD + CG + F^2 + FG. \end{aligned}$$

11. Deduce from Problem 9 that

$$\begin{aligned} A &= 3F + 3G - f - 1, \\ B &= -E - 2F - G + f, \\ C &= -D - F - 2G + f. \end{aligned}$$

12. Deduce from Problems 10 and 11 that

$$\begin{aligned} &E^2 + 5F^2 - 2G^2 - 2DE - 2DF - DG + EF - EG + FG \\ &\qquad = -fD + 2fF - fG + F, \\ &D^2 + 7G^2 + 2DE + 3DF + 3DG + EF + 2EG + 6FG \\ &\qquad = 2fD + fE + 2fF + 5fG + G - f^2. \end{aligned}$$

13. Define integers x, u, v, w by

$$\begin{aligned} x &= 25F + 25G - 10f - 4, \\ u &= -2D - F - 2G + f, \\ v &= -2E - 2F - G + f, \\ w &= -F + G. \end{aligned}$$

Use Problem 12 to show that the integers $x,\ w,\ v,\ u$ satisfy

$$16p = x^2 + 125w^2 + 50v^2 + 50u^2, \quad x \equiv 1 \pmod 5,$$
$$xw = v^2 - 4vu - u^2.$$

(It will be shown after (3.7.35) that this Diophantine system has only four solutions.)

14. Deduce the formulas for the cyclotomic numbers of order 5:

$$\begin{aligned}
25A &= p - 14 + 3x,\\
100B &= 4p - 16 - 3x + 25w + 50v,\\
100C &= 4p - 16 - 3x - 25w + 50u,\\
100D &= 4p - 16 - 3x - 25w - 50u,\\
100E &= 4p - 16 - 3x + 25w - 50v,\\
100F &= 4p + 4 + 2x - 50w,\\
100G &= 4p + 4 + 2x + 50w.
\end{aligned}$$

Deduce from Corollary 2.6.11 that

$$\text{ind}_g 5 \equiv -v - 2u \pmod 5.$$

Determine the sum $\sum_{x=1}^{p-1}(1 + x^5)^{(p-1)/5} \pmod p$ in two different ways, using the method explained in Section 2.3, and deduce that

$$3x + 5w(h - h^2 - h^3 + h^4) + 10v(h - h^4) + 10u(h^2 - h^3) \equiv 0 \pmod p,$$

where $h \equiv g^{(p-1)/5} \pmod p$.

15. Show that the cyclotomic matrix of order 6 is of the form

$$\begin{bmatrix}
A & B & C & D & E & F\\
G & H & I & E & C & I\\
H & J & G & F & I & B\\
A & G & H & A & G & H\\
G & F & I & B & H & J\\
H & I & E & C & I & G
\end{bmatrix},$$

if f is odd, and of the form

$$\begin{bmatrix}
A & B & C & D & E & F\\
B & F & G & H & I & G\\
C & G & E & I & J & H\\
D & H & I & D & H & I\\
E & I & J & H & C & G\\
F & G & H & I & G & B
\end{bmatrix}.$$

if f is even.

16. Show that the cyclotomic matrix of order 7 is of the form

$$\begin{bmatrix} A & B & C & D & E & F & G \\ B & G & H & I & J & K & H \\ C & H & F & K & L & L & I \\ D & I & K & E & J & L & J \\ E & J & L & J & D & I & K \\ F & K & L & L & I & C & H \\ G & H & I & J & K & H & B \end{bmatrix}.$$

17. Show that the cyclotomic matrix of order 8 is of the form

$$\begin{bmatrix} A & B & C & D & E & F & G & H \\ B & H & I & J & K & L & M & I \\ C & I & G & M & N & O & N & J \\ D & J & M & F & L & O & O & K \\ E & K & N & L & E & K & N & L \\ F & L & O & O & K & D & J & M \\ G & M & N & O & N & J & C & I \\ H & I & J & K & L & M & I & B \end{bmatrix},$$

if f is even, and of the form

$$\begin{bmatrix} A & B & C & D & E & F & G & H \\ I & J & K & L & F & D & L & M \\ N & O & N & M & G & L & C & K \\ J & O & O & I & H & M & K & B \\ A & I & N & J & A & I & N & J \\ I & H & M & K & B & J & O & O \\ N & M & G & L & C & K & N & O \\ J & K & L & F & D & L & M & I \end{bmatrix}.$$

if f is odd.

18. Establish the inversion of (2.5.1), namely, for $u, v = 0, 1, \ldots, k-1$

$$J^*(\chi^u, \chi^v) = (-1)^{uf} \sum_{s=0}^{k-1} \sum_{t=0}^{k-1} (s,t)_k \beta^{su+tv},$$

where J^* is defined in Section 2.5. (For an extension to general finite fields F_q, see (11.6.6).)

19. Let $2 \mid k$ and put $E = k/2$. Prove that

$$(s,t)_E = (s,t)_k + (s+E,t)_k + (s,t+E)_k + (s+E,t+E)_k.$$

20. Show that in the notation of Section 2.5,

$$J^*(\chi^u, \chi^v) = (-1)^{vf} \sum_{j=0}^{k-1} A_j(u,v)\beta^j,$$

where, for $0 \le j \le k-1$,

$$A_j(u,v) = \frac{1}{k} \sum_{n=0}^{k-1} (-1)^{nvf} J^*(\chi^{nu}, \chi^{nv})\beta^{-nj}.$$

21. In the notation of Problem 20, prove that, for $j,\ v = 0, 1, \ldots, k-1$,

$$A_j(1,v) = \sum_{s=0}^{k-1} (s, j - vs)_k.$$

22. Suppose that $4 \mid k$. Using the notation of Problems 19 and 20, let $D_j = A_j(1,E) - A_{j+E}(1,E)$. Prove that

(a) $$K(\chi) = \sum_{j=0}^{E-1} D_j \beta^j,$$

(b) $$p = \sum_{j=0}^{E-1} D_j^2 = D_0^2 + D_{E/2}^2 + 2 \sum_{j=1}^{E/2-1} D_j^2,$$

(c) $$D_{E/2} = 0 \text{ if } 8 \mid k.$$

23. Let $k = 2^n$ with $n \ge 3$. Let D_0 be as given in Problem 22. Prove that $D_0 \equiv -1 \pmod 4$.

24. Show that in the notation of Theorem 2.1.14, with $r = 1$,

$$G^k(\chi)O_K = \prod_{\substack{j=1 \\ \gcd(j,k)=1}}^{k} P_{j^{-1}}^{k-j}.$$

(For a generalization, see Theorem 11.2.8.)
Hint: Use Problem 1.

25. Let k be a squarefree positive integer. If the Gauss sums $\tau_k(\chi_1, m_1)$ and $\tau_k(\chi_2, m_2)$ defined in (1.6.1) are associates (i.e., their quotient is a unit), prove that

$\chi_1 = \chi_2$.

Hint: In the case that k is prime, this follows from Problem 24. The general case follows from (1.6.3).

26. Let χ be a character on F_q of order 2ℓ, where $\ell > 1$ is odd. Write $\lambda = 1 - e^{2\pi i/\ell}$. Prove that for even n not divisible by ℓ, $J_r(\chi, \chi^n) \equiv -\chi^n(4) \pmod{\lambda^2}$.

Hint: First show that

$$J_r(\chi^n, \chi) - J_r(\chi^n, \chi^\ell) = \sum_{\substack{\alpha \in F_q \\ \alpha \neq 0,1}} (\chi^n(1-\alpha) - 1)(\chi(\alpha) - \chi^\ell(\alpha)) \equiv 0 \pmod{\lambda^2}.$$

Then use Theorems 2.1.7 and 2.1.4 to show that

$$J_r(\chi^n, \chi) \equiv -\frac{J_r(\chi^n, \chi^\ell)}{J_r(\chi^n, \chi^n)} \equiv -\chi^n(4) \pmod{\lambda^2}.$$

Notes on Chapter 2

The Jacobi sums $J(\chi, \psi)$ over F_p were mentioned by Jacobi [1] in a letter to Gauss dated February 8, 1827. Ten years later, Jacobi [3] published his first account of these sums. Kummer [CW, vol. 2, p. 702] suggested that this work of Jacobi [1], [3] provided the most significant advance in "Kreisteilungstheorie" since Gauss' *Disquisitiones Arithmeticae* (published in 1801). Other early papers on Jacobi sums include Cauchy [1], [2], Gauss [7], Eisenstein [1], and Lebesgue [4]. Accounts of this early work on Jacobi sums have been given by Bachmann [1], Dickson, Mitchell, Vandiver, and Wahlin [1], H. J. S. Smith [1], and Weil [6]. Kummer [4] introduced sums equivalent to Jacobi sums over general finite fields F_q. Accounts of the properties of Jacobi sums have been given by Hasse [2], Ireland and Rosen [1], Joly [1], and Lidl and Niederreiter [1, Chapter 5, §3]. See also the classic papers of Weil [5], [7] and Iwasawa [1]. Jacobi sums play an important role in the Jacobi sum primality test of Adleman, Pomerance, and Rumely [1], which was simplified by Cohen and Lenstra [1]. For good expositions, see the books of H. Cohen [1, §9.1] and Washington [1, §16.1], and the article of A. K. Lenstra [1].

In Section 1.6, we studied Gauss sums over the ring of integers (mod p^n). Wang [1] studied both Gauss and Jacobi sums (mod p^n). As an application, Wang related the number of solutions of the congruence $x_1^p + \cdots + x_r^p \equiv 1 \pmod{p^2}$ to the number of certain real Jacobi sums over rings.

Theorem 2.1.2, as well as the method used in the proof of Theorem 2.1.7, occurs in the work of Jacobsthal [1], [2]. Theorem 2.1.3 gives a basic relation between Gauss and Jacobi sums. A far-reaching multidimensional analogue has been given by Loeser [1]. An analogue for Jacobi sums over finite local rings has been found by Wang [4].

The sum $K_r(\chi)$, which appears in Theorems 2.1.4, 2.1.6, and 2.1.8, was studied in the papers of Berndt and Evans [1], [2]. Theorem 2.1.5 plays a key role in S. Cohen's proof [3, p. 49] that every finite field of order $q \notin \{2,3,7\}$ contains a pair of consecutive primitive roots. Theorem 2.1.11 has been used by Acharya and Katre [1, p. 59] in connection with the determination of cyclotomic numbers. Problem 26 extends a result of Acharya and Katre [1, p. 59]. For a generalization of Problem 26, see Evans [21]. Different proofs of (2.1.6) can be found in Narkiewicz [2, p. 169] and Spearman and Williams [2]. Theorem 2.1.12 is due to Kronecker; see Narkiewicz [2, p. 73]. Other proofs of Theorems 2.1.9 and 2.1.13 are given in the book of Weiss [1, p. 267]. Mitchell [1] gave the prime ideal factorization of Jacobi sums over finite fields F_q, generalizing and correcting the pioneering work of Kummer [5], [6]. We give the special case over F_p in (2.1.9), and the general case in Theorem 11.2.3. The congruence for Jacobi sums over F_q in (2.1.11) follows from the work of Stickelberger [1], but it is already essentially in the work of Jacobi [1, pp. 397–398], [3] in the case $q = p$. Sepanski [1] used the congruence (2.1.11) to determine when certain Lie algebras connected with PSL(2,q) can be decomposed as a direct sum of Cartan subalgebras.

The notion of a cyclotomic number of order k originated in the works of Gauss [2, art. 358], [5], which examine the cases $k = 3, 4$, respectively. The connection between cyclotomic numbers and Jacobi sums was recognized by Kummer [3], [4]. A double finite Fourier series expressing Jacobi sums in terms of cyclotomic numbers (Problem 18) occurs in the work Mitchell [1, eq. (4)] and Vandiver [8, p. 682], for example. The inverted form expressing cyclotomic numbers in terms of Jacobi sums (Theorem 2.5.1) has been used by many authors to evaluate cyclotomic numbers; see, for example, Whiteman [10, p. 96]. Extensions of these Fourier expansions to general finite fields F_q are given in (11.6.5) and (11.6.6). The cyclotomic numbers over F_q are defined just above (11.6.5); for their basic properties, see the book of Storer [1]. It follows directly from (11.6.5) that each cyclotomic number of order k over F_q is asymptotic to q/k^2 as q tend to infinity. This proves a conjecture of Snapper [1, p. 277]; see Katre [1].

Thaine [2, Theorem 1] lists conditions on a set of k^2 integers that ensure that they comprise all the cyclotomic numbers of order k. In related work, Thaine [3] lists conditions characterizing certain Jacobi sums. Thaine [1] applies this work to the study of the orders of the components of the p-Sylow subgroup of the ideal class group of $Q(\cos 2\pi/p)$. Vandiver's conjecture states that this class group has no element of order p; see Washington [1, p. 158]. Jakubec [2] discussed conditions for Jacobi sums $J(\chi,\chi)$ over F_p of prime order ℓ to be uniquely determined up to conjugation and association by the solution $A \in Z[e^{2\pi i/\ell}]$, $A \equiv 1 (\text{mod } 2)$ to the equation $A\bar{A} = p$. In the same paper, Jakubec used these Jacobi sums $J(\chi,\chi)$ to determine the 11th– and 19th–power residuacity of 2 (mod p) in certain cases. Acharya and Katre [1] have given a Diophantine system that uniquely determines the cyclotomic numbers over F_q of order twice an odd prime dividing $(p-1)$, in terms of parameters occurring in Jacobi sums, free from sign ambiguities associated with the choice of a primitive root of F_q^*. Katre and Rajwade [4] had earlier obtained such a result for cyclotomic numbers of odd prime order, extending work of Parnami, Agrawal, and Rajwade [2]. See also the closely related work of Thaine [3]. For references to explicit determinations of cyclotomic numbers of small orders (≤ 24), see the Notes on Chapter 3. Problems 8–14 on cyclotomic numbers of order 5 are based on the work of Rajwade [3] and Whiteman [10]. Du and Hwang [1] have used cyclotomic numbers and Jacobi sums to prove the existence of symmetric skew–balanced starters, which are used in combinatorial designs. D. H. and E. Lehmer [3] have shown that certain sums of products of cyclotomic numbers of order k related to Gauss periods can be calculated by an explicit kth–order recurrence sequence. Rajwade [12] has a comprehensive survey article discussing cyclotomic numbers.

Theorem 2.6.1 is a more precise form of a result of E. Lehmer [2, Lemma 1], which has its origins in the formulation of a theorem of Libri given by Lebesgue [1, paper II]. Corollary 2.6.4 is a famous result of Gauss [2, art. 358]. Jacobi [2, p. 236] stated similar criteria for 3, 5, 7, 11, 13, 17, 19, 23, 29, 31, 37 to be cubes modulo a prime $p \equiv 1 \pmod 3$. Theorem 2.6.5 is also due to Gauss [5, art. 21]. A simple demonstration of Corollary 2.6.6 was given by Dirichlet [5]. Reuschle [1] stated, but did not prove, a similar result for 2 to be an octic residue modulo a prime p. The first proof of Reuschle's criterion was given by Western [3]. (The life and mathematical work of Western is described in Miller [1].) Cunningham [1] conjectured a criterion for the 16th–power residue character of 2. Aigner [1] rediscovered Cunningham's criterion and gave the first proof. Whiteman [7] employed cyclotomy to reprove the criteria of Gauss, Reuschle, and Cunningham. Hasse [1] and Evans [5] have treated the 2^nth–power residue character of 2 modulo a prime p. For extensions, see Hudson and Williams [8]. A number of authors, in particular Aigner [2] and von Lienen [2], have treated the quartic and octic residue characters (mod p) of other small primes. The quartic, octic, and bioctic character of certain fundamental units in quadratic fields has been investigated with the aid of cyclotomic numbers; see, for example, Evans [14] and the references therein. For further references on residuacity, consult Lemmermeyer [3]. We return to this topic in Chapter 7.

Lemma 2.6.7 and Theorem 2.6.8 are due to Muskat [2]. Related ideas are contained in Ankeny [1] and Muskat [1], [3], [5]. Muskat [6, Theorem 4] applied Lemma 2.6.7 to prove that for a prime $p = 15f + 1 = M^2 + 15N^2$, the element $(1+\sqrt{5})/2 \in F_p$ is a cubic residue (mod p) if and only if $3|N$. See also E. Lehmer [10]. Corollary 2.6.9 is formulated in a different but equivalent way in Williams [13, eq. (8)]. Corollary 2.6.10 appears in Jacobi [2].

The ideas in Section 2.7 are based on and extend the work of Alderson [2], who studied the residuacity of 3 modulo a prime p. Thus, for example, a corrected form of Alderson [2, Theorem 2] is the special case $t = 0$ of Theorem 2.7.5. Theorem 2.7.6 is well known and can be found, for example, in E. Lehmer [7, p. 24]. It can be deduced from the work of Pépin [1], but its origin is probably earlier.

Chapter 3

Evaluation of Jacobi Sums over F_p

In this chapter, we evaluate certain Jacobi sums over F_p, where p is an odd prime. Throughout, $p = kf + 1$, g is a primitive root (mod p), $\beta = \exp(2\pi i/k)$, O_K is the ring of algebraic integers in $K = Q(\beta)$, $\sigma_m \in \mathrm{Gal}(K/Q)$ maps β to β^m, and χ is a character (mod p) of order k such that $\chi(g) = \beta$. In the evaluations of Jacobi sums that follow, this last requirement is really no restriction, for given a character χ of order k, there always exists a primitive root g (mod p) such that $\chi(g) = \beta$. A specified value of k is examined in each section. Thus, f equals $(p-1)/6$ in Section 3.1, whereas the same letter f denotes $(p-1)/4$ in Section 3.2. We wish to evaluate the Jacobi sums $J(\chi^i, \chi^j)$, $0 < i, j < k$, for the values $k = 3$, 4, 5, 6, 7, 8, 10, 12, 16, 20, and 24, in terms of parameters occurring in quadratic partitions of p. We claim that it suffices to consider the sums $J(\chi, \chi^j)$ for $1 \le j < k/2$. To see this, consider a sum $J(\chi^m, \chi^n)$ with $\gcd(m, n, k) = 1$. Since each of the values of k under consideration here has at most two distinct prime factors, it follows that at least one of the integers m, n, and $m + n$ is relatively prime to k. Thus, by Theorem 2.1.5, $\pm J(\chi^m, \chi^n)$ is conjugate to a sum of the form $J(\chi, \chi^j)$, where, again by Theorem 2.1.5, we may take $1 \le j \le k/2$. By Theorem 2.1.4, we may further assume that $j < k/2$. For example, when $k = 24$, $J(\chi^2, \chi^3)$ can be expressed in terms of $J(\chi, \chi^9)$ as follows:

$$J(\chi^2, \chi^3) = (-1)^f J(\chi^3, \chi^{19}) = (-1)^f \sigma_{19}(J(\chi, \chi^9)).$$

In Section 2.1, we defined $K(\chi) = \chi(4)J(\chi, \chi)$. We prove several theorems (see, for example, Theorem 3.1.1) showing that $K(\chi) = a + ib\sqrt{d}$, where a, b, d are nonzero integers with d squarefree and positive. The integer d does not depend on the choice of primitive root g and is such that $Q(i\sqrt{d})$ is a quadratic subfield of K. By Theorem 2.1.3(b), $|K(\chi)|^2 = p = a^2 + db^2$. Lemma 3.0.1 below then shows that $|a|$ and $|b|$ are uniquely determined, that is, they do not depend on the choice of the primitive root g. We now examine the dependence of a and b on g. Let g and g' be two primitive roots (mod p). Let χ and χ' be the characters of order k such that $\chi(g) = \chi'(g') = \beta$, and suppose that $K(\chi) = a(g) + ib(g)\sqrt{d}$ and $K(\chi') = a(g') + ib(g')\sqrt{d}$. As g and g' are primitive roots (mod p), there exists an integer u such that $g' \equiv g^u \pmod{p}$ and $\gcd(u, p-1) = 1$. Hence

$$\chi(g) = \chi'(g') = \chi'(g^u) = \chi'(g)^u.$$

As $\gcd(u, k) = 1$, there exists an integer v with $uv \equiv 1 \pmod{k}$. Then $\chi'(g) = \chi(g)^v = \sigma_v(\chi(g))$, where σ_v is the automorphism of K such that $\sigma_v(\beta) = \beta^v$. Thus

$$\begin{aligned} a(g') + ib(g')\sqrt{d} &= K(\chi') = \sigma_v(K(\chi)) = \sigma_v(a(g) + ib(g)\sqrt{d}) \\ &= a(g) + b(g)\sigma_v(i\sqrt{d}) = a(g) + b(g)\epsilon(v)i\sqrt{d}, \end{aligned}$$

where $\epsilon(v) = \pm 1$. Therefore $a(g') = a(g)$ and $b(g') = b(g)\epsilon(v)$. Thus a is independent of the choice of g, but b is dependent on the choice of g. To see this latter assertion, take $g' = g^{-1}$ so that $u = v = -1$ and $b(g') = -b(g)$. We prove several theorems (e.g., Theorem 3.1.2) that show that the sign of $b(g)$ can be computed relatively quickly once $|b(g)|$ and $g^f \pmod p$ are known. It is well known (see, for example, Knuth's book [1, vol. 2, p. 399]) that $g^f \pmod p$ can be computed in $O(\log p)$ operations. In the case $d = -1$, for example, $|b(g)|$ can also be computed in a time polynomial in $\log p$; see the paragraph following the proof of Theorem 4.2.4. For general fixed d, $|b(g)|$ can, of course, be computed in $O(\sqrt{p})$ operations. In contrast, $b(g)$ could be directly computed by summing all of the terms of $K(\chi)$, but this would require $O(p)$ operations.

Lemma 3.0.1 *Fix positive integers p, r, c, d, m such that c, d, m, and p are pairwise coprime and p is prime. Let*

$$t = \begin{cases} 4, & \text{if } p = 2, \\ \gcd(2, m), & \text{if } p > 2. \end{cases}$$

If $cd \geq (m/t)^2$, then there exists at most one pair of positive integers u, v with $p \nmid uv$ such that

$$mp^r = cu^2 + dv^2.$$

In particular, for pairwise coprime positive integers c, d, p with p prime, there is at most one pair of positive integers u, v such that

$$p = cu^2 + dv^2.$$

Proof Write $q = p^r$ for brevity. Suppose that

$$mq = cu^2 + dv^2 = cx^2 + dy^2 \tag{3.0.1}$$

for positive integers u, v, x, y with $p \nmid uvxy$. If $y = v$, then $x = u$, and so assume that $y \neq v$. It remains to show that $y = u$ and $c = d = 1$. Since $(mq - cx^2)v^2 = (mq - cu^2)y^2$ by (3.0.1),

$$0 \neq mq(y^2 - v^2) = c(u^2y^2 - v^2x^2). \tag{3.0.2}$$

By (3.0.1), $u, x < \sqrt{mq/c}$ and $v, y < \sqrt{mq/d}$, and so

$$uy,\ vx < mq/\sqrt{cd} \leq qt. \tag{3.0.3}$$

First, suppose that q is odd. Then by (3.0.2) and the fact that $p \nmid uvxy$, exactly one of $(uy + vx)$, $(uy - vx)$ is divisible by p, and hence by q. In fact,

$$uy \equiv \pm vx \pmod{qt}, \tag{3.0.4}$$

since if m is even (so that $t = 2$), then cd is odd, so that the pairs u, v and x, y each have the same parity.

Next, suppose that q is even. Then $t = 4$ and $mcduvxy$ is odd, since $p = 2$ is prime to $uvxy$. The right side of (3.0.2) must thus be divisible by $8q$. One of $(uy + vx)$, $(uy - vx)$ is congruent to 2 (mod 4), and the other is divisible by $4q = qt$. Thus (3.0.4) holds whether q is even or odd.

If $uy \equiv vx \pmod{qt}$, then by (3.0.3), we have $uy = vx$, which contradicts (3.0.2). Thus $uy \equiv -vx \pmod{q}$, and so by (3.0.3), $uy + vx = qt$. Then by (3.0.1),

$$\begin{aligned} m^2q^2 &= (cx^2 + dy^2)(cu^2 + dv^2) = (cxu - dyv)^2 + cd(uy + vx)^2 \\ &= (cxu - dyv)^2 + cdq^2t^2 \geq (cxu - dyv)^2 + m^2q^2. \end{aligned}$$

Thus $cxu = dyv$. As $\gcd(c, dyv) = \gcd(d, cxu) = 1$, this implies that $c = d = 1$. Therefore $xu = yv$, so that $y/u = x/v = s$ for some $s > 0$. Then by (3.0.1), $mq = u^2 + v^2 = s^2(u^2 + v^2)$, so that $s = 1$ and $y = u$, $x = v$. ■

We remark that if $p = 7$, the equation $4p = u^2 + 3v^2$ has three pairs of solutions u, v with $p \nmid uv$, namely, 1,3; 5,1; and 4,2 (see Problem 2). This, however, does not contradict Lemma 3.0.1, because the hypothesis $cd \geq (m/t)^2$ is not satisfied, since $cd = 3$ while $m/t = 4/2 = 2$.

In the next lemma, we record some properties of $\chi(-1)$, $\chi(2)$, $\chi(3)$, $\chi(4)$ and $\chi(5)$, which will be used throughout this chapter.

Lemma 3.0.2 *Let k be an integer with $k \geq 2$, and let $p = kf + 1$ be a prime. Let g be a primitive root* (mod p), *and let χ be a character* (mod p) *of order k such that $\chi(g) = \beta$, where $\beta = \exp(2\pi i/k)$. Set*

(3.0.5) $$Z = \mathrm{ind}_g 2, \quad T = \mathrm{ind}_g 3, \quad R = \mathrm{ind}_g 5.$$

Then

(a) $\chi(-1) = (-1)^f$.
Moreover, $\chi(-1) = \left(\frac{-1}{p}\right)$, *if* $k \equiv 2 \pmod 4$.

(b) $\chi(2) = \beta^Z$.
Moreover, $Z \equiv 0 \pmod 2$, *if* $k \equiv 0 \pmod 8$, *and* $Z \equiv f \pmod 2$, *if* $k \equiv 4 \pmod 8$.

(c) $\chi(3) = \beta^T$.
Moreover, $T \equiv 0 \pmod 2$, *if* $k \equiv 0 \pmod{12}$.

(d) $\chi(4) = \beta^{2Z}$.

(e) $\chi(5) = \beta^R$.
Moreover, $R \equiv 0 \pmod 2$, *if* $k \equiv 0 \pmod 5$.

Proof (a) By (1.1.3), $\chi(-1) = (-1)^f$. If $k \equiv 2 \pmod 4$ then, as $k/2$ is odd,

$$\chi(-1) = (-1)^f = (-1)^{kf/2} = (-1)^{(p-1)/2} = \left(\frac{-1}{p}\right).$$

(b) We have
$$\chi(2) = \chi(g^{\text{ind}_g 2}) = \chi(g^Z) = \chi(g)^Z = \beta^Z.$$
If $k \equiv 4 \pmod 8$ then $k/4$ is odd and $k^2/8$ is even, so that
$$\begin{aligned}(-1)^f &= (-1)^{(k^2/8)f^2+(k/4)f} = (-1)^{(k^2f^2+2kf)/8} \\ &= (-1)^{((kf+1)^2-1)/8} = (-1)^{(p^2-1)/8} \\ &= \left(\frac{2}{p}\right) = \left(\frac{g^Z}{p}\right) = \left(\frac{g}{p}\right)^Z = (-1)^Z,\end{aligned}$$
and thus $Z \equiv f \pmod 2$. If $k \equiv 0 \pmod 8$ then $p \equiv 1 \pmod 8$ and
$$(-1)^Z = \left(\frac{g}{p}\right)^Z = \left(\frac{g^Z}{p}\right) = \left(\frac{2}{p}\right) = 1,$$
so that $Z \equiv 0 \pmod 2$.

(c) $\chi(3) = \chi(g^T) = \chi(g)^T = \beta^T$.

If $k \equiv 0 \pmod{12}$, then $p \equiv 1 \pmod{12}$, and (by the law of quadratic reciprocity)
$$(-1)^T = \left(\frac{g}{p}\right)^T = \left(\frac{g^T}{p}\right) = \left(\frac{3}{p}\right) = \left(\frac{-1}{p}\right)\left(\frac{p}{3}\right) = 1,$$
showing that $T \equiv 0 \pmod 2$.

(d) $\chi(4) = \chi(2)^2 = \beta^{2Z}$.

(e) $\chi(5) = \chi(g^R) = \chi(g)^R = \beta^R$.

If $k \equiv 0 \pmod 5$, then $p \equiv 1 \pmod 5$, and (by the law of quadratic reciprocity)
$$(-1)^R = \left(\frac{g}{p}\right)^R = \left(\frac{g^R}{p}\right) = \left(\frac{5}{p}\right) = \left(\frac{p}{5}\right) = 1,$$
showing that $R \equiv 0 \pmod 2$. ■

3.1 Cubic and sextic sums

In this section, $k = 6$, $p = 6f + 1$, g is a primitive root (mod p), $Z = \text{ind}_g 2$, $\beta = \exp(2\pi i/6)$, and χ is a character (mod p) of order 6 with $\chi(g) = \beta$. By Lemma 3.0.2(a), $\chi(-1) = \left(\frac{-1}{p}\right)$. We note that $\beta = (1 + i\sqrt{3})/2$ and $\beta^2 = (-1 + i\sqrt{3})/2$.

Theorem 3.1.1 *We have*
$$J(\chi, \chi^2) = \left(\frac{-1}{p}\right) K(\chi) = K(\chi^2) = a_3 + ib_3\sqrt{3},$$
where a_3 and b_3 are integers such that $a_3^2 + 3b_3^2 = p$ and $a_3 \equiv -1 \pmod 3$.

Proof The first and second equalities follow from (2.1.2). Next,

$$K(\chi^2) = \chi^2(4)J(\chi^2,\chi^2) = \chi^2(4)\sum_n \chi^2(n(1-n)) = \sum_{n=2}^{p-1} \chi^2(4n(1-n))$$
$$= 1 + 2\sum_{n=2}^{(p-1)/2} \chi^2(4n(1-n)),$$

since $4(p+1-n)(1-(p+1-n)) \equiv 4n(1-n) \pmod p$ and $4((p+1)/2)(1-((p+1)/2)) \equiv 1 \pmod p$. Now, as each term in the last sum has the value 1 or $(-1\pm i\sqrt{3})/2$, it follows that $K(\chi^2) = a_3 + ib_3\sqrt{3}$, where a_3 and b_3 are integers. Since $|K(\chi^2)|^2 = p$, the integers a_3 and b_3 are such that $p = a_3^2 + 3b_3^2$. Finally, by the multinomial theorem, we see that

$$a_3 \equiv a_3^3 \equiv K(\chi^2)^3 \equiv K(\chi^6) - 2 = p - 2 \equiv -1 \pmod 3. \qquad \blacksquare$$

We emphasize that, by Lemma 3.0.1, a_3 is uniquely determined by $p = a_3^2 + 3b_3^2$ and $a_3 \equiv -1 \pmod 3$. Thus a_3 does not depend on the choice of primitive root g such that $\chi(g) = \beta$. We now determine the sign of b_3.

Theorem 3.1.2 *In the notation of Theorem* 3.1.1,

$$3b_3 \equiv (2g^{(p-1)/3} + 1)a_3 \pmod p.$$

Proof Choose a prime ideal P dividing pO_K such that $\chi = \chi_P$. As $\chi(g) = \beta$ by the definition of χ, we have $\beta \equiv g^{(p-1)/6} \pmod P$. By (2.1.10), $P \mid K(\chi)O_K$. Thus, by Theorem 3.1.1,

$$0 \equiv K(\chi^2) = a_3 + b_3(\beta^2 - \overline{\beta}^2) \equiv a_3 + b_3(g^{(p-1)/3} - g^{2(p-1)/3}) \pmod P.$$

By (2.1.8), the right side is divisible by p, so that

$$(2g^{(p-1)/3} + 1)a_3 \equiv (2g^{(p-1)/3} + 1)(g^{2(p-1)/3} - g^{(p-1)/3})b_3$$
$$\equiv (2 - g^{(p-1)/3} - g^{2(p-1)/3})b_3 \equiv 3b_3 \pmod p. \qquad \blacksquare$$

Example Take $p = 7$ and $g = 3$. Then, by Theorem 3.1.1, $a_3 = 2$ and $|b_3| = 1$. By Theorem 3.1.2, $3b_3 \equiv 3 \pmod 7$, and so $b_3 = 1$.

In the next two theorems, we evaluate $J(\chi,\chi)$ and $J(\chi^2,\chi^2)$ in terms of quadratic partitions of $4p$. By Lemma 3.0.2(b), we have $\chi(2) = \beta^Z$, and so 2 is a cubic residue (mod p) if and only if $Z \equiv 0 \pmod 3$.

Theorem 3.1.3 *We have*

(3.1.1) $$2J(\chi^2,\chi^2) = r_3 + is_3\sqrt{3},$$

where r_3 and s_3 are integers such that

$$r_3^2 + 3s_3^2 = 4p, \ r_3 \equiv 1 \pmod 3, \ s_3 \equiv 0 \pmod 3, \ 3s_3 \equiv (2g^{(p-1)/3} + 1) r_3 \pmod p.$$

Proof By Theorem 3.1.1,

$$2J(\chi^2, \chi^2) = 2\chi(4) K(\chi^2) = 2\beta^{2Z}(a_3 + ib_3\sqrt{3}).$$

Thus (3.1.1) and $r_3^2 + 3s_3^2 = 4p$ hold with

$$\text{(3.1.2)} \qquad \begin{cases} r_3 = 2a_3, \ s_3 = 2b_3, & \text{if } Z \equiv 0 \pmod 3, \\ r_3 = -a_3 - 3b_3, \ s_3 = a_3 - b_3, & \text{if } Z \equiv 1 \pmod 3, \\ r_3 = -a_3 + 3b_3, \ s_3 = -a_3 - b_3, & \text{if } Z \equiv 2 \pmod 3. \end{cases}$$

By (3.1.1) and Theorem 2.1.11,

$$0 \equiv 2J(\chi^2, \chi^2) - 1 = r_3 - 1 + is_3\sqrt{3} \pmod 3,$$

so that

$$r_3 \equiv 1 \pmod 3, \ s_3 \equiv 0 \pmod 3.$$

By (3.1.2), $b_3/a_3 \equiv s_3/r_3 \pmod p$, so by Theorem 3.1.2, $3s_3 \equiv (2g^{(p-1)/3}+1) r_3 \pmod p$. ■

As $a_3 \equiv -1 \pmod 3$ and $s_3 \equiv 0 \pmod 3$ we observe from (3.1.2) that

$$\begin{cases} b_3 = s_3/2 \equiv 0 \pmod 3, & \text{if } Z \equiv 0 \pmod 3, \\ b_3 = a_3 - s_3 \equiv -1 \pmod 3, & \text{if } Z \equiv 1 \pmod 3, \\ b_3 = -a_3 - s_3 \equiv 1 \pmod 3, & \text{if } Z \equiv 2 \pmod 3, \end{cases}$$

so that

$$\text{(3.1.3)} \qquad b_3 \equiv -Z \pmod 3.$$

It is interesting to note that the congruence (3.1.3) determines the sign of b_3 in terms of $Z = \mathrm{ind}_g 2$ when 2 is a cubic nonresidue $\pmod p$ but not when 2 is a cubic residue $\pmod p$.

Theorem 3.1.4 *We have*

$$2\left(\frac{-1}{p}\right) J(\chi, \chi) = u_3 + iv_3\sqrt{3},$$

where u_3 and v_3 are integers such that

$$u_3^2 + 3v_3^2 = 4p, \ u_3 \equiv 1 \pmod 3, \ v_3 \equiv -Z \pmod 3, \ 3v_3 \equiv (2g^{(p-1)/3} + 1) u_3 \pmod p.$$

Proof By Theorem 3.1.1,

$$2J(\chi,\chi) = 2\overline{\chi}(4)K(\chi) = 2\left(\frac{-1}{p}\right)\beta^{-2Z}(a_3 + ib_3\sqrt{3}).$$

Thus $2\left(\frac{-1}{p}\right)J(\chi,\chi) = u_3 + iv_3\sqrt{3}$, where the integers u_3, v_3 satisfy $u_3^2 + 3v_3^2 = 4p$ and

(3.1.4)
$$\begin{cases} u_3 = 2a_3,\ v_3 = 2b_3, & \text{if } Z \equiv 0 \pmod 3, \\ u_3 = 3b_3 - a_3,\ v_3 = -a_3 - b_3, & \text{if } Z \equiv 1 \pmod 3, \\ u_3 = -3b_3 - a_3,\ v_3 = a_3 - b_3, & \text{if } Z \equiv 2 \pmod 3. \end{cases}$$

Recalling that $a_3 \equiv -1 \pmod 3$ and $b_3 \equiv -Z \pmod 3$, we find from (3.1.4) that

(3.1.5)
$$u_3 \equiv 1 \pmod 3, \quad v_3 \equiv -Z \pmod 3.$$

By (3.1.4), $b_3/a_3 \equiv v_3/u_3 \pmod p$, so by Theorem 3.1.2, $3v_3 \equiv (2g^{(p-1)/3}+1)u_3 \pmod p$. ■

From (3.1.2), (3.1.3), and (3.1.4), we see that

(3.1.6)
$$\begin{cases} (r_3 + is_3\sqrt{3})/2 = \beta^{-2b_3}(a_3 + ib_3\sqrt{3}) = \beta^{2Z}(a_3 + ib_3\sqrt{3}), \\ (u_3 + iv_3\sqrt{3})/2 = \beta^{2b_3}(a_3 + ib_3\sqrt{3}) = \beta^{-2Z}(a_3 + ib_3\sqrt{3}). \end{cases}$$

In particular, in the case that 2 is a cubic residue $\pmod p$, $v_3 = s_3$. In the case that 2 is a cubic nonresidue $\pmod p$, v_3 can be distinguished from s_3 by the fact that $3 \mid s_3$ whereas $3 \nmid v_3$. For the prime $p = 6f + 1$, the only solutions (x, y) to $x^2 + 3y^2 = 4p$ in positive integers x, y are $(|r_3|, |s_3|)$, $(|u_3|, |v_3|)$, and $(2|a_3|, 2|b_3|)$, when 2 is a cubic nonresidue $\pmod p$. When 2 is a cubic residue $\pmod p$, the only solutions are $(2|a_3|, 2|b_3|)$, $(|a_3 + 3b_3|, |a_3 - b_3|)$ and $(|a_3 - 3b_3|, |a_3 + b_3|)$; see Problem 2.

For the convenience of the reader we give the values of the Jacobi sums of orders 3 and 6 in tabular form.

Table 3.1.1: Jacobi sums of order 3

Let $p = 3f + 1$ be a prime. Let g be a primitive root $\pmod p$. Let χ be a character $\pmod p$ of order 3 such that $\chi(g) = \beta$, where $\beta = \exp(2\pi i/3) = (-1 + i\sqrt{3})/2$. Then the values of the nine Jacobi sums $J(\chi^m, \chi^n)$ $(m, n = 0, 1, 2)$ of order 3 are given in the table below.

$m\backslash^n$	0	1	2
0	p	0	0
1	0	$\frac{1}{2}(r_3 + is_3\sqrt{3})$	-1
2	0	-1	$\frac{1}{2}(r_3 - is_3\sqrt{3})$

Table 3.1.2 : Jacobi sums of order 6

Let $p = 6f + 1$ be a prime. Let g be a primitive root (mod p). Let χ be a character (mod p) of order 6 such that $\chi(g) = \beta$, where $\beta = \exp(2\pi i/6) = (1+i\sqrt{3})/2$. Then the values of the 36 Jacobi sums $J(\chi^m, \chi^n)$ $(m, n = 0, 1, 2, 3, 4, 5)$ of order 6 are given in the table below.

$m\backslash^n$	0	1	2	3	4	5
0	p	0	0	0	0	0
1	0	$(-1)^f \frac{1}{2}(u_3 + iv_3\sqrt{3})$	$a_3 + ib_3\sqrt{3}$	$(-1)^f(a_3 + ib_3\sqrt{3})$	$\frac{1}{2}(u_3 + iv_3\sqrt{3})$	$-(-1)^f$
2	0	$a_3 + ib_3\sqrt{3}$	$\frac{1}{2}(r_3 + is_3\sqrt{3})$	$a_3 + ib_3\sqrt{3}$	-1	$\frac{1}{2}(u_3 - iv_3\sqrt{3})$
3	0	$(-1)^f(a_3 + ib_3\sqrt{3})$	$a_3 + ib_3\sqrt{3}$	$-(-1)^f$	$a_3 - ib_3\sqrt{3}$	$(-1)^f(a_3 - ib_3\sqrt{3})$
4	0	$\frac{1}{2}(u_3 + iv_3\sqrt{3})$	-1	$a_3 - ib_3\sqrt{3}$	$\frac{1}{2}(r_3 - is_3\sqrt{3})$	$a_3 - ib_3\sqrt{3}$
5	0	$-(-1)^f$	$\frac{1}{2}(u_3 - iv_3\sqrt{3})$	$(-1)^f(a_3 - ib_3\sqrt{3})$	$a_3 - ib_3\sqrt{3}$	$(-1)^f \frac{1}{2}(u_3 - iv_3\sqrt{3})$

The integers r_3 and s_3 (resp. u_3 and v_3) are determined uniquely in terms of a_3 and b_3 by means of (3.1.2) (resp. (3.1.4)); see also (3.1.6).

3.2 Quartic sums

In this section, $k = 4$, $p = 4f + 1$, g is a primitive root (mod p), $\beta = \exp(2\pi i/4) = i$, $K = Q(i)$, $O_K = Z[i]$, and χ is a character (mod p) of order 4 with $\chi(g) = i$. By Lemma 3.0.2(a)(b)(d), $\chi(-1) = \chi(4) = (-1)^f$.

Theorem 3.2.1 *We have*

$$K(\chi) = a_4 + ib_4,$$

where a_4 and b_4 are integers such that

$$a_4^2 + b_4^2 = p, \ a_4 \equiv -\left(\frac{2}{p}\right) \pmod 4.$$

Proof Since $\chi(n) \in Z[i]$ for each integer n, it follows that $K(\chi) = a_4 + ib_4$, where a_4 and b_4 are integers. Moreover, by Theorem 2.1.3(b), a_4 and b_4 satisfy $a_4^2 + b_4^2 = p$. Further, by (2.1.3),

$$a_4 + ib_4 \equiv -p \pmod{2(1-i)}.$$

Thus 8 must divide

$$|(a_4 + p) + ib_4|^2 = a_4^2 + b_4^2 + p^2 + 2a_4 p = p(p + 1 + 2a_4),$$

and so

$$a_4 \equiv -(p+1)/2 \equiv -\left(\frac{2}{p}\right) \pmod 4. \qquad \blacksquare$$

The sign of b_4 is determined in the next theorem.

Theorem 3.2.2 *In the notation of Theorem* 3.2.1,

$$b_4 \equiv a_4 g^{(p-1)/4} \pmod p.$$

Proof Choose a prime ideal P dividing pO_K such that $\chi = \chi_P$. Then $i \equiv g^{(p-1)/4} \pmod P$. By (2.1.10), $P \mid K(\chi)O_K$. Thus, by Theorem 3.2.1,

$$0 \equiv K(\chi) = a_4 + ib_4 \equiv a_4 + g^{(p-1)/4} b_4 \pmod P.$$

Then, by (2.1.8), the right side of this congruence is divisible by p, and the required congruence easily follows. ■

We give an example to illustrate Theorem 3.2.2.

Example Take $p = 5$ and $g = 2$. By Theorem 3.2.1, $a_4 = 1$ and $|b_4| = 2$. Then Theorem 3.2.2 yields $b_4 \equiv 2 \pmod 5$, and so $b_4 = 2$.

We provide for the convenience of the reader the values of the Jacobi sums of order 4 in tabular form.

Table 3.2.1: Jacobi sums of order 4

Let $p = 4f + 1$ be a prime. Let g be a primitive root (mod p). Let χ be a character (mod p) of order 4 such that $\chi(g) = \beta$, where $\beta = \exp(2\pi i/4) = i$. Then the values of the 16 Jacobi sums $J(\chi^m, \chi^n)$ $(m, n = 0, 1, 2, 3)$ of order 4 are given in the table below.

$m \backslash n$	0	1	2	3
0	p	0	0	0
1	0	$(-1)^f(a_4 + ib_4)$	$a_4 + ib_4$	$-(-1)^f$
2	0	$a_4 + ib_4$	-1	$a_4 - ib_4$
3	0	$-(-1)^f$	$a_4 - ib_4$	$(-1)^f(a_4 - ib_4)$

Just as in (2.4.13), the integers a_4 and b_4 are determined uniquely by

$$p = a_4^2 + b_4^2,\ a_4 \equiv -\left(\frac{2}{p}\right) \pmod 4,\ b_4 \equiv a_4 g^{(p-1)/4} \pmod p.$$

3.3 Octic sums

In this section, $k = 8$, $p = 8f + 1$, g is a primitive root (mod p), $\beta = \exp(2\pi i/8) = (1+i)/\sqrt{2}$, $Z = \mathrm{ind}_g 2$, and χ is a character (mod p) of order 8 with $\chi(g) = \beta$. We note that $\beta^2 = i$, $\beta+\beta^3 = i\sqrt{2}$, and $\beta-\beta^3 = \sqrt{2}$. By Lemma 3.0.2(a)(b)(d), $\chi(-1) = (-1)^f$, $Z \equiv 0 \pmod 2$, and $\chi(4) = \beta^{2Z} = i^Z = (-1)^{Z/2} = \pm 1$.

Theorem 3.3.1 *We have*

$$K(\chi) = K(\chi^3) = a_8 + ib_8\sqrt{2},$$

where a_8 and b_8 are integers with

$$a_8^2 + 2b_8^2 = p,\ a_8 \equiv -1 \pmod 4.$$

Proof Observe that $Q(\beta)$ is the biquadratic field $Q(i, \sqrt{2})$ and $Q(i\sqrt{2})$ is the fixed field of the subgroup $< \sigma_3 >$ of $G = \mathrm{Gal}(Q(\beta)/Q)$. Appealing to (2.1.2), we see that $K(\chi) = K(\chi^3) = \sigma_3(K(\chi))$ so that $K(\chi) \in Q(i\sqrt{2})$. Thus $K(\chi) = a_8 + ib_8\sqrt{2}$, where a_8 and b_8 are integers such that $a_8^2 + 2b_8^2 = p$. From (2.1.3) we deduce that

$$a_8 + ib_8\sqrt{2} \equiv -p \pmod{2(1-\beta)}.$$

Thus

$$\prod_{\sigma\in G} \sigma(2-2\beta) = 16(1-\beta)(1-\beta^3)(1-\beta^5)(1-\beta^7) = 32$$

divides

$$\prod_{\sigma\in G} \sigma(a_8 + p + ib_8\sqrt{2}) = \{(a_8+p)^2 + 2b_8^2\}^2,$$

and so

$$(a_8+p)^2 + 2b_8^2 = p(p+1+2a_8) \equiv 0 \pmod 8.$$

It follows that $a_8 \equiv -(p+1)/2 \equiv -1 \pmod 4$. ∎

The sign of b_8 is determined in the next theorem.

Theorem 3.3.2 *In the notation of Theorem* 3.3.1,

$$2b_8 \equiv (g^f + g^{3f})a_8 \pmod p.$$

Proof Choose a prime ideal P dividing pO_K such that $\chi = \chi_P$. Then $\beta \equiv g^{(p-1)/8} \pmod{P}$. By (2.1.10), $P \mid K(\chi)O_K$. Thus, by Theorem 3.3.1,

$$\begin{aligned} 0 &\equiv i\sqrt{2}K(\chi) = i\sqrt{2}(a_8 + ib_8\sqrt{2}) = -2b_8 + i\sqrt{2}a_8 \\ &= -2b_8 + (\beta + \beta^3)a_8 \equiv -2b_8 + (g^f + g^{3f})a_8 \pmod{P}. \end{aligned}$$

By (2.1.8), the right side of the congruence above must be divisible by p, and we obtain $2b_8 \equiv (g^f + g^{3f})a_8 \pmod{p}$. ∎

Example We take $p = 17$ and $g = 3$. By Theorem 3.3.1, $a_8 = 3$ and $|b_8| = 2$. Theorem 3.3.2 then gives $2b_8 \equiv 3(g^2 + g^6) \equiv 4 \pmod{17}$, so that $b_8 = 2$.

As noted in the first paragraph of this chapter, the evaluation of the octic Jacobi sums may be reduced to the evaluation of $J(\chi, \chi)$, $J(\chi, \chi^2)$, and $J(\chi, \chi^3)$. We know from Theorem 3.3.1 that

$$J(\chi, \chi) = \chi(4)(a_8 + ib_8\sqrt{2}).$$

In the next theorem, we evaluate $J(\chi, \chi^2)$, and $J(\chi, \chi^3)$.

Theorem 3.3.3 *We have*

$$J(\chi, \chi^2) = \chi(-4)(a_4 + ib_4) \tag{3.3.1}$$

and

$$J(\chi, \chi^3) = \chi(-1)(a_8 + ib_8\sqrt{2}). \tag{3.3.2}$$

Proof From Theorem 2.1.6, $K(\chi) = \chi(-1)J(\chi, \chi^3)$, and so (3.3.2) follows from Theorem 3.3.1.

The equality $K(\chi) = \chi(-1)J(\chi, \chi^3)$ may be written in the form

$$\frac{\chi(4)G^2(\chi)}{G(\chi^2)} = \frac{\chi(-1)G(\chi)G(\chi^3)}{G(\chi^4)},$$

by Theorem 2.1.3(a). Thus

$$\frac{G(\chi)}{G(\chi^3)} = \frac{\chi(-4)G(\chi^2)}{G(\chi^4)}.$$

Multiplying by $G(\chi^2)$, we obtain

$$J(\chi, \chi^2) = \chi(-4)J(\chi^2, \chi^2),$$

and so (3.3.1) follows in view of Theorem 3.2.1. ∎

3.4 Bioctic sums

In this section, $k = 16$, $p = 16f + 1$, g is a primitive root (mod p), $\beta = \exp(2\pi i/16) = \frac{1}{2}(\sqrt{2+\sqrt{2}} + i\sqrt{2-\sqrt{2}})$, and χ is a character (mod p) of order 16 with $\chi(g) = \beta$. We note that

$$\beta^2 = (1+i)/\sqrt{2}, \qquad \beta^4 = i,$$
$$\beta + \beta^7 = i\sqrt{2-\sqrt{2}}, \qquad \beta - \beta^7 = \sqrt{2+\sqrt{2}},$$
$$\beta^3 + \beta^5 = i\sqrt{2+\sqrt{2}}, \qquad \beta^3 - \beta^5 = \sqrt{2-\sqrt{2}},$$
$$\beta^2 + \beta^6 = i\sqrt{2}, \qquad \beta^2 - \beta^6 = \sqrt{2}.$$

Theorem 3.4.1 *We have*

(3.4.1) $$K(\chi) = K(\chi^7) = a_{16} + b_{16}\sqrt{2} + ic_{16}\sqrt{2-\sqrt{2}} + id_{16}\sqrt{2+\sqrt{2}},$$

where a_{16}, b_{16}, c_{16}, and d_{16} are integers such that

(3.4.2) $$a_{16} \equiv -1 \pmod 8,$$

(3.4.3) $$a_{16}^2 + 2b_{16}^2 + 2c_{16}^2 + 2d_{16}^2 = p,$$

and

(3.4.4) $$2a_{16}b_{16} = c_{16}^2 - 2c_{16}d_{16} - d_{16}^2.$$

Furthermore, b_{16}, c_{16}, and d_{16} are even. The solutions $\pm(a,b,c,d)$ to (3.4.3) and (3.4.4) are "essentially unique" in the sense that the only other integral solutions are the "conjugate solutions" $\pm(a,b,-c,-d)$, $\pm(a,-b,-d,c)$, and $\pm(a,-b,d,-c)$.

Proof The subgroup $<\sigma_7>$ of $G = \text{Gal}(Q(\beta)/Q)$ has fixed field $Q(\beta^3+\beta^5) = Q(i\sqrt{2+\sqrt{2}})$. By (2.1.2), σ_7 fixes $K(\chi)$, and so $K(\chi) \in Q(i\sqrt{2+\sqrt{2}})$. Since $\{1, \sqrt{2}, i\sqrt{2-\sqrt{2}}, i\sqrt{2+\sqrt{2}}\}$ forms a basis for $Q(i\sqrt{2+\sqrt{2}})$ over Q, we conclude that (3.4.1) holds for some rational numbers a_{16}, b_{16}, c_{16}, and d_{16}. By Theorem 2.1.3(b), a_{16}, b_{16}, c_{16}, d_{16} satisfy (3.4.3) and (3.4.4).

By (2.1.4) and (3.4.1), $a_{16} + 1 + b_{16}\sqrt{2} \in 2O_{Q(\sqrt{2})}$. Thus, a_{16} is an odd integer and b_{16} is an even integer. Also, by (2.1.4) and (3.4.1),

$$c_{16}\sqrt{2-\sqrt{2}} + d_{16}\sqrt{2+\sqrt{2}} \in 2O_{Q(\sqrt{2+\sqrt{2}})}.$$

Multiplying the quantity above by $\sqrt{2+\sqrt{2}}$, we see that $2d_{16} + (c_{16}+d_{16})\sqrt{2} \in 2O_{Q(\sqrt{2})}$. Thus c_{16} and d_{16} are integers of the same parity. Since $a_{16}^2 \equiv 1 \pmod 8$, $2b_{16}^2 \equiv 0 \pmod 8$, and $p \equiv 1 \pmod 8$, it follows from (3.4.3) that $c_{16}^2 + d_{16}^2 \equiv 0 \pmod 4$, so that c_{16} and d_{16} are both even.

We now prove (3.4.2). By (2.1.3) we have $1 + K(\chi) \equiv 0 \pmod{2 - 2\beta}$. Hence $2^9 = \prod_{\sigma \in G} \sigma(2 - 2\beta)$ divides

$$\prod_{j=0}^{7} \{1 + K(\chi^{2j+1})\} = \{(p + 1 + 2a_{16})^2 - 8b_{16}^2\}^2.$$

Thus 32 divides $(p+1+2a_{16})^2 - 8b_{16}^2$. As b_{16} is even, it follows that $8 \mid (p+1+2a_{16})$, and hence

$$a_{16} \equiv -1 \pmod 4. \tag{3.4.5}$$

Further, using (3.4.4) and (3.4.5), we find that

$$b_{16}^2 + c_{16}^2 + d_{16}^2 = b_{16}^2 + c_{16}^2 + (c_{16}^2 - 2c_{16}d_{16} - 2a_{16}b_{16}) \equiv b_{16}^2 + 2b_{16} \equiv 0 \pmod 8.$$

Then (3.4.3) gives $a_{16}^2 \equiv p \equiv 1 \pmod{16}$, and so, appealing to (3.4.5), we obtain the congruence (3.4.2).

It remains to prove the uniqueness. We give here an algebraic proof; for an arithmetic proof, see Muskat and Zee [2].

First, by (2.1.10), the prime ideal factorization of $K(\chi)O_K$ is

$$K(\chi)O_K = P_1 P_7 P_{11} P_{13}.$$

Assume that (3.4.3) and (3.4.4) hold for integers a, b, c, d in place of a_{16}, b_{16}, c_{16}, d_{16}, respectively. Then

$$\gamma = a + b\sqrt{2} + ic\sqrt{2 - \sqrt{2}} + id\sqrt{2 + \sqrt{2}} \in O_K$$

satisfies the equality $\gamma\overline{\gamma} = p$. It suffices to show that, for some $\epsilon = \pm 1$, $\epsilon K(\chi)$ and γ are conjugates, i.e., $\epsilon K(\chi) = \sigma_j(\gamma)$ for some odd integer j, $1 \le j \le 15$. Replacing γ by an appropriate conjugate, if necessary, we may assume that $P \mid \gamma O_K$. As σ_7 fixes γ and $\gamma\overline{\gamma} = p$, either $\gamma O_K = P_1 P_7 P_{11} P_{13}$ or $\gamma O_K = P_1 P_7 P_3 P_5$. Thus, $\gamma O_K = K(\chi) O_K$ or $\gamma O_K = K(\chi^3) O_K$. Replacing γ by an appropriate conjugate, if necessary, we may assume that the former equality holds. Thus, there is a unit $\mu \in O_K$ such that $\gamma = \mu K(\chi)$. All conjugates of μ have modulus 1, so by Theorems 2.1.12 and 2.1.13, μ is a 16th root of unity. Since σ_7 fixes γ and $K(\chi)$, σ_7 fixes μ. Thus $\mu = \pm 1$, as desired. ∎

Theorem 3.4.1 shows that a_{16} is uniquely determined by (3.4.2), (3.4.3), and (3.4.4). The next theorem gives two additional congruences which determine b_{16}, c_{16}, and d_{16} uniquely.

Theorem 3.4.2 *Let $h = g^{(p-1)/16}$. Define a_{16}, b_{16}, c_{16}, and d_{16} by (3.4.1). Then a_{16}, b_{16}, c_{16}, and d_{16} are uniquely determined by (3.4.2), (3.4.3), (3.4.4), and the two congruences*

$$a_{16} + b_{16}(h^2 - h^6) + c_{16}(h + h^7) + d_{16}(h^3 + h^5) \equiv 0 \pmod p \tag{3.4.6}$$

and

$$(3.4.7) \qquad 2b_{16}^2 - a_{16}^2 \equiv (c_{16}^2 + 2c_{16}d_{16} - d_{16}^2)(h^2 - h^6) \pmod p.$$

Both members of (3.4.7) *are nonzero* $\pmod p$.

Proof Choose a prime ideal P dividing pO_K such that $\chi = \chi_P$. Then $\beta \equiv h \pmod P$. By Theorem 2.1.14, $K(\chi) \equiv K(\chi^3) \equiv 0 \pmod P$. Thus, by (3.4.1),

$$\begin{aligned} &a_{16} + b_{16}(h^2 - h^6) + c_{16}(h + h^7) + d_{16}(h^3 + h^5) \\ &\quad \equiv a_{16} + b_{16}(\beta^2 - \beta^6) + c_{16}(\beta + \beta^7) + d_{16}(\beta^3 + \beta^5) \equiv K(\chi) \equiv 0 \pmod P \end{aligned}$$

and

$$\begin{aligned} &a_{16} - b_{16}(h^2 - h^6) - d_{16}(h + h^7) + c_{16}(h^3 + h^5) \\ &\quad \equiv a_{16} - b_{16}(\beta^2 - \beta^6) - d_{16}(\beta + \beta^7) + c_{16}(\beta^3 + \beta^5) \equiv K(\chi^3) \equiv 0 \pmod P. \end{aligned}$$

Hence,

$$a_{16} + b_{16}(h^2 - h^6) \equiv -c_{16}(h + h^7) - d_{16}(h^3 + h^5) \pmod p,$$

which is (3.4.6), and

$$-a_{16} + b_{16}(h^2 - h^6) \equiv -d_{16}(h + h^7) + c_{16}(h^3 + h^5) \pmod p.$$

The multiplication of the last two congruences yields (3.4.7).

Now suppose that (a, b, c, d) is a solution to (3.4.2), (3.4.3), (3.4.4), (3.4.6), and (3.4.7). By the previous theorem, in order to prove uniqueness, we must show that none of the other three solutions $(a, b, -c, -d)$, $(a, -b, -d, c)$, and $(a, -b, d, -c)$, to (3.4.2)–(3.4.4) is also a solution to (3.4.6) and (3.4.7). Suppose that (3.4.7) holds with $(a, -b, -d, c)$ in place of (a, b, c, d). Then $2b^2 - a^2 \equiv 0 \pmod p$. By (3.4.3), we have $p > a^2 + 2b^2 \geq |2b^2 - a^2|$. Thus, $2b^2 - a^2 = 0$, which is impossible as a is odd. Similarly, (3.4.7) cannot hold for $(a, -b, d, -c)$. Last, suppose that (3.4.6) holds for $(a, b, -c, -d)$ in place of (a, b, c, d). Then $a + b(h^2 - h^6) \equiv 0 \pmod p$, and so

$$a^2 \equiv b^2(h^2 - h^6)^2 \equiv 2b^2 \pmod p,$$

which yields a contradiction as above. ■

Example We take $p = 17$ and $g = 3$. By Theorem 3.4.1, $a_{16} = -1$, $|b_{16}| = 2$, and $\{|c_{16}|, |d_{16}|\} = \{0, 2\}$. From (3.4.7), we find that $c_{16} = 0$ and $d_{16} = \pm 2$. From (3.4.6), we deduce that $1 + 6b_{16} + 2d_{16} \equiv 0 \pmod{17}$, so that $b_{16} = d_{16} = 2$.

In the next theorem, we evaluate Jacobi sums of order 16 in terms of $K(\chi)$, $K(\chi^2)$, and $J(\chi, \chi^2)$. Since $\{1, \beta, \ldots, \beta^7\}$ is an integral basis for $Q(\beta)$ over Q and $|J(\chi, \chi^2)|^2 = p$, it is not hard to see that

$$J(\chi, \chi^2) = \sum_{j=0}^{7} d_j \beta^j,$$

where $d_0, d_1, \ldots, d_7$ are integers such that $p = \sum_{j=0}^{7} d_j^2$. However, we will not further investigate this quadratic form here.

Theorem 3.4.3 *We have*
(a) $J(\chi, \chi^3) = \chi(4)\sigma_3(J(\chi, \chi^2))$,
(b) $J(\chi, \chi^4) = \overline{\chi}(-4)J(\chi, \chi^2)$,
(c) $J(\chi, \chi^5) = \chi(-1)\sigma_5(J(\chi, \chi^2))$,
(d) $J(\chi, \chi^6) = \overline{\chi}(-4)K(\chi^2)$,
(e) $J(\chi, \chi^7) = \chi(-1)K(\chi)$.

Proof Part (e) is a direct consequence of Theorem 2.1.6. Using Theorem 2.1.3(a), we see that part (c) is equivalent to $G(\chi)G(\chi^{15}) = \chi(-1)G(\chi^6)G(\chi^{10})$, which is obvious by Theorem 1.1.4(a).

By Theorem 2.1.4, $K(\chi) = J(\chi, \chi^8)$. From this and Theorem 2.1.3(a), it follows readily that $\chi(4)J(\chi, \chi^9) = J(\chi^2, \chi^8)$. Applying Theorem 2.1.4 again, we find that $J(\chi, \chi^9) = \overline{\chi}(4)K(\chi^2)$. Next, using Theorem 2.1.5, we conclude that $J(\chi, \chi^6) = \overline{\chi}(-4)K(\chi^2)$, which proves part (d).

By Theorems 3.3.1 and 2.1.5,

$$K(\chi^2) = K(\chi^6) = \chi^6(4)J(\chi^6, \chi^6) = \chi^6(4)J(\chi^4, \chi^6).$$

Thus, by part (d),

$$J(\chi, \chi^6) = \chi(-4)J(\chi^4, \chi^6).$$

Then by Theorem 2.1.5,

$$J(\chi, \chi^9) = \chi(4)J(\chi^4, \chi^6),$$

and so

$$G(\chi)G(\chi^9) = \chi(4)G(\chi^4)G(\chi^6).$$

Thus

$$\frac{G(\chi)G(\chi^3)}{G(\chi^4)} = \frac{\chi(4)G(\chi^3)G(\chi^6)}{G(\chi^9)},$$

which yields part (a).

Now apply σ_{11} to both sides of part (a) to obtain

$$J(\chi, \chi^{11}) = \overline{\chi}(4)J(\chi, \chi^2).$$

By Theorem 2.1.5, $J(\chi, \chi^{11}) = \chi(-1)J(\chi, \chi^4)$, which completes the proof of part (b) and the theorem. ∎

3.5 Duodecic sums

In this section, $k = 12$, $p = 12f + 1$, g is a primitive root (mod p), $\beta = \exp(2\pi i/12) = (\sqrt{3} + i)/2$, $Z = \text{ind}_g 2$, $T = \text{ind}_g 3$, and χ is a character (mod p) of order 12 with $\chi(g) = \beta$. By Lemma 3.0.2(b)(c), $Z \equiv f \pmod 2$ and $T \equiv 0 \pmod 2$. We define the integers a_{12} and b_{12} as follows:

$$\begin{cases} a_{12} = a_4, \ b_{12} = b_4, & \text{if } 3 \nmid a_4, \\ a_{12} = -a_4, \ b_{12} = -b_4, & \text{if } 3 \mid a_4, \end{cases}$$

where a_4 and b_4 are specified in Theorems 3.2.1 and 3.2.2. Since $p = a_4^2 + b_4^2 \equiv 1$ (mod 3), exactly one of a_4 and b_4 is divisible by 3.

The next lemma is a consequence of Theorem 2.7.6 on the quartic residuacity of 3 (mod p).

Lemma 3.5.1 *For $p = 12f + 1$, $3 \mid a_4$ if and only if $f + T/2$ is odd.*

Proof By definition of T, $f + T/2$ is even if and only if either $2 \mid f$ and 3 is a fourth power (mod p) or $2 \nmid f$ and 3 is not a fourth power (mod p). Thus by Theorem 2.7.6, $f + T/2$ is even if and only if $3 \mid b_4$. ■

Theorem 3.5.2 *We have*

$$K(\chi) = K(\chi^5) = (-1)^{f+T/2} K(\chi^3) = a_{12} + ib_{12}.$$

Proof The last equality follows from Theorem 3.2.1 and Lemma 3.5.1. Hence, it suffices to prove that $K(\chi) = K(\chi^5) = a_{12} + ib_{12}$. The subgroup $< \sigma_5 >$ of Gal$(Q(\beta)/Q)$ has the fixed field $Q(i)$. By (2.1.2), σ_5 fixes $K(\chi)$. Thus $K(\chi) \in Q(i)$, and so $K(\chi) = a + ib$, where a and b are integers such that $a^2 + b^2 = p$. By Theorem 2.1.8, a is odd. Since a_4 is odd, it follows from Lemma 3.0.1 that $a = \pm a_4$ and $b = \pm b_4$. Therefore, recalling that $K(\chi) = a + ib$ and $K(\chi^3) = a_4 + ib_4$, we must have

$$K(\chi) = \epsilon K(\chi^3) \quad \text{or} \quad K(\overline{\chi}) = \epsilon K(\chi^3),$$

where $\epsilon = \pm 1$.

Assume, for the purpose of contradiction, that $K(\overline{\chi}) = \epsilon K(\chi^3)$. Then we have

$$\begin{aligned} J^2(\chi, \chi^3) &= \frac{G^2(\chi)G^2(\chi^3)}{G^2(\chi^4)} \\ &= \frac{J(\chi,\chi)G(\chi^2)J(\chi^3,\chi^3)G(\chi^6)}{J(\chi^4,\chi^4)G(\chi^8)} && \text{(by Theorem 2.1.3(a))} \\ &= \frac{J(\chi,\chi)J(\chi^3,\chi^3)J(\chi^2,\chi^6)}{J(\chi^4,\chi^4)} && \text{(by Theorem 2.1.3(a))} \end{aligned}$$

$$= \frac{J(\chi,\chi)J(\chi^3,\chi^3)K(\chi^2)}{J(\chi^4,\chi^4)} \qquad \text{(by Theorem 2.1.4)}$$

$$= \frac{K(\chi)K(\chi^3)K(\chi^2)}{K(\chi^4)} \qquad \text{(by definition of } K(\chi))$$

$$= K(\chi)K(\chi^3) \qquad \text{(by Theorem 3.1.1)}$$

$$= \epsilon K(\chi)K(\overline{\chi}) \qquad \text{(by assumption)}$$

$$= \epsilon p,$$

so that $\pm\sqrt{\epsilon p} = J(\chi,\chi^3) \in Q(e^{2\pi i/12})$, that is $\sqrt{p} \in Q(\sqrt{3}, i)$, which is impossible, as $Q(\sqrt{3})$, $Q(i)$, and $Q(i\sqrt{3})$ are the only quadratic subfields of $Q(\sqrt{3}, i)$. Thus

$$K(\chi) = \epsilon K(\chi^3). \tag{3.5.1}$$

Cubing each side of (3.5.1), and working modulo 3, we obtain

$$K(\chi^3) \equiv K(\chi)^3 = \epsilon K(\chi^3)^3 \equiv \epsilon K(\chi^9) = \epsilon K(\overline{\chi}^3) \pmod 3,$$

so that, by Theorem 3.2.1,

$$a_4 + ib_4 \equiv \epsilon(a_4 - ib_4) \pmod 3.$$

Since exactly one of the integers in the pair a_4, b_4 is divisible by 3, $\epsilon = 1$ or $\epsilon = -1$ according as $3 \nmid a_4$ or $3 \mid a_4$. In view of (3.5.1), this proves that $K(\chi) = a_{12} + ib_{12}$. ■

We next evaluate other Jacobi sums of order 12 in terms of previously determined sums.

Theorem 3.5.3 *We have*

$$J(\chi,\chi^2) = K(\chi^2)/c_{12}, \tag{3.5.2}$$

$$J(\chi,\chi^3) = J(\chi^3,\chi^3)/c_{12}, \tag{3.5.3}$$

$$J(\chi,\chi^4) = J(\chi^4,\chi^4), \tag{3.5.4}$$

and

$$J(\chi,\chi^5) = \chi(-1)K(\chi), \tag{3.5.5}$$

where c_{12} *is uniquely determined by*

$$\begin{cases} c_{12} = \pm 1 & \textit{with} \quad c_{12} \equiv -a_4 \pmod 3, \quad \textit{if } 3 \mid b_4, \\ c_{12} = \pm i & \textit{with} \quad c_{12} \equiv -ib_4 \pmod 3, \quad \textit{if } 3 \mid a_4. \end{cases}$$

Proof We first define c_{12} by (3.5.2). Then, using Theorems 2.1.3–2.1.5, we find that

$$\begin{aligned} c_{12} &= \frac{K(\chi^2)}{J(\chi,\chi^2)} = \frac{J(\chi^2,\chi^6)}{J(\chi,\chi^2)} = \frac{G(\chi^6)G(\chi^3)}{G(\chi^8)G(\chi)} \\ &= \frac{J(\chi^3,\chi^6)}{J(\chi,\chi^8)} = \frac{K(\chi^3)}{(-1)^f J(\chi,\chi^3)} = \frac{J(\chi^3,\chi^3)}{J(\chi,\chi^3)}, \end{aligned}$$

where for the last equality we used the fact that $\chi^3(4) = \chi^6(2) = \left(\frac{2}{p}\right) = (-1)^f$. This establishes (3.5.3).

Squaring the last equality for c_{12}, and appealing to Theorem 2.1.3(a), we find that

$$c_{12}^2 = \frac{G^2(\chi^3)G^2(\chi^4)}{G^2(\chi^6)G^2(\chi)}.$$

Since $K(\chi) = (-1)^{f+T/2}K(\chi^3)$ (Theorem 3.5.2), we obtain

$$G^2(\chi) = \frac{(-1)^{f+T/2}\chi^2(4)G(\chi^2)G^2(\chi^3)}{G(\chi^6)} = \frac{\beta^{-2Z+3T}G(\chi^2)G^2(\chi^3)}{G(\chi^6)},$$

as $(-1)^{f+T/2}\chi^2(4) = \beta^{6f+3T+4Z} = \beta^{-6Z+3T+4Z} = \beta^{-2Z+3T}$. Substituting this expression for $G^2(\chi)$ into the expression for c_{12}^2, we obtain (noting that $3T \equiv -3T \pmod{12}$ as T is even)

$$c_{12}^2 = \frac{\beta^{2Z+3T}G^2(\chi^4)}{G(\chi^6)G(\chi^2)} = \frac{\beta^{2Z+3T}J(\chi^4,\chi^4)}{J(\chi^2,\chi^6)} = \frac{\beta^{2Z+3T}K(\chi^4)}{\chi^4(4)K(\chi^2)} = \frac{\beta^{6Z+3T}K(\chi^4)}{K(\chi^2)},$$

where we used Theorem 2.1.4. Hence, as $K(\chi^2) = K(\chi^4)$ by Theorem 3.1.1, we deduce that $c_{12}^2 = \beta^{6Z+3T} = (-1)^{Z+T/2} = (-1)^{f+T/2}$, so that, by Lemma 3.5.1,

$$c_{12} = \begin{cases} \pm 1, & \text{if } 3 \mid b_4, \\ \pm i, & \text{if } 3 \mid a_4. \end{cases}$$

Next, cubing both sides of (3.5.2), we obtain

$$J^3(\chi,\chi^2) = K^3(\chi^2)/c_{12}^3.$$

Hence as $J^3(\chi,\chi^2) \equiv J(\chi^3,\chi^6) \pmod 3$,

$$J(\chi^3,\chi^6) \equiv K^3(\chi^2)/c_{12}^3 \pmod 3.$$

Now,

$$K^3(\chi^2) = J^3(\chi^2,\chi^2) \equiv J(\chi^6,\chi^6) \pmod 3,$$

and, according to Theorem 2.1.1(c), it follows that

$$J(\chi^6, \chi^6) = -\chi^6(-1) = -1,$$

so that

$$J(\chi^3, \chi^6) \equiv -1/c_{12}^3 = -c_{12} \pmod 3.$$

From Theorems 3.2.1 and 2.1.4, we deduce that

$$a_4 + ib_4 = K(\chi^3) = J(\chi^3, \chi^6) \equiv -c_{12} \pmod 3.$$

This yields the desired evaluation of c_{12} and completes the proof of (3.5.2) and (3.5.3).

Next we prove (3.5.4). We have

$$\frac{J(\chi, \chi^4)}{J(\chi^4, \chi^4)} = \frac{G(\chi)G(\chi^8)}{G(\chi^5)G(\chi^4)} = \frac{J(\chi, \chi^8)}{\sigma_5(J(\chi, \chi^8))}. \tag{3.5.6}$$

Since

$$J(\chi, \chi^8) = (-1)^f J(\chi, \chi^3) = (-1)^f J(\chi^3, \chi^3)/c_{12},$$

by Theorem 2.1.5 and (3.5.3), σ_5 fixes $J(\chi, \chi^8)$. Thus (3.5.4) follows from (3.5.6).

Finally we observe that (3.5.5) follows from (2.1.2). ■

3.6 Biduodecic sums

In this section, $k = 24$, $p = 24f + 1$, g is a primitive root (mod p), $\beta = \exp(2\pi i/24) = (\sqrt{6} + \sqrt{2})/4 + i(\sqrt{6} - \sqrt{2})/4$, $Z = \mathrm{ind}_g 2$, $T = \mathrm{ind}_g 3$, and χ is a character (mod p) of order 24 with $\chi(g) = \beta$. By Lemma 3.0.2(b)(c), $Z \equiv T \equiv 0 \pmod 2$.

Theorem 3.6.1 *We have*

$$K(\chi) = K(\chi^5) = K(\chi^7) = K(\chi^{11}) = a_{24} + ib_{24}\sqrt{6}, \tag{3.6.1}$$

where a_{24} and b_{24} are integers such that

$$a_{24}^2 + 6b_{24}^2 = p, \quad a_{24} \equiv a_8 \pmod 3.$$

Proof By Theorem 2.1.14,

$$K(\chi)O_K = P_1P_5P_7P_{11} = K(\chi^5)O_K.$$

Thus, there is a unit $\mu \in O_K$ for which

$$K(\chi) = \mu K(\chi^5). \tag{3.6.2}$$

By Theorems 2.1.12 and 2.1.13, μ is a 24th root of unity. By (2.1.2), σ_{11} fixes $K(\chi)$ and $K(\chi^5)$. Hence, $\mu^{11} = \mu$, and so $\mu = \pm 1$. Cubing both sides of (3.6.2) and using Theorem 3.3.1, we find that, since $3 \mid b_8$,

$$a_8 \equiv K(\chi^3) \equiv \mu^3 K(\chi^{15}) \equiv \mu a_8 \pmod 3.$$

Thus $\mu = 1$, and so $K(\chi) = K(\chi^5)$. Since σ_{11} fixes $K(\chi)$ and $K(\chi^5)$, this completes the proof of the first three equalities in (3.6.1). It remains to show that $K(\chi) = a_{24}+ib_{24}\sqrt{6}$. As $\sqrt{2} = \beta^3 + \overline{\beta}^3$, $\sqrt{3} = \beta^2 + \overline{\beta}^2$, and $i = \beta^6$, we see that

$$\sigma_5(\sqrt{2}) = -\sqrt{2},\ \sigma_5(\sqrt{3}) = -\sqrt{3},\ \sigma_5(i) = i,$$
$$\sigma_{11}(\sqrt{2}) = -\sqrt{2},\ \sigma_{11}(\sqrt{3}) = \sqrt{3},\ \sigma_{11}(i) = -i.$$

Hence the subgroup $< \sigma_5, \sigma_{11} >$ of $\mathrm{Gal}(Q(\beta)/Q)$ has the fixed field $Q(i\sqrt{6})$. Thus $K(\chi) \in Q(i\sqrt{6})$, and (3.6.1) is established for some integers a_{24} and b_{24} such that $a_{24}^2 + 6b_{24}^2 = p$. Cubing the extreme left and right sides of (3.6.1) and using Theorem 3.3.1 again, we find that

$$a_8 \equiv K(\chi^3) \equiv a_{24}^3 \equiv a_{24} \pmod 3. \qquad \blacksquare$$

The next theorem gives two other determinations of a_{24}, one in terms of T, the other in terms of a_4.

Theorem 3.6.2 *We have*
(a) $a_{24} \equiv -(-1)^{T/2} \pmod 4$,
(b) $a_{24} \equiv \begin{cases} 1 \pmod 4, & \text{if } 3 \mid a_4, \\ -1 \pmod 4, & \text{if } 3 \nmid a_4. \end{cases}$

Proof (a) By Theorem 3.2.1, $a_4 \equiv -1 \pmod 4$. By the definition of a_{12} and Lemma 3.5.1, $a_{12} = (-1)^{T/2} a_4$. Thus we must show that $a_{24} \equiv a_{12} \pmod 4$. Now, by Theorem 2.1.4,

$$K(\chi) - K(\chi^{10}) = J(\chi, \rho) - J(\chi^{10}, \rho) = \sum_n \rho(1-n)\chi(n)(1-\chi^9(n))$$
$$= \sum_n \chi(n)(\rho(1-n) - 1)(1-\chi^9(n)).$$

When $n \not\equiv 0, 1 \pmod p$, $\rho(1-n) = \pm 1$ and $\chi^9(n)$ is a power of β^3. Thus,

$$K(\chi) - K(\chi^{10}) \equiv 0 \pmod{2(1-\beta^3)}.$$

By Theorem 2.1.6, $K(\chi^{10}) = K(\chi^2)$. Hence, by Theorems 3.6.1 and 3.5.2,

$$a_{24} + ib_{24}\sqrt{6} = K(\chi) \equiv K(\chi^2) = a_{12} + ib_{12} \pmod{2(1-\beta^3)}.$$

Since $p = a_{24}^2 + 6b_{24}^2 = a_{12}^2 + b_{12}^2 \equiv 1 \pmod 8$, it is clear that $2 \mid b_{24}$ and $4 \mid b_{12}$. Also, since $(1-\beta^3)(1-\beta^9) = -i\sqrt{2}$, it follows that

$$ib_{24}\sqrt{6} \equiv 0 \equiv ib_{12} \pmod{2(1-\beta^3)}.$$

Thus $a_{24} \equiv a_{12} \pmod{2(1-\beta^3)}$, and so $a_{24} \equiv a_{12} \pmod 4$.

(b) By Lemma 3.5.1, $T/2$ is odd if and only if $3 \mid a_4$. The required congruence now follows from part (a). ■

The next theorem determines the sign of b_{24}.

Theorem 3.6.3 *Let* $h = g^{(p-1)/24}$. *Then*

$$-a_{24} \equiv b_{24}(h + h^5 + h^7 + h^{11}) \pmod p.$$

Proof Note that $i\sqrt{6} = \beta + \beta^5 + \beta^7 + \beta^{11}$. The remainder of the proof is analogous to that of Theorem 3.2.2. ■

In Theorem 3.6.6 we evaluate other Jacobi sums of order 24 in terms of $J(\chi, \chi^2)$ and previously determined Jacobi sums. Because $\{1, \beta, \ldots, \beta^7\}$ is an integral basis for $Q(\beta)$ and $\beta^4 + \overline{\beta}^4 = 1$, it follows that $J(\chi, \chi^2) = \sum_{j=0}^{7} d_j \beta^j$, where $d_0, d_1, \ldots, d_7$ are integers such that $p = \sum_{j=0}^{7} d_j^2 + \sum_{j=0}^{3} d_j d_{j+4}$. We will, however, not investigate this quadratic form here. Two lemmas will be required. The first lemma involves $\text{ind}_g 3$ modulo 8, and so requires a form of octic reciprocity. We therefore delay the proof to Chapter 7; see the end of Section 7.3.

Lemma 3.6.4 *For* $p = 24f + 1$,

$$\begin{cases} a_8 a_4 \equiv (-1)^{f+T/4} \pmod 3, & \text{if } 4 \mid T, \\ a_8 b_4 \equiv (-1)^{f+(T-2)/4} \pmod 3, & \text{if } 4 \nmid T. \end{cases}$$

We remark that for primes $p \equiv 1 \pmod{24}$, Lemma 3.5.1 follows from Lemma 3.6.4. Write $V = b_{24}/2$. Define d_{24} by

$$d_{24} = \pm 1, \; d_{24} \equiv (-1)^{f+(Z/2)+1} a_{24} \pmod 3. \tag{3.6.3}$$

The next lemma characterizes d_{24} in terms of V.

Lemma 3.6.5 *We have* $d_{24} = (-1)^V$.

Proof To simplify notation, put $u = a_{24}$. Let ψ be a character (mod p) of order 4. By (3.6.3), $d_{24} = (-1)^{f+1}\psi(2)\left(\frac{u}{3}\right)$. Now, by the law of quadratic reciprocity (1.2.7), which allows the possibility of negative u,

$$\left(\frac{u}{3}\right) = \left(\frac{3}{u}\right)(-1)^{(u-1)/2} = -\left(\frac{3}{u}\right)\psi(3),$$

where the last equality follows from Theorem 3.6.2(a). Thus,

$$\begin{aligned} d_{24} &= (-1)^f\psi(6)\left(\frac{6}{u}\right)\left(\frac{2}{u}\right) = \psi(6)\left(\frac{6}{u}\right)(-1)^{3f+(1-u^2)/8} \\ &= \psi(6)\left(\frac{6}{u}\right)(-1)^{(p-u^2)/8} = \psi(6)\left(\frac{6}{u}\right)(-1)^{3V^2}. \end{aligned}$$

It remains to show that $\psi(6)\left(\frac{6}{u}\right) = 1$. Clearly

$$\psi(u^2) = \left(\frac{u}{p}\right) = \left(\frac{p}{u}\right) = \left(\frac{u^2+24V^2}{u}\right) = \left(\frac{6}{u}\right).$$

Write $V = 2^a v$, where v is odd. Then

$$\left(\frac{V}{p}\right) = \left(\frac{2^a}{p}\right)\left(\frac{v}{p}\right) = \left(\frac{p}{v}\right) = \left(\frac{u^2+24V^2}{v}\right) = 1.$$

Hence,

$$\psi(6)\left(\frac{6}{u}\right) = \psi(6u^2) = \psi(-144V^2) = \left(\frac{V}{p}\right) = 1. \qquad \blacksquare$$

Theorem 3.6.6 *We have*

$$J(\chi,\chi^3) = (-1)^f\beta^{6Z+3T}\sigma_5(J(\chi,\chi^2)), \tag{3.6.4}$$
$$J(\chi,\chi^4) = (-1)^{f+V}\beta^{-2Z}K(\chi^8), \tag{3.6.5}$$
$$J(\chi,\chi^5) = (-1)^f\beta^{6Z-3T}K(\chi), \tag{3.6.6}$$
$$J(\chi,\chi^6) = \beta^{6Z-3T}K(\chi^6), \tag{3.6.7}$$
$$J(\chi,\chi^7) = (-1)^V J(\chi,\chi), \tag{3.6.8}$$
$$J(\chi,\chi^8) = (-1)^{f+V}\beta^{3T}J(\chi,\chi^2), \tag{3.6.9}$$
$$J(\chi,\chi^9) = \beta^{-3T}\sigma_7(J(\chi,\chi^2)), \tag{3.6.10}$$
$$J(\chi,\chi^{10}) = (-1)^f\beta^{6T-2Z}K(\chi^6), \tag{3.6.11}$$

and

$$J(\chi,\chi^{11}) = (-1)^f K(\chi). \tag{3.6.12}$$

Proof Observe first that (3.6.12) follows from (2.1.2).

In the proofs that follow, we shall frequently employ Theorems 2.1.3–2.1.5 without comment.

First, we prove (3.6.8). By Theorem 3.6.1, $K(\chi) = K(\chi^7)$. Thus $G^2(\chi^7) = G^2(\chi)G(\chi^{14})/G(\chi^2)$, and so

$$\frac{J^2(\chi,\chi^7)}{J^2(\chi,\chi)} = \frac{G^2(\chi^2)G^2(\chi^7)}{G^2(\chi^8)G^2(\chi)} = \frac{G(\chi^2)G(\chi^{14})}{G^2(\chi^8)} = \frac{J(\chi^2,\chi^{14})}{J(\chi^8,\chi^8)} = \frac{J(\chi^2,\chi^8)}{J(\chi^8,\chi^8)} = 1,$$

where the last equality follows from (3.5.4). The equalities above show that

$$J(\chi,\chi^7) = \epsilon J(\chi,\chi), \tag{3.6.13}$$

where $\epsilon = \pm 1$. Cubing both sides of (3.6.13) and using Theorems 2.1.1(c), 3.3.1, and 3.6.1, we find that

$$-(-1)^f \equiv \epsilon\beta^{-6Z}K(\chi^3) \equiv \epsilon\beta^{-6Z}a_8 \equiv \epsilon\beta^{-6Z}a_{24} \pmod 3.$$

By (3.6.3), $\epsilon = d_{24}$, and from Lemma 3.6.5, we see that $d_{24} = (-1)^V$. Thus $\epsilon = (-1)^V$, and (3.6.13) gives (3.6.8).

We next prove (3.6.5). First, by Theorem 3.1.1,

$$\beta^{4Z} = \frac{J(\chi^4,\chi^4)}{K(\chi^4)} = \frac{J(\chi^4,\chi^{16})}{K(\chi^8)} = \frac{J(\chi^4,\chi^{16})}{J(\chi^8,\chi^{12})} = \frac{G(\chi^4)G(\chi^{16})}{G(\chi^8)G(\chi^{12})}. \tag{3.6.14}$$

Also, by (3.6.12) and Theorem 3.6.1,

$$\beta^{4Z} = \frac{\beta^{2Z}K(\chi)}{J(\chi,\chi)} = \frac{\beta^{14Z}(-1)^f J(\chi,\chi^{11})}{J(\chi^7,\chi^7)}.$$

An application of σ_7 to the equality above yields

$$\beta^{4Z} = \beta^{2Z}(-1)^f\frac{G(\chi^7)G(\chi^5)G(\chi^2)}{G(\chi^{12})G^2(\chi)}. \tag{3.6.15}$$

Equating the right sides of (3.6.14) and (3.6.15), and then multiplying both sides by $G(\chi)G(\chi^{12})/G(\chi^5)G(\chi^8)$, we obtain

$$\frac{J(\chi,\chi^4)}{J(\chi^8,\chi^8)} = \beta^{2Z}(-1)^f\frac{J(\chi,\chi^7)}{J(\chi,\chi)}.$$

Employing (3.6.8) above, we deduce (3.6.5).

We next prove (3.6.7). By (3.5.3), (3.6.8) and Lemma 3.6.5,

$$c_{12}d_{24} = c_{12}(-1)^V = \frac{J(\chi^6,\chi^6)J(\chi,\chi^7)}{J(\chi^2,\chi^6)J(\chi,\chi)} = \frac{G(\chi^6)G(\chi^7)}{G(\chi^{12})G(\chi)} = \frac{J(\chi^6,\chi^6)}{J(\chi,\chi^6)}. \tag{3.6.16}$$

Thus,

$$c_{12}d_{24}J(\chi,\chi^6) = K(\chi^6). \tag{3.6.17}$$

To prove (3.6.7), it remains to prove that

$$c_{12}d_{24} = \beta^{6Z+3T}. \tag{3.6.18}$$

Cubing both sides of (3.6.16), we find that

$$(-1)^f J(\chi^3,\chi^3) = J(\chi^3,\chi^{18}) \equiv c_{12}d_{24}K(\overline{\chi}^6) \pmod 3.$$

Then

$$(-1)^f \beta^{6Z} K(\chi^3)K(\chi^6) \equiv c_{12}d_{24} \pmod 3.$$

By Theorems 3.2.1 and 3.3.1, since $3 \mid b_8$, the equality above gives

$$(-1)^f \beta^{6Z}(a_8a_4 + ia_8b_4) \equiv c_{12}d_{24} \pmod 3.$$

Appealing to Lemma 3.6.4, we obtain (3.6.18), which completes the proof of (3.6.7).

We next prove (3.6.6). By (3.6.16),

$$c_{12}d_{24}K(\chi) = J(\chi^6,\chi^7).$$

Then, by (3.6.18),

$$\beta^{6Z+3T}K(\chi) = J(\chi^6,\chi^7).$$

Applying σ_7 to this, we deduce

$$\beta^{6Z-3T}K(\chi^7) = J(\chi^{18},\chi).$$

Then by Theorem 3.6.1,

$$\beta^{6Z-3T}K(\chi) = (-1)^f J(\chi,\chi^5),$$

which proves (3.6.6).

To prove (3.6.9), observe that

$$\begin{aligned}\frac{J(\chi,\chi^8)}{J(\chi,\chi^2)} &= \frac{G(\chi^3)G(\chi^8)}{G(\chi^2)G(\chi^9)} = \frac{(-1)^f G(\chi^3)G(\chi^{15})}{G(\chi^2)G(\chi^{16})}\\ &= \frac{(-1)^f J(\chi^3,\chi^{15})}{J(\chi^2,\chi^{16})} = \frac{J(\chi^3,\chi^6)}{J(\chi^2,\chi^6)} = \frac{(-1)^f \beta^{6Z} J(\chi^6,\chi^6)}{c_{12}^{-1} J(\chi^6,\chi^6)},\end{aligned}$$

where the last numerator and denominator are obtained from (3.3.1) and (3.5.3), respectively. Appealing to (3.6.18) and Lemma 3.6.5, we obtain (3.6.9).

To prove (3.6.11), observe that

$$\frac{J(\chi,\chi^{10})}{J(\chi,\chi^6)} = \frac{G(\chi^7)G(\chi^{10})}{G(\chi^6)G(\chi^{11})} = \sigma_7\left(\frac{J(\chi,\chi)}{J(\chi,\chi^5)}\right) = (-1)^f \beta^{4Z-3T},$$

where the last equality follows from (3.6.6). The equation (3.6.11) now follows from (3.6.7).

To prove (3.6.10), note that

$$\frac{J(\chi,\chi^9)}{\sigma_7(J(\chi,\chi^8))} = \frac{G(\chi)G(\chi^9)G(\chi^{15})}{G(\chi^{10})G(\chi^7)G(\chi^8)} = \frac{(-1)^f J(\chi,\chi^7)}{J(\chi^7,\chi^7)} = \frac{(-1)^f J(\chi,\chi^7)}{J(\chi,\chi)} = (-1)^{f+V},$$

where we have used Theorem 3.6.1 and (3.6.8). Appealing to (3.6.9), we deduce (3.6.10).

Finally, to prove (3.6.4), observe that

$$\frac{J(\chi,\chi^3)}{\sigma_5(J(\chi,\chi^8))} = \frac{G(\chi)G(\chi^3)G(\chi^{21})}{G(\chi^4)G(\chi^5)G(\chi^{16})} = \frac{(-1)^f J(\chi,\chi^4)}{J(\chi^4,\chi^4)} = \frac{(-1)^f J(\chi,\chi^4)}{\beta^{4Z}K(\chi^8)} = (-1)^V\beta^{6Z},$$

where we have used Theorem 3.1.1 and (3.6.5). Then, appealing to (3.6.9), we obtain (3.6.4). ■

It is interesting to note that by Lemma 3.6.5 and (3.6.18), c_{12} can be expressed solely in terms of Z, T, and V in the case $p = 24f + 1$, viz.,

$$c_{12} = (-1)^V\beta^{6Z+3T}.$$

3.7 Quintic and decic sums

In this section, $k = 10$, $p = 10f + 1$, g is a primitive root (mod p), $\beta = \exp(2\pi i/10)$, $K = Q(\beta)$, $Z = \text{ind}_g 2$, and χ is a character (mod p) of order 10 with $\chi(g) = \beta$. By Lemma 3.0.2(a), $\chi(-1) = (-1)^f = \left(\frac{-1}{p}\right)$. We note that

$$\beta = \exp(2\pi i/10) = ((\sqrt{5}+1) + i\sqrt{10-2\sqrt{5}}\,)/4,$$
$$\beta^2 = \exp(2\pi i/5) = ((\sqrt{5}-1) + i\sqrt{10+2\sqrt{5}}\,)/4.$$

Thus $\beta^3 = -\overline{\beta}^2$, $\beta^4 = -\overline{\beta}$, $\beta^5 = -1$ and

(3.7.1) $\quad 2(\beta+\overline{\beta}) = \sqrt{5}+1, \qquad 2(\beta^2+\overline{\beta}^2) = \sqrt{5}-1,$

(3.7.2) $\quad 2(\beta-\overline{\beta}) = i\sqrt{10-2\sqrt{5}}, \qquad 2(\beta^2-\overline{\beta}^2) = i\sqrt{10+2\sqrt{5}},$

(3.7.3) $\quad \beta-\overline{\beta}+\beta^2-\overline{\beta}^2 = i\sqrt{5+2\sqrt{5}}, \qquad \overline{\beta}-\beta+\beta^2-\overline{\beta}^2 = i\sqrt{5-2\sqrt{5}}.$

From (3.7.1)–(3.7.3) we deduce that

$$\sigma_3(\sqrt{5}) = -\sqrt{5},\ \sigma_3(i\sqrt{10-2\sqrt{5}}\,) = i\sqrt{10+2\sqrt{5}},\ \sigma_3(i\sqrt{10+2\sqrt{5}}\,) = -i\sqrt{10-2\sqrt{5}},$$
$$\sigma_3(i\sqrt{5+2\sqrt{5}}) = i\sqrt{5-2\sqrt{5}},\ \sigma_3(i\sqrt{5-2\sqrt{5}}) = -i\sqrt{5+2\sqrt{5}}.$$

Lemma 3.7.1 *Let c and d be rationals such that $\alpha = c\sqrt{5+2\sqrt{5}} + d\sqrt{5-2\sqrt{5}}$ is an algebraic integer. Then 2c and 2d are integers.*

Proof Since $\alpha\sqrt{5+2\sqrt{5}} = 5c+(2c+d)\sqrt{5}$, either $5c$ and $(2c+d)$ are both integers, or $10c$ and $2(2c+d)$ are both odd integers. Moreover, as the norm of $\sqrt{5+2\sqrt{5}}$ is 5, the norm of $10c+2(2c+d)\sqrt{5}$ is a multiple of 5. Thus, $10c$ is a multiple of 5, and so $2c$ is an integer. Consequently, $2d$ is also an integer. ∎

Theorem 3.7.2 *We have*

$$\left(\frac{-1}{p}\right)K(\chi) = K(\chi^4) = a_5 + b_5\sqrt{5} + ic_5\sqrt{5+2\sqrt{5}} + id_5\sqrt{5-2\sqrt{5}}, \tag{3.7.4}$$

where a_5, b_5, c_5, and d_5 are integers such that

$$a_5 \equiv -1 \pmod 5, \tag{3.7.5}$$

$$a_5^2 + 5b_5^2 + 5c_5^2 + 5d_5^2 = p, \tag{3.7.6}$$

and

$$a_5 b_5 = d_5^2 - c_5^2 - c_5 d_5. \tag{3.7.7}$$

The solutions $\pm(a,b,c,d)$ to (3.7.6) and (3.7.7) are "essentially unique" in the sense that the only other solutions in integers are the conjugate solutions $\pm(a,b,-c,-d)$, $\pm(a,-b,-d,c)$, and $\pm(a,-b,d,-c)$.

Proof The first equality in (3.7.4) follows from (2.1.2). Moreover, it is easily seen that $\{1,\ \sqrt{5},\ i\sqrt{5+2\sqrt{5}},\ i\sqrt{5-2\sqrt{5}}\}$ is a basis for K/Q, so that the second equality in (3.7.4) holds for some rational numbers a_5, b_5, c_5, and d_5 satisfying (3.7.6) and (3.7.7).

We now show that a_5, b_5, c_5, and d_5 are integers. By Theorem 2.1.8, either $\mathrm{Re}\{K(\chi)\} - 1 \in 2O_K$ or $\mathrm{Re}\{K(\chi)\} \in 2O_K$. By (3.7.4),

$$\mathrm{Re}\{K(\chi)\} = \left(\frac{-1}{p}\right)(a_5 + b_5\sqrt{5}),$$

so in either case it follows that a_5 and b_5 are integers. Then, by (3.7.4), $c_5\sqrt{5+2\sqrt{5}} + d_5\sqrt{5-2\sqrt{5}}$ is an algebraic integer. Lemma 3.7.1 now implies that $2c_5$ and $2d_5$ are integers, and (3.7.6) shows that both $2c_5$ and $2d_5$ are even. This completes the proof that a_5, b_5, c_5, and d_5 are integers.

Raising each side of (3.7.4) to the fifth power, and appealing to Theorem 2.1.1(c), we obtain

$$-1 = \left(\frac{-1}{p}\right)K(\chi^5) \equiv a_5 \pmod 5,$$

which proves (3.7.5).

We now give a proof of the statement on uniqueness that parallels that in the proof of Theorem 3.4.1. First, by (2.1.10),

$$K(\chi)O_K = P_1P_7. \tag{3.7.8}$$

Assume that (3.7.6) and (3.7.7) hold for integers a, b, c, and d in place of a_5, b_5, c_5, and d_5, respectively. Then

$$\gamma = a + b\sqrt{5} + ic\sqrt{5+2\sqrt{5}} + id\sqrt{5-2\sqrt{5}} \tag{3.7.9}$$

satisfies $\gamma\overline{\gamma} = p$. It suffices to show that $\pm K(\chi)$ and γ are conjugates. As $\gamma\overline{\gamma} = p$, the factorization of γO_K into prime ideals is given by

$$\gamma O_K = P_{\pm 1}P_{\pm 7},$$

for some choice of signs. By (3.7.8), $\gamma O_K = K(\chi^j)O_K$ for some $j \in \{\pm 1, \pm 3\}$. Thus, there is a 10th root of unity $\nu \in O_K$ such that

$$\nu\gamma = K(\chi^j), \tag{3.7.10}$$

for some $j \in \{\pm 1, \pm 3\}$. It remains to show that $\nu = \pm 1$. If not, by applying an appropriate automorphism in (3.7.10), we may assume that $\nu = \pm\beta^2$. Then, noting that

$$\begin{aligned}
\sqrt{10+2\sqrt{5}} &= \sqrt{5+2\sqrt{5}} + \sqrt{5-2\sqrt{5}},\\
\sqrt{5}\sqrt{10+2\sqrt{5}} &= 3\sqrt{5+2\sqrt{5}} - \sqrt{5-2\sqrt{5}},\\
\sqrt{5}\sqrt{5+2\sqrt{5}} &= 2\sqrt{5+2\sqrt{5}} + \sqrt{5-2\sqrt{5}},\\
\sqrt{5}\sqrt{5-2\sqrt{5}} &= \sqrt{5+2\sqrt{5}} - 2\sqrt{5-2\sqrt{5}},\\
\sqrt{10+2\sqrt{5}}\sqrt{5+2\sqrt{5}} &= 5 + 3\sqrt{5},\\
\sqrt{10+2\sqrt{5}}\sqrt{5-2\sqrt{5}} &= 5 - \sqrt{5},
\end{aligned}$$

we obtain

$$\begin{aligned}
4\beta^2\gamma &= (-1+\sqrt{5}+i\sqrt{10+2\sqrt{5}})(a + b\sqrt{5} + ic\sqrt{5+2\sqrt{5}} + id\sqrt{5-2\sqrt{5}})\\
&= (-a+5b-5c-5d) + (a-b-3c+d)\sqrt{5}\\
&\quad + i(a+3b+c+d)\sqrt{5+2\sqrt{5}} + i(a-b+c-3d)\sqrt{5-2\sqrt{5}}.
\end{aligned} \tag{3.7.11}$$

By (3.7.10) and (3.7.11), $4 \mid (-a+5b-5c-5d)$. However, by (3.7.6), $a+b+c+d$ is odd. Hence, $-a+5b-5c-5d$ is odd, and this is the desired contradiction. ■

Theorem 3.7.3 *Let $h = g^{(p-1)/10}$. Define a_5, b_5, c_5, and d_5 by (3.7.4). Then a_5, b_5, c_5, and d_5 are uniquely determined by (3.7.5)-(3.7.7) together with the congruences*

(3.7.12) $$a_5+b_5(2h^2-2h^3+1)+c_5(h+h^2+h^3+h^4)+d_5(h^2+h^3-h-h^4) \equiv 0 \pmod p$$

and

(3.7.13) $$5b_5^2 - a_5^2 \equiv (2h^2-2h^3+1)(c_5^2-d_5^2-4c_5d_5) \pmod p.$$

Proof The proof is analogous to that of Theorem 3.4.2. ■

Example To illustrate Theorem 3.7.3, we choose $p = 11$ and $g = 2$, so $h = g^{(p-1)/10} = 2$. By (3.7.6) and (3.7.5), $a_5 = -1$. Then, from (3.7.6), $b_5^2 + c_5^2 + d_5^2 = 2$. Hence

$$(|b_5|, |c_5|, |d_5|) = (1,1,0),\ (1,0,1) \text{ or } (0,1,1).$$

Then, from (3.7.7),

$$(b_5, c_5, d_5) = (1, \pm 1, 0) \text{ or } (-1, 0, \pm 1).$$

Congruences (3.7.12) and (3.7.13) in this case are

$$\begin{cases} a_5 + 4b_5 + 8c_5 + 5d_5 \equiv 0 \pmod{11}, \\ 5b_5^2 - a_5^2 \equiv 4(c_5^2 - d_5^2 - 4c_5d_5) \pmod{11}. \end{cases}$$

These are satisfied only by $(a_5, b_5, c_5, d_5) = (-1, 1, 1, 0)$.

Applying σ_3 to (3.7.4), we obtain

(3.7.14) $$\left(\frac{-1}{p}\right) K(\chi^3) = K(\chi^2) = a_5 - b_5\sqrt{5} - id_5\sqrt{5+2\sqrt{5}} + ic_5\sqrt{5-2\sqrt{5}}.$$

We now evaluate other Jacobi sums of orders 5 and 10 in terms of $K(\chi)$ and $K(\chi^2)$.

Theorem 3.7.4 *We have*

(3.7.15) $$K(\chi) = \left(\frac{-1}{p}\right) J(\chi, \chi^4) = \overline{\chi}^3(-4) J(\chi, \chi^2) = \overline{\chi}(-4) J(\chi^2, \chi^4)$$

and

(3.7.16) $$K(\chi^2) = \chi(-4) J(\chi, \chi^3).$$

Proof By Theorem 2.1.6, $K(\chi) = \chi(-1)J(\chi, \chi^4)$. Next, by Theorem 2.1.5 and (3.7.4),

$$J(\chi^2, \chi^4) = J(\chi^4, \chi^4) = \overline{\chi}^4(4) K(\chi^4) = \chi(-4) K(\chi).$$

Third, by Theorems 2.1.4 and 2.1.3, we find that

$$\frac{\chi(4)G^2(\chi)}{G(\chi^2)} = \frac{G(\chi)G(\chi^5)}{G(\chi^6)},$$

which implies that

$$\chi(4)J(\chi,\chi^6) = J(\chi^2,\chi^5) = K(\chi^2).$$

By Theorem 2.1.5,

$$J(\chi,\chi^6) = \chi(-1)J(\chi,\chi^3).$$

Combining the last two equalities, we complete the proof of (3.7.16).

Last, applying σ_7 to both sides of (3.7.16), we obtain

$$\overline{\chi}^3(-4)J(\chi,\chi^7) = K(\chi^4).$$

Applying Theorem 2.1.5 and (3.7.4) to the equation above, we find that

$$\overline{\chi}^3(-4)\chi(-1)J(\chi,\chi^2) = \chi(-1)K(\chi),$$

which completes the proof of (3.7.15). ■

In Section 3.1, we showed that (3.1.3) gives a criterion for determining the sign of b_3 in terms of $\text{ind}_g 2$, provided that $\text{ind}_g 2 \not\equiv 0 \pmod 3$. Analogously, we now give a criterion for determining b_5, c_5, and d_5 in terms of $Z = \text{ind}_g 2$, provided that $Z \not\equiv 0 \pmod 5$. Without loss of generality, by choosing an appropriate primitive root $g \pmod p$, we may assume that $Z \equiv 1 \pmod 5$.

Theorem 3.7.5 *Let* $Z \equiv 1 \pmod 5$. *Then* (3.7.4) *holds with* a_5, b_5, c_5, *and* d_5 *uniquely determined by* (3.7.5)–(3.7.7) *together with the conditions* $b_5 \equiv c_5 - 2d_5 \equiv 1 \pmod 5$.

Proof For brevity in this proof, we delete the subscript 5 from each of a_5, b_5, c_5, and d_5. First we have

$$G^5(\chi^4) \equiv \sum_{n=1}^{p-1} \chi^{20}(n)e^{10\pi in/p} = \sum_{n=1}^{p-1} e^{2\pi in/p} = -1 \pmod 5.$$

Next, using the fact $\chi^5(4) = 1$ and Theorem 2.1.3, we find that

$$\frac{p^2K^2(\chi^4)}{\chi(4)K(\chi^2)} = p^2\frac{J^2(\chi^4,\chi^4)}{J(\chi^2,\chi^2)} = p^2\frac{G^4(\chi^4)/G^2(\chi^8)}{G^2(\chi^2)/G(\chi^4)} = p^2\frac{G^5(\chi^4)}{\{G(\chi^2)G(\chi^8)\}^2} = G^5(\chi^4),$$

by Theorem 1.1.4(a). Hence,

$$K^2(\chi^4) \equiv 4\chi(4)K(\chi^2) \pmod 5. \qquad (3.7.17)$$

According to (3.7.14), the right side of (3.7.17) is

$$4\chi(4)(a - b\sqrt{5} - id\sqrt{5+2\sqrt{5}} + ic\sqrt{5-2\sqrt{5}}), \tag{3.7.18}$$

and, by (3.7.4) and (3.7.7), the left side of (3.7.17) is

$$\begin{aligned}(a^2+5b^2-5c^2-5d^2)+4ab\sqrt{5}\\ +i(2ac+4bc+2bd)\sqrt{5+2\sqrt{5}}+i(2ad+2bc-4bd)\sqrt{5-2\sqrt{5}}.\end{aligned} \tag{3.7.19}$$

Since $Z \equiv 1 \pmod 5$, we have $\chi(4) = \beta^2$. By (3.7.11), the expression in (3.7.18) becomes

$$\begin{aligned}(-a-5b-5c+5d)+(a+b+c+3d)\sqrt{5}\\ +i(a-3b+c-d)\sqrt{5+2\sqrt{5}}+i(a+b-3c-d)\sqrt{5-2\sqrt{5}}.\end{aligned} \tag{3.7.20}$$

Equating (3.7.19) and (3.7.20), we easily derive the congruences

$$4ab \equiv a+b+c+3d \pmod 5 \tag{3.7.21}$$

and

$$2ac+4bc+2bd \equiv a-3b+c-d \pmod 5. \tag{3.7.22}$$

By (3.7.5), $a \equiv -1 \pmod 5$, and so (3.7.21) yields

$$c \equiv 2d+1 \pmod 5, \tag{3.7.23}$$

and (3.7.22) and (3.7.23) imply that $b \equiv 1 \pmod 5$.

The assertion that a, b, c, and d are uniquely determined by (3.7.5)–(3.7.7) and the congruences $b \equiv c - 2d \equiv 1 \pmod 5$ follows from the last part of Theorem 3.7.2. ■

The following theorem shows that the congruence $b_5 \equiv c_5 - 2d_5 \equiv Z \pmod 5$ also holds in the case $5 \mid Z$, i.e., in the case when 2 is a quintic residue (mod p).

Theorem 3.7.6 *Let* $5 \mid Z$. *Then* (3.7.4) *holds with* a_5, b_5, c_5, *and* d_5 *satisfying*

$$b_5 \equiv c_5 - 2d_5 \equiv 0 \pmod 5.$$

Proof As in the previous proof, we delete the subscripts from a_5, b_5, c_5, and d_5. Here $\chi(4) = 1$. Equating (3.7.18) and (3.7.19) and then equating the coefficients of $i\sqrt{5+2\sqrt{5}}$ and $i\sqrt{5-2\sqrt{5}}$, we obtain, respectively, the congruences

$$d \equiv 2ac+4bc+2bd \pmod 5$$

and

$$-c \equiv 2ad+2bc-4bd \pmod 5.$$

Adding the latter congruence to twice the former, and, recalling that $a \equiv -1 \pmod 5$, we obtain $-c + 2d \equiv -4c - 2d \pmod 5$, which is equivalent to $c - 2d \equiv 0 \pmod 5$. Thus, $5 \mid b$ by (3.7.7). ∎

The following theorem enables us to evaluate the Jacobi sums $J(\chi, \chi)$ and $J(\chi^2, \chi^2)$ in terms of a quadratic partition of $16p$.

Theorem 3.7.7 *Let j be a given integer with $0 \le j \le 4$. Then there are integers $A = A(j)$, $B=B(j)$, $C = C(j)$, and $D = D(j)$ such that*

$$4\beta^{2j}K(\chi^4) = A + B\sqrt{5} + iC\sqrt{5+2\sqrt{5}} + iD\sqrt{5-2\sqrt{5}} \tag{3.7.24}$$

with

$$16p = A^2 + 5B^2 + 5C^2 + 5D^2, \quad AB = D^2 - C^2 - CD, \tag{3.7.25}$$

and $A \equiv 1 \pmod 5$. If $j = 0$, then A, B, C, and D are multiples of 4; otherwise, A, B, C, and D are odd. The ten solutions $\pm(A, B, C, D)$, $0 \le j \le 4$, to (3.7.25) are "essentially unique" in the sense that the only other solutions in integers are the conjugate solutions $\pm(A, B, -C, -D)$, $\pm(A, -B, -D, C)$, and $\pm(A, -B, D, -C)$.

Proof If $j = 0$, the result (3.7.24) follows from (3.7.4) with

$$A = 4a_5, \quad B = 4b_5, \quad C = 4c_5, \quad D = 4d_5. \tag{3.7.26}$$

If $j = 1$, then by (3.7.9) and (3.7.11) we obtain (3.7.24) with

$$\begin{cases} A = -a_5 + 5b_5 - 5c_5 - 5d_5, & B = a_5 - b_5 - 3c_5 + d_5, \\ C = a_5 + 3b_5 + c_5 + d_5, & D = a_5 - b_5 + c_5 - 3d_5. \end{cases} \tag{3.7.27}$$

Similarly, for $j = 2$, choose

$$\begin{cases} A = -a_5 - 5b_5 - 5c_5 + 5d_5, & B = -a_5 - b_5 - c_5 - 3d_5, \\ C = a_5 + b_5 - 3c_5 - d_5, & D = -a_5 + 3b_5 - c_5 + d_5; \end{cases} \tag{3.7.28}$$

for $j = 3$, choose

$$\begin{cases} A = -a_5 - 5b_5 + 5c_5 - 5d_5, & B = -a_5 - b_5 + c_5 + 3d_5, \\ C = -a_5 - b_5 - 3c_5 - d_5, & D = a_5 - 3b_5 - c_5 + d_5; \end{cases} \tag{3.7.29}$$

and for $j = 4$, choose

$$\begin{cases} A = -a_5 + 5b_5 + 5c_5 + 5d_5, & B = a_5 - b_5 + 3c_5 - d_5, \\ C = -a_5 - 3b_5 + c_5 + d_5, & D = -a_5 + b_5 + c_5 - 3d_5. \end{cases} \tag{3.7.30}$$

Since $a_5 \equiv -1 \pmod 5$, by (3.7.5), $A \equiv 1 \pmod 5$ in all cases. Since $a_5 + b_5 + c_5 + d_5$ is odd, (3.7.26)–(3.7.30) show that A, B, C, and D are multiples of 4 if $j = 0$, and are odd otherwise. Finally, the uniqueness follows by a variant of the argument used to establish the uniqueness in Theorem 3.7.2. ■

Example The solutions to the system (3.7.25) for $p = 11$ are $\pm(-4,4,4,0)$, $\pm(1,-5,3,-1)$, $\pm(-9,-1,-3,3)$, $\pm(1,1,-3,-5)$, $\pm(11,1,-1,3)$, and the corresponding conjugate solutions.

The right side of (3.7.24) can be expressed as

$$q + r\sqrt{5} + is\sqrt{10+2\sqrt{5}} + it\sqrt{10-2\sqrt{5}},$$

where

$$q = A,\ r = B,\ s = (C+D)/2,\ t = (C-D)/2. \tag{3.7.31}$$

The right side of (3.7.24) can also be expressed as

$$x - 5w\sqrt{5} + iv\sqrt{50+10\sqrt{5}} - iu\sqrt{50-10\sqrt{5}},$$

where

$$\begin{cases} x = A,\ w = -B/5,\ v = (3C-D)/10,\ u = -(3D+C)/10, & \text{or} \\ A = x,\ B = -5w,\ C = 3v - u,\ D = -3u - v. \end{cases} \tag{3.7.32}$$

These relations give rise to other quadratic partitions of $16p$, namely,

$$16p = q^2 + 5r^2 + 10s^2 + 10t^2,\ qr = t^2 - s^2 - 4st, \tag{3.7.33}$$

and Dickson's [8, p. 402] well-known partition

$$16p = x^2 + 125w^2 + 50v^2 + 50u^2,\ xw = v^2 - u^2 - 4vu, \tag{3.7.34}$$

which we have already encountered in Problem 13 of Exercises 2. Numerical solutions of (3.7.34) can be deduced from the tables given by Tanner [2]. We just give a solution for each prime $p < 100$ with $p \equiv 1 \pmod 5$.

p	x	w	v	u
11	1	1	1	0
31	11	−1	1	2
41	−9	−1	3	0
61	1	1	−1	4
71	−19	1	−3	−2

We observe that, as $C \equiv D \pmod 2$ for every solution of (3.7.25), (3.7.31) provides a one-to-one correspondence between the integer solutions (A, B, C, D) of (3.7.25) and the integer solutions (q, r, s, t) of (3.7.33). Also, as $C \equiv 2D \pmod 5$ for every solution (A, B, C, D) of (3.7.25) with $5 \mid B$, (3.7.32) gives a one-to-one correspondence between the integer solutions (A, B, C, D) of (3.7.25) for which $5 \mid B$ and the integer solutions (x, w, v, u) of (3.7.34).

Next we investigate the relationship between $Z \pmod 5$ and $B \pmod 5$, where (A, B, C, D) is an integer solution of (3.7.25). Consider first the case $5 \mid Z$. By (3.7.26)–(3.7.30) and Theorem 3.7.6, $5 \mid B$ if $j = 0$, but $B \equiv \pm 1 \pmod 5$ for $1 \leq j \leq 4$. Now consider the case when $5 \nmid Z$. As before, without loss of generality, we may assume that $Z \equiv 1 \pmod 5$. By (3.7.26)–(3.7.30) and Theorem 3.7.5, $5 \mid B$ if $j = 1$ but $B \equiv \pm 1 \pmod 5$ otherwise. We have shown that when $5 \mid Z$ we have $5 \mid B$ only if $j = 0$, and when $Z \equiv 1 \pmod 5$ we have $5 \mid B$ only if $j = 1$. Thus, in view of Theorem 3.7.7 and the statement following (3.7.31), there are integers x, w, v, and u satisfying (3.7.34) such that

$$(3.7.35) \quad \begin{cases} 4J(\chi^4, \chi^4) = x - 5w\sqrt{5} + iv\sqrt{50 + 10\sqrt{5}} - iu\sqrt{50 - 10\sqrt{5}}, & \text{equivalently} \\ 4J(\chi^2, \chi^2) = x + 5w\sqrt{5} + iu\sqrt{50 + 10\sqrt{5}} + iv\sqrt{50 - 10\sqrt{5}}, \end{cases}$$

and the integral solutions (x, w, v, u) to (3.7.34) are "essentially unique"; that is, if (x, w, v, u) is an integral solution of (3.7.34), then all integral solutions are given by

$$\pm(x, w, v, u), \ \pm(x, w, -v, -u), \ \pm(x, -w, -u, v), \ \pm(x, -w, u, -v).$$

Dickson [8, pp. 402–403] gave the original elementary proof of this result.

Example When $p = 11$, the only such solutions are $(x, w, v, u) = \pm(1, 1, 1, 0)$ and the conjugate solutions $\pm(1, 1, -1, 0)$, $\pm(1, -1, 0, 1)$, $\pm(1, -1, 0, -1)$.

We note that (3.7.26), (3.7.27), and (3.7.32) yield the following relations between the tuples (x, w, v, u) in (3.7.35) and (a_5, b_5, c_5, d_5) in (3.7.4): If $Z \equiv 0 \pmod 5$, then

$$(3.7.36) \quad \begin{cases} x = 4a_5, \ w = -4b_5/5, \ v = 2(3c_5 - d_5)/5, \ u = -2(c_5 + 3d_5)/5, \ \text{or} \\ a_5 = x/4, \ b_5 = -5w/4, \ c_5 = (3v - u)/4, \ d_5 = -(3u + v)/4; \end{cases}$$

if $Z \equiv 1 \pmod 5$, then

$$(3.7.37) \quad \begin{cases} x = -a_5 + 5b_5 - 5c_5 - 5d_5, \\ 5w = -a_5 + b_5 + 3c_5 - d_5, \\ 5v = a_5 + 5b_5 + c_5 + 3d_5, \\ 5u = -2a_5 - 2c_5 + 4d_5, \end{cases} \quad \text{or} \quad \begin{cases} 16a_5 = -x - 25w + 10v - 20u, \\ 16b_5 = x + 5w + 10v, \\ 16c_5 = -x + 15w + 2v - 4u, \\ 16d_5 = -x - 5w + 6v + 8u. \end{cases}$$

We observe that $x \equiv 1 \pmod 5$, as $a_5 \equiv -1 \pmod 5$ by (3.7.5). The two congruences $\pmod p$ in Theorem 3.7.3 could be used to determine uniquely the integer solution (x, w, v, u) to (3.7.34) with $x \equiv 1 \pmod 5$. However, the following theorem shows that these two congruences can be replaced by the single congruence

$$F(x, w, v, u) \equiv g^{(p-1)/5} \pmod p,$$

where

$$F(x,w,v,u) := \frac{x^2 - 125w^2 - 10(2xu - xv - 25vw)}{x^2 - 125w^2 + 10(2xu - xv - 25vw)}. \tag{3.7.38}$$

Theorem 3.7.8 *Let $H = g^{(p-1)/5}$. Then there is exactly one 4-tuple of integers (x, w, v, u) satisfying both*

$$\begin{cases} 16p = x^2 + 125w^2 + 50v^2 + 50u^2, \\ xw = v^2 - u^2 - 4uv, \\ x \equiv 1 \pmod 5, \end{cases} \tag{3.7.39}$$

and

$$F(x,w,v,u) \equiv H \pmod p, \tag{3.7.40}$$

and this 4-tuple satisfies (3.7.35). *Moreover, the cyclotomic numbers of order* 5 *are as given in Problem* 14 *of Exercises* 2 *and*

$$\mathrm{ind}_g 5 = -v - 2u \pmod 5.$$

Proof From the statement containing (3.7.35), we see that the only integer solutions to (3.7.39) are the 4-tuple (x, w, v, u) defined by (3.7.35) together with the conjugate solutions $(x, w, -v, -u)$, $(x, -w, u, -v)$, and $(x, -w, -u, v)$. It remains to show that

$$\begin{cases} F(x,w,v,u) \equiv H \pmod p, \\ F(x,-w,u,-v) \equiv H^2 \pmod p, \\ F(x,-w,-u,v) \equiv H^3 \pmod p, \\ F(x,w,-v,-u) \equiv H^4 \pmod p. \end{cases} \tag{3.7.41}$$

It suffices to prove the first congruence in (3.7.41), for then the other three follow by replacing g by the primitive root g^t for values of t coprime with $(p-1)$ for which $t \equiv 2 \pmod 5$, $t \equiv 3 \pmod 5$, and $t \equiv 4 \pmod 5$.

Choose a prime ideal P dividing pO_K such that $\chi = \chi_P$. Then $\beta^2 \equiv H \pmod P$. Define integers R, S, T (mod p) by

$$\begin{cases} R = H - H^2 - H^3 + H^4, \\ S = 2H + 4H^2 - 4H^3 - 2H^4, \\ T = 4H - 2H^2 + 2H^3 - 4H^4. \end{cases} \tag{3.7.42}$$

Note that

$$\begin{cases} R^2 \equiv 5 \pmod p, \\ S^2 \equiv -50 - 10R \pmod p, \\ T^2 \equiv -50 + 10R \pmod p, \\ ST \equiv -20R \pmod p. \end{cases} \tag{3.7.43}$$

By (2.1.10), the Jacobi sums in (3.7.35) are divisible by P, and so

$$\left\{\begin{array}{l} x + 5wR + uS + vT \equiv 0 \pmod{p}, \\ x - 5wR + vS - uT \equiv 0 \pmod{p}. \end{array}\right. \tag{3.7.44}$$

Thus,

$$\begin{aligned} x^2 - 125w^2 &\equiv (x + 5wR)(x - 5wR) \equiv (vT + uS)(vS - uT) \\ &\equiv -20R(v^2 + vu - u^2) \pmod{p}. \end{aligned} \tag{3.7.45}$$

If $x^2 - 125w^2 \equiv 0 \pmod{p}$, then from the first equality in (3.7.39), we would have $50(5w^2 + v^2 + u^2) \equiv 0 \pmod{p}$, which contradicts the fact that $5w^2 + v^2 + u^2 < p$. Thus the members of (3.7.45) are nonzero $\pmod{p}$.

Solving (3.7.44) for x and then for w, we obtain

$$x \equiv -\frac{(u+v)}{2}S + \frac{(u-v)}{2}T \pmod{p}$$

and

$$w \equiv \frac{(v-u)}{10R}S - \frac{(v+u)}{10R}T \pmod{p}.$$

By (3.7.43),

$$S/R \equiv (2T + S)/5 \pmod{p}$$

and

$$T/R \equiv (2S - T)/5 \pmod{p},$$

so that

$$w \equiv -\frac{(v+3u)}{50}S + \frac{(3v-u)}{50}T \pmod{p}.$$

Thus

$$2xu - xv - 25vw \equiv (v^2 + vu - u^2)(S - T) \pmod{p}.$$

Then, by (3.7.38) and (3.7.45),

$$F(x, w, v, u) \equiv \frac{2R + S - T}{2R + T - S} \pmod{p}.$$

Therefore $F(x, w, v, u) \equiv H \pmod{p}$ by (3.7.42). The last assertion of the theorem follows after some calculation from Theorem 2.5.1, (3.7.35), and Corollary 2.6.11. ■

The next theorem classifies the 4-tuples (x, w, v, u) in Theorem 3.7.8 in terms of the value of $Z = \text{ind}_g 2 \pmod{5}$. The proof depends on a result on quintic residuacity that is proved in Chapter 7 (Theorem 7.4.1).

Theorem 3.7.9 *Let (x, w, v, u) be the unique 4-tuple defined in Theorem 3.7.8. Then*

$$\begin{aligned}
Z &\equiv 0 \pmod 5), \quad \text{if } x \equiv 0 \pmod 2,\\
Z &\equiv 1 \pmod 5, \quad \text{if } x \equiv 1 \pmod 2,\ u \equiv 0 \pmod 2,\ x - v + u \equiv 0 \pmod 4,\\
Z &\equiv 2 \pmod 5, \quad \text{if } x \equiv 1 \pmod 2,\ u \equiv 1 \pmod 2,\ x - v + u \equiv 2 \pmod 4,\\
Z &\equiv 3 \pmod 5, \quad \text{if } x \equiv 1 \pmod 2,\ u \equiv 1 \pmod 2,\ x - v + u \equiv 0 \pmod 4,\\
Z &\equiv 4 \pmod 5, \quad \text{if } x \equiv 1 \pmod 2,\ u \equiv 0 \pmod 2,\ x - v + u \equiv 2 \pmod 4.
\end{aligned}$$

Proof By Theorem 7.4.1, x is even if and only if $Z \equiv 0 \pmod 5$. Suppose that x is odd. Then by (3.7.39), w is odd and v, u are of opposite parity, so that $x - v + u$ is even. If u is even, then from (3.7.39),

$$xw = v^2 - u^2 - 4vu \equiv 1 - u^2 \equiv 1 - 2u \pmod 8,$$

so that

$$w \equiv x - 2xu \equiv x - 2u \pmod 8. \tag{3.7.46}$$

If u is odd, then

$$xw = v^2 - u^2 - 4vu \equiv v^2 - 1 \equiv 2v - 1 \pmod 8,$$

so that

$$w \equiv 2xv - x \equiv 2v - x \pmod 8. \tag{3.7.47}$$

We now examine four cases, using the formulas for the cyclotomic numbers $B = (0,1)_5$, $C = (0,2)_5$, $D = (0,3)_5$ and $E = (0,4)_5$ given in Problem 14 of Exercises 2.

Case 1: u even, $x - v + u \equiv 0 \pmod 4$.
By (3.7.46),

$$\begin{aligned}
4B &\equiv 100B = 4p - 16 - 3x + 25w + 50v \equiv 4 - 3x + w + 2v\\
&\equiv 4 - 2(x - v + u) \equiv 4 \pmod 8,
\end{aligned}$$

so that B is odd, and thus by Theorem 2.6.1, $Z \equiv 1 \pmod 5$.

Case 2: u odd, $x - v + u \equiv 2 \pmod 4$.
By (3.7.47),

$$4C \equiv 100C = 4p - 16 - 3x - 25w + 50u \equiv 4 - 3x - w + 2u \equiv 2(x - v + u) \equiv 4 \pmod 8,$$

so that C is odd and thus $Z \equiv 2 \pmod 5$.

Case 3: u odd, $x - v + u \equiv 0 \pmod 4$.
By (3.7.47),

$$4D \equiv 100D = 4p - 16 - 3x - 25w - 50u \equiv 4 - 3x - w - 2u \equiv 4 - 2(x - v + u) \equiv 4 \pmod 8,$$

so that D is odd and thus $Z \equiv 3 \pmod 5$.
Case 4: u even, $x - v + u \equiv 2 \pmod 4$.
By (3.7.46),

$$4E \equiv 100E = 4p - 16 - 3x + 25w - 50v \equiv 4 - 3x + w - 2v \equiv 2(x - v + u) \equiv 4 \pmod 8,$$

so that E is odd and thus $Z \equiv 4 \pmod 5$. ■

Example Suppose that $p = 41$ and $g = 6$. Then $(x, w, v, u) = (-9, -1, 3, 0)$. Here x is odd, u is even, and $x - v + u \equiv 0 \pmod 4$. Thus, by Theorem 3.7.9, $Z \equiv 1 \pmod 5$. It is easy to check directly that $2 \equiv 6^{26} \pmod{41}$, so that $Z = 26 \equiv 1 \pmod 5$.

3.8 Bidecic sums

In this section, $k = 20$, $p = 20f + 1$, g is a primitive root $\pmod p$, $\beta = \exp(2\pi i/20)$, $K = Q(\beta)$, $Z = \text{ind}_g 2$, $R = \text{ind}_g 5$, and χ is a character $\pmod p$ of order 20 with $\chi(g) = \beta$. By Lemma 3.0.2(a)(b)(e), $\chi(-1) = (-1)^f$, $\chi(2) = \beta^Z$, $Z \equiv f \pmod 2$, and $R \equiv 0 \pmod 2$. Note further that

$$\chi^5(-4) = \chi^5(-1)\chi^{10}(2) = (-1)^{f+Z} = 1.$$

In Theorem 7.2.3 a necessary and sufficient condition for 5 to be a quartic residue $\pmod p$ is proved. This condition is used in the proof of the next lemma.

Lemma 3.8.1 *For $p = 20f + 1$, $5 \mid a_4$ if and only if $4 \nmid R$.*

Proof By Theorem 7.2.3, $4 \mid R$ if and only if $5 \mid b_4$. The result now follows since exactly one of a_4, b_4 is divisible by 5. ■

Theorem 3.8.2 *Let χ be a character $\pmod p$ of order 20. Then*

$$K(\chi) = K(\chi^9) = (-1)^{R/2} K(\chi^3) = (-1)^{R/2} K(\chi^7) = (a_{20} + i b_{20}\sqrt{5})/c_{20},$$

where a_{20} and b_{20} are integers such that $a_{20}^2 + 5b_{20}^2 = p$ and $a_{20} \equiv -(-1)^{R/2} \pmod 5$. Moreover, c_{20} is defined by

$$c_{20} = \pm 1, \ c_{20} \equiv -a_4 \pmod 5, \quad \text{if } 4 \mid R, \text{ i.e., if } 5 \mid b_4,$$

and

$$c_{20} = \pm i, \ c_{20} \equiv -i b_4 \pmod 5, \quad \text{if } 4 \nmid R, \text{ i.e., if } 5 \mid a_4.$$

Proof By (2.1.10), $K(\chi)O_K$ and $K(\chi^3)O_K$ have the same factorization into prime ideals, namely,

$$K(\chi)O_K = K(\chi^3)O_K = P_1P_3P_7P_9.$$

Thus there is a unit $\mu \in O_K$ for which

$$K(\chi^3) = \mu K(\chi). \tag{3.8.1}$$

By Theorems 2.1.12 and 2.1.13, μ is a 20th root of unity, say, $\mu = \beta^j$ $(0 \le j \le 19)$. From (2.1.2), $K(\chi) = K(\chi^9)$ and $K(\chi^3) = K(\chi^7)$. Hence

$$\beta^{9j} = \sigma_9(\beta^j) = \sigma_9\left(\frac{K(\chi^3)}{K(\chi)}\right) = \frac{K(\chi^7)}{K(\chi^9)} = \frac{K(\chi^3)}{K(\chi)} = \beta^j,$$

so that $\beta^{8j} = 1$. Thus $j = 5l$ for some $l = 0, 1, 2, 3$, and

$$\mu = \beta^j = \beta^{5l} = i^l.$$

Taking the fifth power of each side of $K(\chi^3) = i^l K(\chi)$ and employing Theorem 3.2.1, we obtain

$$a_4 - ib_4 \equiv i^l(a_4 + ib_4) \pmod 5.$$

Hence, $\mu = i^l = 1$ or -1 according as $5 \mid b_4$ or $5 \mid a_4$, respectively. Appealing to Lemma 3.8.1, we see that $\mu = (-1)^{R/2}$. Then from (3.8.1), $K(\chi^3) = (-1)^{R/2}K(\chi)$.

From the definition of c_{20} we see that $\sigma_3(c_{20}) = (-1)^{R/2}c_{20}$, so that

$$\sigma_3(c_{20}K(\chi)) = (-1)^{R/2}c_{20}K(\chi^3) = c_{20}K(\chi).$$

As the subgroup $<\sigma_3>$ of Gal$(Q(\beta)/Q)$ has fixed field $Q(i\sqrt{5}\,)$,

$$c_{20}K(\chi) = a_{20} + ib_{20}\sqrt{5}, \tag{3.8.2}$$

where a_{20} and b_{20} are certain integers such that $p = a_{20}^2 + 5b_{20}^2$. Raising both sides of (3.8.2) to the fifth power and utilizing Theorem 3.2.1, we obtain

$$c_{20}(a_4 + ib_4) \equiv a_{20} \pmod 5.$$

It now follows from the definition of c_{20} that $a_{20} \equiv -(-1)^{R/2} \pmod 5$. ∎

Theorem 3.8.3 *For $h = g^{(p-1)/20}$,*

$$a_{20} \equiv b_{20}(h^5 - 2h^3 - 2h^7) \pmod p.$$

Proof The proof is analogous to that of Theorem 3.2.2. ∎

Corollary 3.8.4 *Let $p = 20f + 1 = a^2 + b^2 = u^2 + 5v^2$ with a odd. Then $5 \mid a$ if and only if $2 \mid u$.*

Proof By Theorem 2.1.8, Re $K(\chi) \equiv 1 \pmod 2$ and Im $K(\chi) \equiv 0 \pmod 2$. The result now follows from Theorem 3.8.2. ■

In Corollary 8.3.4, Corollary 3.8.4 is reproved and its analogue for primes $p \equiv 9 \pmod{20}$ is established.

In the next theorem, we evaluate other Jacobi sums of order 20 in terms of $J(\chi, \chi^5)$ and previously determined sums. Since $\beta^0, \beta^1, \ldots, \beta^7$ is an integral basis for $Q(\beta)$ over Q, it is not hard to show that $J(\chi, \chi^5) = \sum_{j=0}^{7} d_j \beta^j$, where the integers d_j, $0 \le j \le 7$, are such that $p = \sum_{j=0}^{7} d_j^2 + \sum_{j=0}^{5} d_j d_{j+2}$. We do not study this partition here.

Theorem 3.8.5 *Let* $p = 20f + 1$, *and let* χ *be a character* $\pmod p$ *of order* 20. *Then*

$$J(\chi, \chi^2) = c_{20} J(\chi^2, \chi^2), \tag{3.8.3}$$
$$J(\chi, \chi^3) = J(\chi, \chi)/c_{20}, \tag{3.8.4}$$
$$J(\chi, \chi^4) = c_{20}(-1)^f \sigma_3(J(\chi, \chi^5)), \tag{3.8.5}$$
$$J(\chi, \chi^6) = c_{20} J(\chi^6, \chi^6), \tag{3.8.6}$$
$$J(\chi, \chi^7) = \beta^{6Z} K(\chi)/c_{20}, \tag{3.8.7}$$
$$J(\chi, \chi^8) = \beta^{8Z} K(\chi^2), \tag{3.8.8}$$

and

$$J(\chi, \chi^9) = (-1)^f K(\chi). \tag{3.8.9}$$

Proof We will make frequent use of Theorems 2.1.3–2.1.5. First we observe that

$$K(\chi) = K(\chi^9) = J(\chi^9, \chi^{10}) = \chi^9(-1) J(\chi, \chi^9) = (-1)^f J(\chi, \chi^9),$$

which gives (3.8.9). We now prove (3.8.4). Since $K(\chi) = \beta^{5R} K(\chi^3)$ by Theorem 3.8.2,

$$G^2(\chi^3)/G^2(\chi) = \beta^{5R-4Z} G(\chi^6)/G(\chi^2).$$

Therefore,

$$\frac{J^2(\chi, \chi^3)}{J^2(\chi, \chi)} = \frac{G^2(\chi^3) G^2(\chi^2)}{G^2(\chi^4) G^2(\chi)} = \beta^{5R-4Z} \frac{J(\chi^2, \chi^6)}{J(\chi^4, \chi^4)} = \beta^{5R},$$

where the last equality follows from Theorem 3.7.4. Thus,

$$\mu J(\chi, \chi^3) = J(\chi, \chi), \tag{3.8.10}$$

where $\mu^2 = \beta^{5R} = \pm 1$. Raising both sides of (3.8.10) to the fifth power and using Theorem 3.2.1, we see that

$$\mu \equiv -a_4 - i b_4 \pmod 5.$$

Then, by the definition of c_{20}, we deduce that $\mu = c_{20}$. This proves (3.8.4), and since $J(\chi,\chi)/J(\chi,\chi^3) = J(\chi,\chi^2)/J(\chi^2,\chi^2)$, (3.8.3) also follows. Since

$$J(\chi,\chi^6) = \frac{G(\chi)G(\chi^6)}{G(\chi^7)} = \frac{G(\chi^{13})G(\chi^6)}{G(\chi^{19})} = J(\chi^{13},\chi^6) = \sigma_{13}(J(\chi,\chi^2)),$$

(3.8.6) follows from (3.8.3).

By (3.8.4),

$$c_{20} = \frac{J(\chi,\chi)}{J(\chi,\chi^3)} = \frac{(-1)^f J(\chi,\chi^{18})}{(-1)^f J(\chi^3,\chi^{16})} = \frac{J(\chi,\chi^{15})}{J(\chi^3,\chi^{15})} = \frac{(-1)^f J(\chi,\chi^4)}{\sigma_3(J(\chi,\chi^5))},$$

which proves (3.8.5).

We next prove (3.8.7). By (3.8.9),

$$\begin{aligned}\frac{\beta^{6Z}K(\chi)}{J(\chi,\chi^7)} &= \frac{J(\chi,\chi^9)}{\beta^{4Z}J(\chi,\chi^7)} = \frac{G(\chi^8)G(\chi^9)}{\beta^{4Z}G(\chi^7)G(\chi^{10})}\\ &= \frac{J(\chi^8,\chi^9)}{\beta^{4Z}J(\chi^7,\chi^{10})} = \frac{(-1)^f J(\chi^3,\chi^9)}{\beta^{4Z}K(\chi^7)} = \frac{(-1)^f\sigma_3(J(\chi,\chi^3))}{\beta^{4Z}K(\chi^7)}.\end{aligned}$$

Appealing to (3.8.4), we obtain (recalling that $Z \equiv f \pmod 2$)

$$\begin{aligned}\frac{\beta^{6Z}K(\chi)}{J(\chi,\chi^7)} &= \frac{(-1)^f\sigma_3(J(\chi,\chi)/c_{20})}{\beta^{4Z}K(\chi^7)} = \frac{(-1)^f J(\chi^3,\chi^3)}{\beta^{4Z}K(\chi^7)\sigma_3(c_{20})} = \frac{(-1)^f J(\chi^3,\chi^3)c_{20}}{\beta^{4Z}K(\chi^7)}\\ &= \frac{(-1)^{f+Z}J(\chi^3,\chi^3)c_{20}}{\beta^{14Z}K(\chi^7)} = \frac{J(\chi^3,\chi^3)c_{20}}{\beta^{14Z}K(\chi^7)}.\end{aligned}$$

Further, as $\chi^3(4) = \beta^{6Z}$, and $K(\chi^3) = K(\chi^7)$ by Theorem 3.8.2,

$$\frac{J(\chi^3,\chi^3)c_{20}}{\beta^{14Z}K(\chi^7)} = \frac{\chi^3(4)J(\chi^3,\chi^3)c_{20}}{K(\chi^7)} = \frac{c_{20}K(\chi^3)}{K(\chi^7)} = c_{20},$$

which completes the proof of (3.8.7).

Finally, we prove (3.8.8). As $K(\chi) = J(\chi,\chi^{10})$,

$$G(\chi^{10})/G(\chi^{11}) = \beta^{2Z}G(\chi)/G(\chi^2),$$

and so

$$G(\chi)/G(\chi^9) = \beta^{8Z}G(\chi^2)/G(\chi^{10}).$$

Thus

$$J(\chi,\chi^8) = \beta^{8Z}J(\chi^2,\chi^8) = \beta^{8Z}K(\chi^2),$$

where the last equality follows from Theorem 2.1.6. ∎

3.9 Septic sums

In this section, $k = 7$, $p = 7f + 1$, g is a primitive root (mod p), $\beta = \exp(2\pi i/7)$, and χ is a character (mod p) of order 7 with $\chi(g) = \beta$. Note that f is even and $\chi(-1) = 1$. From Theorem 1.2.4 we have $g(1;7) = \sum_{n=0}^{6} \beta^{n^2} = i\sqrt{7}$, so that

$$\beta + \beta^2 + \beta^4 = \frac{1}{2}(-1 + i\sqrt{7}),$$

and

$$\beta^3 + \beta^5 + \beta^6 = \frac{1}{2}(-1 - i\sqrt{7}).$$

Thus $Q(i\sqrt{7})$ is a quadratic subfield of the cyclotomic field $K = Q(\beta)$.

We first evaluate the Jacobi sum $J(\chi, \chi)$. From this, we also obtain an evaluation of $J(\chi, \chi^3)$, since $J(\chi, \chi^3) = \sigma_3(J(\chi, \chi))$ by Theorem 2.1.5. For an evaluation of $J(\chi, \chi^2)$, see Theorem 3.9.4.

Theorem 3.9.1 *Let $p = 7f + 1$ and let χ be a character* (mod p) *of order* 7. *Then*

$$J(\chi, \chi) = a_1\beta + a_2\beta^2 + a_3\beta^3 + a_4\beta^4 + a_5\beta^5 + a_6\beta^6, \tag{3.9.1}$$

where the integers a_1, a_2, a_3, a_4, a_5, a_6 satisfy

$$p = a_1^2 + a_2^2 + a_3^2 + a_4^2 + a_5^2 + a_6^2 - a_1a_2 - a_2a_3 - a_3a_4 - a_4a_5 - a_5a_6, \tag{3.9.2}$$

and

$$\begin{aligned} a_1a_2 + a_2a_3 + a_3a_4 + a_4a_5 + a_5a_6 &= a_1a_3 + a_2a_4 + a_3a_5 + a_4a_6 + a_6a_1 \\ &= a_1a_4 + a_2a_5 + a_3a_6 + a_5a_1 + a_6a_2. \end{aligned} \tag{3.9.3}$$

Proof As $J(\chi, \chi) \in O_K$ and $\{\beta, \beta^2, \beta^3, \beta^4, \beta^5, \beta^6\}$ is an integral basis for K, there are integers $a_1, a_2, a_3, a_4, a_5, a_6$ satisfying (3.9.1). By Theorem 2.1.3(b),

$$(a_1\beta + a_2\beta^2 + a_3\beta^3 + a_4\beta^4 + a_5\beta^5 + a_6\beta^6)(a_6\beta + a_5\beta^2 + a_4\beta^3 + a_3\beta^4 + a_2\beta^5 + a_1\beta^6) = p,$$

that is,

$$S + X(\beta + \beta^6) + Y(\beta^2 + \beta^5) + Z(\beta^3 + \beta^4) = p, \tag{3.9.4}$$

where

$$\begin{cases} S = a_1^2 + a_2^2 + a_3^2 + a_4^2 + a_5^2 + a_6^2, \\ X = a_1a_2 + a_2a_3 + a_3a_4 + a_4a_5 + a_5a_6, \\ Y = a_1a_3 + a_2a_4 + a_3a_5 + a_4a_6 + a_6a_1, \\ Z = a_1a_4 + a_2a_5 + a_3a_6 + a_5a_1 + a_6a_2. \end{cases} \tag{3.9.5}$$

We can now rewrite (3.9.4) in the following form:

$$(X - S + p)(\beta + \beta^6) + (Y - S + p)(\beta^2 + \beta^5) + (Z - S + p)(\beta^3 + \beta^4) = 0.$$

As $\beta, \beta^2, \beta^3, \beta^4, \beta^5, \beta^6$ are linearly independent over Q, we have

$$X - S + p = Y - S + p = Z - S + p = 0,$$

which gives (3.9.2), and

$$X = Y = Z,$$

which is (3.9.3). ■

Definition *An algebraic integer κ in $K = Q(\beta)$ is said to be normalized if* $\kappa \equiv -1 \pmod{(1-\beta)^2}$.

We now prove some congruences satisfied by the integer coefficients $a_1, \ldots, a_6$ of a normalized element $\kappa = a_1\beta + a_2\beta^2 + \cdots + a_6\beta^6$ such that $\kappa\bar{\kappa} = p$. These congruences must hold for $\kappa = J(\chi, \chi)$, for example, since $J(\chi, \chi)$ is normalized by Theorem 2.1.11.

Theorem 3.9.2 *If $a_1, \ldots, a_6$ are integers for which $\kappa = a_1\beta + a_2\beta^2 + \cdots + a_6\beta^6$ is normalized and $\kappa\bar{\kappa} = p$, then*

(3.9.6) $$a_1 + a_2 + a_3 + a_4 + a_5 + a_6 \equiv -1 \pmod 7,$$

(3.9.7) $$a_1 + 2a_2 + 3a_3 + 4a_4 + 5a_5 + 6a_6 \equiv 0 \pmod 7,$$

(3.9.8) $$a_1 + a_6 \equiv a_2 + a_5 \equiv a_3 + a_4 \equiv 2 \pmod 7,$$

(3.9.9) $$a_1 - a_2 - a_5 + a_6 \equiv 0 \pmod 7,$$

and

(3.9.10) $$a_1 + a_2 - 2a_3 - 2a_4 + a_5 + a_6 \equiv 0 \pmod 7.$$

Proof By definition of κ,

(3.9.11) $$\kappa \equiv \sum_{i=1}^{6} a_i\beta^i \equiv -1 \pmod{(1-\beta)^2}.$$

Now, $\beta^i \equiv 1 \pmod{1-\beta}$, so (3.9.11) implies (3.9.6). Thus (3.9.11) can be written in the form

(3.9.12) $$\sum_{i=1}^{6} a_i(1 - \beta^i) \equiv 0 \pmod{(1-\beta)^2}.$$

Dividing both sides of (3.9.12) by $1 - \beta$, we obtain

(3.9.13) $$\sum_{i=1}^{6} a_i(1 + \beta + \cdots + \beta^{i-1}) \equiv 0 \pmod{1-\beta}.$$

Thus,

$$\sum_{i=1}^{6} ia_i \equiv 0 \pmod{1-\beta},$$

and (3.9.7) follows. It remains to prove (3.9.8), since (3.9.9) and (3.9.10) are immediate consequences thereof. By (3.9.11), there is an integer t such that

$$\kappa \equiv -1 + t(1-\beta)^2 \pmod{(1-\beta)^3}. \tag{3.9.14}$$

Write $w = \kappa + 1$. Then since $\kappa\overline{\kappa} = p$,

$$p = |w-1|^2 = w\overline{w} - (w + \overline{w}) + 1. \tag{3.9.15}$$

Since $(1-\beta)^4$ divides both $p-1$ and $w\overline{w}$, it follows from (3.9.15) that $(1-\beta)^4$ also divides $w + \overline{w}$. However, by (3.9.14),

$$w + \overline{w} \equiv 2t(1-\beta)^2 \pmod{(1-\beta)^3}.$$

Therefore t must be an integral multiple of 7, and so $(1-\beta)^3$ divides w, by (3.9.14). Then 7 divides $w\overline{w}$, and so by (3.9.15), 7 divides $w + \overline{w}$. Now,

$$w + \overline{w} = A(\beta + \beta^6) + B(\beta^2 + \beta^5) + C(\beta^3 + \beta^4), \tag{3.9.16}$$

where

$$A = a_1 + a_6 - 2, \quad B = a_2 + a_5 - 2, \quad C = a_3 + a_4 - 2.$$

Since $\beta, \beta^2, \ldots, \beta^6$ is an integral basis for O_K, it follows from (3.9.16) that $A \equiv B \equiv C \equiv 0 \pmod 7$. This completes the proof of (3.9.8). ∎

Define $a_1, \ldots, a_6$ as in (3.9.1). Congruences (3.9.9) and (3.9.10) show that we can define integers $x_1, \ldots, x_6$ by

$$\begin{cases} x_1 &= -a_1 - a_2 - a_3 - a_4 - a_5 - a_6, \\ x_2 &= a_1 - a_6, \\ x_3 &= a_2 - a_5, \\ x_4 &= a_3 - a_4, \\ 7x_5 &= a_1 + a_2 - 2a_3 - 2a_4 + a_5 + a_6, \\ 7x_6 &= a_1 - a_2 - a_5 + a_6. \end{cases} \tag{3.9.17}$$

Observe that, by (3.9.6),

$$x_1 \equiv 1 \pmod 7. \tag{3.9.18}$$

Solving the system (3.9.17) of linear equations for $a_1,\ a_2,\ a_3,\ a_4,\ a_5,\ a_6$, we obtain

$$(3.9.19)\qquad \begin{cases} 12a_1 = -2x_1 + 6x_2 + 7x_5 + 21x_6, \\ 12a_2 = -2x_1 + 6x_3 + 7x_5 - 21x_6, \\ 12a_3 = -2x_1 + 6x_4 - 14x_5, \\ 12a_4 = -2x_1 - 6x_4 - 14x_5, \\ 12a_5 = -2x_1 - 6x_3 + 7x_5 - 21x_6, \\ 12a_6 = -2x_1 - 6x_2 + 7x_5 + 21x_6. \end{cases}$$

Substituting these values of $a_1,\ a_2,\ a_3,\ a_4,\ a_5,\ a_6$ into (3.9.2) and (3.9.3), we find after a long calculation that the integers $x_1,\ x_2,\ x_3,\ x_4,\ x_5,\ x_6$ satisfy the following system, which is an analogue of Dickson's system (3.7.34) in the quintic case:

$$(3.9.20)\quad 72p = 2x_1^2 + 42(x_2^2 + x_3^2 + x_4^2) + 343(x_5^2 + 3x_6^2),$$

$$(3.9.21)\quad 12x_2^2 - 12x_4^2 + 147x_5^2 - 441x_6^2 + 56x_1x_6 + 24x_2x_3 - 24x_2x_4 + 48x_3x_4 + 98x_5x_6 = 0,$$

$$(3.9.22)\quad 12x_3^2 - 12x_4^2 + 49x_5^2 - 147x_6^2 + 28x_1x_5 + 28x_1x_6 + 48x_2x_3 + 24x_2x_4 + 24x_3x_4 + 490x_5x_6 = 0.$$

A table of solutions of (3.9.20)–(3.9.22) has been given by Williams [12] and Hardy and Williams [2]. Further, replacing $a_1,\ a_2,\ a_3,\ a_4,\ a_5,\ a_6$ in (3.9.1) by the expressions given in (3.9.19), we obtain the following determination of the Jacobi sum $J(\chi,\chi)$ in terms of the integers $x_1, x_2, x_3, x_4,\ x_5, x_6$.

Theorem 3.9.3 *Let $p = 7f + 1$ and let χ be a character* (mod p) *of order* 7 *with $\chi(g) = \beta$. Then*

$$(3.9.23)\quad J(\chi,\chi) = \frac{1}{12}(2x_1 + 6x_2(\beta - \beta^6) + 6x_3(\beta^2 - \beta^5) + 6x_4(\beta^3 - \beta^4) + 7x_5(\beta + \beta^2 - 2\beta^3 - 2\beta^4 + \beta^5 + \beta^6) + 21x_6(\beta - \beta^2 - \beta^5 + \beta^6)),$$

where $x_1,\ x_2,\ x_3,\ x_4,\ x_5,\ x_6$ are integers satisfying (3.9.18), (3.9.20), (3.9.21), *and* (3.9.22).

Before discussing the number of solutions in integers $x_1,\ x_2,\ x_3,\ x_4,\ x_5,\ x_6$ of the system (3.9.20)–(3.9.22), we first determine the Jacobi sum $J(\chi,\chi^2)$.

Theorem 3.9.4 *Let $p = 7f + 1$ and let χ be a character* (mod p) *of order* 7 *with $\chi(g) = \beta$. Then*

$$(3.9.24)\qquad J(\chi,\chi^2) = a_7 + ib_7\sqrt{7},$$

where a_7 and b_7 are integers satisfying

$$(3.9.25)\qquad p = a_7^2 + 7b_7^2,\quad a_7 \equiv -1 \pmod 7.$$

Proof The fixed field of the subgroup $<\sigma_2>$ of the Galois group $\mathrm{Gal}(Q(\beta)/Q)$ is $Q(\beta+\beta^2+\beta^4) = Q(i\sqrt{7})$. As

$$\sigma_2(J(\chi,\chi^2)) = J(\chi^2,\chi^4) = J(\chi,\chi^2),$$

by Theorem 2.1.5, we see that $J(\chi,\chi^2)$ is an integer of $Q(i\sqrt{7})$. Hence there are integers x and y of the same parity such that

$$J(\chi,\chi^2) = \frac{1}{2}(x+iy\sqrt{7}),\ 4p = x^2+7y^2.$$

Taking the equation $4p = x^2+7y^2$ modulo 8, we see that $4 \equiv x^2 - y^2 \pmod 8$. Therefore, x and y must both be even, say, $x = 2a_7$, $y = 2b_7$, and thus

$$J(\chi,\chi^2) = a_7 + ib_7\sqrt{7},\quad p = a_7^2 + 7b_7^2.$$

Finally, by Theorem 2.1.7,

$$J(\chi,\chi^2) \equiv -p \pmod{(1-\beta)^2},$$

so that

$$a_7 + b_7(\beta+\beta^2-\beta^3+\beta^4-\beta^5-\beta^6) \equiv -1 \pmod{(1-\beta)^2}.$$

Taking this congruence modulo $1-\beta$, we see that $a_7 \equiv -1 \pmod{(1-\beta)}$, so that $a_7 \equiv -1 \pmod 7$. ■

Theorem 3.9.5 *Let $h = g^{(p-1)/7}$. Define the integers a_7 and b_7 by (3.9.24). Then a_7 and b_7 are uniquely determined by (3.9.25) and the congruence*

$$a_7 + b_7(h+h^2-h^3+h^4-h^5-h^6) \equiv 0 \pmod p. \tag{3.9.26}$$

Proof The proof is analogous to that of Theorem 3.3.2. ■

We now discuss the nature of the integral solutions $(x_1,x_2,x_3,x_4,x_5,x_6)$ of the system (3.9.20)–(3.9.22). We say that a solution of this system is a trivial solution if $x_5 = x_6 = 0$, and nontrivial otherwise. The next lemma determines the trivial solutions of the system.

Lemma 3.9.6 *There are exactly four trivial solutions of the Diophantine system (3.9.20)–(3.9.22), namely,*

$$(x_1,x_2,x_3,x_4,x_5,x_6) = (6\delta a_7, 2\epsilon b_7, 2\epsilon b_7, -2\epsilon b_7, 0, 0),$$

where a_7 and b_7 are integers satisfying $p = a_7^2 + 7b_7^2$, and $\delta = \pm 1$, $\epsilon = \pm 1$.

Proof Let $(x_1, x_2, x_3, x_4, x_5, x_6)$ be a trivial solution of (3.9.20)–(3.9.22). Then $x_5 = x_6 = 0$, and (3.9.20), (3.9.21) and (3.9.22) reduce to

(3.9.27) $$36p = x_1^2 + 21(x_2^2 + x_3^2 + x_4^2),$$

(3.9.28) $$x_2^2 - x_4^2 + 2x_2x_3 - 2x_2x_4 + 4x_3x_4 = 0,$$

(3.9.29) $$x_3^2 - x_4^2 + 4x_2x_3 + 2x_2x_4 + 2x_3x_4 = 0.$$

Clearly $(x_2, x_3) \neq (0, 0)$. Subtracting (3.9.29) from (3.9.28) we obtain

$$(4x_2 - 2x_3)x_4 = x_2^2 - 2x_2x_3 - x_3^2.$$

Using this in (3.9.28), we obtain, after some simplification,

$$(x_2 - x_3)(x_2^3 + 9x_2^2x_3 - x_2x_3^2 - x_3^3) = 0.$$

We show next that $x_2^3 + 9x_2^2x_3 - x_2x_3^2 - x_3^3 \neq 0$. Suppose on the contrary that for some integers a and b (not both zero),

(3.9.30) $$a^3 + 9a^2b - ab^2 - b^3 = 0.$$

Then (3.9.30) has a solution with $\gcd(a, b) = 1$, but for such a solution, (3.9.30) is impossible (mod 3). Thus $x_2^3 + 9x_2^2x_3 - x_2x_3^2 - x_3^3 \neq 0$, and so we must have $x_2 = x_3$. Then, from (3.9.28) and (3.9.29),

$$(x_2 + x_4)(3x_2 - x_4) = (x_2 + x_4)(5x_2 - x_4) = 0,$$

so that $x_2 = -x_4$. Then (3.9.27) gives $36p = x_1^2 + 63x_2^2$, from which we deduce that $6 \mid x_1$ and $2 \mid x_2$, so that $p = (x_1/6)^2 + 7(x_2/2)^2$. Hence, by Lemma 3.0.1, we have $x_1/6 = \delta a_7$ and $x_2/2 = \epsilon b_7$, where $\delta,\ \epsilon = \pm 1$, and the result follows. ■

Next we show that if $(x_1, x_2, x_3, x_4, x_5, x_6)$ is a solution in integers of (3.9.20)–(3.9.22) given by (3.9.23), then it is a nontrivial solution.

Lemma 3.9.7 *The integers $x_1, x_2, x_3, x_4, x_5, x_6$ given by* (3.9.23) *give a nontrivial solution $(x_1, x_2, x_3, x_4, x_5, x_6)$ of* (3.9.20)–(3.9.22).

Proof Suppose $(x_1, x_2, x_3, x_4, x_5, x_6)$ is a trivial solution of (3.9.20)–(3.9.22). Then, by (3.9.23) and Lemma 3.9.6,

$$\begin{aligned} J(\chi,\chi) &= \frac{1}{12}(12\delta a_7 + 12\epsilon b_7(\beta - \beta^6) + 12\epsilon b_7(\beta^2 - \beta^5) - 12\epsilon b_7(\beta^3 - \beta^4)) \\ &= \pm a_7 \pm i b_7\sqrt{7}, \end{aligned}$$

so that $J(\chi,\chi) \in Q(i\sqrt{7})$. Hence, as $Q(i\sqrt{7})$ is the fixed field of σ_2,

$$J(\chi,\chi) = \sigma_2(J(\chi,\chi)) = J(\chi^2,\chi^2).$$

Then, according to Theorem 2.1.14, we find that

$$P_1P_4P_5 = J(\chi,\chi)O_K = J(\chi^2,\chi^2)O_K = P_1P_2P_3,$$

which contradicts the uniqueness of the prime ideal decomposition, as $P_1,\ldots,P_6$ are distinct prime ideals. ■

Lemma 3.9.8 *There are exactly eight normalized algebraic integers κ in $K = Q(\beta)$ such that $\kappa\bar{\kappa} = p$, namely, $J(\chi,\chi^2)$, $J(\chi^3,\chi^6)$, and $J(\chi^i,\chi^i)$, $i = 1,2,\ldots,6$.*

Proof The eight Jacobi sums listed above are normalized by Theorem 2.1.11, and in O_K, they have the respective prime ideal factorizations $P_1P_2P_4$, $P_3P_5P_6$, $P_1P_4P_5$, $P_1P_2P_3$, $P_1P_3P_5$, $P_2P_4P_6$, $P_4P_5P_6$, and $P_2P_3P_6$, by Theorem 2.1.14. For any $\kappa \in O_K$ with $\kappa\bar{\kappa} = p$, the prime ideal factorization of κO_K must be one of the eight listed above, and so κ must equal a unit times one of the eight listed Jacobi sums. However, two distinct associates with modulus $\sqrt{p}$ cannot both be normalized (see Problem 21), and so the result follows. ■

We are now ready to determine the number of integral solutions of (3.9.20)–(3.9.22).

Theorem 3.9.9 *If p is a prime with $p \equiv 1 \pmod 7$, there are exactly 16 integral solutions $(x_1,x_2,x_3,x_4,x_5,x_6)$ of the system (3.9.20)–(3.9.22). Of these 16 solutions, 4 solutions are trivial and 12 are nontrivial.*

Proof Each integral solution $(x_1,\ldots,x_6)$ of (3.9.20)–(3.9.22), (3.9.18) provides two solutions of (3.9.20)–(3.9.22), namely, $\pm(x_1,\ldots,x_6)$. Conversely, exactly one of the pair of solutions $\pm(x_1,\ldots,x_6)$ of (3.9.20)–(3.9.22) satisfies (3.9.18). Hence, by Lemma 3.9.8, it suffices to demonstrate a one-to-one correspondence between the solutions $(x_1,\ldots,x_6)$ of (3.9.18), (3.9.20)–(3.9.22) and the normalized elements $\kappa \in O_K$ with $\kappa\bar{\kappa} = p$.

Let $a_1,\ldots,a_6$ be rational integers such that $\kappa = a_1\beta + a_2\beta^2 + \cdots + a_6\beta^6$ is a normalized element of O_K with $\kappa\bar{\kappa} = p$. Then by (3.9.9) and (3.9.10), we can define integers $x_1,\ldots,x_6$ by the equations in (3.9.17). These integers $x_1,\ldots,x_6$ satisfy (3.9.20)–(3.9.22) and (3.9.18); see Problem 18.

Conversely, for any solution $(x_1,\ldots,x_6)$ of (3.9.20)–(3.9.22) and (3.9.18), we have (reducing the equations modulo 8, 3, and 7)

$$(3.9.31)\qquad \begin{cases} x_2 + x_3 + x_6 \equiv 0 \pmod 2, \\ 2x_2 + 2x_4 + x_5 + x_6 \equiv 0 \pmod 4, \\ 2x_1 + 2x_2 + x_5 + 3x_6 \equiv 0 \pmod 4, \\ x_1 + x_5 \equiv 0 \pmod 3, \\ x_2 + 2x_3 + 3x_4 \equiv 0 \pmod 7; \end{cases}$$

see Problem 19. The first four congruences in (3.9.31) show that we can define integers

$a_1, a_2, \ldots, a_6$ by the equations in (3.9.19). Finally, we can check from (3.9.18)–(3.9.22) and (3.9.31) that $\kappa = a_1\beta + a_2\beta^2 + \cdots + a_6\beta^6$ is a normalized element of O_K such that $\kappa\bar{\kappa} = p$; see Problem 20. ■

We remark that if $(x_1, \ldots, x_6)$ is one of the six nontrivial solutions in Theorem 3.9.9 with $x_1 \equiv 1 \pmod 7$, then two others are given by

$$(x_1, -x_3, x_4, x_2, (-x_5 - 3x_6)/2, (x_5 - x_6)/2),$$
$$(x_1, -x_4, x_2, -x_3, (-x_5 + 3x_6)/2, (-x_5 - x_6)/2),$$

and the remaining three can be obtained from the previous three above by changing the signs of x_2, x_3, and x_4.

Exercises 3

1. Let p be a prime with $p \equiv 1 \pmod 3$. Show that there exist integers x and y such that $p = x^2 + xy + y^2$. How many such pairs (x, y) of integers are there?

2. Let p be a prime with $p \equiv 1 \pmod 3$. Show that $x^2 + 3y^2 = 4p$ has three pairs of integral solutions (x, y) with x and y positive.

3. (a) If $p \equiv 1 \pmod{12}$, prove that $a_{12} + a_3 \equiv 2 \pmod 4$.

   (b) If $p \equiv 1 \pmod{24}$, prove that $2a_{24} + a_{12} + a_3 \equiv 4 \pmod 8$.

4. Use Theorem 3.1.2 and (3.1.3) to show that, for a prime $p \equiv 1 \pmod 3$,

$$2^{(p-1)/3} \equiv \begin{cases} 1 \pmod p, & \text{if } b_3 \equiv 0 \pmod 3, \\ (-3b_3 - a_3)/2a_3 \pmod p, & \text{if } b_3 \equiv 1 \pmod 3, \\ (3b_3 - a_3)/2a_3 \pmod p, & \text{if } b_3 \equiv 2 \pmod 3. \end{cases}$$

5. Prove the assertion about uniqueness made at the end of the proof of Theorem 3.7.7.

6. Let $p = 7f + 1$, and let χ have order 7.

   (a) Prove that $pJ(\chi, \chi^2) = J(\chi, \chi)J(\chi^2, \chi^2)J(\chi^4, \chi^4)$.

   (b) Use (3.9.23), (3.9.24) and part (a) to relate a_7 and b_7 to $x_1, x_2, x_3, x_4, x_5, x_6$.

   (c) Prove that $b_7 \equiv 3x_2 + 2x_3 \pmod 7$.

7. Let $p = 9f + 1$, $\beta = \exp(2\pi i/9)$, and let χ be a character (mod p) of order 9.

(a) Prove that $J(\chi,\chi) = \mu J(\chi,\chi^3)$, where μ is a cube root of 1.

(b) Prove that

$$\frac{J(\chi,\chi)}{J(\chi,\chi^3)} \equiv \frac{G^3(\chi)}{G(\chi^3)} \pmod{(1-\beta)^4}.$$

(c) Conclude that $\mu = \overline{\chi}^3(3)$.

8. Let $p = 11f + 1$, and let χ have order 11. Prove that either

$$\frac{J(\chi,\chi)J(\chi^4,\chi^4)}{J(\chi,\chi^4)} = A + iB\sqrt{11},$$

where A and B are integers with $p = A^2 + 11B^2$ and $A \equiv -1 \pmod{11}$, or

$$\frac{J(\chi,\chi)J(\chi^4,\chi^4)}{J(\chi,\chi^4)} = \frac{C + iD\sqrt{11}}{2},$$

where C and D are odd integers with $4p = C^2 + 11D^2$ and $C \equiv 9 \pmod{22}$.

9. Let $p = 13f + 1$, and let χ have order 13. Prove that

$$4J(\chi,\chi^3) = A + B\sqrt{13} + iC\sqrt{26 + 6\sqrt{13}} + iD\sqrt{26 - 6\sqrt{13}},$$

where A, B, C, and D are integers such that

(i) $16p = A^2 + 13B^2 + 26C^2 + 26D^2$,

(ii) $AB = 3D^2 - 3C^2 - 4CD$,

and $A \equiv -4 \pmod{13}$. Prove, moreover, that the solutions $\pm(A, B, C, D)$ to (i) and (ii) are essentially unique in the sense that the only other solutions in integers are $\pm(A, B, -C, -D)$, $\pm(A, -B, D, -C)$, and $\pm(A, -B, -D, C)$.

10. Let $p = 15f + 1$, $\beta = \exp(2\pi i/15)$, and let χ have order 15. As in (3.0.5), set $T = \text{ind}_g 3$ and $R = \text{ind}_g 5$, where g is a primitive root (mod p) such that $\chi(g) = \beta$.

(a) Prove that

$$J(\chi,\chi^4) = \beta^{5R}(-a_{15} + ib_{15}\sqrt{15}),$$

where a_{15} and b_{15} are integers such that $a_{15}^2 + 15b_{15}^2 = p$ and $a_{15} \equiv 1 \pmod 3$.

(b) Prove that

$$J(\chi,\chi^2) = c_{15}\beta^{3T}(-a_{15} + ib_{15}\sqrt{15}),$$

where $c_{15} = \pm 1$ and $c_{15} \equiv a_{15} \pmod 5$.

11. Let $p = 21f + 1$, and let χ have order 21. Prove that

$$\overline{\chi}(7)J(\chi, \chi^4) = \mu(a_7 + ib_7\sqrt{7}),$$

where a_7 and b_7 are the integers defined in Theorem 3.9.4 and $\mu = 1$ or -1 according as $3 \mid b_7$ or $3 \mid a_7$.

12. Let $p = 39f + 1$, and let χ have order 39. Prove that there are integers a, b, c, and d such that

$$\chi^{13}(13)J(\chi, \chi^{16}) = \begin{cases} a + ib\sqrt{39}, & \text{if } a_3 \equiv \pm 1 \text{ or } \pm 6 \pmod{13}, \\ ic\sqrt{3} + d\sqrt{13}, & \text{if } a_3 \equiv 0 \text{ or } \pm 2 \pmod{13}, \end{cases}$$

where in the first case $p = a^2 + 39b^2$ with $a \equiv \chi^{13}(13)J(\chi^{13}, \chi^{13}) \pmod{13}$, and in the second case $p = 3c^2 + 13d^2$ with $c \equiv -i\chi^{13}(13)J(\chi^{13}, \chi^{13})/\sqrt{3} \pmod{13}$.

13. Let $p = 40f + 1$, and let χ have order 40. Prove that

$$\frac{J(\chi, \chi^9)K(\chi^{10})}{c_{20}J(\chi^3, \chi^{27})} = -a_{40} + ib_{40}\sqrt{10}\,,$$

where c_{20} is defined in Theorem 3.8.2, and a_{40} and b_{40} are integers such that $a_{40}^2 + 10b_{40}^2 = p$ and $a_{40} \equiv 1 \pmod 5$.

14. Let $p = 60f + 1$, and let χ have order 60. Prove that, in the notation of Theorem 3.8.2,

$$K(\chi)/c_{20} = A + 3B\sqrt{5} + iC\sqrt{15 + 6\sqrt{5}} + iD\sqrt{15 - 6\sqrt{5}},$$

where A, B, C, and D are integers such that

(i) $p = A^2 + 45B^2 + 15C^2 + 15D^2$,

(ii) $AB = D^2 - C^2 - CD$,

and $A \equiv a_{20} \pmod 3$. Prove that the solutions $\pm(A, B, C, D)$ to (i) and (ii) are essentially unique in the sense that the only other solutions in integers are $\pm(A, B, -C, -D)$, $\pm(A, -B, D, -C)$, and $\pm(A, -B, -D, C)$.

15. Prove Theorem 3.8.3.

16. Determine explicit expressions for the cyclotomic numbers of orders 3, 4, 5, 6, 7 and 8 by making use of Theorem 2.5.1 and the values of the Jacobi sums of orders 3, 4, 5, 6, 7 and 8 obtained in this chapter. (For references, see the Notes at the end

of this chapter.) The expressions obtained for orders 3, 4, 5 should agree with those in Theorem 2.3.1, Theorem 2.4.1, and Problem 14 of Exercises 2.

17. Let $K = Q(\beta)$, where $\beta = \exp(2\pi i/7)$. Let p be a prime such that $p \equiv 1 \pmod 7$. Let $\kappa = a_1\beta + \cdots + a_6\beta^6$ be a normalized element of O_K, such that $\kappa\bar{\kappa} = p$. Prove that

$$\sum_{i=1}^{6} i^2 a_i \equiv \sum_{i=1}^{6} i^4 a_i \equiv 0 \pmod 7.$$

18. Prove that the integers $x_1, \ldots, x_6$ defined by (3.9.17) satisfy (3.9.18), (3.9.20), (3.9.21), and (3.9.22).

19. Prove the statement containing (3.9.31).

20. Prove the assertion about κ made at the end of the proof of Theorem 3.9.9.

21. With the notation of Problem 17, prove that, if $\kappa_1 \in O_K$ is such that $\kappa_1\bar{\kappa}_1 = p$, then κ_1 posseses a unique normalized associate κ with $\kappa\bar{\kappa} = p$.

22. In Problem 14 of Exercises 2 it was indicated that the solution (x, w, v, u) of (3.7.39) satisfies a certain congruence modulo p. Does this congruence determine the solution of (3.7.39) uniquely?

Notes on Chapter 3

The literature contains perhaps more material on the relationships between Jacobi sums of a given order than on the explicit determination of Jacobi sums. The relationship between two Jacobi sums of a particular order often involves a certain root of unity that at times is difficult to determine. Dickson [8] established relationships between sums of order k for $k \le 6$, $k = 8$, $k = 10$, and $k = 12$, and Muskat [6] determined the sign of a fourth root of unity that arose in the relationship between two sums of order 12. Dickson [7] found certain relationships for sums of order 15 and for sums of order 24. Muskat [6] completed Dickson's work and extended it to sums of order 30. A special case of one of Muskat's relationships for Jacobi sums of order 30 is the "merkwürdigen Satz" mentioned by Jacobi in [3; CW, vol. 6, p. 260]. Dickson [6] gave certain relationships for sums of orders 14 and 22. Complete results for order 14 were later found by Muskat [4]. Dickson [7] also studied sums of orders 9 and 18, but Baumert and Fredricksen [1] corrected some of his results and removed a sign ambiguity. Relationships and evaluations for sums of order 16 were found by Whiteman [8]. In fact, prior even to the work of Dickson, Western [2] had determined Jacobi sums of order 16 (see the last eight lines in Western [2, p. 290]). His evaluations were however not given in terms of the parameters a_{16}, b_{16}, c_{16}, and d_{16} in (3.4.1). Western also determined octic sums [2, p. 286, lines 12–14] and sums of order 9 (see four lines at the bottom of [2, p. 295]). Muskat and Whiteman [1] established relationships and evaluations for sums of order 20. Zee found relationships for sums of orders 13 and 60 [1] and for sums of order 22 [2]. Relationships for orders 21, 28, 39, 55, and 56 are given in a paper of Muskat and Zee [1]. Sums of orders 3, 4, 6, 8, 12, 20, and 24 were evaluated by Berndt and Evans [1], and sums of orders 5, 10, and 16 were determined by the same authors in [2]. Leonard and Williams [5] determined $K(\chi)$ when χ is octic. Evaluations and relationships for Jacobi sums of orders 3, 4, and 7 were given by Jacobi himself [1] in a letter to Gauss in 1827. In general, it is difficult to determine who deserves credit for first evaluating a particular Jacobi sum.

The evaluation of the quartic Jacobi sum (Theorem 3.2.1) was used by Yamada [3] to compute supplementary difference sets.

E. Lehmer [9] removed sign ambiguities in the Jacobi sums $K(\chi)$ for $k = 3, 4, 5$, by giving evaluations in terms of $\text{ind}_g 2$ in the case that 2 is not a kth–power residue, where g is a primitive root (mod p). Cf. the remark above Theorem 3.1.4. Similarly, Evans [2], [3] has removed ambiguities in $K(\chi)$ for $k = 8, 16, 20$, for certain values of $\text{ind}_g 2$. The systematic method for removing the ambiguous signs that we have given in this chapter (e.g., in Theorems 3.1.2, 3.2.2, and 3.3.2) was also employed by Evans [13]. This method of removing sign ambiguities via congruences (mod p) in some respects solves the problem of sign ambiguities discussed by D. H. Lehmer [1] in his review of Whiteman's paper [10]. Nonetheless, the complete determination of ambiguous signs in terms of $\text{ind}_g 2$ is still an interesting open problem.

Yamada [2] used explicit evaluations of Gauss and Jacobi sums over finite rings algebraic over $Z/4Z$, to classify certain distance–regular digraphs of girth 4. Ishibashi [1] also evaluated Gauss and Jacobi sums over these rings and applied them to classify certain irreducible modules over Terwilliger algebras.

The representation (3.4.1) or the form (3.4.3) has been studied by Whiteman [8], Hasse [1], and Giudici, Muskat, and Robinson [1]. The statement on uniqueness in Theorem 3.4.1 comes from Muskat and Zee [2].

Theorem 3.6.2, Lemma 3.6.5, and Theorem 3.6.6 are due to Muskat [6, p. 496]. The last observation in Section 3.6 is also due to Muskat [6, p. 486].

In connection with Lemma 3.7.1, we remark that the integers of the quartic field $Q(\sqrt{c}, \sqrt{a + b\sqrt{c}})$ with quadratic subfield $Q(\sqrt{c})$ are given in Huard, Spearman, and Williams [2]. The representation of p in (3.7.6) appears to have been first noticed by Giudici, Muskat, and Robinson [1]. The representations for $16p$ in (3.7.33) and (3.7.34) were studied by Dickson [6], [8]. The statement on uniqueness following (3.7.35) is also due to Dickson [8, pp. 402–403].

A very simple proof of Corollary 3.8.4 has been found by E. Lehmer [15], who also gave some generalizations as well. Other proofs have been given by E. Lehmer [13], Muskat and Whiteman [1], and Muskat [8].

The results of Section 3.9 are based on the work of Leonard and Williams [2]. The

Diophantine system (3.9.20)–(3.9.22) is discussed in Williams [10], [12] and in Leonard and Williams [2], [3]. The cyclotomic numbers of orders 7 and 14 have been given in terms of solutions of this system by Leonard and Williams [3] and Buck and Williams [1], respectively.

In connection with Problem 16, we mention that the explicit determination of the cyclotomic numbers of order k has been carried out for $k \leq 6$ (Dickson [8]), $k = 7$ (Leonard and Williams [3]), $k = 8$ (E. Lehmer [4]), $k = 9$ (Baumert and Fredricksen [1]), $k = 10$ (Whiteman [10]), $k = 11$ (Leonard and Williams [4]), $k = 12$ (Whiteman [9]), $k = 14$ (Muskat [4], Buck and Williams [1]), $k = 15$ (Buck, Smith, Spearman, and Williams [1]), $k = 16$ (Whiteman [8], Evans and Hill [1]), $k = 18$ (Baumert and Fredricksen [1]), $k = 20$ (Muskat and Whiteman [1]), and $k = 24$ (Evans [6]). For cyclotomic numbers of order k over F_q, see also Katre and Rajwade [7], [3] for k =4, 5; Friesen, Muskat, Spearman, and Williams [1] for k =15; Acharya [1] for k =10; and Acharya and Katre [1] for $k = 6$ and for $k = \ell$ or 2ℓ, where ℓ is an odd prime dividing $(p-1)$. See also Thaine [2], [3].

Problem 10 is treated in Muskat's paper [6]. Problems 11 and 12 are found in the work of Muskat and Zee [1]. Zee [1] has established the results in Problems 9 and 14. To determine when a prime p has the form $A^2 + 11B^2$ in Problem 8, see a paper of E. Lehmer [15, p. 296].

Chapter 4

Determination of Gauss Sums over F_p

Let k be an integer with $k \geq 2$, and let p be a prime with $p \equiv 1 \pmod{k}$. In Section 1.1, we defined the sums

$$g(k) = \sum_{n=0}^{p-1} e^{2\pi i n^k/p} \tag{4.0.1}$$

and

$$G(\chi) = \sum_{n=0}^{p-1} \chi(n) e^{2\pi i n/p}, \tag{4.0.2}$$

where χ is a character (mod p) of order k. Both $g(k)$ and $G(\chi)$ are called Gauss sums of order k, and by (1.1.4) they are related by

$$g(k) = \sum_{j=1}^{k-1} G(\chi^j). \tag{4.0.3}$$

We showed in Theorem 1.2.4 that

$$g(2) = \begin{cases} \sqrt{p}, & \text{if } p \equiv 1 \pmod{4}, \\ i\sqrt{p}, & \text{if } p \equiv 3 \pmod{4}, \end{cases}$$

that is,

$$g(2) = i^{(p-1)^2/4}\sqrt{p}. \tag{4.0.4}$$

In this chapter we determine $g(k)$ for $k = 3, 4, 6, 8$, and 12 up to certain sign ambiguities. Some of these sign ambiguities remain unresolved to this day.

The following theorem enables us to determine when the Gauss sum $g(k)$ is real.

Theorem 4.0.1 *For $p \equiv 1 \pmod{k}$, the Gauss sum $g(k)$ is real if and only if $p \equiv 1 \pmod{2k}$.*

Proof Let S denote the group of $(p-1)/k$ distinct kth–power residues (mod p). Then

$$g(k) = 1 + k \sum_{r \in S} e^{2\pi i r/p}.$$

If $p \equiv 1 \pmod{2k}$, then $-1 \in S$, and it follows that

$$g(k) = 1 + k \sum_{r \in S} e^{-2\pi i r/p}.$$

This proves that $g(k)$ is real when $p \equiv 1 \pmod{2k}$. Suppose that $p \not\equiv 1 \pmod{2k}$. Then $-1 \notin S$, so that S and $-S$ are disjoint sets. Thus

$$\sum_{r \in S} (e^{2\pi i r/p} - e^{-2\pi i r/p})$$

cannot vanish, since the $p-1$ elements $\exp(2\pi i n/p)$, $1 \le n \le p-1$, are linearly independent over Q. Thus $g(k)$ is nonreal when $p \not\equiv 1 \pmod{2k}$. ■

It is convenient at this point to introduce some notation that will be used throughout this chapter. If χ is a character (mod p) of order 24, we define R_j for $j = 1, 2, \ldots, 11$ by

$$R_j = G(\chi^j) + G(\overline{\chi}^j). \tag{4.0.5}$$

It is easy to check that R_4, R_6, R_8, $R_3 + R_9$, and $R_2 + R_{10}$ are all independent of the choice of the character χ.

4.1 The Gauss sums $g(3)$ and $g(6)$

Let p be a prime with $p \equiv 1 \pmod 6$. Let g be a primitive root (mod p) and let χ be a character (mod p) of order 6 with $\chi(g) = \beta$, where $\beta = e^{2\pi i/6}$. Set $Z = \operatorname{ind}_g 2$, so that

$$\chi(4) = \chi(g^{2\operatorname{ind}_g 2}) = \chi(g)^{2\operatorname{ind}_g 2} = \beta^{2Z}.$$

Lemma 4.1.1 *We have*
(a) $G^3(\chi^2) = pJ(\chi^2, \chi^2)$,
(b) $G(\chi) = \beta^{-2Z} i^{(p-1)^2/4} p^{-1/2} G^2(\chi^2)$.

Proof (a) By Theorems 2.1.3(a) and 1.1.4(a),

$$J(\chi^2, \chi^2) = \frac{G^2(\chi^2)}{G(\chi^4)} = \frac{G^3(\chi^2)}{p},$$

from which assertion (a) follows.

(b) By Theorem 2.1.4, we see that $\chi^2(4)J(\chi^2, \chi^2) = K(\chi^2) = J(\chi^2, \chi^3)$, so that $J(\chi^2, \chi^2) = \chi(4)J(\chi^2, \chi^3)$. Then, by part (a),

$$G^3(\chi^2) = p\chi(4)J(\chi^2, \chi^3) = p\chi(4)\frac{G(\chi^2)G(\chi^3)}{G(\chi^5)},$$

so that

$$\frac{G^2(\chi^2)}{p} = \chi(4)\frac{G(\chi^3)}{G(\chi^5)} = \beta^{2Z}\frac{G(\chi^3)}{G(\chi^5)}.$$

By Theorem 1.1.4(a) and Theorem 1.2.4,

$$\frac{G(\chi^3)}{G(\chi^5)} = \frac{G(\chi)}{G(\chi^3)} = \frac{G(\chi)}{g(2)} = \frac{G(\chi)}{i^{(p-1)^2/4}\sqrt{p}},$$

so that

$$\frac{G^2(\chi^2)}{p} = \frac{\beta^{2Z}G(\chi)}{i^{(p-1)^2/4}\sqrt{p}},$$

from which part (b) readily follows. ■

The next theorem shows that $g(3)$ is one of the three real roots of a certain cubic polynomial.

Theorem 4.1.2 *The Gauss sum $g(3)$ is one of the three real roots of the irreducible cubic polynomial $x^3 - 3px - pr_3$ of discriminant $81p^2s_3^2$, where r_3 and s_3 are the integers defined in Theorem 3.1.3. Moreover, each of the three intervals $(-2\sqrt{p}, -\sqrt{p})$, $(-\sqrt{p}, \sqrt{p})$, and $(\sqrt{p}, 2\sqrt{p})$ contains exactly one root of $x^3 - 3px - pr_3$.*

Proof By (4.0.3),

$$g(3) = G(\chi^2) + G(\chi^4).$$

From Theorem 1.1.4(b),

$$\overline{G(\chi^2)} = G(\overline{\chi}^2) = G(\chi^4),$$

so that

$$g(3) = G(\chi^2) + \overline{G(\chi^2)} \tag{4.1.1}$$

is real. Cubing each side of (4.1.1) and using Theorem 1.1.4(c), Lemma 4.1.1(a), and Theorem 3.1.3, we find that

$$\begin{aligned} g^3(3) &= G^3(\chi^2) + \overline{G^3(\chi^2)} + 3pg(3) \\ &= pJ(\chi^2,\chi^2) + p\overline{J(\chi^2,\chi^2)} + 3pg(3) = pr_3 + 3pg(3), \end{aligned}$$

which shows that $g(3)$ is a root of the cubic polynomial $x^3-3px-pr_3$. This polynomial is p-Eisenstein, so it is irreducible over the rationals. The discriminant of $x^3 - 3px - pr_3$ is $-4(-3p)^3 - 27(-pr_3)^2 = 27p^2(4p - r_3^2) = 81p^2s_3^2$, which is positive, so the cubic polynomial has three distinct real roots.

Set $f(x) = x^3 - 3px - pr_3$, and denote the three (real and distinct) roots of $f(x)$ by m_1, m_2, m_3. Order the roots of $f(x)$ so that $m_1 < m_2 < m_3$. From Theorem 3.1.3 we see that $-2\sqrt{p} < r_3 < 2\sqrt{p}$. Therefore

$$\begin{aligned} f(-2\sqrt{p}) &= -p(2\sqrt{p} + r_3) < 0, \\ f(-\sqrt{p}) &= p(2\sqrt{p} - r_3) > 0, \\ f(\sqrt{p}) &= -p(2\sqrt{p} + r_3) < 0, \\ f(2\sqrt{p}) &= p(2\sqrt{p} - r_3) > 0. \end{aligned}$$

Hence, by the intermediate value theorem, $m_1 \in (-2\sqrt{p}, -\sqrt{p})$, $m_2 \in (-\sqrt{p}, \sqrt{p})$, and $m_3 \in (\sqrt{p}, 2\sqrt{p})$. ■

It follows immediately from Theorem 4.1.2 that the three roots of $x^3 - 3px - pr_3$ are

$$g(3) = \sum_{n=0}^{p-1} \exp(2\pi i n^3/p), \quad \sum_{n=0}^{p-1} \exp(2\pi i n^3 g/p), \quad \sum_{n=0}^{p-1} \exp(2\pi i n^3 g^2/p).$$

No simple criterion is known for determining which of the three intervals $(-2\sqrt{p}, -\sqrt{p})$, $(-\sqrt{p}, \sqrt{p})$, and $(\sqrt{p}, 2\sqrt{p})$ contains the Gauss sum $g(3)$.

The next result gives $G(\chi)$ and $G(\chi^2)$ in terms of $g(3)$.

Lemma 4.1.3 *We have*

$$\text{(a) } G(\chi^2) = \frac{1}{2}(g(3) + i\nu\{4p - g^2(3)\}^{1/2})$$

and

$$\text{(b) } G(\chi) = \frac{1}{2} i^{(p-1)^2/4} \beta^{-2Z} p^{-1/2} (g^2(3) - 2p + i\nu g(3)\{4p - g^2(3)\}^{1/2}),$$

where $\nu = sgn\{s_3(g^2(3) - p)\}$, *and where* s_3 *is defined in Theorem* 3.1.3.

Proof From (4.1.1), $\mathrm{Re}\,(G(\chi^2)) = \frac{1}{2} g(3)$, and since $|G(\chi^2)|^2 = p$, we deduce that $\mathrm{Im}\,(G(\chi^2)) = \pm\frac{1}{2}\{4p - g^2(3)\}^{1/2}$. Thus, (a) holds for some $\nu \in \{-1, 1\}$. By Lemma 4.1.1(a) and Theorem 3.1.3, we then find that

$$\frac{1}{2} p s_3 \sqrt{3} = \mathrm{Im}\,(G^3(\chi^2)) = \frac{1}{2}\nu(g^2(3) - p)\{4p - g^2(3)\}^{1/2},$$

from which the desired expression for ν follows.

Secondly, part (b) follows from part (a) and Lemma 4.1.1(b). ■

The next theorem determines $g(6)$ in terms of $g(3)$.

Theorem 4.1.4 *If* 2 *is a cubic residue* (mod p), *then*

$$g(6) = g(3) + i^{(p-1)^2/4} p^{-1/2} (g^2(3) - p);$$

if 2 *is a cubic nonresidue* (mod p), *then with* ν *as in Lemma* 4.1.3,

$$g(6) = g(3) + \frac{1}{2} i^{(p-1)^2/4} p^{-1/2} (4p - g^2(3) + Z_3 \nu g(3)\{12p - 3g^2(3)\}^{1/2}),$$

where

$$Z_3 = \begin{cases} +1, & \textit{if } Z \equiv 1 \pmod 3, \\ -1, & \textit{if } Z \equiv 2 \pmod 3. \end{cases}$$

Proof By (4.0.3),

$$g(6) = G(\chi) + G(\overline{\chi}) + G(\chi^2) + G(\overline{\chi}^2) + G(\chi^3).$$

Appealing to (4.0.5), (4.1.1), and (4.0.4), we see that

$$g(6) = R_4 + g(3) + i^{(p-1)^2/4}\sqrt{p}, \tag{4.1.2}$$

where $R_4 = G(\chi) + G(\overline{\chi})$.

If 2 is a cubic residue (mod p) then, by Lemma 4.1.3(b) and Theorem 1.1.4(b), $R_4 = i^{(p-1)^2/4}p^{-\frac{1}{2}}(g^2(3) - 2p)$, and the first part of the theorem follows.

Suppose now that 2 is a cubic nonresidue (mod p). Then $\beta^{-2Z} = -(1 + iZ_3\sqrt{3})/2$, and Lemma 4.1.3(b) yields

$$G(\chi) = -\frac{1}{4}i^{(p-1)^2/4}p^{-1/2}(1 + iZ_3\sqrt{3})(g^2(3) - 2p + i\nu g(3)\{4p - g^2(3)\}^{1/2}).$$

Now $\chi(-1) = (-1)^{(p-1)/6} = (-1)^{(p-1)/2} = (-1)^{(p-1)^2/4}$, and so

$$\begin{aligned} G(\overline{\chi}) &= \chi(-1)\overline{G(\chi)} = (-1)^{(p-1)^2/4}\overline{G(\chi)} \\ &= -\frac{1}{4}i^{(p-1)^2/4}p^{-1/2}(1 - iZ_3\sqrt{3})(g^2(3) - 2p - i\nu g(3)\{4p - g^2(3)\}^{1/2}). \end{aligned}$$

Thus,

$$R_4 = G(\chi) + G(\overline{\chi}) = -\frac{1}{2}i^{(p-1)^2/4}p^{-1/2}\{g^2(3) - 2p - Z_3\nu g(3)\{12p - 3g^2(3)\}^{1/2}\}.$$

The evaluation of $g(6)$ now follows from (4.1.2). ■

The following special case of Theorem 4.0.1 also can be seen directly from Theorems 4.1.2 and 4.1.4.

Corollary 4.1.5 *For $p \equiv 1 \pmod 6$, $g(6)$ is real if and only if $p \equiv 1 \pmod{12}$.*

Now let $\wp(z)$ be a Weierstrass $\wp$-function with $\wp'(z)^2 = 4\wp(z)^3 - 1$. Thus

$$\wp(z) = \frac{1}{z^2} + \sum_{0 \neq u \in \mathcal{L}} \left(\frac{1}{(z-u)^2} - \frac{1}{u^2} \right), \tag{4.1.3}$$

where $\mathcal{L} = \{\theta m + \theta n\beta^2 : m, n \in Z\}$ is the lattice of periods of $\wp$, with θ equal to the least positive period. The reader will find basic facts about Weierstrass $\wp$-functions in Cox's book [1, Chapter 3]. Let R be a set of $(p-1)/3$ residues (mod p) such that $R \cup R\omega \cup R\omega^2$ is a reduced residue system (mod p), where ω is a primitive cube root of 1 (mod p), and $\prod_{r \in R} r \equiv -1 \pmod{p}$. (See Problem 9.) Cassels [2] constructed the product

$$H(\chi^2) = -\prod_{r \in R} \wp(r\theta / J(\chi^2, \chi^2)), \tag{4.1.4}$$

which does not depend on the choice of R, and observed that

$$H^3(\chi^2) = J^{-2}(\chi^2, \chi^2). \tag{4.1.5}$$

Making use of (4.1.5), we deduce that

$$(p^{1/3} J(\chi^2, \chi^2) H(\chi^2))^3 = pJ^3(\chi^2, \chi^2) H^3(\chi^2) = pJ^3(\chi^2, \chi^2) J^{-2}(\chi^2, \chi^2) = pJ(\chi^2, \chi^2),$$

and thus, by Lemma 4.1.1(a),

$$(p^{1/3} J(\chi^2, \chi^2) H(\chi^2))^3 = G^3(\chi^2). \tag{4.1.6}$$

It follows from (4.1.6) that

$$G(\chi^2) = \lambda p^{1/3} J(\chi^2, \chi^2) H(\chi^2), \tag{4.1.7}$$

for some cube root λ of 1. Cassels [2, eq. (4.2)] conjectured that λ is always 1, and, in 1978, Matthews [1], [2] succeeded in proving this deep result by making use of some ideas of Cassels. A generalization for cubic Gauss sums over finite rings of integers in $Z[\beta^2]$ has been given by Tsuyoshi Takagi [1]. A completely elementary algorithm for computing the right side of (4.1.7) has been devised by McGettrick [1]; see also Cassels [2, §4]. This algorithm involves little more than counting the lattice points within a triangle whose vertices are determined by $J(\chi^2, \chi^2)$. Combining these facts with Theorem 3.1.3, we deduce the following theorem.

Theorem 4.1.6

$$G(\chi^2) = \frac{1}{2} p^{1/3} (r_3 + is_3\sqrt{3}) H(\chi^2).$$

From Lemma 4.1.1(b) and Theorem 4.1.6, we obtain the following evaluation of $G(\chi)$.

Theorem 4.1.7

$$G(\chi) = \frac{1}{4} \beta^{-2Z} i^{(p-1)^2/4} p^{1/6} ((r_3^2 - 3s_3^2) + 2r_3 s_3 i\sqrt{3}) H^2(\chi^2).$$

From (4.1.1) we see that

$$g(3) = 2\mathrm{Re}(G(\chi^2)).$$

Appealing to Theorem 4.1.6, we obtain the following evaluation of $g(3)$.

Theorem 4.1.8

$$g(3) = p^{1/3}\{r_3\mathrm{Re}(H(\chi^2)) - s_3\sqrt{3}\ \mathrm{Im}(H(\chi^2))\}.$$

One can now also determine $g(6)$ in terms of $\mathrm{Re}(H(\chi^2))$ and $\mathrm{Im}(H(\chi^2))$ from Theorems 4.1.4 and 4.1.8.

By Theorem 4.1.2 we know that $g(3)$ lies in one of the three intervals $J_0 = (\sqrt{p}, 2\sqrt{p})$, $J_1 = (-\sqrt{p}, \sqrt{p})$, and $J_2 = (-2\sqrt{p}, -\sqrt{p})$. The next lemma shows that, if $g(3)$ belongs to the interval J_m $(m = 0, 1, 2)$, then $|\mathrm{Arg}\, G(\chi^2)|$ belongs to the interval $I_m = (m\pi/3, (m+1)\pi/3)$.

Lemma 4.1.9

$$\begin{aligned} g(3) \in (-2\sqrt{p}, -\sqrt{p}) &\iff |\mathrm{Arg}\ G(\chi^2)| \in (2\pi/3, \pi), \\ g(3) \in (-\sqrt{p}, \sqrt{p}) &\iff |\mathrm{Arg}\ G(\chi^2)| \in (\pi/3, 2\pi/3), \\ g(3) \in (\sqrt{p}, 2\sqrt{p}) &\iff |\mathrm{Arg}\ G(\chi^2)| \in (0, \pi/3). \end{aligned}$$

Proof Set

$$X = \frac{1}{2}g(3),\ Y = \frac{1}{2}\nu\{4p - g^2(3)\}^{1/2},$$

so that $X^2 + Y^2 = p$ and, by Lemma 4.1.3(a),

$$G(\chi^2) = X + iY.$$

Let $\theta = \mathrm{Arg}\, G(\chi^2)$, so that

$$\cos\theta = \frac{X}{\sqrt{p}}, \quad \sin\theta = \frac{Y}{\sqrt{p}}, \quad -\pi < \theta \le \pi.$$

If $g(3) \in (-\sqrt{p}, \sqrt{p})$, then $|\cos\theta| = |X/\sqrt{p}| < 1/2$, and so $\pi/3 < |\theta| < 2\pi/3$. If $g(3) \in (\sqrt{p}, 2\sqrt{p})$, then $1/2 < \cos\theta < 1$, so that $0 < |\theta| < \frac{\pi}{3}$. If $g(3) \in (-2\sqrt{p}, -\sqrt{p})$, then $-1 < \cos\theta < -1/2$, so that $2\pi/3 < |\theta| < \pi$. ∎

The first mathematician to focus on the distribution of the values of $g(3)$ was Kummer. Kummer's calculations (Kummer [1], [2]) showed that for the 45 primes p with $p \equiv 1 \pmod 6$, $p < 500$, $|\mathrm{Arg}\, G(\chi^2)|$ falls into the three intervals $I_2 = (2\pi/3, \pi)$, $I_1 = (\pi/3, 2\pi/3)$, and $I_0 = (0, \pi/3)$, respectively, 7, 14, and 24 times. On the basis of this very limited evidence, Kummer speculated that the ratios of the number of occurrences of $|\mathrm{Arg}\, G(\chi^2)|$ in the intervals I_2, I_1, and I_0 would continue to be $1 : 2 : 3$ as p increases. Kummer's tenuous speculation was elevated to what is now called "Kummer's conjecture." Because of the prominence of this conjecture, some authors (for example, Cassels [2]) refer to the cubic Gauss sum as Kummer's sum. Von Neumann and Goldstine [1] extended Kummer's investigations by considering primes p with $p < 10,000$ and $p \equiv 1 \pmod 6$. Their calculations showed that the ratios of the number of occurrences of $|\mathrm{Arg}\, G(\chi^2)|$ in the intervals I_2, I_1, and I_0, respectively, are nearly 2 : 3 : 4 in contrast to 1 : 2 : 3. Still further calculations of Beyer [1], E. Lehmer [6], Cassels [1], Fröberg [1] (see Shanks [2]), and Tabakova [1] (see Cassels [3]) suggest that perhaps the values of $G(\chi^2)p^{-1/2}$ are uniformly distributed on the unit circle as $p \to \infty$ with $p \equiv 1 \pmod 6$, despite the apparent tendency of the initial values to lie in the right half-plane. An unsuccessful attempt to prove the uniform distribution of $\operatorname{Arg} G(\chi^2)$ was made by Vinogradov [1] in 1967. In 1974, Moreno [2] was able to prove the weaker result that the values of $G^3(\chi^2)p^{-3/2}$ are uniformly distributed on the unit circle. Patterson [1]–[3] developed the work of Kubota [1]–[3] and established a cubic analogue of the classical theta function. This enabled Patterson [4] to prove that the arguments of more general cubic Gauss sums (mod α) are uniformly distributed, where α runs through the squarefree algebraic integers in $Q(i\sqrt{3})$ that are congruent to 1 (mod 3). By combining the techniques of Patterson with a powerful method of Vaughan [1] for estimating certain trigonometric sums, Heath-Brown and Patterson [1] proved in 1979 that the values of $G(\chi^2)p^{-1/2}$ are, indeed, uniformly distributed on the unit circle. Thus Kummer's conjecture was finally disproved after more than 130 years. Despite the equidistribution of the values of $G(\chi^2)p^{-1/2}$ on the unit circle, it is possible that the average value of $\operatorname{Re}(G(\chi^2)p^{-1/2}) = g(3)/(2\sqrt{p})$ for $p \le x$ is positive for sufficiently large x. Indeed, a conjecture of Patterson [4, p. 127] supported by numerical evidence implies that

$$\sum_{p\le x} \operatorname{Re}(G(\chi^2)p^{-1/2}) \sim \frac{(2\pi)^{2/3}x^{5/6}}{5\Gamma(2/3)\log x},$$

as x tends to ∞. It is then perhaps not surprising that the calculations of Kummer and later mathematicians showed a tendency for $G(\chi^2)p^{-1/2}$ to lie in the right half-plane.

4.2 The Gauss sum $g(4)$

Let p be a prime with $p \equiv 1 \pmod 4$, let g be a primitive root (mod p), and let χ be a character (mod p) of order 4 with $\chi(g) = i$. We note that

$$\chi(4) = \chi(-1) = \left(\frac{2}{p}\right).$$

Theorem 4.2.1 *For* $p \equiv 1 \pmod 4$,

$$g(4) = \sqrt{p} \pm \left\{2\left(\frac{2}{p}\right)(p + a_4\sqrt{p})\right\}^{1/2},$$

where the integer a_4 *is defined in Theorem* 3.2.1.

Proof By (4.0.3)–(4.0.5),

$$g(4) = \sqrt{p} + R_6,$$

where

$$R_6 = G(\chi) + G(\overline{\chi}). \tag{4.2.1}$$

It remains to show that

$$R_6 = \pm\left\{2\left(\frac{2}{p}\right)(p + a_4\sqrt{p})\right\}^{1/2}. \tag{4.2.2}$$

By (4.2.1), Theorem 1.1.4(a), Theorem 2.1.3(a), and Theorem 3.2.1, we find that

$$\begin{aligned} R_6^2 &= G(\chi^2)J(\chi,\chi) + G(\overline{\chi}^2)J(\overline{\chi},\overline{\chi}) + 2\chi(-1)p \\ &= \sqrt{p}\left(\frac{2}{p}\right)K(\chi) + \sqrt{p}\left(\frac{2}{p}\right)K(\overline{\chi}) + 2\left(\frac{2}{p}\right)p = 2\left(\frac{2}{p}\right)(a_4\sqrt{p} + p), \end{aligned}$$

which gives (4.2.2). ■

Corollary 4.2.2 *The Gauss sum $g(4)$ is a root of the irreducible quartic polynomial*

$$x^4 - 2p\left(1 + 2\left(\frac{2}{p}\right)\right)x^2 - 8\left(\frac{2}{p}\right)a_4px + p\left(\left(1 - 4\left(\frac{2}{p}\right)\right)p + 4b_4^2\right), \tag{4.2.3}$$

which has four distinct real roots if $p \equiv 1 \pmod 8$ and four distinct nonreal roots if $p \equiv 5 \pmod 8$. The discriminant of the quartic polynomial (4.2.3) *is*

$$2^{14}b_4^2p^3\left\{p\left(1 - 2\left(\frac{2}{p}\right)\right) + a_4^2\right\}^2.$$

Proof By Theorem 4.2.1,

$$(g(4) - \sqrt{p})^2 = 2\left(\frac{2}{p}\right)(p + a_4\sqrt{p}),$$

so that

$$g^2(4) + p\left(1 - 2\left(\frac{2}{p}\right)\right) = 2\sqrt{p}\left(g(4) + \left(\frac{2}{p}\right)a_4\right).$$

Squaring and expanding, we obtain the following equation:

$$g^4(4) - 2p\left(1 + 2\left(\frac{2}{p}\right)\right)g^2(4) - 8\left(\frac{2}{p}\right)a_4pg(4) + \left(5p^2 - 4\left(\frac{2}{p}\right)p^2 - 4pa_4^2\right) = 0.$$

Noting that $a_4^2 = p - b_4^2$, we see that $g(4)$ is a root of the quartic polynomial (4.2.3). The four roots x_1, x_2, x_3, x_4 of the quartic polynomial (4.2.3) are

$$\sqrt{p} \pm \left\{2\left(\frac{2}{p}\right)(p + a_4\sqrt{p})\right\}^{1/2}, \quad -\sqrt{p} \pm \left\{2\left(\frac{2}{p}\right)(p - a_4\sqrt{p})\right\}^{1/2}.$$

These are clearly distinct. As $|a_4| < \sqrt{p}$, these roots are real if $\left(\frac{2}{p}\right) = 1$ and nonreal if $\left(\frac{2}{p}\right) = -1$. The quartic polynomial (4.2.3) is p-Eisenstein, so it is irreducible over the rationals.

Finally, a lengthy computation of $\prod_{1\le i<j\le 4}(x_i - x_j)^2$ yields the desired formula for the discriminant. ■

Now let $\wp(z)$ be the Weierstrass $\wp$-function with $\wp'(z)^2 = 4\wp(z)^3 - \wp(z)$, and period lattice $\{\theta(m+in) : m, n \in Z\}$, where θ is the least positive period. Matthews [3] constructed the product

$$h(\chi) = \prod_{r=1}^{(p-1)/2} \wp'(r\theta/J(\chi,\chi)) \tag{4.2.4}$$

and observed that

$$h^2(\chi) = -J^{-3}(\chi,\chi). \tag{4.2.5}$$

Now let us set

$$a = \left(\frac{2}{p}\right)a_4, \quad b = \left(\frac{2}{p}\right)b_4, \tag{4.2.6}$$

so that, by Theorem 3.2.1,

$$J(\chi,\chi) = a + ib, \ p = a^2 + b^2, \ a \equiv -1 \pmod 4). \tag{4.2.7}$$

Clearly

$$p - 5 - 2b \equiv p - 5 - b^2 \equiv a^2 - 5 \equiv 4 \pmod 8,$$

so that $(p-5-2b)/4$ is an odd integer. Thus, from $G(\chi)^2 = \sqrt{p}J(\chi,\chi)$ and (4.2.5), we deduce that

$$G(\chi)^2 = \{i^{(p-5-2b)/4}p^{1/4}h(\chi)J^2(\chi,\chi)\}^2, \tag{4.2.8}$$

so that

$$G(\chi) = B(\chi)i^{(p-5-2b)/4}p^{1/4}h(\chi)J^2(\chi,\chi), \tag{4.2.9}$$

for some $B(\chi) = \pm 1$. Matthews [3] proved, in fact, that $B(\chi)$ is determined by

$$B(\chi) \equiv \frac{a}{b}\left(\frac{p-1}{2}\right)! \pmod p. \tag{4.2.10}$$

We note that (4.2.9) is the quartic analogue of (4.1.7). An equivalent version of (4.2.9)–(4.2.10), expressed in terms of the elliptic sn function instead of the $\wp$-function, was conjectured in 1972 by McGettrick [2]. Using the theory of modular forms, Matthews [3] was able to deduce the following explicit formula for $G(\chi)$, which had been conjectured by Loxton [1], [2] in a slightly different form.

Theorem 4.2.3 *With* a *and* b *defined by* (4.2.6),

$$G(\chi) = C\left(\frac{|b|}{|a|}\right)(-1)^{(b^2+2b)/8} p^{1/4} J^{1/2}(\chi, \chi), \tag{4.2.11}$$

where $\text{Re}(J^{1/2}(\chi,\chi)) > 0$, $\left(\frac{|b|}{|a|}\right)$ *is the Legendre–Jacobi symbol, and* $C = \pm 1$ *is determined by*

$$C \equiv \frac{|b|}{a}\left(\frac{p-1}{2}\right)!\left(\frac{2}{p}\right) \pmod{p}. \tag{4.2.12}$$

From Theorem 4.2.3, we can obtain the Gauss sum $g(4)$ explicitly, as follows.

Theorem 4.2.4 *If* $p \equiv 1 \pmod 8$, *then*

$$g(4) = \sqrt{p} + C\left(\frac{|b|}{|a|}\right)(-1)^{(b^2+2|b|)/8}\sqrt{2p + 2a\sqrt{p}},$$

and, if $p \equiv 5 \pmod 8$,

$$g(4) = \sqrt{p} + iC\left(\frac{|b|}{|a|}\right)(-1)^{(b^2+2|b|)/8}\sqrt{2p - 2a\sqrt{p}},$$

where a *and* b *are defined in* (4.2.6) *and* C *is given in* (4.2.12).

Proof By (4.2.7), $J(\chi,\chi) = a + ib$, so that $J^{1/2}(\chi,\chi)$ with $\text{Re}(J^{1/2}(\chi,\chi)) > 0$ is given by

$$J^{1/2}(\chi,\chi) = \sqrt{\frac{a+\sqrt{p}}{2}} + i\frac{b}{|b|}\sqrt{\frac{-a+\sqrt{p}}{2}}.$$

Hence, by Theorem 4.2.3,

$$G(\chi) = C\left(\frac{|b|}{|a|}\right)(-1)^{(b^2+2b)/8}\left(\sqrt{\frac{p+a\sqrt{p}}{2}} + i\frac{b}{|b|}\sqrt{\frac{p-a\sqrt{p}}{2}}\right).$$

Taking complex conjugates above and noting that $(-1)^{b/2} = \left(\frac{2}{p}\right) = \chi(-1)$, we find that

$$G(\overline{\chi}) = C\left(\frac{|b|}{|a|}\right)(-1)^{(b^2-2b)/8}\left(\sqrt{\frac{p+a\sqrt{p}}{2}} - i\frac{b}{|b|}\sqrt{\frac{p-a\sqrt{p}}{2}}\right).$$

Then, from (4.2.1), we deduce that

$$R_6 = C\left(\frac{|b|}{|a|}\right)(-1)^{(b^2+2b)/8}E,$$

where

$$E = \left(\sqrt{\frac{p+a\sqrt{p}}{2}} + i\frac{b}{|b|}\sqrt{\frac{p-a\sqrt{p}}{2}}\right) + \left(\frac{2}{p}\right)\left(\sqrt{\frac{p+a\sqrt{p}}{2}} - i\frac{b}{|b|}\sqrt{\frac{p-a\sqrt{p}}{2}}\right).$$

Since $g(4) = \sqrt{p} + R_6$, the result now easily follows. ■

We remark that for large $p \equiv 1 \pmod 4$, the quantity C in (4.2.12), and hence $g(4)$, can be computed relatively quickly. A randomized algorithm for computing a and $|b|$ in expected time $O(\log p)$ has been given by Rabin and Shallit [1]. There is also a deterministic algorithm of Schoof [1] for computing a and $|b|$ whose running time is polynomial in log p. A factorial evaluation scheme outlined in the paper of Crandall, Dilcher, and Pomerance [1] for computing $m! \pmod n$, where m and n are integers with $1 < m < n$, runs in time $O(\sqrt{m}(nm)^\epsilon)$ for every $\epsilon > 0$. In particular, this algorithm computes $((p-1)/2)! \pmod p$ in time $O(p^{1/2+\epsilon})$ for every $\epsilon > 0$.

Calculations of E. Lehmer [6] in 1956 tended to support the conjecture that the values of $G(\chi)p^{-1/2}$ are uniformly distributed on the unit circle. This conjecture has been proved as a special case of a more general result. See Patterson [8], [9] for further details. In 1965 Yamamoto [1] made four conjectures about the distribution of $G(\chi)p^{-1/2}$. These conjectures have all been settled by Ito [1], [2] (Conjectures I, II, III are false, and IV is true). In Problem 6, we outline the simple proof of Conjecture IV given by Barkan [1]. Yamamoto [1, Theorem 3] gave a formula for $G(\chi)$ involving both the class number of the real quadratic field $Q(\sqrt{p})$ and the length of the period of the continued fraction expansion of $\omega + [\omega] + 1$, where $\omega = (-1+\sqrt{p})/2$. It remains to reconcile this result with Theorem 4.2.3.

4.3 The Gauss sum $g(8)$

Let p be a prime with $p \equiv 1 \pmod 8$ and let g be a primitive root $\pmod p$. Let $\beta = \exp(2\pi i/8)$ and let $p = 8f+1$. Let ψ denote a character $\pmod p$ of order 8 such that $\psi(g) = \beta$. Put $\chi = \psi^2$, so that χ is a character of order 4. As usual, we set $Z = \text{ind}_g 2$. Note that

$$(-1)^Z = \beta^{4Z} = \psi^4(2) = \chi^2(2) = \left(\frac{2}{p}\right) = 1,$$

so that Z is even. Also $\psi(-1) = (-1)^f$.

Theorem 4.3.1

$$g(8) = g(4) \pm \left\{2(a_8+\sqrt{p})\left((-1)^{Z/2}(g(4)-\sqrt{p}) + 2(-1)^f\sqrt{p}\right)\right\}^{1/2},$$

where the integer a_8 is defined in Theorem 3.3.1.

Proof By (4.0.3),

$$
\begin{aligned}
g(8) &= \sum_{j=1}^{7} G(\psi^j) \\
&= \{G(\psi) + G(\overline{\psi})\} + \{G(\chi) + G(\overline{\chi})\} + \{G(\psi^3) + G(\overline{\psi}^3)\} + G(\chi^2) \\
&= R_3 + R_6 + R_9 + \sqrt{p}.
\end{aligned}
$$

Since $\sqrt{p} + R_6 = g(4)$ from the proof of Theorem 4.2.1,

$$g(8) = g(4) + (R_3 + R_9). \tag{4.3.1}$$

It remains to show that

$$R_3 + R_9 = \pm\{2(a_8 + \sqrt{p})((-1)^{Z/2}(g(4) - \sqrt{p}) + 2(-1)^f\sqrt{p})\}^{1/2}. \tag{4.3.2}$$

First we determine $R_3^2 + R_9^2$. By Theorem 2.1.3(a) and Theorem 1.1.4(a),

$$
\begin{aligned}
R_3^2 &= \{G(\psi) + G(\overline{\psi})\}^2 \\
&= G(\psi^2)J(\psi,\psi) + G(\overline{\psi}^2)J(\overline{\psi},\overline{\psi}) + 2\psi(-1)p \\
&= G(\chi)\overline{\chi}(2)K(\psi) + G(\overline{\chi})\chi(2)K(\overline{\psi}) + 2\psi(-1)p.
\end{aligned}
$$

In a similar manner, we find that

$$R_9^2 = G(\overline{\chi})\chi(2)K(\psi^3) + G(\chi)\overline{\chi}(2)K(\overline{\psi}^3) + 2\psi(-1)p.$$

Since $\chi(2) = \overline{\chi}(2) = \beta^{2Z} = (-1)^{Z/2}$ and $K(\psi) = K(\psi^3)$ (by Theorem 2.1.6),

$$R_3^2 + R_9^2 = (-1)^{Z/2}\{G(\chi) + G(\overline{\chi})\}\{K(\psi) + K(\overline{\psi})\} + 4\psi(-1)p.$$

Appealing to Theorem 3.3.1, we deduce that

$$R_3^2 + R_9^2 = 2(-1)^{Z/2}(g(4) - \sqrt{p})a_8 + 4\psi(-1)p. \tag{4.3.3}$$

Next we determine R_3R_9. By Theorem 2.1.6,

$$G(\psi)G(\psi^3) = G(\psi^4)J(\psi,\psi^3) = \sqrt{p}\,\psi(-1)K(\psi). \tag{4.3.4}$$

Multiplying both sides of (4.3.4) by $G(\overline{\psi})/G(\psi)$, we find that

$$G(\overline{\psi})G(\psi^3) = \frac{G(\overline{\psi})}{G(\psi)}\sqrt{p}\,\psi(-4)\frac{G^2(\psi)}{G(\psi^2)} = (-1)^{Z/2}\sqrt{p}\,G(\overline{\chi}), \tag{4.3.5}$$

by Theorem 1.1.4(a). By (4.3.4) and (4.3.5),

$$
\begin{aligned}
R_3R_9 &= \{G(\psi) + G(\overline{\psi})\}\{G(\psi^3) + G(\overline{\psi}^3)\} \\
&= G(\psi)G(\psi^3) + G(\overline{\psi})G(\psi^3) + G(\psi)G(\overline{\psi}^3) + G(\overline{\psi})G(\overline{\psi}^3) \\
&= \sqrt{p}\,\psi(-1)K(\psi) + (-1)^{Z/2}\sqrt{p}\,G(\overline{\chi}) + (-1)^{Z/2}\sqrt{p}\,G(\chi) + \sqrt{p}\,\psi(-1)K(\overline{\psi}) \\
&= \sqrt{p}\,\{\psi(-1)(K(\psi) + K(\overline{\psi})) + (-1)^{Z/2}(G(\chi) + G(\overline{\chi}))\} \\
&= \sqrt{p}\,\{2\psi(-1)a_8 + (-1)^{Z/2}R_6\} = \sqrt{p}\,\{2\psi(-1)a_8 + (-1)^{Z/2}(g(4) - \sqrt{p})\}.
\end{aligned}
$$

Then, from (4.3.3),

$$\begin{aligned}(R_3+R_9)^2 &= (R_3^2+R_9^2)+2R_3R_9 = 2(-1)^{Z/2}(g(4)-\sqrt{p})a_8+4\psi(-1)p\\ &\quad +2\sqrt{p}\,\{2\psi(-1)a_8+(-1)^{Z/2}(g(4)-\sqrt{p})\}\\ &= 2(a_8+\sqrt{p})((-1)^{Z/2}(g(4)-\sqrt{p})+2(-1)^f\sqrt{p}),\end{aligned}$$

from which (4.3.2) follows. ■

There is no known criterion (analogous to that for $g(4)$) for determining the ambiguous sign for $g(8)$ in Theorem 4.3.1.

4.4 The Gauss sum $g(12)$

Let p be a prime with $p \equiv 1 \pmod{12}$ and let g be a primitive root (mod p). Let $\beta = \exp(2\pi i/12)$ and let $p = 12f+1$. Let ψ denote a character (mod p) of order 12 such that $\psi(g) = \beta$. Put $\chi = \psi^2$, so that χ is a character of order 6. Note that

$$\chi(2) = \psi^2(2) = \psi^2(g^Z) = \psi^{2Z}(g) = \beta^{2Z},\ \chi(4) = \beta^{4Z},$$

$$\psi(-1) = \psi(g^{6f}) = \beta^{6f} = (-1)^f = \left(\frac{2}{p}\right) = \left(\frac{g^Z}{p}\right) = \left(\frac{g}{p}\right)^Z = (-1)^Z = \beta^{6Z},$$

$$\overline{\chi}(2) = \beta^{-2Z} = \beta^{10Z} = \beta^{6Z}\beta^{4Z} = \left(\frac{2}{p}\right)\chi(4).$$

Theorem 4.4.1 *Let $p \equiv 1 \pmod{12}$. Then*

$$g(12) = g(6) + R_6 \pm p^{-\frac{1}{2}} g(3)\left\{2\left(\frac{2}{p}\right)(p + a_{12}\sqrt{p})\right\}^{1/2},$$

where R_6 is given by (4.2.2) and a_{12} is defined in Theorem 3.5.2.

Proof By (4.0.3),

$$\begin{aligned} g(12) &= G(\psi)+G(\psi^2)+\cdots+G(\psi^{11})\\ &= \{G(\psi^2)+G(\psi^4)+G(\psi^6)+G(\psi^8)+G(\psi^{10})\}\\ &\qquad +\{G(\psi)+G(\overline{\psi})\}+\{G(\psi^3)+G(\overline{\psi}^3)\}+\{G(\psi^5)+G(\overline{\psi}^5)\}\\ &= \{G(\chi)+G(\chi^2)+G(\chi^3)+G(\chi^4)+G(\chi^5)\}+R_2+R_6+R_{10}\\ &= g(6)+R_2+R_6+R_{10},\end{aligned}$$

where

$$R_2 = G(\psi)+G(\overline{\psi}) \quad \text{and} \quad R_{10} = G(\psi^5)+G(\overline{\psi}^5).$$

It remains to show that

$$R_2 + R_{10} = \pm p^{-1/2} g(3)\left\{2\left(\frac{2}{p}\right)(p + a_{12}\sqrt{p})\right\}^{1/2}. \tag{4.4.1}$$

First we calculate $R_2^2 + R_{10}^2$. By Theorems 1.1.4(a) and 2.1.3,

$$R_2^2 = G(\psi^2)\overline{\psi}(4)K(\psi) + G(\overline{\psi}^2)\psi(4)K(\overline{\psi}) + 2\psi(-1)p$$

and

$$R_{10}^2 = G(\overline{\psi}^2)\psi(4)K(\psi^5) + G(\psi^2)\overline{\psi}(4)K(\overline{\psi}^5) + 2\psi(-1)p.$$

By Theorem 2.1.6, $K(\psi) = K(\psi^5)$, and so

$$R_2^2 + R_{10}^2 = \{G(\psi^2)\overline{\psi}(4) + G(\overline{\psi}^2)\psi(4)\}\{K(\psi) + K(\overline{\psi})\} + 4\psi(-1)p.$$

Thus, by Theorem 3.5.2,

(4.4.2) $$R_2^2 + R_{10}^2 = 4a_{12}\mathrm{Re}(\overline{\chi}(2)G(\chi)) + 4\psi(-1)p.$$

Next we determine R_2R_{10}. By Theorems 2.1.3–2.1.5,

(4.4.3) $$G(\psi)G(\psi^5) = \sqrt{p}\ J(\psi, \psi^5) = \sqrt{p}\,\psi(-1)K(\psi).$$

Now multiply (4.4.3) by $G(\overline{\psi})/G(\psi)$ to find that

(4.4.4) $$G(\overline{\psi})G(\psi^5) = \sqrt{p}\ \psi(-4)\frac{G^2(\psi)}{G(\psi^2)}\cdot\frac{G(\overline{\psi})}{G(\psi)} = \sqrt{p}\ \psi(4)G(\overline{\psi}^2) = \sqrt{p}\ \chi(2)G(\overline{\chi}).$$

Then, by (4.4.3) and (4.4.4),

$$\begin{aligned} R_2R_{10} &= (G(\psi) + G(\overline{\psi}))(G(\psi^5) + G(\overline{\psi}^5)) \\ &= G(\psi)G(\psi^5) + G(\overline{\psi})G(\psi^5) + G(\psi)G(\overline{\psi}^5) + G(\overline{\psi})G(\overline{\psi}^5) \\ &= \sqrt{p}\ \psi(-1)K(\psi) + \sqrt{p}\ \chi(2)G(\overline{\chi}) + \sqrt{p}\ \overline{\chi}(2)G(\chi) + \sqrt{p}\ \psi(-1)K(\overline{\psi}) \\ &= \sqrt{p}\ \{\psi(-1)(K(\psi) + K(\overline{\psi})) + \chi(2)G(\overline{\chi}) + \overline{\chi}(2)G(\chi)\}. \end{aligned}$$

Thus, by Theorem 3.5.2, we obtain the following equation:

(4.4.5) $$R_2R_{10} = 2\sqrt{p}\ (\psi(-1)a_{12} + \mathrm{Re}\,(\overline{\chi}(2)G(\chi))).$$

By (4.4.2) and (4.4.5), we deduce that

$$(R_2 + R_{10})^2 = (R_2^2 + R_{10}^2) + 2R_2R_{10} = 4(a_{12} + \sqrt{p})\left(\mathrm{Re}(\overline{\chi}(2)G(\chi)) + \left(\frac{2}{p}\right)\sqrt{p}\right).$$

Now, by Lemma 4.1.3(b),

$$\mathrm{Re}\,(\beta^{4Z}G(\chi)) = \frac{1}{2}p^{-1/2}(g^2(3) - 2p),$$

so that

$$2\text{Re}(\overline{\chi}(2)G(\chi)) = 2\text{Re}\left(\left(\frac{2}{p}\right)\chi(4)G(\chi)\right)$$
$$= 2\left(\frac{2}{p}\right)\text{Re}(\beta^{4Z}G(\chi)) = \left(\frac{2}{p}\right)p^{-1/2}(g^2(3) - 2p).$$

Therefore,

$$(R_2 + R_{10})^2 = 2\left(\frac{2}{p}\right)p^{-1/2}g^2(3)(a_{12} + \sqrt{p}\,),$$

from which (4.4.1) follows. ■

There is no known criterion analogous to that for $g(4)$ for the determination of the ambiguous signs for $g(12)$ in Theorem 4.4.1.

Exercises 4

1. Prove that

$$|g(k)| \leq (k-1)\sqrt{p}.$$

Hint: Use (4.0.3).
For generalizations of this inequality, see the Notes at the end of this chapter.

2. Let h, k, and m be integers with $k, m > 0$. Define

$$g(h,k,m) = \sum_{n \pmod m} e^{2\pi i h n^k/m}.$$

Let $m = m_1 m_2$, where $\gcd(m_1, m_2) = 1$ and $m_1, m_2 > 0$. Prove that

$$g(h,k,m_1m_2) = g(hm_2^{k-1}, k, m_1)g(hm_1^{k-1}, k, m_2).$$

3. Let p be a prime > 3. Let a and n be integers with n positive and $p \nmid a$. Prove that

$$\sum_{j=0}^{p^n-1} \cos(2\pi a j^3/p^n) = \begin{cases} p^{2(n-1)/3} \sum_{j=0}^{p-1} \cos(2\pi a j^3/p), & \text{if } n \equiv 1 \pmod 3, \\ p^{(2n-1)/3}, & \text{if } n \equiv 2 \pmod 3, \\ p^{2n/3}, & \text{if } n \equiv 0 \pmod 3. \end{cases}$$

4. Let p be an odd prime. Prove that

$$\sum_{n=0}^{p-1}(-1)^n e^{\pi i n^4/p} = \begin{cases} \sqrt{p} + 2\text{Re}(\chi(2)G(\chi)), & \text{if } p \equiv 1 \pmod 8, \\ -\sqrt{p} + i2\text{Im}(\chi(2)G(\chi)), & \text{if } p \equiv 5 \pmod 8, \\ i\left(\frac{2}{p}\right)\sqrt{p}, & \text{if } p \equiv 3 \pmod 4, \end{cases}$$

where χ is any quartic character (mod p).

5. Let p be a prime with $p \equiv 1 \pmod 3$. Prove that

$$g(3) = p^{1/3}\left(\left(\frac{r_3 + is_3\sqrt{3}}{2}\right)^{1/3}\omega + \left(\frac{r_3 - is_3\sqrt{3}}{2}\right)^{1/3}\omega^2\right)$$

for some cube root of unity ω.

6. Let p be a prime $\equiv 1 \pmod 8$, and write $p = a^2 + b^2$, where a and b are integers of the same sign, with $a \equiv -1 \pmod 4$. For the Gaussian prime $\pi = a + ib$, let χ be the quartic character (mod p) defined by

$$\chi(n) \equiv n^{(p-1)/4} \pmod{\pi}.$$

Let

$$D(\chi) = -\sum_{n=1}^{(p-1)/2} n\chi(n).$$

(a) Berndt [3, Theorem 13.1] gave a short proof that

$$\frac{\pi^2 D(\chi)}{pG(\chi)} = \prod_{t\in S}(1 - \overline{\chi}(t)/t^2)^{-1},$$

where S is the set of odd primes. Use this to show that

$$\left|\mathrm{Arg}\frac{D(\chi)}{G(\chi)}\right| \le \sum_{t\in S}\arctan\left(\frac{1}{t^2}\right) \le \sum_{t\in S}\frac{1}{t^2} < \frac{\pi}{15}.$$

(b) Show that $G^2(\chi) = \sqrt{p}J(\chi,\chi) = \sqrt{p}\pi$.

(c) Use part (b) to show that if $a > 0$, then $G(\chi)$ is in the first or fifth octant, while if $a < 0$, then $G(\chi)$ is in the third or seventh octant.

(d) Use parts (a) and (c) to prove the following conjecture of Yamamoto [1, p. 121]:

If $a > 0$, then $G(\chi)$ is in the first octant if and only if $\mathrm{Re}(D(\chi)) > 0$; *if $a < 0$, then $G(\chi)$ is in the third octant if and only if* $\mathrm{Im}(D(\chi)) > 0$.

7. Let p be a prime, and let r, s, j, k be integers such that $k = p^s j$ with $j \ge 1$, $p \nmid j$, and $s \ge 0$.

(a) Prove that

$$\sum_{\substack{m=1 \\ p \nmid m}}^{p^r} \exp(2\pi i m^k/p^r) = 0$$

whenever

$$r \geq \begin{cases} s+2, & \text{if } p > 2, \\ s+3, & \text{if } p = 2\ . \end{cases}$$

Hint: Write $m = up^{r-s-1} + v$.
(b) Use part (a) to prove that

$$\sum_{\substack{m=1 \\ p \nmid m}}^{p^r} \exp(2\pi i m^k/p^r) = 0$$

whenever either
(i) $r > k$,
(ii) $r \geq 2$, $p \nmid k$,
or
(iii)$r = k \geq 2$,
except in the two cases $p^r = 4$, $k = 2$ and $p^r = 16$, $k = 4$.

8.(a) Use Problem 7 to show that for $r \geq k \geq 2$, the reduction formula

$$\sum_{m=1}^{p^r} \exp(2\pi i m^k/p^r) = p^{k-1} \sum_{m=1}^{p^{r-k}} \exp(2\pi i m^k/p^{r-k})$$

holds except in the two cases $p^r = 4$, $k = 2$ and $p^r = 16$, $k = 4$. (Compare the case $k = 2$ with Problem 4(b) of Exercises 1.)

(b) Write $r = ak + b$ for nonnegative integers a, b with $0 \leq b < k$. Use part (a) to show that for $r \geq k \geq 2$,

$$\sum_{m=1}^{p^r} \exp(2\pi i m^k/p^r) = p^{a(k-1)} \sum_{m=1}^{p^b} \exp(2\pi i m^k/p^b),$$

except in the two cases $p^r = 4$, $k = 2$ and $p^r = 16$, $k = 4$. The right side equals

$$\begin{cases} p^{a(k-1)}, & \text{if } b = 0, \\ p^{a(k-1)} g(k), & \text{if } b = 1, \end{cases}$$

where $g(k)$ is the Gauss sum defined in (4.0.1). If $2 \leq b < k$, show with the aid of Problem 7(b) that the right side equals $p^{a(k-1)+b-1}$ in the case $p \nmid k$.

9. Prove the existence of the set R defined above (4.1.4).

Notes on Chapter 4

Jacobi [3] used the result in Problem 1 of Exercises 2 to evaluate (up to sign ambiguities) Gauss sums $G(\chi)$ of orders 6, 8, and 12.

The determination of $g(3)$ as one of the roots of $x^3 - 3px - pr_3$ (see Theorem 4.1.2) is due to Gauss [2, art. 358, 359]. For later and clearer presentations, consult the books of Bachmann [1, pp. 209–213], Mathews [1, pp. 219–228], Weber [2, vol. 1, pp. 628–631], Fricke [1, vol. 1, pp. 438–440], or Hasse [2, paragraph 20]. The proof in Hasse's book is similar to that given here. There is no simple criterion known for determining the interval in Theorem 4.1.2 that contains $g(3)$, or equivalently, for determining $|\text{Arg } G(\chi)|$, where χ is a cubic character (mod p) (see (4.1.1)). Weil (see Kummer [CW, vol. 1, p. 3]) has expressed pessimism that such a criterion will ever be found.

Prior to the work of Patterson discussed in Section 4.1, Krätzel [1] had developed an asymptotic series for the Kummer sum that is similar in some respects to Patterson's cubic theta functions. Krätzel [1] has also established an analogous formula for quartic Gauss sums. For other work of Patterson in this area, see Patterson [5]–[9]. Patterson [9] considered general Gauss sums of order k over rings of S-integers of an algebraic number field containing all the k–th roots of unity. Contrary to what is implied in Berndt and Evans [4] and Montgomery, Vaughan, and Wooley [1], it is not evident how to deduce results on uniform distribution for fixed $k > 4$ when this number field is replaced by Q. It would be highly interesting if one could show that as $p \to \infty$ with $p \equiv 1 \pmod{k}$, the values $p^{-1/2}G(\chi)$ are uniformly distributed on the unit circle, where χ runs through the characters (mod p) of order k.

The determination of $g(6)$ in Theorem 4.1.4 is given in Berndt and Evans [1]. Hasse's text [2] gives just a brief discussion of $g(6)$. For the explicit monic sextic polynomial over Z satisfied by $g(6)$, see D. H. and E. Lehmer [5].

The determination of the quartic Gauss sum $g(4)$ in Theorem 4.2.1 was essentially first given by Gauss [5, art. 15–22]. However, the quartic polynomial over Z satisfied by $g(4)$ (see Corollary 4.2.2) was perhaps first made explicit by Lebesgue [5]. Expositions of these determinations may be found in the books of Bachmann [1, pp. 224–230], Weber [2, vol. 1, pp. 632–634], Fricke [1, vol. 1, pp. 441–443], and Hasse [2, pp. 489–494]. Our proof of Theorem 4.2.1 is similar to that given by Hasse. For $p \equiv 5 \pmod{8}$, $\text{Re}(g(4))$ is evaluated in E. Lehmer [5]. A special class of quartic Gauss sums is evaluated simply by Berndt and Chowla [1]. The explicit value of $g(4)$ given in Theorem 4.2.4 has been used by Hardy and Williams [3] to determine the sum of the series

$$\sum_{\substack{n=1 \\ \left(\frac{n}{p}\right)=1}}^{\infty} \left(\frac{n}{p}\right)_4 \frac{1}{n},$$

where p is a prime $\equiv 1 \pmod{4}$ and $\left(\dfrac{n}{p}\right)_4$ is $+1$ if n is a quartic residue (mod p) and

-1 if n is a quadratic but not a quartic residue (mod p). Drmota and Skalba [1] used the value of $g(4)$ in analyzing the zeros of the Thue-Morse fractal function.

The evaluation of $g(8)$ given in Theorem 4.3.1 is due to Berndt and Evans [1]. Very special cases had been previously evaluated by Berndt and Chowla [1]. For the octic polynomial over Z satisfied by $g(8)$, see Evans [12].

The quintic polynomial over Z satisfied by $g(5)$ is exhibited in Berndt and Evans [4, eq. (5.2)]. This is a (corrected) form of the polynomial given by E. Lehmer [2, eq. (10)], whose work relied heavily on the results of Tanner [1]. For further references to quintic Gauss sums, see Berndt and Evans [4, p. 119].

The determination of $g(12)$ in Theorem 4.4.1 is due to Berndt and Evans [1]. In that paper, the authors also evaluate $g(24)$, in terms of $g(3)$ and $g(12)$. Evans [4] has evaluated $g(16)$ up to four sign ambiguities.

In summary, there exist evaluations for the Gauss sums $g(k)$ in the cases $k = 2$, 3, 4, 5, 6, 8, 12, 16, and 24, but ambiguities still persist in the determinations, except when $k = 2$, 3, 4, 6 (if $g(3)$ is taken as known).

A straightforward extension of the estimate $|g(k)| \leq (k-1)\sqrt{p}$ in Problem 1 is the estimate $|g(a,k)| \leq (k-1)\sqrt{p}$, where p is a prime not dividing a; see (1.1.4). This has been improved under certain restrictions. Montgomery, Vaughan, and Wooley [1, Theorem 1] give an elementary proof that

$$|g(a,k)| \leq \left(1 - k + k^2/2\right)^{1/2}\sqrt{p},$$

if $p \equiv k+1 \pmod{2k}$ and k is an (even) integer > 0. (Note that $\left(1-k+k^2/2\right)^{1/2} < k-1$ for all $k > 2$.) Another improvement of the estimate $|g(a,k)| \leq (k-1)\sqrt{p}$ is

$$|g(a,k)| \leq \begin{cases} 2k^{7/12}p^{2/3}, & \text{if } p^{2/5} \leq k \leq p^{1/2}, \\ 2k^{1/3}p^{19/24}, & \text{if } p^{1/2} \leq k \leq p^{5/8}; \end{cases}$$

see Mullen and Shparlinski [1, p. 258] and Shparlinski [1]. An estimate that is useful when k is even larger relative to p is due to Konyagin [1, transl., p. 113] who proved that for any $\epsilon > 0$, $k \geq 2$ and $p \geq k(\log k)/(\log(1+\log k))^{1-\epsilon}$,

$$|g(a,k)| \leq p\left(1 - (\log p)^{-1-\epsilon}\right).$$

In the other direction, Konyagin [1, transl., p. 106] proved that for any $A > 0$, there exist infinitely many triples k, p, a with $0 < a < p$ such that $p \geq Ak(\log k)/\log(\log k)$ and

$$|g(a,k)| \geq p\left(1 - (\log k)^{-A}\right).$$

Korobov [1, p. 27, Theorem 6] obtained the estimate

$$|g(a,k,N)| \leq k^{k^6} N^{(k-1)/k}$$

for arbitrary positive integers a, k, N with $\gcd(a, N) = 1$, where $g(a, k, N)$ is the Gauss sum over the ring of integers (mod N) defined by

$$g(a, k, N) = \sum_{x=1}^{N} \exp(2\pi i a x^k / N).$$

Katz and Zheng [1] proved that as $q \to \infty$, the set of Gauss (as well as Jacobi) character sums over F_q lying on the complex circle of radius $\sqrt{q}$ (with the additive and multiplicative characters ranging over all possible nontrivial choices) is uniformly distributed on that circle; furthermore, they obtain good estimates for the discrepancy. If the additive character is kept fixed, the resulting subset of Gauss sums is still uniformly distributed as $q \to \infty$. R. Smith [1, §2] had shown earlier that as $p \to \infty$, the set of Gauss sums $G(\chi)$ (mod p) is uniformly distributed on the circle of radius $\sqrt{p}$, where χ ranges over all the nontrivial characters on F_p. Babaev and Ismoilov [1] showed that as $k \to \infty$, the set of Gauss sums $\tau_k(m, \chi)$ (defined in (1.6.1)) is uniformly distributed on the complex circle of radius $\sqrt{k}$, where χ ranges over all primitive Dirichlet characters (mod k), and $\gcd(m, k) = 1$.

Problem 3 of Exercises 4 is taken from Kummer's paper [2]. A generalization is given in Problem 8.

Many types of character sums, e.g., hypergeometric character sums, can be evaluated in terms of Gauss sums $G(\chi)$; see Evans [16], [18] and the references therein. Thus the determinations of quartic Gauss sums $G(\chi)$ in this chapter, for example, can be applied to evaluate various other quartic character sums. Another application of the evaluations in this chapter is to power residue difference sets; this is the main topic of Chapter 5.

Chapter 5

Difference Sets

The primary objective of this brief chapter is to illustrate how the determinations of Gauss sums can be used to prove or disprove the existence of power residue difference sets. The reader wishing to learn more about difference sets and their applications may consult the books of Baumert [1], Hall [1, Chapter 11], Jungnickel [2, §§6.5, 6.6, 6.9, 7.8], Mann [1], Ryser [1, Chapter 9], and Storer [1]. See also Jungnickel [1].

5.1 Basic definitions

We begin by defining a difference set in an Abelian group.

Definition *A subset H of a finite (additive) Abelian group G is said to be a difference set of G if for some fixed natural number λ, every nonzero element of G can be written as a difference of two elements of H in exactly λ ways.*

The difference set H in G defined above is called an Abelian difference set to emphasize that the group G is Abelian. If G is cyclic, then H is called a cyclic difference set. If $\lambda = 1$, then H is called a planar difference set. For example, $\{3, 6, 7, 12, 14\}$ and $\{7, 9, 14, 15, 18\}$ are each cyclic planar difference sets in the cyclic group of residues (mod 21). It is interesting to note that a difference set in the group G of residues (mod v) always contains a generator of G, i.e., an element coprime with v, except in the two examples above; see S. Cohen [2]. All known planar Abelian difference sets H have the property that $|H| - 1$ is a prime power; see Gordon [1].

We observe that if the subset H of the finite Abelian group G is a difference set, then the numbers of elements in G and H, say, n_G and n_H, respectively, are related by

$$\lambda(n_G - 1) = n_H(n_H - 1). \tag{5.1.1}$$

This equality will be useful in the sequel.

We are interested in the situation when G is the additive group of integers (mod p). The subsets to be considered are H_k and $M_k = H_k \cup \{0\}$, where H_k is the subset of kth-power residues (mod p). Here, k is an integer with $k \geq 2$, and p is a prime $\equiv 1 \pmod{k}$.

Definition *If H_k is a difference set, it is called a power residue difference set. If M_k is a difference set, it is called a modified power residue difference set.*

Definition *For $p = kf + 1$, define the f-nomial period*

$$S(k) = \sum_{r \in H_k} e^{2\pi i r/p}.$$

In the terminology of Gauss, $S(k)$ is called a cyclotomic period. In the notation of Section 10.10, where f-nomial periods are discussed, the sum $S(k)$ is the f-nomial period η_0. The next lemma relates $S(k)$ to the Gauss sum $g(k)$ defined in (4.0.1).

Lemma 5.1.1 *For $p \equiv 1 \pmod k$,*

$$g(k) = kS(k) + 1.$$

Proof We have

$$g(k) - 1 = \sum_{n=1}^{p-1} e^{2\pi i n^k/p} = k \sum_{r \in H_k} e^{2\pi i r/p} = kS(k). \qquad \blacksquare$$

Since the elements $e^{2\pi i n/p}$, $1 \leq n \leq p-1$, are linearly independent over Q, it is easily seen that $S(k)$ has exactly k distinct conjugates over Q, namely, $\sum_{r \in H_k} \exp(2\pi i r g^j/p)$, $0 \leq j \leq k-1$, where g is a primitive root $\pmod p$.

5.2 Necessary and sufficient conditions for power residue difference sets

The theorem below gives a necessary and sufficient condition for each of H_k and M_k to be a difference set, in terms of the size of $g(k)$.

Theorem 5.2.1 *Let $p \equiv 1 \pmod k$ with $p > k + 1$. Then*
(a) *H_k is a difference set if and only if*

(5.2.1) $$|g(k) - 1|^2 = p(k-1) + 1;$$

(b) *M_k is a difference set if and only if*

(5.2.2) $$|g(k) + k - 1|^2 = (p + k - 1)(k - 1).$$

Proof (a) Suppose that H_k is a difference set. Then

$$\begin{aligned} |S(k)|^2 &= S(k)\overline{S(k)} \\ &= \sum_{r,s \in H_k} e^{2\pi i(r-s)/p} = \sum_{\substack{r,s \in H_k \\ r \equiv s \pmod p}} e^{2\pi i(r-s)/p} + \sum_{\substack{r,s \in H_k \\ r \not\equiv s \pmod p}} e^{2\pi i(r-s)/p} \\ &= (p-1)/k + \lambda \sum_{n \not\equiv 0 \pmod p} e^{2\pi i n/p} = (p-1)/k - \lambda, \end{aligned}$$

where, by (5.1.1), λ is determined by

$$\lambda = (p - 1 - k)/k^2 > 0.$$

Hence

$$|S(k)|^2 = \{p(k-1) + 1\}/k^2,$$

from which (5.2.1) follows by Lemma 5.1.1.

Conversely, suppose that (5.2.1) holds. Appealing to Lemma 5.1.1, we obtain $|S(k)|^2 = \{p(k-1)+1\}/k^2$. Then multiplying, expanding, and collecting terms in the product $S(k)\overline{S(k)}$, we see that

$$\sum_{n=1}^{p-1} a_n e^{2\pi i n/p} = \lambda \sum_{n=1}^{p-1} e^{2\pi i n/p},$$

where $\lambda = (p-1-k)/k^2$ and a_n is the number of ways of expressing n as a difference of two elements of H_k. Since the elements $e^{2\pi i n/p}$ $(1 \le n \le p-1)$ are linearly independent over Q, it follows that $a_n = \lambda$ for each n, and so H_k is a difference set.

(b) Suppose that M_k is a difference set. Then, arguing as in part (a), we find that

$$|S(k) + 1|^2 = (p + k - 1)/k - \lambda,$$

where $\lambda = (p + k - 1)/k^2$ and

$$|S(k) + 1|^2 = (p + k - 1)(k - 1)/k^2,$$

from which (5.2.2) follows by Lemma 5.1.1.

Conversely, suppose that (5.2.2) holds. Then by the same type of reasoning as used in the proof of part (a), M_k is a difference set. ∎

The size of $|g(k)|$ determined by (5.2.1) or (5.2.2) should be compared with the upper estimate of $|g(k)|$ given in Problem 1 of Exercises 4, namely, $|g(k)| \le (k-1)\sqrt{p}$.

5.3 Applications of Gauss sums

In this section, we give several applications of Theorem 5.2.1.

Theorem 5.3.1 *If* $p \equiv 1 \pmod{2k}$, *where* $k > 2$, *then neither* H_k *nor* M_k *is a difference set.*

Proof As was noted at the end of Section 5.1, $S(k)$ has k distinct conjugates over Q. Hence, the Galois extension $Q(S(k))$ over Q has degree k. Since $p \equiv 1 \pmod{2k}$, we see that $-1 \in H_k$. It follows that $S(k)$ is real. If H_k (resp. M_k) is a difference set, then $|S(k)|^2$ (resp. $|S(k)+1|^2$) is rational by Lemma 5.1.1 and Theorem 5.2.1. Since

$S(k)$ is real, we see that $S^2(k)$ (resp. $\{S(k)+1\}^2$) is rational. This contradicts the fact that $[Q(S(k)) : Q] = k > 2$. ■

An immediate corollary to Theorem 5.3.1 is the following result, which shows, for example, that neither H_3 nor M_3 is a difference set.

Corollary 5.3.2 *If $p \equiv 1 \pmod{k}$, where k is odd, then neither H_k nor M_k is a difference set.*

We next answer the question "When do the set of quadratic residues (mod p) form a difference set ?"

Theorem 5.3.3 *Let $p > 3$. The set of quadratic residues H_2 and the set M_2 are difference sets if and only if $p \equiv 3 \pmod{4}$.*

Proof By Theorem 5.2.1, H_2 is a difference set if and only if

$$|g(2) - 1|^2 = p + 1.$$

By Theorem 1.2.4, this is true if and only if $p \equiv 3 \pmod{4}$.

The proof for M_2 is similar. ■

Example By Theorem 5.3.3, the quadratic residues 1, 4, 5, 6, 7, 9, 11, 16, and 17 (mod 19) form a difference set. Here $\lambda = 4$ and we see, for example, that

$$5 \equiv 6 - 1 \equiv 9 - 4 \equiv 11 - 6 \equiv 16 - 11 \pmod{19}.$$

The next theorem gives a necessary and sufficient condition for the quartic residues (mod p) to be a difference set.

Theorem 5.3.4 *Let $p \equiv 1 \pmod{4}$ with $p > 5$. Then*
(a) *H_4 is a difference set if and only if $p = 1 + 4a^2$, for some odd integer a;*
(b) *M_4 is a difference set if and only if $p = 9 + 4b^2$, for some odd integer b.*

Proof We shall just prove part (a), as the proof of part (b) is similar.

Suppose that $p = 1 + 4a^2$, where a is an odd integer. Then $p \equiv 5 \pmod{8}$, and so Theorem 4.2.1 gives

(5.3.1) $$g(4) = \sqrt{p} \pm i(2p + 2\sqrt{p})^{1/2}.$$

Hence,

$$|g(4) - 1|^2 = g(4)\overline{g(4)} - (g(4) + \overline{g(4)}) + 1 = (3p + 2\sqrt{p}) - 2\sqrt{p} + 1 = 3p + 1.$$

It now follows from Theorem 5.2.1(a) that H_4 is a difference set.

Conversely, suppose that H_4 is a difference set. Then, by Theorem 5.3.1, we see that $p \equiv 5 \pmod 8$, and, by Theorems 5.2.1(a) and 4.2.1,

$$3p + 1 = |g(4) - 1|^2 = 3p + 1 + 2(a_4 - 1)\sqrt{p}.$$

Therefore $a_4 = 1$, and since $p \equiv 5 \pmod 8$, we have $p = 1 + 4a^2$, where a is odd. ■

Example The first example of Theorem 5.3.4(a) occurs for $p = 37$. Here the set H_4 of biquadratic residues is $\{1, 7, 9, 10, 12, 16, 26, 33, 34\}$. The first example of Theorem 5.3.4(b) is for $p = 13$. Here the set M_4 is $\{0, 1, 3, 9\}$.

We show next that the sextic residues (mod p) do not form a difference set for a prime $p \equiv 1 \pmod 6$.

Theorem 5.3.5 *If* $p \equiv 1 \pmod 6$, *then neither* H_6 *nor* M_6 *is a difference set.*

Proof If $p \equiv 1 \pmod{12}$, the result follows from Theorem 5.3.1. Thus we may suppose that $p \equiv 7 \pmod{12}$. We may clearly assume that $p > 7$. Assume that H_6 or M_6 is a difference set. Then $|g(6) + \alpha|^2 = \beta$, where α and β are certain rational integers determined by Theorem 5.2.1.

First, suppose that 2 is a cubic residue (mod p). Then, appealing to Theorem 4.1.4, we find that

$$\{g(3) + \alpha\}^2 + p^{-1}\{g^2(3) - p\}^2 = \beta.$$

Since $p^{-1}g^4(3) = 3g^2(3) + r_3 g(3)$ by Theorem 4.1.2, we obtain the simplification

$$2g^2(3) + (2\alpha + r_3)g(3) + (p + \alpha^2 - \beta) = 0,$$

which contradicts the fact that $g(3)$ has degree 3 over Q.

Second, suppose that 2 is a cubic nonresidue (mod p). Then, by Theorem 4.1.4, we obtain

$$\{g(3) + \alpha\}^2 + \frac{1}{4p}(4p - g^2(3) + Z_3\nu g(3)\{12p - 3g^2(3)\}^{1/2})^2 = \beta.$$

Expanding the squares on the left side, and using the equalities $\alpha^2 - \beta = -5p$, $g^3(3) = 3pg(3) + pr_3$, and $g^4(3) = 3pg^2(3) + pr_3 g(3)$, we obtain

$$g^2(3) + (4\alpha - r_3)g(3) - 2p = Z_3\nu(r_3 - g(3))(12p - 3g^2(3))^{1/2}.$$

After squaring both sides and eliminating the terms involving $g^3(3)$ and $g^4(3)$, we obtain quadratic polynomials in $g(3)$ on the left and right sides whose constant terms are $4p^2 + 2pr_3(4\alpha - r_3)$ and $18pr_3^2$, respectively. Since these two constant terms must be equal, $p = 5r_3^2 - 2\alpha r_3$. Thus, $r_3 \mid p$, and so $r_3 = \pm 1$. But $r_3 \equiv 1 \pmod 3$, by Theorem 3.1.3, so $r_3 = 1$. Therefore $p = 5 - 2\alpha$. By Theorem 5.2.1, $\alpha = -1$ or 5. Hence $p = 7$ or -5, which is impossible. ■

Our next theorem gives a necessary and sufficient condition for each of H_8 and M_8 to be difference sets.

Theorem 5.3.6 *Let* $p \equiv 1 \pmod 8$. *Then*

(a) H_8 *is a difference set if and only if* $p = 1 + 8c^2 = 9 + 64d^2$ *for some integers* c *and* d;

(b) M_8 *is a difference set if and only if* $p = 49 + 8e^2 = 441 + 64f^2$ *for some integers* e *and* f.

Proof We just prove part (a), as the proof of part (b) is completely analogous. If $p \equiv 1 \pmod{16}$, then by Theorem 5.3.1 H_8 is never a difference set. Thus in the remainder of the proof we may assume that $p \equiv 9 \pmod{16}$.

By Theorem 4.3.1,

$$g(8) = \sqrt{p} + R_6 \pm i\{(a_8 + \sqrt{p})(4\sqrt{p} - 2\beta^{2Z} R_6)\}^{1/2}, \tag{5.3.2}$$

where $R_6 = \pm(2p + 2a_4\sqrt{p})^{1/2}$, $a_4^2 + b_4^2 = a_8^2 + 2b_8^2 = p$, $a_4 \equiv -1 \pmod 4$, and $a_8 \equiv -1 \pmod 4$. Suppose that $p = 1 + 8c^2 = 9 + 64d^2$. Then, by Corollary 2.6.6, 2 is a quartic residue $\pmod p$, and so $\beta^{2Z} = 1$. Clearly $a_8 = -1$ and $a_4 = 3$. Thus, by (5.3.2),

$$g(8) = \sqrt{p} + R_6 \pm i\{(\sqrt{p} - 1)(4\sqrt{p} - 2R_6)\}^{1/2},$$

where $R_6 = \pm(2p + 6\sqrt{p})^{1/2}$. A short calculation gives $|g(8) - 1|^2 = 7p + 1$, and so H_8 is a difference set by Theorem 5.2.1(a).

Conversely, suppose that H_8 is a difference set. By Theorem 5.2.1(a) and (5.3.2),

$$\begin{aligned} 7p + 1 &= |g(8) - 1|^2 \\ &= 7p + 1 + \sqrt{p}\,(4a_8 + 2a_4 - 2) + 2R_6(\sqrt{p} - \beta^{2Z}\sqrt{p} - 1 - \beta^{2Z} a_8). \end{aligned}$$

Since $R_6 \notin Q(\sqrt{p})$, the coefficient of R_6 must vanish. Thus $\beta^{2Z} = 1$ (so that $4 \mid Z$) and $a_8 = -1$. Consequently, $a_4 = 3$. Therefore, $p = 1 + 2b_8^2 = 9 + b_4^2$. As $4 \mid Z$, it follows that $8 \mid b_4$ by Corollary 2.6.6, and the result follows. ∎

Example The smallest prime satisfying the conditions of Theorem 5.3.6(a) is 73. The octic residues (mod 73) are 1, 2, 4, 8, 16, 32, 37, 55, and 64. The second smallest such prime is 104,411,704,393, for which the difference set H_8 contains $(114,243)^2$ elements, and has multiplicity $\lambda = (40,391)^2$. See E. Lehmer [3].

Example The smallest prime satisfying the conditions of Theorem 5.3.6(b) is 26,041. There are no further such primes < 34,352,398,777. See Baumert [1, p. 124].

Theorem 5.3.7 *If* $p \equiv 1 \pmod{12}$, *then neither* H_{12} *nor* M_{12} *is a difference set.*

Proof For brevity, we give a proof only in the case that 2 is a cubic residue $\pmod p$. The proof when 2 is a cubic nonresidue $\pmod p$ is very similar. If $p \equiv 1 \pmod{24}$, neither H_{12} nor M_{12} is a difference set by Theorem 5.3.1. Thus we may suppose that $p \equiv 13 \pmod{24}$. Assume that H_{12} or M_{12} is a difference set. Then

$|g(12)+\alpha|^2=\beta$, where α and β are integers determined by Theorem 5.2.1. Note that $\alpha^2-\beta=-11p$. By Theorems 4.1.4 and 4.4.1,

$$g(12)=g(3)+(g^2(3)-p)/\sqrt{p}\pm i\{(2p+2a_4\sqrt{p})^{1/2}\pm g(3)(2p+2a_{12}\sqrt{p})^{1/2}/\sqrt{p}\},$$

where $a_4^2+b_4^2=p$, $a_4\equiv 1 \pmod 4$, and $a_{12}=a_4$ if $3\nmid a_4$ and $a_{12}=-a_4$ if $3\mid a_4$. Thus

$$g(12)=g(3)+(g^2(3)-p)/\sqrt{p}\pm iV(2p+2a_4\sqrt{p})^{1/2},$$

where

$$V=\begin{cases} 1\pm g(3)/\sqrt{p}, & \text{if } 3\nmid a_4,\\ 1\pm g(3)(p-a_4\sqrt{p})/(pb_4), & \text{if } 3\mid a_4.\end{cases}$$

Hence,

$$(5.3.3)\quad \beta=|g(12)+\alpha|^2=(g(3)+(g^2(3)/\sqrt{p})+\alpha-\sqrt{p})^2+2V^2(p+a_4\sqrt{p}).$$

Note that $2V^2(p+a_4\sqrt{p})$ is a quadratic polynomial in $g(3)$ with coefficients in $Q(\sqrt{p})$ and constant term $2p+2a_4\sqrt{p}$. Thus, after expanding the right side of (5.3.3) and eliminating the terms involving $g^3(3)$ and $g^4(3)$, we obtain a quadratic polynomial in $g(3)$ over $Q(\sqrt{p})$ with constant term $2p+2a_4\sqrt{p}+(\alpha-\sqrt{p})^2+2r_3\sqrt{p}$. Since $g(3)$ has degree 3 over Q and hence over $Q(\sqrt{p})$, it follows from (5.3.3) that $\beta=2p+2a_4\sqrt{p}+(\alpha-\sqrt{p})^2+2r_3\sqrt{p}$. Thus, since $\alpha^2-\beta=-11p$, we deduce that $-8p+2\sqrt{p}\,(a_4-\alpha+r_3)=0$, which is impossible. ■

Exercises 5

1. If H_k is a difference set, prove that $p\equiv k+1 \pmod{k^2}$.

2. If M_k is a difference set, prove that $p\equiv -k+1 \pmod{k^2}$.

3. Provide the details in the last part of the proof of Theorem 5.2.1.

4. Prove Theorems 5.3.4(b) and 5.3.6(b).

5. Prove Theorem 5.3.7 when 2 is a cubic nonresidue (mod p).

6. Let $p=kf+1$ be prime, and let G denote the additive group of integers (mod p). Let S be any subset of G. Show that $\cup_{s\in S}\, sH_k$ can be a difference set of G only if f is odd and k is even. (Cf. Theorems 5.3.1 and 5.3.2.)

7. In the notation of Problem 6, show that $\cup_{s\in S}\, sH_k$ is a difference set in the case $p=31$, $k=6$, and $S=\{1, 3, 27\}$.

8. Let g be a primitive root of both p and $p+2$, where p and $p+2$ are twin primes. Let G denote the group of integers modulo $p(p+2)$. Show that the $(p^2+2p-1)/2$ numbers $1,\ g,\ g^2,\dots,\ g^{(p^2-3)/2};\ 0,\ p+2,\ 2(p+2),\dots,\ (p-1)(p+2)$ in G form a difference set with $\lambda = (p^2+2p-3)/4$, where λ is as defined in Section 5.1.

Notes on Chapter 5

The use of Gauss sums in the study of power residue difference sets (see Theorem 5.2.1) is due to Berndt and Evans [1], and the proofs given here are the same as those in Berndt and Evans [1]. Other proofs in the literature which do not use Gauss sums are generally more computational in nature.

An approach to difference sets via Jacobi sums has been given by Yamamoto [3]. Yamada [3] obtained conditions, involving the vanishing of linear combinations of Jacobi sums of order k, for certain unions of cosets of kth–power residues in F_q^* to be supplementary difference sets, and he gave explicit results for $k=4$. McFarland [1] used the evaluation of the quadratic Gauss sum in proving that no nontrivial difference set exists in an Abelian group of order $4p^2$, for each prime $p>3$.

Theorem 5.3.3 is due to Paley [1]. Theorem 5.3.4(a) is a theorem of S. Chowla [3], while Theorem 5.3.4(b) was first established by E. Lehmer [3]. Proofs of Theorem 5.3.4 may also be found in the books of Mann [1, pp. 91–92] and Storer [1, pp. 49–50].

E. Lehmer's paper [3] significantly advanced the theory of power residue difference sets, by presenting the following necessary and sufficient conditions for H_k and M_k to be difference sets for a prime $p=kf+1$. Let $(s,t)_k$ denote the cyclotomic number of order k defined in Section 2.2. Then H_k is a difference set if and only if k is even, f is odd, and

$$(s,0)_k = (f-1)/k, \text{ for } s = 0,1,2,3,\dots,(k-2)/2,$$

while M_k is a difference set if and only if k is even, f is odd, and $1+(0,0)_k = (s,0)_k = (f+1)/k$ for $s=1,2,\dots,(k-2)/2$. These results formed the groundwork for much of the research in this area which followed, especially the work of Muskat and Whiteman described below. Proofs of E. Lehmer's theorems on H_k and M_k are given in Baumert's book [1, Theorem 5.17].

Theorems 5.3.5 and 5.3.6 are both due to E. Lehmer [3]. For $k=10$, E. Lehmer [3] showed the nonexistence of residue difference sets if 2 is a quintic residue (mod p). Whiteman [10] established the full theorem that H_{10} and M_{10} never form difference sets. Theorem 5.3.7 is also originally due to Whiteman [9]. For $k=14$, Muskat [4] showed the nonexistence of residue difference sets and modified residue difference sets. Whiteman [8] proved that H_{16} and M_{16} do not form difference sets when 2 is an octic residue (mod p); Evans [4] established the same result when 2 is not an octic residue (mod p). The result that H_{18} and M_{18} are never difference sets is due to Baumert and Fredricksen [1]. In cases where 5 is a biquadratic residue (mod p), Muskat and Whiteman [1] showed that H_{20} and M_{20} are not difference sets. The case when 5 is not a biquadratic residue (mod p) is open. Evans [11] has shown that if either 2 is a cubic residue or 3 is a quartic residue (mod p), then H_{24} and M_{24} are not difference sets. The case when 2 is not cubic and 3 is not quartic (mod p) remains open.

Whiteman [11] has found difference sets modulo $v = pq$ for primes p and q, generalizing Problem 8; see Baumert [1, Theorem 5.32]. D. H. and E. Lehmer [2] have related certain difference sets modulo other composite values of v to cyclotomic periods of modulus v. Baumert, Mills, and Ward [1] found all difference sets that are unions of cosets of the kth–power residues in F_q^* when $k \geq 3$, $q = p^r \equiv 1 (\text{mod } k)$, and -1 is a power of p (mod k).

Problems 6 and 7 are discussed in Baumert's book [1, p. 127]. Problem 8 is in Baumert [1, Theorem 5.27], and a generalization is in Jungnickel [2, Theorem 6.5.7].

Many applications of cyclotomic numbers to difference sets may be found in Baumert [1, pp. 119–142] and Jungnickel [1, pp. 276–280].

Chapter 6

Jacobsthal Sums over F_p

Throughout this chapter, let p denote an odd prime. In Theorem 2.1.2, we evaluated the character sum

$$\sum_{m=0}^{p-1} \left(\frac{f(m)}{p}\right), \tag{6.0.1}$$

for a quadratic polynomial $f(m)$ with integral coefficients. For polynomials $f(m)$ of degree > 2, the problem of evaluating the sum (6.0.1) is much more difficult. In fact, even for degree 3, there is no known simple closed-form evaluation of the sum (6.0.1). However, for special classes of cubics, see Stark [2], Joux and Morain [1], and Padma and Venkataraman [1].

A deep estimate for the character sum (6.0.1) has been obtained by Weil [3]. Let $f(x)$ be a polynomial of degree $n \geq 1$ with integral coefficients. Let p be an odd prime and suppose that $f(x)$ is not congruent (mod p) to $cg^2(x)$, where c is an integer and g is a polynomial with integral coefficients. Then Weil has shown that

$$\left|\sum_{m=0}^{p-1} \left(\frac{f(m)}{p}\right)\right| \leq (n-1)\sqrt{p}. \tag{6.0.2}$$

Note that the inequality (6.0.2) is nontrivial for $p > (n-1)^2$. This work has been put on a more elementary basis by Stepanov, Bombieri, and Schmidt, and the reader should consult W. M. Schmidt's book [1] for a lucid account.

The number N_p of solutions $(x, y) \in F_p \times F_p$ of the congruence $y^2 \equiv f(x) \pmod{p}$ is given by

$$N_p = \sum_{x=0}^{p-1} \left(1 + \left(\frac{f(x)}{p}\right)\right) = p + \sum_{x=0}^{p-1} \left(\frac{f(x)}{p}\right).$$

Thus the problem of determining the sum (6.0.1) is equivalent to finding N_p, and many evaluations of (6.0.1) in the literature are given as formulas for N_p.

In this chapter, we evaluate explicitly certain Jacobsthal sums, which are instances of (6.0.1), and we give some applications. As will be seen at the end of Section 6.1, Jacobsthal sums are intimately connected with Jacobi sums. Other instances of (6.0.1) will be examined in Chapter 13.

6.1 Jacobsthal sums and their elementary properties

We begin by defining the Jacobsthal sums $\phi_k(a)$ and $\psi_k(a)$.

Definition *If a is an integer not divisible by the odd prime p, and k is a positive integer, the Jacobsthal sums $\phi_k(a)$ and $\psi_k(a)$ are defined by*

$$\phi_k(a) = \sum_{m=1}^{p-1} \left(\frac{m}{p}\right)\left(\frac{m^k+a}{p}\right) = \sum_{m=0}^{p-1} \left(\frac{m}{p}\right)\left(\frac{m^k+a}{p}\right) \tag{6.1.1}$$

and

$$\psi_k(a) = \sum_{m=1}^{p-1} \left(\frac{m^k+a}{p}\right) = \sum_{m=0}^{p-1} \left(\frac{m^k+a}{p}\right) - \left(\frac{a}{p}\right). \tag{6.1.2}$$

The following result shows that we may restrict attention in (6.1.1)-(6.1.2) to the case where $p \equiv 1 \pmod{k}$.

Proposition 6.1.1 *For $d = \gcd(k, p-1)$, $\phi_k(a) = \phi_d(a)$ and $\psi_k(a) = \psi_d(a)$.*

Proof By (6.1.1),

$$\phi_k(a) = \sum_{m=1}^{p-1} \left(\frac{m^{2k}+a}{p}\right) - \sum_{m=1}^{p-1} \left(\frac{m^k+a}{p}\right). \tag{6.1.3}$$

Since the sequence $\{m^k \pmod{p}\colon 1 \le m \le p-1\}$ is just a permutation of the sequence $\{m^d \pmod{p}\colon 1 \le m \le p-1\}$, the result follows from (6.1.3) and (6.1.2). ∎

For $p \equiv 3 \pmod{4}$, the Jacobsthal sum $\phi_2(a)$ vanishes. This is a special case of the following result.

Proposition 6.1.2 *If $p \equiv k+1 \pmod{2k}$, then $\phi_k(a) = 0$.*

Proof As $(p-1)/k$ is odd, the sequence $\{m^{2k} \pmod{p}\colon 1 \le k \le p-1\}$ is just a permutation of the sequence $\{m^k \pmod{p}\colon 1 \le k \le p-1\}$. The result now follows from (6.1.3). ∎

We next evaluate $\phi_1(a)$ and $\psi_1(a)$. Later in this chapter we shall evaluate $\phi_k(a)$ for $k = 2, 3, 4, 5, 6, 8, 10, 12$, with $p \equiv 1 \pmod{k}$.

Proposition 6.1.3 *We have $\phi_1(a) = -1$.*

Proof By (6.1.1) and Theorem 2.1.2,

$$\phi_1(a) = \sum_{m=0}^{p-1} \left(\frac{m^2+am}{p}\right) = -1,$$

since $p \nmid a$. ∎

Proposition 6.1.4 *We have* $\psi_1(a) = -\left(\frac{a}{p}\right)$.

Proof This is immediate from (6.1.2). ■

In the next several results, we develop the basic properties of the Jacobsthal sums $\phi_k(a)$ for $p \equiv 1 \pmod{k}$. Write $p = kf + 1$.

Proposition 6.1.5 *If g is a primitive root* (mod p), *and n and j are integers, then*

$$\phi_k(g^{kn+j}) = (-1)^{n(k+1)}\phi_k(g^j).$$

Proof Let m^{-1} denote the inverse of $m \in F_p^*$. Then

$$\begin{aligned}
\phi_k(g^{kn+j}) &= \sum_{m=1}^{p-1} \left(\frac{m}{p}\right)\left(\frac{m^k + g^{kn+j}}{p}\right) \\
&= \left(\frac{g^n}{p}\right)^{k+1} \sum_{m=1}^{p-1} \left(\frac{m^{-1}g^n}{p}\right)^{k+1} \left(\frac{1 + (m^{-1}g^n)^k g^j}{p}\right) \\
&= (-1)^{n(k+1)} \sum_{m=1}^{p-1} \left(\frac{m^{-1}}{p}\right)^{k+1} \left(\frac{1 + m^{-k}g^j}{p}\right) \\
&= (-1)^{n(k+1)} \sum_{m=1}^{p-1} \left(\frac{m}{p}\right)\left(\frac{m^k + g^j}{p}\right) = (-1)^{n(k+1)}\phi_k(g^j).
\end{aligned}$$

■

Corollary 6.1.6 *Suppose that $f = (p-1)/k$ is even. Then*

$$\phi_k(-a) = (-1)^{f(k+1)/2}\phi_k(a).$$

Proof Put $n = (p-1)/(2k)$ in Proposition 6.1.5. Then

$$\phi_k(-g^j) = \phi_k(g^{kn+j}) = (-1)^{f(k+1)/2}\phi_k(g^j),$$

and the result follows. ■

Proposition 6.1.7 *We have*

$$\phi_k(a) = \begin{cases} \left(\frac{a}{p}\right)\phi_k(a^{-1}), & \textit{if } k \textit{ is even,} \\ \left(\frac{a}{p}\right)\psi_k(a^{-1}), & \textit{if } k \textit{ is odd.} \end{cases}$$

Proof Replacing m by m^{-1} in (6.1.1), we obtain

$$\phi_k(a) = \sum_{m\neq 0}\left(\frac{m}{p}\right)\left(\frac{m^{-k}+a}{p}\right) = \begin{cases} \left(\frac{a}{p}\right)\sum_{m\neq 0}\left(\frac{m}{p}\right)\left(\frac{a^{-1}+m^k}{p}\right), & \text{if } k \text{ is even,} \\ \left(\frac{a}{p}\right)\sum_{m\neq 0}\left(\frac{a^{-1}+m^k}{p}\right), & \text{if } k \text{ is odd,} \end{cases}$$

and the result follows. ∎

Corollary 6.1.8 *We have*

$$\phi_k(g^j) = \begin{cases} (-1)^{j+1}\phi_k(g^{k-j}), & \text{if } k \text{ is even,} \\ (-1)^{j+1}\psi_k(g^{k-j}), & \text{if } k \text{ is odd.} \end{cases}$$

Proof Taking $n = -1$ in Proposition 6.1.5, we obtain

$$\phi_k(g^{j-k}) = \begin{cases} -\phi_k(g^j), & \text{if } k \text{ is even,} \\ \phi_k(g^j), & \text{if } k \text{ is odd.} \end{cases}$$

By Proposition 6.1.7,

$$\phi_k(g^{j-k}) = \begin{cases} (-1)^j\phi_k(g^{k-j}), & \text{if } k \text{ is even,} \\ (-1)^{j+1}\psi_k(g^{k-j}), & \text{if } k \text{ is odd.} \end{cases}$$

Putting these together, we obtain the assertion of the corollary. ∎

Corollary 6.1.9 *If* $p \nmid ab$, *then*

$$\phi_k(ab^k) = \left(\frac{b}{p}\right)^{k+1}\phi_k(a).$$

Proof This follows from Proposition 6.1.5 with $a = g^j$ and $b = g^n$. ∎

Proposition 6.1.10 *If* $f = (p-1)/k$ *is odd, then* $\phi_k(a) = 0$.

Proof This is a restatement of Proposition 6.1.2. (See also Problem 1.) ∎

Proposition 6.1.11 *For* k *odd,*

$$\sum_{j=0}^{k-1}\phi(g^j) = -k.$$

Proof By Proposition 6.1.5,

$$\begin{aligned} f\sum_{j=0}^{k-1}\phi_k(g^j) &= \sum_{n=0}^{f-1}\sum_{j=0}^{k-1}\phi_k(g^{kn+j}) = \sum_{a=1}^{p-1}\phi_k(a) = \sum_{a=1}^{p-1}\sum_{m=1}^{p-1}\left(\frac{m}{p}\right)\left(\frac{m^k+a}{p}\right) \\ &= \sum_{m=1}^{p-1}\left(\frac{m}{p}\right)\sum_{a=1}^{p-1}\left(\frac{a+m^k}{p}\right) = -\sum_{m=1}^{p-1}\left(\frac{m}{p}\right)\left(\frac{m^k}{p}\right) = -\sum_{m=1}^{p-1}1 = -kf, \end{aligned}$$

and the asserted result follows on dividing by f. ■

We next prove an orthogonality relationship for the Jacobsthal sums $\phi_k(a)$. For an integer n, set

$$S(n) = \sum_{j=0}^{k-1}\phi_k(g^j)\phi_k(g^{j+n}).$$

If f is odd, then $S(n)=0$ by Proposition 6.1.10.

Theorem 6.1.12 *Let $0 \le n \le k-1$. If k is odd (so that f is even), then*

$$S(n) = \begin{cases} k^2p - k(p-1), & \text{if } n = 0, \\ -k(p-1), & \text{if } n \neq 0; \end{cases}$$

if k is even and f is even, then

$$S(n) = \begin{cases} k^2p, & \text{if } n = 0, \\ 0, & \text{if } n \neq 0. \end{cases}$$

Proof By Proposition 6.1.5,

$$\begin{aligned} fS(n) &= \sum_{m=0}^{f-1}\sum_{j=0}^{k-1}\phi_k(g^{km+j})\phi_k(g^{km+j+n}) = \sum_{r=0}^{p-2}\phi_k(g^r)\phi_k(g^{r+n}) \\ &= \sum_{i,\,j}\left(\frac{i}{p}\right)\left(\frac{j}{p}\right)\sum_{r=0}^{p-2}\left(\frac{i^k+g^r}{p}\right)\left(\frac{j^k+g^{r+n}}{p}\right) \\ &= \sum_{i,\,j}\left(\frac{ij}{p}\right)\sum_{r=0}^{p-2}\left(\frac{g^ng^{2r}+(i^kg^n+j^k)g^r+i^kj^k}{p}\right), \end{aligned}$$

so that

$$(6.1.4)\quad fS(n) = -\sum_{i,j\neq 0}\left(\frac{ij}{p}\right)^{k+1} + \left(\frac{g^n}{p}\right)\sum_{i,j\neq 0}\left(\frac{ij}{p}\right)\sum_{m}\left(\frac{m^2+(i^k+j^kg^{-n})m+i^kj^kg^{-n}}{p}\right).$$

By Theorem 2.1.2 the sum over m in (6.1.4) is equal to -1 except when $(i^k - j^kg^{-n})^2 \equiv 0 \pmod p$, that is, when

$$(6.1.5) \qquad (ij^{-1})^k \equiv g^{-n} \pmod{p},$$

in which case the sum is $p-1$. If $0 < n \leq k-1$, there are no solutions to (6.1.5). Hence (6.1.4) gives

$$(6.1.6) \qquad fS(n) = -\sum_{i,j\neq 0}\left(\frac{ij}{p}\right)^{k+1} = \begin{cases} -(p-1)^2, & \text{if } k \text{ is odd, } n \neq 0, \\ 0, & \text{if } k \text{ is even, } n \neq 0, \end{cases}$$

and the theorem follows for $n \neq 0$. Suppose next that $n = 0$. Let j ($j \neq 0$) be fixed and fix a primitive kth root of unity $r \pmod{p}$. Then $i = r^s j$, $0 \leq s \leq k-1$, gives k distinct solutions of (6.1.5). Since f is even,

$$\left(\frac{r}{p}\right) \equiv r^{(p-1)/2} = (r^k)^{f/2} \equiv 1 \pmod{p}.$$

Hence, $\left(\frac{r^s j}{p}\right) = \left(\frac{j}{p}\right)$, and the second term on the right side of (6.1.4) reduces to

$$p\sum_{j}\sum_{s=0}^{k-1}\left(\frac{j}{p}\right)^2 = p(p-1)k.$$

Combining the equality above with the calculation in (6.1.6), we deduce the desired result for $n = 0$. ■

The next theorem gives a simple relationship between the two types of Jacobsthal sums.

Theorem 6.1.13 *For each natural number k,*

$$\psi_{2k}(a) = \psi_k(a) + \phi_k(a).$$

Proof This follows from (6.1.2) and (6.1.3). ■

Because $\psi_{2k}(a)$ is so easily calculated once $\phi_k(a)$ and $\psi_k(a)$ are known, and because $\psi_k(a) = \left(\frac{a}{p}\right)\phi_k(a^{-1})$ when k is odd by Proposition 6.1.7, we shall concentrate primarily on $\phi_k(a)$ in the sequel. The next theorem relates the Jacobsthal sum $\phi_k(a)$ to Jacobi sums in the case that $p \equiv 1 \pmod{2k}$.

Theorem 6.1.14 *Let p be a prime with $p \equiv 1 \pmod{2k}$. Let χ be a character $\pmod{p}$ of order $2k$. Then*

$$\phi_k(a) = \chi(-1)\sum_{j=0}^{k-1}\chi^{k+2j+1}(a)K(\chi^{2j+1}).$$

Proof We have

$$\begin{aligned}\phi_k(a) &= \sum_m \left(\frac{m}{p}\right)\left(\frac{m^k+a}{p}\right) = \sum_m \chi^k(m)\chi^k(m^k+a)\\ &= \sum_m \chi(m^k)\chi^k(m^k+a) = \sum_m \chi(m)\chi^k(m+a)\sum_{j=0}^{k-1}\chi^{2j}(m)\\ &= \chi(-1)\chi^{k+1}(a)\sum_m \chi(m)\chi^k(1-m)\sum_{j=0}^{k-1}\chi^{2j}(am)\\ &= \chi(-1)\chi^{k+1}(a)\sum_{j=0}^{k-1}\chi^{2j}(a)J(\chi^{2j+1},\chi^k) = \chi(-1)\sum_{j=0}^{k-1}\chi^{k+2j+1}(a)K(\chi^{2j+1}),\end{aligned}$$

by Theorem 2.1.4. ■

The next theorem can be proved in exactly the same way as Theorem 6.1.14.

Theorem 6.1.15 *Let p be a prime with $p \equiv 1 \pmod{2k}$, and let χ be a character $\pmod p$ of order $2k$. Then*

$$\psi_k(a) = \left(\frac{a}{p}\right)\sum_{j=1}^{k-1}\chi^{2j}(a)K(\chi^{2j}) - \left(\frac{a}{p}\right).$$

We observe that Theorems 6.1.14 and 6.1.15 yield the estimates

$$|\phi_k(a)| < k\sqrt{p} \tag{6.1.7}$$

and

$$\left|\psi_k(a) + \left(\frac{a}{p}\right)\right| < (k-1)\sqrt{p}\ , \tag{6.1.8}$$

respectively. These are special cases of Weil's estimate (6.0.2).

6.2 Explicit determination of some Jacobsthal sums

In this section, $p = kf + 1$ is an odd prime, and g is a primitive root $\pmod p$. In evaluating Jacobsthal sums $\phi_k(a)$, it is clear from Proposition 6.1.5 that we need only consider values of $\text{ind}_g a$ with $0 \le \text{ind}_g a \le k-1$. Moreover, if k is even, then by Corollary 6.1.8, we may further restrict $\text{ind}_g a$ to $0 \le \text{ind}_g a \le k/2$. Also, if k is even and f is odd, then by Proposition 6.1.10, $\phi_k(a) = 0$, so that we need only consider the case when f is even, in which case $p \equiv 1 \pmod{2k}$. Note that the value of $\phi_1(a)$ is given in Proposition 6.1.3, so we can take $k \ge 2$.

In Theorems 6.2.1 to 6.2.8 below, we evaluate the Jacobsthal sums $\phi_k(a)$ with $p \equiv 1 \pmod{2k}$, for $k = 2, 3, 4, 5, 6, 8, 10$ and 12, respectively. In each of the proofs, we let χ denote a character $\pmod p$ of order $2k$ such that $\chi(g) = \exp(\pi i/k)$. Observe that $\chi(-1) = (-1)^{(p-1)/(2k)}$.

Theorem 6.2.1 *Let $p \equiv 1 \pmod 4$. With the integers a_4 and b_4 as in Theorem 3.2.1, the values of $\phi_2(a)$ are given as follows:*

$\phi_2(a)$	$\text{ind}_g a \pmod 4$
$2a_4(-1)^{(p-1)/4}$	0
$2b_4(-1)^{(p-1)/4}$	1
$2a_4(-1)^{(p+3)/4}$	2
$2b_4(-1)^{(p+3)/4}$	3

Proof By Theorem 6.1.14 with $k = 2$,

$$\phi_2(a) = (-1)^{(p-1)/4}(\overline{\chi}(a)K(\chi) + \chi(a)K(\overline{\chi})).$$

The theorem now easily follows from Theorem 3.2.1. ■

Theorem 6.2.2 *Let $p \equiv 1 \pmod 3$. With the integers a_3 and b_3 as in Theorem 3.1.1, the values of $\phi_3(a)$ are given as follows:*

$\phi_3(a)$	$\text{ind}_g a \pmod 3$
$-1 + 2a_3$	0
$-1 - a_3 + 3b_3$	1
$-1 - a_3 - 3b_3$	2

Proof By Theorem 6.1.14 with $k = 3$, Theorem 3.1.1, and Theorem 2.1.1 (c),

$$\begin{aligned} \phi_3(a) &= -1 + \chi(-1)\chi^4(a)K(\chi) + \chi(-1)\chi^2(a)K(\overline{\chi}) \\ &= -1 + \chi^4(a)K(\chi^2) + \chi^2(a)K(\overline{\chi}^2) = -1 + 2\text{Re}\,(\overline{\chi}^2(a)(a_3 + ib_3\sqrt{3}\,)), \end{aligned}$$

from which the result follows. ■

Theorem 6.2.3 *Let $p \equiv 1 \pmod 8$. With the integers a_8 and b_8 as in Theorem 3.3.1, the values of $\phi_4(a)$ are given as follows:*

$\phi_4(a)$	$\text{ind}_g a \pmod 8$
$4a_8(-1)^{(p-1)/8}$	0
$4b_8(-1)^{(p-1)/8}$	1
0	2

Proof By Theorem 6.1.14 with $k = 4$,

$$\phi_4(a) = (-1)^{(p-1)/8}(\overline{\chi}^3(a)K(\chi) + \overline{\chi}(a)K(\chi^3) + \chi(a)K(\overline{\chi}^3) + \chi^3(a)K(\overline{\chi})).$$

Appealing to Theorem 3.3.1, we deduce that

$$\phi_4(a) = (-1)^{(p-1)/8}\{\overline{\chi}^3(a) + \overline{\chi}(a)\}K(\chi) + \{\chi^3(a) + \chi(a)\}K(\overline{\chi}).$$

The theorem now readily follows from Theorem 3.3.1. ∎

Theorem 6.2.4 *Let* $p \equiv 1 \pmod 5$. *With the integers* a_5, b_5, c_5, *and* d_5 *as defined in Theorem* 3.7.2, *the values of* $\phi_5(a)$ *are given as follows:*

$\phi_5(a)$	$\text{ind}_g a \pmod 5$
$-1 + 4a_5$	0
$-1 - a_5 - 5b_5 + 5c_5 - 5d_5$	1
$-1 - a_5 + 5b_5 - 5c_5 - 5d_5$	2
$-1 - a_5 + 5b_5 + 5c_5 + 5d_5$	3
$-1 - a_5 - 5b_5 - 5c_5 + 5d_5$	4

Proof By Theorem 6.1.14 with $k = 5$, (3.7.4), and (3.7.14),

$$\begin{aligned}\phi_5(a) &= \left(\frac{-1}{p}\right)\{2\text{Re}\,(\overline{\chi}^4(a)K(\chi) + \overline{\chi}^2(a)K(\chi^3)) + K(\chi^5)\} \\ &= 2\text{Re}\left(\overline{\chi}^4(a)\left(a_5 + b_5\sqrt{5} + ic_5\sqrt{5+2\sqrt{5}} + id_5\sqrt{5-2\sqrt{5}}\right)\right. \\ &\quad \left.+\overline{\chi}^2(a)\left(a_5 - b_5\sqrt{5} + ic_5\sqrt{5-2\sqrt{5}} - id_5\sqrt{5+2\sqrt{5}}\right)\right) - 1.\end{aligned}$$

The theorem follows with the computations in the five cases being facilitated by (3.7.26), (3.7.29), (3.7.27), (3.7.30), and (3.7.28), respectively. ∎

Theorem 6.2.5 *Let $p \equiv 1 \pmod{12}$. With the integers a_4 and b_4 as defined in Theorem 3.2.1, the values of $\phi_6(a)$ are given as follows:*

$\phi_6(a)$ *if* $3 \mid a_4$	$\phi_6(a)$ *if* $3 \not\mid a_4$	$\mathrm{ind}_g a \pmod{12}$
$-2a_4(-1)^{(p-1)/4}$	$6a_4(-1)^{(p-1)/4}$	0
0	$4b_4(-1)^{(p-1)/4}$	1
$-4a_4(-1)^{(p-1)/4}$	0	2
$-6b_4(-1)^{(p-1)/4}$	$2b_4(-1)^{(p-1)/4}$	3

Proof By Theorem 6.1.14 with $k = 6$ and by Theorems 3.5.2 and 3.2.1,

$$\begin{aligned}(-1)^{(p-1)/4}\phi_6(a) &= \{\overline{\chi}(a) + \overline{\chi}^5(a)\}K(\chi) + \overline{\chi}^3(a)K(\chi^3)\\ &\quad + \{\chi(a) + \chi^5(a)\}K(\overline{\chi}) + \chi^3(a)K(\overline{\chi}^3)\\ &= 2\mathrm{Re}\,((\overline{\chi}(a) + \overline{\chi}^5(a))(a_{12} + ib_{12})) + 2\mathrm{Re}\,(\overline{\chi}^3(a)(a_4 + ib_4)).\end{aligned}$$

Since

$$\begin{aligned} a_{12} &= a_4, & b_{12} &= b_4, & &\text{if } 3 \not\mid a_4,\\ a_{12} &= -a_4, & b_{12} &= -b_4, & &\text{if } 3 \mid a_4,\end{aligned}$$

the table follows. ■

Theorem 6.2.6 *Let $p \equiv 1 \pmod{16}$. With the integers a_{16}, b_{16}, c_{16}, and d_{16} as defined in Theorem 3.4.1, the values of $\phi_8(a)$ are given as follows:*

$\phi_8(a)$	$\mathrm{ind}_g a \pmod{16}$
$8a_{16}(-1)^{(p-1)/16}$	0
$8c_{16}(-1)^{(p-1)/16}$	1
$8b_{16}(-1)^{(p-1)/16}$	2
$8d_{16}(-1)^{(p-1)/16}$	3
0	4

Proof Applying the automorphism $\sigma_3 \in \mathrm{Gal}(Q(e^{2\pi i/16})/Q)$ to both sides of (3.4.1), we find that

$$K(\chi^3) = K(\chi^5) = a_{16} - b_{16}\sqrt{2} - id_{16}\sqrt{2-\sqrt{2}} + ic_{16}\sqrt{2+\sqrt{2}}. \tag{6.2.1}$$

Then, by Theorem 6.1.14 with $k = 8$,

$$(-1)^{(p-1)/16}\phi_8(a) = 2\mathrm{Re}\,(\overline{\chi}(a)(1+\overline{\chi}^6(a))K(\chi)) + 2\mathrm{Re}\,(\overline{\chi}^3(a)(1+\overline{\chi}^2(a))K(\chi^3)).$$

Substituting (3.4.1) and (6.2.1) into the equation above, we obtain the values given in the table. ■

Theorem 6.2.7 *Let* $p \equiv 1 \pmod{20}$. *With the integers* a_4 *and* b_4 *as defined in Theorem* 3.2.1, *and the integers* a_{20}, b_{20}, *and* c_{20} *as defined in Theorem* 3.8.2, *the values of* $\phi_{10}(a)$ *are given in the following table:*

$\phi_{10}(a)$ *if* $5 \mid a_4$	$\phi_{10}(a)$ *if* $5 \nmid a_4$	$\mathrm{ind}_g a \pmod{20}$
$2a_4(-1)^{(p-1)/4}$	$(2a_4 + 8a_{20}/c_{20})(-1)^{(p-1)/4}$	0
$(2b_4 + i2a_{20}/c_{20})(-1)^{(p-1)/4}$	$(2b_4 + 10b_{20}/c_{20})(-1)^{(p-1)/4}$	1
$(-2a_4 + i10b_{20}/c_{20})(-1)^{(p-1)/4}$	$(-2a_4 + 2a_{20}/c_{20})(-1)^{(p-1)/4}$	2
$(-2b_4 - i2a_{20}/c_{20})(-1)^{(p-1)/4}$	$(-2b_4 + 10b_{20}/c_{20})(-1)^{(p-1)/4}$	3
$(2a_4 + i10b_{20}/c_{20})(-1)^{(p-1)/4}$	$(2a_4 - 2a_{20}/c_{20})(-1)^{(p-1)/4}$	4
$(2b_4 - i8a_{20}/c_{20})(-1)^{(p-1)/4}$	$2b_4(-1)^{(p-1)/4}$	5

Proof By Theorem 6.1.14 with $k = 10$ and Theorem 3.8.2,

(6.2.2)
$$\begin{aligned}(-1)^{(p-1)/4}\phi_{10}(a) &= 2\mathrm{Re}\,(\chi^5(a)K(\overline{\chi}^5) + \chi^9(a)K(\overline{\chi}) \\ &\quad + \chi^7(a)K(\overline{\chi}^3) + \chi^3(a)K(\overline{\chi}^3) + \chi(a)K(\overline{\chi})).\end{aligned}$$

If $5 \mid a_4$, then $K(\chi) = -K(\chi^3)$, by Lemma 3.8.1 and Theorem 3.8.2. Thus, by Theorems 3.2.1 and 3.8.2, (6.2.2) becomes

$$\begin{aligned}(-1)^{(p-1)/4}\phi_{10}(a) &= 2\mathrm{Re}\,(\chi^5(a)(a_4 - ib_4) \\ &\quad + \{\overline{\chi}(a) - \overline{\chi}^3(a) - \overline{\chi}^7(a) + \overline{\chi}^9(a)\}(a_{20} + ib_{20}\sqrt{5})/c_{20}).\end{aligned}$$

The first column of the table now follows, with the calculations facilitated by the facts that for $\beta = \exp(2\pi i/20)$,

$$\beta - \beta^3 - \beta^7 + \beta^9 = -i \quad \text{and} \quad \beta^2 + \beta^4 - \beta^6 - \beta^8 = \sqrt{5}.$$

If $5 \nmid a_4$, then $K(\chi) = K(\chi^3)$ by Lemma 3.8.1 and Theorem 3.8.2. Thus (6.2.2) yields

$$\begin{aligned}(-1)^{(p-1)/4}\phi_{10}(a) &= 2\mathrm{Re}\,(\chi^5(a)(a_4 - ib_4) \\ &\quad + \{\overline{\chi}(a) + \overline{\chi}^3(a) + \overline{\chi}^7(a) + \overline{\chi}^9(a)\}(a_{20} + ib_{20}\sqrt{5})/c_{20}).\end{aligned}$$

The second column of the table now follows in the same fashion as the first column by making use of the facts

$$\beta + \beta^3 + \beta^7 + \beta^9 = i\sqrt{5} \quad \text{and} \quad \beta^2 - \beta^4 + \beta^6 - \beta^8 = 1.$$

■

Theorem 6.2.8 *Let $p \equiv 1 \pmod{24}$. With the integers a_8 and b_8 as defined in Theorem 3.3.1, and the integers a_{24} and b_{24} as defined in Theorem 3.6.1, the values of $\phi_{12}(a)$ are given in the following table:*

$\phi_{12}(a)$	$\text{ind}_g a \pmod{24}$
$(8a_{24} + 4a_8)(-1)^{(p-1)/8}$	0
$(12b_{24} + 4b_8)(-1)^{(p-1)/8}$	1
0	2
$4b_8(-1)^{(p-1)/8}$	3
$(4a_{24} - 4a_8)(-1)^{(p-1)/8}$	4
$(12b_{24} - 4b_8)(-1)^{(p-1)/8}$	5
0	6

Proof By Theorem 6.1.14 with $k = 12$ and Theorems 3.3.1 and 3.6.1,

$$\begin{aligned}(-1)^{(p-1)/8}\phi_{12}(a) = {} & 2\text{Re}\,(\{\chi^3(a) + \chi^9(a)\}(a_8 - ib_8\sqrt{2}\,) \\ & + \{\chi(a) + \chi^5(a) + \chi^7(a) + \chi^{11}(a)\}(a_{24} - ib_{24}\sqrt{6}\,)),\end{aligned}$$

and the results follow. ■

Example We take $p = 73$ and $g = 5$. From Theorem 3.3.1 we see that $a_8 = -1$, $b_8 = \pm 6$. From Theorem 3.3.2, we find that $2b_8 \equiv -(5^9 + 5^{27}) \equiv 12 \pmod{73}$, so that $b_8 = 6$. Next, from Theorem 3.6.1, $a_{24} = -7$, $b_{24} = \pm 2$. Then, from Theorem 3.6.3, $7 \equiv b_{24}(52+52^5+52^7+52^{11}) \equiv b_{24}(52+30+17+7) \equiv 33b_{24} \pmod{73}$, so that $b_{24} = -2$. By Theorem 6.2.8,

$$\begin{aligned}\phi_{12}(5^0) &= \phi_{12}(1) = -(8a_{24} + 4a_8) = -(-56 - 4) = 60, \\ \phi_{12}(5^1) &= \phi_{12}(5) = -(12b_{24} + 4b_8) = -(-24 + 24) = 0, \\ \phi_{12}(5^2) &= \phi_{12}(25) = 0, \\ \phi_{12}(5^3) &= \phi_{12}(52) = -4b_8 = -24,\end{aligned}$$

$$\phi_{12}(5^4) = \phi_{12}(41) = -(4a_{24} - 4a_8) = -(-28+4) = 24,$$
$$\phi_{12}(5^5) = \phi_{12}(59) = -(12b_{24} - 4b_8) = -(-24-24) = 48,$$
$$\phi_{12}(5^6) = \phi_{12}(3) = 0.$$

We conclude this section by reformulating the evaluations of the Jacobsthal sums $\phi_2(n)$ and $\phi_3(n)$. These two sums have received considerable attention in the literature; see, for example, Berndt and Evans [1], S. Chowla [5], Evans [13], Hudson and Williams [3], Jacobsthal [1], [2], E. Lehmer [4], Monzingo [2], and von Schrutka [1]. The formulations below possess the advantage of making no reference to a primitive root modulo p.

Theorem 6.2.9 *Let p denote an odd prime and let n be an integer not divisible by p. If $p \equiv 3 \pmod 4$, then $\phi_2(n) = 0$. If $p \equiv 1 \pmod 4$, then for the integers a and b uniquely defined by*

(6.2.3)
$$p = a^2 + b^2, \qquad a \equiv -1 \pmod 4, \qquad b > 0,$$

we have

$$\phi_2(n) = \begin{cases} 2a, & \text{if } n^{(p-1)/4} \equiv 1 \pmod p, \\ 2b, & \text{if } n^{(p-1)/4} \equiv b/a \pmod p, \\ -2a, & \text{if } n^{(p-1)/4} \equiv -1 \pmod p, \\ -2b, & \text{if } n^{(p-1)/4} \equiv -b/a \pmod p. \end{cases}$$

Proof If $p \equiv 3 \pmod 4$, then $\phi_2(n) = 0$ by Proposition 6.1.2. Thus we may suppose that $p \equiv 1 \pmod 4$. Let g be a primitive root $\pmod p$. Then $(b/a)^2 \equiv -1 \equiv g^{(p-1)/2} \equiv (g^{(p-1)/4})^2 \pmod p$ so that $g^{(p-1)/4} \equiv \pm b/a \pmod p$. Replacing g by g^{-1}, if necessary, we may suppose that $g^{(p-1)/4} \equiv b/a \pmod p$. By Lemma 3.0.1 and Theorem 3.2.1, $a_4 = \left(\frac{2}{p}\right) a = (-1)^{(p-1)/4} a$ and $b_4 = \pm b$. Then by Theorem 3.2.2, $b_4 \equiv a_4 g^{(p-1)/4} \equiv (-1)^{(p-1)/4} b \pmod p$, so that $b_4 = (-1)^{(p-1)/4} b$. For $j \in \{0,1,2,3\}$, $n^{(p-1)/4} \equiv (b/a)^j \pmod p$ if and only if $\text{ind}_g(n) \equiv j \pmod 4$. Finally, by Theorem 6.2.1 and Proposition 6.1.5,

$$\phi_2(n) = \begin{cases} 2a_4(-1)^{(p-1)/4}, & \text{if } \text{ind}_g n \equiv 0 \pmod 4, \\ 2b_4(-1)^{(p-1)/4}, & \text{if } \text{ind}_g n \equiv 1 \pmod 4, \\ -2a_4(-1)^{(p-1)/4}, & \text{if } \text{ind}_g n \equiv 2 \pmod 4, \\ -2b_4(-1)^{(p-1)/4}, & \text{if } \text{ind}_g n \equiv 3 \pmod 4, \end{cases}$$

from which the asserted result follows. ∎

Theorem 6.2.10 *Let p denote a prime with $p > 3$ and let n be an integer not divisible by p. If $p \equiv 2 \pmod 3$, then $\phi_3(n) = -1$. If $p \equiv 1 \pmod 3$, then for the integers a and b uniquely defined by*

(6.2.4)
$$p = a^2 + 3b^2, \qquad a \equiv -1 \pmod 3, \qquad b > 0,$$

we have

$$\phi_3(n) = \begin{cases} -1+2a, & \text{if } n^{(p-1)/3} \equiv 1 \pmod p, \\ -1-a-3b, & \text{if } n^{(p-1)/3} \equiv (a-b)/2b \pmod p, \\ -1-a+3b, & \text{if } n^{(p-1)/3} \equiv (-a-b)/2b \pmod p. \end{cases}$$

Proof If $p \equiv 2 \pmod 3$, then $\phi_3(n) = -1$ by Propositions 6.1.1 and 6.1.3. Thus we may suppose that $p \equiv 1 \pmod 3$. Since $a^2/b^2 \equiv -3 \pmod p$, it is easily checked that 1, $(a-b)/2b$, and $(-a-b)/2b$ are the three distinct cube roots of unity modulo p. Let g be a primitive root $(\text{mod } p)$. As $g^{(p-1)/3}$ is a cube root of unity $(\text{mod } p)$ different from 1, it follows that $g^{(p-1)/3} \equiv (a-b)/2b$ or $(-a-b)/2b \pmod p$. Replacing g by g^{-1}, if necessary, we may suppose that $g^{(p-1)/3} \equiv (a-b)/2b \pmod p$. For $j \in \{0,1,2\}$, $n^{(p-1)/3} \equiv ((a-b)/2b)^j \pmod p$ if and only if $\text{ind}_g(n) \equiv j \pmod 3$. By Theorem 3.1.1 and Lemma 3.0.1, $a_3 = a$ and $b_3 = \pm b$. Then, by Theorem 3.1.2,

$$3b_3 \equiv (2g^{(p-1)/3}+1)a_3 \equiv \left(2\left(\frac{a-b}{2b}\right)+1\right)a \equiv \frac{a^2}{b} \equiv -3b \pmod p,$$

so that $b_3 = -b$. Finally, by Theorem 6.2.2,

$$\phi_3(n) = \begin{cases} -1+2a_3, & \text{if } \text{ind}_g n \equiv 0 \pmod 3, \\ -1-a_3+3b_3, & \text{if } \text{ind}_g n \equiv 1 \pmod 3, \\ -1-a_3-3b_3, & \text{if } \text{ind}_g n \equiv 2 \pmod 3, \end{cases}$$

from which the asserted result follows. ■

Example Take $p = 7$, so that $(p-1)/3 = 2$, $a = 2$, $b = 1$, and $(a-b)/2b \equiv 4 \pmod 7$. As $5^2 \equiv 4 \pmod 7$, Theorem 6.2.10 yields $\phi_3(5) = -1-2-3 = -6$.

6.3 Applications to the distribution of quadratic residues and nonresidues

Let p be an odd prime and let c be an integer not divisible by p. Let m be a positive integer. We are interested in determining the number of integers n with $0 \le n \le p-1$ for which the m Legendre symbols

$$(6.3.1) \qquad \left(\frac{n}{p}\right), \left(\frac{n+c}{p}\right), \left(\frac{n+2c}{p}\right), \ldots, \left(\frac{n+(m-1)c}{p}\right)$$

have a prescribed set of nonzero values. In the next two theorems, we apply Theorems 2.1.2 and 6.2.9 to determine this number explicitly when $m = 2$ and $m = 3$, respectively.

Theorem 6.3.1 *Let p be an odd prime and let c be an integer not divisible by p. Let $\epsilon_1 = \pm 1$ and $\epsilon_2 = \pm 1$. Then the number of integers n such that $0 \le n \le p-1$ and*

$$\left(\frac{n}{p}\right) = \epsilon_1, \qquad \left(\frac{n+c}{p}\right) = \epsilon_2,$$

is

$$\frac{1}{4}\left\{p - 2 - \epsilon_1\left(\frac{-c}{p}\right) - \epsilon_2\left(\frac{c}{p}\right) - \epsilon_1\epsilon_2\right\}.$$

Proof The required number is

$$\begin{aligned}
&\frac{1}{4}\sum_{n\neq 0,-c}\left\{1+\epsilon_1\left(\frac{n}{p}\right)\right\}\left\{1+\epsilon_2\left(\frac{n+c}{p}\right)\right\}\\
&= \frac{1}{4}\sum_{n\neq 0,-c}\left\{1+\epsilon_1\left(\frac{n}{p}\right)+\epsilon_2\left(\frac{n+c}{p}\right)+\epsilon_1\epsilon_2\left(\frac{n^2+cn}{p}\right)\right\}\\
&= \frac{1}{4}\left\{p-2-\epsilon_1\left(\frac{-c}{p}\right)-\epsilon_2\left(\frac{c}{p}\right)-\epsilon_1\epsilon_2\right\},
\end{aligned}$$

by Theorem 2.1.2. ■

Rabin and Shallit [1] used Theorem 6.3.1 in devising a randomizing algorithm for finding the partition $p = a^2 + b^2$ for large primes $p \equiv 1 \pmod 4$ in $O(\log p)$ expected operations.

Theorem 6.3.2 *Let p be an odd prime and let c be an integer not divisible by p. Let $\epsilon_1 = \pm 1$, $\epsilon_2 = \pm 1$, and $\epsilon_3 = \pm 1$. Then the number of integers n such that $0 \le n \le p-1$ and*

$$\left(\frac{n}{p}\right) = \epsilon_1, \qquad \left(\frac{n+c}{p}\right) = \epsilon_2, \qquad \left(\frac{n+2c}{p}\right) = \epsilon_3,$$

is

$$\begin{aligned}
&\frac{1}{8}\left\{p-3-\epsilon_1\left(\frac{-c}{p}\right)-\epsilon_1\left(\frac{-2c}{p}\right)-\epsilon_2\left(\frac{c}{p}\right)-\epsilon_2\left(\frac{-c}{p}\right)-\epsilon_3\left(\frac{2c}{p}\right)-\epsilon_3\left(\frac{c}{p}\right)-\epsilon_1\epsilon_2\right.\\
&\left.-\epsilon_1\epsilon_2\left(\frac{2}{p}\right)-\epsilon_1\epsilon_3-\epsilon_1\epsilon_3\left(\frac{-1}{p}\right)-\epsilon_2\epsilon_3-\epsilon_2\epsilon_3\left(\frac{2}{p}\right)+\epsilon_1\epsilon_2\epsilon_3\left(\frac{c}{p}\right)\left(1+\left(\frac{-1}{p}\right)\right)a_4\right\},
\end{aligned}$$

where the integer a_4 is defined in Theorem 3.2.1.

Proof The required number is

$$\frac{1}{8}\sum_{n\neq 0,-c,-2c}\left\{1+\epsilon_1\left(\frac{n}{p}\right)\right\}\left\{1+\epsilon_2\left(\frac{n+c}{p}\right)\right\}\left\{1+\epsilon_3\left(\frac{n+2c}{p}\right)\right\}$$

$$= \frac{1}{8} \sum_{n \neq 0, -c, -2c} \left\{ 1 + \epsilon_1 \left(\frac{n}{p} \right) + \epsilon_2 \left(\frac{n+c}{p} \right) + \epsilon_3 \left(\frac{n+2c}{p} \right) + \epsilon_1 \epsilon_2 \left(\frac{n^2+cn}{p} \right) \right.$$

$$\left. + \epsilon_1 \epsilon_3 \left(\frac{n^2+2cn}{p} \right) + \epsilon_2 \epsilon_3 \left(\frac{n^2+3cn+2c^2}{p} \right) + \epsilon_1 \epsilon_2 \epsilon_3 \left(\frac{n(n+c)(n+2c)}{p} \right) \right\}$$

$$= \frac{1}{8} \left\{ p - 3 - \epsilon_1 \left(\frac{-c}{p} \right) - \epsilon_1 \left(\frac{-2c}{p} \right) - \epsilon_2 \left(\frac{c}{p} \right) - \epsilon_2 \left(\frac{-c}{p} \right) - \epsilon_3 \left(\frac{2c}{p} \right) - \epsilon_3 \left(\frac{c}{p} \right) - \epsilon_1 \epsilon_2 \right.$$

$$\left. - \epsilon_1 \epsilon_2 \left(\frac{2}{p} \right) - \epsilon_1 \epsilon_3 - \epsilon_1 \epsilon_3 \left(\frac{-1}{p} \right) - \epsilon_2 \epsilon_3 - \epsilon_2 \epsilon_3 \left(\frac{2}{p} \right) + \epsilon_1 \epsilon_2 \epsilon_3 \sum_{n=0}^{p-1} \left(\frac{n(n+c)(n+2c)}{p} \right) \right\}.$$

It remains to evaluate $\sum_{n=0}^{p-1} \left(\frac{n(n+c)(n+2c)}{p} \right)$. This sum equals

$$\sum_{n=0}^{p-1} \left(\frac{(n-c)n(n+c)}{p} \right) = \sum_{n=0}^{p-1} \left(\frac{n(n^2-c^2)}{p} \right) = \phi_2(-c^2) = \left(\frac{c}{p} \right) \phi_2(-1),$$

by Corollary 6.1.9. By Theorem 6.2.9,

$$\phi_2(-1) = \begin{cases} 2a_4, & \text{if } p \equiv 1 \pmod 4, \\ 0, & \text{if } p \equiv 3 \pmod 4, \end{cases}$$

and the asserted result follows. ∎

Example Let $p = 41$, $c = 7$, $\epsilon_1 = \epsilon_2 = \epsilon_3 = 1$. Here $a_4 = -5$. By Theorem 6.3.2, there are six triples $(n, n+7, n+14)$ with $0 \leq n \leq 40$ and $\left(\frac{n}{41} \right) = \left(\frac{n+7}{41} \right) = \left(\frac{n+14}{41} \right) = 1$. They are

$$(2, 9, 16),\ (9, 16, 23),\ (18, 25, 32),\ (25, 32, 39),\ (32, 39, 5),\ (36, 2, 9).$$

If $m \geq 4$, we are unable to obtain an exact formula for the number of integers n with $0 \leq n \leq p-1$ for which the m Legendre symbols in (6.3.1) have a prescribed set of values, because we cannot explicitly determine the values of all the character sums which arise. However, it is easily seen by using the foregoing technique and Weil's character sum estimate (6.0.2) that the number is

$$\frac{1}{2^m} p + E_m(p; c), \tag{6.3.2}$$

where

$$|E_m(p; c)| \leq K(m) \sqrt{p},$$

for some positive constant $K(m)$. By simple heuristic reasoning we would expect the number to be approximately $p/2^m$ for large p. For an estimate of $K(m)$ in certain cases, see the paper of Peralta [1].

We conclude this section with a simple application of Theorem 6.3.1.

Theorem 6.3.3 *For each prime $p > 3$ we can find at least $(p-2+(-1)^{(p-1)/2})/4$ positive integers $m \le p$ such that the values of $x^{(p+1)/2} + mx$ for $x = 1, 2, \ldots, p-1$ are all quadratic nonresidues* (mod p).

Proof By Theorem 6.3.1, the number of integers n such that $0 \le n \le p-1$ and $\left(\frac{n}{p}\right) = 1$, $\left(\frac{n+2}{p}\right) = -1$, is

$$\frac{1}{4}\left\{p-2-\left(\frac{-2}{p}\right)+\left(\frac{2}{p}\right)+1\right\} \ge \frac{1}{4}\left(p-2+\left(\frac{-1}{p}\right)\right) > 0.$$

Let n be one of these integers, and set $m = n+1$, so that

$$\left(\frac{-1+m}{p}\right) = 1 \text{ and } \left(\frac{1+m}{p}\right) = -1.$$

For $x \not\equiv 0 \pmod{p}$, write

$$t = x^{(p+1)/2} + mx \quad \text{and} \quad \epsilon = \left(\frac{x}{p}\right).$$

Then by Euler's criterion,

$$\left(\frac{t}{p}\right) = \left(\frac{x}{p}\right)\left(\frac{\epsilon+m}{p}\right) = \epsilon\left(\frac{\epsilon+m}{p}\right) = -1.$$ ■

An immediate consequence of Theorem 6.3.3 is that for each prime $p > 3$ there is at least one integer m depending on p such that

$$\phi_{(p-1)/2}(m) = -(p-1).$$

For example, $\phi_3(5) = -6$ for $p = 7$, as shown at the end of Section 6.2.

6.4 Congruences for Jacobsthal sums

The main result in this section is the following theorem, which gives a congruence for the Jacobsthal sum $\phi_k(a)$ modulo p.

Theorem 6.4.1 *Let $p = 2nf+1$ be a prime with $n \ge 1$. Then*

$$\phi_n(a) \equiv -\sum_{j=0}^{[(n-1)/2]} \binom{nf}{(2j+1)f} a^{f(n-2j-1)} \pmod{p}.$$

Proof Using Euler's criterion and the binomial theorem, we find that

$$\phi_n(a) = \sum_{m \neq 0} \left(\frac{m}{p}\right)\left(\frac{m^n + a}{p}\right) \equiv \sum_{m=1}^{p-1} (m^{n+1} + ma)^{nf}$$
$$= \sum_{r=0}^{nf} \binom{nf}{r} a^{nf-r} \sum_{m=1}^{p-1} m^{nf+nr} \pmod{p}.$$

Then by (2.1.13), we obtain the congruence

$$\phi_n(a) \equiv - \sum_{\substack{r=0 \\ (p-1)|(nf+nr)}}^{nf} \binom{nf}{r} a^{nf-r} \pmod{p}.$$

Next we observe that $nf + nr$ is divisible by $p-1$ if and only if r is an odd multiple of f, that is, if and only if $r = (2j+1)f$, $0 \le j \le (n-1)/2$. The theorem now follows. ■

We next give an application of Theorem 6.4.1 when $n = 3$. In this case, $p = 6f + 1$ and

$$\phi_3(a) \equiv -1 - \binom{3f}{f} a^{2f} \pmod{p}.$$

Therefore, by Proposition 6.1.7,

$$\psi_3(a) + \left(\frac{a}{p}\right) \equiv -\binom{3f}{f} a^f \pmod{p}.$$

The left side has absolute value $< 2\sqrt{p}$, by (6.1.8). As a consequence, R. M. Robinson [1, §4] noted that the integers in the interval $(-p/2,\ p/2)$ which are the sixth roots of $\binom{3f}{f}^6 \pmod{p}$ each have absolute value $< 2\sqrt{p}$ (so that they are all relatively small). For example, $\binom{3f}{f} \equiv 194$ modulo $p = 9601$, and the sixth roots of $194^6 \pmod{9601}$ are ± 194, ± 73, and ± 121. All six of these roots have absolute value $< 195 < 2\sqrt{p}$.

In the remaining theorems of this section, we apply Theorem 6.4.1 to determine the integers a_3, a_4, a_5, a_8, a_{16}, a_{20}, and a_{24} modulo p, in terms of binomial coefficients. More general results of this kind are presented in Chapter 9.

Theorem 6.4.2 *Let $p = 4f + 1$ be a prime. Then*

$$2(-1)^f a_4 \equiv -\binom{2f}{f} \pmod{p}.$$

Proof By Theorems 6.2.1 (with $a = 1$) and 6.4.1 (with $n = 2$ and $a = 1$),

$$2a_4(-1)^f = \phi_2(1) \equiv -\binom{2f}{f} \pmod{p}.$$ ■

Example We take $p = 17$. Then $f = 4$, $a_4 = -1$, and Theorem 6.4.2 gives $2 \equiv \binom{8}{4} = 70 \pmod{17}$.

Theorem 6.4.3 *Let $p = 6f + 1$ be a prime. Then*

$$2a_3 \equiv -\binom{3f}{f} \pmod{p}.$$

Proof By Theorem 6.2.2 (with $a = 1$) and Theorem 6.4.1 (with $n = 3$ and $a = 1$),

$$-1 + 2a_3 = \phi_3(1) \equiv -1 - \binom{3f}{f} \pmod{p}.$$ ■

Theorem 6.4.4 *Let $p = 8f + 1$ be a prime. Then*

$$2(-1)^f a_8 \equiv -\binom{4f}{f} \pmod{p}.$$

Proof By Theorem 6.2.3 (with $a = 1$) and Theorem 6.4.1 (with $n = 4$ and $a = 1$),

$$4a_8(-1)^f = \phi_4(1) \equiv -\binom{4f}{f} - \binom{4f}{3f} = -2\binom{4f}{f} \pmod{p}.$$ ■

Theorem 6.4.5 *Let $p = 10f + 1$ be a prime. Then*

$$4a_5 \equiv -\binom{5f}{f} - \binom{5f}{3f} \pmod{p}.$$

Proof By Theorem 6.2.4 (with $a = 1$) and Theorem 6.4.1 (with $n = 5$ and $a = 1$),

$$-1 + 4a_5 = \phi_5(1) \equiv -\binom{5f}{f} - \binom{5f}{3f} - \binom{5f}{5f} \pmod{p}.$$ ■

Theorem 6.4.6 *Let $p = 12f + 1$ be a prime. Then*

$$2(-1)^{(f+T/2)}a_4 \equiv -\binom{6f}{f} \pmod{p},$$

where T is given in Lemma 3.5.1.

Proof Let $n = 6$ and $a = 1$ in Theorem 6.4.1. Applying Theorems 6.4.2 and 6.2.5, we find that modulo p,

$$-2\binom{6f}{f} + 2(-1)^f a_4 \equiv -2\binom{6f}{f} - \binom{6f}{3f} \equiv \begin{cases} -2(-1)^f a_4, & \text{if } 3 \mid a_4, \\ 6(-1)^f a_4, & \text{if } 3 \nmid a_4. \end{cases}$$

Then, appealing to Lemma 3.5.1, we obtain the asserted congruence. ■

Theorem 6.4.7 *Let $p = 16f + 1$ be a prime. Then*

$$4(-1)^f a_{16} \equiv -\binom{8f}{f} - \binom{8f}{3f} \pmod{p}.$$

Proof We take $n = 8$ and $a = 1$ in Theorem 6.4.1. Then, applying Theorem 6.2.6, we obtain

$$8(-1)^f a_{16} \equiv -2\binom{8f}{f} - 2\binom{8f}{3f} \pmod{p},$$

which gives the required congruence. ∎

Theorem 6.4.8 *Let $p = 20f + 1$ be a prime. Then*

$$\begin{aligned} 4(-1)^f a_{20}/c_{20} &\equiv -\binom{10f}{f} - \binom{10f}{3f} \pmod{p}, && \text{if } 5 \nmid a_4, \\ \binom{10f}{f} &\equiv -\binom{10f}{3f} \pmod{p}, && \text{if } 5 \mid a_4. \end{aligned}$$

Proof We take $n = 10$ and $a = 1$ in Theorem 6.4.1. Then, appealing to Theorems 6.2.7 and 6.4.2, we deduce that modulo p,

$$-2\binom{10f}{f} - 2\binom{10f}{3f} + 2(-1)^f a_4 \equiv \begin{cases} 2(-1)^f a_4, & \text{if } 5 \mid a_4, \\ (-1)^f(2a_4 + 8(a_{20}/c_{20})), & \text{if } 5 \nmid a_4. \end{cases}$$

The required congruences now follow. ∎

Theorem 6.4.9 *Let $p = 24f + 1$ be a prime. Then*

$$4(-1)^f a_{24} \equiv -\binom{12f}{f} - \binom{12f}{5f} \pmod{p}.$$

Proof We take $n = 12$ and $a = 1$ in Theorem 6.4.1. Then, appealing to Theorems 6.2.8 and 6.4.4, we obtain

$$(-1)^f(8a_{24} + 4a_8) \equiv -2\binom{12f}{f} - 2\binom{12f}{5f} + 4(-1)^f a_8 \pmod{p},$$

from which the asserted congruence follows. ∎

6.5 Double Jacobsthal sums

Let $p = kf + 1$ be a prime, and let a be an integer not divisible by p.

Definition *The double Jacobsthal sum $\xi_k(a)$ is defined by*

$$\xi_k(a) = \sum_{m\neq 0}\sum_{n\neq 0}\left(\frac{m^k+an}{p}\right)\left(\frac{m^k+an^2}{p}\right).$$

For $k = 3$, this sum arose in the work of D. H. Lehmer and E. Lehmer [1] on Kloosterman sums. They showed that $\xi_3(1) = 1 + 4a_3^2 - 2p$ [1, p. 21].

Our next theorem is the analogue of Theorem 6.1.14 for double Jacobsthal sums.

Theorem 6.5.1 *Let χ be a character* (mod p) *of order k. Then*

$$\xi_k(a) = 1 + \sum_{\substack{j=1\\ j\neq k/2}}^{k-1} \chi^j(-a)K(\chi^j)K(\overline{\chi}^{2j}) + \left(\frac{a}{p}\right)(1+(-1)^k)/2.$$

Proof We begin by expressing $\xi_k(a)$ in the form

(6.5.1) $$\xi_k(a) = \sum_{m\neq 0}\sum_{n\neq 0}\left(\frac{m+an}{p}\right)\left(\frac{m+an^2}{p}\right)\sum_{j=0}^{k-1}\chi^j(m) = \sum_{j=0}^{k-1} S_j(\chi),$$

where

$$S_j(\chi) = \sum_{m\neq 0}\sum_{n\neq 0}\left(\frac{m+an}{p}\right)\left(\frac{m+an^2}{p}\right)\chi^j(m).$$

Next, replacing n by mn and then replacing m by m/n^2, we find that

$$\begin{aligned} S_j(\chi) &= \sum_{m\neq 0}\sum_{n\neq 0}\left(\frac{1+an}{p}\right)\left(\frac{1+amn^2}{p}\right)\chi^j(m)\\ &= \sum_{m\neq 0}\sum_{n\neq 0}\left(\frac{1+an}{p}\right)\left(\frac{1+am}{p}\right)\chi^j(m)\overline{\chi}^{2j}(n), \end{aligned}$$

that is,

(6.5.2) $$S_j(\chi) = \left\{\sum_{m\neq 0}\left(\frac{1+am}{p}\right)\chi^j(m)\right\}\left\{\sum_{n\neq 0}\left(\frac{1+an}{p}\right)\overline{\chi}^{2j}(n)\right\}.$$

If $j = 0$, (6.5.2) becomes

$$S_0(\chi) = \left\{\sum_{m\neq 0}\left(\frac{1+am}{p}\right)\right\}\left\{\sum_{n\neq 0}\left(\frac{1+an}{p}\right)\right\} = (-1)(-1) = 1.$$

If k is even and $j = k/2$, (6.5.2) becomes

$$S_{k/2}(\chi) = \left\{\sum_{m\neq 0}\left(\frac{1+am}{p}\right)\left(\frac{m}{p}\right)\right\}\left\{\sum_{n\neq 0}\left(\frac{1+an}{p}\right)\right\} = \left\{-\left(\frac{a}{p}\right)\right\}(-1) = \left(\frac{a}{p}\right).$$

If $j \neq 0$ and $j \neq k/2$, then by Theorem 2.1.4,

$$\begin{aligned} S_j(\chi) &= \left\{ \sum_{m \neq 0} \left(\frac{1-m}{p} \right) \chi^j(-a^{-1}m) \right\} \left\{ \sum_{n \neq 0} \left(\frac{1-n}{p} \right) \overline{\chi}^{2j}(-a^{-1}n) \right\} \\ &= \left\{ \overline{\chi}^j(-a) \sum_{m \neq 0} \left(\frac{1-m}{p} \right) \chi^j(m) \right\} \left\{ \chi^{2j}(-a) \sum_{n \neq 0} \left(\frac{1-n}{p} \right) \overline{\chi}^{2j}(n) \right\} \\ &= \chi^j(-a) J(\chi^j, \rho) J(\overline{\chi}^{2j}, \rho) = \chi^j(-a) K(\chi^j) K(\overline{\chi}^{2j}). \end{aligned}$$

Substituting these values into (6.5.1), we obtain the asserted formula for $\xi_k(a)$. ■

Taking $k = 2, 3, 4$ in Theorem 6.5.1, we obtain the following evaluations of $\xi_k(a)$.

Corollary 6.5.2 *For an odd prime p and an integer a not divisible by p,*

$$\xi_2(a) = 1 + \left(\frac{a}{p} \right).$$

Corollary 6.5.3 *Let p be a prime with $p \equiv 1 \pmod 3$, and a an integer not divisible by p. Let g be a primitive root $\pmod p$. Then we have the following table of values of $\xi_3(a)$.*

$\xi_3(a)$	$\operatorname{ind}_g a \pmod 3$
$1 - 2p + 4a_3^2$	0
$1 + p - 2a_3^2 - 6a_3b_3$	1
$1 + p - 2a_3^2 + 6a_3b_3$	2

Proof Let χ be a character $\pmod p$ of order 3 with $\chi(g) = \exp(2\pi i/3)$. Then, by Theorem 6.5.1,

$$\xi_3(a) = 1 + \chi(a)K^2(\chi) + \overline{\chi}(a)K^2(\overline{\chi}) = 1 + 2\mathrm{Re}\,(\chi(a)K^2(\chi)).$$

By Theorem 3.1.1,

$$\xi_3(a) = 1 + 2\mathrm{Re}\,(\chi(a)(a_3 + ib_3\sqrt{3}\,)^2),$$

from which the asserted values follow. ■

Corollary 6.5.4 *Let $p = 4f + 1$ be a prime, and let a be an integer not divisible by p. Let g be a primitive root $\pmod p$. Then we have the following table of values of $\xi_4(a)$.*

$\xi_4(a)$	$\text{ind}_g a \pmod 4$
$2-2(-1)^f a_4$	0
$2(-1)^f b_4$	1
$2+2(-1)^f a_4$	2
$-2(-1)^f b_4$	3

Proof Let χ be a character (mod p) of order 4 with $\chi(g) = i$. Then, by Theorems 6.5.1 and 3.2.1 (recalling that $K(\chi^2) = -1$),

$$\begin{aligned}\xi_4(a) &= 1 + \left(\frac{a}{p}\right) + \chi(-a)K(\chi)K(\chi^2) + \overline{\chi}(-a)K(\overline{\chi})K(\chi^2)\\ &= 1 + \left(\frac{a}{p}\right) - 2\chi(-1)\text{Re}\,(\chi(a)K(\chi)) = 1 + \left(\frac{a}{p}\right) - 2(-1)^f\text{Re}\,(\chi(a)(a_4 + ib_4)),\end{aligned}$$

from which the asserted results follow. ■

Exercises 6

1. Deduce Proposition 6.1.10 from Corollary 6.1.9.

2. If g is a primitive root (mod p) and n and j are integers, prove that
$$\psi_k(g^{kn+j}) = (-1)^{kn}\psi_k(g^j).$$

3. If $f = (p-1)/k$ is even, show that
$$\psi_k(-a) = (-1)^{(p-1)/2}\psi_k(a).$$

4. Show that, if k is even,
$$\psi_k(a) = \left(\frac{a}{p}\right)\psi_k(a^{-1}).$$

5. For k even, show that
$$\psi_k(g^j) = (-1)^j\psi_k(g^{k-j}).$$

6. Given $p \nmid ab$, show that
$$\psi_k(ab^k) = \left(\frac{b}{p}\right)^k\psi_k(a).$$

7. Show that for a primitive root $g \pmod{p}$,

$$\sum_{j=0}^{k-1}(-1)^j \psi_k(g^j) = -k.$$

8. For each integer n, set

$$T(n) = \sum_{j=0}^{k-1} \psi_k(g^j)\psi_k(g^{j+n}),$$

where g is a primitive root $\pmod{p}$. If k is even, prove that

$$T(n) = \begin{cases} k^2p - 2k(p-1), & \text{if } n = 0, \\ -2k(p-1), & \text{if } n \text{ is even and } n \neq 0, \\ 0, & \text{if } n \text{ is odd;} \end{cases}$$

if k is odd, prove that

$$T(n) = \begin{cases} k^2p - k(p-1), & \text{if } n = 0, \\ -k(p-1), & \text{if } n \text{ is even and } n \neq 0, \\ k(p-1), & \text{if } n \text{ is odd.} \end{cases}$$

9. Let $p = 24f + 1$. Prove that

$$a_{24} = \frac{1}{4}(-1)^f \sum_m \left(\frac{m+2}{p}\right)\left(\frac{m^2-2}{p}\right)\left\{\left(\frac{m^4-4m^2+1}{p}\right) - 1\right\}.$$

10. Many sums of the form (6.0.1) can be evaluated by employing a linear fractional transformation $m = (an+b)/(cn+d)$, where a, b, c, and d are integers such that $ad - bc \neq 0$, to reduce the sum to a more familiar one. For a, b, c, and d as above, prove that if g is any arithmetical function with period p, then

$$\sum_{m \neq a/c} g(m) = \sum_{m \neq -d/c} g\left(\frac{am+b}{cm+d}\right).$$

In addition to Problem 10, the next four problems are also useful in showing that certain sums of type (6.0.1) *can be reduced to more familiar sums. In the following, assume that $a_1, \ldots, a_r$ are pairwise incongruent integers* $\pmod{p}$, *where p is an odd prime. Let*

$$F_r(a_1, \ldots, a_r) = \sum_m \left(\frac{(m+a_1)\cdots(m+a_r)}{p}\right).$$

11. Prove that

$$F_4(a_1, \ldots, a_4) = -1 + \left(\frac{c}{p}\right) F_3(0, 1, d),$$

where

$$c \equiv -(a_1 - a_4)(a_2 - a_3) \pmod{p}$$

and

$$d \equiv \frac{(a_1 - a_3)(a_2 - a_4)}{(a_1 - a_4)(a_2 - a_3)} \pmod{p}.$$

12. In general, show that the problem of evaluating $F_r(a_1, \ldots, a_r)$ can be reduced to evaluating $F_r(0, 1, b_1, \ldots, b_{r-2})$, if r is odd, and $F_{r-1}(0, 1, b_1, \ldots, b_{r-3})$, if r is even.

13. Let $f_1(x) = a_2x^2 + 2a_1x + a_0$ and $f_2(x) = b_2x^2 + 2b_1x + b_0$ with $p \nmid a_2b_2$. Put $\Delta_1 = 2(a_0a_2 - a_1^2)$, $\Delta_2 = 2(b_0b_2 - b_1^2)$, and $\Delta_{12} = a_0b_2 - 2a_1b_1 + a_2b_0$. Prove that

$$\sum_n \left(\frac{f_1(n)f_2(n)}{p}\right) = -\left(\frac{a_2b_2}{p}\right) + \left(\frac{2}{p}\right)\sum_n \left(\frac{n}{p}\right)\left(\frac{\Delta_2 n^2 + 2\Delta_{12}n + \Delta_1}{p}\right).$$

14. If $p \nmid ac$, prove that

$$\sum_n \left(\frac{n^2 + a}{p}\right)\left(\frac{n^2 + ac}{p}\right) = -1 + \left(\frac{a}{p}\right)\sum_n \left(\frac{n}{p}\right)\left(\frac{n+1}{p}\right)\left(\frac{n+c}{p}\right).$$

15. In the notation prior to Problem 11, let

$$S_n = \sum_m F_3^n(0, 1, m).$$

Prove that

$$S_2 = p^2 - 2p - 1.$$

(The values for S_4 and S_6 lie much deeper; see a paper of Yamauchi [1].)

16. Show that, for $p \nmid m$,

$$F_3(0, 1, m) = \left(\frac{m}{p}\right) F_3(0, 1, 1/m)$$

and

$$F_3(0, 1, m) = \left(\frac{-1}{p}\right) F_3(0, 1, 1 - m).$$

17. Prove that, for $p \nmid m$,

$$F_3(0, 1, m^2) = \left(\frac{m}{p}\right) F_3(0, 1, (m+1)^2/(4m)).$$

18. Prove that

$$\sum_m \left(\frac{m - n}{p}\right) F_3(0, 1, m) = 1 + p\left(\frac{n}{p}\right)\left(\frac{n-1}{p}\right).$$

19. Prove that

$$\sum_m F_3(0,1,m)F_3(0,1,m+1) = -1 - \left\{1 + \left(\frac{-1}{p}\right)\right\} p.$$

20. Show that

$$\sum_n \left(\frac{6n^4 + 11n^3 + 6n^2 + n}{p}\right) = \begin{cases} 2a_4 - \left(\frac{6}{p}\right), & \text{if } p \equiv 1 \pmod 4, \\ -\left(\frac{6}{p}\right), & \text{if } p \equiv 3 \pmod 4. \end{cases}$$

21. Show that, if $p \nmid a$,

$$\sum_n \left(\frac{(n^2 + 4an + a^2)(n^2 + 2an)}{p}\right) = \begin{cases} 2a_3 - 1, & \text{if } p \equiv 1 \pmod 3, \\ -1, & \text{if } p \equiv 2 \pmod 3. \end{cases}$$

22. Show that

$$\sum_n \left(\frac{n^4 - 8n^3 + 12n^2 - 16n + 4}{p}\right) = \begin{cases} 2a_4 - 1, & \text{if } p \equiv 1 \pmod 4, \\ -1, & \text{if } p \equiv 3 \pmod 4. \end{cases}$$

23. Show that

$$\sum_n \left(\frac{n^4 - 8n^3 + 12n^2 - 8n + 4}{p}\right) = \begin{cases} 2a_3 - 1, & \text{if } p \equiv 1 \pmod 3, \\ -1, & \text{if } p \equiv 2 \pmod 3. \end{cases}$$

24. Prove that

$$\phi_k(a) \equiv 0 \pmod k.$$

25. Evaluate $\xi_k(a)$ for $k = 5, 6, 8, 12, 16, 20$, and 24, where $\xi_k(a)$ is defined in Section 6.5.

26. Using Theorem 6.5.1, or otherwise, prove that, for $p \equiv 1 \pmod 6$,

$$\sum_{m \neq 0} \sum_{n \neq 0} \left(\frac{m + n^{-1} + 1}{p}\right)\left(\frac{m^{-1} + n + 1}{p}\right) = 4a_3^2.$$

27. Suppose that $p \nmid a$, where $p = kf + 1$. Define another type of double Jacobsthal sum $\zeta_k(a)$ by

$$\zeta_k(a) = \sum_{m \neq 0} \sum_{n \neq 0} \left(\frac{n}{p}\right)\left(\frac{m^k + an}{p}\right)\left(\frac{m^k + an^2}{p}\right).$$

Prove that, for any character $\chi \pmod p$ of order k,

$$\zeta_k(a) = \sum_{\substack{j=0 \\ j \neq k/2}}^{k-1} \chi^j(-a) K(\rho\chi^j) K(\rho\overline{\chi}^{2j}) + \left(\frac{a}{p}\right)(1 + (-1)^k)/2,$$

where ρ denotes the quadratic character $\pmod p$.

28. In the notation of Problem 27, evaluate $\zeta_k(a)$ for $k = 2, 3, 4, 5, 6, 8, 12, 16, 20$, and 24.

Notes on Chapter 6

The basic properties of Jacobsthal sums given in Section 6.1, as well as Problems 2, 7, 8, and 24 can be found in Whiteman's elegant paper [5].

The study of Jacobsthal sums commenced with Jacobsthal's paper [2] in 1907. In this paper it is shown that if p is a prime $\equiv 1 \pmod 4$, r is a quadratic residue $\pmod p$, and n is a quadratic nonresidue $\pmod p$, then $\phi_2(r)/2$ and $\phi_2(n)/2$ are integers such that

$$p = (\phi_2(r)/2)^2 + (\phi_2(n)/2)^2.$$

This result is now well known and can be found in the books of Andrews [3, pp. 135–138], S. Chowla [12, pp. 45–54], Davenport [4, p. 122], Hasse [2, pp. 167–170], and Storer [1, pp. 47–48]. Jacobsthal's formula has been generalized by Rankin [1]. S. Chowla [9] used Jacobsthal's formula to show that, for any given $\epsilon > 0$, there is an integer a depending on ϵ for which there are infinitely many primes p with $|\phi_2(a)| > (2-\epsilon)\sqrt{p}$. This shows that the constant 2 in the inequality $|\phi_2(a)| < 2\sqrt{p}$ (see (6.1.7)) is best possible.

We evaluated $\phi_2(a)$ in Theorems 6.2.1 and 6.2.9. Evaluations of $\phi_2(a)$ (in some cases only up to sign) may be found in the papers of Burde [2], S. Chowla [7], Davenport and Hasse [1], Evans [13], Hudson and Williams [3], Katre and Rajwade [7], E. Lehmer [4], Monzingo [2], Morlaye [1], Rajwade [4], Singh and Rajwade [1], and Whiteman [2], [5].

We evaluated $\phi_3(a)$ in Theorems 6.2.2 and 6.2.10. The determination of $\phi_3(1)$ was first achieved by von Schrutka [1] in 1911. Hasse's text [2, pp. 171–175] contains a proof. Evaluations of $\phi_3(a)$ (in some cases only up to sign) have been given by S. Chowla [5], Davenport and Hasse [1], Hudson and Williams [3], E. Lehmer [4], Rajwade [2], [3], Whiteman [2], [5], and Williams [14].

We evaluated $\phi_4(a)$ in Theorem 6.2.3. Evaluations of $\phi_4(a)$ (in some cases only up to sign) have been given by Hudson and Williams [4], Katre and Rajwade [5], and Whiteman [2], [5]. The determination of $\phi_5(a)$ has been discussed by E. Lehmer [1], [6], [9], Whiteman [2], [5], and Rajwade [6], [9]; see Theorem 6.2.4. The sum $\psi_6(a)$ was considered by Rajwade [7] and the sum $\phi_6(a)$ by Acharya [1]; see Theorem 6.2.5. The Jacobsthal sum $\phi_9(a)$ is discussed in Katre and Rajwade [1]. Jacobsthal sums of prime order have been considered by Katre and Rajwade [6]. See also Whiteman [3].

The proofs of Theorems 6.2.1–6.2.8 given here are essentially those given by Berndt and Evans [1], [2]. The method, based on Theorem 6.1.14, gives a unified procedure for evaluating Jacobsthal sums. A unified method of a different nature has been found by Leonard and Williams [7]. Evans [13] showed how to unambiguously evaluate $\phi_k(a)$

for k =2, 3, 4, 6, 10, and 12.

Most of the proofs mentioned in the preceding paragraphs are elementary, although the methods vary. Some of these proofs of Rajwade, however, are less elementary and are based on a theorem of Deuring [1] in the theory of elliptic curves with complex multiplication. Cf. Stark [2].

The study of consecutive quadratic residues and nonresidues began with a paper in 1896 by Aladov [1], who established Theorem 6.3.1 with $c = \epsilon_1 = \epsilon_2 = 1$. A paper 2 years later by von Sterneck [1] established additional elementary results. Theorem 6.3.2 is due to Jacobsthal [2].

In the late 1920s and early 1930s, several papers were written on distribution problems for quadratic residues and nonresidues (mod p), e.g., Brauer [1], Davenport [1], [2], Salié [2], [3], Dörge [1], and Hopf [1]. Their error term estimates, such as estimates of $E_m(p,c)$ in (6.3.2) for $m = 4$, 5, have for the most part been superseded by Weil's estimates (see (6.0.2)). Brauer [1] showed that for each pair of positive integers k and n, there exist n consecutive kth–power residues modulo some prime p. Moser [1] has a fine discussion of consecutive quadratic residues. Perron [1] discussed pairs of quadratic residues separated by c units, but with proofs different from those presented here. Other papers related to the distribution of residues and nonresidues (mod p) include Buell and Hudson [1], [2], Burde [4], [5], Caragui [1], Fabrykowski [1], Karatsuba [1], Genocchi [2], Monzingo [1], Peralta [1], Walum [1], and Williams [31]. See also the references in Lidl and Niederreiter [1, p. 257]. The distribution of quadratic residues and nonresidues modulo p among certain integer sequences is discussed in Korobov's book [1, p. 65].

Problems 10–15, which offer transformation formulas for character sums of the type in (6.0.1), are taken from the aforementioned papers of Salié, Davenport, and Hopf. A more general transformation formula for character sums has been established by Williams [3]. Special instances of this formula, in particular, special cases of Problems 11, 13, and 14, were employed by Williams [30] to evaluate several sums of the type (6.0.1). Problems 20–23 are found in his paper [30]. Problem 20, in fact, was first stated by S. Chowla [14, paper I] without proof, and first proved by Williams [29]. For sums of the type (6.0.1) that have been evaluated explicitly, see also Joux and Morain [1], Leprévost and Morain [1], Padma and Venkataraman [1], Poulakis [1], Rajwade [11], [13], Rishi, Parnami, and Rajwade [1], Parnami, Agrawal, and Rajwade [3], and Stark [2]. It is interesting to note that while the sum in (6.0.1) has been evaluated only for special cubic polynomials f, Wright [1] has evaluated the sum $\sum_{u\in F_q} \chi(f(u))$ for general cubic polynomials $f(u) \in F_q[u]$, where χ is a cubic character on F_q (so that $q := p^r \equiv 1 \pmod 3$). Wright's elegant evaluation is in terms of the cubic Jacobi sum $J_r(\chi,\chi)$. The sums $\sum \chi(f(u))$ appear in the residues of twists by cubic characters of Shintani's Dirichlet series associated with the space of binary cubic forms.

Theorem 6.4.1 can be found in Whiteman's paper [5]. The congruence of Theorem 6.4.2 was first established by Gauss [5]. The congruence in Theorem 6.4.2 can be rewritten as

$$\binom{(p-1)/2}{(p-1)/4} \equiv 2a \pmod{p},$$

where $p = a^2 + b^2$ is a prime with $p \equiv 1 \pmod 4$ and $a \equiv 1 \pmod 4$. Chowla, Dwork, and Evans [1] have proved more generally that

$$\binom{(p-1)/2}{(p-1)/4} \equiv \left(1 + \frac{2^{p-1}-1}{2}\right)\left(2a - \frac{p}{2a}\right) \pmod{p^2}.$$

We prove this and further congruences of this type in Section 9.4. Theorem 6.4.3 is due to Jacobi [2]. Theorem 6.4.4 is a theorem of Stern [1]. The proofs of Theorems 6.4.2–6.4.4 that we have given are the same as those of Whiteman [5]. In regard to Theorems 6.4.2 and 6.4.4, see also papers of Brewer [2] and Rajwade [8]. Theorem 6.4.8, in a slightly less precise form, was first stated by Brewer [2] and proved by Whiteman [13].

As previously indicated, Theorem 6.5.1 in the case $k = 3$ and $a = 1$ is due to D. H. and E. Lehmer [1]. Their proof is entirely different. They used this result to evaluate the sum of Problem 26, remarking that the evaluation of this sum took them 7 years. Mordell [4] has given a slightly more transparent evaluation. The result in Problem 26 has been generalized by Evans [7, eq. (5)].

Problem 9 is taken from Berndt and Evans [1, p. 389]. Problems 16–19 are taken from Burde [2].

The evaluation of certain Jacobsthal sums over F_{p^2} has been given by Berndt and Evans [2, §6] and over F_{p^r} by Storer [1, pp. 56, 62, 78], Katre and Rajwade [7], and Acharya [1]. Rankin [1] considered more general Jacobsthal sums and used them to deduce results on representations by sums of squares. S. D. Cohen [1] evaluated certain Jacobsthal sums over F_{q^2}, and discussed the application to the construction of projective planes of order q^4. K. Ono [2] used the evaluation of Jacobsthal sums (e.g., Theorem 6.2.1) to simply compute certain Fourier coefficients of newforms associated with elliptic curves. Using linear recurrence relations for these Fourier coefficients, Ono showed that, for every prime p, there are infinitely many rank zero elliptic curves of the form $y^2 = x^3 - D^2x$ whose Tate–Shafarevich groups do not contain an element of order p.

Chapter 7

Residuacity

Let p be a prime of the form $kf+1$, where k and f are positive integers with $k \geq 2$. In this chapter we use Gauss sums and cyclotomic numbers to provide necessary and sufficient conditions for certain primes $q(\neq p)$ to be kth–power residues modulo p. The following result on Gauss sums will be crucial in deriving such criteria for residuacity.

Lemma 7.0.1 *Let χ be a character* (mod p) *of order k. Let q be a prime with $q \neq p$. Then*

$$G^q(\chi) \equiv \overline{\chi}^q(q)G(\chi^q) \pmod{q}.$$

Proof By the multinomial theorem and Theorem 1.1.3,

$$G^q(\chi) \equiv G(q, \chi^q) = \chi^q(q^{-1})G(\chi^q) = \overline{\chi}^q(q)G(\chi^q) \pmod{q}. \qquad \blacksquare$$

When $q \equiv \pm 1 \pmod{k}$, the congruence of Lemma 7.0.1 can be simplified.

Corollary 7.0.2 *In the notation of Lemma* 7.0.1,

$$\begin{array}{ll} G^{q-1}(\chi) \equiv \overline{\chi}(q) \pmod{q}, & \text{if } q \equiv 1 \pmod{k}, \\ G^{q+1}(\chi) \equiv \chi(-q)p \pmod{q}, & \text{if } q \equiv -1 \pmod{k}. \end{array}$$

Proof If $q \equiv 1 \pmod{k}$, then $\chi^q = \chi$, and Lemma 7.0.1 gives $G^q(\chi) \equiv \overline{\chi}(q)G(\chi) \pmod{q}$. Since $p \neq q$, we may divide both sides of the congruence by $G(\chi)$, and the result follows.

If $q \equiv -1 \pmod{k}$, then $\chi^q = \overline{\chi}$, and Lemma 7.0.1 gives $G^q(\chi) \equiv \chi(q)G(\overline{\chi}) \pmod{q}$. Multiplying this congruence by $G(\chi)$, and appealing to Theorem 1.1.4(a), we obtain

$$G^{q+1}(\chi) \equiv \chi(-q)p \pmod{q},$$

as asserted. $\blacksquare$

7.1 Cubic residuacity

Throughout this section $k = 3$, $p = 3f + 1$ is a prime, and χ is a character (mod p) of order 3.

We begin by using Lemma 7.0.1 to give a necessary and sufficient condition for 2 to be a cubic residue (mod p), thus reproving part of Theorem 2.6.3, as well as Corollary 2.6.4.

Theorem 7.1.1 2 *is a cubic residue* (mod p) *if and only if* $2 \mid r_3$.

Proof By Theorem 2.1.3(a), $J(\chi,\chi) = G^2(\chi)/G(\chi^2)$, and, from Theorem 3.1.3, $J(\chi,\chi) = \frac{1}{2}(r_3 + is_3\sqrt{3}\ )$. Hence

$$\begin{aligned}
2 \text{ is a cubic residue (mod } p) &\iff \chi(2) = 1 \iff \overline{\chi}^2(2) = 1 \\
&\iff G^2(\chi) \equiv G(\chi^2) \pmod 2 \qquad \text{(by Lemma 7.0.1)} \\
&\iff J(\chi,\chi) \equiv 1 \pmod 2 \\
&\iff \tfrac{1}{2}(r_3 + is_3\sqrt{3}\ ) \equiv 1 \pmod 2 \\
&\iff r_3 \equiv s_3 \equiv 0 \pmod 2,\ r_3 + s_3 + 2 \equiv 0 \pmod 4 \\
&\iff r_3 \equiv 0 \pmod 2,
\end{aligned}$$

where the last equivalence follows because

$$4 \equiv 4p \equiv r_3^2 + 3s_3^2 \pmod 8. \qquad \blacksquare$$

If $q = k$, it is clear that Lemma 7.0.1 cannot be used to determine when q is a kth–power residue (mod p). Thus, in particular, we cannot deduce a necessary and sufficient condition for 3 to be a cubic residue (mod p) from Lemma 7.0.1. However, we have already given such a condition in Corollary 2.6.10 using the theory of cyclotomic numbers. For completeness, we restate this result here as Theorem 7.1.2.

Theorem 7.1.2 3 *is a cubic residue* (mod p) *if and only if* $9 \mid s_3$.

In the next three theorems, we give necessary and sufficient conditions for the primes 5, 7, and 11 to be cubic residues (mod p).

Theorem 7.1.3 5 *is a cubic residue* (mod p) *if and only if* $5 \mid r_3$ *or* $5 \mid s_3$.

Proof Recalling that $\chi(-1) = 1$, we see from Corollary 7.0.2, with $q = 5$ and $k = 3$, that $\chi(5) = 1$ if and only if $G^6(\chi) \equiv p \pmod 5$. By Lemma 4.1.1(a), we have $G^3(\chi) = pJ(\chi,\chi)$. Hence, $\chi(5) = 1$ if and only if $p^2J^2(\chi,\chi) \equiv p \pmod 5$, and so, by Theorem 2.1.3(b), $\chi(5) = 1$ if and only if $p^2J(\chi,\chi) \equiv J(\overline{\chi},\overline{\chi}) \pmod 5$. Thus, by Theorem 3.1.3 and Euler's criterion, 5 is a cubic residue (mod p) if and only if

$$\left(\frac{p}{5}\right)(r_3 + is_3\sqrt{3}\ ) \equiv r_3 - is_3\sqrt{3} \pmod 5,$$

or equivalently

$$\left\{\left(\frac{p}{5}\right) - 1\right\} r_3 \equiv \left\{\left(\frac{p}{5}\right) + 1\right\} s_3 \equiv 0 \pmod 5. \tag{7.1.1}$$

If $5 \mid r_3$, observe that

$$\left(\frac{p}{5}\right) = \left(\frac{4p}{5}\right) = \left(\frac{3s_3^2}{5}\right) = -1.$$

Similarly, if $5 \mid s_3$, then $\left(\frac{p}{5}\right) = 1$. Hence, the theorem follows from (7.1.1). ■

Example We take $p = 67$. Then $4p = 5^2 + 3 \cdot 9^2$, and so 5 is a cubic residue (mod 67) by Theorem 7.1.3. In fact, $5 \equiv 32^3 \pmod{67}$.

Corollary 7.1.4 *5 is a cubic residue* (mod p) *if and only if*

$$5 \mid a_3,\ 3 \mid b_3 \quad or \quad 15 \mid b_3 \quad or \quad 15 \mid a_3 \pm b_3 \quad or \quad 15 \mid 2a_3 \pm b_3.$$

Proof If $5 \mid a_3$, $3 \mid b_3$, then by (3.1.2) and (3.1.3), $r_3 = 2a_3$, so that $5 \mid r_3$, and, by Theorem 7.1.3, 5 is a cubic residue (mod p).

If $15 \mid b_3$, then by (3.1.2) and (3.1.3), $s_3 = 2b_3$, so that $5 \mid s_3$, and, by Theorem 7.1.3, 5 is a cubic residue (mod p).

If $15 \mid a_3 + b_3$, then, as $a_3 \equiv -1 \pmod 3$, we have $b_3 \equiv 1 \pmod 3$, so that $s_3 = -a_3 - b_3$, and thus $5 \mid s_3$, which proves that 5 is a cubic residue (mod p).

If $15 \mid a_3 - b_3$, then, as $a_3 \equiv -1 \pmod 3$, we have $b_3 \equiv 2 \pmod 3$, so that $s_3 = a_3 - b_3$, and thus $5 \mid s_3$, which proves that 5 is a cubic residue (mod p).

If $15 \mid 2a_3 + b_3$, then, as $a_3 \equiv -1 \pmod 3$, we have $b_3 \equiv 2 \pmod 3$, so that $r_3 = -a_3 - 3b_3$, and thus $5 \mid r_3$, which proves that 5 is a cubic residue (mod p).

If $15 \mid 2a_3 - b_3$, then, as $a_3 \equiv -1 \pmod 3$, we have $b_3 \equiv 1 \pmod 3$, so that $r_3 = -a_3 + 3b_3$, and thus $5 \mid r_3$, which proves that 5 is a cubic residue (mod p).

Conversely, suppose that 5 is a cubic residue (mod p), so that, by Theorem 7.1.3, either $5 \mid r_3$ or $5 \mid s_3$. It is easy to check the following implications from (3.1.2) and (3.1.3), as $a_3 \equiv -1 \pmod 3$:

$$\begin{aligned}
&5 \mid r_3,\ 3 \mid b_3 \Longrightarrow 5 \mid a_3,\\
&5 \mid r_3,\ b_3 \equiv 1 \pmod 3 \Longrightarrow 15 \mid 2a_3 - b_3,\\
&5 \mid r_3,\ b_3 \equiv 2 \pmod 3 \Longrightarrow 15 \mid 2a_3 + b_3,\\
&5 \mid s_3,\ 3 \mid b_3 \Longrightarrow 15 \mid b_3,\\
&5 \mid s_3,\ b_3 \equiv 1 \pmod 3 \Longrightarrow 15 \mid a_3 + b_3,\\
&5 \mid s_3,\ b_3 \equiv 2 \pmod 3 \Longrightarrow 5 \mid a_3 - b_3.
\end{aligned}$$

This completes the proof of Corollary 7.1.4. ■

Example We take $p = 67$. Then $p = 8^2 + 3 \cdot 1^2$, so that $a_3 = 8$, $|b_3| = 1$. As $15 \mid 2a_3 \pm b_3$, it follows from Corollary 7.1.4 that 5 is a cubic residue (mod 67).

Theorem 7.1.5 *7 is a cubic residue* (mod p) *if and only if* $7 \mid r_3$ *or* $7 \mid s_3$.

Proof By Corollary 7.0.2 with $q = 7$ and $k = 3$, and Lemma 4.1.1, we see that $\chi(7) = 1$ if and only if $G^6(\chi) = p^2 J^2(\chi, \chi) \equiv 1 \pmod 7$, or, equivalently, $p^3 J(\chi, \chi) \equiv$

$J(\overline{\chi}, \overline{\chi}) \pmod 7$. It follows from Theorem 3.1.3 and Euler's criterion that 7 is a cubic residue (mod p) if and only if

(7.1.2) $$\left\{\left(\frac{p}{7}\right) - 1\right\} r_3 \equiv \left\{\left(\frac{p}{7}\right) + 1\right\} s_3 \equiv 0 \pmod 7.$$

By an argument similar to that following (7.1.1), it is easily shown that, if $7 \mid r_3$, then $\left(\frac{p}{7}\right) = -1$, and if $7 \mid s_3$, then $\left(\frac{p}{7}\right) = 1$. The theorem now follows from (7.1.2). ■

Theorem 7.1.6 11 *is a cubic residue* (mod p) *if and only if* $11 \mid (r_3^2 - 3s_3^2)$ *or* $11 \mid r_3 s_3$.

Proof By Corollary 7.0.2 with $q = 11$ and $k = 3$, and Lemma 4.1.1, we see that $\chi(11) = 1$ if and only if $G^{12}(\chi) = p^4 J^4(\chi, \chi) \equiv p \pmod{11}$, or, equivalently, $p^5 J^2(\chi, \chi) \equiv J^2(\overline{\chi}, \overline{\chi}) \pmod{11}$. Hence, by Theorem 3.1.3 and Euler's criterion, 11 is a cubic residue (mod p) if and only if

$$\left\{\left(\frac{p}{11}\right) - 1\right\} (r_3^2 - 3s_3^2) \equiv \left\{\left(\frac{p}{11}\right) + 1\right\} r_3 s_3 \equiv 0 \pmod{11}.$$

The proof can now be completed by the same type of argument as in the previous proof. ■

Example We take $p = 61$. Then $4p = 1^2 + 3 \cdot 9^2$. Note that $11 \mid (1^2 - 3 \cdot 9^2)$, and so 11 is a cubic residue (mod 61). In fact, $11 \equiv 32^3 \pmod{61}$.

It is clear that the method employed in the proofs of Theorems 7.1.3, 7.1.5, and 7.1.6 can be utilized to give a necessary and sufficient condition for any prime $q \neq 2, 3, p$ to be a cubic residue (mod p). Our next theorem gives this condition.

Theorem 7.1.7 *Let p be a prime with* $p \equiv 1 \pmod 3$. *Let q be a prime with* $q \neq 2, 3, p$, *so that* $q \equiv \pm 1 \pmod 6$. *Define* $\epsilon = \pm 1$ *by* $q \equiv \epsilon \pmod 6$. *Then q is a cubic residue* (mod p) *if and only if*

$$\left(\frac{r_3 + i s_3 \sqrt{3}}{r_3 - i s_3 \sqrt{3}}\right)^{(q-\epsilon)/6} \equiv \left(\frac{p}{q}\right) \pmod q.$$

Proof From Corollary 7.0.2 with $k = 3$, we deduce that $\chi(q) = 1$ if and only if

(7.1.3) $$G^{q-\epsilon}(\chi) \equiv \begin{cases} 1 \pmod q, & \text{if } q \equiv 1 \pmod 3, \\ p \pmod q, & \text{if } q \equiv 2 \pmod 3. \end{cases}$$

By Lemma 4.1.1(a), $G^3(\chi) = pJ(\chi, \chi)$, so that (7.1.3) is equivalent to

(7.1.4) $$p^{(q-\epsilon)/3} J^{(q-\epsilon)/3}(\chi, \chi) \equiv \begin{cases} 1 \pmod q, & \text{if } q \equiv 1 \pmod 3, \\ p \pmod q, & \text{if } q \equiv 2 \pmod 3. \end{cases}$$

Since $J(\chi,\chi)J(\overline{\chi},\overline{\chi}) = p$, (7.1.4) is equivalent to

$$p^{(q-1)/2} J^{(q-\epsilon)/6}(\chi,\chi) \equiv J^{(q-\epsilon)/6}(\overline{\chi},\overline{\chi}) \pmod{q}.$$

Then, by Euler's criterion and Theorem 3.1.3, we obtain the asserted result. ■

7.2 Quartic residuacity

In this section, $k = 4$, $p = 4f + 1$ is a prime, and χ is a character (mod p) of order 4.

Corollary 2.6.6 gives a necessary and sufficient condition for 2 to be a quartic residue (mod p). We will return to this topic in Section 7.5, where we also give a necessary and sufficient condition for 2 to be an octic residue (mod p), as well as one for 2 to be a bioctic residue (mod p). In the next three theorems, we give necessary and sufficient conditions for -3, 3 and 5 to be quartic residues (mod p). Theorem 7.2.4 then gives a criterion for the quartic residuacity of any odd prime $q \neq p$.

Theorem 7.2.1 *A necessary and sufficient condition for* -3 *to be a quartic residue* (mod p) *is that* $3 \mid b_4$.

Proof By Corollary 7.0.2 with $k = 4$ and $q = 3$, we see that $\chi(-3) = 1$ if and only if $G^4(\chi) \equiv p \pmod{3}$. By Theorem 2.1.3(a),

$$J(\chi,\chi) = \frac{G^2(\chi)}{G(\chi^2)} = \frac{G^2(\chi)}{\sqrt{p}},$$

so that

$$G^4(\chi) = pJ^2(\chi,\chi) = p\frac{K^2(\chi)}{\chi^2(4)} = p\frac{K^2(\chi)}{\chi^4(2)} = pK^2(\chi).$$

Thus the congruence $G^4(\chi) \equiv p \pmod{3}$ can be written as $K^2(\chi) \equiv 1 \pmod{3}$, or, equivalently, as $pK(\chi) \equiv K(\overline{\chi}) \pmod{3}$. Thus, by Theorem 3.2.1, -3 is a quartic residue (mod p) if and only if

(7.2.1) $$a_4(p-1) \equiv b_4(p+1) \equiv 0 \pmod{3}.$$

These congruences can hold only when $3 \mid b_4$, in which case $p \equiv 1 \pmod{3}$. ■

Theorem 7.2.2 *3 is a quartic residue* (mod p) *if and only if either* $3 \mid b_4$ *and* $p \equiv 1 \pmod{8}$ *or* $3 \mid a_4$ *and* $p \equiv 5 \pmod{8}$.

Proof For $p \equiv 1 \pmod{8}$, the result follows immediately from Theorem 7.2.1. In the case that $p \equiv 5 \pmod{8}$, $\chi(-1) = -1$. Proceeding then by the same argument as in the proof of Theorem 7.2.1, we find, in place of the congruence (7.2.1), that 3 is a quartic residue (mod p) if and only if

$$a_4(p+1) \equiv b_4(p-1) \equiv 0 \pmod{3}.$$

These congruences can hold only if $3 \mid a_4$, in which case $p \equiv 1 \pmod{3}$. ■

Theorem 7.2.3 *5 is a quartic residue* (mod p) *if and only if* $5 \mid b_4$.

Proof By Corollary 7.0.2 with $q = 5$ and $k = 4$, we see that $\chi(5) = 1$ if and only if $G^4(\chi) = pK^2(\chi) \equiv 1 \pmod 5$, or, equivalently, $p^2K(\chi) \equiv K(\overline{\chi}) \pmod 5$. Thus, by Theorem 3.2.1 and Euler's criterion, 5 is a quartic residue (mod p) if and only if

$$\left(\frac{p}{5}\right)(a_4 + ib_4) \equiv a_4 - ib_4 \pmod 5,$$

or

$$\left\{\left(\frac{p}{5}\right) - 1\right\} a_4 \equiv \left\{\left(\frac{p}{5}\right) + 1\right\} b_4 \equiv 0 \pmod 5.$$

Since $p = a_4^2 + b_4^2$,

$$5 \mid a_4 \Longrightarrow 5 \nmid b_4 \text{ and } \left(\frac{p}{5}\right) = \left(\frac{b_4^2}{5}\right) = 1$$

and

$$5 \mid b_4 \Longrightarrow 5 \nmid a_4 \text{ and } \left(\frac{p}{5}\right) = \left(\frac{a_4^2}{5}\right) = 1,$$

so that 5 is a quartic residue (mod p) if and only if $5 \mid b_4$. ■

Example Take $p = 269$. Then $p = 13^2 + 10^2$. By Theorem 7.2.3, 5 is a quartic residue (mod 269). In fact, $5 \equiv 113^4 \pmod{269}$.

Theorem 7.2.4 *Let p be a prime with $p \equiv 1 \pmod 4$. Let q be a prime with $q \neq 2, p$. Define $\epsilon = \pm 1$ by $q \equiv \epsilon \pmod 4$. Then $(-1)^{(q-1)/2}q$ is a quartic residue* (mod p) *if and only if*

$$\left(\frac{a_4 + ib_4}{a_4 - ib_4}\right)^{(q-\epsilon)/4} \equiv \left(\frac{p}{q}\right) \pmod q.$$

Proof From Corollary 7.0.2 with $k = 4$, we see that $(-1)^{(q-1)/2}q$ is a quartic residue (mod p) if and only if

$$G^{q-\epsilon}(\chi) \equiv \begin{cases} 1 \pmod q, & \text{if } q \equiv 1 \pmod 4, \\ p \pmod q, & \text{if } q \equiv 3 \pmod 4. \end{cases} \tag{7.2.2}$$

As $G^4(\chi) = pK^2(\chi)$, the congruence (7.2.2) is equivalent to

$$p^{(q-\epsilon)/4}K^{(q-\epsilon)/2}(\chi) \equiv \begin{cases} 1 \pmod q, & \text{if } q \equiv 1 \pmod 4, \\ p \pmod q, & \text{if } q \equiv 3 \pmod 4. \end{cases} \tag{7.2.3}$$

Since $K(\chi)K(\overline{\chi}) = p$, (7.2.3) is equivalent to

$$p^{(q-1)/2}K^{(q-\epsilon)/4}(\chi) \equiv K^{(q-\epsilon)/4}(\overline{\chi}) \pmod q. \tag{7.2.4}$$

The asserted result now follows from Euler's criterion and Theorem 3.2.1. ■

7.3 Octic residuacity

In this section, $k = 8$, $p = 8f + 1$ is a prime, and χ is a character (mod p) of order 8. Recall from Theorems 3.2.1 and 3.3.1 that $K(\chi^2) = a_4 + ib_4$ and $K(\chi) = a_8 + ib_8\sqrt{2}$. We begin this section by giving a criterion for 3 to be an octic residue (mod p). Theorem 7.3.2 then gives a criterion for the octic residuacity of any odd prime $q \neq p$. The octic residuacity of 2 is discussed in Section 7.5.

Theorem 7.3.1 *3 is an octic residue* (mod p) *if and only if* $3 \mid b_4$, $3 \mid b_8$, *and* $(-1)^f a_4 \equiv a_8 \pmod 3$.

Proof From Lemma 7.0.1 with $q = 3$ and $k = 8$,

$$G^3(\chi) \equiv \chi^5(3)G(\chi^3) \pmod 3.$$

By Theorem 3.3.3, we have modulo 3

$$\begin{aligned}(a_4 + ib_4)(a_8 + ib_8\sqrt{2}\,) &= K(\chi)K(\chi^2) = (\chi(4)J(\chi,\chi))(\chi(-4)J(\chi,\chi^2)) \\ &= \chi(-1)J(\chi,\chi)J(\chi,\chi^2) = \chi^5(-1)G^3(\chi)/G(\chi^3) \equiv \chi^5(-3).\end{aligned}$$

Then, as $(a_4 - ib_4)(a_4 + ib_4) = p$,

$$p(a_8 + ib_8\sqrt{2}\,) \equiv \chi^5(-3)(a_4 - ib_4) \pmod 3. \tag{7.3.1}$$

Thus, $\chi(3) = 1$ if and only if

$$0 \equiv pa_8 - (-1)^f a_4 \equiv b_8 \equiv b_4 \pmod 3.$$

Further, $3 \mid b_4$ implies that $p = a_4^2 + b_4^2 \equiv a_4^2 \equiv 1 \pmod 3$, and so

$$3 \mid b_4,\ 3 \mid b_8,\ (-1)^f a_4 \equiv pa_8 \pmod 3 \iff 3 \mid b_4,\ 3 \mid b_8,\ (-1)^f a_4 \equiv a_8 \pmod 3. \ \blacksquare$$

Example Let $p = 313$. Here, $f = 39$, $a_4 = -13$, $a_8 = -5$, and $|b_4| = |b_8| = 12$. Since $3 \mid b_4$, $3 \mid b_8$, and $(-1)^f a_4 = 13 \equiv -5 = a_8 \pmod 3$, 3 is an octic residue (mod 313), by Theorem 7.3.1. In fact, $3 \equiv 14^8 \pmod{313}$.

By the same type of reasoning as in the proof of Theorem 7.3.1, we can obtain a criterion for any odd prime $q \neq p$ to be an octic residue (mod p).

Theorem 7.3.2 *Let* $p = 8f + 1$ *be a prime. Let* q *be a prime with* $q \neq 2, p$. *Then* q *is an octic residue* (mod p) *if and only if*

$$\begin{aligned}
&p^{(q-1)/8}(a_8 + ib_8\sqrt{2}\,)^{(q-1)/2}(a_4 + ib_4)^{(q-1)/4} \equiv 1 \pmod q, && \text{if } q \equiv 1 \pmod 8,\\
&(-1)^f p^{(q-3)/8}(a_8 + ib_8\sqrt{2}\,)^{(q-1)/2}(a_4 + ib_4)^{(q+1)/4} \equiv 1 \pmod q, && \text{if } q \equiv 3 \pmod 8,\\
&p^{(q-5)/8}(a_8 + ib_8\sqrt{2}\,)^{(q+1)/2}(a_4 + ib_4)^{(q-1)/4} \equiv 1 \pmod q, && \text{if } q \equiv 5 \pmod 8,\\
&(-1)^f p^{(q-7)/8}(a_8 + ib_8\sqrt{2}\,)^{(q+1)/2}(a_4 + ib_4)^{(q+1)/4} \equiv 1 \pmod q, && \text{if } q \equiv 7 \pmod 8.
\end{aligned}$$

Proof First we recall from the definition of $K(\chi)$, and Theorems 2.1.4, 2.1.5, and 3.3.3, the following expressions for the Jacobi sums $J(\chi, \chi^j)$ $(j = 1, 2, \ldots, 6)$ in terms of $K(\chi)$ and $K(\chi^2)$:

$$\begin{array}{lll} J(\chi,\chi) = \overline{\chi}(4)K(\chi), & J(\chi,\chi^2) = \overline{\chi}(-4)K(\chi^2), & J(\chi,\chi^3) = \chi(-1)K(\chi), \\ J(\chi,\chi^4) = K(\chi), & J(\chi,\chi^5) = \overline{\chi}(4)K(\chi^2), & J(\chi,\chi^6) = \overline{\chi}(-4)K(\chi). \end{array}$$

We use these expressions to determine the quantities $G^3(\chi)/G(\chi^3)$, $G^5(\chi)/G(\chi^5)$, $G^7(\chi)/G(\chi^7)$, and $G^8(\chi)$ in terms of $K(\chi)$ and $K(\chi^2)$. Since $\chi(-1) = (-1)^f$ and $\chi^2(4) = 1$,

$$\begin{aligned} \frac{G^3(\chi)}{G(\chi^3)} &= \frac{G^2(\chi)}{G(\chi^2)} \frac{G(\chi)G(\chi^2)}{G(\chi^3)} = J(\chi,\chi)J(\chi,\chi^2) = (-1)^f K(\chi)K(\chi^2); \\ \frac{G^5(\chi)}{G(\chi^5)} &= \prod_{j=1}^{4} \frac{G(\chi)G(\chi^j)}{G(\chi^{j+1})} = \prod_{j=1}^{4} J(\chi,\chi^j) = K^3(\chi)K(\chi^2); \\ \frac{G^7(\chi)}{G(\chi^7)} &= \prod_{j=1}^{6} \frac{G(\chi)G(\chi^j)}{G(\chi^{j+1})} = \prod_{j=1}^{6} J(\chi,\chi^j) = (-1)^f K^4(\chi)K^2(\chi^2); \\ G^8(\chi) &= (-1)^f p \frac{G^7(\chi)}{G(\chi^7)} = pK^4(\chi)K^2(\chi^2). \end{aligned}$$

We now examine the four cases $q \equiv 1, 3, 5, 7 \pmod 8$ separately.

Case 1: $\boldsymbol{q \equiv 1 \pmod 8}$. By Corollary 7.0.2,

$$\begin{aligned} & q \text{ is an octic residue} \pmod p \\ & \iff G^{q-1}(\chi) \equiv 1 \pmod q \\ & \iff p^{(q-1)/8} K^{(q-1)/2}(\chi) K^{(q-1)/4}(\chi^2) \equiv 1 \pmod q \\ & \iff p^{(q-1)/8} (a_8 + ib_8\sqrt{2}\,)^{(q-1)/2} (a_4 + ib_4)^{(q-1)/4} \equiv 1 \pmod q. \end{aligned}$$

Case 2: $\boldsymbol{q \equiv 3 \pmod 8}$. By Lemma 7.0.1,

$$\begin{aligned} & q \text{ is an octic residue} \pmod p \\ & \iff G^{q}(\chi) \equiv G(\chi^3) \pmod q \\ & \iff G^{q-3}(\chi) \frac{G^3(\chi)}{G(\chi^3)} \equiv 1 \pmod q \\ & \iff p^{(q-3)/8} K^{(q-3)/2}(\chi) K^{(q-3)/4}(\chi^2) (-1)^f K(\chi) K(\chi^2) \equiv 1 \pmod q \\ & \iff (-1)^f p^{(q-3)/8} K^{(q-1)/2}(\chi) K^{(q+1)/4}(\chi^2) \equiv 1 \pmod q \\ & \iff (-1)^f p^{(q-3)/8} (a_8 + ib_8\sqrt{2}\,)^{(q-1)/2} (a_4 + ib_4)^{(q+1)/4} \equiv 1 \pmod q. \end{aligned}$$

Case 3: $\boldsymbol{q \equiv 5 \pmod 8}$. By Lemma 7.0.1,

q is an octic residue (mod p)

$$\iff G^q(\chi) \equiv G(\chi^5) \pmod q$$
$$\iff G^{q-5}(\chi)\frac{G^5(\chi)}{G(\chi^5)} \equiv 1 \pmod q$$
$$\iff p^{(q-5)/8}K^{(q-5)/2}(\chi)K^{(q-5)/4}(\chi^2)K^3(\chi)K(\chi^2) \equiv 1 \pmod q$$
$$\iff p^{(q-5)/8}K^{(q+1)/2}(\chi)K^{(q-1)/4}(\chi^2) \equiv 1 \pmod q$$
$$\iff p^{(q-5)/8}(a_8 + ib_8\sqrt{2}\,)^{(q+1)/2}(a_4 + ib_4)^{(q-1)/4} \equiv 1 \pmod q.$$

Case 4: $q \equiv 7 \pmod 8$. By Corollary 7.0.2,

q is an octic residue (mod p)

$$\iff G^{q+1}(\chi) \equiv (-1)^f p \pmod q$$
$$\iff p^{(q+1)/8}K^{(q+1)/2}(\chi)K^{(q+1)/4}(\chi^2) \equiv (-1)^f p \pmod q$$
$$\iff (-1)^f p^{(q-7)/8}K^{(q+1)/2}(\chi)K^{(q+1)/4}(\chi^2) \equiv 1 \pmod q$$
$$\iff (-1)^f p^{(q-7)/8}(a_8 + ib_8\sqrt{2}\,)^{(q+1)/2}(a_4 + ib_4)^{(q+1)/4} \equiv 1 \pmod q. \qquad \blacksquare$$

Let p be a prime with $p \equiv 1 \pmod 8$. We now use Theorem 7.3.2 to determine a necessary and sufficient condition for 5 to be an octic residue (mod p), given that it is a quartic residue (mod p). As 5 is a quartic residue (mod p), Theorem 7.2.3 implies that $5 \mid b_4$, and thus $p \equiv a_4^2 \equiv \pm 1 \pmod 5$. Reducing $p = a_8^2 + 2b_8^2$ modulo 5, we obtain the following table of numerical values (mod 5):

p	a_8^2	b_8^2	$(4a_8^2 - p)b_8$	$(4a_8^2 - 3p)a_8^2$
1	1	0	0	1
1	4	1	0	2
4	1	4	0	2
4	4	0	0	1

Then, by Theorem 7.3.2,

5 is an octic residue (mod p)

$$\iff (a_8 + ib_8\sqrt{2}\,)^3(a_4 + ib_4) \equiv 1 \pmod 5 \iff (a_8 + ib_8\sqrt{2}\,)^3 a_4 \equiv 1 \pmod 5$$
$$\iff \{(a_8^3 - 6a_8b_8^2) + i(3a_8^2b_8 - 2b_8^3)\sqrt{2}\,\}a_4 \equiv 1 \pmod 5$$
$$\iff \{(4a_8^2 - 3p)a_8 + i(4a_8^2 - p)b_8\sqrt{2}\,\}a_4 \equiv 1 \pmod 5$$
$$\iff (4a_8^2 - 3p)a_8a_4 \equiv 1 \pmod 5 \iff (4a_8^2 - 3p)a_8^2a_4 \equiv a_8 \pmod 5$$
$$\iff a_8 \equiv a_4 \text{ or } 2a_4 \pmod 5, \text{ according as } p \equiv a_8^2 \text{ or } b_8^2 \pmod 5.$$

A few numerical examples are given below.

p	a_4	a_8	$a_8/a_4 \pmod 5$	
401	-1	3	2	$5 \equiv 73^8 \pmod{401}$
409	3	11	2	$5 \equiv 263^8 \pmod{409}$
449	7	-21	2	$5 \equiv 274^8 \pmod{449}$
521	11	3	3	$5 \equiv 63^4 \pmod{521}$
569	-13	-21	2	$5 \equiv 252^8 \pmod{569}$
761	19	27	3	$5 \equiv 305^4 \pmod{761}$
929	23	27	4	$5 \equiv 409^4 \pmod{929}$

We conclude this section by giving a proof of Lemma 3.6.4.

Proof of Lemma 3.6.4 Let $p = 24f + 1$ and $T = \text{ind}_g 3$ for a primitive root $g \pmod p$. Then $2 \mid T$, $3 \mid a_4 b_4$, $3 \mid b_8$, and $3 \nmid a_8$. First we examine the case $4 \mid T$. By Theorem 7.2.2, $3 \mid b_4$, and thus $3 \nmid a_4$. Further, by Theorem 7.3.1,

$$\begin{array}{ll} 8 \mid T, & \text{if } (-1)^f a_4 \equiv a_8 \pmod 3, \\ 8 \nmid T, & \text{if } (-1)^f a_4 \equiv -a_8 \pmod 3, \end{array}$$

so that

$$a_8 a_4 \equiv (-1)^{f+T/4} \pmod 3, \quad \text{if } 4 \mid T.$$

Second, we examine the case $4 \nmid T$. By Theorem 7.2.2, $3 \nmid b_4$, so that $3 \mid a_4$. For a character χ such that $\chi(g) = \exp(2\pi i/8)$,

$$\chi(3) = i(-1)^{(T-2)/4}, \quad \chi(-3) = i(-1)^{f+(T-2)/4},$$

and so by (7.3.1),

$$a_8 \equiv i(-1)^{f+(T-2)/4}(-ib_4) \pmod 3.$$

Thus

$$a_8 b_4 \equiv (-1)^{f+(T-2)/4} \pmod 3, \quad \text{if } 4 \nmid T.$$

■

7.4 Quintic residuacity

In this section, $k = 5$, $p = 5f + 1$ is a prime, g is a primitive root $\pmod p$, and χ is a character $\pmod p$ of order 5 such that $\chi(g) = \exp(2\pi i/5)$. Let $Z = \text{ind}_g 2$. Recall that the integers x, w, v, and u are defined in (3.7.35) and satisfy Dickson's system (3.7.34).

Theorem 7.4.1 (a) *2 is a quintic residue* $\pmod p$ *if and only if* $2 \mid x$.
(b) *2 is a quintic residue* $\pmod p$ *if and only if* $5 \mid b_5$.

Proof (a) By Theorem 2.6.1 and Problem 14 of Exercises 2,

$$2 \text{ is a quintic residue} \pmod p \iff \operatorname{ind}_g 2 \equiv 0 \pmod 5 \iff (0,0)_5 \equiv 1 \pmod 2$$
$$\iff \frac{1}{25}(p - 14 + 3x) \equiv 1 \pmod 2 \iff x \equiv 0 \pmod 2.$$

(b) If 2 is a quintic residue (mod p), then, by Theorem 3.7.6, we have $b_5 \equiv 0 \pmod 5$. If 2 is not a quintic residue (mod p), then we may suppose without loss of generality that $Z \equiv 1 \pmod 5$, and then, by Theorem 3.7.5, $b_5 \not\equiv 0 \pmod 5$. ■

Theorem 7.4.2 *3 is a quintic residue* (mod p) *if and only if* $3 \mid u$ *and* $3 \mid v$.

Proof By Lemma 7.0.1, 3 is a quintic residue (mod p) if and only if $G^3(\chi) \equiv G(\chi^3) \pmod 3$, which is equivalent to

$$pJ(\chi,\chi) = \frac{pG^2(\chi)}{G(\chi^2)} \equiv \frac{G^2(\chi^3)}{G(\chi)} = J(\chi^3,\chi^3) \pmod 3. \tag{7.4.1}$$

For brevity, set $N_j = 50 - 10(-1)^j\sqrt{5}$, for $j = 1, 2$. By (3.7.35) and (7.4.1), it follows that 3 is a quintic residue (mod p) if and only if

$$p\left(x + 5w\sqrt{5} + iu\sqrt{N_1} + iv\sqrt{N_2}\right) \equiv \left(x - 5w\sqrt{5} - iv\sqrt{N_1} + iu\sqrt{N_2}\right) \pmod 3,$$

or, equivalently, if and only if

$$x(p-1) \equiv w(p+1) \equiv pu + v \equiv pv - u \equiv 0 \pmod 3. \tag{7.4.2}$$

Thus, $u \equiv v \equiv 0 \pmod 3$ is a necessary condition for 3 to be a quintic residue (mod p). If $3 \mid u$ and $3 \mid v$, then, by (3.7.34),

$$p \equiv x^2 + 125w^2 \equiv \begin{cases} 1, & \text{if } 3 \mid w, \\ -1, & \text{if } 3 \mid x. \end{cases}$$

This, with (7.4.2), also shows that it is a sufficient condition. ■

Theorem 7.4.3 *5 is a quintic residue* (mod p) *if and only if* $5 \mid (v + 2u)$.

Proof From Problem 14 of Exercises 2, we have

$$\operatorname{ind}_g 5 \equiv -v - 2u \pmod 5,$$

from which the result follows immediately. ■

Let (x, w, v, u) be one of the solutions of (3.7.34). As noted following (3.7.35), $\pm(x, w, v, u)$, $\pm(x, -w, -u, v)$, $\pm(x, -w, u, -v)$, $\pm(x, w, -v, -u)$ comprise all the solutions of (3.7.34). Since $-u + 2v \equiv 2(v + 2u) \pmod 5$, $u - 2v \equiv 3(v + 2u) \pmod 5$, and $-v - 2u \equiv 4(v + 2u) \pmod 5$, Theorem 7.4.3 is independent of the choice of solution of (3.7.34).

Example We choose $p = 271$. A solution of (3.7.34) is $(x, w, v, u) = (31, -1, -1, 8)$. Since $5 \mid v + 2u$, Theorem 7.4.3 shows that 5 is a quintic residue (mod 271); indeed, $5 \equiv 225^5 \pmod{271}$.

Theorem 7.4.4 *If* $\left(\frac{p}{7}\right) = 1$, *then 7 is a quintic residue* (mod p) *if and only if*

$$v \equiv 2u \equiv \pm 2w \pmod 7 \quad or \quad -2v \equiv u \equiv \pm 2w \pmod 7;$$

if $\left(\frac{p}{7}\right) = -1$, *then 7 is a quintic residue* (mod p) *if and only if*

$$7 \mid uv \quad and \quad u + v \equiv \pm 4x \pmod 7.$$

Proof By Lemma 7.0.1, 7 is a quintic residue (mod p) if and only if $G^7(\chi) \equiv G(\chi^2) \pmod 7$, which is equivalent to

$$pJ^3(\chi,\chi)J(\chi^2,\chi^2) = p\frac{G^6(\chi)}{G^3(\chi^2)}\frac{G^2(\chi^2)}{G(\chi^4)} = \frac{G^7(\chi)}{G(\chi^2)} \equiv 1 \pmod 7. \tag{7.4.3}$$

Multiplying both sides of (7.4.3) by $J^2(\overline{\chi},\overline{\chi})$, we see that 7 is a quintic residue (mod p) if and only if

$$\left(\frac{p}{7}\right)J(\chi,\chi)J(\chi^{?},\chi^{?}) = J^{?}(\overline{\chi},\overline{\chi}) \pmod 7. \tag{7.4.4}$$

Using (3.7.35) and the notation from the proof of Theorem 7.4.2, we see that the congruence (7.4.4) is equivalent to

$$\begin{aligned} &\left(\frac{p}{7}\right)\left(x + 5w\sqrt{5} + iu\sqrt{N_1} + iv\sqrt{N_2}\right)\left(x - 5w\sqrt{5} + iv\sqrt{N_1} - iu\sqrt{N_2}\right) \\ &\equiv \left(x + 5w\sqrt{5} - iu\sqrt{N_1} - iv\sqrt{N_2}\right)^2 \pmod 7. \end{aligned} \tag{7.4.5}$$

It is easily verified that

$$\sqrt{5N_1} = \sqrt{N_1} + 2\sqrt{N_2} \quad \text{and} \quad \sqrt{5N_2} = 2\sqrt{N_1} - \sqrt{N_2}. \tag{7.4.6}$$

Expanding (7.4.5) with the aid of (7.4.6), we find that (7.4.5) is equivalent to

(7.4.7)
$$\begin{aligned}&\left(\frac{p}{7}\right)(x^2-125w^2)+\left(\frac{p}{7}\right)(20u^2-20v^2-20uv)\sqrt{5}\\&+i\left(\frac{p}{7}\right)(xu+xv-5wv-15wu)\sqrt{N_1}+i\left(\frac{p}{7}\right)(xv-xu+15wv-5wu)\sqrt{N_2}\\&\equiv(x^2+125w^2-50u^2-50v^2)+(10v^2-10u^2-40uv+10xw)\sqrt{5}\\&-i(2xu+20wv+10wu)\sqrt{N_1}+i(10wv-20wu-2xv)\sqrt{N_2}\pmod 7.\end{aligned}$$

Equating coefficients (mod 7) of 1, $\sqrt{5}$, $i\sqrt{N_1}$, and $i\sqrt{N_2}$ in (7.4.7), we find that, if $\left(\frac{p}{7}\right)=1$, then 7 is a quintic residue (mod p) if and only if

(7.4.8)
$$\begin{cases}5w^2-v^2-u^2\equiv 0\pmod 7,\\xw+3v^2-3u^2-2vu\equiv 0\pmod 7,\\xv+3xu+15wv-5wu\equiv 0\pmod 7,\\3xv-xu+5wv+15wu\equiv 0\pmod 7;\end{cases}$$

whereas, if $\left(\frac{p}{7}\right)=-1$, then 7 is a quintic residue (mod p) if and only if

(7.4.9)
$$\begin{cases}x^2-25v^2-25u^2\equiv 0\pmod 7,\\xw-v^2+u^2-6vu\equiv 0\pmod 7,\\xv-xu-25wv-25wu\equiv 0\pmod 7,\\xv+xu-25wv+25wu\equiv 0\pmod 7.\end{cases}$$

First we examine the case when $\left(\frac{p}{7}\right)=1$. From (3.7.34) and the second congruence in (7.4.8),

$$4v^2-4u^2-6vu\equiv 0\pmod 7,$$

so that

$$v\equiv 2u \text{ or } 3u\pmod 7.$$

If $v\equiv 2u\pmod 7$, then from the first congruence in (7.4.8), $w^2-u^2\equiv 0\pmod 7$, so that $u\equiv\pm w\pmod 7$, and thus $v\equiv 2u\equiv\pm 2w\pmod 7$. If $v\equiv 3u\pmod 7$, then from the first congruence in (7.4.8), $w^2-2u^2\equiv 0\pmod 7$, so that $w\equiv\pm 4u\pmod 7$, and thus $-2v\equiv u\equiv\pm 2w\pmod 7$.

Conversely, if $v\equiv 2u\equiv\pm 2w\pmod 7$ or $-2v\equiv u\equiv\pm 2w\pmod 7$, it is easy to see that all the congruences in (7.4.8) are satisfied. This completes the proof in the case when $\left(\frac{p}{7}\right)=1$.

Now we turn to the case $\left(\frac{p}{7}\right)=-1$. Replacing xw by v^2-u^2-4vu in the second congruence in (7.4.9), we obtain $7\mid vu$. If $7\mid v$, then the first congruence in (7.4.9) gives $x \equiv \pm 2u \pmod 7$, whereas, if $7 \mid u$, it gives $x \equiv \pm 2v \pmod 7$. Thus, in both cases,

$$\pm 4x \equiv u + v \pmod 7.$$

Conversely, if $7 \mid vu$ and $u+v \equiv \pm 4x \pmod 7$, it is easy to see that all the congruences in (7.4.9) are satisfied. This completes the proof in the case $\left(\frac{p}{7}\right) = -1$. ■

7.5 The quartic, octic, and bioctic character of 2

Theorem 7.5.1 *For a prime $p = 8f + 1$, 2 is a quartic residue* $(\bmod p)$ *if and only if* $8 \mid b_4$.

Proof By the law of quadratic reciprocity and Theorem 3.2.1,

$$\left(\frac{a_4}{p}\right) = \left(\frac{|a_4|}{p}\right) = \left(\frac{p}{|a_4|}\right) = \left(\frac{a_4^2 + b_4^2}{|a_4|}\right) = \left(\frac{b_4^2}{|a_4|}\right) = 1. \tag{7.5.1}$$

Define ϵ by $b_4 \equiv \epsilon a_4 \pmod p$, where $\epsilon^2 \equiv -1 \pmod p$. Since $(a_4+b_4)^2 \equiv 2a_4b_4 \pmod p$, it follows from Euler's criterion and (7.5.1) that

$$\begin{aligned}\left(\frac{a_4 + b_4}{p}\right) &\equiv (a_4 + b_4)^{(p-1)/2} \equiv (2a_4b_4)^{(p-1)/4} \\ &\equiv (2\epsilon)^{(p-1)/4} a_4^{(p-1)/2} \equiv (2\epsilon)^{(p-1)/4} \left(\frac{a_4}{p}\right) = (2\epsilon)^{(p-1)/4} \pmod p.\end{aligned} \tag{7.5.2}$$

On the other hand,

$$\begin{aligned}\left(\frac{a_4 + b_4}{p}\right) &= \left(\frac{p}{|a_4 + b_4|}\right) = \left(\frac{2}{|a_4 + b_4|}\right)\left(\frac{2p}{|a_4 + b_4|}\right) \\ &= (-1)^{\{(a_4+b_4)^2-1\}/8} \left(\frac{(a_4 + b_4)^2 + (a_4 - b_4)^2}{|a_4 + b_4|}\right) = \epsilon^{(p-1)/4+(a_4b_4/2)}.\end{aligned} \tag{7.5.3}$$

Equating (7.5.2) and (7.5.3), we deduce that

$$2^{(p-1)/4} \equiv \epsilon^{a_4b_4/2} \equiv (-1)^{a_4b_4/4} \equiv (-1)^{b_4/4} \pmod p,$$

from which the theorem is immediate. ■

Theorem 7.5.1 is clearly equivalent to Corollary 2.6.6, which for completeness we restate here as Theorem 7.5.2.

Theorem 7.5.2 *For $p = 8f + 1$, 2 is a quartic residue* $(\bmod p)$ *if and only if $p = x^2 + 64y^2$ for some integers x and y.*

The next theorem determines the quartic residuacity of 2 in terms of b_8 rather than b_4.

Theorem 7.5.3 *For* $p = 8f+1$, *2 is a quartic residue* (mod p) *if and only if* $f + (b_8/2)$ *is even.*

Proof Set $d = b_8/2$. By the law of quadratic reciprocity and Theorem 3.3.1,

$$\text{(7.5.4)}\qquad \left(\frac{a_8}{p}\right) = \left(\frac{p}{|a_8|}\right) = \left(\frac{a_8^2+2b_8^2}{|a_8|}\right) = \left(\frac{2}{|a_8|}\right) = (-1)^{(a_8^2-1)/8} = (-1)^{f-d^2} = (-1)^{f+d}.$$

Let $d = 2^\alpha d'$, where d' is odd. Then

$$\text{(7.5.5)}\qquad \left(\frac{d}{p}\right) = \left(\frac{d'}{p}\right) = \left(\frac{p}{d'}\right) = \left(\frac{a_8^2}{d'}\right) = 1.$$

Let ϵ be such that $\epsilon^2 \equiv 2 \pmod{p}$, so that $(a_8+2\epsilon d)^2 \equiv 4\epsilon a_8 d \pmod{p}$. Hence, by (7.5.4) and (7.5.5),

$$2^{(p-1)/4} \equiv \left(\frac{\epsilon}{p}\right) = \left(\frac{4a_8 d}{p}\right) = (-1)^{f+d} \pmod{p}. \qquad \blacksquare$$

Theorem 7.5.3 can be reformulated as follows.

Theorem 7.5.4 *2 is a quartic residue* (mod p) *if and only if there are integers x and y such that* $p = x^2+32y^2$ *when* $p \equiv 1 \pmod{16}$, *or there are odd integers x and y such that* $p = x^2+8y^2$ *when* $p \equiv 9 \pmod{16}$.

By Theorem 3.3.1 we know that $a_8 \equiv -1 \pmod{4}$. Making use of Theorem 7.5.3, we can determine a_8 modulo 16 and in some cases modulo 32.

Theorem 7.5.5 *For* $p = 8f+1$,

$$a_8 \equiv \begin{cases} (p-3)/2 \pmod{16}, & \text{if } \chi(2)=1 \text{ and } 2 \mid f,\\ (23-p)/2 \pmod{32}, & \text{if } \chi(2)=-1 \text{ and } 2 \mid f,\\ (7-p)/2 \pmod{32}, & \text{if } \chi(2)=1 \text{ and } 2 \nmid f,\\ (-p-17)/2 \pmod{16}, & \text{if } \chi(2)=-1 \text{ and } 2 \nmid f,\end{cases}$$

where χ is a character (mod p) *of order* 4.

Proof We just give the details for the case $\chi(2) = -1$ and $2 \mid f$. By Theorem 7.5.3, $b_8 \equiv 2 \pmod{4}$, so that $2b_8^2 \equiv 8 \pmod{64}$, and hence

$$a_8^2 = 8f+1-2b_8^2 \equiv 8f-7 \equiv (4f-11)^2 = \{(23-p)/2\}^2 \pmod{64}.$$

This shows that $a_8 \equiv \pm(23-p)/2 \pmod{32}$, and since $a_8 \equiv -1 \pmod 4$ by Theorem 3.3.1, the plus sign must be taken. ∎

In the next theorem we determine a_4 modulo 64. Of course, from Theorem 3.2.1, we know that $a_4 \equiv -1 \pmod 4$.

Theorem 7.5.6 *For* $p = 8f+1$,

$$a_4 \equiv \begin{cases} (3p-5)/2 + 16(1-\psi(2)) \pmod{64}, & \text{if } \chi(2)=1 \text{ and } 2 \mid f, \\ (3p-53)/2 \pmod{64}, & \text{if } \chi(2)=-1 \text{ and } 2 \mid f, \\ (15-p)/2 + 16(1+\psi(2)) \pmod{64}, & \text{if } \chi(2)=1 \text{ and } 2 \nmid f, \\ (31-p)/2 \pmod{64}, & \text{if } \chi(2)=-1 \text{ and } 2 \nmid f, \end{cases}$$

where χ *is a character* $\pmod p$ *of order* 4 *and* ψ *is a character* $\pmod p$ *of order* 8.

Proof From Theorem 2.5.1 with $k=8$ and Theorems 2.1.1, 3.3.1, and 3.3.3, we find after some calculation that

$$(7.5.6)\quad 64(0,0)_8 = \begin{cases} p-23+18a_4+24a_8, & \text{if } \chi(2)=1 \text{ and } 2 \mid f, \\ p-23-6a_4, & \text{if } \chi(2)=-1 \text{ and } 2 \mid f, \\ p-15+2a_4, & \text{if } \chi(2)=1 \text{ and } 2 \nmid f, \\ p-15+10a_4+8a_8, & \text{if } \chi(2)=-1 \text{ and } 2 \nmid f. \end{cases}$$

By Theorem 2.6.1, the cyclotomic number $(0,0)_8$ is odd if and only if $\psi(2)=1$. We proceed to give the details just for the case when $\chi(2)=1$ and $2 \mid f$, as the other three cases follow similarly. By (7.5.6) and Theorem 7.5.5,

$$\begin{aligned} 32\{1+\psi(2)\} &\equiv p-23+18a_4+12(p-3) = 13p+18a_4-59 \\ &= 16(p+a_4)-59-3p+2a_4 \pmod{128}. \end{aligned}$$

Since $p \equiv 1 \pmod{16}$, it follows from Theorem 3.2.1 that $a_4 \equiv -1 \pmod 8$. Hence $8 \mid (p+a_4)$, and so

$$32\{1+\psi(2)\} \equiv 69-3p+2a_4 \pmod{128},$$

from which we obtain

$$32\{1-\psi(2)\} \equiv 5-3p+2a_4 \pmod{128},$$

which is equivalent to the desired result. ∎

We next discuss the octic character of 2 in the case that 2 is a quartic residue $\pmod p$. Note that in this case we have $b_4 \equiv 0 \pmod 8$ by Theorem 7.5.1.

Theorem 7.5.7 *Let* $p = 8f+1$ *be a prime for which* 2 *is a quartic residue. Then* 2 *is an octic residue* $\pmod p$ *if and only if* $f + (b_4/8)$ *is even.*

Proof We just give the details when 2 is an octic residue (mod p) and $2 \nmid f$. We must show that $b_4/8$ is odd. By Theorem 7.5.6,

$$a_4^2 \equiv \{32 + (15 - p)/2\}^2 = (39 - 4f)^2 \equiv 16f^2 - 56f - 15 \pmod{128},$$

so that

$$b_4^2 = 8f + 1 - a_4^2 \equiv 64\{f + (1 - f^2)/4\} \equiv 64 \pmod{128},$$

and thus $b_4 \equiv 8 \pmod{16}$, as desired. ■

An immediate consequence of Theorem 7.5.7 is the following criterion for 2 to be an octic residue (mod p).

Corollary 7.5.8 *2 is an octic residue* (mod p) *if and only if either* $p = x^2 + 256y^2 \equiv 1 \pmod{16}$ *for some integers* x *and* y, *or* $p = x^2 + 64y^2 \equiv 9 \pmod{16}$ *for some odd integers* x *and* y.

In the remainder of this section, we assume that $p = 16f + 1$.

Lemma 7.5.9 *If* $p = 16f + 1$ *and 2 is a quartic residue* (mod p), *then* $a_8 \equiv a_4 + 4b_8 \pmod{32}$.

Proof From Theorems 7.5.1 and 7.5.3, $8 \mid b_4$ and $4 \mid b_8$, respectively. Since $p = a_4^2 + b_4^2 = a_8^2 + 2b_8^2$, we deduce that $a_4^2 \equiv a_8^2 + 2b_8^2 \pmod{64}$, and hence

$$(a_8 + a_4)(a_8 - a_4) \equiv \begin{cases} 32 \pmod{64}, & \text{if } 4 \parallel b_8, \\ 0 \pmod{64}, & \text{if } 8 \mid b_8. \end{cases}$$

Since $a_4 \equiv a_8 \equiv -1 \pmod 4$, it follows that $2 \parallel (a_8 + a_4)$. Thus

$$a_8 - a_4 \equiv \begin{cases} 16 \pmod{32}, & \text{if } 4 \parallel b_8, \\ 0 \pmod{32}, & \text{if } 8 \mid b_8, \end{cases}$$

from which $a_8 - a_4 \equiv 4b_8 \pmod{32}$ follows. ■

Theorem 7.5.10 *Suppose that 2 is an octic residue of the prime* $p = 16f + 1$. *Then 2 is a bioctic residue* (mod p) *if and only if* $32 \mid (b_4 + 4b_8)$.

Proof Since $16 \mid b_4$ by Theorem 7.5.7,

$$p = a_4^2 + b_4^2 \equiv a_4^2 + 16b_4 \pmod{512}. \tag{7.5.7}$$

From Theorem 7.5.6 we deduce that

$$a_4 \equiv 24f - 1 \pmod{64}. \tag{7.5.8}$$

We now make use of the table of cyclotomic numbers of order 16 computed by Evans and Hill [1]. First, suppose that $2 \nmid f$. Then

$$\begin{aligned} 256(0,0)_{16} &= p - 31 + 18a_4 + 16a_8 \\ &\equiv a_4^2 + 16b_4 - 31 + 18a_4 + 16a_4 + 64b_8 \\ &\equiv (a_4+17)^2 + 16b_4 + 64b_8 - 320 \pmod{512}, \end{aligned}$$

by (7.5.7) and Lemma 7.5.9. In view of (7.5.8),

$$\begin{aligned} a_4 + 17 &\equiv 8(3f+2) \pmod{64}, \\ (a_4+17)^2 &\equiv 64(3f+2)^2 \equiv 64 \pmod{512}, \end{aligned}$$

so that

(7.5.9) $$256(0,0)_{16} \equiv 16b_4 + 64b_8 + 256 \pmod{512}.$$

Next, suppose that $2 \mid f$. Then, again making use of the table of cyclotomic numbers of order 16, we have

$$256\{(0,0)_{16} - 6(1,2)_{16} + 6(2,4)_{16} - 6(3,6)_{16}\} = -5p - 53 + 54a_4 + 144a_8.$$

Hence, by (7.5.7) and Lemma 7.5.9,

$$\begin{aligned} 256(0,0)_{16} &\equiv -5a_4^2 - 80b_4 - 53 + 54a_4 + 144a_4 + 64b_8 \\ &= -5(a_4-71)^2 - 80b_4 + 64b_8 + 64 \pmod{512}. \end{aligned}$$

In view of (7.5.8),

$$\begin{aligned} a_4 - 71 &\equiv 24f - 8 \pmod{64}, \\ (a_4-71)^2 &\equiv (24f-8)^2 = 64(3f-1)^2 \equiv 64 \pmod{512}, \end{aligned}$$

so that

$$256(0,0)_{16} \equiv -80b_4 + 64b_8 + 256 \equiv 16b_4 + 64b_8 + 256 \pmod{512},$$

which is (7.5.9) again. By Theorem 2.6.1, 2 is a bioctic residue (mod p) if and only if $(0,0)_{16}$ is odd. The result now follows from (7.5.9). ∎

The next result is an immediate consequence of Theorem 7.5.10.

Corollary 7.5.11 *Let p be a prime with $p \equiv 1 \pmod{16}$. Then 2 is a bioctic residue* (mod p) *if and only if $p = x^2 + 1024y^2 = a^2 + 128b^2$ for some integers x, y, a, and b, or $p = x^2 + 256y^2 = a^2 + 32b^2$ for some odd integers x, y, a, and b.*

Example The prime $p = 337$ satisfies $p \equiv 1 \pmod{16}$ and $337 = 9^2 + 256 \cdot 1^2 = 7^2 + 32 \cdot 3^2$. Thus, by Corollary 7.5.11, 2 is a bioctic residue (mod 337). Indeed, $2 \equiv 9^{16}$ (mod 337).

Exercises 7

1. Prove that 13 is a cubic residue (mod p), where $p \equiv 1 \pmod 3$, if and only if $13 \mid r_3 s_3$ or $13 \mid (r_3^2 - 3s_3^2)$.

2. Prove that -7 is a quartic residue (mod p), where $p \equiv 1 \pmod 4$, if and only if $7 \mid a_4 b_4$.

3. (a) If $p \equiv 1 \pmod 8$, prove that 7 is a quartic residue (mod p) if and only if $7 \mid a_4 b_4$.

(b) If $p \equiv 5 \pmod 8$, prove that 7 is a quartic residue (mod p) if and only if $7 \mid (a_4^2 - b_4^2)$.

4. Use Theorem 7.3.2 to determine a necessary and sufficient condition for 7 to be an octic residue (mod p), where $p \equiv 1 \pmod 8$, given that it is a quartic residue (mod p).

5. Complete the proof of Theorem 7.5.5 by deriving the three remaining congruences for a_8.

6. Complete the proof of Theorem 7.5.6 by deriving the three remaining congruences for a_4 from (7.5.6) and Theorem 7.5.5.

7. Complete the proof of Theorem 7.5.7.

8. For $p = 16f + 1$, prove that

$$a_4 \equiv \begin{cases} (-p+31)/2 - 16(-1)^f \pmod{128}, & \text{if } \psi(2) = 1, \\ (-5p+99)/2 - (-1)^f(2p+14) \pmod{256}, & \text{if } \psi(2) = -1, \\ -(p+81)/2 + 4(-1)^f(b_4 - 8) \pmod{128}, & \text{if } \chi(2) = -1, \end{cases}$$

where χ and ψ are characters (mod p) of orders 4 and 8, respectively.

9. (a) Let $p \equiv 1 \pmod 8$, and write $a_8 = -(4a+1)$. Using Theorem 7.5.3, show that 2 is a quartic residue (mod p) if and only if a is even.

(b) Prove that Theorems 7.5.1 and 7.5.3 are equivalent, as follows. Define integers b and d by $b_4 = 4b$ and $b_8 = 2d$. Using the equality

$$\{4(a+b)+1\}\{4(a-b)+1\} = a_4^2 - 8d^2,$$

show that $\gcd(a_4, d) = \delta_1\delta_2$, where $\gcd(\delta_1, \delta_2) = 1$, and where $\delta_1^2 \mid \{4(a+b)+1\}$ and $\delta_2^2 \mid \{4(a-b)+1\}$. Deduce that 2 is a quadratic residue (mod D), where

$D = \{4(a+b)+1\}/\delta_1^2$. Hence, $D^2 \equiv 1 \pmod{16}$. Deduce that $a+b \equiv 0 \pmod 2$. Then use part (a) to deduce that $b \equiv f+d \pmod 2$, which is the desired result.

10. Use Theorem 7.5.6 to prove Theorem 7.5.1.

11. Let $p \equiv 1 \pmod{12}$ be a prime. Prove that 3 is a quartic residue (mod p) if and only if $a_3 \equiv -1 \pmod 4$.

Notes on Chapter 7

The first criteria for cubic and quartic residuacity of primes q (mod p) were established by Gauss. However, criteria for $q = 2$, 3, 5, and 7 were already conjectured in the mid 1700s by Euler; see Bickmore [2, p. 7] and Euler [CW, vol. 5]. For example, Corollaries 2.6.4 and 7.1.4 (giving criteria for the cubic residuacity of 2 and 5) and Theorem 7.5.2 (giving a criterion for the quartic residuacity of 2) were conjectured by Euler in [CW, vol. 5, pp. 250, 251 and 258]. The criterion for the cubic residuacity of 2 given in Theorem 7.1.1 is due to Gauss [2]. Theorem 7.1.2 is due to Jacobi [2]. In fact, Jacobi established criteria for the cubic residuacity of all primes $q \equiv 1 \pmod 3$ for $q \leq 37$. Using work of Pépin [1], Cunningham and Gosset [1] slightly extended these criteria to $q \leq 43$. Different proofs for $q = 2$, 3, 5, and 7 were found by Nagell [1]. S. Chowla [4], [6] gave a simple proof of Theorem 7.1.1. Still different proofs of criteria for cubic residuacity for $q = 2$ and 3 were found by Tihanyi [1] and Vaidyanathaswamy [1]. Theorem 7.1.7 is essentially due to Pépin [1]; see also Cunningham and Gosset [1, p. 15]. Some general criteria for cubic residuacity were found by E. Lehmer [7]. The work of Lehmer has been generalized by Llorente [1].

The criteria for the quartic residuacity of 2 (Theorems 7.5.1, 7.5.2, and 7.5.3) are due to Gauss [5]. The simple proof that we have given for Theorem 7.5.1 is due to Dirichlet [5]. Dirichlet's proof is also reproduced in H. J. S. Smith's report [1]. Dirichlet [5] showed the equivalence of Theorems 7.5.1 and 7.5.3 in a very elementary fashion (see Problem 9). Dirichlet's technique has been extended by Muskat [8] and more recently by Muskat, Spearman, and Williams [1]. In a different direction it has been generalized by Hudson and Williams [7]. Theorem 7.5.3 was rediscovered by Barrucand and Cohn [1]. The simple proof that is presented here is due to Williams [21]. Criteria for the quartic residuacity of 3 and 5 (Theorems 7.2.2 and 7.2.3) are also due to Gauss [5]. Criteria for the quartic residuacity of 7 (Problem 3) were first given by Bickmore [1]. Theorem 7.2.4 is due to Pépin [1]; see also Cunningham and Gosset [1, p. 2]. Some general theorems for quartic residuacity have been established by E. Lehmer [7] and Aigner [2]. Other proofs of Gauss' criteria for 2 and 3 have been given by Subbarao [1]. Another criterion for 3 to be a quartic residue (mod p) has been given by Hudson and Williams [1]; see Problem 11.

Theorem 7.5.7 (or Corollary 7.5.8) giving a criterion for the octic residuacity of 2 was first stated without proof by Reuschle [1]. It was established by Perott [1] in 1888 and Western [3] in 1911. Proofs have also been given by Hasse [1], Whiteman [7], and Evans [5]. Halter-Koch [1] developed criteria for the octic residuacity of 3, 5, and 7. Theorem 7.3.2 is due to Western [3]. Using Western's work, von Lienen [2] has established criteria for the octic residuacity of primes ≤ 41. Another criterion for 3 to be an octic residue (mod p) has been established by Hudson and Williams [1]. Evans [17, Theorem 5.1] has given a general criterion for an arbitrary odd prime $q \neq p$ to be an octic residue (mod p).

Theorem 7.5.10, or Corollary 7.5.11, on the bioctic residuacity of 2, was first conjectured by Cunningham [1]. Apparently the first proof is due to Aigner [1]. However, Beeger [2] has shown that Cunningham's conjecture follows from much earlier work of Goldscheider [1]. Other proofs have been given by Whiteman [7], Hasse [1], and Evans [5]. In 1958 Hasse [1] gave a necessary and sufficient condition for 2 to be a 32nd power (mod p). An elementary proof of Hasse's criterion has been given by Evans [5]. Both Hasse [1] and Evans [5] have established criteria for the 2^nth–power residuacity of 2.

Criteria for the quintic character of 2 and 3 were apparently first established by Pépin [1]. Using work of Tanner [2], Bickmore [2] established criteria for the quintic residuacity of 5, as well as for 2 and 3. In terms of Dickson's representation (3.7.34) of $16p$, E. Lehmer [1], [2] established different criteria (Theorems 7.4.1 and 7.4.2) for the quintic residuacity of 2 and 3. Various versions of Theorem 7.4.3 are due to E. Lehmer [11] and Muskat [2]. Equivalent statements of Theorem 7.4.4 giving the quintic character of 7 have also been found by E. Lehmer [12] and Muskat [3]. Criteria for the quintic residuacity of primes $q \leq 19$ have been established by Williams [24]. A yet different criterion for the quintic residuacity of 2 was found by Alderson [1]. E. Lehmer [12] has proved some general theorems leading to criteria for quintic residuacity.

Criteria for the septic residuacity of 2 and 3 were established by Alderson [2]. Leonard and Williams [1] have determined criteria for the septic residuacity of 2, 3, 5, and 7. Parnami, Agrawal, Pall, and Rajwade [1] have given another proof of Leonard and Williams' criterion for 3 to be a seventh power (mod p). Muskat [2] also found a criterion for the septic character of 7. In fact, he proved a general theorem giving a criterion for the kth–power character of k. Williams [11], [16] has determined when 2 and 3 are ninth powers (mod p). Pépin [1] and Leonard, Mortimer, and Williams [1] have established necessary and sufficient conditions for 2 to be an 11th–power (mod p). As we mentioned in the Notes on Chapter 2, Jakubec [2] has given a criterion for 2 to be an 11th power (mod p) as well as one for 2 to be a 19th power (mod p), in certain cases. Jakubec [3] has also given a criterion for 3 to be an 11th power (mod p). Llorente [1] has given a criterion for the kth–power character of 2 and 3. Parnami, Agrawal, Pall, and Rajwade [2] have given a criterion for 2 to be a kth–power (mod p). Evans [5] established a conjecture of Muskat [7] giving an explicit expression for $\text{ind}_g 2$ (mod 2^k). Burde [6] used Gauss sums to determine $\text{ind}_g q$ (mod q^k) for *odd* primes q. Some rather general theorems for kth–power residuacity have been proved by Ankeny

[1] and Muskat [1], [2], [3], [5].

Hasse's book [3] is an excellent source for theorems on residuacity. The papers of Bickmore [1], [2] contain many specific criteria for residuacity and much historical information.

Let p be an odd prime, and let k be an integer greater than or equal to 3 that divides $p-1$. If q is a prime not equal to p, then $q^{(p-1)/k}$ is a kth root of unity modulo p. The problem of determining exactly which root of unity $q^{(p-1)/k}$ is, was first posed by E. Lehmer [8] in 1959, who completely solved the problem for $q = 2$ and $k = 3, 4, 5$, and 8. See also E. Lehmer [20]. Williams [15] solved the problem for $k = 3$ and arbitrary q; see also Williams [20]. In another paper [17] he solved the problem for $k = 5$ and arbitrary q; see also Katre and Rajwade [2]. Nashier and Rajwade [1] have examined the case $q = 2$ and $k = 7$. Hudson and Williams [8] examined the cases $q = 2$ and $k = 16, 32$.

We now give further examples of criteria for residuacity involving Diophantine representations of p with four or more parameters. Evans [2] has shown that for $p = 16f+1$, 2 is a quartic residue (mod p) if and only if $4|b_{16}$, where b_{16} is defined in Theorem 3.4.1. A related criterion in terms of b_{16} for the octic character of 2 is also given in Evans [2, p. 76]. Alternative criteria not involving b_{16} are presented in Theorem 7.5.4 and Corollary 7.5.8. Hudson and Williams [1] have shown that if $p \equiv 1 \pmod{20}$, then 5 is a quintic residue (mod p) if and only if

$$\begin{cases} x \equiv 4 \pmod{8}, & \text{if } 2 \mid x, \\ x \equiv \pm 3w \pmod{8}, & \text{if } 2 \nmid x, \end{cases}$$

where (x, w, v, u) is a solution of Dickson's system (3.7.34). Hudson and Williams [2] have also given a criterion for 7 to be a quartic residue (mod p), where $p \equiv 1$ (mod 28), in terms of the parameters of the Diophantine system (3.9.20)–(3.9.22). General criteria for quartic residuacity (mod p) in terms of a quaternary representation of p have been given by Williams, Friesen, and Howe [1]. Some results on octic residuacity are given in Hudson [1] and Hudson and Williams [6].

The Notes on Chapters 2 and 8 contain further references on residuacity.

Chapter 8

Reciprocity Laws

In Chapter 1 we employed Gauss sums in giving one of Gauss' proofs of the law of quadratic reciprocity (see Theorem 1.2.6). In this chapter, we shall use some of the simpler properties of Gauss and Jacobi sums to prove the laws of cubic and quartic reciprocity. The law of quartic reciprocity is also referred to as the law of biquadratic reciprocity. In Chapter 14, we shall prove a general kth–power reciprocity law, called Eisenstein's reciprocity law. The last section of this chapter is devoted to proving some rational reciprocity laws. These reciprocity laws involve only rational integers, in contrast to the cubic and quartic reciprocity laws, which involve Eisenstein and Gaussian integers, respectively.

8.1 Cubic reciprocity

An Eisenstein integer is an integer of the imaginary quadratic field $Q(i\sqrt{3}\,) = Q(e^{2\pi i/3})$. We denote the ring of Eisenstein integers by $Z[\omega]$, where $\omega = \exp(2\pi i/3)$. It is known (see, e.g., the book of Ireland and Rosen [1, Chapter 1]) that $Z[\omega]$ is a unique factorization domain. Before stating and proving the law of cubic reciprocity, we shall recall several facts about Eisenstein integers and make some definitions.

The units in $Z[\omega]$ are ± 1, $\pm\omega$, and $\pm\omega^2$. The norm $N(\alpha)$ of an element $\alpha \in Z[\omega]$ is $N(\alpha) = \alpha\overline{\alpha}$. The Eisenstein primes in $Z[\omega]$ consist of the associates of the rational primes $q \equiv 2 \pmod 3$, the associates of $1-\omega$, and the associates of those elements $\alpha \in Z[\omega]$ whose norm $N(\alpha)$ is a rational prime $p \equiv 1 \pmod 3$. We make the convention that p denotes a rational prime $\equiv 1 \pmod 3$, $\pi \in Z[\omega]$ denotes any (Eisenstein) prime, and q denotes a rational prime $\equiv 2 \pmod 3$. Note that $N(q) = q^2$ and $N(1-\omega) = 3$. If α, $\pi \in Z[\omega]$, where π is a prime with $\pi \nmid \alpha$, then we have the analogue of Fermat's little theorem:

$$\alpha^{N(\pi)-1} \equiv 1 \pmod{\pi}. \tag{8.1.1}$$

(See Problem 1.) The finite field $Z[\omega]/(\pi)$ has $N(\pi)$ elements.

Proposition 8.1.1 *Let $\pi, \alpha \in Z[\omega]$ with π prime, $N(\pi) \neq 3$, and $\pi \nmid \alpha$. Then there exists a unique integer m, $0 \leq m \leq 2$, such that*

$$\alpha^{(N(\pi)-1)/3} \equiv \omega^m \pmod{\pi}.$$

Proof First observe that, for all primes $\pi \in Z[\omega]$ with $N(\pi) \neq 3$, we have $3|(N(\pi)-1)$. Hence, we can write

$$(8.1.2) \qquad \alpha^{N(\pi)-1} - 1 = (\alpha^{(N(\pi)-1)/3} - 1)(\alpha^{(N(\pi)-1)/3} - \omega)(\alpha^{(N(\pi)-1)/3} - \omega^2).$$

By (8.1.1), π divides the left side of (8.1.2). Hence, π divides at least one of the three factors on the right side of (8.1.2). Suppose that π divides two of the factors on the right side of (8.1.2), say,

$$\pi \mid (\alpha^{(N(\pi)-1)/3} - \omega^r), \quad \pi \mid (\alpha^{(N(\pi)-1)/3} - \omega^s), \quad 0 \le r < s \le 2.$$

Then $\pi \mid \omega^r(1-\omega^{s-r})$, and thus, as ω^r is a unit, $\pi \mid 1-\omega^t$ for $t = 1$ or 2. Hence $N(\pi) | N(1-\omega^t) = 3$, which is impossible as $N(\pi) \neq 1, 3$. ■

We now define the cubic residue symbol.

Definition *Let π be an Eisenstein prime with $N(\pi) \neq 3$, and let $\alpha \in Z[\omega]$. If $\pi \mid \alpha$, set $\left(\frac{\alpha}{\pi}\right)_3 = 0$. If $\pi \nmid \alpha$, let $\left(\frac{\alpha}{\pi}\right)_3$ be the unique power of ω defined by*

$$\alpha^{(N(\pi)-1)/3} \equiv \left(\frac{\alpha}{\pi}\right)_3 \pmod{\pi}.$$

(This definition is unambiguous by Proposition 8.1.1.) Moreover, if $\beta = \pi_1 \cdots \pi_n$, where each π_j ($1 \le j \le n$) is a prime of $Z[\omega]$ with $N(\pi_j) \neq 3$, define

$$\left(\frac{\alpha}{\beta}\right)_3 = \left(\frac{\alpha}{\pi_1}\right)_3 \cdots \left(\frac{\alpha}{\pi_n}\right)_3.$$

Lastly, if β is a unit of $Z[\omega]$, define $\left(\frac{\alpha}{\beta}\right)_3 = 1$ for every $\alpha \in Z[\omega]$, $\alpha \neq 0$, and define $\left(\frac{0}{\beta}\right)_3 = 0$.

When $N(\pi) \neq 3$, it is easy to see that $\left(\frac{\alpha}{\pi}\right)_3 = 1$ if and only if α is a cubic residue (mod π). In particular, if $\pi = q \equiv 2 \pmod 3$ and n is a rational integer such that $q \nmid n$, then $\left(\frac{n}{q}\right)_3 = 1$, since every such integer n is a cubic residue (mod q).

Proposition 8.1.2 *Let $x \in Z$, $\theta \in Z[\omega]$, and let p be a rational prime with $p \equiv 1 \pmod 3$, so that $p = \pi\overline{\pi}$ for an Eisenstein prime π. Then*

(a) *$x \equiv \alpha^3 \pmod p$ for some α in $Z[\omega]$ if and only if $x \equiv a^3 \pmod p$ for some $a \in Z$,*

and

(b) *$\theta \equiv \gamma^3 \pmod p$ for some $\gamma \in Z[\omega]$ if and only if $\theta \equiv \alpha^3 \pmod \pi$ and $\theta \equiv \beta^3 \pmod{\overline{\pi}}$ for some α, $\beta \in Z[\omega]$.*

Proof Since $N(\pi) = p$, we have the field isomorphism

$$Z[\omega]/(\pi) \approx Z/pZ.$$

Thus every $\alpha \in Z[\omega]$ satisfies $\alpha \equiv a \pmod{\pi}$ for some $a \in Z$, and part (a) follows.

Given $\alpha,\ \beta \in Z[\omega]$, there exists $\gamma \in Z[\omega]$ such that

$$\gamma \equiv \alpha \pmod{\pi}, \quad \gamma \equiv \beta \pmod{\overline{\pi}},$$

by the Chinese remainder theorem. Now part (b) easily follows. ∎

The product formula $\left(\frac{\alpha\beta}{\pi}\right)_3 = \left(\frac{\alpha}{\pi}\right)_3 \left(\frac{\beta}{\pi}\right)_3$, where $\alpha, \beta \in Z[\omega]$, follows directly from the definition of the cubic residue symbol. Likewise, it follows easily that if $\alpha \equiv \beta \pmod{\pi}$, then $\left(\frac{\alpha}{\pi}\right)_3 = \left(\frac{\beta}{\pi}\right)_3$. Hence, $\left(\frac{\alpha}{\pi}\right)_3$ is a cubic character on the field $Z[\omega]/(\pi)$ of $N(\pi)$ elements, and we shall frequently write $\chi_\pi(\alpha) = \left(\frac{\alpha}{\pi}\right)_3$.

Proposition 8.1.3 *Let $\pi, \alpha \in Z[\omega]$ with π prime, $N(\pi) \neq 3$, and $\pi \nmid \alpha$. Then $\overline{\chi_\pi}(\alpha) = \chi_{\overline{\pi}}(\overline{\alpha})$.*

Proof Since $\alpha^{(N(\pi)-1)/3} \equiv \chi_\pi(\alpha) \pmod{\pi}$, it follows by complex conjugation that $\overline{\alpha}^{(N(\pi)-1)/3} \equiv \overline{\chi}_\pi(\alpha) \pmod{\overline{\pi}}$. By definition of $\chi_{\overline{\pi}}$, $\overline{\alpha}^{(N(\pi)-1)/3} \equiv \chi_{\overline{\pi}}(\overline{\alpha}) \pmod{\overline{\pi}}$. The last two congruences imply the desired result. ∎

Definition *Let $\alpha \in Z[\omega]$. We say that α is primary if $\alpha \equiv \pm 1 \pmod 3$.*

Note that the only primary units in $Z[\omega]$ are ± 1. If $(1-\omega)$ divides α in $Z[\omega]$, then none of the associates of α are primary. For any other α in $Z[\omega]$, exactly two associates of α are primary, as we show in Proposition 8.1.4 below. The notion of "primary" will be extended to integers of general cyclotomic fields $Q(\beta)$ in Section 14.2.

Proposition 8.1.4 *Let $\alpha \in Z[\omega]$ be such that $(1-\omega) \nmid \alpha$. Then, of the six associates of α, exactly two are primary, and these differ only by sign.*

Proof Put $\alpha = a + b\omega$ with $a,\ b \in Z$. As $(1-\omega) \nmid a + b\omega$, it follows that $a + b \not\equiv 0 \pmod 3$, and the asserted result follows from the following table, where $\epsilon = \pm 1$.

	$a \pmod 3$	$b \pmod 3$	$\pm\alpha \pmod 3$	$\pm\omega\alpha \pmod 3$	$\pm\omega^2\alpha \pmod 3$
(i)	ϵ	0	± 1	$\pm\omega$	$\pm\omega^2$
(ii)	ϵ	ϵ	$\pm\omega^2$	± 1	$\pm\omega$
(iii)	0	ϵ	$\pm\omega$	$\pm\omega^2$	± 1

∎

Proposition 8.1.5 *An Eisenstein integer that is not a unit is primary if and only if it can be factored into a product of primary Eisenstein primes.*

Proof Let α be a primary Eisenstein integer that is not a unit. As $Z[\omega]$ is a unique factorization domain, $\alpha = \pi_1 \cdots \pi_k$, where each π_i $(1 \le i \le k)$ is an Eisenstein prime. Further, as α is primary, none of the π_i is an associate of the prime $1-\omega$. Hence, by Proposition 8.1.4, for $i = 1, \ldots, k$, $\pi_i = \theta_i \lambda_i$, where θ_i is a unit of $Z[\omega]$, and λ_i is a primary Eisenstein prime. Clearly $\theta_1 \cdots \theta_k$ is a unit of $Z[\omega]$, and so $\theta_1 \cdots \theta_k = \pm\omega^r$ for some $r = 0, 1, 2$. Thus $\alpha = \pm\omega^r \lambda_1 \cdots \lambda_k$. Taking this relation modulo 3, as $\alpha \equiv \pm 1 \pmod 3$ and each $\lambda_i \equiv \pm 1 \pmod 3$, we see that $\omega^r \equiv \pm 1 \pmod 3$. But $3 = -\omega^2(1-\omega)^2$, so we must have $r = 0$, and thus $\alpha = \pm\lambda_1 \cdots \lambda_k$, as required.

Conversely, it is clear that a product of primary Eisenstein primes is primary. ■

Lemma 8.1.6 *Let π be a primary Eisenstein prime with $N(\pi) = p \equiv 1 \pmod 3$. Then*

$$J(\chi_\pi, \chi_\pi) = \begin{cases} -\pi, & \textit{if } \pi \equiv 1 \pmod 3, \\ \pi, & \textit{if } \pi \equiv -1 \pmod 3. \end{cases}$$

Proof Put $\chi_\pi = \chi$ and write $J(\chi, \chi) = a + b\omega$. By Theorem 3.1.3, we have $2J(\chi, \chi) = r_3 + is_3\sqrt{3}$, where $r_3 \equiv 1 \pmod 3$ and $s_3 \equiv 0 \pmod 3$. Moreover, as $\omega = (-1+i\sqrt{3}\,)/2$, it is easily seen that $a = (r_3 + s_3)/2$ and $b = s_3$. Hence, we have $a \equiv 2 \pmod 3$ and $b \equiv 0 \pmod 3$. It follows that $J(\chi, \chi)$ is a primary Eisenstein prime, as $N(J(\chi, \chi)) = p$. Thus

$$J(\chi, \chi)\overline{J(\chi, , \chi)} = p = N(\pi) = \pi\overline{\pi}.$$

By Theorem 2.1.14, $J(\chi, \chi) \equiv 0 \pmod \pi$. Hence $J(\chi, \chi) = \pm\pi$. The assertion of the lemma now follows as $J(\chi, \chi) \equiv -1 \pmod 3$. ■

We are now ready to state and prove the law of cubic reciprocity.

Theorem 8.1.7 (*Law of cubic reciprocity.*) *Let α and β be coprime primary Eisenstein integers. Then*

$$\left(\frac{\alpha}{\beta}\right)_3 = \left(\frac{\beta}{\alpha}\right)_3.$$

Proof If α is a unit (so that $\alpha = \pm 1$), it is easily seen that both members of the cubic reciprocity law equal 1. By Proposition 8.1.5 and the multiplicativity of the cubic residue symbol, it suffices to prove the theorem when α and β are coprime primary Eisenstein primes. We shall divide the proof into three cases.

First, suppose that α and β are both rational primary Eisenstein primes. Then, as α and β are coprime,

$$\left(\frac{\alpha}{\beta}\right)_3 = 1 = \left(\frac{\beta}{\alpha}\right)_3,$$

by the remark preceding Proposition 8.1.2.

Second, suppose that $\alpha = q \equiv 2 \pmod 3$ and that $\beta = \pi$ with $N(\pi) = p \equiv 1 \pmod 3$. By Lemmas 4.1.1(a) and 8.1.6,

$$G^3(\chi_\pi) = pJ(\chi_\pi, \chi_\pi) = \pm p\pi,$$

and so

$$G(\chi_\pi)^{q^2-1} = (p\pi)^{(q^2-1)/3} \equiv \chi_q(p\pi) = \chi_q(\pi) \pmod q,$$

which implies that

$$G(\chi_\pi)^{q^2} \equiv \chi_q(\pi) G(\chi_\pi) \pmod q. \tag{8.1.3}$$

On the other hand, by the multinomial theorem and Theorem 1.1.3,

$$\begin{aligned} G(\chi_\pi)^{q^2} &\equiv \sum_n \chi_\pi^{q^2}(n) e^{2\pi i q^2 n/p} = G(q^2, \chi_\pi) \\ &= \overline{\chi}_\pi(q^2) G(\chi_\pi) = \chi_\pi(q) G(\chi_\pi) \pmod q. \end{aligned} \tag{8.1.4}$$

On comparing (8.1.3) and (8.1.4), we find that

$$\chi_q(\pi) = \chi_\pi(q),$$

as desired.

Third, suppose that $\alpha = \pi_1$ with $N(\pi_1) = p_1 \equiv 1 \pmod 3$ and that $\beta = \pi_2$ with $N(\pi_2) = p_2 \equiv 1 \pmod 3$. Since $\overline{\pi}_1$ is also primary,

$$G(\chi_{\overline{\pi}_1})^{p_2-1} = (\pm p_1 \overline{\pi}_1)^{(p_2-1)/3} \equiv \chi_{\pi_2}(p_1 \overline{\pi}_1) \pmod{\pi_2},$$

or

$$G(\chi_{\overline{\pi}_1})^{p_2} \equiv \chi_{\pi_2}(p_1 \overline{\pi}_1) G(\chi_{\overline{\pi}_1}) \pmod{\pi_2}. \tag{8.1.5}$$

On the other hand, by the multinomial theorem and Theorem 1.1.3,

$$G(\chi_{\overline{\pi}_1})^{p_2} \equiv \sum_n \chi_{\overline{\pi}_1}^{p_2}(n) e^{2\pi i n p_2/p_1} = G(p_2, \chi_{\overline{\pi}_1}) = \overline{\chi}_{\overline{\pi}_1}(p_2) G(\chi_{\overline{\pi}_1}) \pmod{p_2}. \tag{8.1.6}$$

From (8.1.5) and (8.1.6),

$$\chi_{\pi_2}(p_1 \overline{\pi}_1) = \chi_{\overline{\pi}_1}(p_2^2). \tag{8.1.7}$$

Now repeat the same argument as in the preceding paragraph but with $\overline{\pi}_1$ replaced by π_2 and π_2 replaced by π_1. Instead of (8.1.7), we obtain

$$\chi_{\pi_1}(p_2 \overline{\pi}_2) = \chi_{\pi_2}(p_1^2). \tag{8.1.8}$$

Next, by Proposition 8.1.3,

$$\chi_{\overline{\pi}_1}(p_2^2) = \overline{\chi}_{\pi_1}(p_2^2) = \chi_{\pi_1}(p_2). \tag{8.1.9}$$

Using (8.1.7), (8.1.9), and (8.1.8) in this order, we obtain

$$\begin{aligned}\chi_{\pi_1}(\pi_2)\chi_{\pi_2}(p_1\overline{\pi}_1) &= \chi_{\pi_1}(\pi_2)\chi_{\overline{\pi}_1}(p_2^2) = \chi_{\pi_1}(\pi_2)\chi_{\pi_1}(p_2) = \chi_{\pi_1}(\pi_2 p_2)\\ &= \chi_{\pi_2}(p_1^2) = \chi_{\pi_2}(p_1\pi_1\overline{\pi}_1) = \chi_{\pi_2}(\pi_1)\chi_{\pi_2}(p_1\overline{\pi}_1).\end{aligned}$$

Hence, on cancellation, it follows that

$$\chi_{\pi_1}(\pi_2) = \chi_{\pi_2}(\pi_1),$$

as required. ■

The criterion for 2 to be a cubic residue (mod p), given in Theorem 7.1.1, can be established with the aid of the law of cubic reciprocity. First we need the following result.

Proposition 8.1.8 *Let π be a nonrational primary Eisenstein prime. Then the congruence $x^3 \equiv 2 \pmod{\pi}$ is solvable for x in $Z[\omega]$ if and only if $\pi \equiv 1 \pmod 2$.*

Proof By the law of cubic reciprocity, we have

$$\chi_\pi(2) = \chi_2(\pi). \tag{8.1.10}$$

Since

$$\pi = \pi^{(N(2)-1)/3} \equiv \chi_2(\pi) \pmod 2, \tag{8.1.11}$$

and since $x^3 \equiv 2 \pmod{\pi}$ is solvable if and only if $\chi_\pi(2) = 1$, the result follows from (8.1.10) and (8.1.11). ■

Second proof of Theorem 7.1.1. Write $p = \pi\overline{\pi}$, where π is an Eisenstein prime. By Lemma 8.1.6, we can choose π so that $\pi = J(\chi_\pi, \chi_\pi)$. Recall again from Theorem 3.1.3 that $2J(\chi_\pi, \chi_\pi) = r_3 + is_3\sqrt{3}$, where $r_3 \equiv 1 \pmod 3$ and $s_3 \equiv 0 \pmod 3$. Thus, by Proposition 8.1.8, 2 is a cubic residue (mod p) if and only if $(r_3 + is_3\sqrt{3}\,)/2 \equiv 1$ (mod 2), that is, if and only if

$$(r_3 + s_3)/2 \equiv 1 \pmod 2 \qquad \text{and} \qquad s_3 \equiv 0 \pmod 6.$$

From these congruences we see immediately that r_3 is even if 2 is a cubic residue (mod p). Conversely, if r_3 is even, then s_3 is even as $4p = r_3^2 + 3s_3^2$, and so we can write $r_3 = 2r$ and $s_3 = 6s$. Then

$$(r_3 + s_3)/2 = r + 3s \equiv r^2 + 3(3s)^2 = p \equiv 1 \pmod 2,$$

so that 2 is a cubic residue (mod p). ■

The fact that $\left(\dfrac{2}{p}\right) = (-1)^{(p^2-1)/8}$, where p is any odd prime, is called the supplement to the law of quadratic reciprocity, because the prime 2 is not covered by the law of quadratic reciprocity. Likewise, the prime $1-\omega$ is not covered by the law of cubic reciprocity. In the next theorem we prove the supplement to the law of cubic reciprocity.

Theorem 8.1.9 (*Supplement to the law of cubic reciprocity.*) *Let π be a primary Eisenstein prime such that (without loss of generality) $\pi \equiv -1 \pmod 3$. If $\pi = q$ for a rational prime q, write $m = (q+1)/3$. If $N(\pi) = p \equiv 1 \pmod 3$, write $\pi = a + b\omega$, $(a,\ b \in Z)$ and $m = (a+1)/3$. Then*

$$\chi_\pi(1-\omega) = \omega^{2m}.$$

Proof Since $(1-\omega)^2 = -3\omega$, it suffices to show that $\chi_\pi(-3\omega) = \omega^m$. In the case $\pi = q$, this follows because

$$\chi_q(-3\omega) = \chi_q(\omega) = \omega^{(N(q)-1)/3} = \omega^{(q^2-1)/3} = \omega^m.$$

Now suppose that $N(\pi) = p \equiv 1 \pmod 3$. By Lemma 8.1.6 and Theorem 3.1.3, $\pi = a + b\omega = J(\chi_\pi, \chi_\pi) = (r_3 + is_3\sqrt{3}/2)$, so that $a = (r_3+s_3)/2$ and $b = s_3$, with $a \equiv -r_3 \equiv -1 \pmod 3$ and $b \equiv s_3 \equiv 0 \pmod 3$. Let g be a primitive root $\pmod p$ such that $\chi_\pi(g) = \omega$. Then since

$$\text{ind}_g 3 \equiv -s_3/3 \equiv -b/3 \pmod 3$$

by Corollary 2.6.9,

$$\chi_\pi(-3) = \chi_\pi(3) = \omega^{-b/3}.$$

Therefore

$$\chi_\pi(-3\omega) = \omega^{(p-1-b)/3}.$$

Since $p \equiv r_3^2/4 \pmod 9$, it is easily checked that $p - 1 - b \equiv a + 1 \pmod 9$. Thus

$$\chi_\pi(-3\omega) = \omega^{(p-1-b)/3} = \omega^{(a+1)/3} = \omega^m,$$

as desired. ■

The following numerical example illustrates manipulation of the cubic residue symbol.

Example The Eisenstein domain $Z[\omega]$ is a Euclidean domain with respect to its norm, that is, if $\alpha \in Z[\omega]$ and $\beta(\neq 0) \in Z[\omega]$, then there exist $\gamma \in Z[\omega]$ and $\delta \in Z[\omega]$ such that

$$\alpha = \beta\gamma + \delta, \quad N(\delta) < N(\beta);$$

see the book of Ireland and Rosen [1, Chapter 1]. By repeatedly using this property and the law of cubic reciprocity or its supplement, we can evaluate the cubic residue symbol $\left(\frac{\alpha}{\beta}\right)_3$. We illustrate the method with $\left(\frac{-9+4\omega}{5+8\omega}\right)_3$. First we note that $5+8\omega = -\omega^2(8+3\omega)$, where $8+3\omega$ is primary. Then

$$-9+4\omega = (8+3\omega)(-1+\omega)+(2+2\omega) = (8+3\omega)(-1+\omega)+(-\omega^2)2,$$

where $N(2+2\omega) < N(8+3\omega)$, $-\omega^2$ is a unit, and 2 is primary. Next

$$8+3\omega = 2(4+\omega)+\omega,$$

where $N(\omega) < N(2)$ and ω is a unit. Hence

$$\left(\frac{-9+4\omega}{5+8\omega}\right)_3 = \left(\frac{-9+4\omega}{8+3\omega}\right)_3 = \left(\frac{(-\omega^2)2}{8+3\omega}\right)_3 = \omega^2\left(\frac{2}{8+3\omega}\right)_3 = \omega^2\left(\frac{8+3\omega}{2}\right)_3$$
$$= \omega^2\left(\frac{\omega}{2}\right)_3 = \omega^2\omega = 1.$$

8.2 Biquadratic reciprocity

A Gaussian integer is an integer of the imaginary quadratic field $Q(\sqrt{-1}\,) = Q(i) = Q(e^{2\pi i/4})$, that is, a complex number of the form $x+iy$, where x and y are integers. The ring of Gaussian integers is $Z[i]$. It is known (see the book of Ireland and Rosen [1, Chapter 1] or the monograph of Pollard and Diamond [1, Chapter 1]) that $Z[i]$ is a unique factorization domain. We briefly review some facts about the Gaussian integers. The units in $Z[i]$ are $\pm 1, \pm i$. The primes in $Z[i]$ consist of the associates of rational primes $q \equiv 3 \pmod 4$, the associates of $1+i$, and the associates of those elements $\pi \in Z[i]$ whose norm $N(\pi)$ is a rational prime $p \equiv 1 \pmod 4$. If $\pi, \alpha \in Z[i]$, where π is prime and $\pi \nmid \alpha$, then $\alpha^{N(\pi)-1} \equiv 1 \pmod \pi$. For a Gaussian prime π, the finite field $Z[i]/(\pi)$ has $N(\pi)$ elements. By employing the same type of argument as in the proof of Proposition 8.1.1, we see that there is a unique integer m, $0 \le m \le 3$, such that $\alpha^{(N(\pi)-1)/4} \equiv i^m \pmod \pi$ whenever α, $\pi \in Z[i]$, π is prime, $\pi \nmid \alpha$, and $N(\pi) \neq 2$. This leads to the following definition of the quartic residue symbol.

Definition *Let π be a Gaussian prime with $N(\pi) \neq 2$ and let $\alpha \in Z[i]$. If $\pi \mid \alpha$ set $\left(\frac{\alpha}{\pi}\right)_4 = 0$. If $\pi \nmid \alpha$, let $\left(\frac{\alpha}{\pi}\right)_4$ be the unique power of i defined by*

$$\alpha^{(N(\pi)-1)/4} \equiv \left(\frac{\alpha}{\pi}\right)_4 \pmod \pi.$$

If $\beta = \pi_1 \cdots \pi_n$, where each π_j $(1 \le j \le n)$ is a Gaussian prime with $N(\pi_j) \neq 2$, define $\left(\frac{\alpha}{\beta}\right)_4$ by

$$\left(\frac{\alpha}{\beta}\right)_4 = \left(\frac{\alpha}{\pi_1}\right)_4 \cdots \left(\frac{\alpha}{\pi_n}\right)_4.$$

Lastly, if β is a unit of $Z[i]$, define $\left(\frac{\alpha}{\beta}\right)_4 = 1$ *for every* $\alpha \in Z[i]$, $\alpha \neq 0$, *and define* $\left(\frac{0}{\beta}\right)_4 = 0$.

When $N\pi \neq 2$, it is easy to see that $\left(\frac{\alpha}{\pi}\right)_4 = 1$ if and only if α is a quartic residue modulo the prime π. From the definition of the quartic residue symbol, it is easily deduced that $\left(\frac{\alpha}{\pi}\right)_4$ is a multiplicative function of α, and that $\left(\frac{\alpha}{\pi}\right)_4 = \left(\frac{\beta}{\pi}\right)_4$ if $\alpha \equiv \beta \pmod{\pi}$. Hence, $\left(\frac{\alpha}{\pi}\right)_4$ is a quartic residue character on the field $Z[i]/(\pi)$ of $N(\pi)$ elements. Write $\chi_\pi(\alpha) = \left(\frac{\alpha}{\pi}\right)_4$.

Definition *A Gaussian integer* $a + ib$ $(a,\ b \in Z)$ *is said to be primary if*

$$a \equiv 1 \pmod 2, \quad b \equiv 0 \pmod 2, \quad a + b \equiv 1 \pmod 4.$$

Equivalently, $a + ib$ *is primary if and only if* $a + ib \equiv 1 \pmod{2 + i2}$. *(See Problem 9.)*

The only unit that is primary is 1. If a Gaussian integer is primary, then necessarily it is not divisible by $1 + i$. A Gaussian integer that is not divisible by $1 + i$ is said to be odd. If α is an odd Gaussian integer, then, of the four associates of α, exactly one is primary; this is the analogue of Proposition 8.1.4. A Gaussian integer that is not a unit is primary if and only if it can be factored into a product of primary Gaussian primes; the proof is essentially the same as that of Proposition 8.1.5.

Lemma 8.2.1 *We have*

(8.2.1) $\left(\frac{\alpha}{\beta}\right)_4 \left(\frac{\overline{\alpha}}{\overline{\beta}}\right)_4 = 1$, *if* $\alpha, \beta \in Z[i]$, $\alpha \neq 0$, $\beta \neq 0$, *and* $\gcd(\alpha, \beta) = 1$;

(8.2.2) $\left(\frac{a}{b}\right)_4 = 1$, *if a and b are nonzero coprime rational integers with b odd;*

and

(8.2.3) $$\left(\frac{i}{a + ib}\right)_4 = i^{(1-a)/2}, \quad \textit{if } a + ib \textit{ is primary.}$$

Proof By an argument similar to the one used in the proof of Proposition 8.1.3,

$$\left(\frac{\overline{\alpha}}{\overline{\beta}}\right)_4 = \overline{\left(\frac{\alpha}{\beta}\right)_4},$$

from which (8.2.1) follows.

It suffices to prove (8.2.2) when b is an odd prime p that does not divide a. By Fermat's little theorem, $a^{p-1} \equiv 1 \pmod p$. If $p \equiv 3 \pmod 4$, then $a^{(p^2-1)/4} = a^{(p-1)(p+1)/4} \equiv 1 \pmod p$, and the desired result follows. If $p \equiv 1 \pmod 4$, then $p = \pi\overline{\pi}$ for some Gaussian prime π, and

$$\left(\frac{a}{p}\right)_4 = \left(\frac{a}{\pi}\right)_4 \left(\frac{a}{\overline{\pi}}\right)_4 = 1,$$

by (8.2.1), which proves (8.2.2).

Let $a + ib$ and $c + id$ be primary Gaussian integers. Since $b = 1 - a + 4j$ and $d = 1 - c + 4k$ for some $j, k \in Z$, it is easily seen that $(1-a)(1-c) \equiv bd \pmod 8$ and hence

$$i^{(1-a)/2} i^{(1-c)/2} = i^{(1-ac+bd)/2}.$$

Thus it suffices to prove (8.2.3) in the case that $a + ib$ is a primary Gaussian prime π. Since $b = 1 - a + 4j$, it follows easily that $a^2 + b^2 - 1 \equiv 2 - 2a \pmod{16}$. Thus, since $N(\pi) = a^2 + b^2$,

$$\left(\frac{i}{a+ib}\right)_4 = i^{(a^2+b^2-1)/4} = i^{(1-a)/2},$$

which proves (8.2.3). ■

The next lemma is the analogue of Lemma 8.1.6 for Gaussian primes.

Lemma 8.2.2 *Let π be a primary Gaussian prime with $N(\pi) = p \equiv 1 \pmod 4$. Then*

$$J(\chi_\pi, \chi_\pi) = (-1)^{(p+3)/4}\pi.$$

Proof By Theorem 2.1.3(b),

$$J(\chi_\pi, \chi_\pi)\overline{J(\chi_\pi, \chi_\pi)} = p = \pi\overline{\pi},$$

and, by Theorem 2.1.14,

$$J(\chi_\pi, \chi_\pi) \equiv 0 \pmod \pi.$$

Hence, $J(\chi_\pi, \chi_\pi) = \theta\pi$ for some unit θ of $Z[i]$. Next, by Theorem 3.2.1,

$$(-1)^{(p-1)/4} J(\chi_\pi, \chi_\pi) = K(\chi_\pi) = a_4 + ib_4,$$

where a_4 and b_4 are integers such that

$$\begin{array}{lll} a_4 \equiv -1 \pmod 4, & b_4 \equiv 0 \pmod 4, & \text{if } p \equiv 1 \pmod 8, \\ a_4 \equiv 1 \pmod 4, & b_4 \equiv 2 \pmod 4, & \text{if } p \equiv 5 \pmod 8, \end{array}$$

so that $-a_4 - ib_4 = (-1)^{(p+3)/4} J(\chi_\pi, \chi_\pi) = (-1)^{(p+3)/4}\theta\pi$ is primary. But π is primary, so that $(-1)^{(p+3)/4}\theta$ must be a primary unit, that is, $(-1)^{(p+3)/4}\theta = 1$, which gives $J(\chi_\pi, \chi_\pi) = (-1)^{(p+3)/4}\pi$, as required. ■

We are now ready to prove the law of biquadratic reciprocity.

Theorem 8.2.3 (*Law of biquadratic reciprocity.*) *Let α and β be coprime primary Gaussian integers. Set $\alpha = a + ib$ and $\beta = c + id$. Then*

$$\left(\frac{\alpha}{\beta}\right)_4 = \left(\frac{\beta}{\alpha}\right)_4 (-1)^{bd/4}. \tag{8.2.4}$$

Proof We first establish the theorem when α is a primary rational integer, that is, we shall show in this case that

$$\left(\frac{\alpha}{\beta}\right)_4 = \left(\frac{\beta}{\alpha}\right)_4. \tag{8.2.5}$$

It suffices to consider those primary α and β for which $\alpha = \pm q$, where $q(\neq 2)$ is a rational prime and $\beta = \pi$ is a Gaussian prime. If π is real, then (8.2.5) follows immediately from (8.2.2). We thus suppose that $N(\pi) = p \equiv 1 \pmod 4$. For brevity, we put $\chi = \chi_\pi$, so that by Lemma 8.2.2, $J(\chi, \chi) = \pm\pi$.

First, assume that $\alpha = -q$, where q is a rational prime with $q \equiv -1 \pmod 4$. By the multinomial theorem,

$$J^q(\chi, \chi) \equiv J(\overline{\chi}, \overline{\chi}) \pmod q,$$

and so

$$\pi^{q+1} = J^{q+1}(\chi, \chi) \equiv p \pmod q. \tag{8.2.6}$$

Also, by the multinomial theorem and Theorem 1.1.3,

$$G^q(\chi) \equiv \overline{\chi}^q(q)G(\chi^q) = \chi(q)G(\overline{\chi}) \pmod q,$$

which implies that

$$G^{q+1}(\chi) \equiv \chi(-q)p \pmod q. \tag{8.2.7}$$

By (8.2.6), (8.2.7), and Theorem 2.1.3,

$$\begin{aligned}\chi(-q)\pi^{q+1} &\equiv \chi(-q)p \equiv G^{q+1}(\chi) = \{G^2(\chi)\}^{(q+1)/2} = \{p^{\frac{1}{2}}J(\chi,\chi)\}^{(q+1)/2}\\ &= \{pJ^2(\chi,\chi)\}^{(q+1)/4} = (p\pi^2)^{(q+1)/4} \equiv \pi^{(q+3)(q+1)/4} \pmod q.\end{aligned}$$

Thus,

$$\left(\frac{-q}{\pi}\right)_4 = \chi(-q) \equiv \pi^{(q^2-1)/4} \equiv \left(\frac{\pi}{-q}\right)_4 \pmod q,$$

and (8.2.5) follows in the case that $\alpha = -q \equiv 1 \pmod 4$.

Now assume that $\alpha = q$, where q is a rational prime with $q \equiv 1 \pmod 4$. Let q have the prime factorization $\gamma\overline{\gamma}$ in $Z[i]$. By the multinomial theorem and Theorem 1.1.3 again,

$$G^q(\chi) \equiv \overline{\chi}^q(q)G(\chi^q) = \overline{\chi}(q)G(\chi) \pmod q,$$

and so

$$G^{q-1}(\chi) \equiv \overline{\chi}(q) \pmod{q}.$$

Thus,

$$\overline{\chi}(q) \equiv \{pJ^2(\chi,\chi)\}^{(q-1)/4} = (\overline{\pi}\pi^3)^{(q-1)/4} \equiv \left(\frac{\overline{\pi}}{\gamma}\right)_4 \left(\frac{\pi}{\gamma}\right)_4^3 = \left(\frac{\overline{\pi}}{\gamma}\right)_4 \overline{\left(\frac{\pi}{\gamma}\right)_4} \pmod{\gamma}.$$

Taking complex conjugates and using (8.2.1), we deduce that

$$\left(\frac{q}{\pi}\right)_4 = \chi(q) \equiv \left(\frac{\pi}{\overline{\gamma}}\right)_4 \left(\frac{\pi}{\gamma}\right)_4 = \left(\frac{\pi}{q}\right)_4 \pmod{\overline{\gamma}},$$

and (8.2.5) follows in the case that $\alpha = q \equiv 1 \pmod 4$. This completes the proof of (8.2.5).

We now prove (8.2.4) for arbitrary coprime primary Gaussian integers $\alpha = a + ib$ and $\beta = c + id$. In view of (8.2.5), we may assume that $\gcd(a,b) = \gcd(c,d) = 1$. Next choose $\epsilon = \pm 1$ so that ϵc is primary, that is, let $\epsilon = 1$ or -1 according to whether $4|d$ or $2 \parallel d$, respectively. Similarly, choose $\delta = \pm 1$ so that δa is primary. Since $c\alpha \equiv ac + bd \pmod{\beta}$,

$$(8.2.8) \quad \left(\frac{\epsilon c}{\beta}\right)_4 \left(\frac{\alpha}{\beta}\right)_4 = \left(\frac{\epsilon(ac+bd) + ib\epsilon(c+id)}{\beta}\right)_4 = \left(\frac{\delta}{\beta}\right)_4 \left(\frac{\delta\epsilon(ac+bd)}{\beta}\right)_4,$$

as $\left(\frac{\delta}{\beta}\right)_4 = \pm 1$ by (8.2.3). By (8.2.5), (8.2.2), and (8.2.3),

$$\left(\frac{\epsilon c}{\beta}\right)_4 = \left(\frac{\beta}{\epsilon c}\right)_4 = \left(\frac{c+id}{\epsilon c}\right)_4 = \left(\frac{id}{\epsilon c}\right)_4 = \left(\frac{i}{\epsilon c}\right)_4 = i^{(1-\epsilon c)/2}.$$

Next, we note that $\delta\epsilon(ac+bd) \equiv \delta\epsilon ac = (\delta a)(\epsilon c) \equiv 1 \pmod 4$, so that $\delta\epsilon(ac+bd)$ is primary. Then, by (8.2.5) and (8.2.8),

$$(8.2.9) \qquad \left(\frac{\alpha}{\beta}\right)_4 = \left(\frac{\delta}{\beta}\right)_4 \left(\frac{\beta}{\delta\epsilon(ac+bd)}\right)_4 i^{(\epsilon c-1)/2}.$$

Similarly,

$$(8.2.10) \qquad \left(\frac{\beta}{\alpha}\right)_4 = \left(\frac{\epsilon}{\alpha}\right)_4 \left(\frac{\alpha}{\delta\epsilon(ac+bd)}\right)_4 i^{(\delta a-1)/2}.$$

We show next that

$$\gcd(ad - bc, ac + bd) = 1.$$

Suppose, on the contrary, that there exists a rational prime p with

$$p \mid ad - bc, \quad p \mid ac + bd.$$

Clearly $ac+bd$ is odd, so that $p \neq 2$. Let π be a Gaussian prime dividing p. Then $\pi \mid \overline{\alpha}\beta$ and $\pi \mid \alpha\overline{\beta}$. But $\gcd(\alpha,\beta) = \gcd(\overline{\alpha},\overline{\beta}) = 1$, so we must have $\pi \mid \alpha$, $\pi \mid \overline{\alpha}$ or $\pi \mid \beta$, $\pi \mid \overline{\beta}$. If the first possibility holds, we have $\pi \mid \alpha + \overline{\alpha}$ and $\pi \mid \alpha\overline{\alpha}$, so that $\pi \mid 2a$, $\pi \mid a^2+b^2$. Hence, as $N(\pi) \neq 2$, we must have $\pi \mid a$, $\pi \mid b$, and so $p \mid a$, $p \mid b$, contradicting $\gcd(a,b) = 1$. The second possibility leads similarly to the contradiction $p \mid c$, $p \mid d$. This completes the proof of the assertion that $ad - bc$ and $ac + bd$ are coprime.

Then, by (8.2.2),

$$(8.2.11) \qquad \left(\frac{\beta\overline{\alpha}}{\delta\epsilon(ac+bd)}\right)_4 = \left(\frac{ac+bd+i(ad-bc)}{\delta\epsilon(ac+bd)}\right)_4 = \left(\frac{i(ad-bc)}{\delta\epsilon(ac+bd)}\right)_4 = \left(\frac{i}{\delta\epsilon(ac+bd)}\right)_4 .$$

Since $\epsilon = i^d$ and $\delta = i^b$, it follows easily from (8.2.3) that

$$\left(\frac{\epsilon}{\alpha}\right)_4 = \begin{cases} \epsilon, & \text{if } \epsilon = 1, \\ \delta, & \text{if } \epsilon = -1, \end{cases}$$

and

$$\left(\frac{\delta}{\beta}\right)_4 = \begin{cases} \delta, & \text{if } \delta = 1, \\ \epsilon, & \text{if } \delta = -1, \end{cases}$$

so that

$$\left(\frac{\epsilon}{\alpha}\right)_4 \left(\frac{\delta}{\beta}\right)_4 = 1.$$

Now take the complex conjugate of both sides of (8.2.10) to obtain (as $\left(\frac{\epsilon}{\alpha}\right)_4 = \pm 1$)

$$\overline{\left(\frac{\beta}{\alpha}\right)_4} = \left(\frac{\epsilon}{\alpha}\right)_4 \left(\frac{\overline{\alpha}}{\delta\epsilon(ac+bd)}\right)_4 i^{(1-\delta a)/2}.$$

From this equality and (8.2.9), by multiplication,

$$\left(\frac{\alpha}{\beta}\right)_4 \overline{\left(\frac{\beta}{\alpha}\right)_4} = \left(\frac{\epsilon}{\alpha}\right)_4 \left(\frac{\delta}{\beta}\right)_4 \left(\frac{\beta\overline{\alpha}}{\delta\epsilon(ac+bd)}\right)_4 i^{(\epsilon c-\delta a)/2} = \left(\frac{\beta\overline{\alpha}}{\delta\epsilon(ac+bd)}\right)_4 i^{(\epsilon c-\delta a)/2}.$$

Appealing to (8.2.11), we deduce that

$$(8.2.12) \qquad \left(\frac{\alpha}{\beta}\right)_4 \overline{\left(\frac{\beta}{\alpha}\right)_4} = \left(\frac{i}{\delta\epsilon(ac+bd)}\right)_4 i^{(\epsilon c-\delta a)/2}.$$

As $\epsilon c \equiv \delta a \equiv 1 \pmod 4$, it follows from (8.2.3) and (8.2.12) that

$$\left(\frac{\alpha}{\beta}\right)_4 \overline{\left(\frac{\beta}{\alpha}\right)_4} = i^{\{1-\delta\epsilon(ac+bd)+\epsilon c-\delta a\}/2} = i^{(1-\delta a)(1+\epsilon c)/2-\delta\epsilon bd/2}$$
$$= i^{-\delta\epsilon bd/2} = (-1)^{-\delta\epsilon bd/4} = (-1)^{bd/4},$$

which proves (8.2.4). ■

The law of biquadratic reciprocity can be used to derive criteria for residuacity. We shall illustrate this by giving another proof of Theorem 7.2.1, which states that for a prime $p \equiv 1 \pmod 4$, -3 is a quartic residue $\pmod p$ if and only if $3 \mid b_4$.

Second proof of Theorem 7.2.1 Let π be a primary prime with $N(\pi) = p \equiv 1 \pmod 4$. By Theorem 8.2.3,

$$\left(\frac{-3}{\pi}\right)_4 = \left(\frac{\pi}{-3}\right)_4.$$

Next we have

$$\left(\frac{\pi}{-3}\right)_4 = \chi_{-3}(\pi) \equiv \pi^{(N(3)-1)/4} = \pi^2 \pmod 3,$$

so that -3 is a quartic residue (mod p) if and only if $\pi^2 \equiv 1 \pmod 3$. By Lemma 8.2.2, $J(\chi_\pi, \chi_\pi) = (-1)^{(p+3)/4}\pi$, so that -3 is a quartic residue (mod p) if and only if $J^2(\chi_\pi, \chi_\pi) \equiv 1 \pmod 3$. Thus -3 is a quartic residue (mod p) if and only if (7.2.1) holds, and the result follows. ■

In Theorem 8.1.9 we gave a supplement to the law of cubic reciprocity for the exceptional Eisenstein prime $1-\omega$. For quartic reciprocity, the prime $1+i$ is exceptional, and we next prove an analogue of Theorem 8.1.9.

Theorem 8.2.4 (*Supplement to the law of biquadratic reciprocity.*) *Let $\alpha = a + ib$ be a primary Gaussian integer. Then*

$$\left(\frac{1+i}{\alpha}\right)_4 = i^{(a-b-1-b^2)/4}. \tag{8.2.13}$$

Proof We first establish (8.2.13) when $\alpha = a$ is a primary rational integer, that is, we shall show that, for $a \equiv 1 \pmod 4$,

$$\left(\frac{1+i}{a}\right)_4 = i^{(a-1)/4}. \tag{8.2.14}$$

If $a_1 \equiv a_2 \equiv 1 \pmod 4$, it is easily checked that

$$(a_1 - 1)/4 + (a_2 - 1)/4 \equiv (a_1 a_2 - 1)/4 \pmod 4. \tag{8.2.15}$$

Thus, it suffices to prove (8.2.14) when $a=p\equiv 1 \pmod 4$ and when $a=-q\equiv 1 \pmod 4$, where p and q are primes.

Suppose first that $a = p = \pi\overline{\pi}$, where $\pi = c+id$ is primary. Then, by (8.2.1),

$$\left(\frac{1+i}{p}\right)_4 = \left(\frac{1+i}{\pi}\right)_4\left(\frac{1+i}{\overline{\pi}}\right)_4 = \left(\frac{1+i}{\pi}\right)_4\left(\frac{i}{\overline{\pi}}\right)_4\left(\frac{1-i}{\overline{\pi}}\right)_4$$
$$= \left(\frac{i}{\overline{\pi}}\right)_4\left(\frac{1+i}{\pi}\right)_4\overline{\left(\frac{1+i}{\pi}\right)_4} = \left(\frac{i}{\overline{\pi}}\right)_4 = i^{(N\pi-1)/4} = i^{(p-1)/4},$$

as desired.

Second, suppose that $a=-q$, where q is a rational prime congruent to 3 modulo 4. Then

$$\left(\frac{1+i}{-q}\right)_4 \equiv (1+i)^{(q^2-1)/4} = (2i)^{(q^2-1)/8} = (2^{(q-1)/2})^{(q+1)/4}i^{(q^2-1)/8}$$
$$\equiv \left(\frac{2}{q}\right)^{(q+1)/4} i^{(q^2-1)/8} = ((-1)^{(q+1)/4})^{(q+1)/4}i^{(q^2-1)/8}$$
$$= (-1)^{-(q+1)^2/16}i^{(q^2-1)/8} = i^{-(q+1)^2/8+(q^2-1)/8} = i^{(-q-1)/4} \pmod q.$$

Hence, (8.2.14) is valid for $a=-q$. This completes the proof of (8.2.14).

By (8.2.14) and (8.2.15), it now suffices to prove (8.2.13) when $a+ib$ is primary with $\gcd(a,b)=1$. We set $a^* = (-1)^{b/2}a$ and note that

$$(8.2.16) \qquad a^* \equiv 1 \pmod 4 \quad\text{and}\quad (a^*-1)/2 \equiv (a-1)/2 + (b^2/4) \pmod 4.$$

By (8.2.2), (8.2.3), Theorem 8.2.3, and (8.2.16),

$$\left(\frac{1+i}{a+ib}\right)_4 = \left(\frac{i}{a^*}\right)_4^3\left(\frac{ib}{a^*}\right)_4\left(\frac{1+i}{a+ib}\right)_4 = i^{(a^*-1)/2}\left(\frac{a+ib}{a^*}\right)_4\left(\frac{1+i}{a+ib}\right)_4$$
$$= i^{(a^*-1)/2}\left(\frac{a^*}{a+ib}\right)_4\left(\frac{1+i}{a+ib}\right)_4 = i^{(a^*-1)/2}\left(\frac{i}{a+ib}\right)_4^b\left(\frac{a+ia}{a+ib}\right)_4$$
$$= i^{(a-1)/2+b^2/4+b^2/2}\left(\frac{i(a-b)}{a+ib}\right)_4 = i^{3b^2/4}\left(\frac{a-b}{a+ib}\right)_4.$$

As $a-b\equiv 1 \pmod 4$, we see that $a-b$ is primary. Hence, by Theorem 8.2.3, (8.2.2), and (8.2.14),

$$\left(\frac{1+i}{a+ib}\right)_4 = i^{-b^2/4}\left(\frac{a+ib}{a-b}\right)_4 = i^{-b^2/4}\left(\frac{a-b+b+ib}{a-b}\right)_4 = i^{-b^2/4}\left(\frac{b}{a-b}\right)_4\left(\frac{1+i}{a-b}\right)_4$$
$$= i^{-b^2/4}\left(\frac{1+i}{a-b}\right)_4 = i^{-b^2/4+(a-b-1)/4}.$$

This completes the proof. ■

The next two theorems give applications of the biquadratic reciprocity law.

Theorem 8.2.5 *Let p be a prime with $p \equiv 1 \pmod 4$ and write $p = a^2 + b^2$ with a odd. If $p \equiv 1 \pmod 8$, then b is a quartic residue $\pmod p$. If $p \equiv 5 \pmod 8$, then $b/2$ is a quartic residue $\pmod p$, provided that the sign of b is chosen such that $b/2 \equiv 1$ $\pmod 4$.*

Proof First suppose that $p \equiv 1 \pmod 8$. Replacing a by $-a$, if necessary, we may suppose without loss of generality that $a \equiv 1 \pmod 4$. As $p \equiv 1 \pmod 8$, we have $b \equiv 0 \pmod 4$. Then $a+b \equiv 1 \pmod 4$, and so $\pi = a + ib$ is primary. By Theorem 8.2.3 and (8.2.2),

$$\left(\frac{a}{\pi}\right)_4 = \left(\frac{\pi}{a}\right)_4 = \left(\frac{ib}{a}\right)_4 = \left(\frac{i}{a}\right)_4,$$

and thus, by (8.2.3),

$$\left(\frac{b}{\pi}\right)_4 = \left(\frac{i}{\pi}\right)_4 \left(\frac{a}{\pi}\right)_4 = \left(\frac{i}{\pi a}\right)_4 = i^{(1-a^2)/2} = 1.$$

Hence b is a quartic residue $\pmod \pi$, so that b is a quartic residue $\pmod p$.

Next suppose that $p \equiv 5 \pmod 8$, so that $b/2$ is odd. Choose the signs of a and b such that $a \equiv -1 \pmod 4$ and $b/2 \equiv 1 \pmod 4$. Then $\pi = a + ib$ and $b/2$ are primary, and so by Theorem 8.2.3 and (8.2.2),

$$\left(\frac{b/2}{\pi}\right)_4 = \left(\frac{\pi}{b/2}\right)_4 = \left(\frac{a}{b/2}\right)_4 = 1.$$

It follows that $b/2$ is a quartic residue $\pmod \pi$. Since $N(\pi) = p$, we have the field isomorphism

$$Z[i]/(\pi) \approx Z/pZ,$$

and so $b/2$ is a quartic residue $\pmod p$ as well. ∎

Let p be a prime such that

(8.2.17) $$p = 4f + 1 = a^2 + b^2 \equiv 5 \pmod 8,$$

and fix the signs of a and b such that

(8.2.18) $$a \equiv -1 \pmod 4, \quad b \equiv 2 \pmod 8.$$

There exists an $\epsilon = \pm 1$ such that

(8.2.19) $$2^f + 1 \equiv \epsilon 2^{(f+1)/2} \pmod p,$$

since $(2^f + 1)^2 \equiv 2^{f+1} \pmod p$. Brillhart [1] asked which sign $\epsilon = \pm 1$ in (8.2.19) is correct. This is answered by the next theorem. For further results of this kind, see the work of Lemmermeyer [3].

Theorem 8.2.6 *In the notation of* (8.2.17)–(8.2.18),

(8.2.20) $$2^f \equiv -b/a \pmod{p}$$

and

(8.2.21) $$2^f + 1 \equiv (-1)^{(b-10)/8} 2^{(f+1)/2} \pmod{p}.$$

Proof Let $\pi = a + ib$. By (8.2.18), π is primary. Write $\chi = \chi_\pi$. By (8.2.13), (8.2.3), and (8.2.18),

(8.2.22) $$\chi(2) = \chi^2(1+i)\overline{\chi}(i) = i^{(a-b-1-b^2)/2} i^{(a-1)/2} = i^{a-1} i^{(-b-b^2)/2} = -i.$$

Since $2^f \equiv \chi(2) \pmod{\pi}$ and $-i \equiv -b/a \pmod{\pi}$, (8.2.20) follows from (8.2.22). It remains to prove (8.2.21). By (8.2.20), (8.2.5), (8.2.13), (8.2.2), and (8.2.3),

(8.2.23) $$\chi(-2^f - 1) = \chi(a-b)\overline{\chi}(-a) = \left(\frac{\pi}{a-b}\right)_4 \left(\frac{\overline{\pi}}{-a}\right)_4 = \left(\frac{b(1+i)}{a-b}\right)_4 \left(\frac{-ib}{-a}\right)_4$$
$$= \left(\frac{1+i}{a-b}\right)_4 \left(\frac{i}{-a}\right)_4 = i^{(a-b-1)/4} i^{(a+1)/2} = i^{(3a+1-b)/4}.$$

By (8.2.22),

(8.2.24) $$\chi(2^{(f+1)/2}) = i^{3(f+1)/2}.$$

Combining (8.2.19), (8.2.23), and (8.2.24), we have

$$\chi(-\epsilon) = i^{(3a-b-6f-5)/4}.$$

Since $b^2 \equiv 4 \pmod{32}$, it follows easily from (8.2.17) and (8.2.18) that

(8.2.25) $$a \equiv \begin{cases} 1 - 2f \pmod{16}, & \text{if } f \equiv 1 \pmod{4}, \\ 9 - 2f \pmod{16}, & \text{if } f \equiv 3 \pmod{4}. \end{cases}$$

Hence $\chi(-\epsilon) = (-1)^{(2-b)/8}$. Since f is odd, $\chi(-\epsilon) = -\epsilon$, and so $\epsilon = (-1)^{(b-10)/8}$, which proves (8.2.21). ∎

We conclude this section with a numerical example to illustrate manipulation of the quartic residue symbol.

Example The Gaussian domain $Z[i]$ is Euclidean with respect to the norm, that is, if $\alpha \in Z[i]$ and $\beta(\neq 0) \in Z[i]$, then there exist $\gamma \in Z[i]$ and $\delta \in Z[i]$ such that

$$\alpha = \beta\gamma + \delta, \ N(\delta) < N(\beta);$$

see the book of Ireland and Rosen [1, Chapter 1]. Note that γ and δ are not necessarily uniquely determined by α and β. To see this observe that

$$2+3i=(1+i)2+i=(1+i)(3+i)-i.$$

Making use of this property, we evaluate $\left(\dfrac{7+2i}{-10+3i}\right)_4$. First, observe that

$$\left(\frac{7+2i}{-10+3i}\right)_4=\left(\frac{7+2i}{3+10i}\right)_4=-\left(\frac{3+10i}{7+2i}\right)_4.$$

Now,

$$3+10i=(7+2i)(1+i)+(-2+i)=(7+2i)(1+i)+(-i)(-1-2i),$$

where $-1-2i$ is primary and $N(-1-2i)<N(7+2i)$. Next,

$$7+2i=(-1-2i)(-2+2i)+1,$$

where 1 is primary and $N(1)<N(-1-2i)$. As 1 is a unit, we do not need to go any further. Thus

$$\begin{aligned}\left(\frac{3+10i}{7+2i}\right)_4&=\left(\frac{(-i)(-1-2i)}{7+2i}\right)_4=-i\left(\frac{-1-2i}{7+2i}\right)_4\\&=i\left(\frac{7+2i}{-1-2i}\right)_4=i\left(\frac{1}{-1-2i}\right)_4=i,\end{aligned}$$

so that

$$\left(\frac{7+2i}{-10+3i}\right)_4=-i.$$

8.3 Rational reciprocity laws

We begin by defining the rational residue symbol $\left(\dfrac{a}{q}\right)_{2^r}$ of order 2^r.

Definition *Let r be a natural number and assume that a is a nonzero 2^{r-1}th-power residue* (mod q), *where q is an odd prime. Then the rational residue symbol $\left(\dfrac{a}{q}\right)_{2^r}$ of order 2^r is defined by*

$$\left(\frac{a}{q}\right)_{2^r}=\begin{cases}1, & \text{if } a \text{ is a } 2^r\text{th-power residue} \pmod q,\\ -1, & \text{otherwise.}\end{cases}$$

When $r = 1$, this definition coincides with the definition of the Legendre symbol. In this section, $\left(\frac{a}{q}\right)_{2^r}$ always denotes the rational residue symbol. Thus, the meaning of $\left(\frac{a}{q}\right)_4$ in this section is not the same as the meaning of $\left(\frac{a}{q}\right)_4$ in Section 8.2.

Let p and q be distinct primes, each congruent to 1 modulo 4. Write

(8.3.1) $$p = a^2 + b^2 \text{ and } q = A^2 + B^2, \quad b \equiv B \equiv 0 \pmod 2).$$

Moreover, if $p \equiv q \equiv 1 \pmod 8$, write

(8.3.2) $$p = c^2 + 2d^2 \text{ and } q = C^2 + 2D^2, \quad d \equiv D \equiv 0 \pmod 2).$$

Further, if $p \equiv q \equiv 1 \pmod{16}$, from (3.4.3) and (3.4.4) we can write

(8.3.3) $$p = x^2 + 2v^2 + 2u^2 + 2w^2, \quad 2xv = u^2 - w^2 - 2uw$$

and

(8.3.4) $$q = X^2 + 2V^2 + 2U^2 + 2W^2, \quad 2XV = U^2 - W^2 - 2UW.$$

(Note that x, v, u, and w do not have the same meaning as in (3.7.34).)

Theorem 8.3.1 (*Rational quartic reciprocity law.*) *Let p and q be distinct primes with $p \equiv q \equiv 1 \pmod 4$ and $\left(\frac{p}{q}\right) = 1$. Then,*

(8.3.5) $$\left(\frac{p}{q}\right)_4 \left(\frac{q}{p}\right)_4 = \left(\frac{aA + bB}{q}\right) = (-1)^{(q-1)/4} \left(\frac{aB + bA}{q}\right),$$

where the values of the symbols above do not depend on the choice of signs of a, b, A, and B.

Proof First we note that

$$\left(\frac{aA + bB}{q}\right)\left(\frac{aA - bB}{q}\right) = \left(\frac{a^2A^2 - b^2B^2}{q}\right) = \left(\frac{a^2A^2 + b^2A^2 - b^2A^2 - b^2B^2}{q}\right)$$
$$= \left(\frac{pA^2 - b^2q}{q}\right) = \left(\frac{pA^2}{q}\right) = 1,$$

so that

$$\left(\frac{aA + bB}{q}\right) = \left(\frac{aA - bB}{q}\right) = \left(\frac{-aA + bB}{q}\right) = \left(\frac{-aA - bB}{q}\right).$$

Thus the value of $\left(\frac{aA + bB}{q}\right)$ is independent of the choice of signs for a, b, A, B. A similar argument shows that the value of $\left(\frac{aB + bA}{q}\right)$ is also independent of the choice of signs of a, b, A, B.

Let χ be a quartic character (mod p). By Theorem 3.2.1, we may choose the signs of a and b so that $K(\chi) = a + ib$. By Corollary 7.0.2 and Theorem 2.1.3, we have (recalling that $G(\chi^2) = p^{1/2}$)

$$1 \equiv \chi(q)G^{q-1}(\chi) = \left(\frac{q}{p}\right)_4 p^{(q-1)/4}K^{(q-1)/2}(\chi) \pmod{q}.$$

Thus, by Euler's criterion,

$$\left(\frac{p}{q}\right)_4\left(\frac{q}{p}\right)_4 \equiv (a+ib)^{(q-1)/2} \pmod{q}. \tag{8.3.6}$$

Next, appealing to the law of quadratic reciprocity, we have

$$A^{(q-1)/2} \equiv \left(\frac{A}{q}\right) = \left(\frac{|A|}{q}\right) = \left(\frac{q}{|A|}\right) = \left(\frac{A^2+B^2}{|A|}\right) = \left(\frac{B^2}{|A|}\right) = 1 \pmod{q}. \tag{8.3.7}$$

Therefore, by (8.3.6) and (8.3.7),

$$\left(\frac{p}{q}\right)_4\left(\frac{q}{p}\right)_4 \equiv (aA + iAb)^{(q-1)/2} = (aA + bB + ib(A+iB))^{(q-1)/2} \pmod{q}.$$

Then, working modulo $A + iB$, we obtain

$$\left(\frac{p}{q}\right)_4\left(\frac{q}{p}\right)_4 \equiv (aA+bB)^{(q-1)/2} \equiv \left(\frac{aA+bB}{q}\right) \pmod{A+iB}.$$

This proves the first equality in (8.3.5). Finally,

$$\begin{aligned}\left(\frac{aA+bB}{q}\right)\left(\frac{aB+bA}{q}\right) &= \left(\frac{abq+ABp}{q}\right) = \left(\frac{2}{q}\right)\left(\frac{2AB}{q}\right)\\ &= \left(\frac{2}{q}\right)\left(\frac{(A+B)^2-q}{q}\right) = \left(\frac{2}{q}\right) = (-1)^{(q-1)/4},\end{aligned}$$

from which the second equality in (8.3.5) follows. ∎

Example We use the rational quartic reciprocity law to determine if the congruence $x^4 \equiv 73 \pmod{137}$ is solvable. Set $p = 73$ and $q = 137$. First,

$$\left(\frac{p}{q}\right) = \left(\frac{73}{137}\right) = \left(\frac{137}{73}\right) = \left(\frac{64}{73}\right) = 1.$$

We can take $a = 3$, $b = 8$, $A = 11$, and $B = 4$, so that

$$\left(\frac{aA+bB}{q}\right) = \left(\frac{65}{137}\right) = \left(\frac{7}{65}\right) = \left(\frac{2}{5}\right)\left(\frac{7}{13}\right) = -\left(\frac{13}{7}\right) = -\left(\frac{-1}{7}\right) = 1.$$

Therefore, by Theorem 8.3.1,

$$\left(\frac{73}{137}\right)_4 = \left(\frac{137}{73}\right)_4 = \left(\frac{64}{73}\right)_4 = \left(\frac{8}{73}\right) = \left(\frac{2}{73}\right) = 1,$$

and so the congruence $x^4 \equiv 73 \pmod{137}$ is solvable. Indeed, $61^4 \equiv 73 \pmod{137}$.

If there are integers m and n for which

$$p = m^2 + qn^2, \tag{8.3.8}$$

then the rational quartic reciprocity law takes on a simpler form, as the next theorem shows.

Theorem 8.3.2 *Let p and q be distinct primes with $p \equiv q \equiv 1 \pmod 4$ and $\left(\frac{p}{q}\right) = 1$. Suppose that there are integers m and n so that (8.3.8) holds. Then*

$$\left(\frac{p}{q}\right)_4 \left(\frac{q}{p}\right)_4 = \begin{cases} 1, & \text{if } q \equiv 1 \pmod 8, \\ (-1)^n, & \text{if } q \equiv 5 \pmod 8. \end{cases}$$

Proof From (8.3.8), we obtain $\left(\frac{p}{q}\right)_4 = \left(\frac{m}{q}\right)$ and

$$\left(\frac{q}{p}\right)_4 = \left(\frac{pn^{-2} - m^2n^{-2}}{p}\right)_4 = \left(\frac{-m^2n^2}{p}\right)_4 = \left(\frac{2mn}{p}\right).$$

Thus,

$$\left(\frac{p}{q}\right)_4 \left(\frac{q}{p}\right)_4 = \left(\frac{m}{pq}\right)\left(\frac{2n}{p}\right). \tag{8.3.9}$$

Next, we put $m = 2^\lambda m_1$ and $n = 2^\nu n_1$, where m_1 and n_1 are odd. Exactly one of λ and ν is 0. Then, by (8.3.8),

$$\left(\frac{2}{p}\right)^\nu \left(\frac{n}{p}\right) = \left(\frac{n_1}{p}\right) = \left(\frac{p}{|n_1|}\right) = \left(\frac{m^2}{|n_1|}\right) = 1$$

and

$$\left(\frac{2}{pq}\right)^\lambda \left(\frac{m}{pq}\right) = \left(\frac{m_1}{pq}\right) = \left(\frac{pq}{|m_1|}\right) = \left(\frac{q^2n^2}{|m_1|}\right) = 1.$$

Then, from (8.3.9), we deduce that

$$\left(\frac{p}{q}\right)_4 \left(\frac{q}{p}\right)_4 = \left(\frac{2}{p}\right)^{\nu+\lambda+1} \left(\frac{2}{q}\right)^\lambda. \tag{8.3.10}$$

First, we suppose that $2 \mid n$, so that $\nu \geq 1$ and $\lambda = 0$. If $p \equiv 1 \pmod 8$, then $\left(\frac{2}{p}\right) = 1$, whereas if $p \equiv 5 \pmod 8$, we see that $\nu = 1$ from (8.3.8). Therefore, the right side of (8.3.10) is equal to 1 in both cases, as desired.

Second, suppose that $2 \nmid n$, so that $\nu = 0$. If $p \not\equiv q \pmod 8$, we see from (8.3.8) that $\lambda = 1$, and if $p \equiv q \pmod 8$, $\left(\frac{2}{p}\right) = \left(\frac{2}{q}\right)$. Therefore, in both cases, the right side of (8.3.10) is equal to $\left(\frac{2}{q}\right)$, as desired. ∎

Corollary 8.3.3 *Let $q = 5, 13,$ or 37 and let p be a prime with $p \equiv 1 \pmod 4$ and $\left(\frac{p}{q}\right) = 1$. Then $p = m^2 + qn^2$ for some integers m and n and*

$$\left(\frac{p}{q}\right)_4 \left(\frac{q}{p}\right)_4 = (-1)^n.$$

Proof It suffices to show that

$$p = m^2 + qn^2$$

for some integers m and n, for then the remainder of the theorem follows immediately from Theorem 8.3.2. We follow the approach of Mordell [7]. Since

$$\left(\frac{-q}{p}\right) = \left(\frac{p}{q}\right) = 1,$$

there is an integer $s \pmod p$ such that $s^2 \equiv -q \pmod p$. Consider the set $\{x - sy : 0 \leq x,\ y < \sqrt{p}\}$. This set contains more than p elements, and so there are two distinct elements in the set congruent to each other $\pmod p$. Taking their difference, we see that $m - sn \equiv 0 \pmod p$ for some integers m and n with $0 < |m| < \sqrt{p},\ 0 < |n| < \sqrt{p}$. Thus

$$m^2 + qn^2 = Mp$$

for some positive integer M with $Mp < (q+1)p$, i.e., $M \leq q$.

We consider now only the case $q = 5$, as the cases $q = 13$ and $q = 37$ are analogous. Since $m^2 \equiv Mp \pmod 5$ and $\left(\frac{p}{5}\right) = 1$, we cannot have $\left(\frac{M}{5}\right) = -1$, and so $M = 1,\ 4,$ or 5. If $M = 1$, then $p = m^2 + qn^2$, as desired. If $M = 5$, then $5 \mid m$ and

$$p = n^2 + 5(m/5)^2,$$

as desired. Finally, if $M = 4$, then both m and n must be even, and

$$p = (m/2)^2 + 5(n/2)^2,$$

as desired. ∎

The next corollary reestablishes Corollary 3.8.4 as well as extending it to primes $p \equiv 9 \pmod{20}$.

Corollary 8.3.4 *Let p be a prime with $p \equiv 1$ or $9 \pmod{20}$, so that (by Corollary 8.3.3) there are integers a, b, u, and v such that*

$$p = a^2 + b^2 = u^2 + 5v^2 \tag{8.3.11}$$

with a odd. If $p \equiv 1 \pmod{20}$, then

$$5 \mid a \text{ if and only if } 2 \mid u;$$

and if $p \equiv 9 \pmod{20}$, then

$$5 \mid a \text{ if and only if } 2 \nmid u.$$

Proof By Corollary 8.3.3,

$$\left(\frac{p}{5}\right)_4 \left(\frac{5}{p}\right)_4 = (-1)^v = (-1)^{u+1}. \tag{8.3.12}$$

Clearly,

$$\left(\frac{p}{5}\right)_4 = \begin{cases} \left(\frac{1}{5}\right)_4 = 1, & \text{if } p \equiv 1 \pmod{20}, \\ \left(\frac{9}{5}\right)_4 = \left(\frac{3}{5}\right) = -1, & \text{if } p \equiv 9 \pmod{20}. \end{cases} \tag{8.3.13}$$

By Theorem 7.2.3 (since exactly one of a and b is divisible by 5),

$$\left(\frac{5}{p}\right)_4 = \begin{cases} 1, & \text{if } b \equiv 0 \pmod 5, \\ -1, & \text{if } a \equiv 0 \pmod 5. \end{cases} \tag{8.3.14}$$

The corollary follows by putting (8.3.12), (8.3.13), and (8.3.14) together. ∎

Theorem 8.3.5 (*Rational octic reciprocity law.*) *Let p and q be distinct primes such that $p \equiv q \equiv 1 \pmod 8$ and $\left(\frac{p}{q}\right)_4 = \left(\frac{q}{p}\right)_4 = 1$. Then*

$$\left(\frac{p}{q}\right)_8 \left(\frac{q}{p}\right)_8 = \left(\frac{aB + bA}{q}\right)_4 \left(\frac{cD + dC}{q}\right), \tag{8.3.15}$$

where the symbols in (8.3.15) do not depend on the choice of signs for a, b, c, d, A, B, C, and D.

Proof By (8.3.5), $\left(\frac{aB+bA}{q}\right) = 1$, and so $\left(\frac{aB+bA}{q}\right)_4$ is well defined. Moreover, the value of the symbol $\left(\frac{aB+bA}{q}\right)_4$ is independent of the choice of signs of a, b, A, and B, because

$$\left(\frac{aB+bA}{q}\right)_4 \left(\frac{aB-bA}{q}\right)_4 = \left(\frac{-pA^2}{q}\right)_4 = \left(\frac{A}{q}\right) = 1,$$

by (8.3.7). A similar argument shows that the value of the Legendre symbol $\left(\frac{cD+dC}{q}\right)$ is independent of the choice of signs of c, d, C, and D.

Now let χ be an octic character (mod p). By Theorems 3.2.1 and 3.3.1, we can choose a, b, c and d so that $K(\chi^2) = a + ib$ and $K(\chi) = c + id\sqrt{2}$, respectively. Recall from the proof of Theorem 7.3.2 that $G^8(\chi) = pK^4(\chi)K^2(\chi^2)$. Therefore, by Corollary 7.0.2,

$$1 \equiv \chi(q)G^{q-1}(\chi) = \left(\frac{q}{p}\right)_8 p^{(q-1)/8} K^{(q-1)/4}(\chi^2) K^{(q-1)/2}(\chi) \pmod{q}.$$

Thus, by Euler's criterion,

$$\left(\frac{p}{q}\right)_8 \left(\frac{q}{p}\right)_8 \equiv (a+ib)^{(q-1)/4}(c+id\sqrt{2}\,)^{(q-1)/2} \pmod{q}. \tag{8.3.16}$$

Let $D = 2^\alpha D_1$, where α is an integer and D_1 is odd. Then

$$D^{(q-1)/2} \equiv \left(\frac{D}{q}\right) = \left(\frac{D_1}{q}\right) = \left(\frac{q}{|D_1|}\right) = \left(\frac{C^2+2D^2}{|D_1|}\right) = 1 \pmod{q}. \tag{8.3.17}$$

Also, by Theorem 8.2.5,

$$B^{(q-1)/4} \equiv 1 \pmod{q}. \tag{8.3.18}$$

Then, appealing to (8.3.16), (8.3.17), and (8.3.18), we obtain

$$\left(\frac{p}{q}\right)_8 \left(\frac{q}{p}\right)_8 \equiv (aB+ibB)^{(q-1)/4}(cD+idD\sqrt{2}\,)^{(q-1)/2} \pmod{q}. \tag{8.3.19}$$

Now let $K = Q(e^{2\pi i/8}) = Q(i, \sqrt{2}\,)$, and let Q be a prime ideal dividing $(A-iB)O_K$. Thus, Q divides $(A+iB)(A-iB)O_K = qO_K = (C+iD\sqrt{2}\,)(C-iD\sqrt{2}\,)O_K$, and we choose the sign of D so that Q divides $(C-iD\sqrt{2}\,)O_K$. Thus, by (8.3.19),

$$\begin{aligned}\left(\frac{p}{q}\right)_8 \left(\frac{q}{p}\right)_8 &\equiv (aB+bA-b(A-iB))^{(q-1)/4}(cD+dC-d(C-iD\sqrt{2}\,))^{(q-1)/2}\\ &\equiv (aB+bA)^{(q-1)/4}(cD+dC)^{(q-1)/2}\\ &\equiv \left(\frac{aB+bA}{q}\right)_4 \left(\frac{cD+dC}{q}\right) \pmod{Q}.\end{aligned}$$

As Q does not divide $2O_K$, (8.3.15) follows. ∎

Example We use the rational octic reciprocity law to determine if $p = 73$ is an eighth power modulo $q = 137$. From the example following Theorem 8.3.1, we know that 73 is a fourth power (mod 137), and that 137 is a fourth power (mod 73). We take $a = 3$, $b = 8$, $A = 11$, $B = 4$, $c = 1$, $d = 6$, $C = 3$, and $D = 8$. Thus,

$$\left(\frac{aB+bA}{q}\right)_4 = \left(\frac{100}{137}\right)_4 = \left(\frac{10}{137}\right) = \left(\frac{5}{137}\right) = \left(\frac{137}{5}\right) = \left(\frac{2}{5}\right) = -1$$

and

$$\left(\frac{cD+dC}{q}\right) = \left(\frac{26}{137}\right) = \left(\frac{13}{137}\right) = -1.$$

Thus, by Theorem 8.3.5,

$$\left(\frac{73}{137}\right)_8 = \left(\frac{137}{73}\right)_8 = \left(\frac{64}{73}\right)_8 = \left(\frac{8}{73}\right)_4 = \left(\frac{2}{73}\right)_4^3.$$

By Theorem 7.5.4, 2 is a quartic residue (mod 73), and so 73 is an eighth power (mod 137). Indeed, $32^8 \equiv 73 \pmod{137}$.

Theorem 8.3.6 (*Rational bioctic reciprocity law.*) *Let p and q be distinct primes such that $p \equiv q \equiv 1 \pmod{16}$ and $\left(\frac{p}{q}\right)_8 = \left(\frac{q}{p}\right)_8 = 1$. Let (x, v, u, w) and (X, V, U, W) be arbitrary solutions to (8.3.3) and (8.3.4), respectively. Set*

$$M = xX + 2vV + 2uU + 2wW$$

and

$$N = xV + vX + wU + uW + wW - uU.$$

Then the signs of a, b, c, d, A, B, C, and D can be chosen such that

$$bd(x^2 - 2v^2) \equiv ac(u^2 - w^2 + 2uw) \pmod{p} \tag{8.3.20}$$

and

$$BD(X^2 - 2V^2) \equiv AC(U^2 - W^2 + 2UW) \pmod{q}, \tag{8.3.21}$$

and

$$\left(\frac{p}{q}\right)_{16}\left(\frac{q}{p}\right)_{16} = \left(\frac{B}{q}\right)_8\left(\frac{D}{q}\right)_4\left(\frac{(aB+bA)(cD+dC)^2}{q}\right)_8\left(\frac{(XAD+VBC)(MAD+NBC)}{q}\right), \tag{8.3.22}$$

where the value of each of the symbols on the right side of (8.3.22) is independent of the choice of solution of (8.3.1) and (8.3.2) satisfying (8.3.20) and (8.3.21).

Proof First, note that the first three symbols on the right side of (8.3.22) are well defined by (8.3.18), (8.3.17), and (8.3.15), respectively. Fix arbitrary solutions x, v, u, and w of (8.3.3) and X, V, U, and W of (8.3.4). Possibly having to negate each of x, v, u, and w, we choose a bioctic character $\chi \pmod p$, by Theorem 3.4.1, so that

$$K(\chi) = x + v\sqrt{2} + iu\sqrt{2-\sqrt{2}} + iw\sqrt{2+\sqrt{2}}. \tag{8.3.23}$$

(Note that the negation of each of x, v, u, and w leaves (8.3.20) and (8.3.22) unchanged.) Similarly, we may assume that there is a bioctic character $\psi \pmod q$ such that

$$K(\psi) = X + V\sqrt{2} + iU\sqrt{2-\sqrt{2}} + iW\sqrt{2+\sqrt{2}}. \tag{8.3.24}$$

Fix the signs of a, b, c, d, A, B, C, and D by defining

$$K(\chi^2) = c + id\sqrt{2}, \qquad K(\chi^4) = a + ib, \tag{8.3.25}$$

$$K(\psi^2) = C - iD\sqrt{2}, \quad K(\psi^4) = A - iB. \tag{8.3.26}$$

Let $\beta = \exp(2\pi i/16)$, and let g be a primitive root $\pmod p$ such that $\chi(g) = \beta$. Set $K = Q(\beta)$. Choose a prime ideal factor P of pO_K such that $\chi = \chi_P$. Then $\beta \equiv g^{(p-1)/16} \pmod P$. By Theorem 2.1.14, P divides $K(\chi)O_K, K(\chi^2)O_K$, and $K(\chi^4)O_K$. From (8.3.25), we deduce that

$$ac \equiv -bd\sqrt{2} = -bd(\beta^2 - \beta^6) \pmod P.$$

Then, by (3.4.7), we see that (8.3.20) holds. Similarly, there is a prime ideal factor Q of qO_K which divides $K(\psi)O_K, K(\psi^2)O_K$, and $K(\psi^4)O_K$. By the same type of reasoning as above, (8.3.21) holds.

The values of the first two symbols on the right side of (8.3.22) are clearly independent of the choice of signs for B and D, respectively. Also, we have

$$\begin{aligned}\left(\frac{(aB+bA)(cD+dC)^2}{q}\right)_8 \left(\frac{(aB-bA)(cD-dC)^2}{q}\right)_8 &= \left(\frac{pB^2(pD^2)^2}{q}\right)_8 \\ &= \left(\frac{B}{q}\right)_4 \left(\frac{D}{q}\right) = 1,\end{aligned}$$

by (8.3.18) and (8.3.17). By Theorem 3.4.2, the members of (8.3.21) are nonzero $\pmod q$, so that if the signs of A, B, C, or D are changed in such a way as to preserve (8.3.21), then AD and BC are either both unchanged or both negated. In either case, the last symbol on the right side of (8.3.22) is unchanged. Thus, it suffices to prove (8.3.22) for the particular choice of a, b, c, d, A, B, C, and D given in (8.3.25) and (8.3.26). Next, we show that

$$G^{16}(\chi) = pK^2(\chi^4)K^4(\chi^2)K^8(\chi).$$

Making use of Theorem 2.1.3(a), and recalling that $G(\chi^8) = \sqrt{p}$, we obtain

$$\begin{aligned} pK^2(\chi^4)K^4(\chi^2)K^8(\chi) &= p\chi^{24}(4)J^2(\chi^4,\chi^4)J^4(\chi^2,\chi^2)J^8(\chi,\chi) \\ &= p\chi^{48}(2)\frac{G^{16}(\chi)}{G^2(\chi^8)} = G^{16}(\chi). \end{aligned}$$

Then, by Corollary 7.0.2,

$$1 \equiv \chi(q)G^{(q-1)}(\chi) = \left(\frac{q}{p}\right)_{16} p^{(q-1)/16}K^{(q-1)/8}(\chi^4)K^{(q-1)/4}(\chi^2)K^{(q-1)/2}(\chi) \pmod q.$$

Thus, by Euler's criterion, (8.3.25), (8.3.18), and (8.3.17),

$$\begin{aligned} \left(\frac{p}{q}\right)_{16}\left(\frac{q}{p}\right)_{16} &\equiv (a+ib)^{(q-1)/8}(c+id\sqrt{2})^{(q-1)/4}K^{(q-1)/2}(\chi) \\ &\equiv B^{(q-1)/8}(aB+bA-b(A-iB))^{(q-1)/8}\, D^{(q-1)/4} \\ &\quad (cD+dC-d(C-iD\sqrt{2}))^{(q-1)/4}K^{(q-1)/2}(\chi) \pmod q. \end{aligned}$$

Then, by (8.3.26),

$$\begin{aligned} \left(\frac{p}{q}\right)_{16}\left(\frac{q}{p}\right)_{16} &\equiv B^{(q-1)/8}(aB+bA)^{(q-1)/8}\, D^{(q-1)/4}(cD+dC)^{(q-1)/4}K^{(q-1)/2}(\chi) \\ &\equiv \left(\frac{B}{q}\right)_8\left(\frac{D}{q}\right)_4\left(\frac{(aB+bA)(cD+dC)^2}{q}\right)_8 K^{(q-1)/2}(\chi) \pmod Q. \end{aligned}$$

It remains to prove that

(8.3.27) $$0 \not\equiv (XAD+VBC)K(\chi) \equiv MAD+NBC \pmod Q.$$

By (8.3.26),

(8.3.28) $$BC \equiv AD\sqrt{2} \pmod Q,$$

and thus, by (8.3.24),

(8.3.29) $$AD\ \mathrm{Re}(K(\psi)) \equiv XAD+VBC \pmod Q.$$

Next, we show that $Q \nmid XAD+VBC$. Suppose, on the contrary, that we have $Q \mid XAD+VBC$. Then, as $q \nmid AD$, it follows from (8.3.29) that $Q \mid \mathrm{Re}(K(\psi))$. But $Q \mid K(\psi)$, and so we must have $Q \mid K(\overline{\psi})$. Hence Q^2 divides $K(\psi)K(\overline{\psi}) = q$, which is impossible. Hence $Q \nmid XAD+VBC$. It follows from (8.3.29) that

(8.3.30) $$\begin{aligned} 2(XAD+VBC)K(\chi) &\equiv ADK(\chi)\{K(\psi)+K(\overline{\psi})\} \equiv ADK(\chi)K(\overline{\psi}) \\ &\equiv AD\{K(\chi)K(\overline{\psi})+K(\overline{\chi})K(\psi)\} \\ &= 2AD\ \mathrm{Re}\,\{K(\chi)K(\overline{\psi})\} \pmod Q. \end{aligned}$$

A brief calculation, using (8.3.23) and (8.3.24), shows that

$$\operatorname{Re}(K(\chi)K(\overline{\psi})) = M + N\sqrt{2}. \tag{8.3.31}$$

Thus, by (8.3.31) and (8.3.28),

$$2AD \operatorname{Re}(K(\chi)K(\overline{\psi})) \equiv 2(MAD + NBC) \pmod{Q}. \tag{8.3.32}$$

Finally, (8.3.27) follows from (8.3.30) and (8.3.32). ■

Example We shall verify Theorem 8.3.6 for $p = 113$ and $q = 97$. Since $p \equiv 16 \pmod{q}$ and $q \equiv -16 \pmod{p}$, it is clear that p and q are 8th powers, but not 16th powers, modulo each other. Thus the left side of (8.3.22) has value 1. We note that $x = 1$, $v = 4$, $u = 6$, and $w = 2$ satisfy (8.3.3), and that $X = -7$, $V = -2$, $U = 4$, and $W = -2$ satisfy (8.3.4). We also take $a = 7$, $b = 8$, $c = 9$, $d = 4$, $A = 9$, $B = 4$, $C = 5$, and $D = 6$. With these values of the parameters, it is easily checked that (8.3.20) and (8.3.21) hold. Next, by Theorem 7.5.1,

$$\left(\frac{B}{q}\right)_8 = \left(\frac{4}{97}\right)_8 = \left(\frac{2}{97}\right)_4 = -1, \tag{8.3.33}$$

and, by (8.3.33) and Theorem 7.2.2,

$$\left(\frac{D}{q}\right)_4 = \left(\frac{6}{97}\right)_4 = -\left(\frac{3}{97}\right)_4 = 1. \tag{8.3.34}$$

By (8.3.34),

$$\left(\frac{(aB - bA)(cD - dC)^2}{q}\right)_8 = \left(\frac{-44 \cdot 34^2}{97}\right)_8 = \left(\frac{-36}{97}\right)_8 = \left(\frac{6}{97}\right)_4 = 1.$$

Short calculations show that $M = 17$ and $N = -62$. Then we find that $XAD+VBC = -418$ and $MAD + NBC = -322$. Thus,

$$\left(\frac{(XAD + VBC)(MAD + NBC)}{q}\right) = \left(\frac{57}{97}\right) = \left(\frac{97}{57}\right) = \left(\frac{97}{19}\right) = \left(\frac{2}{19}\right) = -1.$$

The preceding four calculations show that the product of the four symbols on the right side of (8.3.22) is 1, as desired.

Exercises 8

1. Prove (8.1.1). Hint: Consider $\{\alpha\mu : 0 \neq \mu \in Z[\omega]/(\pi)\}$.

2. Determine if $x^3 \equiv 2 + 3\omega \pmod{11}$ is solvable for $x \in Z[\omega]$, where $\omega = e^{2\pi i/3}$.

3. Determine if $x^3 \equiv 5 + 9\omega \pmod{7}$ is solvable for $x \in Z[\omega]$, where $\omega = e^{2\pi i/3}$.

4. Show that $\left(\dfrac{3+\omega}{5+6\omega}\right)_3 = \omega$.

5. Show that $\left(\dfrac{5+2\omega}{7+4\omega}\right)_3 = \omega$.

6. With the aid of Proposition 8.1.8, show that for a prime $p = a_3^2 + 3b_3^2 \equiv 1 \pmod{3}$, 2 is a cubic residue $\pmod{p}$ if and only if $3 \mid b_3$.

7. Using the law of cubic reciprocity, give a proof Theorem 7.1.3.

8. Use the supplement to the law of cubic reciprocity (Theorem 8.1.9) to prove Gauss' criterion for the cubic residuacity of 3 (Theorem 7.1.2).

9. Let a be an odd integer and let b be an even integer. Show that the two criteria given in Section 8.2 for $a + ib \in Z[i]$ to be primary are equivalent, that is, $a + b \equiv 1 \pmod{4}$ if and only if $a + ib \equiv 1 \pmod{2+2i}$.

10. Prove that the law of biquadratic reciprocity (Theorem 8.2.3) can also be written in the form

$$\left(\frac{\alpha}{\beta}\right)_4 = \left(\frac{\beta}{\alpha}\right)_4 (-1)^{(N(\alpha)-1)(N(\beta)-1)/16}.$$

11. Show that $\left(\dfrac{-1+2i}{1-4i}\right)_4 = -1$.

12. Show that $\left(\dfrac{-8+7i}{7+12i}\right)_4 = 1$.

13. Show that $\left(\dfrac{5+2i}{-2-7i}\right)_4 = -i$.

14. Use the law of biquadratic reciprocity to prove the criterion for the quartic residuacity of 3 given in Theorem 7.2.2.

15. Use the law of biquadratic reciprocity to prove the criterion for the quartic residuacity of 5 given in Theorem 7.2.3.

16. Use the supplement to the law of biquadratic reciprocity (Theorem 8.2.4) to prove Gauss' criterion for the quartic residuacity of 2 (Theorem 7.5.2).

The residue symbols in the problems below are all rational residue symbols, defined in Section 8.3.

17. Let $\pi = a + ib$ be primary with $8 \mid b$ and $\pi\overline{\pi} = p$, where $p = 8f + 1$ is a rational prime. Observe that

$$\left(\frac{2}{p}\right)_8 \equiv (-i(1+i)^2)^f \pmod{\pi}.$$

Using this, together with the supplement to the law of biquadratic reciprocity, establish Reuschle's criterion for the octic residuacity of 2 (Corollary 7.5.8).

18. Let $p = a^2 + b^2 \equiv 1 \pmod 8$ with $a \equiv 1 \pmod 4$. Prove that

$$\left(\frac{a}{p}\right)_4 = (-1)^{(a-1)/4}.$$

For Problems 19–24 below, use rational reciprocity laws (Theorems 8.3.2, 8.3.5, *and* 8.3.6).

19. Show that $\left(\frac{37}{73}\right)_4 = -\left(\frac{73}{37}\right)_4 = 1.$

20. Show that $\left(\frac{53}{113}\right)_4 = -\left(\frac{113}{53}\right)_4 = 1.$

21. Show that $\left(\frac{137}{409}\right)_4 = -\left(\frac{409}{137}\right)_4 = 1.$

22. (a) Prove that 17 and 409 are fourth powers modulo each other.
(b) Prove that 17 and 409 are eighth powers modulo each other.

23. (a) Show that $\left(\frac{97}{241}\right)_4 = \left(\frac{241}{97}\right)_4 = 1.$

(b) Show that $\left(\frac{97}{241}\right)_8 = \left(\frac{241}{97}\right)_8 = -1.$

24. Show that $\left(\frac{433}{449}\right)_{16} = \left(\frac{449}{433}\right)_{16} = -1.$

25. Let p and q be distinct primes each congruent to 1 (mod 4) and such that $\left(\frac{p}{q}\right) = 1$. Write $p = a^2 + b^2$ and $q = A^2 + B^2$ with b and B even. Prove that

$$\left(\frac{q}{p}\right)_4 \equiv \left(\frac{a/b - A/B}{a/b + A/B}\right)^{(q-1)/4} \pmod{q},$$

which was established by Gosset [1] in 1911. Use the congruence above to prove Theorem 8.3.1, the rational quartic reciprocity law.

26. A consequence of Theorem 8.3.1 is the following result:

$$\left(\frac{aA+bB}{p}\right)=\left(\frac{aA+bB}{q}\right).$$

Prove this result directly from the relation

$$pq=(aA+bB)^2+(aB-bA)^2.$$

27. If p is a prime with $p\equiv 1 \pmod 8$ and χ is a character (mod p) of order 8, we established in the proof of Theorem 7.3.2 that

$$G^8(\chi)=pK^2(\chi^2)K^4(\chi).$$

If p is a prime with $p\equiv 1 \pmod{16}$ and χ is a character (mod p) of order 16, we established in the proof of Theorem 8.3.6 that

$$G^{16}(\chi)=pK^2(\chi^4)K^4(\chi^2)K^8(\chi).$$

State and prove a generalization of these two results.

28. Supply the details for the cases $q=13$ and $q=37$ in the proof of Corollary 8.3.3.

29. Can Theorem 8.3.2 be deduced directly from the rational quartic reciprocity law (Theorem 8.3.1)?

Notes on Chapter 8

A weak form of the law of cubic reciprocity was conjectured by Jacobi [2]; see H. J. S. Smith [1, p. 92]. The first published proof of the law of cubic reciprocity is due to Eisenstein [2] in 1844 and is similar to the proof we have given (Theorem 8.1.7). The first published proof of the supplement to the law of cubic reciprocity is due to Eisenstein [3] in 1844. However, Jacobi had formulated the law of cubic reciprocity and its supplement much earlier, and had given essentially the same proofs in his lectures at Königsberg in the winter of 1836–37. In 1846, Jacobi [3], [CW, vol. 6, pp. 261–262] wrote that Eisenstein's proofs of 1844 were the proofs from the widely disseminated copies of Jacobi's lecture notes, and that the proofs had been known to Dirichlet and Kummer for several years. Eisenstein was deeply hurt by Jacobi's insinuation of plagiarism, and in a gracious response the following year, Eisenstein [9, pp. 273–274] wrote that Jacobi's lecture notes had been unavailable to him and that he had been entirely unaware of Jacobi's proofs. For further details, see the paper of Collison [1].

In [7], Eisenstein remarked that another proof of the law of cubic reciprocity could be constructed by using the theory of the integral $\int \frac{dx}{\sqrt{1-x^3}}$. Such a proof would be analogous to one of Gauss' proofs of the law of quadratic reciprocity and to Eisenstein's own proof of the law of biquadratic reciprocity. However, Eisenstein never published this proof of the cubic reciprocity law. Dantscher [1] in 1877 carried out the program indicated by Eisenstein, and 44 years later Koschmieder [1], evidently unaware of Dantscher's paper, again gave the proof suggested by Eisenstein. It is somewhat ironic that both papers were published in *Mathematische Annalen* and that both begin by referring to Eisenstein's paper [7] and the fact that Eisenstein never published his proof. Mertens [1] also gave an analytic proof of the cubic reciprocity law. While the proofs of Dantscher and Koschmieder involve the Weierstrass $\wp$-function, Mertens' proof uses theta functions. Habicht [1] generalized the famed lemma of Gauss used in many proofs of the law of quadratic reciprocity. He then used this generalization to give an elementary proof of the cubic reciprocity law that is similar to, but considerably more complicated than, the most well known of Gauss' proofs of the quadratic reciprocity law. Other proofs of the law of cubic reciprocity have been given by Friesen, Spearman, and Williams [1], Heima Hayashi [1], Shiratani [1], and Skolem [1]. The book of Ireland and Rosen [1] contains two proofs of the cubic reciprocity law. Cooke [1] gives a proof following Eisenstein's original proof.

Our definition of primary numbers in $Z[\omega]$ is the same as that given in H. J. S. Smith's report [1, p. 89]. However, Smith [1, p. 91] remarks that his definition is incorrect and that he should have written $\alpha \equiv -1 \pmod 3$ instead of $\alpha \equiv \pm 1 \pmod 3$. But he further remarks that in theory it is simpler to employ the more general congruence. We concur with this observation of Smith and thus use the "wrong" definition of primary numbers, as does Cooke [1].

Eisenstein [6] established algorithms for computing $\left(\frac{\alpha}{\beta}\right)_3$ and $\left(\frac{\alpha}{\beta}\right)_4$, but again, these algorithms seem to have been known to Jacobi. The numerical examples given at the end of Sections 8.1 and 8.2 illustrate these algorithms.

The supplement to the law of cubic reciprocity is due to Eisenstein [3]. The proof that we have given (Theorem 8.1.9) is due to Williams [13], who has given another very short proof in [27]. Other proofs of the supplement have been given by Friesen, Spearman, and Williams [1] and Watabe [1].

Gauss [4, p. 4] explains that one of the reasons why he gave so many proofs of the law of quadratic reciprocity is that he hoped that one of these proofs would lead to a proof of the law of biquadratic reciprocity. Indeed, Gauss [6, art. 67] was the first to enunciate the quartic reciprocity law, but he never published a proof. There is some debate as to whether Gauss was ever in possession of a complete proof. Shortly after Gauss stated his conjecture, Jacobi evidently found a proof, and he presented it in the aforementioned lectures at Königsberg. The first proof to be published was by Eisenstein [5] in 1844. This proof, another proof by Eisenstein [6], and Jacobi's proof are dependent on the theory of cyclotomy and are fairly similar to the proof that we have given (Theorem 8.2.3). Eisenstein published five proofs of the law of biquadratic reciprocity. The latter three [7], [8], [9] are based on the theory of the lemniscate integral $\int_0^z \frac{d\omega}{\sqrt{1-\omega^4}}$. It is thought that if Gauss had a proof, it probably was based on the lemniscate integral. Other proofs of the law of biquadratic reciprocity have been given by Mertens [1], Kaplan [1], Shiratani [1], and Skolem [1]. Cooke [1] has given a proof of the quartic reciprocity law based on Eisenstein's latter proofs.

The definitions of primary Gaussian integers that we have given in Section 8.2 are due to Gauss [6, art. 36].

The supplement to the law of biquadratic reciprocity is due to Eisenstein [6, p. 242]. The proof that we have given (Theorem 8.2.4) is due to Williams [25]. Another proof has been given by Watabe [2].

Nakahara [1] used the laws of cubic and quartic reciprocity to give simple proofs that the rings of integers in certain cyclic quartic and sextic algebraic number fields are not monogenic.

In 1847, Kummer discovered a general reciprocity law (along with supplementary laws) that includes the law of quintic reciprocity as a special case; see H. J. S. Smith [1, pp. 120–126] and Kummer [4], [5]. Write $O_K = Z[\beta]$, where $\beta = e^{2\pi i/5}$, and set $\lambda = 1 - \beta$. The law of quintic reciprocity states that for coprime elements $\nu, \mu \in O_K$ prime to λ that are congruent to rational integers (mod λ^3),

$$\left(\frac{\nu}{\mu}\right)_5 = \left(\frac{\mu}{\nu}\right)_5,$$

where the quintic power residue symbol is defined in Section 14.3. For prime $\pi \in O_K$ such that $\pi \equiv 1 + a\lambda^3 + b\lambda^4 + c\lambda^5 \pmod{\lambda^6}$ with $a, b, c \in Z/5Z$, we have the supplementary quintic laws

$$\left(\frac{\beta}{\pi}\right)_5 = \beta^{a+b}, \quad \left(\frac{(1+\sqrt{5})/2}{\pi}\right)_5 = \beta^a, \quad \left(\frac{\lambda}{\pi}\right)_5 = \beta^{-c}.$$

Grant [1] has proved the law of quintic reciprocity and the supplementary laws by studying the Jacobian variety of the hyperelliptic curve $y^2 = x^5 + 1/4$. A supplement to a quintic reciprocity law has also been proved by Williams [22].

The sign (root number) in the functional equation for L-functions of certain elliptic curves can be expressed in term of local Gauss sums over rings. For an application of the law of sextic reciprocity to the computation of the root number for the elliptic curve $y^2 = x^3 + D$, see Liverance [1].

In contrast to the aforementioned reciprocity laws, rational reciprocity laws (except for the law of quadratic reciprocity) have a considerably more recent history. In 1969, Burde [1] published his rational quartic reciprocity law (Theorem 8.3.1). In 1972, E. Lehmer [13] showed that a reciprocity law that she [7] had established in 1958 is equivalent to Burde's law. The proof of Theorem 8.3.1 that we have given is essentially the one given by Williams [26]. Theorem 8.3.2 is due independently to Brown [1] and E. Lehmer [13]. For consequences of the biquadratic rational reciprocity law and other formulations of the law in special cases, see two papers of Brown [2], [3]. See also Lemmermeyer [2].

The rational octic reciprocity law (Theorem 8.3.5) is due independently to Williams [23] and Wu [1]. The proof that we have given is an easier, shorter rendition of that given by Williams. Problem 17 is based on Wu's proof [1] of Corollary 7.5.8.

The rational bioctic reciprocity law (Theorem 8.3.6) was established by Leonard and Williams [6]. Our version is a bit more symmetric, and our proof is considerably shorter.

Evans [9] has established a rational 2^rth–power reciprocity law that includes the quartic, octic, and bioctic laws as special cases. A less explicit 2^rth–power reciprocity law had been previously proved by Furuta [1]. See also Helou [1].

A rational cubic reciprocity law has been established by von Lienen [1]. Burde [3] has also proved such a law. See also Z.–H. Sun [1].

The quadratic character of a quadratic surd modulo a prime p is discussed in Aigner [1], Evans [17], E. Lehmer [10], Lemmermeyer [1], and Williams, Hardy, and Friesen [1]. See also the references therein. This subject is intimately connected with rational biquadratic reciprocity laws. The cubic, quartic, and octic character of certain quadratic surds is examined in Aigner [3], Evans [14], Hiramatsu and Ishii [1], Ishii [1]–[4], and E. Lehmer [14], [15].

Reciprocity relations involving power residue characters of Jacobi sums have been given by Helou [2]–[5]. Helou appeals to Eisenstein's reciprocity laws, which we discuss in Chapter 14. Let $\beta_n = e^{2\pi i/n}$, and let $\omega, \pi \in Z[\beta_n]$ be chosen such that their norms $N(\omega)$, $N(\pi)$ are distinct rational primes $\equiv 1 \pmod{n}$. Helou evaluates the following quotient of nth–power residue symbols for certain n:

$$\left(\frac{N(\pi)}{\omega}\right)_n \left(\frac{N(\omega)}{\pi}\right)_n^{-1}.$$

See Helou's papers [3], [2] for the cases $n = 5$ and $n = 3$, 4, 8, respectively.

E. Lehmer [16] has written an excellent expository article delineating the history of rational reciprocity laws and their connections with other reciprocity laws, residuacity, and related aspects of number theory. An excellent source for general reciprocity laws is Hasse's book [3]. See also the comprehensive work of Lemmermeyer [3]. Further references to reciprocity laws are provided in the Notes on Chapter 14.

Chapter 9

Congruences for Binomial Coefficients

Throughout this chapter, k is an integer with $k \geq 2$ and $p = kf + 1$ is a prime. In Section 9.1, we relate binomial coefficients (mod p) to Jacobi sums. We then use this relationship in Section 9.2 to determine the binomial coefficients $\binom{rf}{sf}$ (mod p) for $k = 3, 4, 5, 6, 7, 8, 10$ and 12, where r and s are integers such that $0 < s < r < k$. The earliest result of this kind is due to Gauss, who showed that for $p = 4f + 1 = a^2 + b^2$ with $a \equiv 1 \pmod 4$,

(9.0.1) $$\binom{2f}{f} \equiv 2a \pmod p.$$

The congruence (9.0.1) was proved in Theorem 6.4.2 by using a congruence (mod p) for Jacobsthal sums (Theorem 6.4.1). Chowla, Dwork, and Evans [1] generalized (9.0.1) by giving a congruence for $\binom{2f}{f}$ (mod p^2), namely,

(9.0.2) $$\binom{2f}{f} = \left(2a - \frac{p}{2a}\right)\left(\frac{2^{p-1}+1}{2}\right) \pmod{p^2},$$

where $p = 4f + 1 = a^2 + b^2$, $a \equiv 1 \pmod 4$. In Section 9.3, we use the Gross–Koblitz formula (see (11.2.12)) to relate binomial coefficients (mod p^2) to Jacobi sums. We then apply this relationship in Section 9.4 to give congruences analogous to (9.0.2) for $\binom{rf}{sf}$ (mod p^2) for $k = 2, 3, 4, 6$ and 8, where r and s are integers with $1 \leq s < r \leq k$. These congruences (mod p^2) generalize the corresponding congruences (mod p) obtained in Section 9.2.

9.1 Binomial coefficients and Jacobi sums

The principal tool for determining the binomial coefficients $\binom{rf}{sf}$ (mod p) is Theorem 2.1.14, which we use in the following form.

Theorem 9.1.1 *Let $p = kf+1$ be prime, with $k \geq 3$. Set $\beta = e^{2\pi i/k}$ and $K = Q(\beta)$. Let χ be a character* (mod p) *of order k, so that $\chi = \chi_P$ for some prime ideal P of O_K dividing pO_K. Then for integers r and s such that $0 < s < r < k$,*

(9.1.1) $$\binom{rf}{sf} \equiv -2\mathrm{Re}J(\chi^s, \chi^{r-s}) \pmod P.$$

Proof In (2.1.11), choose $m = k - s$ and $n = r$, so that $m + n > k$, and thus

$$J(\chi^r, \chi^{k-s}) \equiv (-1)^{sf+1} \binom{rf}{sf} \pmod{P}. \tag{9.1.2}$$

Next choose $m = k - r$ and $n = s$ in (2.1.11), so that $m + n < k$, and thus

$$J(\chi^{k-r}, \chi^s) \equiv 0 \pmod{P}. \tag{9.1.3}$$

Combining (9.1.2) and (9.1.3), we have

$$(-1)^{sf+1} \binom{rf}{sf} \equiv J(\chi^r, \chi^{k-s}) + J(\chi^{k-r}, \chi^s) = 2\mathrm{Re} J(\chi^r, \chi^{k-s}) \pmod{P},$$

and (9.1.1) now follows from Theorem 2.1.5. ■

9.2 Binomial coefficients modulo p

Throughout this section, we use the notation of Theorem 9.1.1. We will apply this theorem for $k = 3, 4, 5, 6, 7, 8, 10$, and 12. In view of (9.1.1) and the remarks at the beginning of Chapter 3, it suffices to consider the binomial coefficients $\binom{mf}{f}$, where m is an integer in the range $2 \le m \le [(k+1)/2]$, for each $k \in \{3, 4, 5, 6, 7, 8, 10, 12\}$.

Theorem 9.2.1 *Let $p = 3f + 1$ be a prime. Then*

$$\binom{2f}{f} \equiv -r_3 \equiv \begin{cases} -2a_3 \pmod{p}, & \text{if } b_3 \equiv 0 \pmod{3}, \\ a_3 - 3b_3 \pmod{p}, & \text{if } b_3 \equiv 1 \pmod{3}, \\ a_3 + 3b_3 \pmod{p}, & \text{if } b_3 \equiv 2 \pmod{3}. \end{cases}$$

Proof The first congruence follows from (9.1.1) and Theorem 3.1.3. The second congruence follows from (3.1.2) and (3.1.3). ■

Theorem 9.2.2 *Let $p = 4f + 1$ be a prime. Then*

$$\binom{2f}{f} \equiv 2(-1)^{f+1} a_4 \pmod{p}.$$

Proof This follows from (9.1.1) and Table 3.2.1. ■

Since $(-1)^{f+1} a_4 = -\left(\frac{2}{p}\right) a_4 \equiv 1 \pmod{4}$, Theorem 9.2.2 is equivalent to Gauss' result (9.0.1).

Theorem 9.2.3 *Let $p = 5f + 1$ be a prime. Then*

$$\binom{2f}{f} \equiv \frac{1}{2}\left(-x + \frac{w(x^2 - 125w^2)}{4(xw + 5vu)}\right) \pmod p$$

and

$$\binom{3f}{f} \equiv \frac{1}{2}\left(-x - \frac{w(x^2 - 125w^2)}{4(xw + 5vu)}\right) \pmod p,$$

where (x, w, v, u) is any of the four solutions to (3.7.39).

Proof Observe that each of the 4-tuples (x, w, v, u), $(x, w, -v, -u)$, $(x, -w, -u, v)$, $(x, -w, u, -v)$ gives the same value in the congruences of Theorem 9.2.3. Let $\chi = \chi_P$ be a character $\pmod p$ of order 5, and choose (x, w, v, u) as in (3.7.35). By Theorem 2.1.5, $J(\chi, \chi^2) = J(\chi^2, \chi^2)$. Thus, from (9.1.1) and (3.7.35),

$$\binom{2f}{f} \equiv -(x + 5w\sqrt{5})/2 \pmod P$$

and

$$\binom{3f}{f} \equiv -(x - 5w\sqrt{5})/2 \pmod P.$$

It remains to show that

(9.2.1) $$\sqrt{5} \equiv -\frac{(x^2 - 125w^2)}{20(xw + 5vu)} \pmod P.$$

By (3.7.34), the denominator in (9.2.1) is $20(v^2 + vu - u^2)$. Hence (9.2.1) follows in the same way as (3.7.45). Note that the numerator and denominator in (9.2.1) are nonzero $\pmod p$, by the remarks following (3.7.45). ■

The next corollary is an immediate consequence of Theorem 9.2.3.

Corollary 9.2.4 *If $p = 5f + 1$ is prime, then*

$$\binom{2f}{f} + \binom{3f}{f} \equiv -x \pmod p.$$

Theorem 9.2.5 *If $p = 6f + 1$ is prime, then*

(9.2.2) $$\binom{2f}{f} \equiv (-1)^{f+1} u_3 \equiv \begin{cases} 2(-1)^{f+1} a_3 \pmod p, & \text{if } b_3 \equiv 0 \pmod 3, \\ (-1)^f (a_3 + 3b_3) \pmod p, & \text{if } b_3 \equiv 1 \pmod 3, \\ (-1)^f (a_3 - 3b_3) \pmod p, & \text{if } b_3 \equiv 2 \pmod 3, \end{cases}$$

and

(9.2.3) $$\binom{3f}{f} \equiv -2a_3 \pmod p.$$

Proof Both (9.2.3) and the first congruence in (9.2.2) follow from (9.1.1) and Table 3.1.2. The second congruence in (9.2.2) follows from (3.1.3) and (3.1.4). ■

Now let $k = 7$, $p = 7f + 1$, and $\beta = e^{2\pi i/7}$. We first determine $\binom{3f}{f}$.

Theorem 9.2.6 *Let* $p = 7f + 1$ *be a prime. Then*

$$\binom{3f}{f} \equiv -2t \pmod{p},$$

where t *is defined by* (3.9.25).

Proof This follows from (9.1.1) and Theorem 3.9.4. ■

We now consider $\binom{2f}{f}$ and $\binom{4f}{f}$ (mod p) for $p = 7f + 1$. Let $\chi = \chi_P$ be a character (mod p) of order 7, and choose $(x_1, x_2, \ldots, x_6)$ as in (3.9.23). By (9.1.1) and (3.9.23),

$$\binom{2f}{f} \equiv -2\mathrm{Re}J(\chi, \chi) \equiv -\frac{1}{6}\big(2x_1 + 7x_5E(\beta) + 21x_6F(\beta)\big) \pmod{P}, \tag{9.2.4}$$

where

$$E(\beta) = \beta + \beta^2 - 2\beta^3 - 2\beta^4 + \beta^5 + \beta^6$$

and

$$F(\beta) = \beta - \beta^2 - \beta^5 + \beta^6.$$

Since $J(\chi, \chi^3) = J(\chi^3, \chi^3)$ by Theorem 2.1.5, it follows from (9.1.1) and (9.2.4) that

$$\binom{4f}{f} \equiv -2\mathrm{Re}J(\chi^3, \chi^3) \equiv -\frac{1}{6}\big(2x_1 + 7x_5E(\beta^3) + 21x_6F(\beta^3)\big) \pmod{P}. \tag{9.2.5}$$

Similarly,

$$\binom{4f}{2f} \equiv -2\mathrm{Re}J(\chi^2, \chi^2) \equiv -\frac{1}{6}\big(2x_1 + 7x_5E(\beta^2) + 21x_6F(\beta^2)\big) \pmod{P}. \tag{9.2.6}$$

Adding (9.2.4), (9.2.5), and (9.2.6), and observing that

$$E(\beta) + E(\beta^2) + E(\beta^3) = F(\beta) + F(\beta^2) + F(\beta^3) = 0,$$

we obtain the following result.

Theorem 9.2.7 *Let $p = 7f+1$ be prime. Then*

$$\binom{2f}{f} + \binom{4f}{f} + \binom{4f}{2f} \equiv -x_1 \pmod{p},$$

where x_1 is defined by (3.9.23).

We could obtain congruences $\pmod{p}$ for the binomial coefficients $\binom{2f}{f}$ and $\binom{4f}{f}$ in terms of rational functions of $x_1, x_2, \ldots, x_6$. In order to do this, in view of (9.2.4) and (9.2.5), it suffices to obtain congruences $\pmod{P}$ for $E(\beta)$ and $F(\beta)$ (analogous to (9.2.1)). We will not do this here, but instead refer to the work of Nashier and Rajwade [1].

Theorem 9.2.8 *Let $p = 8f+1$ be a prime. Then*

$$\binom{2f}{f} \equiv (-1)^{1+(b_4/4)} 2a_8 \pmod{p},$$

$$\binom{3f}{f} \equiv (-1)^{f+1+(b_4/4)} 2a_4 \pmod{p},$$

and

$$\binom{4f}{f} \equiv (-1)^{f+1} 2a_8 \pmod{p}.$$

Proof This follows from (9.1.1) and Theorem 3.3.3, since $\chi(4) = (-1)^{b_4/4}$ by Theorem 7.5.1, where χ is any octic character $\pmod{p}$. ■

Theorem 9.2.9 *Let $p = 10f+1$ be prime. If 2 is a quintic residue* $\pmod{p}$, *then*

$$(-1)^f \binom{2f}{f} \equiv \binom{3f}{f} \equiv \binom{5f}{f} \equiv \frac{1}{8}\left(-4x - \frac{w(x^2-125w^2)}{xw+5vu}\right) \pmod{p}$$

and

$$(-1)^f \binom{4f}{f} \equiv \frac{1}{8}\left(-4x + \frac{w(x^2-125w^2)}{xw+5vu}\right) \pmod{p},$$

where (x, w, v, u) is any solution of (3.7.39). *If 2 is a quintic nonresidue* $\pmod{p}$, *then*

$$8(-1)^f \binom{2f}{f} \equiv x - 25w - 20v - 10u - \frac{(x-5w+10u)(x^2-125w^2)}{20(xw+5vu)} \pmod{p},$$

$$8\binom{3f}{f} \equiv x - 25w + 20v + 10u - \frac{(x-5w-10u)(x^2-125w^2)}{20(xw+5vu)} \pmod{p},$$

$$8(-1)^f \binom{4f}{f} \equiv x - 25w + 20v + 10u + \frac{(x-5w-10u)(x^2-125w^2)}{20(xw+5vu)} \pmod{p},$$

$$8\binom{5f}{f} \equiv x + 25w - 10v + 20u + \frac{(x+5w+10v)(x^2-125w^2)}{20(xw+5vu)} \pmod{p},$$

where (x, w, v, u) *is the solution of* (3.7.39) *satisfying*

$$u \equiv 0 \pmod 2, \quad x - v + u \equiv 0 \pmod 4. \tag{9.2.7}$$

Proof Let g be a primitive root $\pmod p$ and let $\chi = \chi_p$ be a character $\pmod p$ of order 10 such that $\chi(g) = \beta = e^{2\pi i/10}$. Define Z by $\chi(2) = \beta^Z$. Then by (3.7.35) and Theorems 3.7.2 and 3.7.4,

$$\begin{aligned}
4J(\chi,\chi) &= (-1)^f \beta^{6Z}\left(x - 5w\sqrt{5} - iu\sqrt{50-10\sqrt{5}} + iv\sqrt{50+10\sqrt{5}}\right),\\
4J(\chi,\chi^2) &= \beta^{4Z}\left(x - 5w\sqrt{5} - iu\sqrt{50-10\sqrt{5}} + iv\sqrt{50+10\sqrt{5}}\right),\\
4J(\chi,\chi^3) &= (-1)^f \beta^{2Z}\left(x + 5w\sqrt{5} + iv\sqrt{50-10\sqrt{5}} + iu\sqrt{50+10\sqrt{5}}\right),
\end{aligned}$$

and

$$4J(\chi,\chi^4) = \beta^{8Z}\left(x - 5w\sqrt{5} - iu\sqrt{50-10\sqrt{5}} + iv\sqrt{50+10\sqrt{5}}\right).$$

First, suppose that 2 is a quintic residue $\pmod p$, i.e., $Z \equiv 0 \pmod 5$. Then, by (9.1.1),

$$(-1)^f\binom{2f}{f} \equiv \binom{3f}{f} \equiv \binom{5f}{f} \equiv \frac{1}{2}(-x + 5w\sqrt{5}) \pmod P$$

and

$$(-1)^f\binom{4f}{f} \equiv \frac{1}{2}(-x - 5w\sqrt{5}) \pmod P.$$

The result now follows in the case $Z \equiv 0 \pmod 5$ from (9.2.1). Note that in this case the formulas for the binomial coefficients do not depend on which solution of (3.7.39) is taken among (x, w, v, u), $(x, w, -v, -u)$, $(x, -w, -u, v)$, $(x, -w, u, -v)$.

Now suppose that 2 is not a quintic residue $\pmod p$, and choose (x, w, v, u) satisfying (3.7.39) and (9.2.7). Then, by Theorem 3.7.9, $Z \equiv 1 \pmod 5$. The result now follows in the same way as before, with the computations facilitated by the formulas

$$20\beta^2 = -5 + 5\sqrt{5} + i\sqrt{50+10\sqrt{5}} + i2\sqrt{50-10\sqrt{5}},$$

and

$$20\beta^4 = -5 - 5\sqrt{5} + i2\sqrt{50+10\sqrt{5}} - i\sqrt{50-10\sqrt{5}}$$

(see the beginning of Section 3.7). ■

Theorem 9.2.10 *Let $p = 12f + 1$ be prime. Then*

$$\binom{2f}{f} \equiv 2(-1)^{f+1}\rho_4^2\rho_3 a_4 \pmod{p},$$
$$\binom{3f}{f} \equiv 2\rho_4 a_3 \pmod{p},$$
$$\binom{4f}{f} \equiv 2(-1)^f \rho_4 a_4 \pmod{p},$$
$$\binom{5f}{f} \equiv -2a_3\rho_3 \equiv -r_3 \pmod{p},$$

and

$$\binom{6f}{f} \equiv 2(-1)^{f+1}\rho_4^2 a_4 \pmod{p},$$

where

$$\rho_3 = \begin{cases} 1, & \text{if } b_3 \equiv 0 \pmod{3}, \\ (-a_3+b_3)/(a_3+b_3), & \text{if } b_3 \equiv 1 \pmod{3}, \\ (a_3+b_3)/(-a_3+b_3), & \text{if } b_3 \equiv 2 \pmod{3}, \end{cases}$$

and

$$\rho_4 = \begin{cases} 1, & \text{if } a_4 \equiv 1 \pmod{3}, \\ -1, & \text{if } a_4 \equiv 2 \pmod{3}, \\ -a_4/b_4, & \text{if } b_4 \equiv 1 \pmod{3}, \\ a_4/b_4, & \text{if } b_4 \equiv 2 \pmod{3}. \end{cases}$$

Proof Let g be a primitive root $\pmod{p}$ and let $\chi = \chi_P$ be a character $\pmod{p}$ of order 12 such that $\chi(g) = \beta = e^{2\pi i/12}$. Write $\omega = \beta^4 = e^{2\pi i/3} = (-1+i\sqrt{3})/2$ and $\chi(2) = \beta^Z$. By (3.1.3), $Z \equiv -b_3 \pmod{3}$, so that

(9.2.8) $$\chi(4) = \chi^2(2) = \left(\frac{2}{p}\right)\overline{\chi}^4(2) = \left(\frac{2}{p}\right)\omega^{b_3} = (-1)^f\omega^{b_3}.$$

We now express the values of the Jacobi sums $J(\chi,\chi^j)$, $1 \le j \le 5$, in terms of the parameters a_4, b_4, a_3, and b_3. By Theorem 3.5.2,

$$K(\chi) = \rho_4^2(a_4 + ib_4),$$

and so, by (9.2.8),

(9.2.9) $$J(\chi,\chi) = (-1)^f\omega^{-b_3}\rho_4^2(a_4 + ib_4).$$

By (3.5.2) and (9.2.8),

$$J(\chi,\chi^2) = c_{12}^3 \omega^{-b_3} J(\chi^2,\chi^2),$$

where c_{12} is defined just after (3.5.5). Hence, by Theorem 3.1.4 and (3.1.6),

$$J(\chi,\chi^2) = c_{12}^3 (a_3 + ib_3\sqrt{3}). \tag{9.2.10}$$

By (3.5.3) and Table 3.2.1,

$$J(\chi,\chi^3) = c_{12}^3 J(\chi^3,\chi^3) = c_{12}^3 (-1)^f (a_4 + ib_4). \tag{9.2.11}$$

By (3.5.4), (3.1.6), and Table 3.1.1,

$$J(\chi,\chi^4) = J(\chi^4,\chi^4) = \frac{1}{2}(r_3 + is_3\sqrt{3}) = \omega^{-b_3}(a_3 + ib_3\sqrt{3}). \tag{9.2.12}$$

Finally, by (3.5.5) and Theorem 3.5.2,

$$J(\chi,\chi^5) = (-1)^f K(\chi) = (-1)^f \rho_4^2 (a_4 + ib_4). \tag{9.2.13}$$

From (2.1.10),

$$J(\chi,\chi) \equiv J(\chi,\chi^2) \equiv 0 \pmod{P},$$

and so, by (9.2.9) and (9.2.10),

$$a_4 + ib_4 \equiv a_3 + ib_3\sqrt{3} \equiv 0 \pmod{P}. \tag{9.2.14}$$

It follows easily from (9.2.14) that

$$\rho_3 \equiv \omega^{b_3} \pmod{P} \quad \text{and} \quad \rho_4 \equiv -c_{12} \pmod{P}. \tag{9.2.15}$$

(In particular, ρ_3 and ρ_4 are third and fourth roots of unity (mod p), respectively.) We proceed to apply (9.1.1) for the Jacobi sums in (9.2.9)–(9.2.13). Multiplying both sides of (9.2.9) by -2 and taking real parts, we obtain

$$\binom{2f}{f} \equiv 2(-1)^{f+1}\rho_4^2\left\{a_4\mathrm{Re}(\omega^{-b_3}) - b_4\mathrm{Im}(\omega^{-b_3})\right\} \pmod{P}.$$

Consequently, since $b_4 \equiv ia_4 \pmod{P}$ by (9.2.14), we have modulo P

$$\binom{2f}{f} \equiv 2(-1)^{f+1}\rho_4^2 a_4\left\{\mathrm{Re}(\omega^{-b_3}) - i\mathrm{Im}(\omega^{-b_3})\right\} = 2(-1)^{f+1}\rho_4^2 a_4 \omega^{b_3} \equiv 2(-1)^{f+1} a_4 \rho_4^2 \rho_3,$$

where the last congruence is due to (9.2.15). This yields the desired congruence (mod p) for $\binom{2f}{f}$.

Next, by (9.2.10),

$$\binom{3f}{f} \equiv -2\mathrm{Re}(c_{12}^3(a_3 + ib_3\sqrt{3})) = -2(a_3\mathrm{Re}(c_{12}^3) - b_3\sqrt{3}\mathrm{Im}(c_{12}^3)) \pmod{P}.$$

Then, since $b_3\sqrt{3} \equiv ia_3 \pmod{P}$ by (9.2.14),

$$\binom{3f}{f} \equiv -2a_3(\mathrm{Re}(c_{12}^3) - i\mathrm{Im}(c_{12}^3)) = -2a_3c_{12} \equiv 2a_3\rho_4 \pmod{P},$$

where the last congruence is due to (9.2.15). By (9.2.11),

$$\begin{aligned}\binom{4f}{f} &\equiv 2(-1)^{f+1}\big(a_4\mathrm{Re}(c_{12}^3) - b_4\mathrm{Im}(c_{12}^3)\big) \equiv 2(-1)^{f+1}a_4\big(\mathrm{Re}(c_{12}^3) - i\mathrm{Im}(c_{12}^3)\big) \\ &= 2(-1)^{f+1}a_4c_{12} \equiv 2(-1)^f a_4\rho_4 \pmod{P}.\end{aligned}$$

By (9.2.12),

$$\binom{5f}{f} \equiv -r_3 \pmod{P},$$

and also

$$\begin{aligned}\binom{5f}{f} &\equiv -2(a_3\mathrm{Re}(\omega^{-b_3}) - b_3\sqrt{3}\,\mathrm{Im}(\omega^{-b_3})) \equiv -2a_3(\mathrm{Re}(\omega^{-b_3}) - i\mathrm{Im}(\omega^{-b_3})) \\ &= -2a_3\omega^{b_3} \equiv -2a_3\rho_3 \pmod{P}.\end{aligned}$$

Finally, by (9.2.13),

$$\binom{6f}{f} \equiv 2(-1)^{f+1}\rho_4^2 a_4 \pmod{P},$$

as desired. ■

9.3 Binomial coefficients, Jacobi sums, and p-adic gamma functions

Let $p = kf + 1$ be an odd prime, where f and k are integers with fixed $k \geq 2$. Write $\beta = e^{2\pi i/k}$, $K = Q(\beta)$, and $O_K = Z[\beta]$. Let g be a primitive root $\pmod{p}$, and let χ be a character $\pmod{p}$ of order k such that $\chi(g) = \beta$. For some prime ideal P in O_K dividing p, we have $\chi = \chi_P$ and $\beta = \chi(g) \equiv g^f \pmod{P}$ (as in Theorem 2.1.14).

In Theorem 9.3.1, we use the Gross–Koblitz formula (11.2.12) to give a congruence for $\binom{af+bf}{af}$ in terms of the Jacobi sum $J(\chi^{k-a}, \chi^{k-b}) \pmod{P^2}$, where a and b are positive integers with $a + b \leq k$.

We proceed to define Morita's p-adic gamma function Γ_p (Morita [1]). Define a function F on the nonnegative integers by

$$F(0) := 1, \quad F(N) := (-1)^N \prod_{\substack{0<j<N \\ p \nmid j}} j \qquad (N \geq 1). \tag{9.3.1}$$

Interpret (9.3.1) so that $F(1) = -1$. If M, N, and n are nonnegative integers such that $0 \leq M < N$, then

$$M \equiv N \pmod{p^n} \text{ implies } F(M) \equiv F(N) \pmod{p^n}. \tag{9.3.2}$$

To see this, note that when $M \equiv N \pmod{p^n}$,

$$F(N)/F(M) = (-1)^{N-M} \prod_{\substack{M \leq j < N \\ p \nmid j}} j \equiv 1 \pmod{p^n},$$

since the product of elements of a reduced residue system $\pmod{p^n}$ is $-1 \pmod{p^n}$, for any odd prime p.

Let Q_p denote the p-adic completion of the rationals Q, and let Z_p denote the ring of p-adic integers in Q_p. Denote the group of units in Z_p by Z_p^*. Morita's p-adic gamma function is the function $\Gamma_p : Z_p \longrightarrow Z_p^*$ defined by

$$\Gamma_p(z) = \lim_{N \longrightarrow z} F(N) \qquad (z \in Z_p), \tag{9.3.3}$$

where N runs through any sequence of positive integers p-adically approaching z. The limit exists, by (9.3.2). Moreover, (9.3.2) also shows that Γ_p is continuous. Observe that $\Gamma_p(N) = F(N)$ for nonnegative integers N, so that, in particular, $\Gamma_p(0) = 1$ and $\Gamma_p(1) = -1$. For $p^n > N > 0$, it follows easily from (9.3.1) that

$$F(N)F(p^n + 1 - N) \equiv (-1)^{E(N)} \pmod{p^n}, \tag{9.3.4}$$

where $E(z)$ is the least positive integer $\leq p$ congruent to $z \pmod{p}$. Letting $N \longrightarrow z$ in (9.3.4) and using the continuity of $\Gamma_p(z)$ and $E(z)$, we deduce the reflection formula

$$\Gamma_p(z)\Gamma_p(1-z) = (-1)^{E(z)}, \quad z \in Z_p. \tag{9.3.5}$$

We next relate Jacobi sums to p-adic gamma functions via the Gross–Koblitz formula. Any element of K can be viewed as an element in the p-adic completion K_P. Since $p \equiv 1 \pmod{k}$, we can identify K_P with its subfield Q_p; see the book of Narkiewicz [2, p. 217]. Let a and b be positive integers with $a + b \leq k$. Since $J(\chi^{k-a}, \chi^{k-b}) \in O_K$, we can view this Jacobi sum as an element of Z_p. In fact, we can view $J(\chi^{k-a}, \chi^{k-b})$ as an element of Z_p^*, since $J(\chi^{k-a}, \chi^{k-b}) \not\equiv 0 \pmod{P}$ by (2.1.11).

A direct consequence of the Gross–Koblitz formula for Gauss sums (11.2.12) is the following formula expressing Jacobi sums in terms of p-adic gamma functions:

$$J(\chi^{k-a}, \chi^{k-b}) = -\frac{\Gamma_p(a/k)\Gamma_p(b/k)}{\Gamma_p((a+b)/k)}, \tag{9.3.6}$$

where both members of (9.3.6) are elements of Z_p^*. By the reflection formula (9.3.5) and the equalities

$$E(a/k) = p - af, \quad E(b/k) = p - bf, \quad E((a+b)/k) = p - (a+b)f,$$

(9.3.6) can be written in the form

$$J(\chi^{k-a}, \chi^{k-b}) = \frac{\Gamma_p(1 - ((a+b)/k))}{\Gamma_p(1 - (a/k))\Gamma_p(1 - (b/k))}. \tag{9.3.7}$$

We now use (9.3.7) to relate $\binom{af+bf}{af}$ to $J(\chi^{k-a}, \chi^{k-b})$. By (9.3.2) and (9.3.3),

$$z_1 \equiv z_2 \pmod{p^n} \text{ implies } \Gamma_p(z_1) \equiv \Gamma_p(z_2) \pmod{p^n}, \tag{9.3.8}$$

and so, by (9.3.7),

$$\begin{aligned} J(\chi^{k-a}, \chi^{k-b}) &\equiv \frac{\Gamma_p(1 + (a+b)(p^2-1)/k)}{\Gamma_p(1 + a(p^2-1)/k)\Gamma_p(1 + b(p^2-1)/k)} \\ &\equiv \frac{F(1 + (a+b)(p^2-1)/k)}{F(1 + a(p^2-1)/k)F(1 + b(p^2-1)/k)} \pmod{p^2}. \end{aligned} \tag{9.3.9}$$

The right member of (9.3.9) equals

$$\begin{aligned} &- \prod_{\substack{j=1 \\ p \nmid j}}^{(a+b)(p^2-1)/k} j \prod_{\substack{j=1 \\ p \nmid j}}^{a(p^2-1)/k} j^{-1} \prod_{\substack{j=1 \\ p \nmid j}}^{b(p^2-1)/k} j^{-1} \\ &\equiv -(p-1)!^{af+bf} \prod_{j=1}^{af+bf} (afp + bfp + j) \\ &\quad \times (p-1)!^{-af} \prod_{j=1}^{af} (afp + j)^{-1} \times (p-1)!^{-bf} \prod_{j=1}^{bf} (bfp + j)^{-1} \\ &\equiv -\binom{af+bf}{af}\Big(1 + pf(a+b) \sum_{j=1}^{af+bf} 1/j \; - \; pfa\sum_{j=1}^{af} 1/j \; - \; pfb\sum_{j=1}^{bf} 1/j\Big) \\ &\equiv \binom{af+bf}{af}\Big(-1 + (p(a+b)/k) \sum_{j=1}^{af+bf} 1/j - (pa/k)\sum_{j=1}^{af} 1/j - (pb/k)\sum_{j=1}^{bf} 1/j\Big), \end{aligned}$$

where the congruences are modulo $p^2 Z_p$. Thus, by (9.3.9),

$$\binom{af+bf}{af} \equiv J(\chi^{k-a}, \chi^{k-b}) \tag{9.3.10}$$

$$\times\Big(-1-(p(a+b)/k)\textstyle\sum_{j=1}^{af+bf}1/j+(pa/k)\sum_{j=1}^{af}1/j+(pb/k)\sum_{j=1}^{bf}1/j\Big) \pmod{P^2},$$

where in (9.3.10), $J(\chi^{k-a}, \chi^{k-b})$ is now viewed simply as an element of O_K rather than Z_p. In the following theorem, we obtain a version of (9.3.10) in closed form.

Theorem 9.3.1 *Fix $k \geq 2$ and let $p = kf+1$ be an odd prime. Let a and b be positive integers with $a+b \leq k$. Write $\beta = e^{2\pi i/k}$. Let χ be a character* (mod p) *of order k, so that $\chi = \chi_P$ for some prime ideal P in $Z[\beta]$ dividing p. Then*

$$\binom{af+bf}{af} \equiv J(\chi^{k-a}, \chi^{k-b})\big(-1+aA_a+bA_b-(a+b)A_{a+b}\big) \pmod{P^2}, \tag{9.3.11}$$

where for any positive integer s,

$$A_s := \frac{1}{k}\sum_{j=1}^{k-1}\beta^{-sj}\big((1-\beta^j)^{p-1}-1\big). \tag{9.3.12}$$

Proof Observe that $P|A_s$, since $(1-\beta^j)^{p-1} \equiv 1 \pmod{P}$ for each j, $1 \leq j \leq k-1$. We have

$$\begin{aligned}
A_s - A_{s-1} &= -\frac{1}{k}\sum_{j=1}^{k-1}\beta^{-sj}(1-\beta^j)\big(1-(1-\beta^j)^{p-1}\big)\\
&= -\frac{1}{k}\sum_{j=0}^{k-1}\beta^{-sj}\big((1-\beta^j)-(1-\beta^j)^p\big)\\
&= -\frac{1}{k}\sum_{j=0}^{k-1}\sum_{m=1}^{p-1}(-1)^{m-1}\binom{p}{m}\beta^{j(m-s)}\\
&= -p\sum_{\substack{m=1\\ m\equiv s \pmod{k}}}^{p-1}\frac{(p-1)(p-2)\cdots(p-m+1)(-1)^{m-1}}{m!}.
\end{aligned}$$

Thus, for $1 \leq s \leq k$, $A_s - A_{s-1}$ is an integral multiple of p such that

$$\begin{aligned}
A_s - A_{s-1} &\equiv -p\sum_{\substack{m=1\\ m\equiv s \pmod{k}}}^{p-1} 1/m = -p\sum_{t=0}^{f-1}1/(s+tk) \equiv -fp\sum_{t=0}^{f-1}1/(sf-t)\\
&= -fp\sum_{j=(s-1)f+1}^{sf}1/j \equiv \frac{p}{k}\sum_{j=(s-1)f+1}^{sf}1/j \pmod{p^2}.
\end{aligned}$$

Thus, for $1 \le a \le k$,

$$A_a - A_0 = \sum_{s=1}^{a}(A_s - A_{s-1}) \equiv \frac{p}{k}\sum_{j=1}^{af} 1/j \pmod{p^2}.$$

The result (9.3.11) now follows from (9.3.10). ■

9.4 Binomial coefficients modulo p^2

Fix $k \ge 2$ and let a and b be positive integers such that $a + b \le k$. We will apply Theorem 9.3.1 to obtain congruences for $\binom{af+bf}{af} \pmod{p^2}$ in the cases $k = 2, 3, 4, 6$, and 8. As in Theorem 9.3.1, $\beta = e^{2\pi i/k}$ and $\chi = \chi_p$ has order k throughout.

Theorem 9.4.1 *Let $k = 2$, so that $p = 2f + 1$. Then*

$$\binom{2f}{f} = \binom{p-1}{(p-1)/2} \equiv (-1)^f(1 + (2^p - 2)) \pmod{p^2}.$$

Proof Since $\beta = -1$, (9.3.12) gives

$$A_1 = \frac{1}{2}\beta^{-1}\left((1-\beta)^{p-1} - 1\right) = -\frac{1}{2}(2^{p-1} - 1)$$

and

$$A_2 = \frac{1}{2}\beta^{-2}\left((1-\beta)^{p-1} - 1\right) = -A_1.$$

For a quadratic character $\chi \pmod p$,

$$J(\chi,\chi) = -\chi(-1) = -(-1)^f.$$

Now (9.3.11) with $a = b = 1$ yields

$$\binom{2f}{f} \equiv -(-1)^f(-1 + 4A_1) = (-1)^f(1 + (2^p - 2)) \pmod{p^2},$$

as desired. ■

Theorem 9.4.2 *Let $k = 3$, so that $p = 3f + 1$. Then*

$$\binom{2f}{f} \equiv -r_3 + \frac{p}{r_3} \pmod{p^2}$$

and

$$\binom{3f}{f} \equiv 1 + (3^p - 3)/2 \pmod{p^2}.$$

Proof Here $\beta = (-1 + i\sqrt{3})/2$ and

$$(1-\beta)^{p-1} = (\beta - \beta^2)^{p-1} = (i\sqrt{3})^{p-1} = (-3)^{(p-1)/2}.$$

Complex conjugation then gives

$$(1-\beta^2)^{p-1} = (-3)^{(p-1)/2}.$$

By Euler's criterion, $(-3)^{(p-1)/2} \equiv 1 \pmod p$, and so $2(-3)^{(p-1)/2} \equiv 3^{p-1} + 1 \pmod{p^2}$. Thus, by the definition (9.3.12) of A_s,

(9.4.1) $$A_3 = -2A_1 \quad\text{and}\quad A_1 = A_2 \equiv \frac{1}{6}(1 - 3^{p-1}) \pmod{p^2}.$$

For the cubic character (mod p), Table 3.1.1 yields

(9.4.2) $$J(\chi,\chi) = \frac{1}{2}(r_3 + is_3\sqrt{3})$$

and

(9.4.3) $$J(\chi,\chi^2) = -1.$$

By (2.1.11), $P \mid J(\chi,\chi)$, so that by (9.4.2),

$$-2ir_3s_3\sqrt{3} \equiv -3s_3^2 + r_3^2 = 2r_3^2 - 4p \pmod{P^2},$$

i.e.,

$$-is_3\sqrt{3} \equiv r_3 - 2p/r_3 \pmod{P^2}.$$

Thus, by (9.4.2),

(9.4.4) $$J(\chi^2,\chi^2) = \frac{1}{2}(r_3 - is_3\sqrt{3}) \equiv r_3 - p/r_3 \pmod{P^2}.$$

Using (9.4.1), (9.4.3), and (9.4.4) in (9.3.11), we obtain the desired results. ∎

Theorem 9.4.3 *Let $k = 4$, so that $p = 4f + 1$. Then*

$$\binom{2f}{f} \equiv (-1)^{f+1}(2a_4 - p/(2a_4))\big(1 + (2^{p-1} - 1)/2\big) \pmod{p^2},$$

$$\binom{3f}{f} \equiv (2a_4 - p/(2a_4))\big(-1 + (2^{p-1} - 1)/2\big) \pmod{p^2},$$

$$\binom{4f}{f} \equiv (-1)^f\big(1 + 3(2^{p-1} - 1)\big) \pmod{p^2},$$

and

$$\binom{4f}{2f} \equiv 1 + (2^p - 2) \pmod{p^2}.$$

Proof The last congruence follows from Theorem 9.4.1 with f replaced by $2f$, and so we restrict attention to the first three. Here $\beta = i$, and

$$(9.4.5)\qquad (1-\beta)^{p-1} = (1-\beta^3)^{p-1} = (-4)^{(p-1)/4} = (-1)^f 2^{(p-1)/2}.$$

Hence by (9.3.12),

$$(9.4.6)\qquad A_2 = (2^{p-1}-1)/4 - ((-4)^{(p-1)/4}-1)/2,$$

$$(9.4.7)\qquad A_4 = (2^{p-1}-1)/4 + ((-4)^{(p-1)/4}-1)/2,$$

and

$$(9.4.8)\qquad A_1 = A_3 = (1-2^{p-1})/4.$$

By Euler's criterion, $2^{(p-1)/2} \equiv (-1)^f \pmod{p}$, and so, by (9.4.5),

$$2(-4)^{(p-1)/4} \equiv 2^{p-1}+1 \pmod{p^2}.$$

Using this congruence in (9.4.6) and (9.4.7), we obtain

$$(9.4.9)\qquad A_2 \equiv 0 \pmod{p^2}, \quad A_4 \equiv (2^{p-1}-1)/2 \pmod{p^2}.$$

For the quartic character $\chi \pmod{p}$, Table 3.2.1 yields

$$(9.4.10)\qquad J(\chi,\chi) = (-1)^f(a_4+ib_4), \quad J(\chi,\chi^2) = a_4+ib_4, \quad J(\chi,\chi^3) = -(-1)^f.$$

By (2.1.11), $P \mid J(\chi,\chi)$, so that

$$-2a_4b_4i \equiv a_4^2 - b_4^2 = 2a_4^2 - p \pmod{P^2},$$

i.e.,

$$-b_4 i \equiv a_4 - p/(2a_4) \pmod{P^2}.$$

Thus,

$$(9.4.11)\qquad J(\chi^3,\chi^3) = (-1)^f(a_4-ib_4) \equiv (-1)^f(2a_4 - p/(2a_4)) \pmod{P^2}$$

and

$$(9.4.12)\qquad J(\chi^3,\chi^2) = a_4 - ib_4 \equiv 2a_4 - p/(2a_4) \pmod{P^2}.$$

Using (9.4.8)–(9.4.12) in (9.3.11), we obtain the desired results. ∎

Theorem 9.4.4 *Let $k = 6$, so that $p = 6f + 1$. Then*

$$\binom{2f}{f} \equiv (-1)^{f+1}(u_3 - p/u_3)(1 + (2^p - 2)/3) \pmod{p^2},$$
$$\binom{3f}{f} \equiv (2a_3 - p/(2a_3))(-1 + (2^p - 2)/3 - (3^p - 3)/4) \pmod{p^2},$$
$$\binom{4f}{f} \equiv (-1)^f(2a_3 - p/(2a_3))(-1 - 2(2^p - 2)/3 + (3^p - 3)/4) \pmod{p^2},$$
$$\binom{5f}{f} \equiv (u_3 - p/u_3)(-1 + 2(2^p - 2)/3) \pmod{p^2},$$
$$\binom{6f}{f} \equiv (-1)^f(1 + (2^p - 2) + (3^p - 3)/2) \pmod{p^2},$$
$$\binom{4f}{2f} \equiv -r_3 + p/r_3 \pmod{p^2},$$
$$\binom{5f}{2f} \equiv (2a_3 - p/(2a_3))(-1 + (2^p - 2)/3 + (3^p - 3)/4) \pmod{p^2},$$
$$\binom{6f}{2f} \equiv 1 + (3^p - 3)/2 \pmod{p^2},$$
$$\binom{6f}{3f} \equiv (-1)^f(1 + (2^p - 2)) \pmod{p^2}.$$

Proof The congruence for $\binom{6f}{3f}$ follows from Theorem 9.4.1 with f replaced by $3f$. The congruences for $\binom{4f}{2f}$ and $\binom{6f}{2f}$ follow from Theorem 9.4.2 with f replaced by $2f$. Hence we concentrate on the six remaining congruences. Here $\beta = (1 + i\sqrt{3})/2$, $1 - \beta = -\beta^2$, and $(1 - \beta^2)^2 = -3\beta^2$. Hence $(1 - \beta)^{p-1} = (1 - \beta^5)^{p-1} = 1$ and $(1 - \beta^2)^{p-1} = (1 - \beta^4)^{p-1} = (-3)^{(p-1)/2}$. Since $2(-3)^{(p-1)/2} \equiv (3^{p-1} + 1) \pmod{p^2}$, $(1 - \beta^2)^{p-1} = (1 - \beta^4)^{p-1} \equiv (3^{p-1} + 1)/2 \pmod{p^2}$. Thus, by (9.3.12),

$$6A_1 = 6A_5 \equiv -(2^{p-1} - 1) - (3^{p-1} - 1)/2 \pmod{p^2}, \tag{9.4.13}$$

$$6A_2 = 6A_4 \equiv (2^{p-1} - 1) - (3^{p-1} - 1)/2 \pmod{p^2}, \tag{9.4.14}$$

$$6A_3 \equiv -(2^{p-1} - 1) + (3^{p-1} - 1) \pmod{p^2}, \tag{9.4.15}$$

and

$$6A_6 \equiv (2^{p-1} - 1) + (3^{p-1} - 1) \pmod{p^2}. \tag{9.4.16}$$

For the sextic character $\chi \pmod{p}$, Table 3.1.2 yields

$$J(\chi, \chi) = (-1)^f(u_3 + iv_3\sqrt{3})/2, \qquad J(\chi, \chi^2) = J(\chi^2, \chi^3) = a_3 + ib_3\sqrt{3},$$
$$J(\chi, \chi^3) = (-1)^f(a_3 + ib_3\sqrt{3}), \qquad J(\chi, \chi^4) = (u_3 + iv_3\sqrt{3})/2,$$

and

$$J(\chi, \chi^5) = -(-1)^f. \tag{9.4.17}$$

By (2.1.11), $J(\chi,\chi) \equiv J(\chi,\chi^2) \equiv 0 \pmod{P}$, and so

$$-2iu_3v_3\sqrt{3} \equiv u_3^2 - 3v_3^2 \equiv 2u_3^2 - 4p \pmod{P^2},$$

i.e.,

$$-iv_3\sqrt{3} \equiv u_3 - 2p/u_3 \pmod{P^2},$$

and

$$-2ia_3b_3\sqrt{3} \equiv a_3^2 - 3b_3^2 \equiv 2a_3^2 - p \pmod{P^2},$$

i.e.,

$$-ib_3\sqrt{3} \equiv a_3 - p/(2a_3) \pmod{P^2}.$$

Thus

$$J(\chi^5,\chi^5) = (-1)^f(u_3 - iv_3\sqrt{3})/2 \equiv (-1)^f(u_3 - p/u_3) \pmod{P^2}, \tag{9.4.18}$$
$$J(\chi^5,\chi^4) = J(\chi^4,\chi^3) = a_3 - ib_3\sqrt{3} \equiv 2a_3 - p/(2a_3) \pmod{P^2}, \tag{9.4.19}$$
$$J(\chi^5,\chi^3) \equiv (-1)^f(2a_3 - p/(2a_3)) \pmod{P^2}, \tag{9.4.20}$$

and

$$J(\chi^5,\chi^2) \equiv (u_3 - p/u_3) \pmod{P^2}. \tag{9.4.21}$$

Using (9.4.13)–(9.4.21) in (9.3.11), we obtain the desired results. ■

Crandall, Dilcher, and Pomerance [1] have applied Theorems 9.4.2–9.4.4 in implementing a practical algorithm to search for Wilson primes p. (A prime p is called a Wilson prime if $(p-1)! \equiv -1 \pmod{p^2}$.) Their search showed that no Wilson primes $p < 4 \times 10^8$ exist, other than the known Wilson primes $p = 5, 13$, and 563.

We conclude this section by considering the case $k = 8$, with $p = 8f + 1$. For such p, 2 is a square $\pmod{p}$. Moreover, by Theorem 7.5.1,

$$\chi(4) = (-1)^{b_4/4}, \tag{9.4.22}$$

where χ is an octic character $\pmod{p}$. Write $w = \sqrt{2}$ and

$$y = w(2-w)^{p-1} - w(2+w)^{p-1}. \tag{9.4.23}$$

Note that y is an integer and that y does not change if w is replaced by $-w$ in (9.4.23).

The congruences in Theorem 9.4.5 below depend on y, but only on the value of $y \pmod{p^2}$. Thus Theorem 9.4.5 is still valid if, instead of taking $w = \sqrt{2}$ in (9.4.23), we take w to be the element in the ring of integers $\pmod{p^2}$ (unique up to sign) satisfying $w^2 \equiv 2 \pmod{p^2}$. For example, for $p = 17$, we could take $w = \pm 45$.

We can find the value of $w \pmod{p^2}$ in terms of a_8, b_8, a_4, and b_4, as follows. By (2.1.11),

$$J(\chi, \chi) \equiv J(\chi, \chi^2) \equiv 0 \pmod{P},$$

and so, by Theorem 3.3.3,

$$a_4 + ib_4 \equiv a_8 + ib_8\sqrt{2} \equiv 0 \pmod{P}.$$

Thus,

(9.4.24) $$-2a_4b_4i \equiv a_4^2 - b_4^2 \equiv 2a_4^2 - p \pmod{P^2}$$

and

(9.4.25) $$-2a_8b_8i\sqrt{2} \equiv a_8^2 - 2b_8^2 \equiv 2a_8^2 - p \pmod{P^2}.$$

Therefore in (9.4.23) we can take

(9.4.26) $$w = \pm(2a_4^2 - p)(2a_8^2 - p)/(4a_4a_8b_4b_8).$$

Theorem 9.4.5 *Let $k = 8$, so that $p = 8f + 1$. Write $w = \sqrt{2}$, $x = 2^{p-1} - 1$, and $y = w(2-w)^{p-1} - w(2+w)^{p-1}$. (Alternatively, define w and y by (9.4.26) and (9.4.23), respectively.) Then*

$$\binom{2f}{f} \equiv (-1)^{b_4/4}(2a_8 - p/(2a_8))(-1 - x/4 + y/8) \pmod{p^2},$$

$$\binom{3f}{f} \equiv (-1)^{f+(b_4/4)}(2a_4 - p/(2a_4))(-1 + x/4 + y/4) \pmod{p^2},$$

$$\binom{4f}{f} \equiv (-1)^{f}(2a_8 - p/(2a_8))(-1 - x - y/8) \pmod{p^2},$$

$$\binom{5f}{f} \equiv (2a_8 - p/(2a_8))(-1 + x + 3y/8) \pmod{p^2},$$

$$\binom{6f}{f} \equiv (-1)^{b_4/4}(2a_4 - p/(2a_4))(-1 - 3x/4 - y/4) \pmod{p^2},$$

$$\binom{7f}{f} \equiv (-1)^{f+(b_4/4)}(2a_8 - p/(2a_8))(-1 + 3x/4 - 3y/8) \pmod{p^2},$$

$$\binom{8f}{f} \equiv (-1)^{f+1}(-1 - 4x + y/2) \pmod{p^2},$$

$$\binom{4f}{2f} \equiv (2a_4 - p/(2a_4))(-1 - x/2) \pmod{p^2},$$

$$\binom{5f}{2f} \equiv (-1)^{f+(b_4/4)}(2a_8 - p/(2a_8))(-1 + x/4 + y/8) \pmod{p^2},$$

$$\binom{6f}{2f} \equiv (2a_4 - p/(2a_4))(-1 + x/2) \pmod{p^2},$$
$$\binom{7f}{2f} \equiv (-1)^{f+(b_4/4)}(2a_4 - p/(2a_4))(-1 + x/4 - 3y/4) \pmod{p^2},$$
$$\binom{8f}{2f} \equiv 1 + 3x \pmod{p^2},$$
$$\binom{6f}{3f} \equiv (-1)^{b_4/4}(2a_8 - p/(2a_8))(-1 - 3x/4 - 3y/8) \pmod{p^2},$$
$$\binom{7f}{3f} \equiv (2a_8 - p/(2a_8))(-1 + x - 5y/8) \pmod{p^2},$$
$$\binom{8f}{3f} \equiv (-1)^{f+1}(-1 - 4x - y/2) \pmod{p^2},$$

and

$$\binom{8f}{4f} \equiv 1 + 2x \pmod{p^2}.$$

Proof The congruences for $\binom{4f}{2f}$, $\binom{6f}{2f}$, $\binom{8f}{2f}$, and $\binom{8f}{4f}$ follow from Theorem 9.4.3 with f replaced by $2f$, and so we concentrate on the remaining 12 congruences. Since $\beta = (1+i)/\sqrt{2}$,

(9.4.27) $$(1-\beta)^{p-1} = ((1-\sqrt{2})(1+i))^{4f} = (1-\sqrt{2})^{4f}(2i)^{2f} = (-1)^f(2-\sqrt{2})^{4f}.$$

By (9.4.27), $(-1)^f(2-\sqrt{2})^{4f} - 1 \equiv 0 \pmod{P}$, so that

(9.4.28) $$(2-\sqrt{2})^{p-1} + 1 \equiv 2(-1)^f(2-\sqrt{2})^{4f} \pmod{P^2}.$$

By (9.4.27) and (9.4.28),

$$(1-\beta)^{p-1} = (1-\beta^7)^{p-1} \equiv ((2-w)^{p-1}+1)/2 \pmod{P^2}.$$

Similarly,

$$(1-\beta^3)^{p-1} = (1-\beta^5)^{p-1} \equiv ((2+w)^{p-1}+1)/2 \pmod{P^2}.$$

By (9.4.5),

$$(1-\beta^2)^{p-1} = (1-\beta^6)^{p-1} = 2^{(p-1)/2} \equiv (2^{p-1}+1)/2 \pmod{P^2},$$

where the congruence follows from

(9.4.29) $$2^{(p-1)/2} - 1 \equiv 0 \pmod{p}.$$

Thus, by (9.3.12),

$$8A_s \equiv ((2-w)^{p-1}-1)\mathrm{Re}(\beta^{-s}) + ((2+w)^{p-1}-1)\mathrm{Re}(\beta^{-3s}) \tag{9.4.30}$$
$$+ (2^{p-1}-1)\mathrm{Re}(\beta^{-2s}) + \beta^{4s}(2^{p-1}-1) \pmod{P^2}.$$

In particular,

(9.4.31) $8A_1 = 8A_7 \equiv \frac{1}{2}w\Big((2-w)^{p-1}-(2+w)^{p-1}\Big) - (2^{p-1}-1) \pmod{P^2}$,

(9.4.32) $8A_3 = 8A_5 \equiv \frac{1}{2}w\Big((2+w)^{p-1}-(2-w)^{p-1}\Big) - (2^{p-1}-1) \pmod{P^2}$,

(9.4.33) $8A_2 = 8A_6 \equiv 0 \pmod{P^2}$,

$$8A_4 \equiv -\Big((2-w)^{p-1}+(2+w)^{p-1}\Big) + 2^p \pmod{P^2},$$

and

$$8A_8 \equiv \Big((2-w)^{p-1}-1\Big) + \Big((2+w)^{p-1}-1\Big) + 2^p - 2 \pmod{P^2}.$$

We have already seen from (9.4.27) that $(-1)^f(2-w)^{4f} \equiv 1 \pmod{P}$, and similarly, from (9.4.27) with β^5 in place of β, $(-1)^f(2+w)^{4f} \equiv 1 \pmod{P}$. Thus

$$(2-w)^{4f} - (2+w)^{4f} \equiv 0 \pmod{P},$$

so that

$$(2-w)^{p-1}+(2+w)^{p-1} \equiv 2(2-w)^{(p-1)/2}(2+w)^{(p-1)/2} \equiv 2\cdot 2^{(p-1)/2} \equiv 2^{p-1}+1 \pmod{P^2},$$

where the last congruence follows from (9.4.29). Therefore,

$$8A_4 \equiv 2^{p-1}-1 \pmod{P^2} \tag{9.4.34}$$

and

$$8A_8 \equiv 3(2^{p-1}-1) \pmod{P^2}. \tag{9.4.35}$$

For the octic character $\chi \pmod{p}$, by (9.4.22) and Theorems 2.1.5, 3.3.1, and 3.3.3,

$$\begin{aligned}
J(\chi,\chi) &= J(\chi^3,\chi^3) = (-1)^{b_4/4}(a_8+ib_8\sqrt{2}),\\
J(\chi,\chi^2) &= J(\chi^2,\chi^5) = (-1)^{f+(b_4/4)}(a_4+ib_4),\\
(-1)^f J(\chi,\chi^3) &= J(\chi,\chi^4) = J(\chi^3,\chi^4) = a_8+ib_8\sqrt{2},\\
J(\chi,\chi^5) &= (-1)^{b_4/4}(a_4+ib_4),\\
J(\chi,\chi^6) &= J(\chi^2,\chi^3) = (-1)^{f+(b_4/4)}(a_8+ib_8\sqrt{2}),
\end{aligned}$$

and

$$J(\chi,\chi^7) = J(\chi^3,\chi^5) = -(-1)^f. \tag{9.4.36}$$

By (9.4.24) and (9.4.25),

$$-b_4 i \equiv a_4 - p/(2a_4) \pmod{P^2}$$

and

$$-b_8 i\sqrt{2} \equiv a_8 - p/(2a_8) \pmod{P^2}.$$

Thus,

$$J(\chi^7,\chi^7) = J(\chi^5,\chi^5) \equiv (-1)^{b_4/4}(2a_8 - p/(2a_8)) \pmod{P^2}, \tag{9.4.37}$$
$$J(\chi^7,\chi^6) = J(\chi^6,\chi^3) \equiv (-1)^{f+(b_4/4)}(2a_4 - p/(2a_4)) \pmod{P^2}, \tag{9.4.38}$$
$$J(\chi^7,\chi^5) \equiv (-1)^{f}(2a_8 - p/(2a_8)) \pmod{P^2}, \tag{9.4.39}$$
$$J(\chi^7,\chi^4) = J(\chi^5,\chi^4) \equiv (2a_8 - p/(2a_8)) \pmod{P^2}, \tag{9.4.40}$$
$$J(\chi^7,\chi^3) \equiv (-1)^{b_4/4}(2a_4 - p/(2a_4)) \pmod{P^2}, \tag{9.4.41}$$

and

$$J(\chi^7,\chi^2) = J(\chi^6,\chi^5) \equiv (-1)^{f+(b_4/4)}(2a_8 - p/(2a_8)) \pmod{P^2}. \tag{9.4.42}$$

Using (9.4.31)–(9.4.42) in (9.3.11), we obtain the desired results. ■

Exercises 9

1. Let p be a prime, and let m and n be nonnegative integers with $m+n=p-1$. Prove that $m!\,n! \equiv (-1)^{m+1} \pmod{p}$.

Hint: Consider the binomial coefficient $\binom{p-1}{m}$.

2. Let $p = 9f+1$ be a prime. Prove that

$$\binom{2f}{f}\binom{5f}{f}\Big/\binom{4f}{2f} \equiv \binom{4f}{f}\binom{5f}{2f}\Big/\binom{3f}{f} \equiv -r_3 \pmod{p}.$$

3. Use the results in Problem 8 of Exercises 3 and Theorem 9.1.1 to show that if $p = 11f+1$ is a prime, then

$$\binom{3f}{f}\binom{6f}{3f}\Big/\binom{4f}{2f} \equiv a \pmod{p},$$

where $4p = a^2 + 11b^2$, $a \equiv 2 \pmod{11}$.

4. Use the results in Problem 9 of Exercises 3 to determine $\binom{4f}{f}$ and $\binom{7f}{2f}$ modulo a prime $p = 13f + 1$ in terms of A, B, C, D.

5. If $p = 14f + 1$ is a prime, show that

$$\binom{5f}{f} \equiv \binom{7f}{3f} \pmod{p},$$
$$\binom{3f}{f} \equiv \binom{7f}{2f} \pmod{p},$$

and

$$\binom{7f}{f} \equiv (-1)^f \binom{8f}{3f} \pmod{p}.$$

6. Use Theorem 9.1.1 and the results in Problem 10 of Exercises 3 to determine $\binom{3f}{f}$ and $\binom{5f}{f}$ modulo a prime $p = 15f + 1$.

7. Use the results in Problems 11, 12, and 14 of Exercises 3 to determine appropriate binomial coefficients modulo p, where $p = 21f+1$, $39f+1$, and $60f+1$, respectively.

8. Prove that for a prime $p = mnf + 1$, with $m \geq 2$ and $n \geq 2$,

$$n^{(p-1)t/m} \equiv (ntf)! \prod_{j=1}^{n-1} (mjf)! \Big/ \prod_{j=0}^{n-1} ((mj+t)f)! \pmod{p}, \quad t = 1, 2, \ldots, m-1.$$

9. Use the result of Problem 8 to prove that

$$n^{(p-1)t/m} \equiv \prod_{j=2}^{n} \binom{jft}{ft} \Big/ \prod_{j=1}^{n-1} \binom{jmf+ft}{ft} \pmod{p}.$$

10. If $p = 12f + 1$ is a prime, use Problem 9 to show that, in the notation of Theorem 9.2.10,

$$3^{(p-1)/4} \equiv 2(-1)^f c_{12}^2 \rho_3 a_3 / r_3 \pmod{p},$$

where c_{12} is defined just after (3.5.5).

11. Let k be an integer with $k \geq 3$. Let $p = kf + 1$ be a prime. If g, h, i are integers satisfying $1 \leq h < g \leq k-1$, $\ 1 \leq h < i \leq k-1$, $\ k - i > g - h$, prove that

$$\binom{gf}{hf}\binom{(k-g)f}{(i-h)f} \equiv (-1)^{(g+i)f} \binom{if}{hf}\binom{(k-i)f}{(g-h)f} \pmod{p}.$$

12. Prove the analogue of Theorem 9.4.5 in the case $k = 12$, i.e., $p = 12f + 1$. For example, prove that

$$\binom{2f}{f} \equiv 4^f(1 - x/6)(2a_{12} - p/(2a_{12}))(-1 - x/6 + y/12) \pmod{p^2},$$

where

$$x = 2^{p-1} - 1,\ y = w(2-w)^{p-1} - w(2+w)^{p-1},\quad w = \pm(2a_3^2 - p)(2a_4^2 - p)/(4a_4b_4a_3b_3).$$

Hint: First show that $w^2 \equiv 3 \pmod{p^2}$. Then use (9.3.12) with $\beta = e^{2\pi i/12} = (\sqrt{3}+i)/2$ to show that, with v defined by $3^{p-1} - 1$,

$$\begin{aligned}
A_1 = A_{11} &\equiv y/24 - v/24 - x/12 \pmod{p^2},\\
A_2 = A_{10} &\equiv -v/24 \pmod{p^2},\\
A_3 = A_9 &\equiv v/12 - x/12 \pmod{p^2},\\
A_4 = A_8 &\equiv x/6 - v/24 \pmod{p^2},\\
A_5 = A_7 &\equiv -y/24 - v/24 - x/12 \pmod{p^2},\\
A_6 &\equiv v/12 \pmod{p^2},
\end{aligned}$$

and

$$A_{12} \equiv v/12 + x/6 \pmod{p^2}.$$

Then show that, for a character $\chi = \chi_P$ of order 12 as in Theorem 3.5.2,

$$\chi(4) \equiv 4^f(1 - x/6) \pmod{P^2}$$

and

$$K(\overline{\chi}) \equiv (2a_{12} - p/(2a_{12})) \pmod{P^2}.$$

13. All parameters in this problem are nonnegative integers. Let p be a prime, and let $m = a_0 + a_1p + \cdots + a_rp^r$, where $0 \le a_i \le p-1$ for $i < r$ and $a_r \ge 0$. Similarly, for $1 \le j \le t$, write $m_j = a_{j0} + \cdots + a_{jr}p^r$, where $0 \le a_{ji} \le p-1$ for $i < r$ and $a_{jr} \ge 0$. Prove the following congruence for multinomial coefficients $\pmod{p}$:

$$\frac{m!}{m_1!\cdots m_t!(m - m_1 - \cdots - m_t)!} \equiv \prod_{i=0}^{r} \frac{a_i!}{a_{1i}!\cdots a_{ti}!(a_i - a_{1i} - \cdots - a_{ti})!} \pmod{p}.$$

Hint: Compare the coefficients of $x_1^{m_1}\cdots x_t^{m_t}$ on both sides of the identity

$$(1 + x_1 + \cdots + x_t)^m = \prod_{i=0}^{r}(1 + x_1^{p^i} + \cdots + x_t^{p^i})^{a_i}$$

in $F_p[x_1, \ldots, x_t]$.

14. Write $p = 20f + 1 = a_{20}^2 + 5b_{20}^2 = a_4^2 + b_4^2$ and define c_{20} as in Theorem 3.8.2. Prove that, modulo p,

$$\binom{10f}{f} \equiv \begin{cases} 2(-1)^{f+1}\dfrac{a_{20}}{c_{20}}, & \text{if } 5 \nmid a_4, \\[2ex] 2(-1)^f \dfrac{a_{20}a_4}{b_4(c_{20}/i)}, & \text{if } 5 \mid a_4, \end{cases}$$

and

$$\binom{10f}{f} \equiv \begin{cases} \dbinom{10f}{3f}, & \text{if } 5 \nmid a_4, \\[2ex] -\dbinom{10f}{3f}, & \text{if } 5 \mid a_4. \end{cases}$$

Cf. Theorem 6.4.8.

Hint: Use Lemma 3.8.1, Theorem 3.8.2, and Theorem 9.1.1.

Notes on Chapter 9

The first congruence in Theorem 9.2.1 is due to Jacobi [3]. Other proofs of this congruence can be found in Whiteman [5], von Schrutka [1], and Hudson and Williams [9]. The second congruence in Theorem 9.2.1 occurs in the work of Hudson and Williams [9]. Theorem 9.2.2 and the congruence (9.0.1) are due to Gauss [5, pp. 89–91]; see also Whiteman [5, p. 95] and Hudson and Williams [9, p. 449]. The congruences in Theorem 9.2.3 are due to E. Lehmer [8]. Our proof of Theorem 9.2.3 is a simpler version of the proof given by Hudson and Williams [9, Theorem 8.1]. Corollary 9.2.4 was rediscovered by Rajwade [10] in 1976. Theorem 9.2.5 is proved in Hudson and Williams [9, Theorem 9.2]. The companion theorem to Theorem 9.2.5 involving the integers r_3 and s_3 is given in Hudson and Williams [9, Theorem 9.1]. The congruence of Theorem 9.2.6 is due to Jacobi [1], [2]. Theorem 9.2.7 can be found in Hudson and Williams [9, Theorem 10.2]. The third congruence in Theorem 9.2.8 is due to Jacobi [3] and Stern [1]; the remaining congruences are given in Hudson and Williams [9, Theorems 11.1 and 11.2]. Theorem 9.2.9 is an improved version of that in Hudson and Williams [9, Theorem 13.1]. Theorem 9.2.10 can be found in Hudson and Williams [9, Theorem 15.1].

Sections 9.1 and 9.2 of this chapter are based on the ideas contained in the paper of Hudson and Williams [9]. This paper contains many other congruences modulo p for binomial coefficients; for example, results are obtained in the cases $k = 9, 11, 13, 14, 15, 16, 20$, and 24. A few of these results can be found by solving Problems 2–6 and Problem 14.

Cauchy [2] showed in 1840 that, for a prime $p = 20f+1$, $\binom{10f}{f} \equiv \pm \binom{10f}{3f} \pmod{p}$, but it was not until 1965 that the ambiguous sign was determined (Whiteman [13, Theorem 3]). A determination was also given by Yamamoto [5, Theorem 11]. The result is presented in Problem 14. A related result for $p = 20f+9$ is given in Problem 20 of Exercises 12. See also Theorems 12.9.7–12.9.9. Congruences of the type $\binom{rf}{sf} \equiv \pm \binom{r'f}{s'f}$ (mod p) that do not follow immediately from the basic properties of binomial coefficients are said to be of Cauchy–Whiteman type. Such congruences are discussed in Hudson and Williams [5], [9]; see also Problem 5. Some linear relations between binomial coefficients modulo p have been given by Parnami, Agrawal, and Rajwade [4]. Ishikawa [2] obtained congruences (mod p) for Fourier coefficients of special products of eta functions, and, in certain cases, these congruences specialize to ones involving binomial coefficients $\binom{2f}{f}$. Granville [1] contains a variety of interesting congruences modulo prime powers involving binomial coefficients.

In the nineteenth century Cauchy [2] and Jacobi [3] studied modulo a prime $p = kf+1$ such products of factorials as $\prod_{\substack{a=1 \\ (\frac{a}{k})=1}}^{k-1} (af)!$ and $\prod_{\substack{b=1 \\ (\frac{b}{k})=-1}}^{k-1} (bf)!$, where k is odd. For example, let k be a prime with $k \equiv 7 \pmod 8$. Set

$$A = \frac{1}{k} \sum_{\substack{a=1 \\ (\frac{a}{k})=1}}^{k-1} a \quad \text{and} \quad B = \frac{1}{k} \sum_{\substack{b=1 \\ (\frac{b}{k})=-1}}^{k-1} b.$$

Then

$$\prod_{\substack{b=1 \\ (\frac{b}{k})=-1}}^{k-1} (bf)! \equiv 2(-1)^B x \pmod p,$$

where

$$p^{B-A} = x^2 + ky^2, \qquad x \equiv 1 \pmod k.$$

In the special case $k = 7$, this yields

$$(3f)!\,(5f)!\,(6f)! \equiv 2x \pmod p,$$

where $p = x^2 + 7y^2 = 7f + 1$ and $x \equiv 1 \pmod 7$, so that

$$\binom{3f}{f} = \frac{(3f)!}{f!(2f)!} \equiv (3f)!\,(5f)!\,(6f)! \equiv 2x \pmod p,$$

which is Theorem 9.2.6. The reader will find such results of Jacobi and Cauchy described in H. J. S. Smith [1, pp. 268–280]. Extensions of these results to products of the type $\prod_b (bf)!$, where b runs through the quartic residues of a prime, have been given by Buell, Hudson, and Williams [1] and Hudson [2]. See also the related work of Yamamoto [5, Theorem 10].

In the late 1980s, a number of authors considered the problem of determining binomial coefficients $\binom{rf}{sf}$ modulo p^2 rather than modulo p. This is the work described in Sections 9.3 and 9.4. The first congruence of this type, (9.0.2), was conjectured by F. Beukers and proved by Chowla, Dwork, and Evans [1]. The proof given here (Theorem 9.4.3) is somewhat simpler and is due to Evans in an unpublished manuscript circa 1985. Parts of Theorems 9.4.2 and 9.4.4 are also proved in this manuscript. The principal tools are the p-adic gamma function (9.3.3) and the Gross–Koblitz formula (11.2.12). Theorem 9.4.1 is a special case of a theorem of Morley [1], who proved that for a prime $p = 2f + 1 > 3$,

$$\binom{2f}{f} \equiv (-16)^f \pmod{p^3}.$$

Theorems 9.4.2–9.4.4 and the congruence (9.3.10) appear in the work of Yeung [1]. The results of Theorem 9.4.5 and Problem 12 are new. Coster [1], [2] has obtained congruences for binomial coefficients modulo higher powers of p that extend many of the congruences in Section 9.4. See also Ishibashi, Sato, and Shiratani [1] for congruences modulo higher powers of p.

Problem 11 appears in Hudson and Williams [9, Lemma 2.1]. (Note that the inequality $e - k < g - h$ in Hudson and Williams [9] should be $e - k > g - h$.) Problems 8 and 9 occur in the work of Huard, Spearman, and Williams [2]. Problem 13 gives Dickson's generalization of a theorem of Lucas [1, §228]; see Conrad [1, pp. 144–145]. Lucas' theorem has been generalized in several directions. An analogue for q-binomial coefficients has been given by Desarmenien [1]. Davis and Webb [1] and Bailey [1] obtain extensions of Lucas' theorem to prime power moduli. For further references along these lines, see Granville [1].

Chapter 10

Diagonal Equations over Finite Fields

We consider a polynomial equation of the type

$$\alpha_1 x_1^{k_1} + \cdots + \alpha_n x_n^{k_n} = \alpha, \tag{10.0.1}$$

where n, $k_1, \ldots, k_n$ are positive integers, $\alpha_1, \ldots, \alpha_n \in F_q^*$, and $\alpha \in F_q$. Such an equation is called a diagonal equation. By the number N of solutions of this equation in $F_q^n := F_q \times F_q \times \cdots \times F_q$ we mean the number of n-tuples $(\gamma_1, \ldots, \gamma_n) \in F_q^n$ for which $\alpha_1\gamma_1^{k_1} + \cdots + \alpha_n\gamma_n^{k_n} = \alpha$. We shall use Jacobi sums to find some explicit formulas for N, as well as upper and lower bounds for N.

A simple situation occurs when each $k_i = 1$. In this case the equation (10.0.1) is

$$\alpha_1 x_1 + \cdots + \alpha_n x_n = \alpha, \tag{10.0.2}$$

and the unknown x_n is determined in terms of $x_1, \ldots, x_{n-1}$ by means of

$$x_n = \alpha_n^{-1}(\alpha - \alpha_1 x_1 - \cdots - \alpha_{n-1}x_{n-1}).$$

Thus, letting each of $x_1, \ldots, x_{n-1}$ take on the q values in F_q, we see that (10.0.2) has q^{n-1} solutions.

When $q = p$ the equation (10.0.1) becomes the congruence

$$\alpha_1 x_1^{k_1} + \cdots + \alpha_n x_n^{k_n} \equiv \alpha \pmod{p}. \tag{10.0.3}$$

If $k_1, \ldots, k_n$ are positive integers satisfying $\gcd(k_i, p-1) = 1$ $(i = 1, \ldots, n)$, then each congruence $x_i^{k_i} \equiv y_i \pmod{p}$ has exactly one solution x_i for each y_i, and since

$$\alpha_1 y_1 + \cdots + \alpha_n y_n \equiv \alpha \pmod{p}$$

has p^{n-1} solutions, so does (10.0.3). We state this result as a theorem.

Theorem 10.0.1 *Let p be a prime. If $k_1, \ldots, k_n$ are positive integers with $\gcd(k_i, p-1) = 1$ $(i = 1, \ldots, n)$, and $A_1, \ldots, A_n$, A are integers with $p \nmid A_1 \cdots A_n$, then the number of solutions of the congruence*

$$A_1 x_1^{k_1} + \cdots + A_n x_n^{k_n} \equiv A \pmod{p} \tag{10.0.4}$$

is p^{n-1}.

By Theorem 10.0.1, the number of solutions of the congruence (10.0.4) with r $(1 \le r \le n-1)$ of the x_i divisible by p is p^{n-r-1}. Thus, by the inclusion–exclusion principle, the number N' of solutions of (10.0.4) with none of the x_i divisible by p is

$$\epsilon(A)+p^{n-1}-\binom{n}{1}p^{n-2}+\binom{n}{2}p^{n-3}-\cdots+(-1)^{n-1}\binom{n}{n-1}=\frac{(p-1)^n-(-1)^n}{p}+\epsilon(A),$$

where

$$\epsilon(A)=\begin{cases}(-1)^n, & \text{if } A\equiv 0 \pmod{p},\\ 0, & \text{if } A\not\equiv 0 \pmod{p}.\end{cases} \tag{10.0.5}$$

We state this result as a theorem.

Theorem 10.0.2 *Let p be a prime. If $k_1,\ldots,k_n$ are positive integers with $\gcd(k_i,p-1)=1$ $(i=1,\ldots,n)$, and $A_1,\ldots,A_n,A$ are integers with $p \nmid A_1\cdots A_n$, then the number N' of solutions of the congruence*

$$A_1x_1^{k_1}+\cdots+A_nx_n^{k_n}\equiv A \pmod{p}$$

with $x_1\cdots x_n\not\equiv 0 \pmod{p}$ is $((p-1)^n-(-1)^n)/p+\epsilon(A)$, where $\epsilon(A)$ is defined by (10.0.5).

10.1 Generalized Jacobi sums

Let $q=p^r$ for a prime p. In order to determine the number N of solutions of (10.0.1) in F_q^n, we introduce a character sum that generalizes the Jacobi sum defined in Chapter 2. We call this a generalized Jacobi sum.

Definition *Let $\chi_1,\ldots,\chi_k$ be characters on F_q. Then the generalized Jacobi sum $J_r(\chi_1,\ldots,\chi_k)$ is defined by*

$$J_r(\chi_1,\ldots,\chi_k)=\sum_{\gamma_1+\cdots+\gamma_k=1}\chi_1(\gamma_1)\cdots\chi_k(\gamma_k), \tag{10.1.1}$$

where the summation is taken over all k-tuples $(\gamma_1,\ldots,\gamma_k)$ of elements of F_q with $\gamma_1+\cdots+\gamma_k=1$.

Clearly the sum contains q^{k-1} terms. We observe that when $k=2$, this definition agrees with the definition of $J_r(\chi_1,\chi_2)$ given in Chapter 2, and that when $k=1$,

$$J_r(\chi_1,\ldots,\chi_k)=J_r(\chi_1)=\sum_{\gamma_1=1}\chi_1(\gamma_1)=\chi_1(1)=1.$$

It is also clear that, if $\{\sigma(1),\ldots,\sigma(k)\}$ is a permutation of $\{1,\ldots,k\}$, then

$$J_r(\chi_{\sigma(1)},\ldots,\chi_{\sigma(k)})=J_r(\chi_1,\ldots,\chi_k).$$

We further note that if $\alpha \in F_q^*$ then

$$\sum_{\gamma_1+\cdots+\gamma_k=\alpha} \chi_1(\gamma_1)\cdots\chi_k(\gamma_k) = \sum_{\gamma_1+\cdots+\gamma_k=1} \chi_1(\alpha\gamma_1)\cdots\chi_k(\alpha\gamma_k) = (\chi_1\cdots\chi_k)(\alpha)J_r(\chi_1,\ldots,\chi_k).$$

It is easy to evaluate the Jacobi sum $J_r(\chi_1,\ldots,\chi_k)$ if some of the characters χ_i are trivial.

Theorem 10.1.1 *If the characters $\chi_1,\ldots,\chi_k$ are all trivial, then*

$$J_r(\chi_1,\ldots,\chi_k) = q^{k-1}.$$

If some but not all of the χ_i are trivial then

$$J_r(\chi_1,\ldots,\chi_k) = 0.$$

Proof If $\chi_1,\ldots,\chi_k$ are all trivial, then

$$J_r(\chi_1,\ldots,\chi_k) = \sum_{\gamma_1+\cdots+\gamma_k=1} \chi_1(\gamma_1)\cdots\chi_k(\gamma_k) = \sum_{\gamma_1+\cdots+\gamma_k=1} 1 = q^{k-1}.$$

If some but not all of the χ_i are trivial, we reorder the χ_i so that $\chi_1,\ldots,\chi_s$ are nontrivial and $\chi_{s+1},\ldots,\chi_k$ are trivial, where $1 \le s \le k-1$. Then

$$\begin{aligned} J_r(\chi_1,\ldots,\chi_k) &= \sum_{\gamma_1+\cdots+\gamma_k=1} \chi_1(\gamma_1)\cdots\chi_k(\gamma_k) = \sum_{\gamma_1+\cdots+\gamma_k=1} \chi_1(\gamma_1)\cdots\chi_s(\gamma_s) \\ &= \sum_{\gamma_1,\ldots,\gamma_s} \chi_1(\gamma_1)\cdots\chi_s(\gamma_s) \sum_{\substack{\gamma_{s+1},\ldots,\gamma_k \\ \gamma_{s+1}+\cdots+\gamma_k = 1-\gamma_1-\cdots-\gamma_s}} 1 \\ &= \sum_{\gamma_1,\ldots,\gamma_s} \chi_1(\gamma_1)\cdots\chi_s(\gamma_s) q^{k-s-1} \\ &= q^{k-s-1}\sum_{\gamma_1}\chi_1(\gamma_1)\cdots\sum_{\gamma_s}\chi_s(\gamma_s) = 0. \end{aligned}$$

■

It is convenient at this point to introduce a character sum that is closely related to the generalized Jacobi sum.

Definition *Let $\chi_1,\ldots,\chi_k$ be characters on F_q. Then $J_0(\chi_1,\ldots,\chi_k)$ is the character sum defined by*

$$J_0(\chi_1,\ldots,\chi_k) = \sum_{\gamma_1+\cdots+\gamma_k=0} \chi_1(\gamma_1)\cdots\chi_k(\gamma_k), \tag{10.1.2}$$

where the summation is taken over all k-tuples $(\gamma_1,\ldots,\gamma_k)$ of elements of F_q with $\gamma_1+\cdots+\gamma_k=0$.

When $k = 1$, this sum is

$$J_0(\chi_1, \ldots, \chi_k) = J_0(\chi_1) = \sum_{\gamma_1 = 0} \chi_1(\gamma_1) = \chi_1(0) = \begin{cases} 1, & \text{if } \chi_1 \text{ is trivial,} \\ 0, & \text{if } \chi_1 \text{ is nontrivial.} \end{cases}$$

It is also clear that $J_0(\chi_1, \ldots, \chi_k)$ contains q^{k-1} terms and that

$$J_0(\chi_{\sigma(1)}, \ldots, \chi_{\sigma(k)}) = J_0(\chi_1, \ldots, \chi_k)$$

for any permutation $\{\sigma(1), \ldots, \sigma(k)\}$ of $\{1, \ldots, k\}$.

The next theorem gives a relationship between $J_0(\chi_1, \ldots, \chi_k)$ and $J_r(\chi_1, \ldots, \chi_k)$.

Theorem 10.1.2 *Let $\chi_1, \ldots, \chi_k$ be characters on F_q. Then*

$$J_0(\chi_1, \ldots, \chi_k) = \begin{cases} q^{k-1}, & \textit{if } \chi_1, \ldots, \chi_k \textit{ are all trivial,} \\ -(q-1)J_r(\chi_1, \ldots, \chi_k), & \textit{if } \chi_1, \ldots, \chi_k \textit{ are all nontrivial} \\ & \quad \textit{and } \chi_1 \cdots \chi_k \textit{ is trivial,} \\ 0, & \textit{otherwise.} \end{cases}$$

Proof We have

$$\begin{aligned} &\sum_{\gamma \in F_q} \sum_{\gamma_1 + \cdots + \gamma_k = \gamma} \chi_1(\gamma_1) \cdots \chi_k(\gamma_k) \\ &= \sum_{\gamma_1 + \cdots + \gamma_k = 0} \chi_1(\gamma_1) \cdots \chi_k(\gamma_k) + \sum_{\gamma \in F_q^*} \sum_{\gamma_1 + \cdots + \gamma_k = \gamma} \chi_1(\gamma_1) \cdots \chi_k(\gamma_k) \\ &= J_0(\chi_1, \ldots, \chi_k) + J_r(\chi_1, \ldots, \chi_k) \sum_{\gamma \in F_q^*} (\chi_1 \cdots \chi_k)(\gamma) \\ &= \begin{cases} J_0(\chi_1, \ldots, \chi_k) + J_r(\chi_1, \ldots, \chi_k)(q-1), & \text{if } \chi_1 \cdots \chi_k \text{ is trivial,} \\ J_0(\chi_1, \ldots, \chi_k), & \text{if } \chi_1 \cdots \chi_k \text{ is nontrivial.} \end{cases} \end{aligned}$$

On the other hand,

$$\begin{aligned} \sum_{\gamma \in F_q} \sum_{\gamma_1 + \cdots + \gamma_k = \gamma} \chi_1(\gamma_1) \cdots \chi_k(\gamma_k) &= \sum_{\gamma_1} \chi_1(\gamma_1) \cdots \sum_{\gamma_k} \chi_k(\gamma_k) \\ &= \begin{cases} q^k, & \text{if } \chi_1, \ldots, \chi_k \text{ are all trivial,} \\ 0, & \text{if at least one of } \chi_1, \ldots, \chi_k \text{ is nontrivial.} \end{cases} \end{aligned}$$

Hence

$$\begin{array}{ll} J_0(\chi_1, \ldots, \chi_k) + (q-1)J_r(\chi_1, \ldots, \chi_k) = q^k, & \text{if } \chi_1, \ldots, \chi_k \text{ are all trivial,} \\ J_0(\chi_1, \ldots, \chi_k) = 0, & \text{if } \chi_1 \cdots \chi_k \text{ is nontrivial,} \\ J_0(\chi_1, \ldots, \chi_k) + (q-1)J_r(\chi_1, \ldots, \chi_k) = 0, & \text{if at least one of } \chi_1, \ldots, \chi_k \text{ is non-} \\ & \quad \text{trivial and } \chi_1 \cdots \chi_k \text{ is trivial.} \end{array}$$

Appealing to Theorem 10.1.1, we deduce the assertion of the theorem. ■

In the next theorem we give a different evaluation of $J_0(\chi_1, \ldots, \chi_k)$ in the case when $\chi_1, \ldots, \chi_k$ are all nontrivial and $\chi_1 \cdots \chi_k$ is trivial. This evaluation is accomplished by performing the summation over γ_k in (10.1.2).

Theorem 10.1.3 *If $\chi_1, \ldots, \chi_k$ are nontrivial characters on F_q with $\chi_1 \cdots \chi_k$ trivial, then*

$$J_0(\chi_1, \ldots, \chi_k) = \chi_k(-1)(q-1)J_r(\chi_1, \ldots, \chi_{k-1}).$$

Proof We have

$$\begin{aligned}
J_0(\chi_1, \ldots, \chi_k) &= \sum_{\gamma_1+\cdots+\gamma_k=0} \chi_1(\gamma_1)\cdots\chi_k(\gamma_k) \\
&= \sum_{\gamma_k} \left(\sum_{\gamma_1+\cdots+\gamma_{k-1}=-\gamma_k} \chi_1(\gamma_1)\cdots\chi_{k-1}(\gamma_{k-1}) \right) \chi_k(\gamma_k) \\
&= \sum_{\gamma_k\neq 0} \left(\sum_{\gamma_1+\cdots+\gamma_{k-1}=-\gamma_k} \chi_1(\gamma_1)\cdots\chi_{k-1}(\gamma_{k-1}) \right) \chi_k(\gamma_k) \\
&= \sum_{\gamma_k\neq 0} ((\chi_1\cdots\chi_{k-1})(-\gamma_k)J_r(\chi_1, \ldots, \chi_{k-1}))\,\chi_k(\gamma_k) \\
&= (\chi_1\cdots\chi_{k-1})(-1)J_r(\chi_1, \ldots, \chi_{k-1}) \sum_{\gamma_k\neq 0} (\chi_1\cdots\chi_k)(\gamma_k) \\
&= \chi_k(-1)J_r(\chi_1, \ldots, \chi_{k-1})(q-1),
\end{aligned}$$

as $\chi_1 \cdots \chi_k$ is trivial. ■

Comparing Theorems 10.1.2 and 10.1.3, we obtain the next result.

Theorem 10.1.4 *If $\chi_1, \ldots, \chi_k$ are nontrivial characters on F_q with $\chi_1 \cdots \chi_k$ trivial, then*

$$J_r(\chi_1, \ldots, \chi_k) = -\chi_k(-1)J_r(\chi_1, \ldots, \chi_{k-1}).$$

We remark that Theorem 2.1.1(c) is the special case $k = 2$ of Theorem 10.1.4.

10.2 A reduction formula for generalized Jacobi sums

In this section we illustrate how the value of the generalized Jacobi sum $J_r(\chi_1, \ldots, \chi_k)$, where $\chi_1, \ldots, \chi_k$ are nontrivial characters on F_q, can be determined from the values of those Jacobi sums $J_r(\chi_1 \cdots \chi_{i-1}, \chi_i)$ $(i = 2, \ldots, k)$ for which $\chi_1 \cdots \chi_{i-1}$ is nontrivial. This determination is achieved by means of a reduction formula that relates the generalized Jacobi sum $J_r(\chi_1, \ldots, \chi_k)$ to the sum $J_r(\chi_1, \ldots, \chi_{k-1})$.

Theorem 10.2.1 (*Reduction formula for generalized Jacobi sums.*) *If $\chi_1, \ldots, \chi_k$ are nontrivial characters on F_q, where $k \geq 2$, then*

$$J_r(\chi_1, \ldots, \chi_k) = \begin{cases} -qJ_r(\chi_1, \ldots, \chi_{k-1}), & \text{if } \chi_1\cdots\chi_{k-1} \text{ is trivial,} \\ J_r(\chi_1\cdots\chi_{k-1}, \chi_k)J_r(\chi_1, \ldots, \chi_{k-1}), & \text{if } \chi_1\cdots\chi_{k-1} \text{ is nontrivial.} \end{cases}$$

Proof For $k \geq 2$,

$$\begin{aligned}
&J_r(\chi_1, \ldots, \chi_k) \\
&= \sum_{\gamma_1+\cdots+\gamma_k=1} \chi_1(\gamma_1)\cdots\chi_k(\gamma_k) \\
&= \sum_{\substack{\gamma_1+\cdots+\gamma_k=1 \\ \gamma_k=1}} \chi_1(\gamma_1)\cdots\chi_k(\gamma_k) + \sum_{\substack{\gamma_1+\cdots+\gamma_k=1 \\ \gamma_k\neq 1}} \chi_1(\gamma_1)\cdots\chi_k(\gamma_k) \\
&= \sum_{\gamma_1+\cdots+\gamma_{k-1}=0} \chi_1(\gamma_1)\cdots\chi_{k-1}(\gamma_{k-1}) + \sum_{\gamma_k\neq 1} \chi_k(\gamma_k) \sum_{\gamma_1+\cdots+\gamma_{k-1}=1-\gamma_k} \chi_1(\gamma_1)\cdots\chi_{k-1}(\gamma_{k-1}) \\
&= J_0(\chi_1, \ldots, \chi_{k-1}) + \sum_{\gamma_k\neq 1} (\chi_1\cdots\chi_{k-1})(1-\gamma_k)\chi_k(\gamma_k) J_r(\chi_1, \ldots, \chi_{k-1}) \\
&= J_0(\chi_1, \ldots, \chi_{k-1}) + (J_r(\chi_1\cdots\chi_{k-1}, \chi_k) - (\chi_1\cdots\chi_{k-1})(0)) J_r(\chi_1, \ldots, \chi_{k-1}).
\end{aligned}$$

By Theorem 10.1.2,

$$J_0(\chi_1, \ldots, \chi_{k-1}) = \begin{cases} -(q-1)J_r(\chi_1, \ldots, \chi_{k-1}), & \text{if } \chi_1\cdots\chi_{k-1} \text{ is trivial,} \\ 0, & \text{if } \chi_1\cdots\chi_{k-1} \text{ is nontrivial.} \end{cases}$$

Then recalling that

$$J_r(\chi_1\cdots\chi_{k-1}, \chi_k) = 0, \quad \text{if } \chi_1\cdots\chi_{k-1} \text{ is trivial,}$$

and

$$(\chi_1\cdots\chi_{k-1})(0) = \begin{cases} 1, & \text{if } \chi_1\cdots\chi_{k-1} \text{ is trivial,} \\ 0, & \text{if } \chi_1\cdots\chi_{k-1} \text{ is nontrivial,} \end{cases}$$

we obtain the assertion of the theorem. ■

In the next two theorems we use the reduction formula to obtain the values of $J_r(\rho, \ldots, \rho)$ and $J_0(\rho, \ldots, \rho)$, where ρ is the quadratic character on F_q (so that q is necessarily odd). These values will be used to determine explicitly the number N of solutions of (10.0.1) in two important special cases.

Theorem 10.2.2 *Let ρ denote the character of order 2 on F_q. For $k \geq 1$,*

$$J_r(\rho, \ldots, \rho) = \begin{cases} -\rho((-1)^{k/2})q^{(k-2)/2}, & \text{if } k \equiv 0 \pmod 2, \\ \rho((-1)^{(k-1)/2})q^{(k-1)/2}, & \text{if } k \equiv 1 \pmod 2, \end{cases}$$

where the generalized Jacobi sum $J_r(\rho, \ldots, \rho)$ contains k ρ's.

Proof We set $J_k = J_r(\rho, \ldots, \rho)$, where there are k ρ's. Then

$$J_1 = J_r(\rho) = 1,$$

and, by Theorem 2.1.1(c),

$$J_2 = J_r(\rho, \rho) = -\rho(-1).$$

From the reduction formula for generalized Jacobi sums (Theorem 10.2.1), we deduce for $n \ge 1$ that

$$J_{2n+1} = -qJ_{2n}, \quad J_{2n} = J_2 J_{2n-1}.$$

Solving these recurrence relations, we obtain, for $m \ge 0$,

$$J_{4m+1} = q^{2m}, \quad J_{4m+2} = -\rho(-1)q^{2m}, \quad J_{4m+3} = \rho(-1)q^{2m+1}, \quad J_{4m+4} = -q^{2m+1},$$

from which the theorem follows. ■

The next theorem is the analogue of Theorem 10.2.2 for $J_0(\rho, \ldots, \rho)$.

Theorem 10.2.3 *For* $k \ge 1$,

$$J_0(\rho, \ldots, \rho) = \begin{cases} \rho((-1)^{k/2})(q-1)q^{(k-2)/2}, & \text{if } k \equiv 0 \pmod 2, \\ 0, & \text{if } k \equiv 1 \pmod 2, \end{cases}$$

where $J_0(\rho, \ldots, \rho)$ *contains* k ρ*'s.*

Proof By Theorems 10.1.2 and 10.1.3,

$$J_0(\rho, \ldots, \rho) = \begin{cases} \rho(-1)(q-1)J_r(\rho, \ldots, \rho), & \text{if } k \equiv 0 \pmod 2, \\ 0, & \text{if } k \equiv 1 \pmod 2, \end{cases}$$

where there are k ρ's in $J_0(\rho, \ldots, \rho)$ and $k-1$ ρ's in $J_r(\rho, \ldots, \rho)$. The theorem now follows by appealing to Theorem 10.2.2. ■

In the next theorem we determine explicitly the value of the generalized Jacobi sum $J(\chi, \ldots, \chi, \chi^2, \ldots, \chi^2)$, where χ is a cubic character (mod p).

Theorem 10.2.4 *Let* $p = 3f + 1$ *be a prime, and let* g *be a primitive root of* p. *Let* χ *be the cubic character* (mod p) *given by* $\chi(g) = e^{2\pi i/3}$. *Let* a_3 *and* b_3 *be the integers defined in Theorem* 3.1.1, *and set*

$$\pi = \chi(2)(a_3 + ib_3\sqrt{3}\,), \quad \overline{\pi} = \chi^2(2)(a_3 - ib_3\sqrt{3}\,).$$

Let n_1 *and* n_2 *be nonnegative integers with* $n_1 + n_2 \ge 1$. *Set*

$$J_{n_1,n_2} = J(\chi, \ldots, \chi, \chi^2, \ldots, \chi^2),$$

where there are n_1 χ*'s and* n_2 χ^2*'s. Then*

$$(10.2.1) \quad J_{n_1,n_2} = \begin{cases} -\pi^{(2n_1+n_2-3)/3}\,\overline{\pi}^{(n_1+2n_2-3)/3}, & \text{if } n_1 + 2n_2 \equiv 0 \pmod 3, \\ \pi^{(2n_1+n_2-2)/3}\,\overline{\pi}^{(n_1+2n_2-1)/3}, & \text{if } n_1 + 2n_2 \equiv 1 \pmod 3, \\ \pi^{(2n_1+n_2-1)/3}\,\overline{\pi}^{(n_1+2n_2-2)/3}, & \text{if } n_1 + 2n_2 \equiv 2 \pmod 3. \end{cases}$$

Proof By Theorem 10.2.1, for positive integers n_1 and n_2,

$$(10.2.2) \qquad J_{n_1,n_2} = \begin{cases} -J_{n_1-1,n_2}, & \text{if } n_1+2n_2 \equiv 0 \pmod 3, \\ -pJ_{n_1-1,n_2}, & \text{if } n_1+2n_2 \equiv 1 \pmod 3, \\ -pJ_{n_1,n_2-1}, & \text{if } n_1+2n_2 \equiv 2 \pmod 3. \end{cases}$$

Since $p = \pi\overline{\pi}$, (10.2.1) is easily seen to follow by induction from (10.2.2), provided that we can show that (10.2.1) holds in the cases where n_1 or n_2 equals 0. We first consider $J_{n,0}$ for $n > 0$. For $n = 1$, $J_{1,0} = 1$, in agreement with (10.2.1). Thus let $n > 1$. By Theorem 10.2.1,

$$(10.2.3) \qquad J_{n,0} = \begin{cases} -J_{n-1,0}, & \text{if } n \equiv 0 \pmod 3, \\ -pJ_{n-1,0}, & \text{if } n \equiv 1 \pmod 3, \\ J(\chi,\chi)J_{n-1,0}, & \text{if } n \equiv 2 \pmod 3. \end{cases}$$

Since $J(\chi,\chi) = \pi$ by Theorem 3.1.1, it follows by induction from (10.2.3) that (10.2.1) gives the correct values of $J_{n,0}$. Finally, since $J_{0,n} = \overline{J}_{n,0}$, we see that (10.2.1) also gives the correct values of $J_{0,n}$. ∎

In the next theorem we give explicitly the value of the generalized Jacobi sum $J(\chi,\ldots,\chi,\chi^2,\ldots,\chi^2,\chi^3,\ldots,\chi^3)$ for a quartic character $\chi \pmod p$. The proof, which is not given, is exactly analogous to that of Theorem 10.2.4.

Theorem 10.2.5 *Let $p = 4f+1$ be a prime and let g be a primitive root of p. Let χ be the character* (mod p) *given by $\chi(g) = i$. Let a_4 and b_4 be the integers defined in Theorem* 3.2.1, *and set*

$$\pi = (-1)^f(a_4 + ib_4).$$

Let n_1, n_2, and n_3 be nonnegative integers with $n_1+n_2+n_3 \geq 1$. Set

$$J_{n_1,n_2,n_3} = J(\chi,\ldots,\chi,\chi^2,\ldots,\chi^2,\chi^3,\ldots,\chi^3),$$

where there are n_1 χ's, n_2 χ^2's and n_3 χ^3's. Then the value of J_{n_1,n_2,n_3} is

$$\begin{array}{ll} -(-1)^{n_1f}\pi^{(3n_1+2n_2+n_3-4)/4}\,\overline{\pi}^{(n_1+2n_2+3n_3-4)/4}, & \text{if } n_1+2n_2+3n_3 \equiv 0 \pmod 4, \\ (-1)^f(-1)^{n_1f}\pi^{(3n_1+2n_2+n_3-3)/4}\,\overline{\pi}^{(n_1+2n_2+3n_3-1)/4}, & \text{if } n_1+2n_2+3n_3 \equiv 1 \pmod 4, \\ (-1)^{n_1f}\pi^{(3n_1+2n_2+n_3-2)/4}\,\overline{\pi}^{(n_1+2n_2+3n_3-2)/4}, & \text{if } n_1+2n_2+3n_3 \equiv 2 \pmod 4, \\ (-1)^{n_1f}\pi^{(3n_1+2n_2+n_3-1)/4}\,\overline{\pi}^{(n_1+2n_2+3n_3-3)/4}, & \text{if } n_1+2n_2+3n_3 \equiv 3 \pmod 4. \end{array}$$

10.3 Generalized Jacobi sums and Gauss sums

In this section we show how the generalized Jacobi sum $J_r(\chi_1,\ldots,\chi_k)$, where $\chi_1,\ldots,\chi_k$ are nontrivial characters on F_q, can be expressed in terms of Gauss sums. This relationship generalizes Theorem 2.1.3(a) and is easily proved using the reduction formula for generalized Jacobi sums (Theorem 10.2.1).

Theorem 10.3.1 *Let $\chi_1, \ldots, \chi_k$ be nontrivial characters on F_q. Then*

$$(10.3.1)\quad J_r(\chi_1, \ldots, \chi_k) = \begin{cases} G_r(\chi_1)\cdots G_r(\chi_k)/G_r(\chi_1\cdots\chi_k), & \text{if } \chi_1\cdots\chi_k \text{ is nontrivial,} \\ -G_r(\chi_1)\cdots G_r(\chi_k)/q, & \text{if } \chi_1\cdots\chi_k \text{ is trivial.} \end{cases}$$

Proof We prove the assertion (10.3.1) by induction on the number k of nontrivial characters χ_i. If $k = 1$, then

$$J_r(\chi_1) = 1 = G_r(\chi_1)/G_r(\chi_1),$$

and (10.3.1) holds in this case. If $k = 2$ and $\chi_1\chi_2$ is nontrivial, then, by Theorem 2.1.3(a),

$$J_r(\chi_1, \chi_2) = G_r(\chi_1)G_r(\chi_2)/G_r(\chi_1\chi_2),$$

which proves (10.3.1) in this case. If $k = 2$ and $\chi_1\chi_2$ is trivial, then, by Theorem 1.1.4(a) and Theorem 2.1.1(c),

$$J_r(\chi_1, \chi_2) = J_r(\chi_1, \overline{\chi}_1) = -\chi_1(-1) = -G_r(\chi_1)G_r(\chi_2)/q,$$

which proves (10.3.1) in this case.

Let N be an integer with $N \geq 3$ and assume that (10.3.1) holds for any choice of k nontrivial characters $\chi_1, \ldots, \chi_k$, where $1 \leq k \leq N-1$. We consider three cases:

(a) $\chi_1\cdots\chi_{N-1}$ trivial,
(b) $\chi_1\cdots\chi_{N-1}$ nontrivial, $\chi_1\cdots\chi_N$ trivial,
(c) $\chi_1\cdots\chi_{N-1}$ nontrivial, $\chi_1\cdots\chi_N$ nontrivial.

(a) We have

$$\begin{aligned} J_r(\chi_1, \ldots, \chi_N) &= -qJ_r(\chi_1, \ldots, \chi_{N-1}) && \text{(by Theorem 10.2.1)} \\ &= -q(-G_r(\chi_1)\cdots G_r(\chi_{N-1})/q) && \text{(by inductive hypothesis)} \\ &= \frac{G_r(\chi_1)\cdots G_r(\chi_{N-1})G_r(\chi_N)}{G_r(\chi_1\cdots\chi_N)}, && \text{(as } \chi_1\cdots\chi_{N-1} \text{ is trivial)} \end{aligned}$$

which proves (10.3.1) in this case.

(b) We have

$$\begin{aligned} J_r(\chi_1, \ldots, \chi_N) &= J_r(\chi_1\cdots\chi_{N-1}, \chi_N)J_r(\chi_1, \ldots, \chi_{N-1}) && \text{(by Theorem 10.2.1)} \\ &= -\chi_N(-1)\frac{G_r(\chi_1)\cdots G_r(\chi_{N-1})}{G_r(\chi_1\cdots\chi_{N-1})} && \text{(by Theorem 2.1.1(c) and inductive hypothesis)} \\ &= -\chi_N(-1)\frac{G_r(\chi_1)\cdots G_r(\chi_{N-1})G_r(\chi_N)}{G_r(\overline{\chi}_N)G_r(\chi_N)} && \text{(as } \chi_1\cdots\chi_N \text{ is trivial)} \\ &= -\chi_N(-1)\frac{G_r(\chi_1)\cdots G_r(\chi_N)}{\chi_N(-1)q} && \text{(by Theorem 1.1.4(a))} \\ &= -G_r(\chi_1)\cdots G_r(\chi_N)/q, \end{aligned}$$

which proves (10.3.1) in this case.

(c) By Theorem 10.2.1, Theorem 2.1.3(a), and the inductive hypothesis,

$$\begin{aligned} J_r(\chi_1, \ldots, \chi_N) &= J_r(\chi_1 \cdots \chi_{N-1}, \chi_N) J_r(\chi_1, \ldots, \chi_{N-1}) \\ &= \frac{G_r(\chi_1 \cdots \chi_{N-1}) G_r(\chi_N)}{G_r(\chi_1 \cdots \chi_N)} \cdot \frac{G_r(\chi_1) \cdots G_r(\chi_{N-1})}{G_r(\chi_1 \cdots \chi_{N-1})} \\ &= \frac{G_r(\chi_1) \cdots G_r(\chi_N)}{G_r(\chi_1 \cdots \chi_N)}, \end{aligned}$$

which proves (10.3.1) in this case. ■

From Theorem 1.1.4(c) and Theorem 10.3.1 we are able to deduce the value of $|J_r(\chi_1, \ldots, \chi_k)|$ for nontrivial characters $\chi_1, \ldots, \chi_k$.

Theorem 10.3.2 *If $\chi_1, \ldots, \chi_k$ are nontrivial characters on F_q, then*

$$|J_r(\chi_1, \ldots, \chi_k)| = \begin{cases} q^{(k-1)/2}, & \text{if } \chi_1 \cdots \chi_k \text{ is nontrivial,} \\ q^{(k-2)/2}, & \text{if } \chi_1 \cdots \chi_k \text{ is trivial.} \end{cases}$$

Proof If $\chi_1 \cdots \chi_k$ is nontrivial, then, by Theorem 10.3.1 and Theorem 1.1.4(c),

$$|J_r(\chi_1, \ldots, \chi_k)| = \frac{|G_r(\chi_1)| \cdots |G_r(\chi_k)|}{|G_r(\chi_1 \cdots \chi_k)|} = \frac{(q^{\frac{1}{2}})^k}{q^{\frac{1}{2}}} = q^{(k-1)/2}.$$

If $\chi_1 \cdots \chi_k$ is trivial,

$$|J_r(\chi_1, \ldots, \chi_k)| = \frac{|G_r(\chi_1)| \cdots |G_r(\chi_k)|}{q} = \frac{(q^{\frac{1}{2}})^k}{q} = q^{(k-2)/2}.$$ ■

10.4 Number of solutions of the equation $\alpha_1 x_1^{k_1} + \cdots + \alpha_n x_n^{k_n} = \alpha$

In Theorem 10.4.2 we give an explicit formula for the number N of solutions of the equation (10.0.1) in terms of generalized Jacobi sums. We will need the following formula for the number $N(x^k = \gamma)$ of solutions x in F_q of $x^k = \gamma$, where k is a positive integer and $\gamma \in F_q$.

Lemma 10.4.1 *Let $\gamma \in F_q$, and let k be a positive integer. Let χ be a character on F_q of order $d = \gcd(k, q-1)$. Then the number of solutions $x \in F_q$ of $x^k = \gamma$ is*

$$N(x^k = \gamma) = \sum_{j=0}^{d-1} \chi^j(\gamma).$$

Proof If $\gamma = 0$, then the equation $x^k = \gamma$ has exactly one solution, namely, $x = 0$, and $\sum_{j=0}^{d-1} \chi^j(\gamma) = \sum_{j=0}^{d-1} \chi^j(0) = 1$. If $\gamma \neq 0$ and γ is not a kth power in F_q, then $x^k = \gamma$ has no solutions, $\chi(\gamma) \neq 1$, and

$$\sum_{j=0}^{d-1} \chi^j(\gamma) = \frac{1 - \chi^d(\gamma)}{1 - \chi(\gamma)} = \frac{1 - 1}{1 - \chi(\gamma)} = 0.$$

If $\gamma \neq 0$ and γ is a kth power in F_q, then $x^k = \gamma$ has d solutions, $\chi(\gamma) = 1$, and $\sum_{j=0}^{d-1} \chi^j(\gamma) = \sum_{j=0}^{d-1} 1 = d$. ∎

Theorem 10.4.2 *Let $k_1, \ldots, k_n$ be positive integers. Let $\alpha_1, \ldots, \alpha_n \in F_q^*$ and $\alpha \in F_q$. Set $d_i = \gcd(k_i, q-1)$, and let χ_i be a character on F_q of order d_i, $i = 1, \ldots, n$. Then the number N of solutions of the equation $\alpha_1 x_1^{k_1} + \cdots + \alpha_n x_n^{k_n} = \alpha$ is given by*

$$N = q^{n-1} + \sum_{j_1=1}^{d_1-1} \cdots \sum_{j_n=1}^{d_n-1} \chi_1^{j_1}(\alpha\alpha_1^{-1}) \cdots \chi_n^{j_n}(\alpha\alpha_n^{-1}) J_r(\chi_1^{j_1}, \ldots, \chi_n^{j_n}),$$

if $\alpha \neq 0$, and by

$$N = q^{n-1} - (q-1) \sum_{\substack{j_1=1 \\ \chi_1^{j_1} \cdots \chi_n^{j_n} \text{ trivial}}}^{d_1-1} \cdots \sum_{j_n=1}^{d_n-1} \chi_1^{j_1}(\alpha_1^{-1}) \cdots \chi_n^{j_n}(\alpha_n^{-1}) J_r(\chi_1^{j_1}, \ldots, \chi_n^{j_n}),$$

if $\alpha = 0$.

Proof For $\alpha \in F_q^*$,

$$\begin{aligned}
N &= \sum_{\gamma_1 + \cdots + \gamma_n = \alpha} N(\alpha_1 x_1^{k_1} = \gamma_1) \cdots N(\alpha_n x_n^{k_n} = \gamma_n) \\
&= \sum_{\gamma_1 + \cdots + \gamma_n = \alpha} N(x_1^{k_1} = \alpha_1^{-1}\gamma_1) \cdots N(x_n^{k_n} = \alpha_n^{-1}\gamma_n) \\
&= \sum_{\gamma_1 + \cdots + \gamma_n = \alpha} \left(\sum_{j_1=0}^{d_1-1} \chi_1^{j_1}(\alpha_1^{-1}\gamma_1) \right) \cdots \left(\sum_{j_n=0}^{d_n-1} \chi_n^{j_n}(\alpha_n^{-1}\gamma_n) \right) \qquad \text{(by Lemma 10.4.1)} \\
&= \sum_{j_1=0}^{d_1-1} \cdots \sum_{j_n=0}^{d_n-1} \chi_1^{j_1}(\alpha_1^{-1}) \cdots \chi_n^{j_n}(\alpha_n^{-1}) \sum_{\gamma_1 + \cdots + \gamma_n = \alpha} \chi_1^{j_1}(\gamma_1) \cdots \chi_n^{j_n}(\gamma_n) \\
&= \sum_{j_1=0}^{d_1-1} \cdots \sum_{j_n=0}^{d_n-1} \chi_1^{j_1}(\alpha\alpha_1^{-1}) \cdots \chi_n^{j_n}(\alpha\alpha_n^{-1}) J_r(\chi_1^{j_1}, \ldots, \chi_n^{j_n}) \\
&= q^{n-1} + \sum_{j_1=1}^{d_1-1} \cdots \sum_{j_n=1}^{d_n-1} \chi_1^{j_1}(\alpha\alpha_1^{-1}) \cdots \chi_n^{j_n}(\alpha\alpha_n^{-1}) J_r(\chi_1^{j_1}, \ldots, \chi_n^{j_n}),
\end{aligned}$$

by Theorem 10.1.1.

For $\alpha = 0$, we obtain exactly as above

$$N = \sum_{j_1=0}^{d_1-1} \cdots \sum_{j_n=0}^{d_n-1} \chi_1^{j_1}(\alpha_1^{-1}) \cdots \chi_n^{j_n}(\alpha_n^{-1}) \sum_{\gamma_1+\cdots+\gamma_n=0} \chi_1^{j_1}(\gamma_1) \cdots \chi_n^{j_n}(\gamma_n)$$
$$= \sum_{j_1=0}^{d_1-1} \cdots \sum_{j_n=0}^{d_n-1} \chi_1^{j_1}(\alpha_1^{-1}) \cdots \chi_n^{j_n}(\alpha_n^{-1}) J_0(\chi_1^{j_1}, \ldots, \chi_n^{j_n})$$
$$= q^{n-1} - (q-1) \sum_{\substack{j_1=1 \\ \chi_1^{j_1}\cdots\chi_n^{j_n} \text{ trivial}}}^{d_1-1} \cdots \sum_{j_n=1}^{d_n-1} \chi_1^{j_1}(\alpha_1^{-1}) \cdots \chi_n^{j_n}(\alpha_n^{-1}) J_r(\chi_1^{j_1}, \ldots, \chi_n^{j_n}),$$

by Theorem 10.1.2. ■

We note that Theorem 10.4.2 shows that N does not depend directly on the exponents k_i, but only on the integers $d_i = \gcd(k_i, q-1)$.

10.5 Number of solutions of the equation $\alpha_1 x_1^2 + \cdots + \alpha_n x_n^2 = \alpha$

As a first application of Theorem 10.4.2, we determine the number of solutions of the equation $\alpha_1 x_1^2 + \cdots + \alpha_n x_n^2 = \alpha$, when q is odd. Let ρ denote the quadratic character on F_q.

Theorem 10.5.1 *Let q be odd. The number N of solutions of the equation $\alpha_1 x_1^2 + \cdots + \alpha_n x_n^2 = \alpha$, where $\alpha_1, \ldots, \alpha_n \in F_q^*$ and $\alpha \in F_q$, is given as follows:*

$$\begin{aligned}
&N = q^{n-1} - \rho((-1)^{n/2}\alpha_1 \cdots \alpha_n) q^{(n-2)/2}, && \text{if } n \equiv 0 \pmod 2 \text{ and } \alpha \neq 0,\\
&N = q^{n-1} + \rho((-1)^{(n-1)/2}\alpha\alpha_1 \cdots \alpha_n) q^{(n-1)/2}, && \text{if } n \equiv 1 \pmod 2 \text{ and } \alpha \neq 0,\\
&N = q^{n-1} + \rho((-1)^{n/2}\alpha_1 \cdots \alpha_n)(q^{n/2} - q^{(n-2)/2}), && \text{if } n \equiv 0 \pmod 2 \text{ and } \alpha = 0,\\
&N = q^{n-1}, && \text{if } n \equiv 1 \pmod 2 \text{ and } \alpha = 0.
\end{aligned}$$

Proof We take $k_1 = \cdots = k_n = 2$ in Theorem 10.4.2, so that $d_1 = \cdots = d_n = 2$ and $\chi_1 = \cdots = \chi_n = \rho$. For $\alpha \neq 0$,

$$N = q^{n-1} + \rho(\alpha^n \alpha_1 \cdots \alpha_n) J_r(\rho, \ldots, \rho),$$

and, for $\alpha = 0$,

$$N = \begin{cases} q^{n-1} - (q-1)\rho(\alpha_1 \cdots \alpha_n) J_r(\rho, \ldots, \rho), & \text{if } n \equiv 0 \pmod 2,\\ q^{n-1}, & \text{if } n \equiv 1 \pmod 2. \end{cases}$$

By Theorem 10.2.2,

$$J_r(\rho, \ldots, \rho) = \begin{cases} -\rho((-1)^{n/2}) q^{(n-2)/2}, & \text{if } n \equiv 0 \pmod 2,\\ \rho((-1)^{(n-1)/2}) q^{(n-1)/2}, & \text{if } n \equiv 1 \pmod 2, \end{cases}$$

which gives the asserted formulas. ∎

We now show that Theorem 10.5.1 implies the law of quadratic reciprocity. Let p and p' denote distinct odd primes. We consider the number N of solutions of the congruence

(10.5.1) $$x_1^2 + \cdots + x_{p'}^2 \equiv p' \pmod{p}.$$

There are two solutions of (10.5.1) with $x_1 \equiv \cdots \equiv x_{p'} \pmod{p}$, namely, $\pm(1, \ldots, 1)$. Each solution $(x_1, \ldots, x_{p'}) \neq \pm(1, \ldots, 1)$ of (10.5.1) gives rise to p' distinct solutions

$$(x_1, x_2, \ldots, x_{p'-1}, x_{p'}),\ (x_2, x_3, \ldots, x_{p'}, x_1), \ldots, (x_{p'}, x_1, \ldots, x_{p'-2}, x_{p'-1}).$$

In this way the $N-2$ solutions of (10.5.1) different from $\pm(1, \ldots, 1)$ can be placed into $(N-2)/p'$ groups of p' distinct solutions. Hence

$$N \equiv 2 \pmod{p'}.$$

Now, by Theorem 10.5.1,

$$N = p^{p'-1} + \left(\frac{(-1)^{(p'-1)/2}p'}{p}\right) p^{(p'-1)/2},$$

and thus

$$N \equiv 1 + \left(\frac{(-1)^{(p'-1)/2}p'}{p}\right)\left(\frac{p}{p'}\right) \pmod{p'}.$$

Hence we deduce that

$$\left(\frac{(-1)^{(p'-1)/2}p'}{p}\right)\left(\frac{p}{p'}\right) = 1,$$

so that

$$\left(\frac{p}{p'}\right)\left(\frac{p'}{p}\right) = \left(\frac{-1}{p}\right)^{(p'-1)/2} = (-1)^{(p-1)(p'-1)/4},$$

which completes the proof of the law of quadratic reciprocity.

The ideas of this proof can also be used to prove the two supplements to the law of quadratic reciprocity, namely, $\left(\frac{-1}{p}\right) = (-1)^{(p-1)/2}$ and $\left(\frac{2}{p}\right) = (-1)^{(p^2-1)/8}$, where p is an odd prime. This time we consider the number N of solutions of the congruence

(10.5.2) $$x_1^2 + x_2^2 \equiv 1 \pmod{p}.$$

There are four solutions of (10.5.2) with $x_1 \equiv 0 \pmod{p}$ or $x_2 \equiv 0 \pmod{p}$, and there are $2\left(1 + \left(\frac{2}{p}\right)\right)$ solutions with $x_1 \equiv \pm x_2 \not\equiv 0 \pmod{p}$. The remaining solutions of

(10.5.2) can be put in groups of eight distinct solutions $\pm(x_1, x_2)$, $\pm(x_1, -x_2)$, $\pm(x_2, x_1)$, $\pm(x_2, -x_1)$. Thus,

$$N \equiv 4 + 2\left(1 + \left(\frac{2}{p}\right)\right) \pmod 8,$$

and, by Theorem 10.5.1,

$$N = p - \left(\frac{-1}{p}\right),$$

so that

$$p - \left(\frac{-1}{p}\right) \equiv 6 + 2\left(\frac{2}{p}\right) \pmod 8. \tag{10.5.3}$$

Taking the congruence (10.5.3) modulo 4, we obtain

$$\left(\frac{-1}{p}\right) \equiv p \equiv (-1)^{(p-1)/2} \pmod 4,$$

so that

$$\left(\frac{-1}{p}\right) = (-1)^{(p-1)/2}. \tag{10.5.4}$$

Then, from (10.5.3) and (10.5.4),

$$\left(\frac{2}{p}\right) \equiv \frac{1}{2}\left(p - (-1)^{(p-1)/2}\right) - 3 \pmod 4,$$

from which $\left(\frac{2}{p}\right) = (-1)^{(p^2-1)/8}$ follows by a consideration of cases depending on the value of p modulo 8.

10.6 Number of solutions of the congruence $A_1x_1^3 + \cdots + A_nx_n^3 \equiv A \pmod p$

In this section we use Theorems 10.2.4 and 10.4.2 to determine the number of solutions of the congruence

$$A_1x_1^3 + \cdots + A_nx_n^3 \equiv A \pmod p,$$

where $p = 3f + 1$ is a prime and $A, A_1, \ldots, A_n$ are integers with $A_1 \cdots A_n \not\equiv 0 \pmod p$.

Theorem 10.6.1 *Let $p = 3f + 1$ be a prime. Let g be a primitive root of p. Let χ be a cubic character* (mod p) *such that $\chi(g) = \exp(2\pi i/3)$. Let a_3 and b_3 be the integers uniquely determined by*

$$a_3^2 + 3b_3^2 = p,\ a_3 \equiv -1 \pmod 3,\ 3b_3 \equiv (2g^{(p-1)/3} + 1)a_3 \pmod p,$$

as in Theorems 3.1.1 *and* 3.1.2. *Set*

$$\pi = \chi(2)(a_3 + ib_3\sqrt{3}\ ),\ \ \overline{\pi} = \chi^2(2)(a_3 - ib_3\sqrt{3}\ ).$$

Let $A_1, \ldots, A_n, A$ *be integers with* $A_1 \cdots A_n \not\equiv 0 \pmod p$. *Set*

$$\begin{aligned} \textstyle\sum_0 &= 1, \\ \textstyle\sum_1 &= \chi(A_1) + \chi(A_2) + \cdots + \chi(A_n), \\ \textstyle\sum_2 &= \chi(A_1A_2) + \chi(A_1A_3) + \cdots + \chi(A_{n-1}A_n), \\ &\cdots \\ \textstyle\sum_n &= \chi(A_1A_2\cdots A_n). \end{aligned}$$

Let N *denote the number of solutions of the congruence* $A_1x_1^3+\cdots+A_nx_n^3 \equiv A \pmod p$. *Then*

$$N = \begin{cases} p^{n-1} + E_1 + E_2 + E_3, & \text{if } A \not\equiv 0 \pmod p, \\ p^{n-1} - (p-1)E_1, & \text{if } A \equiv 0 \pmod p, \end{cases}$$

where

$$E_1 = -\chi(A_1\cdots A_n) \sum_{\substack{j=0 \\ j \equiv -n \pmod 3}}^{n} \pi^{(j+n-3)/3}\, \overline{\pi}^{(-j+2n-3)/3} \textstyle\sum_j,$$

$$E_2 = \chi(AA_1\cdots A_n) \sum_{\substack{j=0 \\ j \equiv -n-1 \pmod 3}}^{n} \pi^{(j+n-2)/3}\, \overline{\pi}^{(-j+2n-1)/3} \textstyle\sum_j,$$

$$E_3 = \chi(A^2A_1\cdots A_n) \sum_{\substack{j=0 \\ j \equiv -n+1 \pmod 3}}^{n} \pi^{(j+n-1)/3}\, \overline{\pi}^{(-j+2n-2)/3} \textstyle\sum_j.$$

Proof In Theorem 10.4.2 we choose $q = p$, $k_1 = \cdots = k_n = 3$ (so that $d_1 = \cdots = d_n = 3$) and $\chi_1 = \cdots = \chi_n = \chi$, where χ is a cubic character (mod p) with $\chi(g) = \exp(2\pi i/3)$.

For $A \not\equiv 0 \pmod p$,

$$N = p^{n-1} + \sum_{j_1,\ldots,j_n=1}^{2} \chi^{j_1+\cdots+j_n}(A)\,\chi^{j_1}(A_1^{-1})\cdots\chi^{j_n}(A_n^{-1})J(\chi^{j_1},\ldots,\chi^{j_n}).$$

For each pair of integers (n_1, n_2) with

$$0 \le n_1 \le n,\ 0 \le n_2 \le n,\ n_1 + n_2 = n,$$

we collect together in the sum over $j_1, \ldots, j_n$ all those terms having n_1 of the j_i's equal to 1 and n_2 of the j_i's equal to 2. For these terms,

$$j_1 + \cdots + j_n = n_1 + 2n_2$$

and

$$J(\chi^{j_1},\ldots,\chi^{j_n}) = J(\underbrace{\chi,\ldots,\chi}_{n_1},\underbrace{\chi^2,\ldots,\chi^2}_{n_2}) = J_{n_1,n_2},$$

so that

$$N = p^{n-1} + \sum_{\substack{n_1,\ n_2=0 \\ n_1+n_2=n}}^{n} \chi^{n_1+2n_2}(A) J_{n_1,n_2} \sum_{\substack{j_1,\ldots,j_n=1 \\ n_1 \text{ of } j_i\text{'s}=1 \\ n_2 \text{ of } j_i\text{'s}=2}}^{2} \chi^{-j_1}(A_1)\cdots\chi^{-j_n}(A_n).$$

The inner sum is $\chi(A_1\cdots A_n)\Sigma_{n_1}$, and so the formula for N becomes

$$N = p^{n-1} + \chi(A_1\cdots A_n) \sum_{\substack{n_1,\ n_2=0 \\ n_1+n_2=n}}^{n} \chi^{n_1+2n_2}(A) J_{n_1,n_2} \sum\nolimits_{n_1}.$$

Next, appealing to Theorem 10.2.4 for the value of J_{n_1,n_2}, we obtain, considering the three cases $n_1+2n_2 \equiv 0,1,2 \pmod 3$ separately,

$$\sum_{\substack{n_1,\ n_2=0 \\ n_1+n_2=n \\ n_1+2n_2\equiv 0 \ (\mathrm{mod}\ 3)}}^{n} \chi^{n_1+2n_2}(A) J_{n_1,n_2} \sum\nolimits_{n_1} = -\sum_{\substack{n_1=0 \\ n_1\equiv 2n \ (\mathrm{mod}\ 3)}}^{n} \pi^{(n_1+n-3)/3}\,\overline{\pi}^{(-n_1+2n-3)/3} \sum\nolimits_{n_1}$$

$$= \chi^{-1}(A_1\cdots A_n)E_1;$$

$$\sum_{\substack{n_1,\ n_2=0 \\ n_1+n_2=n \\ n_1+2n_2\equiv 1 \ (\mathrm{mod}\ 3)}}^{n} \chi^{n_1+2n_2}(A) J_{n_1,n_2} \sum\nolimits_{n_1} = \chi(A) \sum_{\substack{n_1=0 \\ n_1\equiv 2n-1 \ (\mathrm{mod}\ 3)}}^{n} \pi^{(n_1+n-2)/3}\,\overline{\pi}^{(-n_1+2n-1)/3} \sum\nolimits_{n_1}$$

$$= \chi^{-1}(A_1\cdots A_n)E_2;$$

and

$$\sum_{\substack{n_1,\ n_2=0 \\ n_1+n_2=n \\ n_1+2n_2\equiv 2 \ (\mathrm{mod}\ 3)}}^{n} \chi^{n_1+2n_2}(A) J_{n_1,n_2} \sum\nolimits_{n_1} = \chi^2(A) \sum_{\substack{n_1=0 \\ n_1\equiv 2n-2 \ (\mathrm{mod}\ 3)}}^{n} \pi^{(n_1+n-1)/3}\,\overline{\pi}^{(-n_1+2n-2)/3} \sum\nolimits_{n_1}$$

$$= \chi^{-1}(A_1\cdots A_n)E_3;$$

so that $N = p^{n-1} + E_1 + E_2 + E_3$, as asserted.

It remains to examine the possibility $A \equiv 0 \pmod p$. By Theorem 10.4.2,

$$N = p^{n-1} - (p-1) \sum_{\substack{j_1,\ldots,j_n=1 \\ j_1+\cdots+j_n\equiv 0 \ (\mathrm{mod}\ 3)}}^{2} \chi^{j_1}(A_1^{-1})\cdots\chi^{j_n}(A_n^{-1}) J(\chi^{j_1},\ldots,\chi^{j_n}).$$

In the sum over $j_1,\ldots,j_n$ we collect together those terms having n_1 of the j_i's equal to 1 and n_2 of the j_i's equal to 2, where $0 \le n_1 \le n$, $0 \le n_2 \le n$, $n_1+n_2=n$, $n_1+2n_2 \equiv$

0 (mod 3). Then the formula for N becomes

$$N = p^{n-1} - (p-1) \sum_{\substack{n_1,\, n_2 = 0 \\ n_1 + n_2 = n \\ n_1 + 2n_2 \equiv 0 \pmod 3}}^{n} J_{n_1,n_2} \sum_{\substack{j_1,\ldots,j_n = 1 \\ n_1 \text{ of } j_i\text{'s}=1 \\ n_2 \text{ of } j_i\text{'s}=2}}^{2} \chi^{-j_1}(A_1)\cdots\chi^{-j_n}(A_n),$$

so that

$$N = p^{n-1} - (p-1)\chi(A_1\cdots A_n) \sum_{\substack{n_1,\, n_2 = 0 \\ n_1 + n_2 = n \\ n_1 + 2n_2 \equiv 0 \pmod 3}}^{n} J_{n_1,n_2} \sum\nolimits_{n_1}.$$

Now, appealing to Theorem 10.2.4 for the value of J_{n_1,n_2}, we obtain

$$\begin{aligned} N &= p^{n-1} + (p-1)\chi(A_1\cdots A_n) \sum_{\substack{n_1,\, n_2 = 0 \\ n_1 + n_2 = n \\ n_1 + 2n_2 \equiv 0 \pmod 3}}^{n} \pi^{(2n_1+n_2-3)/3}\,\overline{\pi}^{(n_1+2n_2-3)/3} \sum\nolimits_{n_1} \\ &= p^{n-1} + (p-1)\chi(A_1\cdots A_n) \sum_{\substack{j=0 \\ j \equiv -n \pmod 3}}^{n} \pi^{(j+n-3)/3}\,\overline{\pi}^{(-j+2n-3)/3} \sum\nolimits_{j} \\ &= p^{n-1} - (p-1)E_1, \end{aligned}$$

as asserted. ■

We examine the special cases $n = 2$ and $n = 3$ of Theorem 10.6.1 in the next two corollaries.

Corollary 10.6.2 (a) *With the notation of Theorem 10.6.1, the number N of solutions of the congruence*

$$A_1x_1^3 + A_2x_2^3 \equiv A \pmod p$$

is

$$\begin{cases} p + \pi\chi^2(AA_1A_2) + \overline{\pi}\chi(AA_1A_2) - \chi(A_1)\chi^2(A_2) - \chi^2(A_1)\chi(A_2), & \text{if } A \not\equiv 0 \pmod p, \\ p + (p-1)(\chi(A_1)\chi^2(A_2) + \chi^2(A_1)\chi(A_2)), & \text{if } A \equiv 0 \pmod p. \end{cases}$$

(b) *If $A \not\equiv 0 \pmod p$, then*

$$N \geq p - \max(|r_3|, 3|s_3|) - 2,$$

where r_3 and s_3 are defined in Theorem 3.1.3.

(c) *Set $\omega = e^{2\pi i/3}$. Then the number of solutions of the congruence*

$$x_1^3 + x_2^3 \equiv A \pmod p$$

is

$$\begin{cases} p + r_3 - 2, & \text{if } \chi(A) = 1, \\ (2p - r_3 + 3s_3 - 4)/2, & \text{if } \chi(A) = \omega, \\ (2p - r_3 - 3s_3 - 4)/2, & \text{if } \chi(A) = \omega^2. \end{cases}$$

Proof (a) When $n = 2$,

$$E_1 = -\chi(A_1A_2)(\chi(A_1) + \chi(A_2)),\ E_2 = \overline{\pi}\chi(AA_1A_2),\ E_3 = \pi\chi^2(AA_1A_2),$$

and the assertion of (a) follows from Theorem 10.6.1.

(b) Since $\chi(A_1)\chi^2(A_2) + \chi^2(A_1)\chi(A_2)$ equals either $\omega + \omega^2 = -1$ or $1 + 1 = 2$,

$$-\chi(A_1)\chi^2(A_2) - \chi^2(A_1)\chi(A_2) \geq -2.$$

By (3.1.3) and (3.1.6), $\pi = \chi(2)(a_3 + ib_3\sqrt{3}\,) = (r_3 + is_3\sqrt{3}\,)/2$, and so

$$\begin{aligned}\pi\chi^2(AA_1A_2) + \overline{\pi}\chi(AA_1A_2) &= \pi + \overline{\pi},\ \ \pi\omega + \overline{\pi}\omega^2,\ \ \text{or}\ \ \pi\omega^2 + \overline{\pi}\omega \\ &= r_3,\ \ (-r_3 - 3s_3)/2,\ \ \text{or}\ \ (-r_3 + 3s_3)/2.\end{aligned}$$

Thus

$$\pi\chi^2(AA_1A_2) + \overline{\pi}\chi(AA_1A_2) \geq -\max(|r_3|, 3|s_3|).$$

Hence from (a) we deduce that

$$N \geq p - \max(|r_3|, 3|s_3|) - 2.$$

(c) This follows by taking $A_1 = A_2 = 1$ in (a). ■

Corollary 10.6.3 *With the notation of Theorem* 10.6.1, *the number of solutions of the congruence* $A_1x_1^3 + A_2x_2^3 + A_3x_3^3 \equiv A \pmod{p}$ *is*

$$\begin{cases} p^2 + p\chi(AA_1A_2A_3)(\chi(A_1A_2) + \chi(A_2A_3) + \chi(A_3A_1)) \\ \quad +p\chi(A^2A_1A_2A_3)(\chi(A_1) + \chi(A_2) + \chi(A_3)) \\ \quad\quad -\overline{\pi}\chi(A_1A_2A_3) - \pi\chi^2(A_1A_2A_3), & \text{if } A \not\equiv 0 \pmod{p}, \\ p^2 + (p-1)(\overline{\pi}\chi(A_1A_2A_3) + \pi\chi^2(A_1A_2A_3)), & \text{if } A \equiv 0 \pmod{p}. \end{cases}$$

Proof Since $\pi\overline{\pi} = p$, it follows that for $n = 3$,

$$\begin{aligned} E_1 &= -\overline{\pi}\chi(A_1A_2A_3) - \pi\chi^2(A_1A_2A_3), \\ E_2 &= p\chi(AA_1A_2A_3)(\chi(A_1A_2) + \chi(A_1A_3) + \chi(A_2A_3)), \\ E_3 &= p\chi(A^2A_1A_2A_3)(\chi(A_1) + \chi(A_2) + \chi(A_3)). \end{aligned}$$ ■

Example We use Corollary 10.6.2(a) to determine the number of solutions of the congruence

$$3x_1^3 + 5x_2^3 \equiv 7 \pmod{19}.$$

As 2 is a primitive root (mod 19), we may take $g = 2$. Let χ be the cubic character (mod 19) given by $\chi(2) = \omega$, where $\omega = \exp(2\pi i/3) = (-1 + i\sqrt{3}\,)/2$. The integers a_3 and b_3 are determined by

$$a_3^2 + 3b_3^2 = 19,\ a_3 \equiv -1 \pmod{3},\ 3b_3 \equiv 15a_3 \pmod{19},$$

that is

$$a_3 = -4, \quad b_3 = -1. \tag{10.6.1}$$

Thus

$$\pi = \chi(2)(a_3 + ib_3\sqrt{3}\,) = \omega(-4 - i\sqrt{3}\,) = (7 - i3\sqrt{3}\,)/2, \quad \pi\omega = (1 + i5\sqrt{3}\,)/2.$$

Further,

$$\begin{aligned}\chi(3) &= \chi(2^{13}) = \omega^{13} = \omega,\\ \chi(5) &= \chi(2^{16}) = \omega^{16} = \omega,\\ \chi(7) &= \chi(2^{6}) = \omega^{6} = 1,\end{aligned}$$

so that by Corollary 10.6.2 the required number N of solutions of $3x_1^3 + 5x_2^3 \equiv 7 \pmod{19}$ is

$$N = 19 + \pi\omega + \overline{\pi\omega} - 2 = 17 + 2\ \mathrm{Re}\ (\pi\omega) = 17 + 2 \cdot \frac{1}{2} = 18.$$

Indeed, a simple computer search yields the following 18 solutions:

$$\begin{aligned}(x_1, x_2) = {} & (2,2),\ (2,3),\ (2,14),\ (3,2),\ (3,3),\ (3,14),\ (4,1),\ (4,7),\ (4,11),\\ & (6,1),\ (6,7),\ (6,11),\ (9,1),\ (9,7),\ (9,11),\ (14,2),\ (14,3),\ (14,14).\end{aligned}$$

It is important to stress that in applying Theorem 10.6.1 and its corollaries, it is not enough to select a_3 and b_3 to be integers satisfying $a_3^2 + 3b_3^2 = p$. They must be chosen to satisfy the congruences $a_3 \equiv -1 \pmod 3$ and $3b_3 \equiv (2g^{(p-1)/3} + 1)a_3 \pmod p$ as well. If we took $a_3 = -4$, $b_3 = 1$ in place of (10.6.1), then

$$\begin{aligned}&\pi = \chi(2)(a_3 + ib_3\sqrt{3}\,) = \omega(-4 + i\sqrt{3}\,) = (1 - i5\sqrt{3}\,)/2,\\ &\pi\omega = (7 + i3\sqrt{3}\,)/2, \quad \mathrm{Re}(\pi\omega) = 7/2,\end{aligned}$$

giving

$$N = 19 + (\pi\omega + \overline{\pi}\omega^2) - 2 = 17 + 2\mathrm{Re}(\pi\omega) = 24,$$

which is incorrect.

Our next result is a modification of the special case $A_1 = \cdots = A_n = 1$ of Theorem 10.6.1. It gives the number of solutions of the congruence $x_1^3 + \cdots + x_n^3 \equiv A \pmod p$ in terms of Gauss sums.

Corollary 10.6.4 *Let $p = 3f + 1$ be a prime. Let g be a primitive root of p. Let χ be a cubic character* (mod p) *with $\chi(g) = \omega$, where $\omega = \exp(2\pi i/3) = (-1 + i\sqrt{3}\,)/2$. Set $\tau = G(\chi)$. Then the number N of solutions of the congruence*

$$x_1^3 + x_2^3 + \cdots + x_n^3 \equiv A \pmod p$$

is given by

$$N = p^{n-1} + \frac{\chi(A)}{3\tau}\left((\tau+\overline{\tau})^n + \omega^2(\omega\tau+\omega^2\overline{\tau})^n + \omega(\omega^2\tau+\omega\overline{\tau})^n\right)$$
$$+ \frac{\chi^2(A)}{3\overline{\tau}}\left((\tau+\overline{\tau})^n + \omega(\omega\tau+\omega^2\overline{\tau})^n + \omega^2(\omega^2\tau+\omega\overline{\tau})^n\right)$$
$$- \frac{1}{3p}\left((\tau+\overline{\tau})^n + (\omega\tau+\omega^2\overline{\tau})^n + (\omega^2\tau+\omega\overline{\tau})^n\right), \qquad \text{if } A \not\equiv 0 \pmod{p},$$

and

$$N = p^{n-1} + \frac{(p-1)}{3p}\left((\tau+\overline{\tau})^n + (\omega\tau+\omega^2\overline{\tau})^n + (\omega^2\tau+\omega\overline{\tau})^n\right), \qquad \text{if } A \equiv 0 \pmod{p}.$$

Proof With $A_1 = \cdots = A_n = 1$, we see from Theorem 10.6.1 that

$$\sum\nolimits_i = \binom{n}{i}, \quad i = 0, 1, \ldots, n.$$

Hence, as $p = \pi\overline{\pi}$,

$$E_1 = -\frac{1}{p} \sum_{\substack{j=0 \\ j \equiv -n \pmod 3}}^{n} \binom{n}{j} \pi^{(j+n)/3}\, \overline{\pi}^{(-j+2n)/3},$$

$$E_2 = \frac{\chi(A)}{p} \sum_{\substack{j=0 \\ j \equiv -n-1 \pmod 3}}^{n} \binom{n}{j} \pi^{(j+n+1)/3}\, \overline{\pi}^{(-j+2n+2)/3},$$

and

$$E_3 = \frac{\chi^2(A)}{p} \sum_{\substack{j=0 \\ j \equiv -n+1 \pmod 3}}^{n} \binom{n}{j} \pi^{(j+n+2)/3}\, \overline{\pi}^{(-j+2n+1)/3}.$$

Next, we recall from Theorem 1.1.4, Theorem 3.1.1, and Lemma 4.1.1 that

$$G(\chi)\overline{G(\chi)} = p, \quad G(\chi)^3 = pJ(\chi,\chi) = p\chi(2)(a_3 + ib_3\sqrt{3}\,) = p\pi.$$

Now, by the binomial theorem,

$$(G(\chi) + \overline{G(\chi)})^n = \sum_{j=0}^{n} \binom{n}{j} G(\chi)^j\, \overline{G(\chi)}^{n-j},$$

$$(\omega G(\chi) + \omega^2\overline{G(\chi)})^n = \sum_{j=0}^{n} \binom{n}{j} G(\chi)^j\, \overline{G(\chi)}^{n-j} \omega^{2(n+j)},$$

and

$$(\omega^2 G(\chi) + \omega\overline{G(\chi)})^n = \sum_{j=0}^{n} \binom{n}{j} G(\chi)^j\, \overline{G(\chi)}^{n-j} \omega^{n+j}.$$

Thus,

$$\begin{aligned}
&(G(\chi)+\overline{G(\chi)})^n + (\omega G(\chi)+\omega^2\overline{G(\chi)})^n + (\omega^2 G(\chi)+\omega\overline{G(\chi)})^n \\
&= \sum_{j=0}^{n}\binom{n}{j} G(\chi)^j\overline{G(\chi)}^{n-j}(1+\omega^{n+j}+\omega^{2(n+j)}) = 3\sum_{\substack{j=0\\ j\equiv -n \pmod 3}}^{n}\binom{n}{j} G(\chi)^j\overline{G(\chi)}^{n-j} \\
&= \frac{3}{(G(\chi)\overline{G(\chi)})^n}\sum_{\substack{j=0\\ j\equiv -n \pmod 3}}^{n}\binom{n}{j} G(\chi)^{j+n}\overline{G(\chi)}^{2n-j} \\
&= \frac{3}{p^n}\sum_{\substack{j=0\\ j\equiv -n \pmod 3}}^{n}\binom{n}{j}(p\pi)^{(j+n)/3}(p\overline{\pi})^{(2n-j)/3} \\
&= 3\sum_{\substack{j=0\\ j\equiv -n \pmod 3}}^{n}\binom{n}{j}\pi^{(j+n)/3}\overline{\pi}^{(2n-j)/3} = -3pE_1.
\end{aligned}$$

Similarly,

$$(G(\chi)+\overline{G(\chi)})^n + \omega(\omega G(\chi)+\omega^2\overline{G(\chi)})^n + \omega^2(\omega^2 G(\chi)+\omega\overline{G(\chi)})^n = 3\chi(A)\overline{G(\chi)}E_3$$

and

$$(G(\chi)+\overline{G(\chi)})^n + \omega^2(\omega G(\chi)+\omega^2\overline{G(\chi)})^n + \omega(\omega^2 G(\chi)+\omega\overline{G(\chi)})^n = 3\chi^2(A)G(\chi)E_2.$$

The result now follows from Theorem 10.6.1. ∎

10.7 Number of solutions of the congruence $A_1x_1^4 + \cdots + A_nx_n^4 \equiv A \pmod p$

In this section we use Theorem 10.2.5 and Theorem 10.4.2 to determine an explicit formula for the number N of solutions of the congruence $A_1x_1^4 + \cdots + A_nx_n^4 \equiv A \pmod p$, where p is a prime with $p \equiv 1 \pmod 4$ and $A_1, \ldots, A_n, A$ are integers with $A_1 \cdots A_n \not\equiv 0 \pmod p$.

Theorem 10.7.1 *Let $p = 4f+1$ be a prime. Let g be a primitive root $\pmod p$, and choose χ to be the quartic character $\pmod p$ such that $\chi(g) = i$. Let a_4 and b_4 be the integers uniquely determined by*

$$a_4^2 + b_4^2 = p, \quad a_4 \equiv -\left(\frac{2}{p}\right) \pmod 4, \quad b_4 \equiv a_4 g^{(p-1)/4} \pmod p,$$

as in Theorems 3.2.1 and 3.2.2. Set

$$\pi = (-1)^f(a_4 + ib_4).$$

Let $A_1, \ldots, A_n, A$ be integers with $A_1 \cdots A_n \not\equiv 0 \pmod p$. For each triple of nonnegative integers (n_1, n_2, n_3) with $n_1 + n_2 + n_3 = n$, set

$$E(n_1, n_2, n_3) = \sum_{k_1,\ldots,k_n=0}^{2}{}^{*} \chi^{k_1}(A_1) \cdots \chi^{k_n}(A_n),$$

where the asterisk (*) indicates that the sum is taken over those n-tuples $(k_1, \ldots, k_n)$ having n_1 of the k_i's equal to 2, n_2 of the k_i's equal to 1, and n_3 of the k_i's equal to 0. Further we set

$$F_0 = -\sum_{\substack{n_1, n_2, n_3 = 0 \\ n_1 + n_2 + n_3 = n \\ n_1 + 2n_2 + 3n_3 \equiv 0 \pmod 4}}^{n} (-1)^{n_1 f} \pi^{(3n_1+2n_2+n_3-4)/4} \overline{\pi}^{(n_1+2n_2+3n_3-4)/4} E(n_1, n_2, n_3),$$

$$F_1 = (-1)^f \sum_{\substack{n_1, n_2, n_3 = 0 \\ n_1 + n_2 + n_3 = n \\ n_1 + 2n_2 + 3n_3 \equiv 1 \pmod 4}}^{n} (-1)^{n_1 f} \pi^{(3n_1+2n_2+n_3-3)/4} \overline{\pi}^{(n_1+2n_2+3n_3-1)/4} E(n_1, n_2, n_3),$$

$$F_2 = \sum_{\substack{n_1, n_2, n_3 = 0 \\ n_1 + n_2 + n_3 = n \\ n_1 + 2n_2 + 3n_3 \equiv 2 \pmod 4}}^{n} (-1)^{n_1 f} \pi^{(3n_1+2n_2+n_3-2)/4} \overline{\pi}^{(n_1+2n_2+3n_3-2)/4} E(n_1, n_2, n_3),$$

$$F_3 = \sum_{\substack{n_1, n_2, n_3 = 0 \\ n_1 + n_2 + n_3 = n \\ n_1 + 2n_2 + 3n_3 \equiv 3 \pmod 4}}^{n} (-1)^{n_1 f} \pi^{(3n_1+2n_2+n_3-1)/4} \overline{\pi}^{(n_1+2n_2+3n_3-3)/4} E(n_1, n_2, n_3).$$

Then the number N of solutions to $A_1 x_1^4 + \cdots + A_n x_n^4 \equiv A \pmod p$ is

$$N = \begin{cases} p^{n-1} + \chi(A_1 \cdots A_n)(F_0 + \chi(A)F_1 + \chi^2(A)F_2 + \chi^3(A)F_3), & \text{if } A \not\equiv 0 \pmod p, \\ p^{n-1} - (p-1)\chi(A_1 \cdots A_n)F_0, & \text{if } A \equiv 0 \pmod p. \end{cases}$$

Proof In Theorem 10.4.2 we choose $q = p$, $k_1 = \cdots = k_n = 4$ (so that $d_1 = \cdots = d_n = 4$), and $\chi_1 = \cdots = \chi_n = \chi$. For $A \not\equiv 0 \pmod p$,

$$N = p^{n-1} + \sum_{j_1,\ldots,j_n=1}^{3} \chi^{j_1+\cdots+j_n}(A) \chi^{j_1}(A_1^{-1}) \cdots \chi^{j_n}(A_n^{-1}) J(\chi^{j_1}, \ldots, \chi^{j_n}).$$

For each triple of nonnegative integers (n_1, n_2, n_3) with $n_1 + n_2 + n_3 = n$, we collect together in the sum over $j_1, \ldots, j_n$ all those terms having n_1 of the j_i's equal to 1, n_2 of the j_i's equal to 2, and n_3 of the j_i's equal to 3. For these terms,

$$j_1 + \cdots + j_n = n_1 + 2n_2 + 3n_3$$

and

$$J(\chi^{j_1}, \ldots, \chi^{j_n}) = J(\underbrace{\chi, \ldots, \chi}_{n_1}, \underbrace{\chi^2, \ldots, \chi^2}_{n_2}, \underbrace{\chi^3, \ldots, \chi^3}_{n_3}).$$

This generalized Jacobi sum is the one evaluated in Theorem 10.2.5, that is, J_{n_1,n_2,n_3}. Hence,

$$N = p^{n-1} + \sum_{\substack{n_1,n_2,n_3=0 \\ n_1+n_2+n_3=n}}^{n} \chi^{n_1+2n_2+3n_3}(A) J_{n_1,n_2,n_3} \sideset{}{^\dagger}\sum_{j_1,\ldots,j_n=1}^{3} \chi^{-j_1}(A_1)\cdots\chi^{-j_n}(A_n),$$

where the dagger (†) indicates that the n-tuples $(j_1,\ldots,j_n)$ are restricted to those having n_1 1's, n_2 2's, and n_3 3's. The inner sum is

$$\begin{aligned}\sideset{}{^\dagger}\sum_{j_1,\ldots,j_n=1}^{3} \chi^{4-j_1}(A_1)\cdots\chi^{4-j_n}(A_n) &= \chi(A_1)\cdots\chi(A_n) \sideset{}{^\dagger}\sum_{j_1,\ldots,j_n=1}^{3} \chi^{3-j_1}(A_1)\cdots\chi^{3-j_n}(A_n) \\ &= \chi(A_1\cdots A_n) \sideset{}{^*}\sum_{k_1,\ldots,k_n=0}^{2} \chi^{k_1}(A_1)\cdots\chi^{k_n}(A_n) \\ &= \chi(A_1\cdots A_n)E(n_1,n_2,n_3),\end{aligned}$$

and so the formula for N becomes

$$N = p^{n-1} + \chi(A_1\cdots A_n) \sum_{\substack{n_1,n_2,n_3=0 \\ n_1+n_2+n_3=n}}^{n} \chi^{n_1+2n_2+3n_3}(A) J_{n_1,n_2,n_3} E(n_1,n_2,n_3).$$

We now appeal to Theorem 10.2.5 for the value of J_{n_1,n_2,n_3}. Considering the four possibilities $n_1 + 2n_2 + 3n_3 \equiv 0,1,2,3 \pmod{4}$ separately, we obtain

$$\sum_{\substack{n_1,n_2,n_3=0 \\ n_1+n_2+n_3=n \\ n_1+2n_2+3n_3\equiv 0 \pmod 4}} \chi^{n_1+2n_2+3n_3}(A) J_{n_1,n_2,n_3} E(n_1,n_2,n_3) = F_0,$$

$$\sum_{\substack{n_1,n_2,n_3=0 \\ n_1+n_2+n_3=n \\ n_1+2n_2+3n_3\equiv 1 \pmod 4}} \chi^{n_1+2n_2+3n_3}(A)\, J_{n_1,n_2,n_3}\, E(n_1,n_2,n_3) = \chi(A)F_1,$$

$$\sum_{\substack{n_1,n_2,n_3=0 \\ n_1+n_2+n_3=n \\ n_1+2n_2+3n_3\equiv 2 \pmod 4}} \chi^{n_1+2n_2+3n_3}(A)\, J_{n_1,n_2,n_3}\, E(n_1,n_2,n_3) = \chi^2(A)F_2,$$

and

$$\sum_{\substack{n_1,n_2,n_3=0 \\ n_1+n_2+n_3=n \\ n_1+2n_2+3n_3\equiv 3 \pmod 4}} \chi^{n_1+2n_2+3n_3}(A)\, J_{n_1,n_2,n_3}\, E(n_1,n_2,n_3) = \chi^3(A)F_3.$$

This completes the proof of the asserted formula for N when $A \not\equiv 0 \pmod{p}$.

When $A \equiv 0 \pmod p$, Theorem 10.4.2 yields

$$N = p^{n-1} - (p-1) \sum_{\substack{j_1,\ldots,j_n=1 \\ j_1+\cdots+j_n \equiv 0 \pmod 4}}^{3} \chi^{-j_1}(A_1)\cdots\chi^{-j_n}(A_n) J(\chi^{j_1},\ldots,\chi^{j_n}).$$

Exactly as in the case $A \not\equiv 0 \pmod p$, we obtain

$$\begin{aligned} N &= p^{n-1} - (p-1) \sum_{\substack{n_1,n_2,n_3=0 \\ n_1+n_2+n_3=n \\ n_1+2n_2+3n_3 \equiv 0 \pmod 4}}^{n} J_{n_1,n_2,n_3} \sum_{j_1,\ldots,j_n=1}^{3}{}^{\dagger} \chi^{-j_1}(A_1)\cdots\chi^{-j_n}(A_n) \\ &= p^{n-1} - (p-1)\chi(A_1\cdots A_n) \sum_{\substack{n_1,n_2,n_3=0 \\ n_1+n_2+n_3=n \\ n_1+2n_2+3n_3 \equiv 0 \pmod 4}}^{n} J_{n_1,n_2,n_3} E(n_1,n_2,n_3) \\ &= p^{n-1} - (p-1)\chi(A_1\cdots A_n)F_0. \end{aligned}$$

■

The next corollary is the special case $n = 2$ of Theorem 10.7.1.

Corollary 10.7.2 (a) *With the notation of Theorem* 10.7.1, *the number N of solutions of the congruence*

$$A_1x_1^4 + A_2x_2^4 \equiv A \pmod p$$

is

$$\begin{aligned} &p - (-1)^f(\chi(A_1)\chi^3(A_2) + (-1)^f\chi^2(A_1A_2) + \chi^3(A_1)\chi(A_2)) \\ &\quad +\chi(A)(\chi(A_1)\chi^2(A_2) + \chi^2(A_1)\chi(A_2))(a_4 - ib_4) \\ &\quad +(-1)^f\chi^2(A)(\chi(A_1A_2)(a_4 - ib_4) + \chi^3(A_1A_2)(a_4 + ib_4)) \\ &\quad +\chi^3(A)(\chi^2(A_1)\chi^3(A_2) + \chi^3(A_1)\chi^2(A_2))(a_4 + ib_4), \qquad \text{if } A \not\equiv 0 \pmod p, \end{aligned}$$

and

$$p + (-1)^f(p-1)\{\chi(A_1)\chi^3(A_2) + (-1)^f\chi^2(A_1A_2) + \chi^3(A_1)\chi(A_2)\}, \text{ if } A \equiv 0 \pmod p.$$

In particular, the number of solutions of the congruence

$$x_1^4 + x_2^4 \equiv A \pmod p$$

is

$$\begin{array}{ll} p + 6a_4 - 3, & \text{if } \chi(A) = 1 \text{ and } 2 \mid f, \\ p + 2a_4 + 1, & \text{if } \chi(A) = 1 \text{ and } 2 \nmid f, \\ p - 2a_4 + 4b_4 - 3, & \text{if } \chi(A) = i \text{ and } 2 \mid f, \\ p + 2a_4 + 4b_4 + 1, & \text{if } \chi(A) = i \text{ and } 2 \nmid f, \\ p - 2a_4 - 3, & \text{if } \chi(A) = -1 \text{ and } 2 \mid f, \\ p - 6a_4 + 1, & \text{if } \chi(A) = -1 \text{ and } 2 \nmid f, \\ p - 2a_4 - 4b_4 - 3, & \text{if } \chi(A) = -i \text{ and } 2 \mid f, \\ p + 2a_4 - 4b_4 + 1, & \text{if } \chi(A) = -i \text{ and } 2 \nmid f. \end{array}$$

(b) *If* $A \not\equiv 0 \pmod p$, *then*

$$N \geq p - 6 \max(|a_4|, |b_4|) - 3.$$

Proof Part (a) follows easily from Theorem 10.7.1 with $n = 2$. It remains to deduce part (b) from part (a). First set

$$Z = X + iY = \chi(A)\chi(A_1)\chi^2(A_2) + \chi(A)\chi^2(A_1)\chi(A_2) + (-1)^f\chi^2(A)\chi(A_1A_2),$$

where X and Y are real. As $X + iY = i^\alpha + i^\beta + i^\gamma$, for some $\alpha, \beta, \gamma = 0, 1, 2, 3, 4$, we deduce that $|X| + |Y| \leq 3$. By Corollary 10.7.2,

$$\begin{aligned}
N - (p-3) &\geq N - (p - (-1)^f(\chi(A_1)\chi^3(A_2) + (-1)^f\chi^2(A_1A_2) + \chi^3(A_1)\chi(A_2))) \\
&= (Z + \overline{Z})a_4 + i(-Z + \overline{Z})b_4 \\
&= 2Xa_4 + 2Yb_4 \\
&\geq -2|X||a_4| - 2|Y||b_4| \\
&\geq -2 \max(|a_4|, |b_4|)(|X| + |Y|) \\
&\geq -6 \max(|a_4|, |b_4|),
\end{aligned}$$

as desired. ∎

Example We use Corollary 10.7.2 to determine the number N of solutions of

$$x_1^4 + 2x_2^4 \equiv 9 \pmod{13}.$$

Here $p = 13$, $f = 3$, $A_1 = 1$, $A_2 = 2$, and $A = 9$. We may take $g = 2$. Thus $a_4 = -3$ and $b_4 = 2$. Also $\chi(A_1) = \chi(1) = 1$, $\chi(A_2) = \chi(2) = i$, and $\chi(A) = \chi(9) = \chi(2^8) = 1$. By Corollary 10.7.2(a),

$$N = 13 + 1 + (-1+i)(-3-2i) - (i(-3-2i) - i(-3+2i)) + (-1-i)(-3+2i) = 20.$$

As a check, a simple computer program yields the 20 solutions

$$\begin{aligned}
(x_1, x_2) = {} & (2,2),\ (2,3),\ (2,10),\ (2,11),\ (3,2),\ (3,3),\ (3,10), \\
& (3,11),\ (4,0),\ (6,0),\ (7,0),\ (9,0),\ (10,2),\ (10,3), \\
& (10,10),\ (10,11),\ (11,2),\ (11,3),\ (11,10),\ (11,11).
\end{aligned}$$

10.8 Bounds for the number of solutions

In this section we use Theorem 10.3.2 in conjunction with Theorem 10.4.2 to obtain bounds for the number N of solutions of the equation $\alpha_1 x_1^{k_1} + \cdots + \alpha_n x_n^{k_n} = \alpha$ in F_q^n. We make use of the notation of Theorem 10.4.2 throughout this section. Let χ be a generator of the cyclic group of characters on F_q. Take $\chi_i = \chi^{(q-1)/d_i}$ $(i = 1, \ldots, n)$ in Theorem 10.4.2. Observe that the character

$$\chi_1^{j_1} \cdots \chi_n^{j_n} = \chi^{(j_1(q-1)/d_1) + \cdots + (j_n(q-1)/d_n)}$$

is trivial if and only if

$$(j_1/d_1) + \cdots + (j_n/d_n) \in Z.$$

It is convenient to set

$$\text{(10.8.1)} \qquad t(d_1, \ldots, d_n) = \begin{array}{l} \text{number of } n\text{-tuples of integers } (j_1, \ldots, j_n) \\ \text{with } 1 \leq j_i \leq d_i - 1 \ (i = 1, \ldots, n) \text{ such that} \\ (j_1/d_1) + \cdots + (j_n/d_n) \text{ is an integer.} \end{array}$$

Clearly

$$\text{(10.8.2)} \qquad t(d_1, \ldots, d_n) \leq (d_1 - 1) \cdots (d_n - 1).$$

Theorem 10.8.1 *Let $k_1, \ldots, k_n$ be positive integers. Let $\alpha_1, \ldots, \alpha_n \in F_q^*$ and $\alpha \in F_q$. Set $d_i = \gcd(k_i, q-1)$. Then the number N of solutions in F_q^n of the equation $\alpha_1 x_1^{k_1} + \cdots + \alpha_n x_n^{k_n} = \alpha$ satisfies*

$$|N - q^{n-1}| \leq \left((d_1 - 1) \cdots (d_n - 1) - (1 - q^{-\frac{1}{2}}) t(d_1, \ldots, d_n)\right) q^{(n-1)/2}, \quad \text{if } \alpha \neq 0,$$

and

$$|N - q^{n-1}| \leq t(d_1, \ldots, d_n)(q-1) q^{(n-2)/2}, \qquad \text{if } \alpha = 0.$$

Proof If $\alpha = 0$, then, by Theorems 10.3.2 and 10.4.2, with the choice of χ_i as above,

$$\begin{aligned} |N - q^{n-1}| &= (q-1) \left| \sum_{\substack{j_1=1 \\ \chi_1^{j_1} \cdots \chi_n^{j_n} \text{ trivial}}}^{d_1-1} \cdots \sum_{j_n=1}^{d_n-1} \chi_1^{j_1}(\alpha_1^{-1}) \cdots \chi_n^{j_n}(\alpha_n^{-1}) J_r(\chi_1^{j_1}, \ldots, \chi_n^{j_n}) \right| \\ &\leq (q-1) \sum_{\substack{j_1=1 \\ \chi_1^{j_1} \cdots \chi_n^{j_n} \text{ trivial}}}^{d_1-1} \cdots \sum_{j_n=1}^{d_n-1} |J_r(\chi_1^{j_1}, \ldots, \chi_n^{j_n})| \\ &\leq (q-1) q^{(n-2)/2} \sum_{\substack{j_1=1 \\ \chi_1^{j_1} \cdots \chi_n^{j_n} \text{ trivial}}}^{d_1-1} \cdots \sum_{j_n=1}^{d_n-1} 1 \\ &= (q-1) q^{(n-2)/2} t(d_1, \ldots, d_n). \end{aligned}$$

If $\alpha \neq 0$,

$$\begin{aligned}
|N - q^{n-1}| &= \left| \sum_{j_1=1}^{d_1-1} \cdots \sum_{j_n=1}^{d_n-1} \chi_1^{j_1}(\alpha\alpha_1^{-1}) \cdots \chi_n^{j_n}(\alpha\alpha_n^{-1}) J_r(\chi_1^{j_1}, \ldots, \chi_n^{j_n}) \right| \\
&\leq \sum_{j_1=1}^{d_1-1} \cdots \sum_{j_n=1}^{d_n-1} |J_r(\chi_1^{j_1}, \ldots, \chi_n^{j_n})| \\
&= \sum_{\substack{j_1=1 \\ \chi_1^{j_1} \cdots \chi_n^{j_n} \text{ trivial}}}^{d_1-1} \cdots \sum_{j_n=1}^{d_n-1} |J_r(\chi_1^{j_1}, \ldots, \chi_n^{j_n})| + \sum_{\substack{j_1=1 \\ \chi_1^{j_1} \cdots \chi_n^{j_n} \text{ nontrivial}}}^{d_1-1} \cdots \sum_{j_n=1}^{d_n-1} |J_r(\chi_1^{j_1}, \ldots, \chi_n^{j_n})| \\
&\leq q^{(n-2)/2} t(d_1, \ldots, d_n) + q^{(n-1)/2}((d_1 - 1) \cdots (d_n - 1) - t(d_1, \ldots, d_n)) \\
&= q^{(n-1)/2} \left((d_1 - 1) \cdots (d_n - 1) - (1 - q^{-1/2}) t(d_1, \ldots, d_n) \right). \qquad \blacksquare
\end{aligned}$$

The following corollary is a straightforward consequence of Theorem 10.8.1.

Corollary 10.8.2 *Let $n > 1$ and $\alpha \in F_q^*$. With the notation of Theorem 10.8.1, the equation $\alpha_1 x_1^{k_1} + \cdots + \alpha_n x_n^{k_n} = \alpha$ is solvable in F_q^n if*

$$q > \left\{ (d_1 - 1) \cdots (d_n - 1) - \left(1 - \frac{1}{\sqrt{q}}\right) t(d_1, \ldots, d_n) \right\}^{2/(n-1)}.$$

We wish to apply Corollary 10.8.2 to the congruence

$$A_1 x_1^{m_1} + A_2 x_2^{m_2} \equiv A \pmod{p}, \tag{10.8.3}$$

where p is a prime and m_1, m_2, A_1, A_2, A are integers with $m_1 \geq 1$, $m_2 \geq 1$, $p \nmid A_1 A_2 A$. In order to do this, we must determine $t(d_1, d_2)$.

Theorem 10.8.3 $t(d_1, d_2) = \gcd(d_1, d_2) - 1$.

Proof By (10.8.1), $t(d_1, d_2)$ is the number of pairs of integers (a, b) with $0 < a < d_1$ and $0 < b < d_2$ such that

$$a d_2 + b d_1 = d_1 d_2. \tag{10.8.4}$$

Let $d = \gcd(d_1, d_2)$, $d_2' = d_2/d$, and $d_1' = d_1/d$. Then (10.8.4) holds if and only if

$$a d_2' + b d_1' = d d_1' d_2'. \tag{10.8.5}$$

Since d_1' and d_2' are coprime, (10.8.5) holds if and only if $a = j d_1'$ and $b = (d - j) d_2'$ for some integral value of j with $1 \leq j \leq d - 1$. Since there are $d - 1$ such values of j, $t(d_1, d_2) = d - 1$. ■

Theorem 10.8.4 *Let A_1, A_2, A be integers not divisible by the prime p. Let m_1 and m_2 be positive integers. Set $d_1 = \gcd(m_1, p-1)$ and $d_2 = \gcd(m_2, p-1)$. If*

$$p > \left\{ (d_1 - 1)(d_2 - 1) - \left(1 - \frac{1}{\sqrt{p}}\right)(\gcd(d_1, d_2) - 1) \right\}^2,$$

then the congruence

$$A_1 x_1^{m_1} + A_2 x_2^{m_2} \equiv A \pmod{p}$$

is solvable.

Proof This follows immediately from Corollary 10.8.2 with $n = 2$ and Theorem 10.8.3. ■

In the next two theorems we examine the cases $m_1 = m_2 = 3$ and $m_1 = m_2 = 4$ in more detail.

Theorem 10.8.5 *Let p be a prime such that $p \not| AA_1A_2$. Then the congruence $A_1 x_1^3 + A_2 x_2^3 \equiv A \pmod{p}$ is always solvable unless $p = 7$.*

Proof Choose $m_1 = m_2 = 3$ in Theorem 10.8.4. If $p = 3$ or $p \equiv 2 \pmod{3}$, then $d_1 = d_2 = 1$, and Theorem 10.8.4 guarantees that the congruence $A_1 x_1^3 + A_2 x_2^3 \equiv A \pmod{p}$ is always solvable. Let $p \equiv 1 \pmod{3}$. Then $d_1 = d_2 = 3$, and, by Theorem 10.8.4, the congruence is solvable if

$$p > \left\{ 4 - 2\left(1 - \frac{1}{\sqrt{p}}\right) \right\}^2. \tag{10.8.6}$$

If $p \geq 13$, then the right side of (10.8.6) is < 7, and so (10.8.6) holds. Finally, if $p = 7$, the congruence $x_1^3 + x_2^3 \equiv 3 \pmod{p}$ is not solvable. ■

Theorem 10.8.6 *Let p be a prime such that $p \not| AA_1A_2$. Then the congruence $A_1 x_1^4 + A_2 x_2^4 \equiv A \pmod{p}$ is always solvable unless $p \in \{5, 13, 17, 29\}$*

Proof If $p = 2$, then A_1, A_2, and A are odd, and the congruence is $x_1^4 + x_2^4 \equiv 1 \pmod{2}$, which has the solution $(x_1, x_2) = (1, 0)$. Choose $m_1 = m_2 = 4$ in Theorem 10.8.4. If $p \equiv 3 \pmod{4}$, then $d_1 = d_2 = 2$, and Theorem 10.8.4 assures us that there is always a solution of the congruence $A_1 x_1^4 + A_2 x_2^4 \equiv A \pmod{p}$. If $p \equiv 1 \pmod{4}$, then $d_1 = d_2 = 4$, and there is a solution provided $p > (6 + 3/\sqrt{p})^2$.

It is easily checked that each of the four congruences

$$x_1^4 + x_2^4 \equiv 3 \pmod{5}, \quad x_1^4 + x_2^4 \equiv 7 \pmod{13},$$
$$x_1^4 + x_2^4 \equiv 6 \pmod{17}, \quad x_1^4 + x_2^4 \equiv 4 \pmod{29},$$

is insolvable. Thus let $p \geq 37$. Then since $(6 + 3/\sqrt{p})^2 < 43$, it remains to consider the case $p = 37$. A simple computer program shows that $A_1 x_1^4 + A_2 x_2^4 \equiv A \pmod{37}$

is always solvable when $37 \nmid AA_1A_2$. ■

10.9 Generalized cyclotomic numbers and the congruence $A_1x_1^k + \cdots + A_nx_n^k \equiv A \pmod{p}$

Let k be an integer with $k \geq 2$ and let p be a prime with $p \equiv 1 \pmod{k}$, say, $p = kf+1$. Let $A_1, \ldots, A_n, A$ be integers, none of which is divisible by p. The first theorem of this section shows that the number N' of solutions of the congruence

$$(10.9.1) \qquad A_1x_1^k + \cdots + A_nx_n^k \equiv A \pmod{p}$$

with

$$x_1 \cdots x_n \not\equiv 0 \pmod{p}$$

can be expressed in terms of generalized cyclotomic numbers $(s_1, \ldots, s_n)_{n,k}$ of length n and order k, defined in Section 2.7.

Theorem 10.9.1 *With the notation above,*

$$N' = k^n(\text{ind}_g(A_1/A), \ldots, \text{ind}_g(A_n/A))_{n,k},$$

where g is a primitive root $\pmod{p}$.

Proof Set $s_i = \text{ind}_g(A_i/A)$, $i = 1, \ldots, n$. Then the congruence (10.9.1) can be written in the form

$$g^{k\,\text{ind}_g x_1 + s_1} + \cdots + g^{k\,\text{ind}_g x_n + s_n} \equiv 1 \pmod{p}.$$

Hence

$$\begin{aligned} N' &= \text{card}\{0 \leq u_1, \ldots, u_n < p-1 : \ g^{ku_1+s_1} + \cdots + g^{ku_n+s_n} \equiv 1 \pmod{p}\} \\ &= k^n \ \text{card}\{0 \leq u_1, \ldots, u_n < f : \ g^{ku_1+s_1} + \cdots + g^{ku_n+s_n} \equiv 1 \pmod{p}\}, \end{aligned}$$

as $ku_i' \equiv ku_i \pmod{p-1}$ if and only if $u_i' \equiv u_i \pmod{f}$. Finally, by the definition of generalized cyclotomic numbers,

$$N' = k^n(s_1, \ldots, s_n)_{n,k},$$

as asserted. ■

The next theorem relates the number of solutions of the congruence $x_1^k + x_2^k \equiv 1 \pmod{p}$ to the cyclotomic number $(0,0)_k$.

Theorem 10.9.2 *Let $p = kf+1$ be a prime. Then the number N' of solutions of*

$$x_1^k + x_2^k \equiv 1 \pmod{p}, \quad x_1x_2 \not\equiv 0 \pmod{p},$$

is $N' = k^2(0,0)_k$. The number N of all solutions of $x_1^k + x_2^k \equiv 1 \pmod{p}$ is $N = 2k + k^2(0,0)_k$.

Proof By Theorem 10.9.1 and Lemma 2.7.1,

$$N' = k^2(0,0)_{2,k} = k^2(0,0)_k.$$

Clearly

$$N = 2k + N' = 2k + k^2(0,0)_k,$$

as the congruence $x^k \equiv 1 \pmod{p}$ has k solutions. ■

Since we know explicitly the values of $(0,0)_3$ (Theorem 2.3.1), $(0,0)_4$ (Theorem 2.4.1), and $(0,0)_5$ (Problem 14, Exercises 2), we obtain immediately from Theorem 10.9.2 the following result.

Theorem 10.9.3 (a) *For a prime $p = 3f + 1$, the number N of solutions of the congruence*

$$x_1^3 + x_2^3 \equiv 1 \pmod{p}$$

is

$$N = p + r_3 - 2.$$

(b) *For a prime $p = 4f + 1$, the number N of solutions of the congruence*

$$x_1^4 + x_2^4 \equiv 1 \pmod{p}$$

is

$$N = \begin{cases} p + 6a_4 - 3, & \text{if } 2 \mid f, \\ p + 2a_4 + 1, & \text{if } 2 \nmid f. \end{cases}$$

(c) *For a prime $p = 5f + 1$, the number N of solutions of the congruence*

$$x_1^5 + x_2^5 \equiv 1 \pmod{p}$$

is

$$N = p + 3x - 4,$$

where x is given uniquely by

$$\begin{aligned} 16p &= x^2 + 125w^2 + 50v^2 + 50u^2, \\ xw &= v^2 - u^2 - 4uv, \\ x &\equiv 1 \pmod{5}. \end{aligned}$$

Example When $p = 11$, the four solutions of

$$\begin{aligned} 176 &= x^2 + 125w^2 + 50v^2 + 50u^2, \\ xw &= v^2 - u^2 - 4uv, \\ x &\equiv 1 \pmod{5}, \end{aligned}$$

are

$$(x, w, v, u) = (1,1,1,0),\ (1,1,-1,0),\ (1,-1,0,1),\ (1,-1,0,-1),$$

so that $x = 1$. Thus the number of solutions of $x_1^5 + x_2^5 \equiv 1 \pmod{11}$ is $11 + 3 - 4 = 10$. A computer search yields the 10 solutions $(x_1, x_2) = (1,0), (3,0), (4,0), (5,0), (9,0), (0,1), (0,3), (0,4), (0,5), (0,9)$.

In Corollary 10.6.2 we counted the number N of solutions of the congruence $A_1x_1^3 + A_2x_2^3 \equiv A \pmod p$. In the next theorem, we use Theorem 10.9.1 to count the number N' of solutions of this congruence with $x_1x_2 \not\equiv 0 \pmod p$. We leave it to the reader to obtain this result directly from Corollary 10.6.2 (Problem 14).

Theorem 10.9.4 *Let $p = 3f + 1$ be a prime and let A_1, A_2, A be integers that are not divisible by p. Let g be a primitive root* $\pmod p$. *Then the number N' of solutions of the congruence*

$$A_1x_1^3 + A_2x_2^3 \equiv A \pmod p$$

with $x_1x_2 \not\equiv 0 \pmod p$ is given by

$$\begin{array}{ll} p + r_3 - 8, & \text{if } \operatorname{ind}_g(A_1A_2A) \equiv 0 \pmod 3,\ \operatorname{ind}_g(A_1) \equiv \operatorname{ind}_g(A) \pmod 3, \\ (2p - r_3 + 3s_3 - 4)/2, & \text{if } \operatorname{ind}_g(A_1A_2A) \equiv 1 \pmod 3, \\ (2p - r_3 - 3s_3 - 4)/2, & \text{if } \operatorname{ind}_g(A_1A_2A) \equiv 2 \pmod 3, \\ p + r_3 + 1, & \text{if } \operatorname{ind}_g(A_1A_2A) \equiv 0 \pmod 3,\ \operatorname{ind}_g(A_1) \not\equiv \operatorname{ind}_g(A) \pmod 3. \end{array}$$

Moreover,

$$N' \geq p - \max(|r_3|, 3|s_3|) - 8.$$

Proof From Theorem 10.9.1,

$$N' = 9\,(\operatorname{ind}_g(A_1/A), \operatorname{ind}_g(A_2/A))_{2,3}\,.$$

Then, from Lemma 2.7.1,

$$N' = 9\,(\operatorname{ind}_g(A_2/A_1), \operatorname{ind}_g(A/A_1))_3\,.$$

If $\operatorname{ind}_g(A_1A_2A) \equiv 0 \pmod 3$ and $\operatorname{ind}_g(A_1) \equiv \operatorname{ind}_g(A) \pmod 3$, then

$$\operatorname{ind}_g(A/A_1) \equiv \operatorname{ind}_g(A) - \operatorname{ind}_g(A_1) \equiv 0 \pmod 3$$

and

$$\begin{aligned} \operatorname{ind}_g(A_2/A_1) &\equiv \operatorname{ind}_g(A_2) - \operatorname{ind}_g(A_1) \equiv \operatorname{ind}_g(A_2) + \operatorname{ind}_g(A) + \operatorname{ind}_g(A_1) \\ &\equiv \operatorname{ind}_g(AA_1A_2) \equiv 0 \pmod 3, \end{aligned}$$

so that $(\operatorname{ind}_g(A_2/A_1), \operatorname{ind}_g(A/A_1))_3 = (0,0)_3$.

If $\operatorname{ind}_g(A_1A_2A) \equiv 1 \pmod 3$, then

$$\operatorname{ind}_g(A_2/A_1) + \operatorname{ind}_g(A/A_1) \equiv \operatorname{ind}_g(A_2A/A_1^2) \equiv \operatorname{ind}_g(A_1A_2A) \equiv 1 \pmod 3,$$

so that, as $(0,1)_3 = (1,0)_3 = (2,2)_3$, we have $(\text{ind}_g(A_2/A_1), \text{ind}_g(A/A_1))_3 = (0,1)_3$.

Similarly, if $\text{ind}_g(A_1A_2A) \equiv 2 \pmod 3$, then $(\text{ind}_g(A_2/A_1), \text{ind}_g(A/A_1))_3 = (0,2)_3$, and in the remaining case, $(\text{ind}_g(A_2/A_1), \text{ind}_g(A/A_1))_3 = (1,2)_3$.

The values of N' now follow from Theorem 2.3.1.

In the first and last cases,

$$N' \geq p - |r_3| - 8 \geq p - \max(|r_3|, 3|s_3|) - 8,$$

and, in the second and third cases,

$$\begin{aligned} N' &\geq \tfrac{1}{2}(2p - |r_3| - 3|s_3| - 4) \geq \tfrac{1}{2}(2p - 2\max(|r_3|, 3|s_3|) - 4) \\ &= p - \max(|r_3|, 3|s_3|) - 2 > p - \max(|r_3|, 3|s_3|) - 8. \end{aligned}$$

■

In the next theorem we use Theorem 10.9.4 to obtain the analogue of Theorem 10.8.5 for solutions of $A_1x_1^3 + A_2x_2^3 \equiv A \pmod p$ with $x_1x_2 \not\equiv 0 \pmod p$.

Theorem 10.9.5 *Let p be a prime with $p \nmid AA_1A_2$. The congruence $A_1x_1^3 + A_2x_2^3 \equiv A \pmod p$ is always solvable with $x_1 \not\equiv 0 \pmod p$ and $x_2 \not\equiv 0 \pmod p$ unless $p \in \{2, 7, 13\}$.*

Proof If $p \equiv 1 \pmod 3$, the number N' of solutions of the congruence with $x_1x_2 \not\equiv 0 \pmod p$ satisfies, by Theorem 10.9.4,

$$N' \geq p - \max(|r_3|, 3|s_3|) - 8.$$

In particular, for $p = 19$, $N' \geq 19 - \max(7, 9) - 8 = 2 > 0$. As $r_3^2 + 3s_3^2 = 4p$, it follows that $\max(|r_3|, 3|s_3|) < 2\sqrt{3}\sqrt{p}$, so that

$$N' > p - 2\sqrt{3}\sqrt{p} - 8.$$

Hence $N' > 0$ for $p \geq 19$, $p \equiv 1 \pmod 3$.

If $p = 3$ or $p \equiv 2 \pmod 3$, then, by Theorem 10.0.2,

$$N' = ((p-1)^2 - 1)/p = p - 2,$$

so that $N' = 0$ for $p = 2$ and $N' > 0$ otherwise. When $p = 7$ the congruence $x_1^3 + x_2^3 \equiv 3 \pmod 7$ is insolvable. Finally, when $p = 13$ the congruence $x_1^3 + x_2^3 \equiv 1 \pmod{13}$ has only the solutions $(x_1, x_2) = (1,0), (3,0), (9,0), (0,1), (0,3), (0,9)$. ■

In the next theorem, we use Theorem 10.9.1 to determine the number of solutions of the congruence $x_1^3 + x_2^3 + x_3^3 \equiv 1 \pmod p$ with $x_1x_2x_3 \not\equiv 0 \pmod p$ for a prime $p \equiv 1 \pmod 3$.

Theorem 10.9.6 *Let p be a prime with $p \equiv 1 \pmod 3$. The number N' of solutions of*

$$x_1^3 + x_2^3 + x_3^3 \equiv 1 \pmod p$$

with $x_1x_2x_3 \not\equiv 0 \pmod p$ is $N' = p^2 + 3p - 4r_3 + 15$.

Proof By Theorem 10.9.1,

$$N' = 27(0,0,0)_{3,3}.$$

By Theorem 2.7.2 and (2.2.2),

$$(0,0,0)_{3,3} = \frac{1}{3}(p-1) + \sum_{h=0}^{2} (0,h)_3{}^2.$$

Next, appealing to Theorem 2.3.1, we deduce that

$$\begin{aligned}
&(0,0)_3^2 + (0,1)_3^2 + (0,2)_3^2 \\
&= \frac{1}{324}\left\{(2p+2r_3-16)^2 + (2p-r_3+3s_3-4)^2 + (2p-r_3-3s_3-4)^2\right\} \\
&= \frac{1}{324}\left\{12p^2 + 6r_3^2 + 18s_3^2 - 96p - 48r_3 + 288\right\} \\
&= \frac{1}{324}\left\{12p^2 - 72p - 48r_3 + 288\right\} = \frac{1}{27}\left\{p^2 - 6p - 4r_3 + 24\right\}.
\end{aligned}$$

Finally,

$$N' = 9(p-1) + (p^2 - 6p - 4r_3 + 24) = p^2 + 3p - 4r_3 + 15. \qquad \blacksquare$$

10.10 f-nomial periods and the period polynomial

In this section we define the f-nomial periods associated with F_p for a prime $p = kf+1$, and, after developing some of their properties, show how they provide an alternative method of finding the number of solutions of (10.0.1) over F_p when all the k_i are equal. This was the approach taken by Lebesgue, Libri, and others in the nineteenth century. We discuss the extension of periods from F_p to F_q in Section 12.10.

Let $p = kf + 1$ be a prime, and let g be a primitive root (mod p).

Definition *The f-nomial periods are the complex numbers*

$$\eta_s = \eta_s(g) = \sum_{r=0}^{f-1} \exp(2\pi i g^{kr+s}/p), \quad s = 0, 1, \ldots, k-1.$$

The polynomial

$$P_k(x) = \prod_{s=0}^{k-1} (x - \eta_s)$$

is called the period polynomial of degree k.

It is not difficult to show that the coefficients of $P_k(x)$ are rational integers independent of the choice of primitive root g, and that $P_k(x)$ is irreducible over the integers. Thus $\eta_0, \ldots, \eta_{k-1}$ are algebraic conjugates of degree k over Q, and for each s, $0 \le s \le k-1$, $Q(\eta_0) = Q(\eta_s)$ is an Abelian extension of Q with $[Q(e^{2\pi i/p}) : Q(\eta_s)] = f$.

Definition *The reduced f-nomial periods are the complex numbers*

$$\eta_s^* = \eta_s^*(g) = 1 + k\eta_s(g), \quad s = 0, 1, \ldots, k-1,$$

and the reduced period polynomial is

$$P_k^*(x) = \prod_{s=0}^{k-1} (x - \eta_s^*).$$

The Gauss sum $g(k)$ defined in (4.0.1) is the reduced f-nomial period η_0^*, as

$$\begin{aligned} g(k) &= \sum_{n=0}^{p-1} \exp(2\pi i n^k/p) = 1 + \sum_{n=1}^{p-1} \exp(2\pi i n^k/p) = 1 + \sum_{m=0}^{p-2} \exp(2\pi i g^{mk}/p) \\ &= 1 + k \sum_{m=0}^{f-1} \exp(2\pi i g^{mk}/p) = 1 + k\eta_0 = \eta_0^*. \end{aligned}$$

The two period polynomials are related by

$$P_k^*(x) = k^k P_k((x-1)/k),$$

so that the coefficients of $P_k^*(x)$ are also integers that do not depend on g. Moreover $P_k^*(x)$ is also irreducible over the integers.

The next lemma shows that the coefficient of x^{k-1} in $P_k(x)$ is 1, while that in $P_k^*(x)$ is 0.

Lemma 10.10.1 (a) $\sum_{s=0}^{k-1} \eta_s = -1.$

(b) $\sum_{s=0}^{k-1} \eta_s^* = 0.$

Proof We have

$$\sum_{s=0}^{k-1} \eta_s = \sum_{s=0}^{k-1} \sum_{r=0}^{f-1} \exp(2\pi i g^{kr+s}/p) = \sum_{t=0}^{p-2} \exp(2\pi i g^t/p) = \sum_{u=1}^{p-1} \exp(2\pi i u/p) = -1,$$

so that

$$\sum_{s=0}^{k-1} \eta_s^* = \sum_{s=0}^{k-1} (1 + k\eta_s) = k + k \sum_{s=0}^{k-1} \eta_s = k - k = 0.$$

■

For an arbitrary integer s_1, it is convenient to set $\eta_{s_1} = \eta_s$ and $\eta^*_{s_1} = \eta^*_s$, where $s_1 \equiv s \pmod{k}$, $0 \le s < k$.

The next lemma gives a simple connection between the f-nomial periods η_s $(s = 0, 1, \ldots, k-1)$ and the cyclotomic numbers of order k. For a generalization to F_q, see Problem 23 in Exercises 12.

Lemma 10.10.2 *Let $p = kf + 1$ be an odd prime. For any integers m and n,*

$$\eta_m \eta_{m+n} = \sum_{h=0}^{k-1} (n,h)_k \eta_{m+h} + \begin{cases} f, & \text{if } n \equiv kf/2 \pmod{k}, \\ 0, & \text{otherwise.} \end{cases}$$

Proof We may suppose that $0 \le m \le k-1$ and $0 \le n \le k-1$. By definition of η_m,

$$\eta_m \eta_{m+n} = \sum_{r=0}^{f-1} \sum_{s=0}^{f-1} \exp(2\pi i (g^{kr+m} + g^{ks+m+n})/p).$$

For a fixed r, we may replace the summation variable s by $s + r$, which ranges with s over a complete set of residues modulo f. This yields

$$\eta_m \eta_{m+n} = \sum_{r=0}^{f-1} \sum_{s=0}^{f-1} \exp(2\pi i g^{kr+m}(1 + g^{ks+n})/p). \tag{10.10.1}$$

First we consider those s $(0 \le s \le f-1)$ for which $1 + g^{ks+n} \equiv 0 \pmod{p}$. Since

$$0 \le ks + n \le k(f-1) + (k-1) = kf - 1 = p - 2,$$

we must have

$$ks + n = (p-1)/2 = kf/2.$$

Hence the number of s $(0 \le s \le f-1)$ for which $1 + g^{ks+n} \equiv 0 \pmod{p}$ is

$$\begin{cases} 1, & \text{if } n \equiv kf/2 \pmod{k}, \\ 0, & \text{otherwise,} \end{cases}$$

and the contribution to the sum (10.10.1) is f times this number.

Next we consider those s for which $1 + g^{ks+n} \not\equiv 0 \pmod{p}$. In this case $1 + g^{ks+n} \equiv g^{kt+h} \pmod{p}$ for some integers t and h with $0 \le t \le f-1$ and $0 \le h \le k-1$, and the contribution to the sum (10.10.1) is

$$\begin{aligned}
\sum_{\substack{t=0 \\ 1+g^{ks+n} \equiv g^{kt+h} \pmod{p}}}^{f-1} \sum_{h=0}^{k-1} \sum_{r=0}^{f-1} \sum_{s=0}^{f-1} \exp(2\pi i g^{kr+m+kt+h}/p) &= \sum_{\substack{t=0 \\ 1+g^{ks+n} \equiv g^{kt+h} \pmod{p}}}^{f-1} \sum_{h=0}^{k-1} \sum_{u=0}^{f-1} \sum_{s=0}^{f-1} \exp(2\pi i g^{ku+m+h}/p) \\
&= \sum_{\substack{h=0 \\ 1+g^{ks+n} \equiv g^{kt+h} \pmod{p}}}^{k-1} \left\{ \sum_{s=0}^{f-1} \sum_{t=0}^{f-1} 1 \right\} \left\{ \sum_{u=0}^{f-1} \exp(2\pi i g^{ku+m+h}/p) \right\}
\end{aligned}$$

$$= \sum_{h=0}^{k-1} (n,h)_k \eta_{m+h}.$$

Putting the two contributions together, we obtain the assertion of the lemma. ■

Lemmas 10.10.1 and 10.10.2 allow us to give an explicit (but cumbersome) formula for the period polynomial $P_k(x)$ in terms of cyclotomic numbers of order k.

Theorem 10.10.3 $P_k(x) = (-1)^{k-1} \det(a_{ij})$,
where

$$\begin{aligned}
a_{11} &= 1 + x, && \\
a_{1j} &= 1, && \text{if } 2 \le j \le k, \\
a_{i1} &= f\theta_{i-1} + (i-1,0)_k x, && \text{if } 2 \le i \le k, \\
a_{ii} &= (i-1,i-1)_k - x, && \text{if } 2 \le i \le k, \\
a_{ij} &= (i-1,j-1)_k, && \text{if } 2 \le i \le k,\ 2 \le j \le k,\ i \ne j,
\end{aligned}$$

and, for $i = 0, 1, \ldots, k-1$,

$$\theta_i = \begin{cases} 1, & \text{if either } f \text{ is even and } i = 0 \text{ or } f \text{ is odd and } i = k/2, \\ 0, & \text{otherwise.} \end{cases}$$

Proof Let $m \in \{0, 1, \ldots, k-1\}$. From Lemma 10.10.1(a) and from Lemma 10.10.2 with $n = 1, 2, \ldots, k-1$,

$$\begin{aligned}
&(1 + \eta_m) + \eta_{m+1} + \eta_{m+2} + \cdots + \eta_{m+k-1} = 0, \\
&(f\theta_1 + (1,0)_k \eta_m) + ((1,1)_k - \eta_m)\eta_{m+1} + (1,2)_k \eta_{m+2} + \cdots + (1,k-1)_k \eta_{m+k-1} = 0, \\
&(f\theta_2 + (2,0)_k \eta_m) + (2,1)_k \eta_{m+1} + ((2,2)_k - \eta_m)\eta_{m+2} + \cdots + (2,k-1)_k \eta_{m+k-1} = 0, \\
&\cdots \\
&(f\theta_{k-1} + (k-1,0)_k \eta_m) + (k-1,1)_k \eta_{m+1} + (k-1,2)_k \eta_{m+2} \\
&\qquad + \cdots + ((k-1,k-1)_k - \eta_m)\eta_{m+k-1} = 0.
\end{aligned}$$

Regarding these k equalities as a homogeneous system of k linear equations in the k quantities $1, \eta_{m+1}, \ldots, \eta_{m+k-1}$, we see that the determinant

$$\begin{vmatrix}
1 + \eta_m & 1 & \cdots & 1 \\
f\theta_1 + (1,0)_k \eta_m & (1,1)_k - \eta_m & \cdots & (1,k-1)_k \\
\vdots & \vdots & \vdots & \vdots \\
f\theta_{k-1} + (k-1,0)_k \eta_m & (k-1,1)_k & \cdots & (k-1,k-1)_k - \eta_m
\end{vmatrix} = 0.$$

Thus the polynomial

$$A(x) = \begin{vmatrix}
1 + x & 1 & \cdots & 1 \\
f\theta_1 + (1,0)_k x & (1,1)_k - x & \cdots & (1,k-1)_k \\
\vdots & \vdots & \vdots & \vdots \\
f\theta_{k-1} + (k-1,0)_k x & (k-1,1)_k & \cdots & (k-1,k-1)_k - x
\end{vmatrix}$$

has the roots $x = \eta_0, \ldots, \eta_{k-1}$. The coefficient of x^k in $A(x)$ is $(-1)^{k-1}$, so that $P_k(x) = (-1)^{k-1}A(x)$. ■

Theorem 10.10.3 enables us to determine the period polynomial $P_k(x)$ whenever the cyclotomic numbers of order k are known. The cyclotomic numbers of order 2 are given in Theorem 2.2.2, so for $k = 2$, by Theorem 10.10.3,

$$\begin{aligned} P_2(x) &= -\begin{vmatrix} 1+x & 1 \\ f\theta_1 + (1,0)_2 x & (1,1)_2 - x \end{vmatrix} \\ &= x^2 + ((1,0)_2 - (1,1)_2 + 1)x + (f\theta_1 - (1,1)_2) \\ &= x^2 + x + \frac{1}{4}(1 - (-1)^{(p-1)/2}p), \end{aligned}$$

as $f = (p-1)/2$ and $\theta_1 = (1-(-1)^f)/2$. We have shown the following theorem.

Theorem 10.10.4 *Let p be an odd prime. Then*

$$P_2(x) = x^2 + x + \frac{1}{4}(1 - (-1)^{(p-1)/2}p)$$

and

$$P_2^*(x) = x^2 - (-1)^{(p-1)/2}p.$$

Since P_2^* is the irreducible polynomial of $g(2)$ over the integers, Theorem 10.10.4 also follows from Lemma 1.2.1.

In a similar manner, making use of the formula for the cyclotomic numbers of order 3 given in Theorem 2.3.1, and those for the cyclotomic numbers of order 4 given in Theorem 2.4.1, we obtain the period polynomials of degrees 3 and 4.

Theorem 10.10.5 *Let $p = 3f + 1$ be a prime. Then*

$$P_3(x) = x^3 + x^2 - \frac{1}{3}(p-1)x - \frac{1}{27}(r_3 p + 3p - 1)$$

and

$$P_3^*(x) = x^3 - 3px - r_3 p.$$

Theorem 10.10.6 *Let $p = 4f + 1$ be a prime. Then*

$$\begin{aligned} P_4(x) &= x^4 + x^3 - \frac{1}{8}\left(2\left(\frac{2}{p}\right)p + p - 3\right)x^2 - \frac{1}{16}\left(2\left(\frac{2}{p}\right)a_4 p + 2\left(\frac{2}{p}\right)p + p - 1\right)x \\ &\quad - \frac{1}{256}\left(4\left(\frac{2}{p}\right)p^2 - 5p^2 + 4a_4^2 p + 8\left(\frac{2}{p}\right)a_4 p + 4\left(\frac{2}{p}\right)p + 2p - 1\right) \end{aligned}$$

and

$$P_4^*(x) = x^4 - 2p\left(1 + 2\left(\frac{2}{p}\right)\right)x^2 - 8\left(\frac{2}{p}\right)a_4 px + p\left(\left(5 - 4\left(\frac{2}{p}\right)\right)p - 4a_4^2\right).$$

Of course, the polynomials $P_3^*(x)$ and $P_4^*(x)$ are precisely those that we have seen already in Theorem 4.1.2 and Corollary 4.2.2.

When $p \equiv 1 \pmod 8$, we note that

$$P_4^*(x) = (x^2 - p)^2 - 4p(x + a_4)^2,$$

and when $p \equiv 5 \pmod 8$,

$$P_4^*(x) = (x^2 + 3p)^2 - 4p(x - a_4)^2.$$

These forms of $P_4^*(x)$ were given by Lebesgue [1] in 1838.

We now show how to use f-nomial periods to determine the number of solutions of (10.0.1) over F_p when all the k_i are equal. We examine the case $A \equiv 0 \pmod p$ and leave the case $A \not\equiv 0 \pmod p$ to the reader (Problem 15). For an extension to F_q, see Problem 22 in Exercises 12.

Theorem 10.10.7 *Let $p = kf + 1$ be a prime. Let $A_1, \ldots, A_n$ be integers with $p \nmid A_1 \cdots A_n$. Let g be a primitive root* $\pmod p$, *and set $a_i = \mathrm{ind}_g A_i$ $(i = 1, \ldots, n)$. Then the number N of solutions of the congruence*

$$A_1 x_1^k + \cdots + A_n x_n^k \equiv 0 \pmod p$$

is

$$N = p^{n-1} + \frac{f}{p} \sum_{s=0}^{k-1} \eta_{a_1+s}^* \cdots \eta_{a_n+s}^*.$$

Proof As

$$\frac{1}{p} \sum_{t=0}^{p-1} \exp(2\pi i a t/p) = \begin{cases} 1, & \text{if } a \equiv 0 \pmod p, \\ 0, & \text{otherwise,} \end{cases}$$

we have

$$N = \frac{1}{p} \sum_{x_1, \ldots, x_n = 0}^{p-1} \sum_{t=0}^{p-1} \exp(2\pi i (A_1 x_1^k + \cdots + A_n x_n^k) t/p).$$

Separating out the term with $t = 0$ and changing the order of summation, we deduce that

$$N = p^{n-1} + \frac{1}{p} \sum_{t=1}^{p-1} \left\{ \sum_{x_1=0}^{p-1} \exp(2\pi i A_1 x_1^k t/p) \right\} \cdots \left\{ \sum_{x_n=0}^{p-1} \exp(2\pi i A_n x_n^k t/p) \right\}.$$

Next we express $\sum_{x=0}^{p-1} \exp(2\pi i A x^k t/p)$ as a reduced f-nomial period. Setting $A \equiv g^a \pmod p$ $(0 \le a \le p-2)$, $t \equiv g^s \pmod p$ $(0 \le s \le p-2)$, and, for $x \ne 0$,

$x \equiv g^y \pmod{p}$ $(0 \le y \le p-2)$, we obtain

$$\sum_{x=0}^{p-1} \exp(2\pi i A x^k t/p) = 1 + \sum_{x=1}^{p-1} \exp(2\pi i A x^k t/p) = 1 + \sum_{y=0}^{p-2} \exp(2\pi i g^{ky+a+s}/p)$$
$$= 1 + k \sum_{y=0}^{f-1} \exp(2\pi i g^{ky+a+s}/p) = \eta^*_{a+s}.$$

Thus

$$N = p^{n-1} + \frac{1}{p}\sum_{s=0}^{p-2} \eta^*_{a_1+s} \cdots \eta^*_{a_n+s} = p^{n-1} + \frac{f}{p}\sum_{s=0}^{k-1} \eta^*_{a_1+s} \cdots \eta^*_{a_n+s},$$

as asserted. ■

The next theorem shows that each reduced period η^*_s is a linear combination of kth-order Gauss sums $G(\chi), \ldots, G(\chi^{k-1})$, where the coefficients are kth roots of unity.

Theorem 10.10.8 *Let $p = kf+1$ be a prime, where k is an integer ≥ 2. Let g be a primitive root* $\pmod{p}$, *and let χ be a character* $\pmod{p}$ *of order k with $\chi(g) = \beta$, where $\beta = e^{2\pi i/k}$. Then for each integer s,*

$$\eta^*_s = \sum_{r=1}^{k-1} G(\chi^r)\beta^{-sr}.$$

Proof For each integer s,

$$\sum_{r=1}^{k-1} G(\chi^r)\beta^{-sr} = \sum_{n=1}^{p-1} e^{2\pi i n/p} \sum_{r=1}^{k-1} \chi^r(n)\beta^{-sr} = 1 + \sum_{n=1}^{p-1} e^{2\pi i n/p} \sum_{r=0}^{k-1} (\chi(n)\beta^{-s})^r$$
$$= 1 + k \sum_{\substack{n=1 \\ \chi(n)=\beta^s}}^{p-1} e^{2\pi i n/p} = 1 + k\sum_{u=0}^{f-1} \exp(2\pi i g^{ku+s}/p) = \eta^*_s.$$
■

Example We use Theorems 10.10.7 and 10.10.8 to determine the number N of solutions to the congruence

$$x_1^4 + \cdots + x_n^4 \equiv 0 \pmod{p},$$

where $p = 4f+1$. Let g be a primitive root $\pmod{p}$, and let χ be a quartic character $\pmod{p}$ such that $\chi(g) = i$. Write $\tau = G(\chi)$. Since $G(\chi^3) = (-1)^f\overline{\tau}$ and $G(\chi^2) = \sqrt{p}$, it follows from Theorems 10.10.7 and 10.10.8 that

$$N = p^{n-1} + \frac{(p-1)}{4p}\sum_{s=0}^{3} (\eta^*_s)^n = p^{n-1} + \frac{(p-1)}{4p}\left\{(\tau + \sqrt{p} + (-1)^f\overline{\tau})^n\right.$$
$$\left. + (-i\tau - \sqrt{p} + (-1)^f i\overline{\tau})^n + (-\tau + \sqrt{p} - (-1)^f\overline{\tau})^n + (i\tau - \sqrt{p} - (-1)^f i\overline{\tau})^n\right\}.$$

Exercises 10

1. What is wrong with the following argument?

$$\begin{aligned} J_0(\chi_1, \chi_2) &= \sum_{\gamma_1+\gamma_2=0} \chi_1(\gamma_1)\chi_2(\gamma_2) = \sum_{\gamma_1} \chi_1(\gamma_1)\chi_2(-\gamma_1) \\ &= \chi_2(-1)\sum_{\gamma_1} \chi_1(\gamma_1)\chi_2(\gamma_1) = \chi_2(-1)\sum_{\gamma_1} \chi_1\chi_2(\gamma_1) \\ &= \begin{cases} \chi_2(-1)q, & \text{if } \chi_1\chi_2 \text{ is trivial,} \\ 0, & \text{if } \chi_1\chi_2 \text{ is nontrivial.} \end{cases} \end{aligned}$$

(This does not agree with Theorem 10.1.2 when $k = 2$.)

2. Determine the number of solutions of each of the following congruences:

(a) $x_1^3 - 2x_2^3 \equiv 1 \pmod{37}$,
(b) $x_1^4 + x_2^4 + x_3^4 + x_4^4 \equiv 4 \pmod{13}$,
(c) $x_1^5 + 2x_2^5 \equiv 3 \pmod{11}$,
(d) $x_1^8 + x_2^8 + x_3^8 \equiv 0 \pmod{17}$.

3. Find an analogue of Theorem 2.5.1 for generalized cyclotomic numbers and generalized Jacobi sums.

4. Determine a formula for the number of solutions of the congruence $x_1^6 + x_2^6 \equiv 1 \pmod{p}$, where p is a prime with $p \equiv 1 \pmod{6}$.

5. Determine a formula for the number of solutions of the congruence $x_1^7 + x_2^7 \equiv 1 \pmod{p}$, where p is a prime with $p \equiv 1 \pmod{7}$.

6. Use Corollary 10.6.2 to prove Corollary 2.6.9.

7. Find an analogue of Theorem 10.9.4 for the congruence $A_1x_1^4 + A_2x_2^4 \equiv A \pmod{p}$.

8. Let $p = kf + 1$ be an odd prime with primitive root g. Let $0 \leq r, s < k$. Prove that the number of solutions of the congruence

$$g^r x_1^k - g^s x_2^k \equiv -1 \pmod{p}$$

is

$$\begin{array}{ll} k^2(r,s)_k, & \text{if } s \neq 0 \text{ and } r \not\equiv kf/2 \pmod{k}, \\ k^2(r,s)_k + k, & \text{if either } s = 0 \text{ and } r \not\equiv kf/2 \pmod{k} \\ & \text{or } s \neq 0 \text{ and } r \equiv kf/2 \pmod{k}, \\ k^2(r,s)_k + 2k, & \text{if } s = 0 \text{ and } r \equiv kf/2 \pmod{k}. \end{array}$$

9. Let $p = kf + 1$ be an odd prime with primitive root g. Let $0 \leq r, s < k$. Prove that, if f is even, the number of solutions of

$$g^r x_1^k + g^s x_2^k \equiv -1 \pmod{p}$$

is the same as the number of solutions in Problem 8, whereas, if f is odd, the number is

$$\begin{array}{ll} k^2(r, s+\frac{1}{2}k)_k, & \text{if } r \neq \frac{1}{2}k,\ s \neq \frac{1}{2}k, \\ k^2(r, s+\frac{1}{2}k)_k + k, & \text{if either } r = \frac{1}{2}k,\ s \neq \frac{1}{2}k \text{ or } r \neq \frac{1}{2}k,\ s = \frac{1}{2}k, \\ k^2(r, s+\frac{1}{2}k)_k + 2k, & \text{if } r = s = \frac{1}{2}k. \end{array}$$

10. Let $p = 4f + 1$ be a prime. Determine a_4 modulo 8 and then show directly from Theorem 10.10.6 that the coefficient of x in $P_4(x)$ is an integer.

11. Let $p = 4f + 1$ be a prime. Determine a_4 modulo an appropriate power of 2, and use this to verify directly from Theorem 10.10.6 that $P_4(0)$ is an integer.

12. Prove that

$$t(d_1, d_2, d_3) = \frac{d_1 d_2 d_3}{\operatorname{lcm}(d_1, d_2, d_3)} - \gcd(d_1, d_2) - \gcd(d_2, d_3) - \gcd(d_3, d_1) + 2.$$

13. Let p be a prime with $p \equiv 1 \pmod{3}$. Write $L = r_3$, $M = s_3/3$. Assume that 2 is not a cubic residue (mod p).

(a) Prove that $L \equiv M \equiv 1 \pmod{2}$ and $L \equiv \pm M \pmod{4}$.

(b) If $L \equiv M \pmod{4}$, show that the number N of solutions of the congruence

$$x^3 + y^3 + 2z^3 \equiv 0 \pmod{p}$$

is

$$N = p^2 - \frac{1}{2}(p-1)(L - 9M).$$

(c) What is the number N if $L \equiv -M \pmod{4}$?

14. Prove Theorem 10.9.4 directly from Corollary 10.6.2.

15. Find an analogue of Theorem 10.10.7 for the congruence $A_1x_1^k + \cdots + A_nx_n^k \equiv A \pmod{p}$ when $A \not\equiv 0 \pmod{p}$. Deduce a formula for the number of solutions of the congruence $A_1x_1^2 + \cdots + A_nx_n^2 \equiv A \pmod{p}$.

16. Let $p = kf + 1$ be an odd prime. Prove that for any integer l,

$$\sum_{j=0}^{k-1} \eta_j\eta_{j+l} = \begin{cases} p - f, & \text{if } l \equiv fk/2 \pmod{k}, \\ -f, & \text{otherwise.} \end{cases}$$

17. With $p = 13,\ k = 4,\ f = 3,\ g = 2$, the f-nomial periods are

$$\eta_0 = r + r^3 + r^9, \qquad \eta_1 = r^2 + r^5 + r^6,$$
$$\eta_2 = r^4 + r^{10} + r^{12}, \qquad \eta_3 = r^7 + r^8 + r^{11},$$

where $r = \exp(2\pi i/13)$. Express each of η_1, η_2, η_3 as a cubic polynomial in η_0 with rational coefficients.

18. Let p be a prime with $p \equiv 1 \pmod{3}$. Show that the number of solutions of the congruence

$$x_1^3 + x_2^3 + x_3^3 + x_4^3 \equiv 0 \pmod{p}$$

is

$$p^3 + 6p^2 - 6p.$$

19. Let $p = 3f + 1$ be a prime. Let N_n denote the number of solutions of the congruence $x_1^3 + \cdots + x_n^3 \equiv 0 \pmod{p}$. Prove that the numbers $u_n = N_n - p^{n-1} (n \geq 1)$ satisfy the recurrence relation

$$u_n - 3pu_{n-2} - pr_3u_{n-3} = 0 \ (n \geq 4),$$
$$u_1 = 0,\ u_2 = 2p - 2,\ u_3 = (p-1)r_3.$$

20. Deduce from Problem 19 that $\sum_{n=1}^{\infty} N_nx^n$ is a rational function of x.

21. Let p be a prime of the form $p = 4f + 1$. Let g be a primitive root $\pmod{p}$. Let χ be the quartic character $\pmod{p}$ such that $\chi(g) = i$. Let N denote the number of solutions of the congruence

$$x_1^4 + x_2^4 \equiv 2 \pmod{p}.$$

Let N_1 denote the number of solutions of

$$x_1^4 + x_2^4 \equiv 2 \pmod{p}, \quad x_1x_2 \equiv 0 \pmod{p}.$$

Let N_2 denote the number of solutions of

$$x_1^4 + x_2^4 \equiv 2 \pmod{p}, \quad x_1^4 \equiv x_2^4 \pmod{p}.$$

Prove that

$$\begin{aligned}&\text{(a)}\ \ N_1 = 2(1+\chi(2)+\chi^2(2)+\chi^3(2)),\\&\text{(b)}\ \ N_2 = 16,\\&\text{(c)}\ \ N \equiv N_1 + N_2 \pmod{32}.\end{aligned}$$

Now use parts (a), (b), (c), and Corollary 10.7.2 to deduce that

$$\text{ind}_g 2 \equiv b_4/2 \pmod 4.$$

22. In the notation of Theorem 2.1.14, prove that for each $t \geq 2$,

$$J_r(\bar{\chi}^{a_1}, \ldots, \bar{\chi}^{a_t}) \equiv (-1)^{t+1} \frac{(a_1 + \cdots + a_t)!}{a_1! \cdots a_t!} \pmod P,$$

where $0 < a_i < k$ for $1 \leq i \leq t$. Show that this generalizes (2.1.11).

Notes on Chapter 10

Generalized Jacobi sums (defined in Section 10.1) were introduced by Eisenstein [6] to obtain a proof of the law of biquadratic reciprocity. Properties of these Jacobi sums, with applications to solutions of diagonal equations over finite fields, are given in the books of Ireland and Rosen [1, Chapter 8] and Lidl and Niederreiter [1, Chapter 6]. See also Small [1, Chapter 6].

Kida and Ono [1] have shown that if χ is a character (mod p) of order m, where p is an odd prime, (m, k)=1, and $p \not| (k-1)$, then the k-fold Jacobi sum $J(\chi, \ldots, \chi)$ has degree $\phi(m)$ over Q. Further results characterizing the field $Q(J_r(\chi_1, \ldots, \chi_k))$ have been given by Gyoja and Ono [1], T. Ono [2], and Aoki [1].

Let $\chi_1, \ldots, \chi_t$ and $\chi_1 \cdots \chi_t$ be characters on F_q of prime order $k > 3$, where $q = p^r$. Shioda [1] and Yui [1] have made conjectures on the power of primes dividing the norm

$$N(1 \pm J_r(\chi_1, \ldots, \chi_t)q^{(1-t)/2}).$$

For progress on such conjectures, see Miki [3].

Koblitz [3] used the prime ideal factorization of Jacobi sums $J_r(\chi, \ldots, \chi)$ (see Theorem 11.2.3) to investigate which algebraic conjugates of these Jacobi sums are distinct. This has applications to discrete log cryptosystems.

Special cases of the equation (10.0.1) as well as related equations have been studied by many authors from various points of view over the years, e.g., Carlitz [1]–[11], [13], [14], [16], [20], I. Chowla [1]–[4], S. Chowla [8], Dickson [2]–[10], Hua and Vandiver [3], [4], Hull [1], Hurwitz [1], Mordell [3], Vandiver [1]–[14], and Whiteman [6]. More recent references are discussed below. The reader interested in the historical development of the study of (10.0.1) should consult the excellent overview given by Lidl and Niederreiter [1] in their notes on Chapter 6. A major paper in the subject is that of Weil [4], to whom Theorem 10.4.2 is due. This result was also obtained independently and almost simultaneously by Hua and Vandiver [1], [2]. See also Furtado Gomide [1]. Extensions of this work are given, for example, in Koblitz [2] and Lang and Weil [1].

Theorem 10.5.1 for an arbitrary finite field goes back to Dickson [1], whereas for the prime field case, the result is due to Jordan [1]; see also Lebesgue [7]. The proof of the supplement to the law of quadratic reciprocity given at the end of Section 10.5 is due to Williams [18].

The congruence $A_1x_1^3+\cdots+A_nx_n^3 \equiv A \pmod p$ was examined by Williams [1]. The present treatment (Theorem 10.6.1) is an improved version of Williams' paper [1]. The congruence $A_1x_1^3+\cdots+A_nx_n^3 \equiv A \pmod m$ for an arbitrary modulus m was examined by E. Cohen [2] in the cases $n = 2$ and $n = 3$. Corollaries 10.6.2 and 10.6.3 are both special cases of the results of Williams [1] and E. Cohen [1], [2]. The formulas given in Corollary 10.6.2 and Theorem 10.9.4 for the number of solutions of the congruence $A_1x_1^3 + A_2x_2^3 \equiv A \pmod p$ have their origins in the work of Gauss [3] on the cubic cyclotomic periods. Corollary 10.6.4 can be found in Williams [1].

The formula (Corollary 10.7.2) for the number of solutions of the congruence $A_1x_1^4+A_2x_2^4 \equiv A \pmod p$ for a prime $p = 4f + 1$ is implicit in the work of Gauss [5], and was used by him to obtain the biquadratic character of 2 (mod p); see Problem 21. Kaplan [1] (see also Williams [19]) used the formula (Theorem 10.7.1) for the number of solutions of the congruence $x_1^4 + \cdots + x_q^4 \equiv q \pmod p$, where p and q are primes with $p \neq q$ and $p \equiv 1 \pmod 4$, to obtain a proof of the law of biquadratic reciprocity.

The congruence $A_1x_1^5+\cdots+A_nx_n^5 \equiv A \pmod p$ has been studied by H. S. Hayashi [1].

Bounds of the type given in Theorem 10.8.1 were given by Weil in his classic paper [1]. The number $t(d_1,\ldots,d_n)$ defined in (10.8.1) is discussed in Lidl and Niederreiter [1, pp. 291–293] and in Z.–W. Sun [1, p. 178], where it is shown that

$$t(d_1,\ldots,d_n) = (-1)^n + \sum_{r=1}^{n}(-1)^{n-r} \sum_{1\le i_1<i_2<\cdots<i_r\le n} \frac{d_{i_1}\cdots d_{i_r}}{\operatorname{lcm}[d_{i_1},\ldots,d_{i_r}]}.$$

Theorem 10.8.3 is the special case $n = 2$ of this result, and Problem 12 is the special case $n = 3$. Sun and Wan [1] have shown that $t(d_1,\ldots,d_n) = 0$ if and only if either $\gcd(d_i, \prod_{j\neq i} d_j) = 1$ for some i or k is odd, $e_1/2,\ldots,e_k/2$ are pairwise coprime, and each e_j is coprime with any odd number in $\{d_1,\ldots,d_n\}$, where $\{e_1,\ldots,e_k\}$ is the set of even integers among $d_1,\ldots,d_n$. Sun and Wan [2] have shown that $t(d_1,\ldots,d_n) = 1$ if and only if $2|n$, $d_1/2,\ldots,d_n/2$ are pairwise coprime, and at least $(n-1)$ of the $d_i/2$ are odd. In the cases when $t(d_1,\ldots,d_n) \le 1$, Sun and Yuan [1] have given explicit formulas for the number of solutions N in Theorem 10.8.1 in the case $\alpha = 0$. Granville, Li, and Sun [1] have obtained upper and lower bounds for $t(d_1,\ldots,d_n)$. See also Q. Sun [1].

Theorem 10.8.5 was proved by Dunton [1] using the algebra of cubic residues. Theorem 10.8.6 does not appear to have been stated before. Theorem 10.9.2 is given in Dickson [8] together with some related results, which are given as Problems 8 and 9. Parts (a) and (b) of Theorem 10.9.3 are due to Gauss [3], [5], and part (c) can be deduced from the work of Dickson [8]. Theorem 10.9.5 is due to Dunton [1]. Her proof uses only the elementary properties of cubic residues. In connection with Theorem 10.9.6, we mention that various cases of the equation $ax^3 + by^3 + cz^3 = d$ over a finite field have been examined by Chowla, Cowles, and Cowles [1]–[4]. Problem 13 is based on their work [3].

The notions of "f-nomial period" and "period polynomial", which provide the foundations of the theory of cyclotomy, are due to Gauss [7]. These periods are also called cyclotomic periods or Gauss periods. Basic results about f-nomial periods can be found, for example, in Mathews [1, §177], where it is shown that the period polynomial $P_k(x)$ has rational integral coefficients and is irreducible. The irreducibility of $P_k(x)$ was proved by Eisenstein [11] in 1850, and by Lebesgue [5] in 1859. The reducibility of $P_k(x)$ modulo an odd integer has been considered by Wang [3].

As indicated in the Notes on Chapter 4, the period polynomials of degrees 2, 3, and 4 given in Theorems 10.10.4, 10.10.5, and 10.10.6, respectively, were found by Gauss. See also Lebesgue [1], Cayley [1], and Sylvester [1]. Early work on the determination of $P_5^*(x)$ was done by Burnside [1], Carey [1], Cayley [2], [3], Glashan [1], Legendre [1, vol. 2], Scott [1], and Tanner [1]. The polynomial $P_5(x)$ was determined by E. Lehmer [2] in 1951 (see also Rajwade [5] and Berndt and Evans [4, eq. (5.2)]), the polynomial $P_6(x)$ by D. H. Lehmer and E. Lehmer [5] in 1984, and the polynomial $P_8(x)$ by Evans [12] in 1983. Lazarus [3] evaluated sextic period polynomials for periods defined in terms of roots of unity of composite order.

Lemma 10.10.2 is very well known and can be found in Bachmann [1, p. 201]; see also Kummer [5, pp. 95, 100], [CW, p. 212]. The approach given in Theorem 10.10.3 for finding $P_k(x)$ is that of Dickson [8]. The ideas behind Theorem 10.10.7 go back to the early work of Libri [1], [2] and Lebesgue [1]. Our approach is based on that of Williams [2].

Schoof and Washington [1] showed that, for certain primes $p \equiv 1 \pmod 5$, certain integer translations of quintic periods (found by E. Lehmer [19]) generate the unit group of the quintic subfield of $Q(e^{2\pi i/p})$. For related work on translations of periods and on translations of differences of periods, see Lazarus [1], [2], and D. H. and E. Lehmer [7]. Darmon [1] has connected E. Lehmer's translations of quintic periods with coverings of certain modular curves.

Gauss periods with small values of f are considered in D. H. and E. Lehmer [3]. Periods defined in terms of roots of unity of composite order have been studied by Evans [10], [17] and D. H. and E. Lehmer [2]. The powers of p dividing the coefficients of the reduced period polynomial have been studied by Wan [1] and Evans [15].

The period polynomial can be used to obtain results on residuacity of primes q (mod p) with $\gcd(q, p) = 1$. For example, q is a kth–power residue of p if and only if the period polynomial $P_k(x)$ is a product of k (not necessarily distinct) linear factors modulo q. If furthermore q does not divide the discriminant of $P_k(x)$, then q is a kth-power residue of p if and only if $P_k(x)$ is a product of k distinct linear factors modulo q. These results are proved in Evans [17, pp. 1076–1077] in a more general setting where the periods are defined in terms of roots of unity of order p with p squarefree instead of just prime. For the history of these and related results, see Kummer [3], Glaisher's editorial footnote in H. J. S. Smith [CW, vol. 1, p. 105], and E. Lehmer [7], [12]. For extensions, see papers of Evans [10, Theorem 7] and Llorente [1]. Such criteria for residuacity were used in conjunction with explicit formulas for $P_3(x)$ and $P_4(x)$ by E. Lehmer [7] to give conditions for certain small primes q to be cubic or quartic residues for a prime p. Williams [24] used $P_5(x)$ to give explicit conditions for quintic residuacity. For related papers, see D. H. and E. Lehmer [2], [5], [6] and E. Lehmer [17], [18].

Jakubec [1] proved a congruence for a certain Gauss period, and gave several applications. One of these applications is that for a prime $p \equiv 1 \pmod 3$,

$$r_3 \equiv ((p-1)/3)!^{-3} \pmod p.$$

This is equivalent to the first congruence of Theorem 9.2.1.

Gao, von zur Gathen, and Panario [1] and von zur Gathen and Shparlinski [1] showed that certain types of Gauss periods embedded in finite fields generate normal bases and have high multiplicative order. Gao et al. pointed out that since Gauss periods in finite fields often have high multiplicative orders and are easy to exponentiate, they are useful in certain cryptosystems and in exponential pseudorandom number generators.

For an application of Gauss periods to factorization, see a paper of Bach and Shallit [1].

For further references to Gauss periods, see the Notes on Chapter 12.

We now discuss some recent results related to the equation (10.0.1). Wan [2] has shown that the number of solutions N of the equation $\alpha_1 x_1^{d_1} + \cdots + \alpha_n x_n^{d_n} = \alpha$ in Theorem 10.4.2 is divisible by q^b, where b is the greatest integer less than $1/d_1 + \cdots + 1/d_n$. This generalizes a result of Joly [1]. Wan's proof involves an application of Stickelberger's congruence for Gauss sums (Theorem 11.2.1). See also Wan [3] for related results on systems of polynomial equations.

For a prime p and positive integer k, let $\gamma(k,p)$ denote the smallest s such that, for every integer y, $x_1^k + \cdots + x_s^k \equiv y \pmod p$ has a solution. Konyagin [1], using his upper estimates for the Gauss sums $\sum_{x=1}^{p} \exp(2\pi i a x^k/p)$ discussed in the Notes on Chapter 4, has obtained a strong upper bound for $\gamma(k,p)$. For related results over F_{2^m} in place of F_p, see Shparlinski [1].

Ishikawa [1] has given a $(k+1)$ - term linear recursion formula for the number of solutions $N_k(n)$ of the equation $x_1^k + \cdots + x_n^k = 0$ over F_p ($N_k(n)$ is the N in Theorem 10.10.7 in the case where all $A_i = 1$). The k starting values are computed in terms of sums of Jacobi sums, and explicit results are given in the cases $k = 5$, 7 by using cyclotomy of orders $k = 5, 7$. Such results had been obtained earlier by Myerson [1] for $k = 3, 4$. In this connection, see Problem 19. Problems 18–20 are taken from Chowla, Cowles, and Cowles [1]. Their results have been extended from F_p to F_q by Myerson [1]. Richman [1] found the generating function for $N_k(n)$ in terms of the period polynomial.

Koblitz [2] gave formulas for the number of points on certain projective hypersurfaces over F_q in terms of Gauss and Jacobi sums, and made connections with hypergeometric functions. For example, Koblitz considered deformations of diagonal hypersurfaces of the form $x_1^d + \cdots + x_n^d = d\lambda x_1^{h_1} \cdots x_n^{h_n}$, where $d|(q-1), d = h_1 + \cdots + h_n$, and $\gcd(d, h_1, \ldots, h_n)=1$.

If -1 is a power of $p \pmod k$, then the Gauss sums $G_r(\chi)$ of order k can all be simply evaluated via Theorem 11.6.3. Then, in view of (1.1.4), the reduced periods in Problem 22 of Exercises 12 can be simply evaluated as well, and consequently an explicit formula can be given for the number of solutions N of the equation $A_1x_1^k + \cdots + A_nx_n^k = A$ over F_q. Such a formula was given in Wolfmann [2] in 1992. In 1994, Wolfmann [3] used the relationship between Gauss sums and weights of words in irreducible cyclic codes (Theorem 11.7.2) to compute an explicit formula for the number of solutions of the equation $x_1^k + \cdots + x_n^k = 0$ over F_q.

Lam and Leung [1] have shown that, apart from some trivial cases, the equation $x_1^d + \cdots + x_n^d = 0$ always has a solution with $x_i \in F_q^*$ $(1 \le i \le n)$ if $n > d$.

Let c_s denote the number of solutions of the congruence $f(x_1, \ldots, x_n) := A_1x_1^{k_1} + \cdots + A_nx_n^{k_n} \equiv 0 \pmod{p^s}$, $s \ge 0$. The generating function $\sum_{s=0}^{\infty} c_s t^s$ is said to be the Poincaré series of $f(x_1, \ldots, x_n)$, and has been determined as a rational function of t by Wang [2].

Skinner [1] obtained a good lower bound on s such that the equation $a_1x_1^k + \cdots + a_sx_s^k = 0$ has a nontrivial solution over a finite extension K of the p-adic rationals, where $a_1, \ldots, a_s \in K$ and $k \ge 2$.

Ozeki [1] obtained a formula in terms of Gauss sums for the number of solutions of a certain quadratic matrix congruence modulo large powers of an odd prime p.

Fulton [1] used Carlitz's evaluations [24] of Gauss sums attached to quadratic forms over F_{2^n} to find the number of solutions of certain quadratic systems of equations over F_{2^n}.

Montgomery, Vaughan, and Wooley [1, p. 22] have given a criterion for the existence of a nontrivial p-adic solution to a system of t diagonal equations $a_{i1}x_1^l + \cdots + a_{in}x_n^l = 0$, $1 \le i \le t$, where $a_{ij} \in Z$.

Brüdern [1] has proved that if $n = 14$, then the system of two diagonal cubic equations

$$\sum_{i=1}^{n} a_i x_i^3 = 0, \qquad \sum_{i=1}^{n} b_i x_i^3 = 0 \qquad (a_i, b_i \in Z)$$

has a nontrivial integral solution, provided that it has a nontrivial solution in the 7-adic number field. This system is known to have nontrivial integer solutions if $n \ge 16$, but for $n = 15$, there exists such a system which even has no nontrivial solution in the 7-adic number field. For other references concerning pairs of diagonal equations, see, for example, Corson [1], Spackman [1], [2], and Tietäväinen [1]. See also Lidl and Niederreiter [1, p. 329] for references concerning systems of diagonal equations.

Chapter 11

Gauss Sums over F_q

In Section 11.1, we give the prime ideal factorization of p in certain cyclotomic fields, which prepares us for the proof of Stickelberger's theorem in Section 11.2. Stickelberger's theorem (Theorem 11.2.1) gives a congruence for the Gauss sums $G_r(\chi)$ modulo a certain prime ideal factor of p. This congruence is one of the most important and useful results about Gauss sums. As consequences, we give prime ideal factorizations of Gauss and Jacobi sums in Section 11.2 (Theorems 11.2.2 and 11.2.3). In Section 11.3, we use Stickelberger's theorem to prove the Davenport–Hasse formula for products of Gauss sums (Theorem 11.3.5). This formula generalizes the relation $K_r(\chi) = J_r(\chi, \rho)$ proved in Theorem 2.1.4. No elementary proof of the Davenport–Hasse product formula is known in general, but we show in Section 11.3 how to give an elementary proof in the case that the character ψ in Theorem 11.3.5 has an order that is a power of 2.

In Section 11.5, we prove the Davenport–Hasse theorem on lifted Gauss sums, which gives the relationship between the Gauss sum $G_r(\chi)$ and the Gauss sum $G_{rs}(\chi')$ for the lifted character

$$\chi' = \chi \circ N_{F_{p^{rs}}/F_{p^r}}$$

defined in Section 11.4. This is then used to evaluate the quadratic Gauss sum $G_r(\rho)$ in Theorem 11.5.4.

A Gauss or Jacobi sum is said to be pure if some positive integral power of it is real. In Section 11.6, we prove that if χ is a character of order k on F_{p^r}, then $G_r(\chi^j)$ is pure for all integers j if and only if -1 is a power of p (mod k). In Section 11.7, we find explicitly the weights of code words in certain irreducible cyclic codes, using results on pure Gauss sums as well as many of the earlier results in this chapter, such as Stickelberger's theorem and the Davenport–Hasse lifting theorem.

11.1 Prime ideal factorization of p

Fix an integer $k > 1$, and let p be a prime not dividing k. Let r be the smallest positive integer for which $p^r \equiv 1 \pmod{k}$, that is, r is the order of p (mod k). Set $q = p^r$ and $f = (q-1)/k$, so that $p^r = kf + 1$. Then write

$$\beta = e^{2\pi i/k}, \quad \beta_p = e^{2\pi i/p}, \quad \pi = \beta_p - 1.$$

Define the cyclotomic fields

$$K = Q(\beta), \quad M = K(\beta_p) = Q(\beta, \beta_p),$$

and let O_K and O_M denote their respective rings of integers. Let R denote the group of $\phi(k)$ reduced residues (mod k). For $j \in R$, define $\sigma_j \in \mathrm{Gal}(M/Q(\beta_p))$ by $\sigma_j(\beta) = \beta^j$. Let P be a prime ideal of O_K dividing (i.e., containing) the principal ideal pO_K. Then for some prime ideal $\wp \subseteq O_M$,

$$PO_M = \wp^{p-1}, \quad \wp \cap O_K = P, \tag{11.1.1}$$

(i.e., P ramifies totally in O_M); see the book of Narkiewicz [2, p. 423]. Write

$$P_j = \sigma_j(P), \quad \wp_j = \sigma_j(\wp),$$

so that

$$P_jO_M = \wp_j^{p-1}.$$

Let T denote a set of coset representatives of the multiplicative quotient group

$$T = R/\{1, p, p^2, \ldots, p^{r-1}\},$$

so that T has $\phi(k)/r$ elements. By the remarks following (2.1.8),

$$qO_K = p^rO_K = \prod_{j \in T} P_j^r, \tag{11.1.2}$$

and the prime ideals P_j $(j \in T)$ are distinct. By (11.1.2), pO_K has the prime ideal factorization

$$pO_K = \prod_{j \in T} P_j. \tag{11.1.3}$$

Multiplying (11.1.3) by O_M and using (11.1.1), we obtain the prime ideal factorization

$$pO_M = \prod_{j \in T} \wp_j^{p-1}. \tag{11.1.4}$$

Setting $x = 1$ in the identity

$$x^{p-1} + \cdots + x + 1 = \prod_{n=1}^{p-1} (x - \beta_p^n),$$

we see that

$$p = (-\pi)^{p-1} \prod_{n=1}^{p-1} \left(\frac{1 - \beta_p^n}{1 - \beta_p} \right). \tag{11.1.5}$$

The product on n in (11.1.5) is a unit in O_M by Corollary 2.1.10 with $k = p$. Thus (11.1.5) gives

$$pO_M = \pi^{p-1}O_M. \tag{11.1.6}$$

Comparing (11.1.4) and (11.1.6), we see that

$$\pi O_M = \prod_{j \in T} \wp_j. \tag{11.1.7}$$

The principal ideal (π) in $Z[\beta_p]$ is prime, since π has prime norm p, by Theorem 2.1.9. Formula (11.1.7) tells us how this prime ideal (π) splits in O_M.

11.2 Stickelberger's congruence for Gauss sums

We use the notation of the previous section. As in Section 2.1, define the character χ_P of order k on the finite field O_K/P (identified with F_q) by letting $\chi_P(\alpha+P)$ denote the (unique) power of $\beta = e^{2\pi i/k}$ such that

$$\chi_P(\alpha+P) \equiv \alpha^{(q-1)/k} = \alpha^f \pmod{P}, \tag{11.2.1}$$

when $\alpha \in O_K$, $\alpha \notin P$. When $\alpha \in P$, define $\chi_P(\alpha+P)=0$. Any character χ of order k on F_q can be identified with χ_P. Observe that for $\sigma \in \mathrm{Gal}(K/Q)$,

$$\chi_{\sigma(P)} = \sigma\chi_P\sigma^{-1}. \tag{11.2.2}$$

For any integer $a \not\equiv 0 \pmod{k}$, let $L(a)$ denote the least positive integer congruent to $a \pmod{k}$. Write the base p expansion of $L(a)f$ as

$$L(a)f = a_0 + a_1p + \cdots + a_{r-1}p^{r-1}, \qquad 0 \le a_i \le p-1. \tag{11.2.3}$$

Define

$$s(a) = a_0 + a_1 + \cdots + a_{r-1}, \quad t(a) = a_0!a_1!\cdots a_{r-1}!. \tag{11.2.4}$$

Observe that

$$s(a) \equiv L(a)f \equiv af \pmod{p-1}. \tag{11.2.5}$$

Above Theorem 1.1.3, we defined the Gauss sums of the form

$$G_r(\chi_P^{-a}) = \sum_{\alpha\in O_K/P} \chi_P^{-a}(\alpha)e(\alpha), \tag{11.2.6}$$

where $e(\alpha) = \exp(2\pi i\,\mathrm{tr}(\alpha)/p)$. Note that these Gauss sums are in O_M. Stickelberger's theorem gives a congruence $(\bmod\ \wp^{s(a)+1})$ for the Gauss sum in (11.2.6).

Theorem 11.2.1 (*Stickelberger's congruence for Gauss sums.*) *For an integer* $a \not\equiv 0 \pmod{k}$,

$$G_r(\chi_P^{-a}) \equiv \frac{-\pi^{s(a)}}{t(a)} \pmod{\wp^{s(a)+1}}.$$

In particular, by (11.1.7),

$$\wp^{s(a)}\,||\,G_r(\chi_P^{-a})O_M.$$

Proof We prove only the case $k=q-1$, i.e., $f=1$, as the proof for general k is a straightforward consequence; see Problem 1. The proof is by induction on a, for $0<a<k$. First let $a=1$, so that by (11.2.4), $s(a)=t(a)=1$ (since $f=1$). By (11.2.6),

$$G_r(\chi_P^{-1}) = \sum_{0\not\equiv\alpha \pmod{P}} \chi_P^{-1}(\alpha)\beta_p^{\mathrm{tr}(\alpha)}.$$

Viewing $\operatorname{tr}(\alpha)$ as an integer between 0 and $p-1$, we have

$$\beta_p^{\operatorname{tr}(\alpha)} = (1+\pi)^{\operatorname{tr}(\alpha)} \equiv 1 + \operatorname{tr}(\alpha)\pi \pmod{\pi^2}.$$

Thus, by (11.1.7),

$$G_r(\chi_P^{-1}) \equiv \sum_{0 \not\equiv \alpha \pmod P} \chi_P^{-1}(\alpha)\operatorname{tr}(\alpha)\pi \pmod{\wp^2}.$$

To complete the proof for $a = 1$, we must show that

$$\sum_{0 \not\equiv \alpha \pmod P} \chi_P^{-1}(\alpha)\operatorname{tr}(\alpha) \equiv -1 \pmod P.$$

By (11.2.1), the left member above is congruent $\pmod P$ to

$$\sum_{0 \not\equiv \alpha \pmod P} \alpha^{-1}(\alpha + \alpha^p + \cdots + \alpha^{p^{r-1}}) \equiv \sum_{0 \not\equiv \alpha \pmod P} 1 = q - 1 \equiv -1 \pmod P,$$

by (2.1.13). This completes the proof for the case $a = 1$.

Let $a > 1$. First suppose that $a = pb$ for some integer b. By Theorem 1.1.4(d),

$$G_r(\chi_P^{-a}) = G_r(\chi_P^{-bp}) = G_r(\chi_P^{-b}). \tag{11.2.7}$$

Since also $s(a) = s(b)$ and $t(a) = t(b)$ by the definitions in (11.2.4), the result follows by the inductive hypothesis. Now suppose that $p \nmid a$. Then by (11.2.3) and (11.2.4),

$$a_0 \geq 1, \quad s(a) = s(a-1) + 1, \text{ and } t(a) = a_0 t(a-1). \tag{11.2.8}$$

Since the result has been proved for $a = 1$,

$$-\pi^{s(a)-1} G_r(\chi_P^{-1}) \equiv \pi^{s(a)} \pmod{\wp^{s(a)+1}}.$$

Since the result also holds for $a - 1$ in place of a, it follows from (11.2.8) that

$$-\pi^{s(a)-1} G_r(\chi_P^{-1}) \equiv t(a-1) G_r(\chi_P^{-1}) G_r(\chi_P^{1-a}) \pmod{\wp^{s(a)+1}}.$$

The two congruences above yield

$$t(a-1) G_r(\chi_P^{-1}) G_r(\chi_P^{1-a}) \equiv \pi^{s(a)} \pmod{\wp^{s(a)+1}}.$$

By Theorem 2.1.3(a) and (11.2.8),

$$t(a) J_r(\chi_P^{-1}, \chi_P^{1-a}) G_r(\chi_P^{-a}) \equiv a_0 \pi^{s(a)} \pmod{\wp^{s(a)+1}}.$$

Since $a \equiv a_0 \pmod p$, it remains to show that

$$J_r(\chi_P^{-1}, \chi_P^{1-a}) \equiv -a \pmod P.$$

This follows from (2.1.11) with $k = q-1$, $f = 1$, $m = q-2$, and $n = q - a$. ■

In the next two theorems, we use Stickelberger's theorem to factor Gauss and Jacobi sums into prime ideals of O_M and O_K, respectively.

Theorem 11.2.2 (*Prime ideal factorization of Gauss sums.*) *For an integer* $a \not\equiv 0 \pmod{k}$,

$$G_r(\chi_p^{-a})O_M = \prod_{j\in T} \wp_{j^{-1}}^{s(aj)}, \qquad G_r(\chi_p^{a})O_M = \prod_{j\in T} \wp_{j^{-1}}^{r(p-1)-s(aj)},$$

where j^{-1} *denotes the inverse of* $j \pmod{k}$.

Proof By Theorem 11.2.1,

$$\wp^{s(aj)} || G_r(\chi_p^{-aj}), \quad \text{for each } j \in T.$$

Apply $\sigma_{j^{-1}}$ to see that

$$\wp_{j^{-1}}^{s(aj)} || G_r(\chi_p^{-a}), \quad \text{for each } j \in T.$$

By Theorem 1.1.4(c),

$$G_r(\chi_p^{-a})G_r(\chi_p^{a}) = \pm q, \tag{11.2.9}$$

and so the only prime ideal divisors of $G_r(\chi_p^{-a})$ in O_M are the ideals $\wp_j$ ($j \in T$), by (11.1.4). The first equality thus follows. The second equality then follows with the aid of (11.2.9) and (11.1.4). ■

Theorem 11.2.3 (*Prime ideal factorization of Jacobi sums.*) *Let m, n, and $m+n$ be integers not divisible by k. Then*

$$J_r(\chi_p^{-m}, \chi_p^{-n})O_K = \prod_{j\in T} P_{j^{-1}}^{(s(mj)+s(nj)-s(mj+nj))/(p-1)}.$$

Proof By Theorem 11.2.2 and (11.1.1),

$$J_r(\chi_p^{-m}, \chi_p^{-n})^{p-1}O_M = \prod_{j\in T} P_{j^{-1}}^{s(mj)+s(nj)-s(mj+nj)}O_M.$$

By (11.2.5), $s(mj)+s(nj)-s(mj+nj)$ is an integral multiple of $(p-1)$. Taking $(p-1)$th roots of both sides, we obtain the result. ■

Corollary 11.2.4 *Let m, n, and $m+n$ be integers not divisible by k. Then*

$$J_r(\chi_p^{m}, \chi_p^{n})O_K = \prod_{j\in T} P_{j^{-1}}^{r+(s(mj+nj)-s(mj)-s(nj))/(p-1)}.$$

Proof This follows from Theorem 11.2.3 and (11.1.2), in view of Theorem 2.1.3(b). ■

The next theorem gives a congruence for $((p-1)/2)!$ modulo π.

Theorem 11.2.5 *For an odd prime p,*

$$((p-1)/2)!\,i^{(p-1)^2/4}\sqrt{p} \equiv -\pi^{(p-1)/2} \pmod{\pi^{(p+1)/2}}.$$

Proof This follows from Theorem 11.2.1 with $k=2$, $r=1$, $q=p$, $f=(p-1)/2$, $a=1$, $K=Q$, $M=Q(\beta_p)$, $P=(p)$, $\wp=(\pi)$, in view of Theorem 1.2.4. ■

Another factorization formula for Gauss sums is the following.

Theorem 11.2.6 *For an integer* $a \not\equiv 0 \pmod{k}$,

$$G_r^k(\chi_P^{-a}) \in O_K$$

and

$$G_r^k(\chi_P^{-a})O_K = \prod_{j\in T} P_{j-1}^{ks(aj)/(p-1)}.$$

Proof Write $\chi = \chi_P^{-a}$ for brevity, and let m denote the order of χ. From Problem 1 of Exercises 2,

$$G_r^m(\chi) = \chi(-1)q\prod_{j=1}^{m-2} J_r(\chi,\chi^j),$$

so that $G_r^k(\chi) \in O_K$, since $m \mid k$. By Theorem 11.2.2 and (11.1.1),

$$G_r(\chi)^{k(p-1)}O_M = \prod_{j\in T} P_{j-1}^{ks(aj)}O_M.$$

By (11.2.5), $ks(aj)$ is an integral multiple of $(p-1)$. Taking the $(p-1)$th root of each side, we obtain the result, since $G_r(\chi)^k \in O_K$. ■

The next theorem gives an alternative to the formula for $s(a)$ in (11.2.4).

Theorem 11.2.7 *Let a be an integer not divisible by k, and let $L(x)$ denote the least positive integer congruent to x* (mod *k*). *Then*

$$s(a) = \frac{p-1}{k}\sum_{i=0}^{r-1} L(ap^i).$$

Proof We may assume that $0 < a < k$. By (11.2.3),

$$\begin{aligned} af &= a_0 + a_1p + a_2p^2 + \cdots + a_{r-1}p^{r-1}, \\ afp &\equiv a_{r-1} + a_0p + a_1p^2 + \cdots + a_{r-2}p^{r-1} \pmod{q-1}, \\ afp^2 &\equiv a_{r-2} + a_{r-1}p + a_0p^2 + \cdots + a_{r-3}p^{r-1} \pmod{q-1}, \end{aligned}$$

and so on. Each of the right members of these congruences is thus a positive integral multiple of f that is less than $kf = q - 1$, since $0 \le a_i \le p - 1$ for each i. Therefore, dividing by f, we see that

$$\begin{aligned} L(a) &= \frac{1}{f}(a_0 + a_1 p + a_2 p^2 + \cdots + a_{r-1} p^{r-1}), \\ L(ap) &= \frac{1}{f}(a_{r-1} + a_0 p + a_1 p^2 + \cdots + a_{r-2} p^{r-1}), \\ L(ap^2) &= \frac{1}{f}(a_{r-2} + a_{r-1} p + a_0 p^2 + \cdots + a_{r-3} p^{r-1}), \end{aligned}$$

and so on. Addition of these identities yields

$$\sum_{i=0}^{r-1} L(ap^i) = \frac{1}{f} s(a) \frac{q-1}{p-1},$$

which gives the desired result. ■

The special case $r = 1$ of the following theorem was presented as Problem 24 of Exercises 2.

Theorem 11.2.8 *For $a \not\equiv 0 \pmod{k}$,*

$$G_r^k(\chi_p^{-a}) O_K = \prod_{\substack{j=1 \\ \gcd(j,k)=1}}^{k} P_{j^{-1}}^{L(ja)}$$

and

$$G_r^k(\chi_p^{a}) O_K = \prod_{\substack{j=1 \\ \gcd(j,k)=1}}^{k} P_{j^{-1}}^{k-L(ja)}.$$

In particular,

$$G_r^k(\chi_p^{-1}) O_K = \prod_{\substack{j=1 \\ \gcd(j,k)=1}}^{k} P_{j^{-1}}^{j}.$$

Proof Since $\sigma_{p^i}(P_j) = P_j$ for each i, $0 \le i \le r-1$, it follows from (11.1.2) that

$$G_r^k(\chi_p^{-a}) G_r^k(\chi_p^{a}) O_K = q^k O_K = \prod_{\substack{j=1 \\ \gcd(j,k)=1}}^{k} P_{j^{-1}}^{k}.$$

Thus it suffices to prove the first part of the theorem. Since $T = R/\{1, p, \ldots, p^{r-1}\}$,

$$\begin{aligned}\prod_{\substack{j=1\\ \gcd(j,k)=1}}^{k} P_{j^{-1}}^{L(ja)} &= \prod_{j\in R} P_{j^{-1}}^{L(ja)} = \prod_{j\in T} P_{j^{-1}}^{L(ja)+L(pja)+\cdots+L(p^{r-1}ja)}\\ &= \prod_{j\in T} P_{j^{-1}}^{ks(ja)/(p-1)} && \text{(by Theorem 11.2.7)}\\ &= G_r^k(\chi_P^{-a})O_K. && \text{(by Theorem 11.2.6)}\end{aligned}$$

■

The next theorem gives the exponents of the prime ideals in the decomposition of Jacobi sums (Theorem 11.2.3) in two different forms.

Theorem 11.2.9 *If* m, n, *and* $m+n$ *are integers not divisible by* k, *then*

$$\begin{aligned}(s(m)+s(n)-s(m+n))/(p-1) &= \operatorname{card}\{i:\ 0\le i\le r-1,\ L(mp^i)+L(np^i)>k\}\\ &= \sum_{i=0}^{r-1}\Big\{\Big[\frac{mp^i+np^i}{k}\Big]-\Big[\frac{mp^i}{k}\Big]-\Big[\frac{np^i}{k}\Big]\Big\}.\end{aligned}$$

Proof By Theorem 11.2.7,

$$\begin{aligned}(s(m)+s(n)-s(m+n))/(p-1) &= \frac{1}{k}\sum_{i=0}^{r-1}\Big(L(mp^i)+L(np^i)-L(mp^i+np^i)\Big)\\ &= \sum_{i=0}^{r-1}\Big\{\Big[\frac{mp^i+np^i}{k}\Big]-\Big[\frac{mp^i}{k}\Big]-\Big[\frac{np^i}{k}\Big]\Big\},\end{aligned}$$

since $L(x) = x - k[x/k]$ for integers $x \not\equiv 0 \pmod k$. The expression in braces equals 1 or 0 according as $L(mp^i)+L(np^i)$ is greater than or less than k, and the result follows.

■

We close this section by examining the Gauss sum $G_r(\chi^{-a}) = G_r(\chi_P^{-a})$ (with $a \not\equiv 0 \pmod k$) embedded in the $\wp$ adic completion $M_\wp$ of M. The necessary background in $\wp$-adic theory may be found in the book of Koblitz [1, Chapter 3].

Stickelberger's congruence (Theorem 11.2.1) can be viewed as a congruence in $M_\wp$ upon embedding M into $M_\wp$. Let Q_p denote the field of p-adic rationals. The Gauss sum $G_r(\chi^{-a})$ in $M_\wp$ is in the subfield $Q_p(\beta_p) \subset M_\wp$, since by (11.2.7), it is fixed by each of the r elements of $\mathrm{Gal}(M_\wp/Q_p(\beta_p))$. Now let λ be the prime element in $Q_p(\beta_p)$ such that

(11.2.10) $$\lambda^{p-1} = -p \quad\text{and}\quad \lambda \equiv \pi \pmod{\pi^2}.$$

Stickelberger's congruence can be written as

$$G_r(\chi^{-a}) \equiv \frac{-\lambda^{s(a)}}{t(a)} \pmod{\lambda^{s(a)+1}} \quad \text{in } Q_p(\beta_p),$$

that is,

(11.2.11) $$G_r(\chi^{-a})\lambda^{-s(a)} \equiv -1/t(a) \pmod{\lambda} \quad \text{in } Q_p(\beta_p).$$

We now show that the left side of (11.2.11) is in Q_p, by showing that it is fixed by each of the $p-1$ elements $\sigma_m \in \mathrm{Gal}(Q_p(\beta_p)/Q_p)$, where σ_m is defined by $\sigma_m(\beta_p) = \beta_p^m$, $1 \le m \le p-1$. By (11.2.10),

$$\sigma_m(\lambda) \equiv \sigma_m(\pi) = \beta_p^m - 1 \equiv m(\beta_p - 1) = m\pi \equiv m\lambda \pmod{\pi^2}.$$

Taking the $s(a)$ powers, we obtain, by (11.2.5),

$$\sigma_m(\lambda^{s(a)}) \equiv \lambda^{s(a)} m^{s(a)} \equiv \lambda^{s(a)} m^{af} \pmod{\pi^{s(a)+1}}.$$

Since m is an integer, $\chi(m)$ is a $(p-1)$th root of unity in Q_p, and hence by (11.2.1),

$$\chi(m) \equiv m^f \pmod{p} \quad \text{in } Q_p.$$

Therefore,

$$\sigma_m(\lambda^{s(a)}) \equiv \lambda^{s(a)} \chi^a(m) \pmod{\pi^{s(a)+1}}.$$

Since the $(p-1)$th powers of the left and right sides of this congruence are equal, the congruence must be an equality, i.e.,

$$\sigma_m(\lambda^{s(a)}) = \lambda^{s(a)} \chi^a(m).$$

Also, by Theorem 1.1.3,

$$\sigma_m(G_r(\chi^{-a})) = G_r(\chi^{-a})\chi^a(m),$$

and so the proof that $G_r(\chi^{-a})\lambda^{-s(a)} \in Q_p$ is complete. Thus (11.2.11) becomes

$$G_r(\chi^{-a})\lambda^{-s(a)} \equiv -1/t(a) \pmod{p} \quad \text{in } Q_p(\beta_p).$$

We have therefore proved the following result.

Theorem 11.2.10

$$G_r(\chi^{-a}) \equiv -\frac{\lambda^{s(a)}}{t(a)} \pmod{\lambda^{p-1+s(a)}} \quad \textit{in } Q_p(\beta_p).$$

The case $k = q-1$ of Theorem 11.2.10 is proved in the paper of Washio, Shimaura, and Shiratani [1, Theorem 1]. Theorem 11.2.10 may also be deduced from the deep formula of Gross and Koblitz [1, Theorem 1.7] which connects $G_r(\chi^{-a})$ and the p-adic gamma function Γ_p defined in (9.3.3). The Gross–Koblitz formula is

(11.2.12) $$G_r(\chi^{-a}) = -\lambda^{s(a)} \prod_{i=0}^{r-1} \Gamma_p\left(\frac{L(ap^i)}{k}\right) \quad \text{in } Q_p(\beta_p).$$

Coleman [1], [2] has given an elegant proof of the Gross–Koblitz formula (11.2.12) which, unlike several other proofs, works for $p = 2$. Lang [1, Chapter 15] has given a relatively elementary proof for odd primes p. For other proofs, see the paper of Young [1] and the references therein.

11.3 The Davenport–Hasse product formula

In this section, we use the notation of Section 11.1 with $k = q - 1$. For example, $K = Q(e^{2\pi i/(q-1)})$, where $q = p^r$, p is a prime, and r is a positive integer. Fix a character ψ on F_q of order $l > 1$ (so that $l \mid (q-1)$) and let χ be any character on F_q. Define the Gauss sum

(11.3.1) $$\tau(\chi) = \sum_{\alpha \in F_q^*} \chi(\alpha) e(\alpha).$$

Thus

(11.3.2) $$\tau(\chi) = \begin{cases} G_r(\chi), & \text{if } \chi \text{ is nontrivial,} \\ -1, & \text{if } \chi \text{ is trivial.} \end{cases}$$

Define

(11.3.3) $$N(l, \chi) := \frac{\chi^l(l)\tau(\chi)}{\tau(\chi^l)} \prod_{i=1}^{l-1} \frac{\tau(\chi\psi^i)}{\tau(\psi^i)}.$$

If χ is a power of ψ, i.e., χ^l is trivial, then $N(l, \chi) = 1$. By (11.3.3),

(11.3.4) $$N(l, \chi) = \chi^l(l) \prod_{i=1}^{l-1} \frac{J_r(\chi, \chi^i)}{J_r(\chi, \psi^i)}, \quad \text{if } \chi \text{ is nontrivial.}$$

Thus

(11.3.5) $$N(l, \chi) \in O_K.$$

The Davenport–Hasse product formula asserts that $N(l, \chi) = 1$ for all characters χ on F_q. We will prove this in Theorem 11.3.5 using Stickelberger's theorem (Theorem 11.2.1). It would be desirable to have a more elementary proof. There is an elementary proof in the case $l = 2$ since, in view of (11.3.4), Theorem 2.1.4 implies that $N(2, \chi) = 1$ for all χ. The proof of Theorem 11.3.1 below shows that the problem of providing an elementary proof that $N(l, \chi) = 1$ for all χ is reduced to the (so far unsolved) problem of providing an elementary proof that $N(t, \chi) = 1$ for all χ and all primes t dividing l. In particular, Theorem 11.3.1 yields an elementary proof that $N(l, \chi) = 1$ for all χ when l is a power of 2.

Theorem 11.3.1 *If $N(t, \chi) = 1$ for all χ on F_q and all primes t dividing l, then $N(l, \chi) = 1$ for all χ.*

Proof The proof is by induction on l. Write $l = tu$, where t is the smallest prime factor of l. If $u = 1$, the proof is complete. Assume that $u > 1$. We have

(11.3.6) $$\prod_{m=0}^{l-1} \tau(\chi\psi^m) = \prod_{j=0}^{u-1} \prod_{i=0}^{t-1} \tau(\chi\psi^j\psi^{ui}).$$

Since $N(t, \chi\psi^j) = 1$, it follows from (11.3.3) and (11.3.6) that

$$\text{(11.3.7)} \qquad \prod_{m=0}^{l-1} \tau(\chi\psi^m) = \prod_{j=0}^{u-1} \left\{ \overline{\chi}^t(t)\overline{\psi}^{tj}(t)\tau(\chi^t\psi^{tj}) \prod_{i=1}^{t-1} \tau(\psi^{ui}) \right\}$$
$$= \overline{\chi}^l(t)\overline{\psi}(t)^{l(u-1)/2} \prod_{i=1}^{t-1} \tau(\psi^{ui})^u \prod_{j=0}^{u-1} \tau(\chi^t\psi^{tj}).$$

Since $N(u, \chi^t) = 1$ by the inductive hypothesis,

$$\text{(11.3.8)} \qquad \prod_{j=0}^{u-1} \tau(\chi^t\psi^{tj}) = \overline{\chi}^l(u)\tau(\chi^l) \prod_{j=1}^{u-1} \tau(\psi^{tj}).$$

Thus, (11.3.7) and (11.3.8) yield

$$\prod_{m=0}^{l-1} \tau(\chi\psi^m) = \overline{\chi}^l(l)\tau(\chi^l)\overline{\psi}(t)^{l(u-1)/2} \prod_{i=1}^{t-1} \tau(\psi^{ui})^u \prod_{j=1}^{u-1} \tau(\psi^{tj}).$$

By (11.3.3), it remains to prove that

$$\text{(11.3.9)} \qquad \prod_{n=1}^{l-1} \tau(\psi^n) = \overline{\psi}(t)^{l(u-1)/2} \prod_{i=1}^{t-1} \tau(\psi^{ui})^u \prod_{j=1}^{u-1} \tau(\psi^{tj}).$$

For any nontrivial character λ on F_q, $\tau(\lambda)\tau(\overline{\lambda}) = \lambda(-1)q$. In particular, for the quadratic character ρ on F_q (which exists when q is odd), $\tau(\rho) = i_0 q^{1/2}$, where $i_0^2 = \rho(-1)$. (The value of i_0 will be given in Theorem 11.5.4, but it is not needed in this proof.) Consequently, if λ has order c,

$$\text{(11.3.10)} \qquad q^{(1-c)/2} \prod_{n=1}^{c-1} \tau(\lambda^n) = \begin{cases} 1, & \text{if } c \text{ is odd,} \\ i_0(\lambda(-1))^{(c^2-2c)/8}, & \text{if } c \text{ is even.} \end{cases}$$

By (11.3.10), both sides of (11.3.9) equal $q^{(l-1)/2}$ when l is odd, which completes the proof for odd l. Suppose now that l is even, so that $t = 2$, $l = 2u$, and q is odd. By (11.3.9) and (11.3.10), it remains to show that

$$\text{(11.3.11)} \qquad i_0(\psi(-1))^{(u^2-u)/2} = \begin{cases} i_0^u \rho(2)^{u-1}, & \text{if } u \text{ is odd,} \\ i_0^{u+1} \rho(2)^{u-1}, & \text{if } u \text{ is even.} \end{cases}$$

If u is odd, then $\rho(2)^{u-1} = 1$ and $i_0^{u-1} = (\rho(-1))^{(u-1)/2} = (\psi(-1))^{u(u-1)/2}$, and (11.3.11) follows. Finally, suppose that u is even, so that $l \equiv 0 \pmod 4$ and $q \equiv 1 \pmod 4$. Then $i_0^2 = \rho(-1) = (\psi(-1))^u = 1$, and so it remains to show that

$$\text{(11.3.12)} \qquad (\psi(-1))^{u/2} = \rho(2), \quad \text{when } 2 \mid u.$$

Let γ be a generator of F_q^*, so that $g = \gamma^{(q-1)/(p-1)}$ is a primitive root of p. Let $Z = \text{ind}_g 2$. Then

$$\rho(2) = \rho(g^Z) = \rho(\gamma)^{Z(q-1)/(p-1)} = (-1)^{Zr},$$

since $(q-1)/(p-1) \equiv r \pmod 2$. Also,

$$(\psi(-1))^{u/2} = \psi^u(\gamma^{(q-1)/4}) = \rho(\gamma)^{(q-1)/4} = (-1)^{(q-1)/4}.$$

Thus (11.3.12) is equivalent to

(11.3.13) $$Zr \equiv (q-1)/4 \pmod 2.$$

If r is even, then q is an odd square, so that $8 \mid (q-1)$, and (11.3.13) holds. If r is odd, then

$$q \equiv 1 \pmod 8 \iff p \equiv 1 \pmod 8 \iff Z \equiv 0 \pmod 2,$$

and (11.3.13) again holds. ■

We noted beneath (11.3.3) that $N(l, \chi) = 1$ whenever χ^l is trivial. In order to prove that $N(l, \chi) = 1$ for the remaining χ, we assume without loss of generality that

(11.3.14) $$\psi = \chi_p^v, \quad \text{where } v = (q-1)/l, \text{ and } \chi = \chi_p^b \quad (0 < b < v),$$

where χ_p is the character of order $q-1$ on F_q defined by (11.2.1). We will need the following three lemmas.

Lemma 11.3.2 *For integers $b \geq 0$, define*

$$D(b) = s(bl) - \sum_{i=0}^{l-1} \{s(b+vi) - s(vi)\},$$

where $v = (q-1)/l$. Then $D(b) = 0$.

Proof Since $D(0) = 0$, it suffices to show for $0 < b < v$ that $D(b) = D(b-1)$, i.e.,

(11.3.15) $$\sum_{i=0}^{l-1} \{s(b+vi) - s(b-1+vi)\} = \sum_{i=0}^{l-1} \{s(bl-i) - s(bl-i-1)\}.$$

Fix i, $0 \leq i \leq l-1$. Define $e \in \{0, 1, \ldots, r-1\}$ by $p^e \,\|\, (b+vi)$. Then $p^e \,\|\, (bl-i)$, and so we have the base p expansions beginning

(11.3.16) $$b + vi = Ap^e + \cdots, \quad bl - i = Bp^e + \cdots,$$

with

(11.3.17) $$1 \leq A, B \leq p-1, \quad Al \equiv B \pmod p,$$

and

$$(11.3.18)\qquad b+vi-1 \;=\; \sum_{h=0}^{e-1}(p-1)p^h + (A-1)p^e + \cdots,$$

$$(11.3.19)\qquad bl-i-1 \;=\; \sum_{h=0}^{e-1}(p-1)p^h + (B-1)p^e + \cdots.$$

Thus the summand corresponding to i on each side of (11.3.15) equals $1-(p-1)e$, and so (11.3.15) holds. ■

Lemma 11.3.3 *For some integer j,*

$$N(l,\chi) = \pm e^{2\pi ij/(q-1)}.$$

Proof By (11.3.5), $N(l,\chi) \in O_K$. Then (11.3.4), (11.3.14), and Theorem 11.2.3 give

$$N(l,\chi)O_K = \prod_{j\in T} P_{j-1}^{D(bj)}.$$

Since each $D(bj)=0$ by Lemma 11.3.2, $N(l,\chi)$ is a unit of O_K. By (11.3.4), all the conjugates of $N(l,\chi)$ in O_K have absolute value 1. Therefore, by Theorem 2.1.12, $N(l,\chi)$ is a root of unity in O_K. The result now follows from Theorem 2.1.13. ■

Lemma 11.3.4 *For each χ, $N(l,\chi)-1 \in P$.*

Proof First apply Theorem 1.1.4(a) to each Gauss sum in (11.3.3). Then apply Theorem 11.2.1 and Lemma 11.3.2 to obtain

$$\chi^l(l)N(l,\chi)\prod_{i=0}^{l-1} t(b+vi) \equiv t(bl)\prod_{i=0}^{l-1} t(vi) \pmod P.$$

It remains to show that

$$l^{bl}\prod_{i=0}^{l-1} t(b+vi) \equiv t(bl)\prod_{i=0}^{l-1} t(vi) \pmod p.$$

We prove this by induction on b. Since it holds for $b=0$, it suffices to show that for $0<b<v$,

$$(11.3.20)\qquad \prod_{i=0}^{l-1}\frac{lt(b+vi)}{t(b+vi-1)} \equiv \prod_{i=0}^{l-1}\frac{t(bl-i)}{t(bl-i-1)} \pmod p.$$

Fix i, $0\le i\le l-1$. By (11.3.16)–(11.3.19), the factor corresponding to i on each side of (11.3.20) is congruent to

$$(-1)^e Al \equiv (-1)^e B \pmod p,$$

and so (11.3.20) holds. ■

Theorem 11.3.5 (*Davenport–Hasse product formula.*) *Let ψ be a character on F_q of order $l > 1$. For every character χ on F_q,*

$$N(l,\chi): \quad = \frac{\chi^l(l)\tau(\chi)}{\tau(\chi^l)} \prod_{i=1}^{l-1} \frac{\tau(\chi\psi^i)}{\tau(\psi^i)} = 1,$$

where the Gauss sums τ are defined by (11.3.2).

Proof By Lemma 11.3.3, $N(l,\chi)$ is a root of unity of order dividing $2(q-1)$. Since $p \in P$, no positive integer relatively prime to p can lie in P. Hence by Lemma 11.3.4 and Theorem 2.1.9, the order of the root of unity $N(l,\chi)$ is p^a for some nonnegative integer a. Then $p^a \mid 2(q-1)$, and so $p^a = 1$ or $p^a = 2$. Thus either $N(l,\chi) = 1$ or else $p = 2$ and $N(l,\chi) = \pm 1$. It remains only to prove the result for $p = 2$. Let $p = 2$, so that l is odd. By Theorem 11.3.1, it suffices to consider the case when l is prime. Let $\lambda = 1 - e^{2\pi i/l}$, so that λ has norm l by Theorem 2.1.9. Since $\psi(\alpha) \equiv 1 \pmod{\lambda}$ for each $\alpha \in F_q^*$, (11.3.3) gives

$$\overline{\chi}^l(l)\tau(\chi^l)N(l,\chi) \equiv \tau(\chi)^l \pmod{\lambda}.$$

Since

$$\overline{\chi}^l(l)\tau(\chi^l) = \sum_{\alpha\in F_q^*} \chi^l(\alpha)e(\alpha l) \equiv \Big(\sum_{\alpha\in F_q^*} \chi(\alpha)e(\alpha)\Big)^l = \tau(\chi)^l \pmod{\lambda},$$

it follows that

$$\tau(\chi)^l(1 - N(l,\chi)) \equiv 0 \pmod{\lambda}.$$

If $N(l,\chi) = -1$, then λ divides $2\tau(\chi)^l$, so that, taking norms, we obtain the contradiction that l divides a power of 2. Thus $N(l,\chi) = 1$. ∎

Example We use the Davenport–Hasse product formula to relate two Jacobi sums of order 9. Let $q = p^r \equiv 1 \pmod 9$, and let χ be a character on F_q of order 9. By Theorem 11.3.5 with $l - 3$, and (11.3.4),

$$\chi^3(3)J_r(\chi,\chi)J_r(\chi,\chi^2) = J_r(\chi,\chi^3)J_r(\chi,\chi^6).$$

By Theorem 2.1.5, $J_r(\chi,\chi^2) = J_r(\chi,\chi^6)$, and so

$$\chi^3(3)J_r(\chi,\chi) = J_r(\chi,\chi^3).$$

11.4 Restrictions and lifts of characters

Throughout this section, p is a prime, r and s are positive integers, $q = p^r$, and $E = F_{q^s}$ is an extension field of F_q. When a character χ on a finite field is restricted to a subfield, it defines a character on that subfield which we denote by χ^*.

Proposition 11.4.1 *If χ is a character on $E = F_{q^s}$ of order k, then the order of the restricted character χ^* on F_q is $k/\gcd(k,(q^s-1)/(q-1))$.*

Proof Let γ be a generator of E^*, so that γ^v is a generator of F_q^*, where $v = (q^s-1)/(q-1)$. Since the root of unity $\chi(\gamma)$ has order k, the root of unity $\chi^*(\gamma^v) = \chi(\gamma)^v$ has order $k/\gcd(k,v)$. Thus χ^* has order $k/\gcd(k,v)$. ∎

The following corollary is an immediate consequence of Proposition 11.4.1.

Corollary 11.4.2 *If χ is a character on $E = F_{q^s}$ of order k, then the restricted character χ^* on F_q is trivial if and only if $(q^s-1)/(q-1) \equiv 0 \pmod{k}$, while χ^* has order k if and only if $\gcd(k,(q^s-1)/(q-1)) = 1$.*

Definition *If $\alpha \in E = F_{q^s}$, the norm of α from E to F_q is*

$$N_{E/F_q}(\alpha) = \prod_{j=0}^{s-1} \alpha^{q^j} = \alpha^{(q^s-1)/(q-1)}.$$

The norm $N_{E/F_q}(\alpha)$ is the product of the s conjugates of $\alpha \in E$ over F_q. The basic properties of the norm are given in the next theorem. We omit the proof, which is completely straightforward. (The counterpart, Theorem 11.4.6, for the trace function is a bit less straightforward.)

Theorem 11.4.3 *Let $\alpha, \beta \in E = F_{q^s}$ and $\delta \in F_q$. Then*
(a) $N_{E/F_q}(\alpha) \in F_q$,
(b) $N_{E/F_q}(\alpha\beta) = N_{E/F_q}(\alpha)N_{E/F_q}(\beta)$,
(c) $N_{E/F_q}(\delta\alpha) = \delta^s N_{E/F_q}(\alpha)$,
(d) N_{E/F_q} *maps E onto F_q and E^* onto F_q^*,*
(e) $N_{E/F_q}(\alpha) = (-1)^s a_0$, *where a_0 is the constant term of the characteristic polynomial of α over F_q,*
(f) $N_{E/F_q}(\alpha) = N_{K/F_q}(N_{E/K}(\alpha))$, *where K is an intermediate field between F_q and E.*

Definition *Let χ be a character on F_q. The lift χ' of the character χ from F_q to the extension field E is given by*

$$\chi'(\alpha) = \chi(N_{E/F_q}(\alpha)), \quad \alpha \in E.$$

The basic properties of the lift χ' of χ from F_q to $E = F_{q^s}$ are given in the next theorem.

Theorem 11.4.4 *Let χ be a character on F_q and let χ' denote the lift of χ from F_q to $E = F_{q^s}$. For a character ψ on E, let ψ^* denote the restriction of ψ to F_q. Then*
(a) *χ' is a character on E,*

(b) $(\chi')^* = \chi^s$ *and* $(\psi^*)' = \psi^v$, *where* $v = (q^s - 1)/(q-1)$, $(\chi')^*$ *is the restriction of* χ' *to* F_q, *and* $(\psi^*)'$ *is the lift of* ψ^* *from* F_q *to* E,
(c) *a character* ψ *on* E *equals the lift* χ' *of some character* χ *on* F_q *if and only if the order of* ψ *divides* $(q-1)$,
(d) *if* $\chi_1' = \chi_2'$, *then* $\chi_1 = \chi_2$,
(e) χ' *and* χ *have the same order,*
(f) $(\chi_1\chi_2)' = \chi_1'\chi_2'$.

Proof (a) For α, $\beta \in E^*$, Theorem 11.4.3(b) gives

$$\chi'(\alpha\beta) = \chi(N_{E/F_q}(\alpha\beta)) = \chi(N_{E/F_q}(\alpha))\chi(N_{E/F_q}(\beta)) = \chi'(\alpha)\chi'(\beta),$$

so that χ' is a character on E.

(b) For $\delta \in F_q^*$, Theorem 11.4.3(c) gives

$$\chi'(\delta) = \chi(N_{E/F_q}(\delta)) = \chi(\delta^s) = \chi^s(\delta),$$

so that $(\chi')^* = \chi^s$. For $\alpha \in E^*$,

$$(\psi^*)'(\alpha) = (\psi^*)(N_{E/F_q}(\alpha)) = (\psi^*)(\alpha^v) = \psi(\alpha^v) = \psi^v(\alpha),$$

so that $(\psi^*)' = \psi^v$.

(c) If $\psi = \chi'$, then for $\alpha \in E^*$, $\psi^{q-1}(\alpha) = \chi^{q-1}(N_{E/F_q}(\alpha)) = 1$, so that the order of ψ divides $(q-1)$. Conversely, suppose that ψ^{q-1} is the trivial character on E. Let γ be a generator of the cyclic group E^*, and write $c = \gamma^v$, where $v = (q^s-1)/(q-1)$. Then c is a generator of F_q^*. We can define a character χ on F_q by $\chi(c) = \psi(\gamma)$, since $\psi(\gamma)$ is a $(q-1)$th root of unity. Then

$$\chi'(\gamma) = \chi(\gamma^v) = \chi(c) = \psi(\gamma), \tag{11.4.1}$$

so that $\chi' = \psi$.

(d) If $\chi_1' = \chi_2'$, then by (11.4.1), $\chi_1(c) = \chi_1'(\gamma) = \chi_2'(\gamma) = \chi_2(c)$, so that $\chi_1 = \chi_2$.

(e) The orders of χ' and χ are the orders of the roots of unity $\chi'(\gamma)$ and $\chi(c)$, which are equal by (11.4.1).

(f) By (11.4.1),

$$(\chi_1\chi_2)'(\gamma) = \chi_1\chi_2(\gamma^v) = \chi_1(\gamma^v)\chi_2(\gamma^v) = \chi_1'(\gamma)\chi_2'(\gamma) = \chi_1'\chi_2'(\gamma). \qquad \blacksquare$$

An immediate consequence of Theorem 11.4.4 is the following.

Corollary 11.4.5 *The lift of a trivial character is also a trivial character, and, for* $p > 2$, *the lift of a quadratic character is also a quadratic character.*

Definition *Let χ be a character on $F_q = F_{p^r}$ and let χ' be the lift of χ from F_q to $E = F_{q^s} = F_{p^{rs}}$. Then the lift of the Gauss sum*

$$G_r(\chi) = \sum_{\alpha \in F_q} \chi(\alpha) e(\alpha) \tag{11.4.2}$$

from F_q to E is the Gauss sum

$$G_{rs}(\chi') = \sum_{\alpha \in E} \chi'(\alpha) e(\alpha). \tag{11.4.3}$$

Definition *For $\alpha \in E = F_{q^s}$, the trace of α from E to F_q is*

$$\mathrm{tr}_{E/F_q}(\alpha) = \sum_{j=0}^{s-1} \alpha^{q^j}.$$

Note that this reduces to the definition of the trace function given in Section 1.1, in the case that the ground field is F_p. For example, the factor $e(\alpha)$ in (11.4.3) is

$$e(\alpha) = \exp(2\pi i \, \mathrm{tr}(\alpha)/p) = \exp(2\pi i \, \mathrm{tr}_{E/F_p}(\alpha)/p).$$

The trace $\mathrm{tr}_{E/F_q}(\alpha)$ is the sum of the s conjugates of $\alpha \in E$ over F_q. The basic properties of the trace function are given in the next theorem. We omit the proof, as it is essentially the same as the proof of Proposition 1.1.1.

Theorem 11.4.6 *Let α, $\beta \in E = F_{q^s}$ and $\delta \in F_q$. Then*
(a) $\mathrm{tr}_{E/F_q}(\alpha) \in F_q$,
(b) $\mathrm{tr}_{E/F_q}(\alpha + \beta) = \mathrm{tr}_{E/F_q}(\alpha) + \mathrm{tr}_{E/F_q}(\beta)$,
(c) $\mathrm{tr}_{E/F_q}(\delta\alpha) = \delta \mathrm{tr}_{E/F_q}(\alpha)$,
(d) tr_{E/F_q} *maps E onto F_q,*
(e) $\mathrm{tr}_{E/F_q}(\alpha) = -a_{s-1}$, *where $x^s + a_{s-1}x^{s-1} + \cdots + a_0$ is the characteristic polynomial of α over F_q,*
(f) $\mathrm{tr}_{E/F_q}(\alpha) = \mathrm{tr}_{K/F_q}(\mathrm{tr}_{E/K}(\alpha))$, *where K is an intermediate field between F_q and E.*

11.5 The Davenport–Hasse theorem on lifted Gauss sums

Throughout this section, p is a prime, r and s are positive integers, $q = p^r$, $E = F_{q^s}$, $\mathcal{M}$ denotes the set of monic polynomials in $F_q[x]$, $\mathcal{M}_n$ denotes the subset of $\mathcal{M}$ consisting of polynomials of degree n, and I denotes the subset of $\mathcal{M}$ consisting of irreducible polynomials over F_q. The following result will be used in the proof of the Davenport–Hasse theorem on lifted Gauss sums (Theorem 11.5.2).

Theorem 11.5.1 *Let λ be a complex function on $\mathcal{M}$ such that*

(11.5.1) $$\lambda(1) = 1,$$

(11.5.2) $$\lambda(fg) = \lambda(f)\lambda(g), \quad \text{for all } f,\, g \in \mathcal{M},$$

(11.5.3) $$|\lambda(f)| \le 1, \quad \text{for all } f \in \mathcal{M},$$

and for some nonnegative integer N

(11.5.4) $$\sum_{f \in \mathcal{M}_N} \lambda(f) \ne 0,$$

(11.5.5) $$\sum_{f \in \mathcal{M}_n} \lambda(f) = 0, \quad \text{for all } n > N.$$

Then, for $m = 1, 2, 3, \ldots,$

(11.5.6) $$\sum_{\substack{f \in I \\ \deg f \mid m}} (\deg f)\lambda(f)^{m/\deg f} = -\sum_{i=1}^{N} \omega_i^{-m},$$

where $\omega_1, \omega_2, \ldots, \omega_N$ are the N complex zeros of the polynomial

(11.5.7) $$W(z) = \sum_{n=0}^{N} \Big(\sum_{f \in \mathcal{M}_n} \lambda(f) \Big) z^n.$$

Proof By (11.5.4) and (11.5.7), the leading coefficient of the polynomial $W(z)$ is

$$A = \sum_{f \in \mathcal{M}_N} \lambda(f) \ne 0,$$

so that

$$W(z) = A(z - \omega_1) \cdots (z - \omega_N).$$

By (11.5.1) and (11.5.7),

$$1 = W(0) = A(-1)^N \omega_1 \omega_2 \cdots \omega_N.$$

Thus we can write

$$W(z) = \prod_{i=0}^{N} \left(1 - \frac{z}{\omega_i}\right).$$

If $|z| < |\omega_i|$ for all i, $1 \le i \le N$, then logarithmic differentiation yields

(11.5.8) $$\frac{zW'(z)}{W(z)} = -\sum_{i=1}^{N} \frac{z/\omega_i}{1 - (z/\omega_i)} = -\sum_{m=1}^{\infty} \Big(\sum_{i=1}^{N} \omega_i^{-m} \Big) z^m.$$

On the other hand, appealing to (11.5.3), (11.5.5), and (11.5.7), we have, for $|z| < 1/q$,

$$W(z) = \sum_{n=0}^{\infty} \Big(\sum_{f \in \mathcal{M}_n} \lambda(f) \Big) z^n = \sum_{f \in \mathcal{M}} \lambda(f) z^{\deg f}.$$

Since $F_q[x]$ is a unique factorization domain and $\lambda(f) z^{\deg f}$ is a multiplicative function of f by (11.5.2), it follows that for $|z| < 1/q$,

$$W(z) = \prod_{f \in I} \sum_{j=0}^{\infty} \lambda(f^j) z^{\deg f^j} = \prod_{f \in I} \sum_{j=0}^{\infty} \lambda(f)^j z^{j \deg f} = \prod_{f \in I} (1 - \lambda(f) z^{\deg f})^{-1}.$$

Thus, for $|z| < 1/q$, logarithmic differentiation yields

$$\begin{aligned} \frac{zW'(z)}{W(z)} &= \sum_{f \in I} \frac{\lambda(f)(\deg f) z^{\deg f}}{1 - \lambda(f) z^{\deg f}} = \sum_{f \in I} (\deg f) \sum_{r=1}^{\infty} \lambda(f)^r z^{r \deg f} \\ &= \sum_{f \in I} (\deg f) \sum_{\substack{m=1 \\ \deg f \mid m}}^{\infty} \lambda(f)^{m/\deg f} z^m, \end{aligned}$$

that is,

$$\frac{zW'(z)}{W(z)} = \sum_{m=1}^{\infty} \Big(\sum_{\substack{f \in I \\ \deg f \mid m}} (\deg f) \lambda(f)^{m/\deg f} \Big) z^m. \tag{11.5.9}$$

Equating the coefficients of z^m in (11.5.8) and (11.5.9), we obtain the desired result. ■

Theorem 11.5.2 (*The Davenport–Hasse theorem on lifted Gauss sums.*) *For a nontrivial character χ on $F_q = F_{p^r}$, the Gauss sums in (11.4.2) and (11.4.3) are related by*

$$G_{rs}(\chi') = (-1)^{s-1} G_r^s(\chi).$$

Proof Recall that $\mathcal{M}$ denotes the set of monic polynomials in $F_q[x]$. Define a complex function λ on $\mathcal{M}$ by

$$\lambda(f) = \begin{cases} 1, & \text{if } \deg f = 0, \\ \chi(a_d) e(a_1), & \text{if } \deg f \geq 1, \end{cases}$$

where $f(x) = x^d - a_1 x^{d-1} + \cdots + (-1)^d a_d$, $d = \deg f$. We now show that λ has properties (11.5.1)–(11.5.5). By definition of λ, $\lambda(x^0) = 1$, and so (11.5.1) holds. Let

$$g(x) = x^c - b_1 x^{c-1} + \cdots + (-1)^c b_c \in \mathcal{M}.$$

If $\deg f = 0$ or $\deg g = 0$, then (11.5.2) holds, and so assume that $d \geq 1$, $c \geq 1$. Then

$$f(x)g(x) = x^{c+d} - (a_1 + b_1)x^{c+d-1} + \cdots + (-1)^{c+d}a_d b_c,$$

so that

$$\lambda(fg) = \chi(a_d b_c)e(a_1 + b_1) = \chi(a_d)e(a_1)\chi(b_c)e(b_1) = \lambda(f)\lambda(g),$$

so that (11.5.2) holds. Since $|\lambda(f)| = 1$, (11.5.3) holds. Next, for $n = 2, 3, \ldots$,

$$\sum_{f \in \mathcal{M}_n} \lambda(f) = \sum_{a_1 \in F_q} \cdots \sum_{a_n \in F_q} \chi(a_n)e(a_1) = q^{n-2} \sum_{a_1 \in F_q} e(a_1) \sum_{a_n \in F_q} \chi(a_n) = 0.$$

Thus (11.5.5) holds with $N = 1$. Moreover, (11.5.4) holds for $N = 1$, since

$$\sum_{f \in \mathcal{M}_1} \lambda(f) = \sum_{a_1 \in F_q} \lambda(x - a_1) = \sum_{a_1 \in F_q} \chi(a_1)e(a_1) = G_r(\chi) \neq 0,$$

by Theorem 1.1.4.

By (11.5.7),

$$W(z) = 1 + G_r(\chi)z \quad \text{and} \quad \omega_1 = -1/G_r(\chi).$$

It thus follows from Theorem 11.5.1 that

$$L := \sum_{\substack{f \in I \\ \deg f | s}} (\deg f)\lambda(f)^{s/\deg f} = (-1)^{s-1}G_r^s(\chi).$$

It remains to show that $L = G_{rs}(\chi')$.

The contribution from $f(x) = x$ to L is

$$\lambda(x)^s = (\chi(0)e(0))^s = 0,$$

since χ is nontrivial and $s > 0$. Every other $f \in I$ with $\deg f \mid s$ has $\deg f$ distinct nonzero zeros $y \in E = F_{q^s}$. The characteristic polynomial over F_q of such y equals

$$f(x)^{s/\deg f} = x^s - A_1 x^{s-1} + \cdots + (-1)^s A_s$$

for some $A_i \in F_q$. By Theorems 11.4.3(e) and 11.4.6(e),

$$N_{E/F_q}(y) = A_s \quad \text{and} \quad \mathrm{tr}_{E/F_q}(y) = A_1.$$

Therefore,

$$\lambda(f)^{s/\deg f} = \lambda(f^{s/\deg f}) = \chi(A_s)e(A_1) = \chi(N_{E/F_q}(y))e(\mathrm{tr}_{E/F_q}(y)) = \chi'(y)e(y),$$

by Theorem 11.4.6(f). It follows that

$$L = \sum_{\substack{f \in I \\ \deg f | s \\ f(x) \neq x}} \sum_{\substack{y \in E \\ f(y) = 0}} \chi'(y)e(y) = \sum_{y \in E^*} \chi'(y)e(y) = G_{rs}(\chi'). \qquad \blacksquare$$

The following corollary gives a corresponding result for lifted Jacobi sums. For a generalization, see Problem 7.

Corollary 11.5.3 *Let χ_1, χ_2, and $\chi_1\chi_2$ be nontrivial characters on F_q. Then*

$$J_{rs}(\chi_1', \chi_2') = (-1)^{s-1} J_r(\chi_1, \chi_2)^s.$$

Proof This follows from Theorems 11.5.2, 11.4.4(e)(f), and 2.1.3(a). ■

The Davenport–Hasse theorem on lifted Gauss sums enables us to evaluate quadratic Gauss sums over F_q, thus extending Theorem 1.2.4 from F_p to F_q.

Theorem 11.5.4 *Let p be an odd prime and let r be a positive integer. Write $q = p^r$. Let ρ be the quadratic character on F_q. Then*

$$G_r(\rho) = \begin{cases} (-1)^{r-1} q^{1/2}, & \text{if } p \equiv 1 \pmod 4, \\ (-1)^{r-1} i^r q^{1/2}, & \text{if } p \equiv 3 \pmod 4. \end{cases}$$

Proof By Theorem 1.2.4, the quadratic Gauss sum over F_p equals $i^{(p-1)^2/4} p^{1/2}$. By Corollary 11.4.5, $G_r(\rho)$ is the lift of that Gauss sum from F_p to F_q, and so Theorem 11.5.2 with $s = r$ gives

$$G_r(\rho) = (-1)^{r-1} i^{r(p-1)^2/4} q^{1/2},$$

as desired. ■

11.6 Pure Gauss sums

Definition *A Gauss or Jacobi sum over a finite field F_q is called pure if some positive integral power of it is real.*

Let $q = p^r$ and let χ, χ_1, and χ_2 be characters on F_q. If χ_1 or χ_2 or $\chi_1\chi_2$ is trivial, then $J_r(\chi_1, \chi_2)$ is real, by Theorem 2.1.1. Similarly, $G_r(\chi)$ is pure for trivial χ. In all other cases, $U = p^{-r/2} G_r(\chi)$ and $V = p^{-r/2} J_r(\chi_1, \chi_2)$ have modulus 1, and then $G_r(\chi)$ (resp. $J_r(\chi_1, \chi_2)$) is pure if and only if U (resp. V) is a root of unity.

Let χ be a character of order k on F_q. If $G_r(\chi^j)$ is pure for all integers j, then $J_r(\chi^u, \chi^v)$ is pure for all integers u and v, by Theorem 1.1.4(a). The converse is also true, since

$$G_r^k(\chi) = \chi(-1) q \prod_{v=1}^{k-2} J_r(\chi, \chi^v), \tag{11.6.1}$$

by Problem 1 of Exercises 2.

Theorem 11.5.4 shows that the quadratic Gauss sum $G_r(\rho)$ is pure. We showed in Theorem 1.6.1 that Gauss sums $G_r(\chi)$ of order $k > 2$ are never pure, in the case $r = 1$. However, pure Gauss sums of order $k > 2$ can exist when $r \geq 2$. In fact, Theorem 11.6.3 below shows that if χ is a character of order k on F_{p^r}, then $G_r(\chi^j)$ is pure for all integers j whenever -1 is a power of $p \pmod{k}$. For example, if $p = 3$ and $r = 6$, then $G_6(\psi)$ is pure for all characters ψ on F_{3^6} of order 28. In Theorem 11.6.4, we prove that, conversely, if $G_r(\chi^j)$ is pure for all integers j, then -1 must be a power of $p \pmod{k}$.

Our first theorem determines explicitly the values of certain pure Gauss sums. This theorem will be used to evaluate certain Eisenstein sums; see the proof of Theorem 12.1.6.

Theorem 11.6.1 *Let χ be a character on $F_{p^{2t}}$ of order m. Suppose that $p^t \equiv -1 \pmod{m}$ for some positive integer t. Then*

$$p^{-t}G_{2t}(\chi) = \begin{cases} 1, & \text{if } p = 2, \\ (-1)^{(p^t+1)/m}, & \text{if } p > 2. \end{cases}$$

Proof Write $q = p^t$. Fix a generator γ of the cyclic group $F_{q^2}^*$. Then $c = \gamma^{q+1}$ is a generator of F_q^*. Every $\alpha \in F_{q^2}^*$ can be written uniquely in the form

$$\alpha = \gamma^{(q+1)u+v} = c^u\gamma^v, \quad 0 \le u < q-1, \ 0 \le v < q+1.$$

Note that $\chi(c) = \chi(\gamma^{q+1}) = 1$, as the order of χ divides $q+1$. Hence $\chi(\alpha) = \chi^v(\gamma)$, and so

$$G_{2t}(\chi) = \sum_{\alpha \in F_{q^2}^*} \chi(\alpha)e(\alpha) = \sum_{u=0}^{q-2}\sum_{v=0}^{q} \chi^v(\gamma)e(c^u\gamma^v) = \sum_{v=0}^{q} \chi^v(\gamma) \sum_{h \in F_q^*} e(h\gamma^v),$$

where

$$e(\alpha) = \exp\left(2\pi i \operatorname{tr}_{F_{q^2}/F_p}(\alpha)/p\right).$$

By Theorem 11.4.6(c)(f), for $h \in F_q^*$,

$$\operatorname{tr}_{F_{q^2}/F_p}(h\gamma^v) = \operatorname{tr}_{F_q/F_p}\left(\operatorname{tr}_{F_{q^2}/F_q}(h\gamma^v)\right) = \operatorname{tr}_{F_q/F_p}(hw),$$

where

$$w = \operatorname{tr}_{F_{q^2}/F_q}(\gamma^v) \in F_q.$$

Thus

$$\sum_{h \in F_q^*} e(h\gamma^v) = \sum_{h \in F_q^*} \exp(2\pi i \operatorname{tr}(hw)/p) = \begin{cases} -1, & \text{if } w \neq 0, \\ q-1, & \text{if } w = 0. \end{cases}$$

As $w = \gamma^v + \gamma^{vq}$, we see that

$$w = 0 \iff \gamma^{v(q-1)} = -1 \iff \begin{cases} v = 0, & \text{if } p = 2 \\ v = (q+1)/2, & \text{if } p > 2. \end{cases}$$

Suppose first that $p = 2$. Then

$$G_{2t}(\chi) = -\sum_{v=1}^{q} \chi^v(\gamma) + (q-1)\chi^0(\gamma) = q - \sum_{v=0}^{q} \chi^v(\gamma) = q,$$

as desired. Now suppose that $p > 2$. Then similar reasoning yields

$$G_{2t}(\chi) = q\chi^{(q+1)/2}(\gamma).$$

Since $\chi(\gamma) = \exp(2\pi ij/m)$ for some j relatively prime to m,

$$p^{-t}G_{2t}(\chi) = \chi^{(q+1)/2}(\gamma) = e^{\pi ij(q+1)/m} = (-1)^{j(q+1)/m} = (-1)^{(q+1)/m},$$

where the last equality holds because when j is even, we must have m odd and $(q+1)/m$ even. ■

The next theorem gives the order of $p \pmod m$ when -1 is a power of $p \pmod m$.

Theorem 11.6.2 *Let $m > 2$ be an integer. Suppose that there is a positive integer t such that $-1 \equiv p^t \pmod m$, with t chosen minimal. Then $2t$ is the order of $p \pmod m$.*

Proof Let N denote the order of $p \pmod m$. Since $p^{2t} \equiv 1 \pmod m$, $N \mid 2t$. Assume that $N \neq 2t$, so that $N \leq t$. Since $-1 \equiv p^{t-N} \pmod m$, we must have $t - N > 0$, which contradicts the minimality of t. ■

The next theorem extends Theorem 11.6.1; it essentially reduces to Theorem 11.6.1 when $v = 2t$.

Theorem 11.6.3 *Let $m > 2$ be an integer. Suppose that there is a positive integer t such that $p^t \equiv -1 \pmod m$, with t chosen minimal. Let ψ be a character of order m on F_{p^v} for some positive integer v. Then $v = 2ts$ for some positive integer s, and*

$$p^{-v/2}G_v(\psi) = \begin{cases} (-1)^{s-1}, & if \;\; p = 2, \\ (-1)^{s-1+(p^t+1)s/m}, & if \;\; p > 2. \end{cases}$$

Proof By Theorem 11.6.2, $2t$ is the order of $p \pmod m$, and since $p^v \equiv 1 \pmod m$, we have $v = 2ts$ for some positive integer s. Since $p^{2t} \equiv 1 \pmod m$, ψ is the lift of some character λ of order m on $F_{p^{2t}}$, by Theorem 11.4.4(c)(e). By the Davenport–Hasse lifting theorem (Theorem 11.5.2),

$$G_v(\psi) = (-1)^{s-1}G_{2t}^s(\lambda).$$

By Theorem 11.6.1,

$$G_{2t}(\lambda) = \begin{cases} p^t, & \text{if } p = 2, \\ p^t(-1)^{(p^t+1)/m}, & \text{if } p > 2, \end{cases}$$

and the result follows on taking the sth power. ■

A consequence of Theorem 11.6.3 is that, if χ is a character of order k on F_{p^r}, then $G_r(\chi^j)$ is pure for all integers j whenever -1 is a power of p (mod k). We prove the converse in the next theorem. As preparation, we first discuss cyclotomic numbers over F_q and their relations to Jacobi sums over F_q, as we did in Section 2.5 in the case $q = p$.

Let $q = p^r$ and let χ be a character of order k on F_q. For integers u and v (mod k), define the Jacobi sum

(11.6.2)
$$J_r^*(\chi^u, \chi^v) = \sum_{\substack{\alpha \in F_q \\ \alpha \neq 0,1}} \chi^u(\alpha)\chi^v(1-\alpha).$$

For $v \not\equiv 0$ (mod k),

(11.6.3)
$$J_r^*(\chi^0, \chi^0) = q-2, \quad J_r^*(\chi^0, \chi^v) = -1, \quad J_r^*(\chi^{-v}, \chi^v) = -\chi^v(-1),$$

by Theorem 2.1.1. All of the other Jacobi sums $J_r^*(\chi^u, \chi^v)$ have modulus $\sqrt{q}$. These sums of modulus $\sqrt{q}$ are the ones for which u, v, and $u+v$ are not divisible by k, and they satisfy

(11.6.4)
$$J_r^*(\chi^u, \chi^v) = J_r(\chi^u, \chi^v) = G_r(\chi^u)G_r(\chi^v)/G_r(\chi^{u+v}).$$

Write $\beta = e^{2\pi i/k}$, and let γ be a generator of F_q such that $\chi(\gamma) = \beta$. For integers a and b (mod k), define the cyclotomic number

$$C_r(a,b) = C_{r,k}(a,b)$$

of order k over F_q to be the number of $\alpha \in F_q^*$ for which

$$\chi(\alpha/\gamma^a) = \chi((\alpha+1)/\gamma^b) = 1.$$

The cyclotomic numbers $C_r(a,b)$ are related to the Jacobi sums $J_r^*(\chi^u, \chi^v)$ by the following double Fourier series relations:

(11.6.5)
$$k^2 C_r(a,b) = \sum_{u=0}^{k-1}\sum_{v=0}^{k-1} \chi^u(-1) J_r^*(\chi^u, \chi^v)\beta^{-au-bv},$$

(11.6.6)
$$\chi^u(-1)J_r^*(\chi^u, \chi^v) = \sum_{a=0}^{k-1}\sum_{b=0}^{k-1} C_r(a,b)\beta^{au+bv}.$$

The proofs of (11.6.5) and (11.6.6) are essentially the same as those for Theorem 2.5.1 and Problem 18 of Exercises 2. ∎

Theorem 11.6.4 *Let $q = p^r$ and let χ be a character of order k on F_q. If $G_r(\chi^j)$ is pure for all integers j, then -1 is a power of p (mod k).*

Proof If $k \le 2$, then -1 is a power of $p \pmod{k}$, so assume that $k > 2$. Define the following sets of ordered pairs with entries $\pmod{k}$:

$$W = \{(u,v): \ u, \ v, \ u+v \not\equiv 0 \pmod{k} \ \},$$

$$B = \{(a,b): \ a, \ b, \ a-b \not\equiv 0 \pmod{k} \ \},$$

and

$$C = \{(a,b): \ (a,b) \notin B, \ (a,b) \neq (0,0) \ \}.$$

By the remarks at the beginning of this section, the hypothesis is equivalent to the assertion that $p^{-r/2} J_r(\chi^u, \chi^v)$ is a root of unity for each ordered pair $(u,v) \in W$. Throughout this proof, we consider p and k fixed. Replacing r by an integral multiple of r if necessary, and using Theorem 11.5.2, we may assume that there is a number $\theta \in \{1,-1\}$ independent of u and v for which

$$J_r(\chi^u, \chi^v) = \theta p^{r/2}, \quad \text{for all } (u,v) \in W, \tag{11.6.7}$$

and that $r \ge 1$ is chosen minimal for which (11.6.7) holds for some $\theta \in \{1,-1\}$ independent of u and v.

Assume for the purpose of contradiction that -1 is not a power of $p \pmod{k}$. We will show that there exists an integer n, a character ψ of order k on F_{p^n}, and a number $\theta_1 \in \{1,-1\}$ independent of u and v for which $r = 2n$ and $J_n(\psi^u, \psi^v) = \theta_1 p^{n/2}$ for all $(u,v) \in W$. This will contradict the minimality of r.

Write $\beta = e^{2\pi i/k}$. By (11.6.7) and Lemma 1.2.1,

$$p^{r/2} \in Q(\beta) \cap Q(e^{2\pi i/(4p)}),$$

so that by (2.1.6), $p^{r/2} \in Q(i)$. Thus $p^{r/2}$ is an integer, and so $r = 2n$ for some integer n. By (11.6.3), (11.6.5), and (11.6.7),

$$k^2 C_r(1,2) = p^r + 1 + 2\theta p^{r/2}.$$

Therefore $((p^n + \theta)/k)^2 = C_r(1,2)$ is an integer, and since -1 is not a power of $p \pmod{k}$, we must have $\theta = -1$ and $p^n \equiv 1 \pmod{k}$. Thus there exists a character ψ of order k on F_{p^n}. By (11.6.7) and Theorem 11.5.2,

$$J_n^2(\psi^u, \psi^v) = p^n \quad \text{for all } (u,v) \in W,$$

and so

$$J_n(\psi^u, \psi^v) = \epsilon(u,v) p^{n/2} \quad \text{for all } (u,v) \in W, \tag{11.6.8}$$

where $\epsilon(u,v) \in \{1,-1\}$. It remains to show that all the $\epsilon(u,v)$ in (11.6.8) are equal.

By (11.6.8) and Lemma 1.2.1, $p^{n/2} \in Q(i)$, so that n is even. By (11.6.8) and (11.6.1), $G_n^k(\psi) = \pm p^{nk/2}$. Thus $G_n(\psi) p^{-n/2}$ is a $2k$th root of unity.

Let ζ be a complex primitive $2kp$th root of unity. Define an integer $c \pmod{2kp}$ by $c \equiv 1 \pmod{2k}$ and $c \equiv g \pmod{p}$, where g is a primitive root of p. The map in $\text{Gal}(Q(\zeta)/Q)$ defined by $\zeta \longrightarrow \zeta^c$ maps $G_n(\psi)$ to $\overline{\psi}(g)G_n(\psi)$, yet it must fix $G_n(\psi)$ since $G_n(\psi)$ equals $p^{n/2}$ times a $2k$th root of unity. Thus $\psi(g) = 1$. This implies that $\psi(-1) = 1$.

By (11.6.3), (11.6.5), and (11.6.7),

$$k^2 C_n(a,b) = p^n + 1 + p^{n/2} Y(a,b), \quad \text{for all } (a,b) \in B \tag{11.6.9}$$

and

$$k^2 C_n(a,b) = p^n + 1 - k + p^{n/2} Y(a,b), \quad \text{for all } (a,b) \in C, \tag{11.6.10}$$

where

$$Y(a,b) = \sum_{(u,v)\in W} \epsilon(u,v)\beta^{-au-bv}. \tag{11.6.11}$$

These three equalities show that the algebraic integer $Y(a,b)$ is, in fact, a rational integer with $|Y(a,b)| < k^2$.

Suppose first that $(a,b) \in B$. By (11.6.9), the $Y(a,b)$ with $(a,b) \in B$ are all congruent to each other $\pmod{k^2}$. Moreover, they all have the same parity, since for $(a,b) \in B$,

$$Y(a,b) \equiv \sum_{(u,v)\in W} \beta^{-au-bv} = 2 \equiv 0 \pmod{2}. \tag{11.6.12}$$

Thus, if k is odd, these $Y(a,b)$ are congruent to each other $\pmod{2k^2}$, and so they are all equal. We now suppose that k is even and show again that these $Y(a,b)$ are all equal. As $(a,b) \in B$, either $(a+(k/2), b) \in B$ or $(a, b+(k/2)) \in B$, or $(a+(k/2), b+(k/2)) \in B$. In the case $(a+(k/2), b) \in B$,

$$\sum_{\substack{(u,v)\in W \\ 2 \nmid u}} \epsilon(u,v)\beta^{-au-bv} = \frac{1}{2}\Big(Y(a,b) - Y(a+(k/2), b)\Big) \equiv 0 \pmod{k^2/2},$$

and since the sum on the left has fewer than $k^2/2$ terms, it vanishes. Thus,

$$|Y(a,b)| = \Big|\sum_{\substack{(u,v)\in W \\ 2 \mid u}} \epsilon(u,v)\beta^{-au-bv}\Big| < k^2/2.$$

Similarly, if $(a, b+(k/2))$ or $(a+(k/2), b+(k/2))$ is in B, $|Y(a,b)| < k^2/2$. Therefore, all the $Y(a,b)$ with $(a,b) \in B$ are equal, since they are congruent to each other $\pmod{k^2}$.

Next suppose that $(a,b) \in C$. The counterpart to (11.6.12) for $(a,b) \in C$ is

$$Y(a,b) \equiv \sum_{(u,v)\in W} \beta^{-au-bv} = 2 - k \equiv k \pmod{2}.$$

Then, proceeding as in the case $(a,b) \in B$, we see that the $Y(a,b)$ with $(a,b) \in C$ are all equal. Thus, all the $C_n(a,b)$ in (11.6.10) are equal, and all the $C_n(a,b)$ in (11.6.9) are equal. Since $\psi(-1) = 1$, it now follows from (11.6.6) that all the $J_n(\psi^u, \psi^v)$ with $(u,v) \in W$ are equal, since for $(u,v) \in W$,

$$\sum_{(a,b)\in C} \beta^{au+bv} = -3, \quad \sum_{(a,b)\in B} \beta^{au+bv} = 2.$$

Consequently, $\epsilon(u,v)$ in (11.6.8) is independent of u and v, as desired. ■

11.7 Irreducible cyclic codes

Let p be a prime, and let $k > 1$ be an integer not divisible by p. For every $h(x) \in F_p[x]$, there is a unique vector $(c_0, \ldots, c_{k-1}) \in F_p^k$ such that

$$h(x) \equiv \sum_{i=0}^{k-1} c_i x^i \pmod{x^k - 1}.$$

The coset $\overline{h}(x)$ of $h(x)$ in the quotient ring

$$A := F_p[x]/(x^k - 1)F_p[x]$$

will be identified with the vector $(c_0, \ldots, c_{k-1}) \in F_p^k$. This gives an isomorphism of the k-dimensional vector spaces A and F_p^k over F_p.

The quotient ring A also has the structure of a principal ideal ring, i.e., all ideals of A have the form $A\overline{h}(x)$. Write

$$x^k - 1 = F(x)g(x), \tag{11.7.1}$$

where $F(x)$ and $g(x)$ are polynomials in $F_p[x]$ of degrees r and $k-r$, respectively, with $0 < r < k$. Define C to be the principal ideal

$$C := A\overline{g}(x). \tag{11.7.2}$$

Then C is an r-dimensional (proper) subspace of the k-dimensional vector space A over F_p, with basis $\overline{g}, \overline{x}\overline{g}, \overline{x}^2\overline{g}, \ldots, \overline{x}^{r-1}\overline{g}$. Such a subspace is called a (linear) (k,r)-code in A over F_p, where k and r refer to the dimensions of the vector spaces A and C over F_p, respectively. The p^r elements $(c_0, \ldots, c_{k-1}) \in C$ are called code words, and k is the length of these code words. Since $\overline{x}C = C$, we have $(c_{k-1}, c_0, c_1, \ldots, c_{k-2}) \in C$ whenever $(c_0, \ldots, c_{k-1}) \in C$, i.e., cyclic shifts of code words are still code words. Thus C is called a cyclic code. If $h(x) \in F_p[x]$ and $h_1(x) = \gcd(h(x), x^k - 1)$, then

$$A\overline{h}(x) = A\overline{h_1}(x).$$

Hence all ideals of A have the form $A\overline{h}(x)$, where $h(x) \mid (x^k - 1)$. The subcodes of $C = A\overline{g}(x)$ are the codes $A\overline{h}(x)$ with $g(x) \mid h(x)$ and $h(x) \mid (x^k - 1)$. If $F(x)$ is irreducible over F_p, then C has no proper subcodes, so that C is called an irreducible cyclic code.

The number of nonzero entries in a word $(c_0, \ldots, c_{k-1}) \in A$ is called the weight of that word. Our goal is to find the weights of code words in certain irreducible cyclic codes $C = A\overline{g}(x)$ specified by (11.7.4) below.

Suppose for the remainder of this section that r is the order of $p \pmod k$, and write $q = p^r = 1 + kf$. Let γ generate the cyclic group F_q^*. Choose

(11.7.3)
$$F(x) = \prod_{i=0}^{r-1} (x - \gamma^{fp^i}) \in F_p[x].$$

Note that $F(x)$ is the irreducible polynomial of γ^f over F_p, and that the polynomial $g(x)$ defined by (11.7.1) is

(11.7.4)
$$g(x) = \prod_{\substack{l=1 \\ l \not\equiv \text{power of } p \pmod k}}^{k} (x - \gamma^{fl}) \in F_p[x].$$

The code $C = A\overline{g}(x)$ defined by (11.7.4) is an irreducible cyclic (k, r)-code. For certain classes of these codes C, we will determine explicitly the weights of the code words; see Theorems 11.7.4–11.7.7 and 11.7.9. One reason it is useful to know these weights is that if all nonzero code words have weight $\geq 2t+1$, then C is a t-error correcting code, in the following sense. If a code word gets transmitted as a word $w = (c_0, \ldots, c_{k-1}) \in A$ that is not in C due to at most t bit transmission errors, then we can correct those bit errors and recover the code word $c \in C$ from which w came, because such $c \in C$ is unique. To see this uniqueness, suppose for the purpose of contradiction that w could have arisen from two different code words $c_1, c_2 \in C$. As $w - c_1$ and $w - c_2$ each have weight $\leq t$, their difference $c_2 - c_1$ has weight $\leq 2t$, a contradiction, since the nonzero code word $c_2 - c_1$ has weight $\geq 2t + 1$ by hypothesis.

Theorem 11.7.1 below gives a formula for the $q = p^r$ code words of C in terms of the trace function $\mathrm{tr} = \mathrm{tr}_{F_q/F_p}$. We will use this in Theorem 11.7.2 to express the weights of the code words in terms of Gauss sums.

Theorem 11.7.1 *The q words in the code C (defined by (11.7.2) and (11.7.4)) are*

(11.7.5)
$$(\mathrm{tr}(\alpha), \mathrm{tr}(\alpha\gamma^{-f}), \mathrm{tr}(\alpha\gamma^{-2f}), \ldots, \mathrm{tr}(\alpha\gamma^{(1-k)f})), \ \alpha \in F_q.$$

Proof Let

$$h_\alpha(x) = \sum_{j=0}^{k-1} \mathrm{tr}(\alpha\gamma^{-jf}) x^j, \ \alpha \in F_q.$$

For any γ^{fl} occurring in (11.7.4),

$$h_\alpha(\gamma^{fl}) = \sum_{i=0}^{r-1} \alpha^{p^i} \sum_{j=0}^{k-1} \gamma^{(l-p^i)fj} = 0,$$

as the inner sum vanishes. Thus, by (11.7.4), $g \mid h_\alpha$ in $F_p[x]$, i.e.,

$$\overline{h_\alpha}(x) \in A\overline{g}(x) = C, \ \alpha \in F_q.$$

Since C has q elements, it remains to show that the polynomials $h_\alpha(x)$ $(\alpha \in F_q)$ are distinct. Suppose that they are not distinct. Then for some $\alpha \in F_q^*$,

$$\mathrm{tr}(\alpha) = \mathrm{tr}(\alpha\gamma^f) = \mathrm{tr}(\alpha\gamma^{2f}) = \cdots = \mathrm{tr}(\alpha\gamma^{(k-1)f}) = 0.$$

Therefore, for all choices of $a_j \in F_p$,

$$0 = \sum_{j=0}^{r-1} a_j \,\mathrm{tr}(\alpha\gamma^{jf}) = \mathrm{tr}(\alpha \sum_{j=0}^{r-1} a_j\gamma^{jf}).$$

Since γ^f has degree r over F_p by (11.7.3), $\{1, \gamma^f, \gamma^{2f}, \ldots, \gamma^{(r-1)f}\}$ is a basis for F_q over F_p. Thus tr maps F_q to 0, contradicting Theorem 11.4.6(d). ∎

For $\alpha \in F_q$, let $w(\alpha)$ denote the weight of the code word in (11.7.5). Throughout the remainder of this section, we use the notation

(11.7.6) $$d = \gcd(f, (q-1)/(p-1)), \quad \beta_d = \exp(2\pi i/d).$$

The next theorem shows how to evaluate $w(\alpha)$ in terms of Gauss sums $G_r(\chi^i)$, where χ is the character of order d on F_q defined by

(11.7.7) $$\chi(\gamma) = \beta_d.$$

Theorem 11.7.2 *For $\alpha \in F_q^*$, the weight $w(\alpha)$ of the code word in (11.7.5) is*

$$w(\alpha) = \frac{q(p-1)}{pf} - \frac{(p-1)}{pf}\sum_{i=1}^{d-1} \overline{\chi}^i(\alpha) G_r(\chi^i),$$

where χ is defined by (11.7.6) and (11.7.7).

Proof If $f = 1$, then the code word in (11.7.5) has length $q-1$ and weight $w(\alpha) = q(p-1)/p$, since tr is a homomorphism from the group F_q onto F_p whose kernel has q/p elements. Thus the result follows, since $d = 1$ when $f = 1$. Now assume that $f > 1$. Let ψ be the character of order f on F_q such that $\psi^{f/d} = \chi$, and let $m \in F_p^*$, $\alpha \in F_q^*$. Then

$$\sum_{i=1}^{f-1} \overline{\psi}^i(m\alpha) G_r(\psi^i) = \sum_{u \in F_q^*} e(u) \sum_{i=1}^{f-1} \overline{\psi}^i(m\alpha)\psi^i(u).$$

The inner sum equals $f-1$ if $u = m\alpha\gamma^{jf}$ for some j, $0 \le j \le k-1$, and otherwise the inner sum equals -1. Thus

$$\sum_{i=1}^{f-1} \overline{\psi}^i(m\alpha) G_r(\psi^i) = 1 + f \sum_{j=0}^{k-1} e(m\alpha\gamma^{jf}).$$

Summing on m, we obtain

$$\sum_{i=1}^{f-1} \overline{\psi}^i(\alpha) G_r(\psi^i) \sum_{m=1}^{p-1} \overline{\psi}^i(m) = (p-1) + f \sum_{j=0}^{k-1} \sum_{m=1}^{p-1} e(\alpha\gamma^{jf})^m. \tag{11.7.8}$$

The inner sum on the right equals $p-1$ if $\operatorname{tr}(\alpha\gamma^{jf}) = 0$, and the inner sum equals -1, otherwise. Thus the right side of (11.7.8) equals

$$p - 1 - kf + fp(k - w(\alpha)) = q(p-1) - fpw(\alpha). \tag{11.7.9}$$

The inner sum on the left of (11.7.8) equals $p-1$ if the character $(\psi^i)^*$ restricted to F_p is trivial; otherwise, the inner sum vanishes. Thus, by Corollary 11.4.2, the left side of (11.7.8) equals

$$(p-1) \sum_{\substack{i=1 \\ (p-1) \mid ik}}^{f-1} \overline{\psi}^i(\alpha) G_r(\psi^i). \tag{11.7.10}$$

Equating (11.7.9) and (11.7.10), we see that it remains to prove that

$$\sum_{\substack{i=1 \\ (p-1) \mid ik}}^{f-1} \overline{\psi}^i(\alpha) G_r(\psi^i) = \sum_{i=1}^{d-1} \overline{\chi}^i(\alpha) G_r(\chi^i). \tag{11.7.11}$$

Since $(p-1)$ divides ik if and only if f/d divides i, (11.7.11) follows. ∎

The following corollary shows that C is a t-error correcting code for a large value of t, when q is large relative to f.

Corollary 11.7.3 *For $\alpha \in F_q^*$,*

$$w(\alpha) \ge \frac{(p-1)}{pf}(q - (d-1)q^{1/2}).$$

Proof This follows from Theorem 11.7.2 and the fact that $|G_r(\chi^i)| = q^{1/2}$ for each i, $1 \le i \le d-1$. ■

In the next theorem, we examine a class of codes C for which there are only two nonzero weights among the code words, and we give these weights explicitly.

Theorem 11.7.4 *Suppose that $d > 2$ and -1 is a power of p (mod d). Write*

$$p^u \equiv -1 \pmod{d}, \tag{11.7.12}$$

where the positive integer u is taken to be minimal. Then $r = 2ua$ for some integer a, and for $\alpha \in F_q^*$,

$$w(\alpha) = \frac{(p-1)}{pf}(q + (-1)^a q^{1/2}(d-1)),$$

if either
(a) α is a dth power
or
(b) α is a $d/2$th power with d even, $(p^u+1)/d$ odd, and a odd;
and for all other $\alpha \in F_q^*$,

$$w(\alpha) = \frac{(p-1)}{pf}(q - (-1)^a q^{1/2}).$$

Proof By (11.7.12) and Theorem 11.6.2, $2u$ is the order of p (mod d). Since $p^r \equiv 1 \pmod{d}$, there is an integer a for which $r = 2ua$.

Fix i, $1 \le i \le d-1$. By (11.7.7), χ^i has order

$$m := d/\gcd(d,i) > 1.$$

Assume first that $m > 2$. Let t be the minimal positive integer for which

$$p^t \equiv -1 \pmod{m}. \tag{11.7.13}$$

By Theorem 11.6.2, $2t$ is the order of p (mod m). Since $p^{2u} \equiv 1 \pmod{m}$ by (11.7.12), there is an integer b for which $u = bt$. Thus $r = 2st$, where $s = ab$. Raising both sides of (11.7.13) to the bth power, we see from (11.7.12) that b is odd. Thus a and s have the same parity. By Theorem 11.6.3,

$$p^{-r/2}G_r(\chi^i) = (-1)^{a-1},$$

unless d is even, $(p^u+1)/d$ is odd, and a is odd, in which case

$$p^{-r/2}G_r(\chi^i) = (-1)^{a-1+i} = (-1)^i.$$

Next suppose that $m = 2$, so that d is even and $i = d/2$. It can be checked from Theorem 11.5.4 that the formulas for $p^{-r/2}G_r(\chi^i)$ above hold in the case $m = 2$ as well.

The result now easily follows from Theorem 11.7.2. ∎

Example Consider the code C for which $p = 5$, $k = 208$. Then $r = 4$ and $f = d = 3$, and so $u = 1$ and $a = 2$. Theorem 11.7.4 now gives, for $\alpha \in F_{5^4}^*$,

$$w(\alpha) = \begin{cases} \frac{4}{15}(5^4 + 2 \cdot 5^2) = 180, & \text{if } \alpha \text{ is a cube,} \\ \frac{4}{15}(5^4 - 5^2) = 160, & \text{otherwise.} \end{cases}$$

In the next theorem, we find the weights $w(\alpha)$ of the code words of C in the case that $d = 2$.

Theorem 11.7.5 *Suppose that $d = 2$. Then $r = 2s$ for some integer s, and for $\alpha \in F_q^*$,*

$$w(\alpha) = \begin{cases} \dfrac{q(p-1)}{pf} + \dfrac{(p-1)}{pf}\left(\dfrac{-1}{p}\right)^s q^{1/2}, & \text{if } \alpha \text{ is a square,} \\ \dfrac{q(p-1)}{pf} - \dfrac{(p-1)}{pf}\left(\dfrac{-1}{p}\right)^s q^{1/2}, & \text{otherwise.} \end{cases}$$

Proof By Theorem 11.7.2, for $\alpha \in F_q^*$,

$$w(\alpha) = \frac{q(p-1)}{pf} - \frac{(p-1)}{pf}\overline{\chi}(\alpha)G_r(\chi),$$

where χ is the quadratic character on F_q. Since $\gcd(f, (q-1)/(p-1)) = d = 2$, we see that p is an odd prime and r is even, say $r = 2s$. Then by Theorem 11.5.4,

$$G_r(\chi) = -\left(\frac{-1}{p}\right)^s q^{1/2},$$

and the result follows. ∎

Example Consider the code C for which $p = 5$ and $k = 3$. Then $r = 2$, $f = 8$, $d = 2$, and $s = 1$. Theorem 11.7.5 now gives

$$w(\alpha) = \begin{cases} \frac{5}{2} + \frac{1}{2} = 3, & \text{if } \alpha \text{ is a square,} \\ \frac{5}{2} - \frac{1}{2} = 2, & \text{otherwise.} \end{cases}$$

In the next theorem, we find the weights $w(\alpha)$ of the code words of C when $d = 3$. The case where $p \equiv 2 \pmod 3$ is covered by Theorem 11.7.4; hence we assume that $p \equiv 1 \pmod 3$.

Theorem 11.7.6 *Suppose that $d = 3$ and p is a prime $\equiv 1 \pmod 3$. Then $r = 3t$ for some integer t. Define r_3 and s_3 as in Theorem 3.1.3, and write*

$$((-r_3 - is_3\sqrt{3})/2)^t = (-R_3(t) - i\sqrt{3}S_3(t))/2.$$

Then, for $\alpha \in F_q^*$,

$$w(\alpha) = \frac{q(p-1)}{pf} - \frac{(p-1)}{pf}p^t R_3(t), \qquad \textit{if } \alpha \textit{ is a cube,}$$

and

$$w(\alpha) = \frac{q(p-1)}{pf} + \frac{(p-1)}{2pf}p^t(R_3(t) \pm 3S_3(t)), \qquad \textit{if } \alpha \textit{ is not a cube,}$$

where the plus sign is taken for exactly half of the noncubic values of α.

Proof By Theorem 11.7.2, for $\alpha \in F_q^*$,

$$w(\alpha) = \frac{q(p-1)}{pf} - \frac{2(p-1)}{pf}\text{Re}\big(\overline{\chi}(\alpha)G_r(\chi)\big),$$

where χ is a cubic character on F_q. Since $\gcd(f, (q-1)/(p-1)) = d = 3$ and $p \equiv 1 \pmod 3$, we must have $r \equiv (q-1)/(p-1) \equiv 0 \pmod 3$, so that $r = 3t$ for some integer t. By Theorem 11.5.2 (the Davenport–Hasse lifting theorem),

$$G_r(\chi) = G_{3t}(\chi) = -(-G^3(\chi))^t.$$

Then by Lemma 4.1.1 and Theorem 3.1.3,

$$G_r(\chi) = -(-p(r_3 + is_3\sqrt{3})/2)^t = p^t(R_3(t) + iS_3(t)\sqrt{3})/2.$$

Since $\chi(\alpha) = (-1 \pm i\sqrt{3})/2$ when α is not a cube, and $\chi(\alpha) = 1$ when α is a cube, the result follows. ∎

Example Consider the code C for which $p = 7$ and $k = 19$. Then $r = 3$, $f = 18$, $d = 3$, and $t = 1$. Also, $r_3 = 1$ and $s_3 = \pm 3$. Theorem 11.7.6 now gives, for $\alpha \in F_{7^3}^*$,

$$\begin{aligned} w(\alpha) &= \frac{49}{3} - \frac{1}{3} = 16, \text{ for the 114 cubes } \alpha, \\ w(\alpha) &= \frac{49}{3} + \frac{5}{3} = 18, \text{ for 114 noncubes } \alpha, \end{aligned}$$

and

$$w(\alpha) = \frac{49}{3} - \frac{4}{3} = 15, \text{ for the other 114 noncubes } \alpha.$$

It is not difficult to prove analogues of Theorem 11.7.6 for other small values of d; see Problem 15.

Generally speaking, the smaller the degree of the algebraic number field containing the Gauss sums $G_r(\chi^i)$ in Theorem 11.7.2, the easier it is to compute these sums explicitly. Since χ has order d, these sums are in $Q(\beta_p, \beta_d)$, where

$$\beta_d = e^{2\pi i/d}, \quad \beta_p = e^{2\pi i/p};$$

however, the following lemma shows that these sums are actually in the smaller field $Q(\beta_d)$.

Lemma 11.7.7 *The Gauss sums $G_r(\chi^i)$ in Theorem* 11.7.2 *are in $Q(\beta_d)$.*

Proof For an integer m, $1 \leq m \leq p-1$, define $\tau_m \in \mathrm{Gal}(Q(\beta_p, \beta_d)/Q(\beta_d))$ by $\tau_m(\beta_p) = \beta_p^m$. We must show that τ_m fixes $G_r(\chi^i)$. We have

$$\tau_m(G_r(\chi^i)) = \sum_{u \in F_q^*} \chi^i(u)e(um) = \overline{\chi}^i(m)G_r(\chi^i).$$

Hence it suffices to show that $\chi(m) = 1$. This follows because the restriction χ^* of χ to F_p is trivial, by Corollary 11.4.2, as χ has order $d = \gcd(f, (q-1)/(p-1))$. ∎

We finish this section by focusing on a particular choice of f and p for which the Gauss sums $G_r(\chi^j)$ in Lemma 11.7.7 are in a quadratic extension of Q (so that we will be able to evaluate them). Take $f = c^n$, where n is a positive integer and $c > 3$ is a prime congruent to 3 (mod 4). Let R_f denote the reduced residue system (mod f). Then R_f is cyclic, with generator h, say. For each integer m with $1 \leq m \leq n$, h generates the reduced residue system (mod c^m), and the even powers of h are the squares in a reduced residue system (mod c^m). Choose a prime p with $p \equiv h^2 \pmod{f}$. Since $c > 3$, $p \not\equiv 1 \pmod{c}$, so that $d = f = c^n$. Let $< p >$ denote the cyclic subgroup of $R_f = R_d$ generated by p. The elements of $< p >$ are the $\phi(d)/2$ squares in R_d. Since $c \equiv 3 \pmod 4$, R_d is the union of the two cosets $< p >$ and $- < p >$. Further, as $< p >$ has index 2 in R_d, the fixed field of the automorphism $\sigma_p : \beta_d \longrightarrow \beta_d^p$ in $\mathrm{Gal}(Q(\beta_d)/Q)$ is a quadratic extension of Q. The quadratic Gauss sum (mod c) equals $i\sqrt{c}$ by Theorem 1.2.4, and this Gauss sum is fixed by σ_p, since p is a square (mod c). Thus $Q(i\sqrt{c})$ is the fixed field of σ_p. By (11.2.7), σ_p fixes the Gauss sum $G_r(\chi^j)$ (which is in $Q(\beta_d)$ by Lemma 11.7.7); thus

(11.7.14) $$G_r(\chi^j) \in Q(i\sqrt{c}), \quad 1 \leq j \leq d-1.$$

For each integer m with $1 \leq m \leq n$, write

(11.7.15) $$\psi_m = \chi^{c^{n-m}}.$$

Thus ψ_m is a character on F_q of order c^m. We proceed to evaluate the Gauss sum $G_r(\psi_m) \in Q(i\sqrt{c})$ for each m, $1 \le m \le n$; the result is stated in Theorem 11.7.8.

Fix m, $1 \le m \le n$, and let $\nu = \nu_m = \phi(c^m)/2$, so that ν is the order of $p \pmod{c^m}$. Since $p^r \equiv 1 \pmod{c^m}$, $\nu \mid r$. The character ψ_m on F_{p^r} is the lift of a character μ_m on F_{p^ν}, by Theorem 11.4.4(c). Write

$$(p^\nu - 1)/c^m = \sum_{i=0}^{\nu-1} d_i p^i, \quad 0 \le d_i \le p-1,$$

so that

$$(c^m - 1)(p^\nu - 1)/c^m = \sum_{i=0}^{\nu-1} (p - 1 - d_i)p^i.$$

Since the reduced residue system $\pmod{c^m}$ is the union of its cosets $< p >$ and $- < p >$, it thus follows from (11.1.3) and Theorem 11.2.6 that the highest integral power of p dividing $G_\nu(\mu_m)$ is p^z, where

$$z = z_m = \frac{1}{p-1} \min \left(\sum_{i=0}^{\nu-1} d_i, \ \sum_{i=0}^{\nu-1} (p - 1 - d_i) \right).$$

If the two sums on i just above were equal, we would have $z = \nu/2$, which contradicts the fact that $\nu = \phi(c^m)/2$ is odd. Thus $z < \nu/2$.

It is a rather deep fact that

$$z = z_m = \frac{1}{p-1} \sum_{i=0}^{\nu-1} d_i, \tag{11.7.16}$$

where the d_i are the p-adic digits of $(p^\nu - 1)/c^m$. To prove (11.7.16), it suffices to show that

$$\frac{1}{p-1} \sum_{i=0}^{\nu-1} d_i < \nu/2 = \phi(c^m)/4.$$

In view of Theorem 11.2.7, this is equivalent to

$$\sum_{i=0}^{\nu-1} L(p^i) < c^m \phi(c^m)/4,$$

where $L(p^i)$ denotes the least positive integer congruent to $p^i \pmod{c^m}$. Thus (11.7.16) is equivalent to the assertion that the sum of the squares $\pmod{c^m}$ strictly between 0 and c^m is $< c^m\phi(c^m)/4$. Let $s_1, \ldots, s_{(c-1)/2}$ denote the squares $\pmod{c}$ between 0 and c. Then the sum of the squares $\pmod{c^m}$ between 0 and c^m is

$$\sum_{j=0}^{c^{m-1}-1} \ \sum_{i=1}^{(c-1)/2} (s_i + jc) = c^{m-1} \sum_{i=1}^{(c-1)/2} s_i + \frac{c^m(c-1)(c^{m-1}-1)}{4}.$$

Thus to prove (11.7.16), it suffices to show that

$$\sum_{i=1}^{(c-1)/2} s_i < \frac{c(c-1)}{4}.$$

Since

$$\sum_{i=1}^{(c-1)/2} s_i - \sum_{i=1}^{(c-1)/2} (c - s_i) = 2\sum_{i=1}^{(c-1)/2} s_i - \frac{c(c-1)}{2},$$

and the $c - s_i$ are the nonsquares (mod c) between 0 and c, it suffices to show that

$$\sum_{j=1}^{c-1} j\left(\frac{j}{c}\right) < 0.$$

A short proof of this inequality has been given by Berndt [3, Theorem 3.1]. Alternatively, this follows from Dirichlet's class number formula for the field $K_c = Q(i\sqrt{c})$, which states that for each prime $c \equiv 3 \pmod 4$,

$$-\frac{1}{c}\sum_{j=1}^{c-1} j\left(\frac{j}{c}\right) = \frac{2h}{w},$$

where h is the class number of K_c and w is the number of roots of unity in K_c. (We note in passing that Stark [1] has given a simple proof showing that the left side of this formula is the value of a Dirichlet L-function at 0, namely, $L(0,\chi)$, where $\chi(n) = \left(\frac{n}{c}\right)$.) The proof of (11.7.16) is now complete.

By the Davenport–Hasse lifting theorem (Theorem 11.5.2), $p^{zr/\nu}$ is the highest power of p dividing $G_r(\psi_m)$. Since $G_r(\psi_m) \in Q(i\sqrt{c})$ by (11.7.14), we can write

(11.7.17) $$G_r(\psi_m) = p^{zr/\nu}(a_m + ib_m\sqrt{c})/2,$$

where a_m and b_m are integers of the same parity, and $p \nmid (a_m + ib_m\sqrt{c})/2$. Taking the square of the modulus of each side, we obtain

(11.7.18) $$p^{(\nu-2z)(r/\nu)} = (a_m^2 + cb_m^2)/4.$$

The left side is divisible by p, since $z < \nu/2$. Thus, since $p \nmid (a_m + ib_m\sqrt{c})/2$, we can have neither $a_m/2 \equiv 0 \pmod p$ nor $b_m/2 \equiv 0 \pmod p$. Since $c > 3$, it follows from Lemma 3.0.1 that for each fixed m with $1 \le m \le n$, (11.7.18) determines the integers a_m and b_m uniquely up to sign. We can determine the sign of a_m in (11.7.17) by the congruence

(11.7.19) $$a_m \equiv -2p^{-zr/\nu} \pmod c,$$

since for $\lambda = 1 - \exp(2\pi i/c^m)$,

$$G_r(\psi_m) = \sum_{u\in F_q^*} \psi_m(u)e(u) \equiv \sum_{u\in F_q^*} e(u) = -1 \pmod \lambda,$$

and λ has norm c by Theorem 2.1.9.

If $a \in R_d$ is a square $\pmod{c^m}$, then $G_r(\psi_m^a) = G_r(\psi_m)$, since $a \in \langle p \rangle$ and $G_r(\psi_m^p) = G_r(\psi_m)$ by (11.2.7). If $b \in R_d$ is a nonsquare $\pmod{c^n}$, then $G_r(\psi_m^b) = \overline{G_r(\psi_m)}$, since $b \in -\langle p \rangle$. We have therefore proved the following theorem.

Theorem 11.7.8 *Let p be a prime, and let r be the order of $p \pmod{k}$, where $k > 1$. Assume that $q = p^r = kf + 1$ with $f = c^n$, where $n \geq 1$ and $c > 3$ is a prime such that $c \equiv 3 \pmod 4$ and p is congruent $\pmod{c^n}$ to the square of a primitive root of c^n. Fix m, $1 \leq m \leq n$, and let ψ be any character of order c^m on F_q. Let $\nu = \nu_m = \phi(c^m)/2$, define d by (11.7.6), and define $z = z_m$ by (11.7.16). Then $d = f$ and*

$$G_r(\psi) = p^{zr/\nu}(a_m + ib_m\sqrt{c})/2,$$

where a_m and b_m are integers of the same parity such that neither $a_m/2 \equiv 0 \pmod p$ nor $b_m/2 \equiv 0 \pmod p$, and a_m and b_m are determined uniquely up to sign by (11.7.18). The sign of b_m depends on the choice ψ, but the sign of a_m is uniquely determined by (11.7.19).

Example Take $p = 2$ and $k = 9$. Then $r = 6$ and $q = 64 = kf + 1$ with $f = c = d = 7$, $n = 1$, $\nu_1 = \phi(7)/2 = 3$, and $z_1 = 1$. Further, by (11.7.18), $16 = 4p^2 = a_1^2 + 7b_1^2$, so that $|a_1| = 3$, $|b_1| = 1$. By (11.7.19),

$$a_1 \equiv -2p^{-2} \equiv -\frac{1}{2} \equiv 3 \pmod 7,$$

so that $a_1 = 3$, $b_1 = \pm 1$. Thus, for a character ψ of order 7 on F_{2^6},

$$G_6(\psi) = 2(3 \pm i\sqrt{7}).$$

In the next theorem, we give the weights $w(\alpha)$ of the code words (11.7.5) of C when C is the irreducible cyclic code defined by (11.7.4) with r, f, k, q, n, and p as given in Theorem 11.7.8. It turns out that the code words of C can assume at most $2n+1$ nonzero weights. We will use the following notation. For $\alpha \in F_q^*$, let $M = M_\alpha$ be the maximal integer with $0 \leq M \leq n$ such that α is a c^Mth power in F_q^*. Note that, for each fixed integer N with $0 \leq N \leq n$, there are $k\phi(c^{n-N})$ values of $\alpha \in F_q^*$ for which $M_\alpha = N$. For brevity, write

$$H_m := p^{z_m r/\nu_m}, \tag{11.7.20}$$

where ν_m and z_m are defined in Theorem 11.7.8.

Theorem 11.7.9 *Let p be a prime, and let r be the order of $p \pmod{k}$, where $k > 1$. Assume that $q = p^r = kf + 1$ with $f = c^n$, where n is a positive integer and*

$c > 3$ is a prime such that $c \equiv 3 \pmod 4$ and p is congruent $\pmod{c^n}$ to the square of some primitive root of c^n. Then for each $\alpha \in F_q^*$,

$$w(\alpha) = \frac{q(p-1)}{pf} - \frac{(p-1)}{2pf}\left\{\sum_{m=0}^{N} c^m(a_m H_m - a_{m+1}H_{m+1}) \pm b_{N+1}H_{N+1}c^{N+1}\right\},$$

where $N = M_\alpha$. For $N < n$, the plus sign is taken for half of the $k\phi(c^{n-N})$ values of α for which $M_\alpha = N$, and the minus sign is taken for the other half. For $1 \le m \le n$, a_m and $|b_m|$ are uniquely determined by (11.7.18) and (11.7.19), and $a_m = b_m = 0$ for all other m. In particular, $w(\alpha)$ assumes at most $2n+1$ nonzero values as α runs through F_q^*.

Proof Theorem 11.7.2 gives, for $\alpha \in F_q^*$,

$$w(\alpha) = \frac{q(p-1)}{pf} - \frac{(p-1)}{pf}\sum_{i=1}^{f-1} \overline{\chi}^i(\alpha)G_r(\chi^i),$$

where χ is a character of order f on F_q. Then with ψ_m defined by (11.7.15),

$$w(\alpha) = \frac{q(p-1)}{pf} - \frac{(p-1)}{pf}\sum_{m=1}^{n} \sum_{\substack{j=1 \\ \gcd(j,c)=1}}^{c^m-1} \overline{\psi}_m^j(\alpha)G_r(\psi_m^j). \tag{11.7.21}$$

Let ρ_m denote the quadratic Dirichlet character $\pmod{c^m}$ defined on the integers by

$$\rho_m(a) = \begin{cases} 0, & \text{if } c \mid a, \\ 1, & \text{if } a \text{ is a square } \pmod{c^m},\ c \nmid a, \\ -1, & \text{if } a \text{ is a nonsquare } \pmod{c^m},\ c \nmid a. \end{cases}$$

By Theorem 11.7.8 and the paragraph preceding it, (11.7.21) becomes

$$w(\alpha) = \frac{q(p-1)}{pf} - \frac{(p-1)}{2pf}\sum_{m=1}^{n} H_m \sum_{\substack{j=1 \\ \gcd(j,c)=1}}^{c^m-1} \overline{\psi}_m^j(\alpha)(a_m + i\rho_m(j)b_m\sqrt{c}), \tag{11.7.22}$$

where a_m and $|b_m|$ are uniquely determined by (11.7.18) and (11.7.19). Let $S_m(\alpha)$ denote the inner sum on j in (11.7.22), so that

$$S_m(\alpha) = a_m \sum_{\substack{j=1 \\ \gcd(j,c)=1}}^{c^m-1} \overline{\psi}_m^j(\alpha) + ib_m\sqrt{c}\sum_{j=1}^{c^m-1} \rho_m(j)\overline{\psi}_m^j(\alpha).$$

Write $M_\alpha = N$, for fixed N, $0 \le N \le n$. If $N < n$, α has the form $\alpha = \gamma^{bc^N}$, where $b = b_\alpha$ is one of the $k\phi(c^{n-N})$ integers for which $1 \le b \le kc^{n-N}$, $c \nmid b$. Thus, for any $\alpha \in F_q^*$

$$S_m(\alpha) = a_m U_m(\alpha) + \epsilon_\alpha i b_m \sqrt{c} V_m(\alpha), \tag{11.7.23}$$

where

$$U_m(\alpha) = \sum_{\substack{j=1 \\ \gcd(j,c)=1}}^{c^m-1} \exp(2\pi i j c^N/c^m), \tag{11.7.24}$$

$$V_m(\alpha) = \sum_{j=1}^{c^m-1} \rho_m(j)\exp(2\pi i j c^N/c^m), \tag{11.7.25}$$

and $\epsilon_\alpha = \rho_m(b_\alpha) = \pm 1$. In fact, for each fixed N with $0 \le N < n$, $\epsilon_\alpha = 1$ for $k\phi(c^{n-N})/2$ of the values of $\alpha \in F_q^*$ for which $M_\alpha = N$, and $\epsilon_\alpha = -1$ for the other $k\phi(c^{n-N})/2$ values of $\alpha \in F_q^*$ for which $M_\alpha = N$.

By (1.6.4) and (11.7.24),

$$U_m(\alpha) = \begin{cases} c^m - c^{m-1}, & \text{if } m \le N, \\ -c^N, & \text{if } m = N+1, \\ 0, & \text{otherwise.} \end{cases} \tag{11.7.26}$$

Now, the quadratic character ρ_m (mod c^m) has conductor c, so that, by (1.6.5) and (11.7.25),

$$V_m(\alpha) = \begin{cases} c^{m-1}\sum_{b=0}^{c-1} \left(\frac{b}{c}\right)\exp(2\pi i b c^{N+1-m}/c), & \text{if } m \le N+1, \\ 0, & \text{otherwise.} \end{cases}$$

Thus, by Theorem 1.2.4,

$$V_m(\alpha) = \begin{cases} c^N(i\sqrt{c}), & \text{if } m = N+1, \\ 0, & \text{otherwise.} \end{cases} \tag{11.7.27}$$

By (11.7.26),

$$\sum_{m=1}^{n} H_m a_m U_m(\alpha) = \sum_{m=1}^{N} H_m a_m (c^m - c^{m-1}) - H_{N+1} a_{N+1} c_N = \sum_{m=0}^{N} c^m (a_m H_m - a_{m+1} H_{m+1}),$$

where we define $a_0 = a_{n+1} = 0$. By (11.7.27),

$$\sum_{m=1}^{n} H_m b_m V_m(\alpha) = H_{N+1} b_{N+1} c^N (i\sqrt{c}),$$

where we define $b_{n+1} = 0$. Therefore, (11.7.23) gives

$$\sum_{m=1}^{n} H_m S_m(\alpha) = \sum_{m=0}^{N} c^m (a_m H_m - a_{m+1} H_{m+1}) - \epsilon_\alpha b_{N+1} H_{N+1} c^{N+1}.$$

The result now follows from (11.7.22). ∎

Corollary 11.7.10 *Let p be a prime and let r be the order of $p \pmod k$, where $k > 1$. Assume that $q = p^r = kf + 1$ with $f = c$, where $c > 3$ is a prime $\equiv 3 \pmod 4$ such that p is the square of a primitive root $\pmod c$. Let $\alpha \in F_q^*$. If α is a cth power, then*

$$w(\alpha) = \frac{q(p-1)}{pc} + \frac{(p-1)}{2pc} a_1 H_1 (1-c).$$

For the remaining $q - 1 - k$ values of $\alpha \in F_q^*$,

$$w(\alpha) = \frac{q(p-1)}{pc} + \frac{(p-1)}{2pc}(a_1 H_1 \pm b_1 H_1 c),$$

where the plus sign is taken for half of these $q - 1 - k$ values of α, and the minus sign is taken for the other half.

Proof This is the case $n = 1$ of Theorem 11.7.9. ∎

Example Take $p = 2$, $k = 9$. Then $r = 6$, $f = c = d = 7$, $n = 1$, $q = 64$, $\nu_1 = 3$, $z_1 = 1$, $H_1 = 4$, $a_1 = 3$, and $b_1 = \pm 1$. If $\alpha \in F_{64}^*$ is a seventh power, then

$$w(\alpha) = \frac{64}{14} - \frac{3 \cdot 4 \cdot 6}{28} = 2.$$

Of the remaining 54 values of $\alpha \in F_{64}^*$,

$$w(\alpha) = \frac{64}{14} + \frac{3 \cdot 4 + 4 \cdot 7}{28} = 6$$

for 27 of these α, and

$$w(\alpha) = \frac{64}{14} + \frac{3 \cdot 4 - 4 \cdot 7}{28} = 4$$

for the other 27.

Exercises 11

1. Show that the special case $k = q-1$ of Stickelberger's theorem (Theorem 11.2.1) implies the theorem for general k.

Hint: Let $K' = Q(e^{2\pi i/(q-1)})$ and $M' = K'(\beta_p)$. Let $\wp'$ be a prime ideal dividing $\wp O_{M'}$ and let $P' = \wp' \cap O_{K'}$, so that $\wp' \cap O_M = \wp$ and $P' \cap O_K = P$. Show that $O_{K'}/P' = \{\alpha + P' : \alpha \in O_K/P\}$ and that the character $\chi_{P'}$ of order $q-1$ on $O_{K'}/P'$ satisfies

$$\chi_{P'}^f(\alpha + P') = \chi_P(\alpha + P) \quad (\alpha \in O_K),$$

so that

$$G_r(\chi_{P'}^{-af}) = G_r(\chi_P^{-a}) \in O_M.$$

2. In the notation of Section 11.1, let $p = 11$ and $k = 48$, so that $r = 4$. Show that $J_4(\chi_p, \chi_p^2)O_K = P_1^3 P_5^4 P_{13}$ and $pO_K = P_1 P_5 P_{13} P_{17}$.

3. Let $K = Q(\beta)$ for $\beta = e^{2\pi i/k}$, where k is an integer > 2. Let p be a prime not dividing k, and let r denote the order of $p \pmod k$. We saw in (2.1.11) that $P \mid K_r(\chi_p)$, where P is a prime ideal of O_K dividing p. Show that in the case $r = 2$, $P \parallel K_2(\chi_p)$ if and only if $L(p) > k/2$, where $L(p)$ denotes the least positive residue of $p \pmod k$.

4. If ρ is the quadratic character on F_q, show that $\rho(2) = (-1)^{(q^2-1)/8}$.

5. Determine all odd prime powers q such that the quadratic character ρ on F_q satisfies $\rho(-3) = 1$.

6. Let $q = p^r$, where p is an odd prime and r is even. Let ρ and χ be characters on F_q of orders 2 and 3, respectively. If $p^{r/2} \equiv 5 \pmod 6$, prove that $J_r(\rho, \chi) = p^{r/2}$.

7. Let $\chi_1, \ldots, \chi_k$ be characters on F_q, not all of which are trivial. Suppose that $\chi_1, \ldots, \chi_k$ are lifted to characters $\chi_1', \ldots, \chi_k'$, respectively, of the extension field E of degree s over F_q. Prove that, if $q = p^r$,

$$J_{rs}(\chi_1', \ldots, \chi_k') = (-1)^{(s-1)(k-1)} J_r(\chi_1, \ldots, \chi_k)^s.$$

(This generalizes Corollary 11.5.3.)

8. Let χ be a character on F_q, where $q = p^r$ for an odd prime p. Let ρ be the quadratic character on F_q. Prove that

$$J_r(\chi, \chi\rho) = \overline{\chi}(4) J_r(\chi^2, \rho) = \chi(4) J_r(\chi^2, \chi^2).$$

9. Let p be a prime, and let χ be a character on F_q, where $q = p^r \equiv 1 \pmod 3$. Let ψ be a cubic character on F_q. Prove that

$$J_r(\chi, \chi\psi) = \overline{\chi}^3(-3) J_r(\chi^3, \overline{\chi}\psi).$$

10. Let $q = p^r$ for a prime p, and suppose that χ is a character of order k on F_q, where k is a prime power. Show that $G_r(\chi)$ is pure if and only if -1 is a power of $p \pmod k$.

11. Define

$$((x)) = \begin{cases} x - [x] - 1/2, & \text{if } x \notin Z, \\ 0, & \text{if } x \in Z. \end{cases}$$

Let p be a prime, and let k and r be positive integers such that $p^r \equiv 1 \pmod k$. Show that -1 is a power of $p \pmod k$ if and only if

$$\sum_{v=1}^{r} ((bp^v/k)) = 0 \quad \text{for all } b \in Z.$$

12. Let $q = p^r$ for a prime p, and suppose that χ is a nontrivial character of order k on F_q. Suppose that $G_r(\chi)$ is pure. Show that $2(q-1)/(k(p-1))$ is an integer with the same parity as r, and use this to show that the restriction χ^* to F_p is trivial or quadratic according as r is even or odd.

13. Consider the irreducible cyclic code C defined by (11.7.4) with $p = 2$, $k = 19065$, $r = 20$, $q = 2^{20}$, $f = 55$. Show that $w(\alpha) = 9472$ for $30k$ values of $\alpha \in F^*_{2^{20}}$, $w(\alpha) = 9600$ for $24k$ values of α, and $w(\alpha) = 9728$ for k values of α, where $w(\alpha)$ denotes the weight of the code word in (11.7.5).

14. If $w(\alpha)$ denotes the weight of the code word in (11.7.5), show that

$$\sum_{\alpha \in F_q} w(\alpha) = (p-1)kq/p.$$

15. Prove analogues of Theorem 11.7.6 for primes $p \equiv 1 \pmod{d}$ in the cases $d = 4$, $d = 6$, and $d = 8$.

Hint: Use Problem 1 of Exercises 2.

16. Extend Theorem 11.7.9 to include the case $d = 3^n$, where n is a positive integer. (See Theorem 11.7.6 for the case $n = 1$.)

17. Let $q = p^r$ for a prime p, and let $a, b \in F_q^*$. For a character χ on F_q, define the Kloosterman sum

$$K_r(\chi; a, b) = \sum_{\alpha \in F_q^*} \chi(\alpha) e(a\alpha + b/\alpha),$$

where as usual, $e(\alpha) = \exp(2\pi i \mathrm{tr}(\alpha)/p)$. (If a or b were 0, $K_r(\chi; a, b)$ would reduce to a Gauss sum.) Show that for the quadratic character ρ on F_q with q odd, $K_r(\rho; a, b)$ is related to the quadratic Gauss sum $G_r(\rho)$ (evaluated in Theorem 11.5.4) by

$$K_r(\rho; a, b) = \begin{cases} \rho(a) G_r(\rho)(e(2c) + e(-2c)), & \text{if } c = \sqrt{ab} \in F_q^*, \\ 0, & \text{if } \rho(ab) = -1. \end{cases}$$

Hint:

$$\begin{aligned} G_r(\rho) K_r(\rho; a, b) &= \sum_{y \in F_q} \rho(y) e(y) \sum_{x \in F_q^*} \rho(x) e(ax + b/x) \\ &= \sum_{y \in F_q} e(y) \sum_{x \in F_q^*} \rho(x) \rho(y - ax - b/x) = \sum_{y \in F_q} e(y) \sum_{x \in F_q} \rho(-ax^2 + yx - b). \end{aligned}$$

The sum on x can be evaluated from Theorem 2.1.2.

18. Let $N > 1$ be an odd number, and let a, b be integers coprime with N. Show that the Salié sum over the ring of integers (mod N) defined by

$$S(a,b,N) = \sum_{\substack{x (\mathrm{mod}\ N) \\ \gcd(x,N)=1}} \left(\frac{x}{N}\right) e^{2\pi i(ax+b/x)/N}$$

satisfies $S(a,b,N) = \left(\frac{a}{N}\right) S(1,ab,N)$. Then prove the evaluations $S(1,a,N) = 0$, if $\left(\frac{a}{N}\right) = -1$, and

$$S(1,a^2,N) = i^{(N-1)^2/4}\sqrt{N} \sum_{\substack{x (\mathrm{mod}\ N) \\ x^2 \equiv 1 (\mathrm{mod}\ N)}} e^{4\pi i x a/N}.$$

Show that this agrees with Problem 17 in the case that N equals an odd prime p.

Notes on Chapter 11

Stickelberger's congruence for Gauss sums (Theorem 11.2.1) was proved by Stickelberger [1] in 1890. For other proofs, see Coates [1], Conrad [1], Gras [1], Lang [1, Chapter 1], C. J. Moreno [3], and Yamamoto [2]. Conrad's proof makes use of Problem 13 of Exercises 9, and shows that Theorem 11.2.1 is equivalent to the congruence in Problem 22 of Exercises 10. The prime ideal factorization of Gauss sums (Theorem 11.2.2) is due to Stickelberger [1]. The prime ideal factorization of Jacobi sums (Theorem 11.2.3) is, of course, a consequence of Stickelberger's factorization of Gauss sums, but Mitchell [1] gave a different derivation in 1916, based on the work of Kummer. The factorizations of Gauss and Jacobi sums had already been given in special cases by Kummer [5], [6] in the mid nineteenth century. The prime ideal factorization of Gauss sums over F_p has been deduced by Brinkhuis [1] as a consequence of four basic properties of Gauss sums.

Theorem 11.2.5 was stated by Carlitz [21, eq. (4)], who used it to evaluate the quadratic Gauss sum over F_p (Theorem 1.2.4).

The Davenport–Hasse product formula over F_q (Theorem 11.3.5) was first stated and proved in 1934 by Davenport and Hasse in their classic paper [1, eq. (0.9)]. Their proof made use of Stickelberger's theorem (Theorem 11.2.1). Remarkably, a special case of the Davenport–Hasse product formula over F_p had already been stated without proof by Jacobi [3], [CW, p. 257] in 1837. The simplest nontrivial case of Jacobi's assertion (the case l=2 of Theorem 11.3.5 for prime q) was proved by Mitchell, whose proof is given in Dickson [8, p. 407]. A simpler proof for l=2 is given in Hasse [2, p. 465]. A proof of the Davenport–Hasse product formula over F_p using only arithmetic results has been given by Huard, Spearman, and Williams [1]. Other proofs of this formula over F_p may be found in Yamamoto [2] and Jakubec [4]. For other proofs of the general formula over F_q, see Gras [1] and Lang [1, Chapter 2]. Theorem 11.3.1 shows that the Davenport–Hasse product formula follows elementarily for each l from an elementary proof for all primes dividing l. This has been observed by Gras [1, p. 23]. Consequently, since an elementary proof exists for $l = 2$ (Theorem 2.1.4), there is an elementary proof whenever l is a power of 2. Our proof of Theorem 11.3.1 follows that in Berndt and Evans [2, p. 435]. A proof of the case $l = 3$ over F_p has been given by Greene and Stanton [1] using hypergeometric series over F_p.

Hasse [2, p. 465] conjectured that Theorem 1.1.4(a) and Theorem 11.3.5 are essentially the only multiplicative relations connecting Gauss sums over F_p. Yamamoto [2] proved Hasse's conjecture when the Gauss sums are viewed as ideals, and he gave a simple example to show it is false when the Gauss sums are considered as numbers. Other counterexamples have been given by Muskat [6] and Muskat and Whiteman [1]. Yamamoto [4] showed that any relation

$$\prod_{a=1}^{k-1} G(\chi^a)^{2c_a} = \eta^2,$$

where χ is a character (mod p) of order $k > 2$, $c_a \in Z$, and η is a unit, is a direct consequence of Theorem 1.1.4(a) and Theorem 11.3.5.

The notion of the lift of a character can be found in Davenport and Hasse's paper [1, eq. (0.7)]. The relationship between a Gauss sum and its lift (Theorem 11.5.2) is due to Davenport and Hasse [1, eq. (0.8)]. Other proofs of their theorem on lifted Gauss sums can be found, for example, in the books by Andrews, Askey, and Roy [1], Ireland and Rosen [1, Chapter 11], Jungnickel [2, p. 283], Lang [1, Chapter 1], Lidl and Niederreiter [1, Chapter 5], C. J. Moreno [3, Chapter 4], W. Schmidt [1, Chapter 2], and Small [1, Chapter 7]. Also, elementary arguments proving Theorem 11.5.2 have been given by McEliece and Rumsey [1], Schmid [1], and Stepanov [1]. Our proof of Theorem 11.5.2 via Theorem 11.5.1 follows that of Lidl and Niederreiter [1, Chapter 5]. The ideas behind Theorem 11.5.1 are due to Weil [4]; for directions in which they might lead, see Niederreiter [1]. The counterpart of Theorem 11.5.2 for Jacobi sums over F_q (Corollary 11.5.3) was given by Mitchell [1, p. 176] in 1916. Counterparts for other character sums such as Kloosterman sums have been given by Autuore [1] and Curtis and Shinoda [1].

The evaluation of the quadratic Gauss sum over F_q (Theorem 11.5.4) may be found in the books of Jungnickel [2, p. 286] and Lidl and Niederreiter [1, p. 199].

Theorem 11.6.1 is due to Stickelberger [1, p. 340]. Other proofs of this theorem may be found in Carlitz [14], Jungnickel [2, p. 290], and Lidl and Niederreiter [1, p. 202]. Theorem 11.6.4 is due to Baumert, Mills, and Ward [1], and the proof we have given is due to Evans [8]. Problems 10–12 are taken from Evans [8]; see Theorem 2, Theorem 1, and eqs. (8), (9) of [8], respectively. See also Evans [1] for related results on pure Gauss and Jacobi sums. Gouvêa and Yui [1, Remark 1.11] have given an example of a pure generalized Jacobi sum of order 9. Their example can be justified with the aid of Problem 9. Akiyama [1] used the prime ideal factorization of Gauss sums (Theorem

11.2.2) to prove that the Jacobi sum $J_2(\chi, \psi)$ with ψ of order 2 and χ of order $n \geq 3$ on F_{p^2} satisfies $J_2(\chi, \psi) = \pm p$ (and is hence pure), if either (a) $n|(p+1)$, (b) $2(p-1)/n$ is an odd integer, (c) $n = 24$ with $p \equiv 17$ or 19 (mod 24), or (d) n =60 with $p \equiv 41$ or 49 (mod 60). Moreover, Akiyama proved that if none of (a)–(d) holds, then $J_2(\chi, \psi)$ is not pure. For related work which makes use of the Gross–Koblitz formula (11.2.12), see the paper of Shiratani and Yamada [1].

Section 11.7 is based largely on a paper of McEliece [1]. However, McEliece considered only the case n =1 of Theorems 11.7.8 and 11.7.9. Theorem 11.7.8 for general n was proved by Langevin [3]. A version of these theorems, where $p = 2$ and f is a product of two primes p_1, p_2 with $\gcd(p_1 - 1, p_2 - 1)$=2, has been proved by van der Vlugt [1]. Problem 13 is taken from that paper.

Langevin and Zanotti [1] used Gauss sums to determine those irreducible cyclic codes over finite fields which have the same number of code words for each nonzero weight of the code. Langevin [2] used explicit evaluations of Gauss sums of prime order over finite fields to construct 2–weight and 3–weight irreducible cyclic codes. Lachaud [1] used Theorem 11.6.1 to give examples of projective algebraic curves of genus g over F_q for which the number of rational points attains the upper bound $1 + q + g[2\sqrt{q}]$ of Weil–Serre. For example, if p is a prime with $p = a^2 + b^2 \equiv 1 \pmod 4$, $a \equiv 3 \pmod 4$, and $b^2 \leq a$, then the curve $x^4 + y^4 = (\frac{2}{p})$ has $p + 1 + 3[2\sqrt{p}]$ F_p-rational points. Wolfmann [1] gave similar examples where the Weil–Serre bound is attained by curves associated with irreducible cyclic codes.

The sums over finite fields and rings of the type defined in Problems 17 and 18 were considered by Kloosterman [1] in his work on positive integers represented by $ax^2 + by^2 + cz^2 + dt^2$ for positive integers a, b, c, d, and by Salié [1]. The sums $K_r(\rho; a, b)$ and $S(a, b, N)$ in Problems 17 and 18 are called Salié sums. Problem 18 is taken from Sarnak's book [1, p. 91]. Simple evaluations of some basic Kloosterman and Salié sums have been given by Williams [4], [6], [7], [9]. See also Estermann [2], Carlitz [18], [25], S. Chowla [17], and Whiteman [1], [4]. A formula for the Kloosterman sum $K_r(\chi; a, b)$ in Problem 17 for cubic χ has been given by Duke and Iwaniec [1]; this is a character sum analogue of a classical formula for Airy's integral involving the Bessel function $K_{1/3}$. Using the Davenport–Hasse product formula (Theorem 11.3.5), Duke [1] has evaluated an n-dimensional Salié sum over F_q which reduces for $n = 2$ to the evaluation of $K_r(\rho; a, b)$ given in Problem 17. For applications of Kloosterman sums $K_r(\chi; a, b)$ in graph theory, see Evans [18, §5], [19].

R. Smith [1] related Gauss sums to multiple Kloosterman sums, and established uniform upper bounds for these multiple sums. Fisher [1], using finite Fourier transforms of Kloosterman sums, obtained relations between Gauss sums over finite fields of small characteristic. With the aid of Stickelberger's theorem on Gauss sums, Wan [4] has improved some of the results in Fisher's work [1], [2] on the distinctness and the distribution of Kloosterman sums. Curtis and Shinoda [1] have expressed Kloosterman sums as values of Dickson polynomials (defined in (13.1.2)). For a sheaf–theoretic approach to Gauss and Kloosterman sums, see Katz [1]. For surveys giving overviews on

Kloosterman sums, see Dabrowski [1] and Hurt [1].

In Problem 17, we evaluated a Kloosterman sum in terms of Gauss sums. Many other types of character sums can be evaluated in terms of Gauss and Jacobi sums; see, for example, Evans [7], [16], [18], [20], Greene [1]–[3], Greene and Stanton [2], and van Wamelen [1]. See also the references in Evans [18].

There is a vast literature on exponential sums and character sums, which of course we can scarcely touch on. Two short classic papers in this subject are those of Weil [1], [2]. A few other papers on such sums are Adolphson and Sperber [1], Berndt [1]–[3], Carlitz [17], [19], [23], [26]–[30], S. Chowla [15], [16], Chung [1], Lahtonen [1], Leprévost and Morain [1], Li [2], Niederreiter [1], Rajwade [1], Rajwade and Parnami [1], R. Smith [2], Stepanov [2], Williams [5], [8], and Yamauchi [1].

We proceed to discuss further references on Gauss and Jacobi sums over F_q, particularly those connected to the theorems of Stickelberger and Davenport–Hasse. Weil [5] used Stickelberger's factorization of Gauss sums to construct Hecke characters (see Hecke [1]) for $Q(e^{2\pi i/m})$ from Jacobi sums $J_r(\chi_P^{a_1}, \ldots, \chi_P^{a_t})$, where χ_P has order m. Weil [5] showed that such Hecke characters have conductor dividing m^2. For the precise conductors of these Hecke characters for any prime power m, see Miki [2] for odd m and Miki [4] for even m. In 1974, Weil [7] constructed Hecke characters for arbitrary Abelian extensions of Q from Gauss sums. This construction was extended by Kubert and Lichtenbaum [1]. See also Kubert [1]. The conductors of general Hecke characters constructed from Gauss sums are determined in Abe [1].

The prime ideal factorization of Gauss sums (Theorem 11.2.2) yields the Stickelberger ideal in the group ring $Z[\mathrm{Gal}(K/Q)]$, where K is an Abelian extension of Q. The Stickelberger ideal annihilates the ideal class group of K; see Brinkhuis [1] and Washington [1, Chapter 6]. There is extensive literature on Stickelberger elements and ideals, as may be seen, for example, from Washington's book [1].

The Brumer–Stark conjecture generalizes Stickelberger's theorem. Sands [1] used Stickelberger elements and Jacobi sum Hecke characters to prove the Brumer–Stark conjecture for field extensions K/k, where K is a finite Abelian extension of Q.

The Gross–Koblitz formula (11.2.12) is a deep generalization of Stickelberger's theorem. Young [2, p. 130] has used the Gross–Koblitz formula to obtain congruences relating Jacobi sums over finite fields to quotients of multinomial coefficients. These congruences are employed to obtain p-adic analogues of classical summation theorems for hypergeometric series such as Kummer's evaluation of ${}_2F_1(-1)$. Ishibashi, Sato, and Shiratani [1] have used the Gross–Koblitz formula to determine the Hasse invariant of the elliptic curve $y^2 = x^3 + ax + b$ over F_p in terms of ${}_2F_1$ hypergeometric series over F_p.

C.–G. Schmidt gave a simpler proof of a theorem of Gross and Koblitz [1] on the value of the Frobenius automorphism on certain algebraic products of gamma functions.

Gurak [1] used the prime ideal factorization of Gauss sums to determine explicitly in certain cases the relative norms of the Gauss sums defined in (11.2.6).

Lakkis [1] computed a product of Gauss sums $\prod_{\chi \text{ on } G/H} \tau(\chi)$, where H is a subgroup of finite index in $G = (O_K/P)^*$, for a finite extension K of the p-adic rationals with associated maximal ideal P, $P \nmid 2$.

Let $T : F_{p^{2ab}} \to F_p$ denote the trace map, and let $p^a \equiv -1 \pmod{k}$. For fixed $h \in F_p$, Han [1], using theorems of Stickelberger and Davenport–Hasse, computed explicitly the number of solutions $y \in F_{p^{ab}}$ to $T(y^k) = h$. This proves and generalizes a conjecture of O. Moreno [1]. The result has applications in algebraic coding theory.

For product identities for Gauss sums, see Coleman [1], Boyarsky [1], Evans [7], Kubert [1], and Kubert and Lichtenbaum [1]. Some of these formulas (see, e.g., Coleman [1, Theorem 11]) generalize the Davenport–Hasse product formula (Theorem 11.3.5).

Neumann [4] used the factorization of Gauss sums to prove the Kronecker–Weber theorem without class field theory.

Brinkhuis [2] used Gauss sums to examine the existence of normal integral bases for the ring of integers of $Q\left(e^{2\pi i/p}\right)$ over the ring of integers of a subfield of index n, where p is an odd prime $\equiv 1 \pmod{n}$. For example, if $n = 4$, no normal integral basis exists.

Miki [1, Theorems 1, 2] has given explicit formulas for the l-adic expansions of certain Gauss and Jacobi sums of order l^n, where l is prime. The formula for Jacobi sums has several important applications, and it generalizes congruences for Jacobi sums given by Iwasawa [1, Theorem 1] and Ihara [1, Corollary to Theorem 7], when $l > 2$.

Uehara [1] gave l-adic congruences connecting cyclotomic units with Gauss and Jacobi sums over finite fields, and gave several applications to algebraic number theory.

Rubin [1] gave an exposition on Kolyvagin's study of the χ-component of the p-part of the ideal class group of $Q\left(e^{2\pi i/p}\right)$, which utilizes the prime ideal factorization of Gauss sums in Abelian extensions of $Q\left(e^{2\pi i/p}\right)$.

Chapter 12

Eisenstein Sums

Let p be a prime, r a positive integer, and $q = p^r$. Write tr for the trace map from F_q to F_p. Let χ be a character on F_q. The restriction of χ to F_p will be denoted by χ^*. In particular, if χ is the trivial character on F_q, then χ^* is the trivial character on F_p. In this chapter we evaluate Eisenstein sums, which are defined as follows.

Definition *The Eisenstein sum* $E_r(\chi)$ *is defined by*

$$E_r(\chi) = \sum_{\substack{\alpha \in F_q^* \\ \mathrm{tr}(\alpha) = 1}} \chi(\alpha).$$

We begin by noting some of the elementary properties of the sum $E_r(\chi)$.

Lemma 12.0.1 (a) $E_r(\chi) = 1$, *if* $r = 1$.
(b) $E_r(\chi) = p^{r-1}$, *if* χ *is trivial.*

Proof (a) If $r = 1$, then

$$E_r(\chi) = \chi(1) = 1.$$

(b) For $a \in F_p$ define the set T_a by $T_a = \{\alpha \in F_q \mid \mathrm{tr}(\alpha) = a\}$. By Proposition 1.1.1, tr is a homomorphism from the group F_q onto the group F_p, and so the kernel T_0 has $q/p = p^{r-1}$ elements. Since $T_1 = \{\alpha + t : t \in T_0\}$ for any fixed $\alpha \in T_1$, it follows that T_1 has p^{r-1} elements as well, and so $E_r(\chi) = p^{r-1}$. ∎

Our next result relates the sum $\sum_{\substack{\alpha \in F_q^* \\ \mathrm{tr}(\alpha) = a}} \chi(\alpha)$ to the Eisenstein sum $E_r(\chi)$ for each $a \in F_p$.

Lemma 12.0.2 *For* $a \in F_p$,

$$\sum_{\substack{\alpha \in F_q^* \\ \mathrm{tr}(\alpha) = a}} \chi(\alpha) = \begin{cases} \chi(a)E_r(\chi), & \text{if } a \in F_p^*, \\ p^{r-1} - 1, & \text{if } a = 0 \text{ and } \chi \text{ is trivial}, \\ 0, & \text{if } a = 0 \text{ and both } \chi \text{ and } \chi^* \text{ are nontrivial}, \\ -(p-1)E_r(\chi), & \text{if } a = 0,\ \chi \text{ is nontrivial, and } \chi^* \text{ is trivial}. \end{cases}$$

Proof If $a \in F_p^*$, then the linear mapping $\alpha \longrightarrow a\alpha$ is a bijection on F_q^*, and we obtain

$$\sum_{\substack{\alpha \in F_q^* \\ \mathrm{tr}(\alpha) = a}} \chi(\alpha) = \sum_{\substack{\alpha \in F_q^* \\ \mathrm{tr}(a\alpha) = a}} \chi(a\alpha) = \chi(a) \sum_{\substack{\alpha \in F_q^* \\ \mathrm{tr}(\alpha) = 1}} \chi(\alpha) = \chi(a)E_r(\chi),$$

by Proposition 1.1.1(c). Now suppose that $a = 0$. Then

$$\begin{aligned}\sum_{\substack{\alpha \in F_q^* \\ \operatorname{tr}(\alpha)=0}} \chi(\alpha) &= \sum_{a \in F_p} \sum_{\substack{\alpha \in F_q^* \\ \operatorname{tr}(\alpha)=a}} \chi(\alpha) - \sum_{a \in F_p^*} \sum_{\substack{\alpha \in F_q^* \\ \operatorname{tr}(\alpha)=a}} \chi(\alpha) \\ &= \sum_{\alpha \in F_q^*} \chi(\alpha) - \sum_{a \in F_p^*} \chi(a) E_r(\chi) = \sum_{\alpha \in F_q^*} \chi(\alpha) - E_r(\chi) \sum_{a \in F_p^*} \chi^*(a) \\ &= \begin{cases} (q-1) - p^{r-1}(p-1), & \text{if } \chi \text{ is trivial}, \\ 0 - E_r(\chi)0, & \text{if both } \chi \text{ and } \chi^* \text{ are nontrivial}, \\ 0 - E_r(\chi)(p-1), & \text{if } \chi \text{ is nontrivial and } \chi^* \text{ is trivial}, \end{cases}\end{aligned}$$

which establishes the asserted values. ■

Lemma 12.0.3 $E_r(\chi^p) = E_r(\chi)$.

Proof As $\alpha \longrightarrow \alpha^p$ is a bijection on F_q^* and $\operatorname{tr}(\alpha) = \operatorname{tr}(\alpha^p)$,

$$E_r(\chi^p) = \sum_{\substack{\alpha \in F_q^* \\ \operatorname{tr}(\alpha)=1}} \chi^p(\alpha) = \sum_{\substack{\alpha \in F_q^* \\ \operatorname{tr}(\alpha^p)=1}} \chi(\alpha^p) = \sum_{\substack{\alpha \in F_q^* \\ \operatorname{tr}(\alpha)=1}} \chi(\alpha) = E_r(\chi).$$ ■

If the character χ on F_q is of order k, then the Eisenstein sum $E_r(\chi)$ is said to be of order k. The next lemma shows that Eisenstein sums $E_r(\chi)$ of order k are integers in the field $Q(e^{2\pi i/k})$.

Lemma 12.0.4 *If χ is of order k then $E_r(\chi) \in O_K$, where $K = Q(\beta)$ and $\beta = \exp(2\pi i/k)$.*

Proof If χ has order k, then each term of $E_r(\chi)$ is a power of β, and so $E_r(\chi) \in O_K$. ■

A main goal of this chapter is to determine explicitly Eisenstein sums of small order k. In order to do this, the necessary theory of such sums is developed in Sections 12.1 and 12.2. Then the evaluations of $E_r(\chi)$ for $k = 3, 4, 5, 6$, and 8 are carried out in Sections 12.3, 12.4, 12.5, 12.6, and 12.7, respectively. When $k = 7$ the Eisenstein sum $E_r(\chi)$ is determined for primes $p \equiv 2$ or $4 \pmod 7$ and positive integers $r \equiv 0 \pmod 3$ in Section 12.8. These evaluations for $k = 7$ and 8 are used in Section 12.9, in conjunction with Stickelberger's theorem, to prove three congruences of Eisenstein for binomial coefficients given in Theorems 12.9.7, 12.9.8, and 12.9.9. In Section 12.10, we express the Gauss sum $g_r(k)$ defined in Section 1.1 as a linear combination of Eisenstein sums, and apply this formula to evaluate $g_r(3)$ and $g_r(4)$. Also in Section 12.10, we briefly discuss f-nomial periods associated with F_{p^r} (the case $r = 1$ was discussed in Section 10.10). In Sections 12.11–12.13, we evaluate certain Eisenstein sums of orders 20, 16, and 12, respectively. These evaluations will be useful in the determination of Brewer sums in Chapter 13.

12.1 Properties of the Eisenstein sum $E_r(\chi)$

In this section we explore the relationship between Eisenstein and Gauss sums over F_q, where $q = p^r$. We show that an Eisenstein sum is generally a quotient of two Gauss sums.

Theorem 12.1.1 *Let χ be a nontrivial character on F_q. Let χ^* be the restriction of χ to F_p. Then*

$$E_r(\chi) = \begin{cases} G_r(\chi)/G(\chi^*), & \text{if } \chi^* \text{ is nontrivial,} \\ -G_r(\chi)/p, & \text{if } \chi^* \text{ is trivial.} \end{cases}$$

Proof For $\alpha \in F_q$, write $e(\alpha) = \exp(2\pi i \text{tr}(\alpha)/p)$. As χ is nontrivial,

$$G_r(\chi) = \sum_{\alpha\in F_q^*} \chi(\alpha)e(\alpha) = \sum_{a\in F_p} \sum_{\substack{\alpha \in F_q^* \\ \text{tr}(\alpha) = a}} \chi(\alpha)e(\alpha) = \sum_{\substack{\alpha \in F_q^* \\ \text{tr}(\alpha) = 0}} \chi(\alpha) + \sum_{a\in F_p^*} \sum_{\substack{\alpha \in F_q^* \\ \text{tr}(\alpha) = a}} \chi(\alpha)e(\alpha).$$

By Lemma 12.0.2,

$$\sum_{\substack{\alpha \in F_q^* \\ \text{tr}(\alpha) = 0}} \chi(\alpha) = \begin{cases} 0, & \text{if } \chi^* \text{ is nontrivial,} \\ -(p-1)E_r(\chi), & \text{if } \chi^* \text{ is trivial.} \end{cases}$$

Also, by Lemma 12.0.2,

$$\sum_{a\in F_p^*} \sum_{\substack{\alpha \in F_q^* \\ \text{tr}(\alpha) = a}} \chi(\alpha)e(\alpha) = \sum_{a\in F_p^*} \exp(2\pi i a/p) \sum_{\substack{\alpha \in F_q^* \\ \text{tr}(\alpha) = a}} \chi(\alpha) = \sum_{a\in F_p^*} \exp(2\pi i a/p)\chi(a)E_r(\chi)$$

$$= E_r(\chi) \sum_{a\in F_p^*} \chi^*(a) \exp(2\pi i a/p) = E_r(\chi)(G(\chi^*) - \chi^*(0)).$$

Thus, if χ^* is nontrivial, as $\chi^*(0) = 0$,

$$G_r(\chi) = E_r(\chi)G(\chi^*),$$

whereas, if χ^* is trivial, then since $\chi^*(0) = 1$ and $G(\chi^*) = 0$,

$$G_r(\chi) = -(p-1)E_r(\chi) + (0-1)E_r(\chi) = -pE_r(\chi).$$ ■

We next determine the value of the product $E_r(\chi)E_r(\overline{\chi})$.

Theorem 12.1.2 *Let χ be a nontrivial character on F_q. Let χ^* be the restriction of χ to F_p. Then*

$$E_r(\chi)E_r(\overline{\chi}) = \begin{cases} p^{r-1}, & \text{if } \chi^* \text{ is nontrivial,} \\ p^{r-2}, & \text{if } \chi^* \text{ is trivial.} \end{cases}$$

Proof As χ is nontrivial, Theorem 1.1.4(a) gives

$$G_r(\chi)G_r(\overline{\chi}) = \chi(-1)p^r.$$

Thus, if χ^* is the trivial character on F_p, then, by Theorem 12.1.1,

$$E_r(\chi)E_r(\overline{\chi}) = G_r(\chi)G_r(\overline{\chi})/p^2 = \chi(-1)p^{r-2} = p^{r-2}.$$

On the other hand, if χ^* is nontrivial on F_p, then, by Theorem 12.1.1,

$$E_r(\chi)E_r(\overline{\chi}) = \frac{G_r(\chi)G_r(\overline{\chi})}{G(\chi^*)G(\overline{\chi^*})} = \frac{\chi(-1)p^r}{\chi(-1)p} = p^{r-1}. \qquad \blacksquare$$

We next use the Davenport–Hasse theorem on lifted Gauss sums (Theorem 11.5.2) to give us valuable information about Eisenstein sums. Suppose that χ is a nontrivial character on F_q of order k so that $k \mid p^r - 1$. Let l be the least positive integer such that $k \mid p^l - 1$. Thus $l \mid r$ and F_{p^r} contains the subfield F_{p^l}. As $k \mid p^l - 1$, Theorem 11.4.4(c) shows that the character χ is equal to the lift of some character ψ on F_{p^l}, that is, $\psi' = \chi$. By Theorem 11.4.4(b), the restriction of the character $\psi' = \chi$ to F_{p^l} is the character $\psi^{r/l}$. By the Davenport–Hasse theorem on lifted Gauss sums, with $s = r/l$,

$$G_r(\chi) = G_r(\psi') = (-1)^{s-1}G_l^s(\psi).$$

The restriction of χ to F_p is χ^*, the restriction of ψ to F_p is ψ^*, and $\chi^* = (\psi^s)^* = (\psi^*)^s$. If ψ^* is trivial, then χ^* is trivial, and, by Theorem 12.1.1,

$$E_r(\chi) = -G_r(\chi)/p, \quad E_l(\psi) = -G_l(\psi)/p,$$

so that

$$E_r(\chi) = -G_r(\chi)/p = (-1)^s G_l^s(\psi)/p = (-1)^s(-pE_l(\psi))^s/p = p^{s-1}E_l^s(\psi).$$

If ψ^* is nontrivial but χ^* is trivial, then, by Theorem 12.1.1,

$$E_r(\chi) = -G_r(\chi)/p, \quad E_l(\psi) = G_l(\psi)/G(\psi^*),$$

so that

$$E_r(\chi) = -G_r(\chi)/p = (-1)^s G_l^s(\psi)/p = (-1)^s G^s(\psi^*)E_l^s(\psi)/p.$$

If χ^* is nontrivial, then ψ^* is nontrivial, and Theorem 12.1.1 gives

$$E_r(\chi) = G_r(\chi)/G(\chi^*), \quad E_l(\psi) = G_l(\psi)/G(\psi^*),$$

so that

$$E_r(\chi) = G_r(\chi)/G(\chi^*) = (-1)^{s-1}G_l^s(\psi)/G(\chi^*) = (-1)^{s-1}(G^s(\psi^*)/G(\psi^{*s}))E_l^s(\psi).$$

We have thus proved the following result.

Theorem 12.1.3 *Let χ be a nontrivial character of order k on F_{p^r}. Let l be the least positive integer such that $k \mid p^l - 1$. Then $l \mid r$, χ is the lift of some character ψ on F_{p^l}, and, for $s = r/l$,*

$$\frac{E_r(\chi)}{E_l^s(\psi)} = \begin{cases} p^{s-1}, & \text{if } \psi^* \text{ is trivial,} \\ (-1)^s G^s(\psi^*)/p, & \text{if } \psi^* \text{ is nontrivial and } \chi^* \text{ is trivial,} \\ (-1)^{s-1} G^s(\psi^*)/G(\psi^{*s}), & \text{if } \chi^* \text{ is nontrivial,} \end{cases}$$

where ψ^* and χ^* are the restrictions of ψ and χ to F_p, respectively.

An important special case of Theorem 12.1.3 occurs when k divides $p-1$.

Corollary 12.1.4 *Let $p \equiv 1 \pmod{k}$. Let χ be a nontrivial character of order k on F_{p^r}, which is the lift of a character ψ on F_p, i.e., $\psi' = \chi$. Then*

$$E_r(\chi) = \begin{cases} (-1)^r G^r(\psi)/p, & \text{if } \psi^r \text{ is trivial,} \\ (-1)^{r-1} G^r(\psi)/G(\psi^r), & \text{if } \psi^r \text{ is nontrivial.} \end{cases}$$

Proof As $p \equiv 1 \pmod{k}$, we have $l = 1$, $\psi = \psi^*$, and $\chi^* = \psi^r$. By Lemma 12.0.1(a), $E_1(\psi) = 1$. Since χ is nontrivial, so is ψ. The result now follows from Theorem 12.1.3. ■

Our next theorem gives the value of the Eisenstein sum of order 2.

Theorem 12.1.5 *Let χ be the quadratic character on F_q, where $q = p^r$, p is an odd prime, and r is a positive integer. Then*

$$E_r(\chi) = \begin{cases} (-1)^{(p-1)r/4} p^{(r/2)-1}, & \text{if } r \text{ is even,} \\ (-1)^{(p-1)(r-1)/4} p^{(r-1)/2}, & \text{if } r \text{ is odd.} \end{cases}$$

Proof Since χ is the lift of the quadratic character ρ on F_p, Corollary 12.1.4 gives

$$E_r(\chi) = \begin{cases} G^r(\rho)/p, & \text{if } r \text{ is even,} \\ G^{r-1}(\rho), & \text{if } r \text{ is odd.} \end{cases}$$

By Theorem 1.2.4,

$$G(\rho) = i^{(p-1)^2/4} p^{1/2}.$$

The asserted result now follows as

$$i^{(p-1)^2 r/4} = (-1)^{(p-1)^2 r/8} = (-1)^{(p-1)r/4}, \quad \text{if } r \text{ is even,}$$

and

$$i^{(p-1)^2(r-1)/4} = (-1)^{(p-1)^2(r-1)/8} = (-1)^{(p-1)(r-1)/4}, \quad \text{if } r \text{ is odd.}$$ ■

In the next theorem we evaluate the Eisenstein sum $E_r(\chi)$ when the order k of the character χ divides $p^s + 1$ for some integer s. In view of Theorem 12.1.5 we may assume that $k \geq 3$.

Theorem 12.1.6 *Let χ be a character of order $k(\geq 3)$ on F_q, where $q = p^r$, p is a prime, and r is a positive integer. Suppose that $p^l \equiv -1 \pmod{k}$ for some positive integer l, which we take to be minimal. Then $r \equiv 0 \pmod{2l}$ and*

$$E_r(\chi) = (-1)^{r(p^l+k+1)/(2lk)} p^{(r/2)-1}, \quad \text{if } p > 2,$$

while

$$E_r(\chi) = (-1)^{r/(2l)} p^{(r/2)-1}, \quad \text{if } p = 2.$$

Proof Let n be the least positive integer such that $p^n \equiv 1 \pmod{k}$. By Theorem 11.6.2, $n = 2l$, so that $2l \mid r$. By Theorem 12.1.3, χ is the lift of some character ψ on $F_{p^{2l}}$. Moreover, by Theorem 11.4.4(e), χ and ψ have the same order, namely, k. Let ψ^* be the restriction of ψ to F_p. Since $k \mid (p^l + 1)$, we see that $k \mid (p^l + 1)(p^l - 1)/(p - 1)$, that is $k \mid (p^{2l} - 1)/(p - 1)$. Hence, by Corollary 11.4.2, the character ψ^* is trivial. Then, by Theorem 12.1.3,

$$E_r(\chi) = p^{r/(2l)-1} E_{2l}^{r/(2l)}(\psi).$$

Next, by Theorem 12.1.1,

$$E_{2l}(\psi) = -G_{2l}(\psi)/p,$$

so that

$$E_r(\chi) = (-1)^{r/(2l)} G_{2l}^{r/(2l)}(\psi)/p.$$

But, by Theorem 11.6.3,

$$G_{2l}(\psi) = \begin{cases} p^l, & \text{if } p = 2, \\ (-1)^{(p^l+1)/k} p^l, & \text{if } p > 2, \end{cases}$$

and the asserted result follows. ∎

We next give a useful congruence for a sum of Eisenstein sums.

Theorem 12.1.7 *Let χ be a character of order k on F_{p^r}, where r is a prime. Then $\sum_{j=0}^{k-1} E_r(\chi^j)$ is a rational integer, which satisfies the congruence*

$$\sum_{j=0}^{k-1} E_r(\chi^j) \equiv \begin{cases} k \pmod{kr}, & \text{if } \chi(r) = 1, \\ 0 \pmod{kr}, & \text{if } \chi(r) \neq 1. \end{cases}$$

Proof Write $q = p^r$ and define

$$N_1 = \sum_{\substack{\alpha \in F_p \\ \mathrm{tr}(\alpha) = 1 \\ \chi(\alpha) = 1}} 1, \quad N_2 = \sum_{\substack{\alpha \in F_q, \alpha \notin F_p \\ \mathrm{tr}(\alpha) = 1 \\ \chi(\alpha) = 1}} 1.$$

Suppose $\alpha \in F_p$ is such that $\mathrm{tr}(\alpha) = 1$ and $\chi(\alpha) = 1$. The former equality implies $r\alpha = 1$. The latter equality then gives

$$1 = \chi(1) = \chi(r\alpha) = \chi(r)\chi(\alpha) = \chi(r).$$

Hence

$$N_1 = \begin{cases} 1, & \text{if } \chi(r) = 1, \\ 0, & \text{if } \chi(r) \neq 1. \end{cases}$$

Now suppose that $\alpha \in F_q - F_p$ is such that $\mathrm{tr}(\alpha) = 1$ and $\chi(\alpha) = 1$. Then, for $j = 0, 1, \ldots, r-1$, $\alpha^{p^j} \in F_q - F_p$, $\mathrm{tr}(\alpha^{p^j}) = \mathrm{tr}(\alpha) = 1$, and $\chi(\alpha^{p^j}) = \chi^{p^j}(\alpha) = 1$. As r is prime, the elements $\alpha, \alpha^p, \ldots, \alpha^{p^{r-1}}$ of $F_q - F_p$ are distinct, so that $N_2 \equiv 0 \pmod{r}$. Since

$$\sum_{j=0}^{k-1} E_r(\chi^j) = \sum_{j=0}^{k-1} \sum_{\substack{\alpha \in F_q \\ \mathrm{tr}(\alpha) = 1}} \chi^j(\alpha) = \sum_{\substack{\alpha \in F_q \\ \mathrm{tr}(\alpha) = 1}} \sum_{j=0}^{k-1} \chi^j(\alpha)$$

and

$$\sum_{j=0}^{k-1} \chi^j(\alpha) = \begin{cases} k, & \text{if } \chi(\alpha) = 1, \\ 0, & \text{if } \chi(\alpha) \neq 1, \end{cases}$$

we have

$$\sum_{j=0}^{k-1} E_r(\chi^j) = k \sum_{\substack{\alpha \in F_q \\ \mathrm{tr}(\alpha) = 1 \\ \chi(\alpha) = 1}} 1 = k(N_1 + N_2) \equiv \begin{cases} k \pmod{kr}, & \text{if } \chi(r) = 1, \\ 0 \pmod{kr}, & \text{if } \chi(r) \neq 1, \end{cases}$$

as asserted. ■

We now show how Theorem 12.1.7 can be used to deduce the law of quadratic reciprocity. Let ρ be the quadratic character on F_{p^r}, where p and r are distinct odd primes. By Theorem 12.1.5,

$$E_r(\rho) = (-1)^{(p-1)(r-1)/4} p^{(r-1)/2},$$

and, by Lemma 12.0.1(b),

$$E_r(\rho^0) = p^{r-1}.$$

Hence, by Theorem 12.1.7,

$$p^{r-1} + (-1)^{(p-1)(r-1)/4} p^{(r-1)/2} = E_r(\rho^0) + E_r(\rho) \equiv \begin{cases} 2 \pmod{2r}, & \text{if } \rho(r) = 1, \\ 0 \pmod{2r}, & \text{if } \rho(r) = -1. \end{cases}$$

Now $p^{r-1} \equiv 1 \pmod{r}$, $p^{(r-1)/2} \equiv \left(\frac{p}{r}\right) \pmod{r}$, and $\rho(r) = \left(\frac{r}{p}\right)$, so that

$$1 + (-1)^{(p-1)(r-1)/4} \left(\frac{p}{r}\right) \equiv 1 + \left(\frac{r}{p}\right) \pmod{r},$$

and thus

$$\left(\frac{r}{p}\right) = \left(\frac{p}{r}\right)(-1)^{(p-1)(r-1)/4},$$

which is the law of quadratic reciprocity.

12.2 An Eisenstein sum over F_{p^2}

Let p be an odd prime. Let χ be a character on F_{p^2} of order $k \geq 1$. Fix a generator γ of the cyclic group $F_{p^2}^*$, and set $\tau = \gamma^{(p+1)/2}$ and $g = \tau^2 = \gamma^{p+1}$. We observe that g is a primitive root (mod p) and that $\tau^p = -\tau$. Since $\chi^{p-1}(g) = \chi(g^{p-1}) = \chi(1) = 1$, we have $\chi^p(g) = \chi(g)$. Also $F_{p^2} = \{a + b\tau \mid a, b \in F_p\}$. The restriction of χ to F_p is denoted by χ^*. We may assume that $\chi(\gamma) = \beta$, where $\beta = e^{2\pi i/k}$, so that

$$\chi^*(g) = \chi(g) = \chi(\gamma^{p+1}) = \chi^{p+1}(\gamma) = \beta^{p+1}.$$

Thus χ^* is trivial if and only if $k \mid p+1$, and χ^* is quadratic if $k \mid 2(p+1)$ but $k \nmid p+1$.

Berndt and Evans [2, §2] defined the sum

$$E(\chi) = \sum_{b=0}^{p-1} \chi(1 + b\tau). \tag{12.2.1}$$

This is a slight variant of $E_2(\chi)$, as the next theorem shows. The purpose of this section is to examine the properties of $E(\chi)$.

Theorem 12.2.1 *If χ is a character on F_{p^2}, then*

$$E(\chi) = \chi(2)E_2(\chi).$$

Proof Let a, $b \in F_p$. Since

$$\mathrm{tr}(a + b\tau) = (a + b\tau) + (a + b\tau)^p = (a + b\tau) + (a - b\tau) = 2a,$$

it follows from Lemma 12.0.2 with $a = 2$ that

$$E(\chi) = \sum_{b=0}^{p-1} \chi(1 + b\tau) = \sum_{\mathrm{tr}(\alpha)=2} \chi(\alpha) = \chi(2)E_2(\chi).$$ ∎

We now use the results of Section 12.1 to derive some properties of the sum $E(\chi)$.

Theorem 12.2.2 *If χ is a character on F_{p^2}, then*

$$E(\chi^p) = E(\chi).$$

Proof By Lemma 12.0.3 and Theorem 12.2.1,

$$E(\chi^p) = \chi^p(2)E_2(\chi^p) = \chi(2)E_2(\chi) = E(\chi). \qquad \blacksquare$$

Theorem 12.2.3 *Let χ be a nontrivial character on F_{p^2} of order k, where $k \mid p+1$. Then*

$$E_2(\chi) = E(\chi) = -(-1)^{(p+1)/k}.$$

Proof As χ is nontrivial, $k > 1$. By Theorem 12.1.5 for $k = 2$ and Theorem 12.1.6 for $k \geq 3$,

$$E_2(\chi) = -(-1)^{(p+1)/k}.$$

Since χ^* is trivial, $\chi(2) = \chi^*(2) = 1$. Hence, by Theorem 12.2.1,

$$E(\chi) = \chi(2)E_2(\chi) = E_2(\chi) = -(-1)^{(p+1)/k}. \qquad \blacksquare$$

Theorem 12.2.4 *Let χ be a nontrivial character on F_{p^2} of order k. Then*

$$G_2(\chi) = \begin{cases} \overline{\chi}(2)E(\chi)G(\chi^*), & \text{if } k \nmid p+1, \\ (-1)^{(p+1)/k}p, & \text{if } k \mid p+1. \end{cases}$$

Proof If $k \nmid p+1$, then χ^* is nontrivial, and, by Theorems 12.1.1 and 12.2.1,

$$G_2(\chi) = E_2(\chi)G(\chi^*) = \overline{\chi}(2)E(\chi)G(\chi^*).$$

If $k \mid p+1$, then χ^* is trivial, and, by Theorems 12.1.1, 12.2.1 and 12.2.3,

$$G_2(\chi) = -pE_2(\chi) = -\overline{\chi}(2)pE(\chi) = -\overline{\chi^*}(2)p(-(-1)^{(p+1)/k}) = (-1)^{(p+1)/k}p. \qquad \blacksquare$$

Theorem 12.2.5 *Let χ be a character on F_{p^2} of order $k = 2n$ with $k \nmid p+1$. Then*

$$E(\chi^{n-1}) = (-1)^{(p+1)/2}\frac{E(\chi)}{E(\chi^2)}K(\chi^*).$$

Proof We begin by observing that $(\chi^*)^n$ is trivial, since

$$\chi^n(g) = \chi^n(\gamma^{p+1}) = \chi^{n(p+1)}(\gamma) = \beta^{n(p+1)} = (-1)^{p+1} = 1.$$

Thus, by Theorem 12.2.1, it suffices to prove the result with E_2 in place of E. Let $R = E_2(\chi^{n-1})E_2(\chi^2)/E_2(\chi)$.

Suppose that $n \nmid p+1$. By Theorem 12.2.4,

$$E_2(\chi) = G_2(\chi)/G(\chi^*), \qquad E_2(\chi^2) = G_2(\chi^2)/G(\chi^{*2}),$$

and

$$E_2(\chi^{n-1}) = G_2(\chi^{n-1})/G(\chi^{*n-1}).$$

Thus

$$\begin{aligned}
R &= \frac{G_2(\chi^{n-1})}{G(\chi^{*n-1})} \cdot \frac{G_2(\chi^2)}{G(\chi^{*2})} \cdot \frac{G(\chi^*)}{G_2(\chi)} = \frac{G_2(\chi^{n-1})G_2(\chi^2)}{G_2(\chi)} \cdot \frac{G(\chi^*)}{G(\chi^{*n-1})G(\chi^{*2})} \\
&= \frac{G_2(\chi^{n-1})G_2(\chi^{n+1})G_2(\chi^2)}{G_2(\chi)G_2(\chi^{n+1})} \cdot \frac{G(\chi^*)G(\chi^*)}{G(\chi^{*n-1})G(\chi^*)G(\chi^{*2})} \\
&= \frac{\chi^{n-1}(-1)p^2G_2(\chi^2)}{G_2(\chi)G_2(\chi^{n+1})} \frac{G^2(\chi^*)}{\chi^{*n-1}(-1)pG(\chi^{*2})} \\
&= p\frac{G_2(\chi^2)}{G_2(\chi)G_2(\chi^{n+1})}J(\chi^*,\chi^*) = \overline{\chi}^2(2)p\frac{G_2(\chi^2)}{G_2(\chi)G_2(\chi^{n+1})}K(\chi^*).
\end{aligned}$$

By Theorem 2.1.4,

$$\chi(4)J_2(\chi,\chi) = K_2(\chi) = J_2(\chi,\rho) = J_2(\chi,\chi^n),$$

so that, by Theorem 2.1.3(a),

$$\chi^2(2)\frac{G_2^2(\chi)}{G_2(\chi^2)} = \frac{G_2(\chi)G_2(\chi^n)}{G_2(\chi^{n+1})},$$

and thus, by Theorem 11.5.4,

$$\frac{G_2(\chi^2)}{G_2(\chi)G_2(\chi^{n+1})} = \frac{\chi^2(2)}{G_2(\chi^n)} = \frac{\chi^2(2)}{p\,(-1)^{(p+1)/2}}.$$

Hence

$$R = (-1)^{(p+1)/2}K(\chi^*),$$

as desired.

Now suppose that $n \mid p+1$. As $2n \nmid p+1$, we have $p+1 \equiv n \pmod{2n}$. Thus $\chi^{n-1} = \chi^p$, and so $E_2(\chi^{n-1}) = E_2(\chi)$ by Lemma 12.0.3. Hence

$$R = \frac{E_2(\chi^{n-1})E_2(\chi^2)}{E_2(\chi)} = E_2(\chi^2) = -(-1)^{(p+1)/n} = 1,$$

by Theorem 12.2.3. The restriction character χ^* has order $\dfrac{k}{\gcd(k,p+1)} = \dfrac{2n}{\gcd(2n,n)} =$ 2, and so, by Theorem 2.1.1(c),

$$K(\chi^*) = \chi^*(4)J(\chi^*,\chi^*) = J(\chi^*,\chi^*) = -\chi^*(-1) = -(-1)^{(p-1)/2} = (-1)^{(p+1)/2}.$$

Hence

$$(-1)^{(p+1)/2}\frac{E_2(\chi)}{E_2(\chi^2)}K(\chi^*) = \frac{E_2(\chi)}{E_2(\chi^2)} = \frac{E_2(\chi^{n-1})}{R} = E_2(\chi^{n-1}),$$

which completes the proof of the theorem. ∎

Theorem 12.2.6 *Let χ be a character on F_{p^2} of order k with $k \not| p+1$. Then*

$$E(\chi^{p+1}) = -K(\chi^*).$$

Proof We have

$$\begin{aligned}
E(\chi^{p+1}) &= \sum_{b\in F_p} \chi^{p+1}(1+b\tau) = \sum_{b\in F_p} \chi^p(1+b\tau)\chi(1+b\tau) \\
&= \sum_{b\in F_p} \chi(1-b\tau)\chi(1+b\tau) = \sum_{b\in F_p} \chi(1-b^2 g) = \sum_{b\in F_p} \chi^*(1-b^2 g) \\
&= \sum_{n\in F_p} \chi^*(n) \sum_{\substack{b\in F_p \\ 1-b^2 g = n}} 1 = \sum_{n\in F_p} \chi^*(n)\{1+\rho(g^{-1}(1-n))\} \\
&= \sum_{n\in F_p} \chi^*(n)\rho(g^{-1}(1-n)) \qquad \text{(as } \chi^* \text{ is nontrivial)} \\
&= -\sum_{n\in F_p} \chi^*(n)\rho(1-n) = -J(\chi^*,\rho) = -K(\chi^*),
\end{aligned}$$

by Theorem 2.1.4. ■

Theorem 12.2.7 *Let χ be a character on F_{p^2} of order k with $k \not| (p+1)$. Then*

$$K_2(\chi) = E^2(\chi)K(\chi^*)/E(\chi^2).$$

Proof If $k \not| 2(p+1)$, then by Theorems 2.1.3(a) and 12.2.4,

$$\begin{aligned}
K_2(\chi) &= \chi(4)J_2(\chi,\chi) = \chi(4)G_2^2(\chi)/G_2(\chi^2) = \chi(4)\frac{\overline{\chi}^2(2)E^2(\chi)G^2(\chi^*)}{\overline{\chi}^2(2)E(\chi^2)G(\chi^{*2})} \\
&= \frac{E^2(\chi)}{E(\chi^2)}\chi^*(4)J(\chi^*,\chi^*) = \frac{E^2(\chi)}{E(\chi^2)}K(\chi^*).
\end{aligned}$$

We leave the remaining case as an exercise (Problem 5). ■

Theorem 12.2.8 *Let χ be a character of order k on F_{p^2}. Then $\sum_{j=0}^{k-1} E(\chi^j)$ is a rational integer, which satisfies*

$$\sum_{j=0}^{k-1} E(\chi^j) \equiv k \pmod{2k}.$$

Proof We have

$$\sum_{j=0}^{k-1} E(\chi^j) = \sum_{j=0}^{k-1}\sum_{b=0}^{p-1} \chi^j(1+b\tau) = \sum_{b=0}^{p-1}\sum_{j=0}^{k-1} \chi^j(1+b\tau) = k \sum_{\substack{b=0 \\ \chi(1+b\tau)=1}}^{p-1} 1 = kN,$$

where N is a rational integer. Clearly N is odd, since if $\chi(1+b\tau)=1$, then $\chi(1-b\tau)=\chi^p(1+b\tau)=1$. ∎

A more general Eisenstein sum $E_r(\chi)$ can defined by

$$E_r(\chi)=\sum_{\substack{\alpha\in F_{p^{tr}}\\ \mathrm{tr}_{F_{p^{tr}}/F_{p^t}}(\alpha)=1}}\chi(\alpha),$$

where t is a positive integer and χ is a multiplicative character on $F_{p^{tr}}$. In the case $t=1$, this reduces to our definition at the beginning of this chapter. All of our results on Eisenstein sums up to this point remain valid for the more general Eisenstein sums, with the obvious modifications: replace p by p^t, q by p^{tr}, F_p by F_{p^t}, $G_r(\chi)$ by $G_{tr}(\chi)$, $G(\chi)$ by $G_t(\chi)$, $K(\chi^*)$ by $K_t(\chi^*)$ (where χ^* now denotes the restriction of χ to F_{p^t}), and $K_2(\chi)$ by $K_{2t}(\chi)$.

Evans [19, pp. 41–42] used the general version of Theorem 12.2.3 (with p replaced by p^t) to give the following evaluation of a Soto–Andrade character sum:

$$\sum_{\substack{\alpha\in F_{p^{2t}}\\ N(\alpha)=1}} W(\alpha)\rho(\alpha-2+1/\alpha)=1,$$

where N denotes the norm from $F_{p^{2t}}$ to F_{p^t}, ρ is the quadratic character on F_{p^t}, and W is a multiplicative character on $F_{p^{2t}}$ whose order does not divide $2(p^t-1)$. This approach via Eisenstein sums is more elementary than that of Kuang [1], who evaluated essentially this Soto–Andrade sum by using the Weil representation of $SL_2(F_{p^t})$.

12.3 Eisenstein sums of order 3

Let p be an odd prime and r a positive integer. Set $q=p^r$, and suppose that $3\mid q-1$. Let γ be a generator of the cyclic group F_q^*, and set $g=\gamma^{(q-1)/(p-1)}$. Let χ be the character on F_q of order 3 given by $\chi(\gamma)=\beta$, where $\beta=e^{2\pi i/3}$. As $3\mid p^r-1$, either (a) $p\equiv 1 \pmod 3$ or (b) $p\equiv 2 \pmod 3$, $r\equiv 0 \pmod 2$.

Case (a): $p\equiv 1 \pmod 3$. Define the cubic character ψ on F_p by $\psi(g)=\beta$. By Corollary 12.1.4,

$$E_r(\chi)=\begin{cases}(-1)^r G^r(\psi)/p, & \text{if } r\equiv 0 \pmod 3,\\ (-1)^{r-1}G^r(\psi)/G(\psi^r), & \text{if } r\not\equiv 0 \pmod 3.\end{cases}$$

Thus, as $G(\psi)G(\psi^2)=p$ by Theorem 1.1.4(a),

$$E_r(\chi)=\begin{cases}(-1)^{r-1}G^{r-1}(\psi), & \text{if } r\equiv 1 \pmod 3,\\ (-1)^{r-1}G^{r+1}(\psi)/p, & \text{if } r\equiv 2 \pmod 3.\end{cases}$$

Now, by Lemma 4.1.1(a) and Table 3.1.1,

$$G^3(\psi)=p\,J(\psi,\psi)=\frac{p}{2}(r_3+is_3\sqrt{3}).$$

Hence, if $r \equiv 0 \pmod 3$,

$$E_r(\chi) = (-1)^r \left(\frac{p}{2}\left(r_3 + is_3\sqrt{3}\right)\right)^{r/3} \Big/ p = (-1)^r p^{(r/3)-1} \left(\frac{1}{2}\left(r_3 + is_3\sqrt{3}\right)\right)^{r/3};$$

if $r \equiv 1 \pmod 3$,

$$E_r(\chi) = (-1)^{r-1} \left(\frac{p}{2}\left(r_3 + is_3\sqrt{3}\right)\right)^{(r-1)/3} = (-1)^{r-1} p^{(r-1)/3} \left(\frac{1}{2}\left(r_3 + is_3\sqrt{3}\right)\right)^{(r-1)/3};$$

and if $r \equiv 2 \pmod 3$,

$$E_r(\chi) = (-1)^{r-1} \left(\frac{p}{2}\left(r_3 + is_3\sqrt{3}\right)\right)^{(r+1)/3} \Big/ p = (-1)^{r-1} p^{(r-2)/3} \left(\tfrac{1}{2}\left(r_3 + is_3\sqrt{3}\right)\right)^{(r+1)/3}.$$

Case (b): $p \equiv 2 \pmod 3$, $r \equiv 0 \pmod 2$. By Theorem 12.1.6,

$$E_r(\chi) = (-1)^{r(p+4)/6} p^{(r/2)-1} = (-1)^{r/2} p^{(r/2)-1}.$$

We have established the following theorem.

Theorem 12.3.1 *Let p be an odd prime and r a positive integer. Set $q = p^r$. Let γ be a generator of the cyclic group F_q^*. Let χ be the character on F_q of order 3 given by $\chi(\gamma) = \beta$, where $\beta = \exp(2\pi i/3)$. Let g be the primitive root of p given by $g = \gamma^{(q-1)/(p-1)}$. If $p \equiv 1 \pmod 3$, then*

$$E_r(\chi) = \begin{cases} (-1)^r\, p^{(r/3)-1} \left(\tfrac{1}{2}(r_3 + is_3\sqrt{3})\right)^{r/3}, & \text{if } r \equiv 0 \pmod 3, \\ (-1)^{r-1}\, p^{(r-1)/3} \left(\tfrac{1}{2}(r_3 + is_3\sqrt{3})\right)^{(r-1)/3}, & \text{if } r \equiv 1 \pmod 3, \\ (-1)^{r-1}\, p^{(r-2)/3} \left(\tfrac{1}{2}(r_3 + is_3\sqrt{3})\right)^{(r+1)/3}, & \text{if } r \equiv 2 \pmod 3, \end{cases}$$

where the integers r_3, s_3 are given in Theorem 3.1.3.

If $p \equiv 2 \pmod 3$, then $r \equiv 0 \pmod 2$ and

$$E_r(\chi) = (-1)^{r/2} p^{(r/2)-1}.$$

Example We choose $p = 7$ and $r = 2$, with $F_7 = \{0,1,2,3,4,5,6\}$ and $F_{49} = \{a + bx \,|\, a, b \in F_7,\ x^2 = 5\}$. The cyclic group F_{49}^* is generated by $\gamma = 1 + x$. Let χ be the cubic character on F_{49} given by $\chi(1+x) = \omega$, where $\omega = \exp(2\pi i/3)$. Now $g = \gamma^{(q-1)/(p-1)} = (1+x)^8 = (6+2x)^4 = (3x)^2 = 3$. By (2.3.9), the integers r_3 and s_3 are determined by

$$28 = r_3^2 + 3s_3^2,\ r_3 \equiv 1 \pmod 3,\ s_3 \equiv 0 \pmod 3,\ 3s_3 \equiv (2\cdot 3^{(7-1)/3} + 1) r_3 \equiv 5r_3 \pmod 7,$$

so that $r_3 = 1$ and $s_3 = -3$. By Theorem 12.3.1,

$$E_2(\chi) = -\frac{1}{2}(r_3 + is_3\sqrt{3}\,) = \frac{1}{2}(-1 + i3\sqrt{3}\,).$$

12.4 Eisenstein sums of order 4

Let p be an odd prime and r a positive integer. Set $q = p^r$ and suppose that $4 \mid q-1$. Let γ be a generator of the cyclic group F_q^*. Let $g = \gamma^{(q-1)/(p-1)}$. Let χ be the character on F_q of order 4 given by $\chi(\gamma) = i$. As $4 \mid p^r - 1$, either (a) $p \equiv 1 \pmod 4$ or (b) $p \equiv 3 \pmod 4$ and $r \equiv 0 \pmod 2$. The value of $E_r(\chi)$ is given in the next theorem.

Theorem 12.4.1 (a) *If* $p \equiv 1 \pmod 4$, *then*

$$E_r(\chi) = \begin{cases} p^{(r/4)-1}(a_4 + ib_4)^{r/2}, & \text{if } r \equiv 0 \pmod 4, \\ p^{(r-1)/4}(a_4 + ib_4)^{(r-1)/2}, & \text{if } r \equiv 1 \pmod 4, \\ (-1)^{(p+3)/4} p^{(r-2)/4}(a_4 + ib_4)^{r/2}, & \text{if } r \equiv 2 \pmod 4, \\ (-1)^{(p-1)/4} p^{(r-3)/4}(a_4 + ib_4)^{(r+1)/2}, & \text{if } r \equiv 3 \pmod 4, \end{cases}$$

where the integers a_4 *and* b_4 *are defined in Theorem* 3.2.1.

(b) *If* $p \equiv 3 \pmod 4$ *and* $r \equiv 0 \pmod 2$, *then*

$$E_r(\chi) = (-1)^{r(p-3)/8} p^{(r/2)-1}.$$

Proof (a) Define the quartic character ψ on F_p by $\psi(g) = i$. By Corollary 12.1.4,

$$E_r(\chi) = \begin{cases} G^r(\psi)/p, & \text{if } r \equiv 0 \pmod 4, \\ (-1)^{r-1} G^r(\psi)/G(\psi^r), & \text{if } r \not\equiv 0 \pmod 4. \end{cases}$$

Thus, as $G(\psi)G(\psi^3) = \psi(-1)p$, by Theorem 1.1.4(a), and $G(\psi^2) = p^{1/2}$, by Theorem 1.2.4,

$$E_r(\chi) = \begin{cases} G^{r-1}(\psi), & \text{if } r \equiv 1 \pmod 4, \\ -G^r(\psi)/p^{1/2}, & \text{if } r \equiv 2 \pmod 4, \\ (-1)^{(p-1)/4} G^{r+1}(\psi)/p, & \text{if } r \equiv 3 \pmod 4, \end{cases}$$

as $\psi(-1) = (-1)^{(p-1)/4}$ by Lemma 3.0.2(a). By Theorem 2.1.3(a) and Table 3.2.1,

$$\frac{G^2(\psi)}{p^{1/2}} = \frac{G^2(\psi)}{G(\psi^2)} = J(\psi, \psi) = (-1)^{(p-1)/4}(a_4 + ib_4),$$

so that

$$G^2(\psi) = (-1)^{(p-1)/4} p^{1/2}(a_4 + ib_4).$$

If $r \equiv 0 \pmod 4$, then

$$E_r(\chi) = \left((-1)^{(p-1)/4} p^{1/2}(a_4 + ib_4)\right)^{r/2} \Big/ p = p^{(r/4)-1}(a_4 + ib_4)^{r/2}.$$

If $r \equiv 1 \pmod 4$, then

$$E_r(\chi) = \left((-1)^{(p-1)/4} p^{1/2}(a_4 + ib_4)\right)^{(r-1)/2} = p^{(r-1)/4}(a_4 + ib_4)^{(r-1)/2}.$$

If $r \equiv 2 \pmod 4$, then

$$E_r(\chi) = -\left((-1)^{(p-1)/4}p^{1/2}(a_4+ib_4)\right)^{r/2} \Big/ p^{1/2} = (-1)^{(p+3)/4}p^{(r-2)/4}(a_4+ib_4)^{r/2}.$$

If $r \equiv 3 \pmod 4$, then

$$\begin{aligned} E_r(\chi) &= (-1)^{(p-1)/4}\left((-1)^{(p-1)/4}p^{1/2}(a_4+ib_4)\right)^{(r+1)/2} \Big/ p \\ &= (-1)^{(p-1)/4}p^{(r-3)/4}(a_4+ib_4)^{(r+1)/2}. \end{aligned}$$

(b) By Theorem 12.1.6 with $k=4$ and $l=1$,

$$E_r(\chi) = (-1)^{r(p+5)/8}p^{(r/2)-1} = (-1)^{r(p-3)/8}p^{(r/2)-1}.$$ ■

Example We choose $p = 5$ and $r = 2$, with $F_5 = \{0,1,2,3,4\}$ and $F_{25} = \{a+bx \mid a,b \in F_5,\ x^2 = 2\}$. The cyclic group F_{25}^* is generated by $\gamma = 2+x$. Let χ be the quartic character on F_{25} given by $\chi(2+x) = i$. Now $g = \gamma^{(q-1)/(p-1)} = (2+x)^6 = (1+4x)^3 = (1+4x)(3+3x) = 2$. By Theorems 3.2.1 and 3.2.2, $a_4 = 1$ and $b_4 = 2$. Hence, by Theorem 12.4.1, $E_2(\chi) = a_4 + ib_4 = 1+2i$.

12.5 Eisenstein sums of order 5

Let p be an odd prime and r a positive integer. Set $q = p^r$ and suppose that $5 \mid q-1$. Let γ be a generator of the cyclic group F_q^*. Let $g = \gamma^{(q-1)/(p-1)}$. Let χ be the character on F_q of order 5 given by $\chi(\gamma) = \beta$, where $\beta = e^{2\pi i/5}$. As $5 \mid p^r - 1$, (a) $p \equiv 1 \pmod 5$, (b) $p \equiv 2$ or $3 \pmod 5$ and $r \equiv 0 \pmod 4$, or (c) $p \equiv 4 \pmod 5$ and $r \equiv 0 \pmod 2$. The value of $E_r(\chi)$ is given in the next theorem.

Theorem 12.5.1 (a) *If $p \equiv 1 \pmod 5$, let (x, w, v, u) be the unique 4-tuple of integers defined in Theorem 3.7.8. Set*

$$\tau = \tau(x,w,v,u) = \frac{1}{4}\left(x + 5w\sqrt{5} + iv\sqrt{50-10\sqrt{5}} + iu\sqrt{50+10\sqrt{5}}\right),$$

and

$$\sigma = \tau(x,-w,-u,v) = \frac{1}{4}\left(x - 5w\sqrt{5} - iu\sqrt{50-10\sqrt{5}} + iv\sqrt{50+10\sqrt{5}}\right).$$

Then

$$E_r(\chi) = \epsilon p^{\alpha}\tau^{\beta}\sigma^{\delta},$$

where

$$\alpha = \begin{cases} (r/5) - 1, & \text{if } r \equiv 0 \pmod 5, \\ [r/5], & \text{if } r \not\equiv 0 \pmod 5, \end{cases}$$

$$\beta = \begin{cases} [2r/5], & \text{if } r \equiv 0, 1, 3 \pmod 5, \\ [2r/5] + 1, & \text{if } r \equiv 2, 4 \pmod 5, \end{cases}$$

$$\delta = \begin{cases} [r/5], & \text{if } r \equiv 0, 1, 2 \pmod 5, \\ [r/5] + 1, & \text{if } r \equiv 3, 4 \pmod 5, \end{cases}$$

$$\epsilon = \begin{cases} (-1)^r, & \text{if } r \equiv 0 \pmod 5, \\ (-1)^{r-1}, & \text{if } r \not\equiv 0 \pmod 5. \end{cases}$$

(b) *If* $p \equiv 2$ *or* $3 \pmod 5$ *and* $r \equiv 0 \pmod 4$, *then*

$$E_r(\chi) = (-1)^{r/4} p^{(r/2)-1}.$$

(c) *If* $p \equiv 4 \pmod 5$ *and* $r \equiv 0 \pmod 2$, *then*

$$E_r(\chi) = (-1)^{r/2} p^{(r/2)-1}.$$

Proof (a) Define the quintic character ψ on F_p by $\psi(g) = \beta$. By Corollary 12.1.4,

$$E_r(\chi) = \begin{cases} (-1)^r G^r(\psi)/p, & \text{if } r \equiv 0 \pmod 5, \\ (-1)^{r-1} G^r(\psi)/G(\psi^r), & \text{if } r \not\equiv 0 \pmod 5. \end{cases}$$

Thus, appealing to Theorem 1.1.4(a) and Theorem 2.1.3(a), we see that

$$E_r(\chi) = \begin{cases} (-1)^{r-1} G^{r-1}(\psi), & \text{if } r \equiv 1 \pmod 5, \\ (-1)^{r-1} G^{r-2}(\psi) J(\psi, \psi), & \text{if } r \equiv 2 \pmod 5, \\ (-1)^{r-1} G^{r-3}(\psi) J(\psi, \psi) J(\psi, \psi^2), & \text{if } r \equiv 3 \pmod 5, \\ (-1)^{r-1} G^{r+1}(\psi)/p, & \text{if } r \equiv 4 \pmod 5. \end{cases}$$

Now, by Theorem 3.7.8,

$$J(\psi, \psi) = \frac{1}{4}\left(x + 5w\sqrt{5} + iv\sqrt{50 - 10\sqrt{5}} + iu\sqrt{50 + 10\sqrt{5}}\right) = \tau.$$

By Theorem 2.1.5 and (3.7.35),

$$J(\psi, \psi^2) = J(\psi^2, \psi^2) = \frac{1}{4}\left(x - 5w\sqrt{5} - iu\sqrt{50 - 10\sqrt{5}} + iv\sqrt{50 + 10\sqrt{5}}\right) = \sigma.$$

Thus

$$G^5(\psi) = G(\psi^2)G(\psi^3) \cdot \frac{G^4(\psi)}{G^2(\psi^2)} \cdot \frac{G(\psi)G(\psi^2)}{G(\psi^3)} = pJ^2(\psi, \psi)J(\psi, \psi^2) = p\tau^2\sigma.$$

If $r \equiv 0 \pmod 5$, then

$$E_r(\chi) = (-1)^r (p\tau^2\sigma)^{r/5}/p = (-1)^r p^{(r/5)-1}\tau^{2r/5}\sigma^{r/5}.$$

If $r \equiv 1 \pmod 5$, then

$$E_r(\chi) = (-1)^{r-1}(p\tau^2\sigma)^{(r-1)/5} = (-1)^{r-1}p^{(r-1)/5}\tau^{2(r-1)/5}\sigma^{(r-1)/5}.$$

If $r \equiv 2 \pmod 5$, then

$$E_r(\chi) = (-1)^{r-1}(p\tau^2\sigma)^{(r-2)/5}\,\tau = (-1)^{r-1}p^{(r-2)/5}\tau^{(2r+1)/5}\sigma^{(r-2)/5}.$$

If $r \equiv 3 \pmod 5$, then

$$E_r(\chi) = (-1)^{r-1}(p\tau^2\sigma)^{(r-3)/5}\,\tau\sigma = (-1)^{r-1}p^{(r-3)/5}\tau^{(2r-1)/5}\sigma^{(r+2)/5}.$$

If $r \equiv 4 \pmod 5$, then

$$E_r(\chi) = (-1)^{r-1}(p\tau^2\sigma)^{(r+1)/5}/p = (-1)^{r-1}p^{(r-4)/5}\tau^{2(r+1)/5}\sigma^{(r+1)/5}.$$

(b) By Theorem 12.1.6 with $k = 5$ and $l = 2$,

$$E_r(\chi) = (-1)^{r(p^2+6)/20}p^{(r/2)-1} = (-1)^{r/4}p^{(r/2)-1},$$

as $(p^2+6)/5$ is an odd integer.

(c) By Theorem 12.1.6 with $k = 5$ and $l = 1$,

$$E_r(\chi) = (-1)^{r(p+6)/10}p^{(r/2)-1} = (-1)^{r/2}p^{(r/2)-1},$$

as $(p+6)/5$ is an odd integer. ■

12.6 Eisenstein sums of order 6

Let p be an odd prime and r a positive integer. Set $q = p^r$ and suppose that $6 \mid q-1$. Let γ be a generator of the cyclic group F_q^*. Let $g = \gamma^{(q-1)/(p-1)}$. Let χ be the character on F_q of order 6 given by $\chi(\gamma) = \beta$, where $\beta = e^{2\pi i/6} = (1+i\sqrt{3}\,)/2$. Write $\omega = \beta^2 = (-1+i\sqrt{3})/2$ and $Z = \text{ind}_g 2$. As $6 \mid q-1$, either (a) $p \equiv 1 \pmod 6$ or (b) $p \equiv 5 \pmod 6$ and $r \equiv 0 \pmod 2$. The next theorem gives the value of the Eisenstein sum $E_r(\chi)$ of order 6.

Theorem 12.6.1 (a) *If* $p \equiv 1 \pmod 6$, *say*, $p = 6f+1$, *then*

$$E_r(\chi) = \epsilon p^\alpha \pi^\beta,$$

where

$$\pi = \frac{1}{2}(r_3 + is_3\sqrt{3}\,),$$

$$\alpha = \begin{cases} (r/6) - 1, & \text{if } r \equiv 0 \pmod 6, \\ [r/6], & \text{if } r \not\equiv 0 \pmod 6, \end{cases}$$

$$\beta = \begin{cases} 2r/3, & \text{if } r \equiv 0 \pmod 3, \\ (2r-2)/3, & \text{if } r \equiv 1 \pmod 6, \\ (2r-1)/3, & \text{if } r \equiv 2 \pmod 6, \\ (2r+1)/3, & \text{if } r \equiv 4 \pmod 6, \\ (2r+2)/3, & \text{if } r \equiv 5 \pmod 6, \end{cases}$$

and

$$\epsilon = \begin{cases} (-1)^{fr/6}, & \text{if } r \equiv 0 \pmod 6, \\ (-1)^{f(r-1)/6}, & \text{if } r \equiv 1 \pmod 6, \\ (-1)^{(f(r+4)/6)+1}\omega^{Z}, & \text{if } r \equiv 2 \pmod 6, \\ (-1)^{f(r+3)/6}, & \text{if } r \equiv 3 \pmod 6, \\ (-1)^{(f(r-4)/6)+1}\omega^{-Z}, & \text{if } r \equiv 4 \pmod 6, \\ (-1)^{f(r-5)/6}, & \text{if } r \equiv 5 \pmod 6. \end{cases}$$

(b) *If* $p \equiv 5 \pmod 6$ *and* $r \equiv 0 \pmod 2$, *then*

$$E_r(\chi) = (-1)^{r(p-5)/4} p^{(r/2)-1}.$$

Proof (a) Define the character ψ on F_p of order 6 by $\psi(g) = \beta = e^{2\pi i/6}$. By Lemma 3.0.2(a), $\psi(-1) = (-1)^f$. By Theorem 1.2.4,

$$G(\psi^3) = i^{(p-1)^2/4} p^{1/2} = i^{f^2} p^{1/2}.$$

Corollary 12.1.4 gives

$$E_r(\chi) = \begin{cases} G^r(\psi)/p, & \text{if } r \equiv 0 \pmod 6, \\ (-1)^{r-1} G^r(\psi)/G(\psi^r), & \text{if } r \not\equiv 0 \pmod 6. \end{cases}$$

Appealing to Theorems 1.1.4(a) and 2.1.3(a), we see that

$$E_r(\chi) = \begin{cases} G^{r-1}(\psi), & \text{if } r \equiv 1 \pmod 6, \\ -G^{r-2}(\psi) J(\psi, \psi), & \text{if } r \equiv 2 \pmod 6, \\ G^r(\psi)/(i^{f^2} p^{1/2}), & \text{if } r \equiv 3 \pmod 6, \\ -G^{r-4}(\psi) J^2(\psi, \psi) J(\psi^2, \psi^2), & \text{if } r \equiv 4 \pmod 6, \\ (-1)^f G^{r+1}(\psi)/p, & \text{if } r \equiv 5 \pmod 6. \end{cases}$$

From (3.1.3), (3.1.6), and Table 3.1.2, we deduce that

$$\begin{aligned} J(\psi, \psi) &= (-1)^f \frac{1}{2}(u_3 + iv_3\sqrt{3}\,) = (-1)^f \beta^{2b_3}(a_3 + ib_3\sqrt{3}\,) = (-1)^f \beta^{4b_3}\pi \\ &= (-1)^f \omega^{2b_3}\pi = (-1)^f \omega^{-b_3}\pi = (-1)^f \omega^{Z}\pi, \end{aligned}$$

and

$$J(\psi^2, \psi^2) = \pi.$$

By Lemma 4.1.1,

$$G^3(\psi) = (-1)^f i^{f^2} p^{1/2}\pi^2 \text{ and } G^6(\psi) = (-1)^f p\pi^4.$$

If $r \equiv 0 \pmod 6$, then

$$E_r(\chi) = ((-1)^f p\pi^4)^{r/6}/p = (-1)^{fr/6} p^{(r/6)-1}\pi^{2r/3}.$$

If $r \equiv 1 \pmod 6$, then

$$E_r(\chi) = ((-1)^f p\pi^4)^{(r-1)/6} = (-1)^{f(r-1)/6} p^{(r-1)/6}\pi^{(2r-2)/3}.$$

If $r \equiv 2 \pmod 6$, then

$$E_r(\chi) = -((-1)^f p\pi^4)^{(r-2)/6}((-1)^f \omega^Z \pi) = (-1)^{(f(r+4)/6)+1}\omega^Z p^{(r-2)/6}\pi^{(2r-1)/3}.$$

If $r \equiv 3 \pmod 6$, then

$$E_r(\chi) = ((-1)^f i^{f^2} p^{1/2}\pi^2)^{r/3}/(i^{f^2} p^{1/2}) = (-1)^{f(r+3)/6} p^{(r-3)/6}\pi^{2r/3}.$$

If $r \equiv 4 \pmod 6$, then

$$E_r(\chi) = -((-1)^f p\pi^4)^{(r-4)/6}((-1)^f \omega^Z \pi)^2\pi = (-1)^{(f(r-4)/6)+1}\omega^{-Z} p^{(r-4)/6}\pi^{(2r+1)/3}.$$

If $r \equiv 5 \pmod 6$, then

$$E_r(\chi) = (-1)^f((-1)^f p\pi^4)^{(r+1)/6}/p = (-1)^{f(r-5)/6} p^{(r-5)/6}\pi^{(2r+2)/3}.$$

(b) From Theorem 12.1.6 with $k = 6$ and $l = 1$,

$$E_r(\chi) = (-1)^{r(p+7)/12} p^{(r/2)-1} = (-1)^{r(p-5)/4} p^{(r/2)-1}.$$

12.7 Eisenstein sums of order 8

Let p be an odd prime and r a positive integer. Set $q = p^r$ and suppose that $8 \mid q-1$. Let γ be a generator of the cyclic group F_q^*. Let $g = \gamma^{(q-1)/(p-1)}$ and $Z = \text{ind}_g 2$. Let χ be the character on F_q of order 8 given by $\chi(\gamma) = \beta$, where $\beta = e^{2\pi i/8}$. As $8 \mid q-1$, either (a) $p \equiv 1 \pmod 8$, (b) $p \equiv 3 \pmod 8$, $r \equiv 0 \pmod 2$, (c) $p \equiv 5 \pmod 8$, $r \equiv 0 \pmod 2$, or (d) $p \equiv 7 \pmod 8$, $r \equiv 0 \pmod 2$.

Theorem 12.7.1 (a) *If* $p \equiv 1 \pmod 8$, *say*, $p = 8f + 1$, *then*

$$E_r(\chi) = \epsilon p^{\alpha}(a_4 + ib_4)^{\beta}(a_8 + ib_8\sqrt{2})^{\delta},$$

where

$$\epsilon = \begin{cases} 1, & \text{if } r \equiv 0, 1, 5 \pmod 8, \\ -1, & \text{if } r \equiv 4 \pmod 8, \\ (-1)^f, & \text{if } r \equiv 3 \pmod 4, \\ -i^Z, & \text{if } r \equiv 2 \pmod 4, \end{cases}$$

$$\alpha = \begin{cases} (r/8) - 1, & \text{if } r \equiv 0 \pmod 8, \\ [r/8], & \text{if } r \not\equiv 0 \pmod 8, \end{cases}$$

$$\beta = \begin{cases} r/4, & \text{if } r \equiv 0 \pmod 4, \\ (r-1)/4, & \text{if } r \equiv 1 \pmod 4, \\ (r-2)/4, & \text{if } r \equiv 2 \pmod 8, \\ (r+2)/4, & \text{if } r \equiv 6 \pmod 8, \\ (r+1)/4, & \text{if } r \equiv 3 \pmod 4, \end{cases}$$

$$\delta = \begin{cases} r/2, & \text{if } r \equiv 0 \pmod 2, \\ (r-1)/2, & \text{if } r \equiv 1, 3 \pmod 8, \\ (r+1)/2, & \text{if } r \equiv 5, 7 \pmod 8. \end{cases}$$

(b) *If* $p \equiv 3 \pmod 8$ *and* $r \equiv 0 \pmod 2$, *then*

$$E_r(\chi) = \begin{cases} (-1)^{r/4} p^{(r/4)-1}(a_8 + ib_8\sqrt{2}\,)^{r/2}, & \text{if } r \equiv 0 \pmod 4, \\ (-1)^{(r+2)/4} p^{(r-2)/4}(a_8 + ib_8\sqrt{2}\,)^{r/2}, & \text{if } r \equiv 2 \pmod 4, \end{cases}$$

where the integers a_8 *and* b_8 *satisfy*

$$p = a_8^2 + 2b_8^2, \quad a_8 \equiv (-1)^{(p-3)/8} \pmod 4,$$

with the sign of b_8 *determined by*

$$b_8 \equiv \frac{a_8}{2}(\gamma^{(q-1)/8} - \gamma^{(1-q)/8}) \pmod p.$$

(*The right side of this congruence* $\pmod p$ *is in* F_p^*.)

(c) *If* $p \equiv 5 \pmod 8$ *and* $r \equiv 0 \pmod 2$, *then*

$$E_r(\chi) = \begin{cases} p^{(3r/8)-1}(a_4 + ib_4)^{r/4}, & \text{if } r \equiv 0 \pmod 8, \\ i(-1)^{(Z-1)/2} p^{(3r-6)/8}(a_4 + ib_4)^{(r+2)/4}, & \text{if } r \equiv 2 \pmod 8, \\ -p^{(3r-4)/8}(a_4 + ib_4)^{r/4}, & \text{if } r \equiv 4 \pmod 8, \\ i(-1)^{(Z-1)/2} p^{(3r-2)/8}(a_4 + ib_4)^{(r-2)/4}, & \text{if } r \equiv 6 \pmod 8. \end{cases}$$

(d) *If* $p \equiv 7 \pmod 8$ *and* $r \equiv 0 \pmod 2$, *then*

$$E_r(\chi) = (-1)^{r(p-7)/16} p^{(r/2)-1}.$$

Proof (a) In this case Z is even. Define the octic character ψ on F_p by $\psi(g) = \beta$. By Corollary 12.1.4,

$$(12.7.1) \qquad E_r(\chi) = \begin{cases} G^r(\psi)/p, & \text{if } r \equiv 0 \pmod 8, \\ (-1)^{r-1} G^r(\psi)/G(\psi^r), & \text{if } r \not\equiv 0 \pmod 8. \end{cases}$$

Appealing to Theorem 2.1.3(a), we deduce that

$$(12.7.2) \quad E_r(\chi) = \begin{cases} G^{r-1}(\psi), & \text{if } r \equiv 1 \pmod 8, \\ -G^{r-2}(\psi)J(\psi,\psi), & \text{if } r \equiv 2 \pmod 8, \\ G^{r-3}(\psi)J(\psi,\psi)J(\psi,\psi^2), & \text{if } r \equiv 3 \pmod 8, \\ -G^r(\psi)/G(\psi^4), & \text{if } r \equiv 4 \pmod 8, \\ (-1)^{(p-1)/8}G^{r-5}(\psi)J^2(\psi,\psi)J(\psi,\psi^2)J(\psi^3,\psi^3), & \text{if } r \equiv 5 \pmod 8, \\ -G^{r-6}(\psi)J^3(\psi,\psi)J(\psi^2,\psi^2)J(\psi^2,\psi^4), & \text{if } r \equiv 6 \pmod 8, \\ (-1)^{(p-1)/8}G^{r+1}(\psi)/p, & \text{if } r \equiv 7 \pmod 8. \end{cases}$$

The evaluation of $E_r(\chi)$ can now be deduced from (12.7.1), (12.7.2), and the values

$$\begin{aligned} J(\psi,\psi) &= J(\psi^3,\psi^3) = i^Z(a_8 + ib_8\sqrt{2}\,), \\ J(\psi,\psi^2) &= (-1)^{(p-1)/8} i^Z (a_4 + ib_4), \\ J(\psi^2,\psi^2) &= J(\psi^2,\psi^4) = a_4 + ib_4, \\ G(\psi^4) &= p^{1/2}, \\ G(\psi)^4 &= J^2(\psi,\psi)J(\psi^2,\psi^2)G(\psi^4) = (a_4+ib_4)(a_8+ib_8\sqrt{2}\,)^2 p^{1/2}, \end{aligned}$$

which follow from Theorems 3.3.1, 3.3.3, 3.2.1, and 1.2.4.

(b) In this case $Z \equiv 1 \pmod 2$. Set $p = 8f+3$. Write $\gamma_1 = \gamma^{(p^r-1)/(p^2-1)}$ and $g = \gamma_1^{1+p}$. Then χ is the lift of the octic character ψ on F_{p^2} defined by $\psi(\gamma_1) = \beta$. The restriction ψ^* of ψ to F_p is given by $\psi^*(g) = \beta^{1+p} = \beta^{8f+4} = \beta^4 = -1$, so that ψ^* is the quadratic character on F_p. In particular, $\psi^*(2) = \psi^*(g^Z) = (-1)^Z = -1$. The restriction χ^* of χ to F_p is given by $\chi^*(g) = \beta^{1+p+\cdots+p^{r-1}} = \beta^{1+3+1+3+\cdots+1+3} = \beta^{4+4+\cdots+4} = \beta^{4r/2} = (-1)^{r/2}$, so that χ^* is trivial if $r \equiv 0 \pmod 4$ and χ^* is quadratic if $r \equiv 2 \pmod 4$. By Theorem 12.1.3 with $k=8$ and $l=2$,

$$E_r(\chi) = \begin{cases} G^{r/2}(\psi^*)E_2^{r/2}(\psi)/p, & \text{if } r \equiv 0 \pmod 4, \\ G^{(r/2)-1}(\psi^*)E_2^{r/2}(\psi), & \text{if } r \equiv 2 \pmod 4. \end{cases}$$

Further, by Theorem 1.2.4, $G(\psi^*) = ip^{1/2}$, so that

$$E_r(\chi) = \begin{cases} (-1)^{r/4} p^{(r/4)-1} E_2^{r/2}(\psi), & \text{if } r \equiv 0 \pmod 4, \\ (-1)^{(r-2)/4} p^{(r-2)/4} E_2^{r/2}(\psi), & \text{if } r \equiv 2 \pmod 4. \end{cases}$$

We next show that $E_2(\psi) = -(a_8 + ib_8\sqrt{2}\,)$. By Lemma 12.0.3, $E_2(\psi) = E_2(\psi^3)$ so that $E_2(\psi)$ is fixed under the automorphism $\sigma_3 : \beta \longrightarrow \beta^3$ of $Q(\beta)$. Hence $E_2(\psi)$

belongs to the field $Q(\beta+\beta^3)=Q(i\sqrt{2})$. As $E_2(\psi)$ is an algebraic integer, $E_2(\psi)=E_2(\psi^3)=-(a_8+ib_8\sqrt{2})$, for some integers a_8 and b_8. Also $E_2(\psi^5)=E_2(\psi^7)=E_2(\overline{\psi})=\overline{E_2(\psi)}=-(a_8-ib_8\sqrt{2})$. By Theorem 12.1.2, $p=a_8^2+2b_8^2$. By Lemma 12.0.1(b), Theorem 12.4.1(b), and Theorem 12.1.5,

$$E_2(\psi^m)=\begin{cases} p, & \text{if } m=0,\\ 1, & \text{if } m=2,\\ -1, & \text{if } m=4,\\ 1, & \text{if } m=6,\end{cases}$$

so that

$$\sum_{m=0}^{7} E_2(\psi^m)=p+1-4a_8.$$

On the other hand, by Theorem 12.1.7,

$$\sum_{m=0}^{7} E_2(\psi^m)\equiv 0 \pmod{16},$$

so that

$$p+1-4a_8\equiv 0 \pmod{16},$$

and thus

$$a_8\equiv \frac{1}{4}(p+1)\equiv 2f+1\equiv (-1)^f \pmod 4.$$

It remains to prove the congruence for $b_8 \pmod p$. We defer the proof until Theorem 12.9.6.

(c) In this case $Z\equiv 1 \pmod 2$. Set $p=8f+5$. Write $\gamma_1=\gamma^{(p^r-1)/(p^2-1)}$ and $g=\gamma_1^{1+p}$. Then χ is the lift of the octic character ψ on F_{p^2} defined by $\psi(\gamma_1)=\beta$. The restriction ψ^* of ψ to F_p is given by $\psi^*(g)=\beta^{1+p}=\beta^{8f+6}=\beta^6=-i$, so that ψ^* is a quartic character on F_p. In particular, $\psi^*(-1)=\psi^*(g^{(p-1)/2})=(-i)^{(p-1)/2}=(-i)^2=-1$, and $\psi^*(2)=\psi^*(g^Z)=(-i)^Z=-i^Z=i(-1)^{(Z+1)/2}$. The restriction χ^* of χ to F_p is given by $\chi^*(g)=\beta^{1+p+\cdots+p^{r-1}}=\beta^{1+5+1+5+\cdots+1+5}=\beta^{6+6+\cdots+6}=\beta^{6r/2}=(-i)^{r/2}=(-1)^{r/2}i^{r/2}$, so that χ^* is trivial if $r\equiv 0 \pmod 8$, χ^* is quadratic if $r\equiv 4 \pmod 8$, and χ^* is quartic if $r\equiv 2$ or $6 \pmod 8$.

First we show that

$$E_2(\psi)=i(-1)^{(Z-1)/2}(a_4+ib_4).$$

In view of Theorem 12.2.1, and the fact that $\psi(2)=i(-1)^{(Z+1)/2}$, this is equivalent to showing that

$$E(\psi)=a_4+ib_4.$$

From Theorem 12.2.2 we see that $E(\psi)=E(\psi^5)$, so that $E(\psi)$ is fixed under the automorphism $\sigma_5:\beta\longrightarrow\beta^5$. Hence $E(\psi)$ belongs to the field $Q(\beta^2)=Q(i)$. As $E(\psi)$ is an algebraic integer, $E(\psi)=E(\psi^5)=a+ib$, where a and b are integers. Complex conjugation yields $E(\psi^3)=E(\psi^7)=E(\overline{\psi})=\overline{E(\psi)}=a-ib$. Moreover, by Theorems

12.1.2 and 12.2.1, $a^2 + b^2 = E(\psi)E(\psi^7) = p$, so that $(a, b) = \pm(a_4, \pm b_4)$ or $\pm(b_4, \pm a_4)$. As $\psi^*(g) = -i$, Theorem 3.2.1 gives

$$K(\psi^*) = a_4 - ib_4.$$

Then, by Theorem 12.2.6,

$$E(\psi^6) = E(\psi^{p+1}) = -K(\psi^*) = -(a_4 - ib_4),$$

so that

$$E(\psi^2) = \overline{E(\psi^6)} = -(a_4 + ib_4).$$

From Theorems 12.1.5 and 12.2.1,

$$E_2(\psi^4) = 1, \quad E(\psi^4) = \psi^4(2)E_2(\psi^4) = E_2(\psi^4) = 1.$$

Hence

$$\sum_{j=0}^{7} E(\psi^j) = p + 1 + 4a - 2a_4.$$

On the other hand, by Theorem 12.2.8,

$$\sum_{j=0}^{7} E(\psi^j) \equiv 8 \pmod{16},$$

so that

$$p + 1 + 4a - 2a_4 \equiv 8 \pmod{16}.$$

As $p \equiv 5 \pmod 8$, $a_4 \equiv 1 \pmod 4$ and $b_4 \equiv 2 \pmod 4$, so that

$$p = a_4^2 + b_4^2 \equiv a_4^2 + 4 \equiv 2a_4 + 3 \pmod{16}.$$

Hence $a \equiv 1 \pmod 4$, $a = a_4$, and $b = \pm b_4$. Thus $E(\psi) = a_4 \pm ib_4$. Then, by Theorems 12.2.5 and 12.2.6,

$$a - ib = E(\psi^3) = K(\psi^*) = -E(\psi^6) = a_4 - ib_4,$$

so that $b = b_4$ and $E(\psi) = a_4 + ib_4$, as asserted.

Next we determine $G^2(\psi^*)$. We have

$$G^2(\psi^*) = \frac{G^2(\psi^*)}{G(\psi^{*2})}G(\psi^{*2}) = J(\psi^*, \psi^*)p^{1/2} = -\psi^*(4)J(\psi^*, \psi^*)p^{1/2} = -K(\psi^*)p^{1/2},$$

so that

$$G^2(\psi^*) = -p^{1/2}(a_4 - ib_4).$$

Now, by Theorem 12.1.3, we obtain the following equations:

$$E_r(\chi) = \begin{cases} G^{r/2}(\psi^*)E_2^{r/2}(\psi)/p, & \text{if } r \equiv 0 \pmod 8, \\ (-1)^{(r/2)-1}G^{r/2}(\psi^*)E_2^{r/2}(\psi)/G(\psi^{*r/2}), & \text{if } r \equiv 2,4,6 \pmod 8. \end{cases}$$

Hence, if $r \equiv 0 \pmod 8$,

$$E_r(\chi) = (-p^{1/2}(a_4 - ib_4))^{r/4}(i(-1)^{(Z-1)/2}(a_4 + ib_4))^{r/2}/p = p^{(3r/8)-1}(a_4 + ib_4)^{r/4};$$

if $r \equiv 2 \pmod 8$,

$$\begin{aligned} E_r(\chi) &= (-p^{1/2}(a_4 - ib_4))^{(r-2)/4}(i(-1)^{(Z-1)/2}(a_4 + ib_4))^{r/2} \\ &= (-1)^{(Z-1)/2}ip^{(3r-6)/8}(a_4 + ib_4)^{(r+2)/4}; \end{aligned}$$

if $r \equiv 4 \pmod 8$,

$$\begin{aligned} E_r(\chi) &= -(-p^{1/2}(a_4 - ib_4))^{r/4}(i(-1)^{(Z-1)/2}(a_4 + ib_4))^{r/2}/p^{1/2} \\ &= -p^{(3r-4)/8}(a_4 + ib_4)^{r/4}; \end{aligned}$$

and, if $r \equiv 6 \pmod 8$,

$$\begin{aligned} E_r(\chi) &= -(-p^{1/2}(a_4 - ib_4))^{(r+2)/4}(i(-1)^{(Z-1)/2}(a_4 + ib_4))^{r/2}/p \\ &= i(-1)^{(Z-1)/2}p^{(3r-2)/8}(a_4 + ib_4)^{(r-2)/4}. \end{aligned}$$

(d) In this case the result follows from Theorem 12.1.6 with $k = 8$ and $l = 1$. ∎

12.8 Some Eisenstein sums of order 7

Let p be an odd prime and r a positive integer. Set $q = p^r$ and suppose that $7 \mid q - 1$. Let γ be a generator of the cyclic group F_q^*, and let $g = \gamma^{(q-1)/(p-1)}$. Let χ be the character on F_q of order 7 given by $\chi(\gamma) = \beta$, where $\beta = e^{2\pi i/7}$. As $7 \mid p^r - 1$, either (a) $p \equiv 1 \pmod 7$, (b) $p \equiv 2$ or $4 \pmod 7$ and $r \equiv 0 \pmod 3$, (c) $p \equiv 3$ or $5 \pmod 7$ and $r \equiv 0 \pmod 6$, or (d) $p \equiv 6 \pmod 7$ and $r \equiv 0 \pmod 2$.

Using results in Section 3.9, we can examine case (a) in a manner similar to case (a) of Theorem 12.5.1, and using Theorem 12.1.6, we can easily examine cases (c) and (d). We therefore focus only on case (b). The evaluation of the Eisenstein sum in this case is given in the next theorem.

Theorem 12.8.1 *Let p be an odd prime with $p \equiv 2$ or $4 \pmod 7$. Let r be a positive integer with $r \equiv 0 \pmod 3$. Then*

$$E_r(\chi) = p^{(r/3)-1}(a_7 + ib_7\sqrt{7}\,)^{r/3},$$

where a_7 and b_7 are integers satisfying

$$p = a_7^2 + 7b_7^2, \quad a_7 \equiv \begin{cases} 4 \pmod 7, & \text{if } p \equiv 2 \pmod 7, \\ 2 \pmod 7, & \text{if } p \equiv 4 \pmod 7, \end{cases}$$

with the sign of b_7 determined by

$$b_7 \equiv \frac{a_7}{7} \sum_{m=1}^{6} \left(\frac{m}{7}\right) \gamma^{m(q-1)/7} \pmod p,$$

where $\left(\frac{m}{7}\right)$ denotes the Legendre symbol. (The right side of this congruence (mod p) *is in F_p^*.)*

Proof Write $\gamma_1 = \gamma^{(p^r-1)/(p^3-1)}$ and $g = \gamma_1^{1+p+p^2}$. Then χ is the lift of the septic character ψ on F_{p^3} defined by $\psi(\gamma_1) = \beta$. The restriction ψ^* of the character ψ to F_p is given by $\psi^*(g) = \beta^{1+p+p^2} = 1$, as $1 + p + p^2 \equiv 0 \pmod 7$. Thus ψ^* is the trivial character. Hence, by Theorem 12.1.3,

$$E_r(\chi) = p^{(r/3)-1} E_3^{r/3}(\psi).$$

We now evaluate $E_3(\psi)$. Appealing to Lemma 12.0.3, we have

$$E_3(\psi) = E_3(\psi^2) = E_3(\psi^4), \quad E_3(\psi^3) = E_3(\psi^5) = E_3(\psi^6),$$

so that $E_3(\psi)$ is fixed under the automorphism $\sigma_2 : \beta \longrightarrow \beta^2$. Hence $E_3(\psi)$ belongs to the field $Q(\beta + \beta^2 + \beta^4) = Q(i\sqrt{7}\,)$. As $E_3(\psi)$ is an algebraic integer,

$$E_3(\psi) = \frac{1}{2}(G + iH\sqrt{7}\,),$$

for some integers G and H satisfying $G \equiv H \pmod 2$. From Theorem 12.1.2,

$$E_3(\psi)\overline{E_3(\psi)} = p,$$

so that

$$4p = G^2 + 7H^2.$$

If $G \equiv H \equiv 1 \pmod 2$, then $4p \equiv 1^2 + 7 \cdot 1^2 \equiv 0 \pmod 8$, which is a contradiction since p is odd. Hence $G \equiv H \equiv 0 \pmod 2$, say, $G = 2a_7$, $H = 2b_7$, and thus,

$$E_3(\psi) = a_7 + ib_7\sqrt{7}\,,$$

where $p = a_7^2 + 7b_7^2$.

We next determine a_7 modulo 7. By Lemma 12.0.1 and Theorem 12.1.7,

$$p^2 + 3(a_7 + ib_7\sqrt{7}\,) + 3(a_7 - ib_7\sqrt{7}\,) \equiv 0 \pmod 7,$$

so that

$$a_7 \equiv -6a_7 \equiv p^2 \equiv \begin{cases} 4 \pmod{7}, & \text{if } p \equiv 2 \pmod{7}, \\ 2 \pmod{7}, & \text{if } p \equiv 4 \pmod{7}, \end{cases}$$

as asserted.

It remains to prove the congruence for $b_7 \pmod{p}$. We defer the proof until Theorem 12.9.5. ■

12.9 Congruences of Eisenstein for binomial coefficients

The main goal of this section is to prove three congruences for binomial coefficients due to Eisenstein; see Theorems 12.9.7–12.9.9. We will use Stickelberger's theorem and the evaluations of the Eisenstein sums of orders 7 and 8 obtained in the previous two sections. Our proofs are shorter than Eisenstein's original proofs (see the description by H. J. S. Smith [1, pp. 280–283]).

Also in this section, we complete the proofs of Theorems 12.8.1 and 12.7.1 (using Theorems 12.9.5 and 12.9.6, respectively).

Lemma 12.9.1 *Let k be an integer > 2, and let p be a prime with $k \mid (1+p+p^2)$. Define u by*

$$p \equiv u \pmod{k}, \qquad 1 \le u \le k-2.$$

Then the representation of $(p^3-1)/k$ in base p is

$$\left(\frac{p(k-1)-up-1}{k}\right) + \left(\frac{u(p+1)-(k-1)}{k}\right)p + \left(\frac{p-u}{k}\right)p^2. \tag{12.9.1}$$

Proof Since the sum of the three p-adic digits in (12.9.1) is $(p-1)$, it suffices to show that each of the three digits is a nonnegative integer. Each digit is an integer because $p \equiv u \pmod{k}$ and $u^2+u+1 \equiv 0 \pmod{k}$, so it remains to show that each is nonnegative. First, $p-u \ge 0$, since $u = p$ when $p < k$, while $p > k > u$ when $p > k$. Next, $u(p+1) \ge (k-1)$, since $u(p+1) \ge u(u+1) = (u^2+u+1)-1 \ge k-1$. Finally, $p(k-1) > up+1$, since $p(k-1) > (k-2)p+1 \ge up+1$. ■

The following theorem gives a congruence $\pmod{P}$ for certain Eisenstein sums $E_3(\psi)$.

Theorem 12.9.2 *Let k be an integer > 3, and let p be a prime such that $k \mid (1+p+p^2)$. Let χ_P be the character on F_q of order k defined by (11.2.1), where $q = p^r = p^3$. Then*

$$E_3(\chi_P) = -G_3(\chi_P)/p \equiv 0 \pmod{P}.$$

Proof Write $\psi = \chi_P$, $\beta = e^{2\pi i/k}$, and $g = \gamma^{(q-1)/(p-1)}$, where γ is a generator of the cyclic group F_q^* such that $\psi(\gamma) = \beta$. The restriction ψ^* of ψ to F_p satisfies

$$\psi^*(g) = \beta^{1+p+p^2} = 1.$$

Hence, by Theorem 12.1.1,

$$E_3(\psi) = -G_3(\psi)/p. \tag{12.9.2}$$

Define the ideal $\wp$ as in (11.1.1). By Theorem 11.2.2 with $a = 1$ and $r = 3$,

$$\wp^{2(p-1)} \mid\mid G_3(\psi),$$

since the expression $s(1)$ in Theorem 11.2.2 equals $(p-1)$ by (12.9.1). Since $\wp^{p-1} \mid\mid p$ by (11.1.4), it follows that $\wp^{p-1} \mid\mid E_3(\psi)$, so that by (11.1.1), $P \mid\mid E_3(\psi)$. ■

The next two results are variations of the previous two.

Lemma 12.9.3 *Let k be a positive integer divisible by 4, and let p be a prime $\equiv -1 + (k/2) \pmod{k}$. Then the representation of $(p^2-1)/k$ in base p is*

$$\left(\frac{(kp/2) - p - 1}{k}\right) + p\left(\frac{p + 1 - (k/2)}{k}\right). \tag{12.9.3}$$

Proof Since the sum of the two p-adic digits in (12.9.3) is $(p-1)/2$, it suffices to show that each of these two digits is a nonnegative integer. Each is an integer since $p + 1 \equiv k/2 \pmod{k}$. As $(k/2) \mid p+1$, we have $p+1 = ku/2$ for some positive integer u, so that $p = (ku/2) - 1 \geq (k/2) - 1$, that is, $p + 1 - (k/2) \geq 0$. Thus the rightmost p-adic digit in (12.9.3) is nonnegative. Finally, $kp/2 \geq 2p > p+1$. ■

Theorem 12.9.4 *Let $k > 4$ be an integer divisible by 4, and let p be a prime $\equiv -1 + k/2 \pmod{k}$. Let χ_P be the character on F_q of order k defined by (11.2.1), where $q = p^r = p^2$. Then*

$$E_2(\chi_P) = G_2(\chi_P) i^{-(p-1)^2/4} p^{-1/2} \equiv 0 \pmod{P}.$$

Proof Write $\psi = \chi_P$, $\beta = e^{2\pi i/k}$, and $g = \gamma^{(q-1)/(p-1)}$, where γ is a generator of F_q^* such that $\psi(\gamma) = \beta$. The restriction ψ^* of ψ to F_p satisfies $\psi^*(g) = \beta^{1+p} = -1$. Hence, by Theorems 12.1.1 and 1.2.4,

$$E_2(\psi) = G_2(\psi) i^{-(p-1)^2/4} p^{-1/2}. \tag{12.9.4}$$

Define $\wp$ as in (11.1.1). By Theorem 11.2.2 with $a = 1$ and $r = 2$,

$$\wp^{3(p-1)/2} \mid\mid G_2(\psi),$$

as the expression $s(1)$ in Theorem 11.2.2 equals $(p-1)/2$ by (12.9.3). Since $\wp^{p-1} \mid\mid p$ by (11.1.4), it follows that $\wp^{p-1} \mid\mid E_2(\psi)$, so that by (11.1.1), $P \mid\mid E_2(\psi)$. ■

The following theorem completes the proof of Theorem 12.8.1.

Theorem 12.9.5 *Let p be an odd prime with $p \equiv 2$ or $4 \pmod 7$, and write $q = p^3$. Let γ generate F_q^*, and let ψ be the character on F_q of order 7 such that $\psi(\gamma) = \beta$, where $\beta = e^{2\pi i/7}$. Then*

$$E_3(\psi) = a_7 + ib_7\sqrt{7},$$

where a_7 and b_7 are integers with $p = a_7^2 + 7b_7^2$ whose signs are determined by

$$a_7 \equiv 6 - p \pmod 7$$

and

$$b_7 \equiv \frac{a_7}{7} \sum_{m=1}^{6} \left(\frac{m}{7}\right) \gamma^{m(q-1)/7} \pmod p,$$

where $\left(\frac{m}{7}\right)$ is the Legendre symbol. (The right side above is an element of F_p^*.)

Proof Once we prove the last congruence $\pmod p$, this follows from Theorem 12.8.1 with $r = 3$. We wish to apply Theorem 12.9.2 with $k = 7$. The character χ_P in Theorem 12.9.2 equals ψ for some choice of P, and for this P,

$$a_7 + ib_7\sqrt{7} = E_3(\psi) \equiv 0 \pmod P. \tag{12.9.5}$$

By Theorem 1.2.4,

$$i\sqrt{7} = \sum_{m=1}^{6} \left(\frac{m}{7}\right) \beta^m.$$

By (11.2.1),

$$\beta^m = \psi(\gamma^m) \equiv \gamma^{m(q-1)/7} \pmod P,$$

and so

$$7b_7/a_7 \equiv i\sqrt{7} \equiv \sum_{m=1}^{6} \left(\frac{m}{7}\right) \gamma^{m(q-1)/7} \pmod P.$$

As p is a quadratic residue $\pmod 7$, the right side is fixed by the automorphism of F_q defined by $\gamma \longrightarrow \gamma^p$, and hence the right side lies in F_p^*. This yields the desired congruence for $b_7 \pmod p$. ■

The following theorem completes the proof of Theorem 12.7.1(b).

Theorem 12.9.6 *Let p be a prime with $p \equiv 3 \pmod 8$, and write $q = p^2$. Let γ generate F_q^*, and let ψ be the character on F_q of order 8 such that $\psi(\gamma) = \beta$, where $\beta = e^{2\pi i/8}$. Then*

$$E_2(\psi) = -(a_8 + ib_8\sqrt{2}),$$

where a_8 and b_8 are integers with $p = a_8^2 + 2b_8^2$ whose signs are determined by

$$a_8 \equiv (-1)^{(p-3)/8} \pmod 4$$

and

$$b_8 \equiv \frac{a_8}{2}(\gamma^{(q-1)/8} - \gamma^{(1-q)/8}) \pmod{p}.$$

(The right side above is in F_p^*.)

Proof Once we prove the last congruence $\pmod{p}$, this follows from Theorem 12.7.1(b) with $r = 2$. We wish to apply Theorem 12.9.4 with $k = 8$. The character χ_P in Theorem 12.9.4 equals ψ for some choice of P, and for this P,

$$a_8 + ib_8\sqrt{2} = -E_2(\psi) \equiv 0 \pmod{P}. \tag{12.9.6}$$

Since $\beta = (1+i)/\sqrt{2}$,

$$i\sqrt{2} = \beta - \beta^{-1}.$$

By (11.2.1),

$$\beta = \psi(\gamma) \equiv \gamma^{(q-1)/8} \pmod{P},$$

and so

$$2b_8/a_8 \equiv i\sqrt{2} \equiv \gamma^{(q-1)/8} - \gamma^{(1-q)/8} \pmod{P}.$$

Since $p \equiv 3 \pmod 8$, the right side is fixed by the automorphism of F_q defined by $\gamma \longrightarrow \gamma^p$, and hence the right side lies in F_p^*. This yields the desired congruence for $b_8 \pmod{p}$. ∎

We proceed to prove the three congruences for binomial coefficients due to Eisenstein.

Theorem 12.9.7 *Let p be a prime $\equiv 3 \pmod 8$, say, $p = 8f + 3$. Then*

$$\binom{4f+1}{f} \equiv -2a_8 \pmod{p},$$

where a_8 is the integer determined uniquely by

$$p = a_8^2 + 2b_8^2, \quad a_8 \equiv (-1)^f \pmod 4.$$

Proof By Theorem 12.9.6, it suffices to show that

$$\binom{4f+1}{f} \equiv E_2(\psi) + E_2(\overline{\psi}) \pmod{P},$$

where P is chosen as in the proof of Theorem 12.9.6. As $P \mid E_2(\psi)$ by (12.9.6), we must show that

$$\binom{4f+1}{f} \equiv E_2(\overline{\psi}) \pmod{P}.$$

Using (12.9.4), we obtain the following equation:

$$E_2(\overline{\psi}) = G_2(\overline{\psi})/(i\sqrt{p}).$$

By Theorem 11.2.1 and (12.9.3) with $k = 8$,

$$G_2(\overline{\psi}) \equiv \frac{-\pi^{(p-1)/2}}{(3f+1)!\,f!} \pmod{\wp^{(p+1)/2}},$$

where $\pi = e^{2\pi i/p} - 1$ and $\wp$ is as in (11.1.1). The numerator on the right equals

$$(4f+1)!\,i\sqrt{p} + \delta\pi^{(p+1)/2}$$

for some algebraic integer δ, by Theorem 11.2.5. Thus, in view of (11.1.6) and (11.1.7),

$$E_2(\overline{\psi}) = G_2(\overline{\psi})/(i\sqrt{p}) \equiv \frac{(4f+1)!}{(3f+1)!\,f!} + \frac{\delta\pi^{(p+1)/2}}{i\sqrt{p}(3f+1)!\,f!} \pmod{\wp}.$$

The rightmost term is divisible by π and hence by $\wp$, so that

$$E_2(\overline{\psi}) \equiv \binom{4f+1}{f} \pmod{\wp},$$

and in view of (11.1.1), this congruence also holds $\pmod{P}$. ∎

Theorem 12.9.8 *Let p be an odd prime $\equiv 2 \pmod 7$, say, $p = 7f + 2$. Then*

$$\binom{3f}{f} \equiv -2a_7 \pmod{p},$$

where a_7 is the integer uniquely determined by

$$p = a_7^2 + 7b_7^2, \quad a_7 \equiv 4 \pmod 7.$$

Proof By Theorem 12.9.5, it suffices to show that

$$\binom{3f}{f} \equiv -E_3(\psi) - E_3(\overline{\psi}) \pmod{P},$$

where P is chosen as in the proof of Theorem 12.9.5. As $P \mid E_3(\psi)$ by (12.9.5), we must show that

$$\binom{3f}{f} \equiv -E_3(\overline{\psi}) \pmod{P}.$$

By (12.9.2),

$$-E_3(\overline{\psi}) = G_3(\overline{\psi})/p.$$

By Theorem 11.2.1 and (12.9.1) with $k = 7$,

$$G_3(\overline{\psi}) \equiv \frac{-\pi^{p-1}}{(4f+1)!(2f)!f!} \pmod{\wp^p},$$

where $\pi = e^{2\pi i/p} - 1$ and $\wp$ is as in (11.1.1). The numerator on the right equals $p + \delta\pi p$ for some algebraic integer δ, by (11.1.5) and Wilson's theorem. Thus, in view of (11.1.4),

$$-E_3(\overline{\psi}) = G_3(\overline{\psi})/p \equiv \frac{1}{(4f+1)!(2f)!f!} + \frac{\delta\pi}{(4f+1)!(2f)!f!} \pmod{\wp}.$$

By (11.1.7), the rightmost term is divisible by $\wp$, so that

$$-E_3(\overline{\psi}) \equiv \frac{1}{(4f+1)!(2f)!f!} \pmod{\wp}.$$

By (11.1.1), this congruence also holds $\pmod{P}$, and so it remains to prove that

$$\frac{1}{(4f+1)!(2f)!f!} \equiv \binom{3f}{f} \pmod{p}.$$

This is equivalent to the easy congruence

$$(3f)!(p-1-3f)! \equiv 1 \pmod{p},$$

which is a special case of Problem 1 in Exercises 9. ∎

Theorem 12.9.9 *Let p be a prime $\equiv 4 \pmod 7$, say $p = 7f + 4$. Then*

$$\binom{3f+1}{f} \equiv 2a_7 \pmod{p},$$

where a_7 is the integer uniquely determined by

$$p = a_7^2 + 7b_7^2, \quad a_7 \equiv 2 \pmod 7.$$

Proof The proof is completely analogous to that of Theorem 12.9.8. ∎

We close this section with a congruence (viz., (12.9.7)) for Eisenstein sums $E_r(\chi)$ that generalizes those congruences in Theorems 12.9.2 and 12.9.4.

Theorem 12.9.10 *Let $k > 2$ be an integer and let p be a prime of order $r \pmod k$. Let χ_P be the character on F_q of order k defined by (11.2.1), where $q = p^r$. Then*

(12.9.7) $$E_r(\chi_P) \equiv 0 \pmod{P}$$

if and only if

(12.9.8) $$p \not\equiv 1 \pmod{k} \quad \textit{and} \quad p \not\equiv -1 \pmod{k}.$$

Proof Write $\psi = \chi_P$. If $p \equiv 1 \pmod{k}$, then $r = 1$ and $E_r(\psi) = 1$ by Lemma 12.0.1(a). If $p \equiv -1 \pmod{k}$, then $r = 2$ and $E_r(\psi) = -(-1)^{(p+1)/k}$, by Theorem 12.2.3.

Conversely, assume that (12.9.8) holds. Write $v = (q-1)/(p-1)$. If $k \nmid v$, then the restriction ψ^* of ψ to F_p is nontrivial by Corollary 11.4.2, so that by Theorem 12.1.1,

$$E_r(\psi) = G_r(\psi)/G(\psi^*), \quad \text{if } k \nmid v. \tag{12.9.9}$$

If $k \mid v$, then Theorem 12.1.1 yields

$$E_r(\psi) = -G_r(\psi)/p, \quad \text{if } k \mid v. \tag{12.9.10}$$

Case 1: $k \nmid v$. The lift of ψ^* from F_p to F_q is ψ^v by Theorem 11.4.4(b). Thus, by the Davenport–Hasse lifting theorem (Theorem 11.5.2),

$$G_r(\psi^v) = (-1)^{r-1} G^r(\psi^*).$$

Therefore, by (12.9.9),

$$(-1)^{r-1} E_r^r(\psi) = G_r^r(\psi)/G_r(\psi^v).$$

By Theorem 11.2.2 with $a = 1$,

$$\wp^{(r^2-r)(p-1)-rs(1)+s(v)} \,\|\, G_r^r(\psi)/G_r(\psi^v),$$

and so in view of (11.1.1), it remains to show that

$$(r^2 - r)(p-1) - rs(1) + s(v) > 0.$$

Write

$$f = (q-1)/k = a_0 + a_1 p + \cdots + a_{r-1} p^{r-1}, \quad 0 \le a_i \le p-1,$$

so that by (11.2.4),

$$s(1) = a_0 + \cdots + a_{r-1}.$$

By (11.2.5), $s(1) \equiv f \pmod{p-1}$. Since $k \nmid v$, we see that $(p-1) \nmid f$, and so $s(1) \equiv f \equiv w \pmod{p-1}$ for some integer w with $0 < w < p-1$. Since

$$vf \equiv w + wp + \cdots + wp^{r-1} \pmod{q-1},$$

we have $s(v) = rw$. Thus

$$(r^2 - r)(p-1) - rs(1) + s(v) = (r^2 - r)(p-1) - r(s(1) - w),$$

and so it remains to show that $(r-1)(p-1) > s(1) - w$. By Theorem 11.2.7 with $a = 1$,

$$\frac{k}{(p-1)} s(1) = 1 + \sum_{i=1}^{r-1} L(p^i) < 1 + (r-1)(k-1) \le (r-1)k, \tag{12.9.11}$$

since $L(p) < k-1$ for $p \not\equiv -1 \pmod{k}$. Therefore

$$s(1) < (r-1)(p-1) < (r-1)(p-1) + w.$$

This completes the proof in the case $k \nmid v$.

Case 2: $k \mid v$. By Theorem 11.2.2 with $a = 1$, $\wp^{r(p-1)-s(1)} \parallel G_r(\psi)$, and by (11.1.4), $\wp^{p-1} \parallel p$. Therefore, in view of (12.9.10), it remains to show that $r(p-1)-s(1) > (p-1)$, that is, $s(1) < (r-1)(p-1)$. This follows from (12.9.11). ■

12.10 Gauss sums and f-nomial periods over F_q

Let p be a prime, and let r be a positive integer. Set $q = p^r$, and let k be a positive integer. As in Section 1.1, define the Gauss sum $g_r(k)$ by

$$g_r(k) = \sum_{\alpha \in F_q} e^{2\pi i \operatorname{tr}(\alpha^k)/p} = \sum_{\alpha \in F_q} e(\alpha^k).$$

Since

$$\{\alpha^k : \alpha \in F_q\} = \{\alpha^d : \alpha \in F_q\},$$

where $d = \gcd(k, q-1)$, we see that

(12.10.1) $$g_r(k) = g_r(d).$$

Thus, in particular, if $\gcd(k, q-1) = 1$, then

(12.10.2) $$g_r(k) = g_r(1) = \sum_{\alpha \in F_q} e^{2\pi i \operatorname{tr}(\alpha)/p} = 0,$$

by Proposition 1.1.2(c). Since $d \mid q-1$, equality (12.10.1) shows that the evaluation of $g_r(k)$ reduces to the case when $k \mid q-1$. In this case we show that $g_r(k)$ can be expressed as a linear combination of Eisenstein sums, and we apply this result to evaluate $g_r(3)$ and $g_r(4)$.

Theorem 12.10.1 *Let k be a divisor of $q-1$, and let χ be a character of order k on F_q. Let χ^* denote the restriction of the character χ to F_p. Set*

$$k^* = \frac{k}{\gcd(k, (q-1)/(p-1))}.$$

Then

$$g_r(k) = \sum_{\substack{j=1 \\ j \not\equiv 0 \pmod{k^*}}}^{k-1} G(\chi^{*j}) E_r(\chi^j) - p \sum_{\substack{j=1 \\ j \equiv 0 \pmod{k^*}}}^{k-1} E_r(\chi^j).$$

Proof By Proposition 11.4.1, χ^* has order k^*. Hence $(\chi^j)^* = (\chi^*)^j$ is trivial if and only if $k^* \mid j$ Thus, by Theorem 12.1.1,

$$\sum_{\substack{j=1 \\ j \not\equiv 0 \pmod{k^*}}}^{k-1} G(\chi^{*j})E_r(\chi^j) - p \sum_{\substack{j=1 \\ j \equiv 0 \pmod{k^*}}}^{k-1} E_r(\chi^j) = \sum_{j=1}^{k-1} G_r(\chi^j).$$

As α runs through F_q^*, α^k runs through the kth powers in F_q^* k times. Thus

$$(12.10.3) \quad g_r(k) = 1 + \sum_{\alpha \in F_q^*} e(\alpha^k)$$

$$= 1 + \sum_{\alpha \in F_q^*} e(\alpha)\{1 + \chi(\alpha) + \cdots + \chi^{k-1}(\alpha)\} = \sum_{j=1}^{k-1} G_r(\chi^j). \quad \blacksquare$$

In the next three theorems, we evaluate $g_r(2)$, $g_r(3)$, and $g_r(4)$.

Theorem 12.10.2 *For an odd prime p,*

$$g_r(2) = \begin{cases} (-1)^{((p-1)r/4)+1} p^{r/2}, & \text{if } r \equiv 0 \pmod 2, \\ i^{(p-1)^2/4}(-1)^{(p-1)(r-1)/4} p^{r/2}, & \text{if } r \equiv 1 \pmod 2. \end{cases}$$

Proof By (12.10.3), $g_r(2) = G_r(\rho)$, where ρ is the quadratic character on F_q. The result now follows from Theorem 11.5.4. $\blacksquare$

Theorem 12.10.3

(a) *If $p \equiv 1 \pmod 3$ and $r \equiv 0 \pmod 3$, then*

$$g_r(3) = 2(-1)^{r+1} p^{r/3} \mathrm{Re} \left(\frac{1}{2}(r_3 + i s_3 \sqrt{3}\,)\right)^{r/3}.$$

(b) *If $p \equiv 1 \pmod 3$ and $r \equiv 1 \pmod 3$, then*

$$g_r(3) = 2(-1)^{r+1} p^{r/3} \mathrm{Re} \left(H(\chi^*) \left(\frac{1}{2}(r_3 + i s_3 \sqrt{3}\,)\right)^{(r+2)/3}\right),$$

where H is defined in (4.1.4).

(c) *If $p \equiv 1 \pmod 3$ and $r \equiv 2 \pmod 3$, then*

$$g_r(3) = 2(-1)^{r+1} p^{r/3} \mathrm{Re} \left(H^{-1}(\chi^{*2}) \left(\frac{1}{2}(r_3 + i s_3 \sqrt{3}\,)\right)^{(r-2)/3}\right).$$

(d) *If $p \equiv 2 \pmod 3$ and $r \equiv 0 \pmod 2$, then*

$$g_r(3) = 2(-1)^{(r+2)/2} p^{r/2}.$$

(e) *If $p \equiv 2 \pmod 3$ and $r \equiv 1 \pmod 2$, then*

$$g_r(3) = 0.$$

Proof Suppose first that $p \equiv 1 \pmod 3$. In the notation of Theorem 12.10.1 with $k = 3$,

$$k^* = \frac{3}{\gcd(3,(p^r-1)/(p-1))} = \frac{3}{\gcd(3,1+p+\cdots+p^{r-1})} = \begin{cases} 1, & \text{if } r \equiv 0 \pmod 3, \\ 3, & \text{if } r \not\equiv 0 \pmod 3. \end{cases}$$

(a) If $r \equiv 0 \pmod 3$, then by Theorems 12.10.1 and 12.3.1,

$$\begin{aligned} g_r(3) &= -p(E_r(\chi)+E_r(\chi^2)) = -2p\text{Re}\,(E_r(\chi)) \\ &= -2p\ \text{Re}\left((-1)^r p^{(r/3)-1}\left(\frac{1}{2}(r_3+is_3\sqrt{3}\,)\right)^{r/3}\right) \\ &= 2(-1)^{r+1}p^{r/3}\text{Re}\left(\frac{1}{2}(r_3+is_3\sqrt{3}\,)\right)^{r/3}. \end{aligned}$$

(b) If $r \equiv 1 \pmod 3$, then appealing to Theorems 12.10.1, 4.1.6, and 12.3.1, we deduce that

$$\begin{aligned} g_r(3) &= G(\chi^*)E_r(\chi)+G(\chi^{*2})E_r(\chi^2) = 2\ \text{Re}\,(G(\chi^*)E_r(\chi)) \\ &= 2\text{Re}\left(\left\{p^{1/3}\frac{1}{2}(r_3+is_3\sqrt{3})H(\chi^*)\right\}\left\{(-1)^{r-1}p^{(r-1)/3}\left(\frac{1}{2}(r_3+is_3\sqrt{3})\right)^{(r-1)/3}\right\}\right) \\ &= 2(-1)^{r+1}p^{r/3}\text{Re}\left(H(\chi^*)\left(\frac{1}{2}(r_3+is_3\sqrt{3}\,)\right)^{(r+2)/3}\right). \end{aligned}$$

(c) If $r \equiv 2 \pmod 3$, then, by Theorems 12.10.1, 4.1.6, and 12.3.1,

$$\begin{aligned} g_r(3) &= G(\chi^*)E_r(\chi)+G(\chi^{*2})E_r(\chi^2) \\ &= 2\,\text{Re}\left(\left\{p^{1/3}\frac{1}{2}(r_3-is_3\sqrt{3})H(\chi^*)\right\}\left\{(-1)^{r-1}p^{(r-2)/3}\left(\frac{1}{2}(r_3+is_3\sqrt{3}\,)\right)^{(r+1)/3}\right\}\right) \\ &- 2(-1)^{r+1}p^{(r+2)/3}\text{Re}\left(H(\chi^*)\left(\frac{1}{2}(r_3+is_3\sqrt{3}\,)\right)^{(r-2)/3}\right). \end{aligned}$$

Now, from (4.1.5),

$$H(\chi^*)H(\chi^{*2}) = p^{-2/3},$$

so that

$$g_r(3) = 2(-1)^{r+1}p^{r/3}\text{Re}\left(H^{-1}(\chi^{*2})\left(\frac{1}{2}(r_3+is_3\sqrt{3})\right)^{(r-2)/3}\right).$$

Now suppose that $p \equiv 2 \pmod 3$.

(d) If $r \equiv 0 \pmod 2$, then

$$k^* = \frac{3}{\gcd(3,1-1+1-\cdots-1)} = \frac{3}{\gcd(3,0)} = 1,$$

and, by Theorems 12.10.1 and 12.3.1, we then find that

$$\begin{aligned} g_r(3) &= -p(E_r(\chi)+E_r(\chi^2)) \\ &= -p\left((-1)^{r/2}p^{(r/2)-1}+(-1)^{r/2}p^{(r/2)-1}\right) = 2(-1)^{(r+2)/2}p^{r/2}. \end{aligned}$$

(e) If $r \equiv 1 \pmod 2$, then $\gcd(k, q-1) = 1$, and $g_r(3) = 0$ by (12.10.2). ∎

Theorem 12.10.4

(a) *If* $p \equiv 1 \pmod 4$ *and* $r \equiv 0 \pmod 4$, *then*

$$g_r(4) = -p^{r/2} - p^{r/4}\left((a_4+ib_4)^{r/2} + (a_4-ib_4)^{r/2}\right).$$

(b) *If* $p \equiv 1 \pmod 4$ *and* $r \equiv 1 \pmod 4$, *then*

$$g_r(4) = p^{r/2} + p^{(r-1)/4}\left(G(\chi^*)(a_4+ib_4)^{(r-1)/2} + G(\chi^{*3})(a_4-ib_4)^{(r-1)/2}\right),$$

where the value of $G(\chi^*)$ *is given in Theorem* 4.2.3.

(c) *If* $p \equiv 1 \pmod 4$ *and* $r \equiv 2 \pmod 4$, *then*

$$g_r(4) = -p^{r/2} + (-1)^{(p+3)/4}p^{r/4}\left((a_4+ib_4)^{r/2} + (a_4-ib_4)^{r/2}\right).$$

(d) *If* $p \equiv 1 \pmod 4$ *and* $r \equiv 3 \pmod 4$, *then*

$$g_r(4) = p^{r/2} + (-1)^{(p-1)/4}p^{(r-3)/4}\left(G(\chi^*)(a_4+ib_4)^{(r+1)/2} + G(\chi^{*3})(a_4-ib_4)^{(r+1)/2}\right),$$

where the value of $G(\chi^{*3})$ *is given in Theorem* 4.2.3.

(e) *If* $p \equiv 3 \pmod 4$ *and* $r \equiv 0 \pmod 2$, *then*

$$g_r(4) = -(2(-1)^{(p-3)r/8} + (-1)^{r/2})p^{r/2}.$$

(f) *If* $p \equiv 3 \pmod 4$ *and* $r \equiv 1 \pmod 2$, *then*

$$g_r(4) = i(-1)^{(r-1)/2}p^{r/2}.$$

Proof We first examine the case $p \equiv 1 \pmod 4$. In the notation of Theorem 12.10.1 with $k = 4$,

$$k^* = \frac{4}{\gcd(4,(q-1)/(p-1))} = \frac{4}{\gcd(4, 1+p+\cdots+p^{r-1})} = \begin{cases} 1, & \text{if } r \equiv 0 \pmod 4, \\ 2, & \text{if } r \equiv 2 \pmod 4, \\ 4, & \text{if } r \equiv 1 \pmod 2. \end{cases}$$

(a) If $r \equiv 0 \pmod 4$, then by Theorems 12.10.1, 12.4.1, and 12.1.5,

$$\begin{aligned} g_r(4) &= -p(E_r(\chi)+E_r(\chi^2)+E_r(\chi^3)) \\ &= -p\left(p^{(r/4)-1}(a_4+ib_4)^{r/2}+p^{(r/2)-1}+p^{(r/4)-1}(a_4-ib_4)^{r/2}\right) \\ &= -p^{r/2}-p^{r/4}\left((a_4+ib_4)^{r/2}+(a_4-ib_4)^{r/2}\right). \end{aligned}$$

(b) If $r \equiv 1 \pmod 4$, then by Theorems 12.10.1, 12.4.1, 12.1.5, and 1.2.4,

$$\begin{aligned} g_r(4) &= G(\chi^*)E_r(\chi)+G(\chi^{*2})E_r(\chi^2)+G(\chi^{*3})E_r(\chi^3) \\ &= G(\chi^*)p^{(r-1)/4}(a_4+ib_4)^{(r-1)/2}+p^{r/2}+G(\chi^{*3})p^{(r-1)/4}(a_4-ib_4)^{(r-1)/2} \\ &= p^{r/2}+p^{(r-1)/4}\{G(\chi^*)(a_4+ib_4)^{(r-1)/2}+G(\chi^{*3})(a_4-ib_4)^{(r-1)/2}\}, \end{aligned}$$

where the value of $G(\chi^*)$ is given by Theorem 4.2.3.

(c) If $r \equiv 2 \pmod 4$, then χ^* has order $k^*=2$, so that by Theorems 12.10.1, 1.2.4, 12.4.1, and 12.1.5,

$$\begin{aligned} g_r(4) &= G(\chi^*)E_r(\chi)+G(\chi^{*3})E_r(\chi^3)-pE_r(\chi^2) \\ &= p^{1/2}(-1)^{(p+3)/4}p^{(r-2)/4}(a_4+ib_4)^{r/2} \\ &\qquad +p^{1/2}(-1)^{(p+3)/4}p^{(r-2)/4}(a_4-ib_4)^{r/2}-p^{r/2} \\ &= -p^{r/2}+(-1)^{(p+3)/4}p^{r/4}((a_4+ib_4)^{r/2}+(a_4-ib_4)^{r/2}). \end{aligned}$$

(d) If $r \equiv 3 \pmod 4$, then by Theorems 12.10.1, 12.4.1, 12.1.5, and 1.2.4,

$$\begin{aligned} g_r(4) &= G(\chi^*)E_r(\chi)+G(\chi^{*2})E_r(\chi^2)+G(\chi^{*3})E_r(\chi^3) \\ &= G(\chi^*)(-1)^{(p-1)/4}p^{(r-3)/4}(a_4+ib_4)^{(r+1)/2}+p^{r/2} \\ &\qquad +G(\chi^{*3})(-1)^{(p-1)/4}p^{(r-3)/4}(a_4-ib_4)^{(r+1)/2} \\ &= p^{r/2}+(-1)^{(p-1)/4}p^{(r-3)/4}\{G(\chi^*)(a_4+ib_4)^{(r+1)/2}+G(\chi^{*3})(a_4-ib_4)^{(r+1)/2}\}, \end{aligned}$$

where the value of $G(\chi^{*3})$ is given in Theorem 4.2.3.

We now turn to the case $p \equiv 3 \pmod 4$. In this case,

$$k^* = \frac{4}{\gcd(4, 1-1+1-\cdots+(-1)^{r-1})} = \begin{cases} 1, & \text{if } r \equiv 0 \pmod 2, \\ 4, & \text{if } r \equiv 1 \pmod 2. \end{cases}$$

(e) If $r \equiv 0 \pmod 2$, then by Theorems 12.10.1, 12.4.1, and 12.1.5,

$$\begin{aligned} g_r(4) &= -p(E_r(\chi)+E_r(\chi^2)+E_r(\chi^3)) \\ &= -p((-1)^{(p-3)r/8}p^{(r/2)-1}+(-1)^{(p-1)r/4}p^{(r/2)-1}+(-1)^{(p-3)r/8}p^{(r/2)-1}) \\ &= -p^{r/2}(2(-1)^{(p-3)r/8}+(-1)^{r/2}). \end{aligned}$$

(f) If $r \equiv 1 \pmod 2$, then $\gcd(k, q-1) = \gcd(4,2) = 2$, so that, by (12.10.1) and Theorem 12.10.2,

$$g_r(4) = g_r(2) = i(-1)^{(r-1)/2}p^{r/2}.$$ ■

The evaluation of $g_r(k)$ for other small values of k such as $k=6$ and $k=8$ can be obtained in a similar manner (Problem 12).

We conclude this section by extending the concepts of f-nomial periods and period polynomials, defined in Section 10.10, from F_p to F_q. As in the case with F_p, the f-nomial periods are intimately connected with Gauss sums over F_q; see (12.10.8) below. However, the period polynomial behaves quite differently in the case of F_q.

Write $q = p^r$. When k divides $q-1$, say, $q = kf+1$, we define the f-nomial periods of F_q with respect to the generator γ of F_q^* to be the complex numbers

(12.10.4) $$\eta_{s,r} = \eta_{s,r}(\gamma) = \sum_{v=0}^{f-1} e^{2\pi i \,\mathrm{tr}(\gamma^{kv+s})/p}, \quad s = 0, 1, \ldots, k-1.$$

The period polynomial of degree k for F_q is the polynomial

(12.10.5) $$P_{k,r}(x) = \prod_{s=0}^{k-1}(x - \eta_{s,r}).$$

The reduced f-nomial periods of F_q are the complex numbers

(12.10.6) $$\eta_{s,r}^* = \eta_{s,r}^*(\gamma) = 1 + k\eta_{s,r}(\gamma), \quad s = 0, 1, \ldots, k-1,$$

and the reduced period polynomial of degree k for F_q is

(12.10.7) $$P_{k,r}^*(x) = \prod_{s=0}^{k-1}(x - \eta_{s,r}^*).$$

From the definitions at the beginning of Section 10.10 and from (12.10.4)–(12.10.7), we see that $\eta_{s,1} = \eta_s$, $P_{k,1}(x) = P_k(x)$, $\eta_{s,1}^* = \eta_s^*$, and $P_{k,1}^*(x) = P_k^*(x)$. Exactly as in Section 10.10, we find that

(12.10.8) $$g_r(k) = \eta_{0,r}^*, \qquad \sum_{s=0}^{k-1} \eta_{s,r}^* = 0.$$

The polynomials $P_{k,r}(x)$ and $P_{k,r}^*(x)$ have integral coefficients and are independent of the choice of generator γ. When $r=1$ the polynomials $P_{k,r}(x) = P_k(x)$ and $P_{k,r}^*(x) = P_k^*(x)$ are irreducible over Q, but this is not generally the case when $r > 1$. Indeed $P_{k,r}(x)$ splits over the rationals into $k/k^* = \gcd(k, (q-1)/(p-1))$ factors $P^{(w)}(x)$, $1 \le w \le k/k^*$, where each $P^{(w)}(x) = \prod_{j=0}^{k^*-1}(x - \eta_{w+jk/k^*,r})$, and each of these factors is irreducible or a power of an irreducible polynomial. Moreover, $P_{k,r}(x)$ and

$P^*_{k,r}(x)$ are irreducible over the rationals if and only if $\gcd(k,(q-1)/(p-1)) = 1$. For a proof of these facts, see the paper of Myerson [1]. For information on the factors $P^{(w)}(x)$, see Gurak [2]–[6].

We now examine the period polynomial of degree 2. Let $k = 2$ and $q = 2f+1$. By (12.10.3), $g_r(2) = G_r(\rho)$ for the quadratic character ρ on F_q. Thus, by Theorem 1.1.4(a) and (12.10.8),

$$\eta^*_{0,r} = g_r(2) = \epsilon p^{r/2},$$

where ϵ is a power of i such that $\epsilon^2 = (-1)^f$. As $\eta^*_{0,r} + \eta^*_{1,r} = 0$, we deduce that

$$\eta^*_{1,r} = -\eta^*_{0,r} = -\epsilon p^{r/2}.$$

Thus

$$P^*_{2,r}(x) = (x-\eta^*_{0,r})(x-\eta^*_{1,r}) = (x-\epsilon p^{r/2})(x+\epsilon p^{r/2}) = x^2 - \epsilon^2 p^r = x^2 - (-1)^f q.$$

Taking $r = 1$, we recover Theorem 10.10.4. Clearly $P^*_{2,r}(x) = (x-p^{r/2})(x+p^{r/2})$, if r is even, and $P^*_{2,r}$ is irreducible, if r is odd. Since $k^* = 1$ if r is even and $k^* = 2$ if r is odd, this is in agreement with the remarks above.

We next examine the period polynomial of degree 3. Let $k = 3$ and define integers c and d by

$$\frac{1}{2}(c + id\sqrt{3}\,) = \begin{cases} (-1)^{r+1}\left(\frac{1}{2}(r_3 + is_3\sqrt{3}\,)\right)^r, & \text{if } p \equiv 1 \pmod 3, \\ (-1)^{(r+2)/2} p^{r/2}, & \text{if } p \equiv 2 \pmod 3, \end{cases}$$

so that

$$4p^r = c^2 + 3d^2, \quad c \equiv 1 \pmod 3, \quad d \equiv 0 \pmod 3,$$

with

$$p \nmid c, \qquad \text{if } p \equiv 1 \pmod 3.$$

These conditions determine c uniquely. This follows from Lemma 3.0.1 when $p \equiv 1 \pmod 3$. In the case $p \equiv 2 \pmod 3$, it follows because $c^2 + 3d^2 \equiv 0 \pmod p$ has no solutions c, d with $p \nmid c$. Using (4.1.5), we deduce from Theorem 12.10.3 that $g_r(3)$ is a root of the cubic polynomial $x^3 - 3p^r x - cp^r$. Indeed, it is not difficult to show that

$$P^*_{3,r}(x) = x^3 - 3p^r x - cp^r$$

and

$$P_{3,r}(x) = x^3 + x^2 - ((p^r-1)/3)x - ((c+3)p^r - 1)/27,$$

which agrees with Theorem 10.10.5 when $r = 1$. The factorization of $P^*_{3,r}(x)$ is as follows:

irreducible, if $p \equiv 1 \pmod 3$ and $r \not\equiv 0 \pmod 3$,

$(x + 2(-1)^{r/2}p^{r/2})(x - (-1)^{r/2}p^{r/2})^2$, if $p \equiv 2 \pmod 3$ and $r \equiv 0 \pmod 2$,

$(x - c_1 p^{r/3})(x + \frac{1}{2}(c_1 + 9d_1)p^{r/3})(x + \frac{1}{2}(c_1 - 9d_1)p^{r/3})$,

if $p \equiv 1 \pmod 3$ and $r \equiv 0 \pmod 3$,

where $4p^{r/3} = c_1^2 + 27d_1^2$, $c_1 \equiv 1 \pmod 3$, $p \nmid c_1$.

The determination and the factorization of the period polynomial $P^*_{4,r}(x)$ can be found in a similar manner (Problems 13 and 14).

12.11 An Eisenstein sum of order 20

The following theorem will be needed for the evaluation of the Brewer sum Λ_5 in Section 13.4.

Theorem 12.11.1 *Let p be a prime with $p \equiv 9 \pmod{20}$, so that $q = p^2 = 20f+1$ for an integer f. Let γ be a generator of F_q^*, and let ψ be the character of order 20 on F_q given by $\psi(\gamma) = \beta$, where $\beta = e^{2\pi i/20}$. Then*

$$E_2(\psi) = \begin{cases} \left(\frac{2}{p}\right)(a_{20} + ib_{20}\sqrt{5}), & if \;\; 5 \nmid a_4, \\ \left(\frac{2}{p}\right)i(a_{20} + ib_{20}\sqrt{5}), & if \;\; 5 \mid a_4, \end{cases}$$

where a_{20} and b_{20} are integers such that

$$p = a_{20}^2 + 5b_{20}^2 \tag{12.11.1}$$

whose signs are determined by

$$a_{20} \equiv \begin{cases} -a_4 \pmod 5, & if \;\; 5 \nmid a_4, \\ -b_4 \pmod 5, & if \;\; 5 \mid a_4, \end{cases} \tag{12.11.2}$$

and

$$b_{20} \equiv \frac{a_{20}}{5}(\gamma^f + \gamma^{3f} - \gamma^{-f} - \gamma^{-3f}) \pmod p. \tag{12.11.3}$$

(The right side of (12.11.3) *is an element of F_p^*.)*

Proof By Proposition 11.4.1, the restriction ψ^* of ψ to F_p is the quadratic character. Hence, by Theorem 12.2.1,

$$E_2(\psi) = \left(\frac{2}{p}\right)E(\psi),$$

and by Theorem 12.1.2,

$$E(\psi)E(\overline{\psi}) = p.$$

Define $\sigma_j \in \mathrm{Gal}(Q(\beta)/Q)$ by $\sigma_j(\beta) = \beta^j$. Then σ_{17} generates a subgroup $< \sigma_{17} >$ of this Galois group whose fixed field is $Q(i)$. By Theorem 12.2.2, $E(\psi) = E(\psi^9)$. Hence

$$\sigma_{17}(E(\psi^3)/E(\psi)) = E(\psi^{11})/E(\psi^{17}) = E(\psi^3)/E(\psi^9) = E(\psi^3)/E(\psi),$$

so that $\epsilon := E(\psi^3)/E(\psi) \in Q(i)$. By Theorem 12.2.4,

$$E(\psi^3) = \psi^3(2)G_2(\psi^3)/G(\psi^{*3})$$

and

$$E(\psi) = \psi(2)G_2(\psi)/G(\psi^*).$$

Since $\psi^* = (\psi^*)^3$,

$$G_2(\psi^3)/G_2(\psi) = E(\psi^3)/E(\psi) = \epsilon.$$

Set $M = Q(\beta, e^{2\pi i/p})$. We now use the prime ideal factorization of Gauss sums in O_M (Theorem 11.2.2) to show that $\epsilon = G_2(\psi^3)/G_2(\psi)$ is a unit of O_M. By (12.9.3),

$$f = \left(\frac{9p-1}{20}\right) + p\left(\frac{p-9}{20}\right),$$

and thus

$$3f = \left(\frac{7p-3}{20}\right) + p\left(\frac{3p-7}{20}\right)$$

and

$$7f = \left(\frac{3p-7}{20}\right) + p\left(\frac{7p-3}{20}\right).$$

It follows from (11.2.4) that

$$s(1) = s(3) = s(7) = (p-1)/2$$

and

$$s(-1) = s(-3) = s(-7) = 3(p-1)/2.$$

Thus, by Theorem 11.2.2, $G_2(\psi)$ and $G_2(\psi^3)$ have exactly the same prime ideal decomposition in O_M. Thus ϵ is a unit of O_M. We must have $\epsilon = \pm 1$ or $\epsilon = \pm i$, since $\epsilon \in Q(i)$. Raising each side of $E(\psi^3) = \epsilon E(\psi)$ to the fifth power, and reducing modulo 5, we obtain, by Theorems 12.2.1 and 12.4.1,

$$-(a_4 - ib_4) = E(\psi^{15}) \equiv E^5(\psi^3) = \epsilon^5 E^5(\psi) \equiv \epsilon E(\psi^5) = -\epsilon(a_4 + ib_4) \pmod 5,$$

so that

$$\epsilon = \begin{cases} 1, & \text{if } 5 \nmid a_4, \\ -1, & \text{if } 5 \mid a_4. \end{cases}$$

Suppose first that $5 \nmid a_4$. Then $\sigma_3(E(\psi)) = E(\psi^3) = E(\psi)$, and so $E(\psi) \in Q(i\sqrt{5})$, the fixed field of $< \sigma_3 > \subset \mathrm{Gal}(Q(\beta)/Q)$. Hence,

(12.11.4) $$E(\psi) = a_{20} + ib_{20}\sqrt{5}, \quad \text{if } 5 \nmid a_4,$$

for integers a_{20} and b_{20} such that $p = a_{20}^2 + 5b_{20}^2$. Raising each side of (12.11.4) to the fifth power and reducing modulo 5, we obtain $-a_4 \equiv a_{20} \pmod 5$. Now suppose that $5 \mid a_4$. Then

$$\sigma_3(iE(\psi)) = -iE(\psi^3) = iE(\psi),$$

so that $iE(\psi) \in Q(i\sqrt{5})$. Hence, we find that

$$E(\psi) = i(a_{20} + ib_{20}\sqrt{5}), \quad \text{if} \quad 5 \mid a_4, \tag{12.11.5}$$

for integers a_{20} and b_{20} such that $p = a_{20}^2 + 5b_{20}^2$. Raising each side of (12.11.5) to the fifth power and reducing modulo 5, we obtain $-b_4 \equiv a_{20} \pmod 5$. It remains to prove (12.11.3). We apply Theorem 12.9.4 with $k = 20$. The character χ_P in Theorem 12.9.4 equals ψ for some choice of P, and for this P,

$$a_{20} + ib_{20}\sqrt{5} \equiv 0 \pmod P. \tag{12.11.6}$$

Note that

$$i\sqrt{5} = \beta + \beta^3 - \overline{\beta} - \overline{\beta}^3. \tag{12.11.7}$$

By (11.2.1),

$$\beta = \psi(\gamma) \equiv \gamma^f \pmod P,$$

and so

$$5b_{20}/a_{20} \equiv i\sqrt{5} \equiv \gamma^f + \gamma^{3f} - \gamma^{-f} - \gamma^{-3f} \pmod P.$$

Since $p \equiv 9 \pmod{20}$, the right side is fixed by the automorphism of F_q defined by $\gamma \longrightarrow \gamma^p$, and hence the right side lies in F_p^*. This completes the proof of (12.11.3). ■.

12.12 An Eisenstein sum of order 16

The following theorem will be needed for the evaluation of the Brewer sum Λ_8 in Section 13.5.

Theorem 12.12.1 *Let p be a prime with $p \equiv 7 \pmod{16}$, so that $q = p^2 = 16f + 1$ for an integer f. Let γ be a generator of F_q^*, and let ψ be the character of order 16 on F_q given by $\psi(\gamma) = \beta$, where $\beta = e^{2\pi i/16}$. Then*

$$E_2(\chi) = a_{16} + b_{16}\sqrt{2} + ic_{16}\sqrt{2 - \sqrt{2}} + id_{16}\sqrt{2 + \sqrt{2}}, \tag{12.12.1}$$

where a_{16}, b_{16}, c_{16}, and d_{16} are odd integers such that

$$\begin{cases} p = a_{16}^2 + 2b_{16}^2 + 2c_{16}^2 + 2d_{16}^2, \\ 2a_{16}b_{16} = c_{16}^2 - 2c_{16}d_{16} - d_{16}^2, \\ a_{16} \equiv (-1)^{(p-7)/16} \pmod 8. \end{cases} \tag{12.12.2}$$

Moreover, the solutions (a, b, c, d) to (12.12.2) are "essentially unique" in the sense that the only other integral solutions are given by the "conjugate solutions" $(a, b, -c, -d)$, $(a, -b, -d, c)$, and $(a, -b, d, -c)$. In particular, a_{16} is uniquely determined by (12.12.2).

Proof By Proposition 11.4.1, the restriction ψ^* of ψ to F_p is the quadratic character. Hence, by Theorems 12.1.2 and 12.2.1,

$$E_2(\psi) = E(\psi)$$

and

$$E(\psi)E(\overline{\psi}) = p.$$

Define $\sigma_j \in \mathrm{Gal}(Q(\beta)/Q)$ by $\sigma_j(\beta) = \beta^j$. Then σ_7 generates a subgroup $< \sigma_7 >$ of $\mathrm{Gal}(Q(\beta)/Q)$ with fixed field $Q(\beta^3 + \beta^5) = Q(i\sqrt{2+\sqrt{2}})$. Since σ_7 fixes $E(\psi)$ by Theorem 12.2.2, we deduce (as in the proof of Theorem 3.4.1) that $E(\psi)$ has the representation given in (12.12.1) for rational numbers a_{16}, b_{16}, c_{16}, and d_{16} satisfying the two equations in (12.12.2). For brevity, write $a = a_{16}$, $b = b_{16}$, $c = c_{16}$, and $d = d_{16}$. As $i\sqrt{2-\sqrt{2}} = \beta + \beta^7$ and $\sqrt{2} = \beta^2 - \beta^6$,

(12.12.3) $$E(\psi) = E(\psi^7) = a + b\sqrt{2} + ic\sqrt{2-\sqrt{2}} + id\sqrt{2+\sqrt{2}},$$

(12.12.4) $$E(\psi^3) = E(\psi^5) = a - b\sqrt{2} - id\sqrt{2-\sqrt{2}} + ic\sqrt{2+\sqrt{2}},$$

(12.12.5) $$E(\psi^9) = E(\psi^{15}) = a + b\sqrt{2} - ic\sqrt{2-\sqrt{2}} - id\sqrt{2+\sqrt{2}},$$

and

(12.12.6) $$E(\psi^{11}) = E(\psi^{13}) = a - b\sqrt{2} + id\sqrt{2-\sqrt{2}} - ic\sqrt{2+\sqrt{2}}.$$

By Theorem 12.2.3,

(12.12.7) $$E(\psi^2) = E(\psi^6) = E(\psi^{10}) = E(\psi^{14}) = 1$$

and

(12.12.8) $$E(\psi^4) = E(\psi^8) = E(\psi^{12}) = -1.$$

Adding the $E(\psi^j)$ in (12.12.3)–(12.12.8), we obtain

$$\sum_{j=0}^{15} E(\psi^j) = p + 1 + 8a.$$

It then follows from Theorem 12.2.8 that $p+1+8a \equiv 16 \pmod{32}$, so that a is an odd integer satisfying

(12.12.9) $$a \equiv (-1)^{(p-7)/16} \pmod 4.$$

Adding (12.12.3) and (12.12.5), we obtain, by (12.2.1),

$$\begin{aligned} a+b\sqrt{2} &= (E(\psi)+E(\psi^9))/2 \\ &= \frac{1}{2}\sum_{m=0}^{p-1}\psi(1+m\tau)(1+\psi^8(1+m\tau)) = \sum_{\substack{m=0 \\ \psi^8(1+m\tau)=1}}^{p-1} \psi(1+m\tau). \end{aligned}$$

Thus $a+b\sqrt{2}$ is an algebraic integer in $Q(\sqrt{2})$, so that b is an integer. Then from (12.12.3), $c\sqrt{2-\sqrt{2}}+d\sqrt{2+\sqrt{2}}$ is an algebraic integer. Multiplying by $\sqrt{2-\sqrt{2}}$, we see that $2c+(d-c)\sqrt{2}$ is an integer in $Q(\sqrt{2})$. This tells us that either c and d are both integers or else $2c$ and $2d$ are both odd integers. The latter is impossible, since $8ab \equiv (2c)^2 - 2(2c)(2d) - (2d)^2 \pmod 8$ by (12.12.2). Thus a, b, c, and d are integers. Since $7 \equiv a^2+2b^2+2c^2+2d^2 \pmod{16}$ by (12.12.2), the integers a, b, c, and d are all odd, and moreover, $a^2 \equiv 1 \pmod{16}$. Hence the congruence in (12.12.9) holds modulo 8.

Let A, B, C, and D be (odd) integers for which

$$\begin{aligned} p &= A^2+2B^2+2C^2+2D^2, \\ 2AB &= C^2-2CD-D^2, \end{aligned}$$

and

$$A \equiv (-1)^{(p-7)/16} \pmod 8.$$

Write

$$\mu = A+B\sqrt{2}+iC\sqrt{2-\sqrt{2}}+iD\sqrt{2+\sqrt{2}}.$$

Thus $\mu\bar{\mu}=p$, and $\mu \in O_K$, where $K=Q(\beta)$. To prove the "essential uniqueness", it suffices to show that μ and $E(\psi)$ are conjugates, i.e., $\sigma_j(\mu)=E(\psi)$ for some odd integer j, $1 \le j \le 15$. The character χ_P in Theorem 12.9.4 (with $k=16$) equals ψ for some choice of P, and for this P, Theorem 12.9.4 yields

$$G_2(\psi)/(i\sqrt{p}) = E(\psi) \equiv 0 \pmod P.$$

From the prime ideal factorizations of $G_2(\psi)$ (Theorem 11.2.2) and $i\sqrt{p}$ (formula (11.1.4)), we see with the aid of (12.9.3) and (11.1.1) that

$$E(\psi)O_K = P_1P_{13}.$$

Replacing μ by a conjugate if necessary, we may assume that $P \mid \mu$. Since

$$\mu\bar{\mu}O_K = pO_K = P_1P_3P_{13}P_{15} = P_1P_{13}\overline{P_1}\,\overline{P_{13}},$$

either $\mu O_K = P_1P_{13}$ or $\mu O_K = P_1P_3$. Replacing μ by $\sigma_{13}(\mu)$ if necessary, we may assume that $\mu O_K = P_1P_{13}$. Thus

$$E(\psi) = \delta\mu$$

for some unit $\delta \in O_K$ of absolute value 1. By Theorem 2.1.13, $\delta = \pm\beta^j$ for some integer j. Since σ_7 fixes both $E(\psi)$ and μ, σ_7 fixes δ. Thus $\delta = \pm 1$, and as $A \equiv a_{16} \pmod 8$, we have $\delta = 1$, so that $\mu = E(\psi)$. ■

The integers a_{16}, b_{16}, c_{16}, d_{16} in Theorem 12.12.1 can be determined uniquely; see Problem 15.

12.13 An Eisenstein sum of order 12

The following theorem will be needed for the evaluation of the generalized Brewer sum $\Lambda_3(a)$ in Section 13.7.

Theorem 12.13.1 *Let p be a prime with $p \equiv 5 \pmod{12}$, so that $q = p^2 = 12f+1$ for an integer f. Let γ be a generator of F_q^*, and let ψ be the character of order 12 on F_q given by $\psi(\gamma) = \beta$, where $\beta = e^{2\pi i/12}$. Then*

$$E_2(\psi) = i\left(\frac{2}{p}\right)\delta(a_4 + ib_4),$$

where δ is defined by

$$\delta = \pm 1, \quad \delta \equiv a_4 b_4 \pmod 3.$$

Proof By Proposition 11.4.1, the restriction ψ^* of ψ to F_p is the quadratic character. Hence, by Theorems 12.1.2 and 12.2.1,

$$E_2(\psi) = \left(\frac{2}{p}\right)E(\psi)$$

and

$$E_2(\psi)E_2(\overline{\psi}) = p.$$

Define $\sigma_j \in \mathrm{Gal}(Q(\beta)/Q)$ by $\sigma_j(\beta) = \beta^j$. Then σ_5 generates a subgroup $< \sigma_5 >$ of $\mathrm{Gal}(Q(\beta)/Q)$ with fixed field $Q(i)$. Since σ_5 fixes $E(\psi)$ by Theorem 12.2.2, $E(\psi) \in Q(i)$. Thus $E(\psi) = A + iB$ for integers A and B such that $p = A^2 + B^2$. We have

$$(12.13.1) \qquad E(\psi) = E(\psi^5) = A + iB, \quad E(\psi^7) = E(\psi^{11}) = A - iB.$$

By Theorem 12.2.3,

$$(12.13.2) \qquad E(\psi^2) = E(\psi^6) = E(\psi^{10}) = 1, \quad E(\psi^4) = E(\psi^8) = -1.$$

By Theorem 12.4.1,

$$(12.13.3) \qquad E(\psi^3) = -a_4 - ib_4, \quad E(\psi^9) = -a_4 + ib_4.$$

Adding the $E(\psi^j)$ in (12.13.1)–(12.13.3), we obtain

$$\sum_{j=0}^{11} E(\psi^j) = p + 1 - 2a_4 + 4A.$$

It then follows from Theorem 12.2.8 that

$$p + 1 - 2a_4 + 4A \equiv 12 \pmod{24}. \tag{12.13.4}$$

Reducing this modulo 3, we deduce that

$$A \equiv -a_4 \pmod 3. \tag{12.13.5}$$

From the proof of Theorem 3.2.1, $p + 1 + 2a_4 \equiv 0 \pmod 8$, so by (12.13.4), A is even. Since $A^2 + B^2 = p = a_4^2 + b_4^2$, it follows from (12.13.5) that $A = -\delta b_4$, where $\delta = \pm 1$, $\delta \equiv a_4 b_4 \pmod 3$. Then $B = \mu a_4$ for some $\mu = \pm 1$. It remains to show that $\mu = \delta$. By (12.13.3),

$$-a_4 - ib_4 = E(\psi^3) \equiv E(\psi)^3 \equiv (A + iB)^3 \equiv A - iB = -\delta b_4 - \mu i a_4 \pmod 3.$$

Comparing imaginary parts, we see that $b_4 \equiv \mu a_4 \pmod 3$, so that $\mu = \delta$. ■

Exercises 12

1. Prove that the sum $E(\chi)$ defined in (12.2.1) is independent of the choice of generator γ of $F^*_{p^2}$.

2. Prove Theorem 12.2.2 directly from (12.2.1).

3. Deduce Theorem 12.2.3 from the equality

$$\sum_{b=0}^{p-1} \chi(b\tau) + \sum_{a=1}^{p-1} \sum_{b=0}^{p-1} \chi(a + b\tau) = \sum_{a,b=0}^{p-1} \chi(a + b\tau) = \sum_{\alpha \in F_{p^2}} \chi(\alpha) = 0$$

by determining explicitly the two sums on the left side.

4. Prove that if χ is a character of order k on F_{p^2}, where $k \nmid p+1$, then $|E(\chi)| = \sqrt{p}$.

5. If χ is a character of order k on F_{p^2} with $k \nmid p+1$ and $k \mid 2(p+1)$, prove that

$$K_2(\chi) = \frac{E^2(\chi)}{E(\chi^2)} K(\chi^*).$$

6. Let p be a prime with $p = 15f + 4$. Prove that

$$5^{(p-1)/3}\binom{5f+1}{f} \equiv 2c \pmod{p},$$

where the integer c is given uniquely by

$$p = c^2 + 15d^2, \quad c \equiv -1 \pmod{3}.$$

7. Determine the values of the Eisenstein sums of orders 10 and 12.

8. Prove Theorem 12.9.9.

9. Let $p = 3f + 1$ be a prime, and let χ be a cubic character on F_q, where $q = p^3$. Prove that $E_3(\chi) = -J(\psi, \psi)$, where χ is the lift of the cubic character $\psi \pmod{p}$.

10. Let $p = 4f + 1$ be a prime, and let χ be a quartic character on F_q, where $q = p^2$. Prove that $E_2(\chi) = -J(\psi, \psi)$, where χ is the lift of the quartic character $\psi \pmod{p}$.

11. Use Theorems 12.5.1 and 12.10.1 to determine $g_r(5)$ when $p \equiv 1 \pmod{5}$ and $r \equiv 0 \pmod{5}$.

12. Determine $g_r(6)$ and $g_r(8)$.

13. Show that

$$P^*_{4,r}(x) = \begin{cases} (x^2 - p^r)^2 - 4p^r(x-s)^2, & \text{if } f \equiv 0 \pmod{2}, \\ (x^2 + 3p^r)^2 - 4p^r(x-s)^2, & \text{if } f \equiv 1 \pmod{2}, \end{cases}$$

where $p^r = s^2 + 4t^2 = 4f + 1$, $s \equiv 1 \pmod{4}$, with $p \nmid s$ if $p \equiv 1 \pmod{4}$.

14. Obtain the factorization of $P^*_{4,r}(x)$ into irreducible polynomials over Q.

15. Obtain uniqueness in Theorem 12.12.1 by proving an analogue of Theorem 3.4.2 for $p \equiv 7 \pmod{16}$.

16. Let p be a prime with $p \equiv 11 \pmod{24}$, so that $q = p^2 = 24f + 1$ for an integer f. Let γ be a generator of F_q^*, and let ψ be the character of order 24 on F_q given by $\psi(\gamma) = \beta$, where $\beta = e^{2\pi i/24}$. Prove that

$$E_2(\psi) = \left(\frac{2}{p}\right)(e_{24}\sqrt{3} + i f_{24}\sqrt{2}),$$

where e_{24} and f_{24} are integers such that $p = 3e_{24}^2 + 2f_{24}^2$ whose signs are determined by

$$f_{24} \equiv b_8 \pmod{3},$$

where b_8 is defined in Theorem 12.7.1(b) and

$$e_{24} \equiv f_{24}(\gamma^{3f} - \gamma^{-3f})/(\gamma^{2f} + \gamma^{-2f}) \pmod{p}.$$

Hint: For the last congruence, see Theorem 12.9.10.

17. Let p be a prime with $p \equiv 5 \pmod{24}$, so that $q = p^2 = 24f + 1$ for an integer f. Let γ be a generator of F_q^*, and let ψ be the character of order 24 on F_q given by $\psi(\gamma) = \beta$, where $\beta = e^{2\pi i/24}$. Prove that

$$E(\psi) = \beta^{6+9\delta}(e_{24}\sqrt{3} + i f_{24}\sqrt{2}),$$

where e_{24} and f_{24} are integers such that $p = 3e_{24}^2 + 2f_{24}^2$ whose signs are determined by

$$f_{24} \equiv a_4 \pmod{3},$$

and

$$e_{24} \equiv f_{24}(\gamma^{3f} - \gamma^{-3f})/(\gamma^{2f} + \gamma^{-2f}) \pmod{p},$$

with δ defined as in Theorem 12.13.1.

18. Let p be a prime with $p \equiv 7 \pmod{24}$, so that $q = p^2 = 24f + 1$ for an integer f. Let γ be a generator of F_q^*, and let ψ be the character of order 24 on F_q given by $\psi(\gamma) = \beta$, where $\beta = e^{2\pi i/24}$. Prove that

$$E(\psi) = a_{24} + i b_{24}\sqrt{6},$$

where a_{24} and b_{24} are integers such that $p = a_{24}^2 + 6b_{24}^2$ whose signs are determined by

$$a_{24} \equiv (-1)^{(p-7)/24} \pmod{3}$$

and

$$b_{24} \equiv -a_{24}/(\gamma^{5f} - \gamma^{-5f} + \gamma^{f} - \gamma^{-f}) \pmod{p}.$$

19. Let p be a prime with $p \equiv 9 \pmod{16}$, so that $q = p^2 = 16f + 1$ for an integer f. Let ψ be a character of order 16 on F_q given by $\psi(\gamma) = \beta$, where $\beta = e^{2\pi i/16}$ and γ is a generator of F_q^*. Prove that

$$E(\psi) = -a_8 - i b_8\sqrt{2},$$

where a_8 and b_8 are integers such that $p = a_8^2 + 2b_8^2$ whose signs are determined by

$$a_8 \equiv -1 \pmod{4}$$

and

$$2b_8 \equiv (\gamma^{2f} + \gamma^{6f})a_8 \pmod{p}.$$

20. Write $p = 20f + 9 = a_4^2 + b_4^2 = a_{20}^2 + 5b_{20}^2$, as in Theorem 12.11.1. Prove that, modulo p,

$$\binom{10f+4}{f} \equiv \begin{cases} 2(-1)^f a_{20}, & \text{if } 5 \nmid a_4, \\ 2(-1)^{f+1} a_{20} b_4 / a_4, & \text{if } 5 | a_4, \end{cases}$$

and

$$\binom{10f+4}{f} \equiv \begin{cases} \binom{10f+4}{3f+1}, & \text{if } 5 \nmid a_4, \\ -\binom{10f+4}{3f+1}, & \text{if } 5 | a_4. \end{cases}$$

Cf. Problem 14 in Exercises 9.

21. Prove that for $p = 8f + 5 = a_4^2 + b_4^2$,

$$\binom{6f+3}{f} \equiv a_4/4^f \pmod{p}.$$

22. Define the reduced f-nomial periods $\eta_{s,r}^*$ of F_q with respect to a generator γ of F_q^* as in (12.10.6), where $q = p^r = kf + 1$. Let $A_1, \ldots, A_n \in F_q^*$, and write $A_i = \gamma^{a_i}$, $1 \le i \le n$. Let $A \in F_q$. Show that the number N of solutions $(x_1, \cdots, x_n) \in F_q^n$ of the equation

$$A_1 x_1^k + \cdots + A_n x_n^k = A$$

is

$$N = q^{n-1} + \frac{f}{q} \sum_{s=0}^{k-1} \beta_p^{\operatorname{tr}(-A\gamma^s)} \eta_{a_1+s,r}^* \cdots \eta_{a_n+s,r}^*,$$

where $\beta_p = \exp(2\pi i/p)$. (This extends Theorem 10.10.7. See also Problem 15 of Exercises 10.)

23. Let $q = p^r = kf + 1$, where p is a prime. In the notation of (12.10.4) and (11.6.5), prove that for any integers m and n,

$$\eta_{m,r}\eta_{m+n,r} = \sum_{h=0}^{k-1} C_r(n,h)\eta_{m+h,r} + N_n,$$

where

$$N_n = \begin{cases} f, & \text{if either } p = 2 \text{ and } k|n \text{ or } p > 2 \text{ and } n \equiv kf/2 \pmod{k}, \\ 0, & \text{otherwise.} \end{cases}$$

(This generalizes Lemma 10.10.2.)

Also prove that the cyclotomic numbers $C_r(n, h)$ of order k can be expressed in terms of periods by

$$qC_r(n,h) = f^2 + \sum_{i=0}^{k-1} \eta_{i,r}\eta_{i+n,r}\eta_{i+h,r}, \qquad \text{if either } p = 2 \text{ or } p > 2,\ 2|f,$$

and

$$qC_r(n,h) = f^2 + \sum_{i=0}^{k-1} \eta_{i,r}\eta_{i+n,r}\eta_{i+h+k/2,r}, \qquad \text{if } p > 2 \text{ and } f \text{ is odd.}$$

Notes on Chapter 12

The mathematical literature contains surprisingly little about Eisenstein sums. These sums were introduced by Eisenstein in his classic paper [10] of 1848 in order to examine the representations of primes p by positive–definite, binary quadratic forms of discriminant Δ when $p \not\equiv 1 \pmod{|\Delta|}$, in a manner analogous to the use of Jacobi sums when $p \equiv 1 \pmod{|\Delta|}$. This work of Eisenstein was described by H. J. S. Smith in Part V of his *Report on the Theory of Numbers* [1] published in 1863. Smith [1, p. 283] concludes his discussion of Eisenstein's results with the statement "There can be no doubt that the principles of this method are capable of many other applications; but nothing has as yet been added to these researches of Eisenstein." In 1890, Stickelberger [1, §3] carried out a systematic development of the properties of Eisenstein sums. The subject does not appear to have received much further attention until it was applied in the study of Brewer sums by Whiteman [14, p. 66] and Giudici, Muskat, and Robinson [1]. Berndt and Evans [2] explicitly evaluated Eisenstein sums of orders $3, 4, 5, 6, 8, 10, 12, 16, 20,$ and 24 over F_{p^2}. Williams, Hardy, and Spearman [1] evaluated Eisenstein sums of orders ≤ 8 over arbitrary F_q. We follow their approach in Sections 12.3–12.8, with simplifications incorporated from the work of Berndt and Evans [2]. These simplifications enable us to avoid the use of Stickelberger's congruence (see Williams, Hardy, and Spearman [1, eq. (2.4)]). Lidl and Niederreiter [1, pp. 263–265] develop properties of Eisenstein sums in a series of exercises, but they give few explicit evaluations.

The basic properties of Eisenstein sums given in Lemmas 12.0.1–12.0.4 and Theorems 12.1.1–12.1.2 are implicit in the work of Stickelberger [1]. Theorem 12.1.2 was given in Jungnickel [2, eq. (7.5.9)], where it is shown how it can be applied to prove the existence of a primitive element in a finite field with prescribed trace. Theorems 12.1.3–12.1.6 can found in Williams, Hardy, and Spearman [1]. Theorem 12.1.7 is an extension of a result given by Berndt and Evans [2]. All of the results of Section 12.2 are taken from Berndt and Evans [2]. Theorems 12.9.7–12.9.9 are due to Eisenstein [10]. Our proofs are simpler than Eisenstein's proofs, which have been summarized by H. J. S. Smith [1, pp. 280–283].

The Gauss sums $g_r(2)$, $g_r(3)$, and $g_r(4)$ examined in Section 12.10 were evaluated for $r = 2$ by Berndt and Evans [2], who also evaluated other Gauss sums not considered here, for example $g_2(6)$, $g_2(8)$, and $g_2(12)$. Theorem 12.10.2 was given by Myerson [1, Proposition 10], who also gave special cases of Theorems 12.10.3 and 12.10.4 [1, Theorems 22 and 24]. The evaluations of the Eisenstein sums E_2 in Sections 11, 12, and 13 are due to Berndt and Evans [2, Theorems 4.22, 4.28, and 4.10].

Problems 4–5 and 16–19 are taken from Berndt and Evans [2], Problem 6 from Stickelberger [1, p. 365], Problems 13 and 14 from Myerson [1], Problems 20–21 from Yamamoto [5], and Problem 23 from McEliece and Rumsey [1, p. 309]. The result of Problem 22 is equivalent to a proposition in Wolfmann [3, Prop. 2.1].

Langevin [1] used Theorem 12.1.1 (with p replaced by p^t; see the remarks following Theorem 12.2.8) to investigate the autocorrelation properties of sequences of the form

$$\lambda\left(\operatorname{tr}_{F_{p^{tr}}/F_{p^t}}(\alpha^k)\right), \qquad k = 1, 2, \ldots, n,$$

where α is a primitive nth root of unity in $F_{p^{tr}}$ and λ is a nontrivial multiplicative character on F_{p^t}. For more on autocorrelation, see Jungnickel's book [2, p. 221].

Yamamoto [5] used relative Gauss sums (which are essentially Eisenstein sums, in view of Theorem 12.1.1) to prove congruences such as those in Problem 20. Yamamoto and Yamada [1] used Eisenstein sums of order $2(q+1)$ attached to quadratic extensions F_{q^2}/F_q, where $q \equiv 1 \pmod 4$, in their investigation of Hadamard matrices of order $2(q+1)$. Yamada [1] used Eisenstein sums to construct Hadamard matrices of order $4q$ when q and $q-2$ are prime powers with $q \equiv 1 \pmod 4$.

McEliece and Rumsey [1] and Baumert and McEliece [1] applied the periods defined in (12.10.4) in their study of the weight distribution of irreducible cyclic codes. The results in Problem 23 relating periods and cyclotomic numbers, taken from McEliece and Rumsey [1, p. 309], were essentially known to Kummer [5, pp. 95, 100], [CW, pp. 212–213].

Gurak [3, Prop. 4] showed that the period $\eta_{s,r}$ of F_q (see (12.10.4)) is an integral translate of a certain sum of periods η_i of F_p of order k^*, where k^* is defined in Theorem 12.10.1. The reducibility of the factors $P^{(w)}(x)$ of the period polynomial (defined below (12.10.8)) has been studied by Gurak [2], [3]. Gurak [4], [5] found explicit expressions for the beginning coefficients of the "last factor" $P^{(k/k^*)}(x)$ in certain cases, and also found situations where this last factor is irreducible [5, p. 96].

Chapter 13

Brewer Sums

For each natural number n and each nonzero complex number a the Dickson polynomial $V_n(x,a)$ is defined by

$$\text{(13.0.1)} \qquad \begin{aligned} &\text{(i) } V_{n+1}(x,a) = xV_n(x,a) - aV_{n-1}(x,a), \ n \geq 2, \\ &\text{(ii) } V_1(x,a) = x, \ V_2(x,a) = x^2 - 2a. \end{aligned}$$

We put $V_n(x) = V_n(x,1)$. For example,

$$\text{(13.0.2)} \qquad \begin{aligned} V_3(x,a) &= x^3 - 3ax, \\ V_4(x,a) &= x^4 - 4ax^2 + 2a^2, \\ V_5(x,a) &= x^5 - 5ax^3 + 5a^2x, \\ V_6(x,a) &= x^6 - 6ax^4 + 9a^2x^2 - 2a^3, \\ V_7(x,a) &= x^7 - 7ax^5 + 14a^2x^3 - 7a^3x, \\ V_8(x,a) &= x^8 - 8ax^6 + 20a^2x^4 - 16a^3x^2 + 2a^4. \end{aligned}$$

For material on Dickson polynomials, see, for example, Lidl and Niederreiter's book [1, p. 355] or Lidl, Mullen, and Turnwald's book [1].

For an odd prime p and an integer a not divisible by p, the generalized Brewer sum $\Lambda_n(a)$ is defined for a natural number n by

$$\text{(13.0.3)} \qquad \Lambda_n(a) = \sum_{x=0}^{p-1} \left(\frac{V_n(x,a)}{p} \right),$$

where the summands are Legendre symbols. The ordinary Brewer sum Λ_n is

$$\text{(13.0.4)} \qquad \Lambda_n = \Lambda_n(1) = \sum_{x=0}^{p-1} \left(\frac{V_n(x)}{p} \right).$$

See the papers of Brewer [2], [3]. In this chapter we develop the basic properties of Brewer sums, and use them, in conjunction with our knowledge of Gauss, Jacobi, Jacobsthal, and Eisenstein sums, to determine $\Lambda_n(a)$ for certain small values of n.

13.1 Dickson polynomials

In this section we develop the basic properties of Dickson polynomials. The following identity is the key to this development.

Proposition 13.1.1 *For $n \geq 1$,*

$$V_n(y+ay^{-1},a) = y^n + a^n y^{-n}. \tag{13.1.1}$$

Proof For $n = 1$ the result (13.1.1) is immediate from (13.0.1)(ii). For $n = 2$, we have $V_2(y+ay^{-1},a) = (y+ay^{-1})^2 - 2a = y^2 + a^2y^{-2}$ by (13.0.1)(ii). Assume that (13.1.1) is true for $n = 1, 2, \ldots, N$, where $N \geq 2$. Then, by (13.0.1)(i),

$$\begin{aligned} V_{N+1}(y+ay^{-1},a) &= (y+ay^{-1})V_N(y+ay^{-1},a) - aV_{N-1}(y+ay^{-1},a) \\ &= (y+ay^{-1})(y^N + a^N y^{-N}) - a(y^{N-1} + a^{N-1}y^{-(N-1)}) \\ &= y^{N+1} + a^{N+1}y^{-(N+1)}. \end{aligned}$$

The asserted result now follows by the principle of mathematical induction. ■

The next result gives an explicit expression for $V_n(x,a)$.

Proposition 13.1.2 *For $n \geq 1$,*

$$V_n(x,a) = \left(\frac{1}{2}(x+\sqrt{x^2-4a}\,)\right)^n + \left(\frac{1}{2}(x-\sqrt{x^2-4a}\,)\right)^n.$$

Proof Let $y = \frac{1}{2}(x+\sqrt{x^2-4a}\,)$. Then

$$ay^{-1} = \frac{2a}{x+\sqrt{x^2-4a}} = \frac{x-\sqrt{x^2-4a}}{2},$$

so that

$$y+ay^{-1} = \frac{1}{2}(x+\sqrt{x^2-4a}\,) + \frac{1}{2}(x-\sqrt{x^2-4a}\,) = x.$$

Hence, by Proposition 13.1.1,

$$\begin{aligned} V_n(x,a) &= V_n(y+ay^{-1},a) = y^n + (ay^{-1})^n \\ &= \left(\frac{1}{2}(x+\sqrt{x^2-4a}\,)\right)^n + \left(\frac{1}{2}(x-\sqrt{x^2-4a}\,)\right)^n, \end{aligned}$$

as asserted. ■

The next result gives $V_n(x,a)$ explicitly as a polynomial of degree n in x.

Proposition 13.1.3 *For $n \geq 1$,*

$$V_n(x,a) = \sum_{j=0}^{[n/2]} \frac{n}{n-j}\binom{n-j}{j}(-1)^j a^j x^{n-2j}. \tag{13.1.2}$$

Proof Formula (13.1.2) is easily checked for $n = 1$ and $n = 2$. Assume that (13.1.2) holds for $n = 1, 2, \ldots, N$, where $N \geq 2$. Then by (13.0.1),

$$V_{N+1}(x,a) = \sum_j \frac{N}{N-j}\binom{N-j}{j}(-a)^j x^{N-2j+1} + \sum_j \frac{N-1}{N-1-j}\binom{N-1-j}{j}(-a)^{j+1}x^{N-1-2j},$$

where the sums are over all integers j. Replacing j by $j-1$ in the second sum, we obtain

$$V_{N+1}(x,a) = \sum_j \left\{\frac{N}{N-j}\binom{N-j}{j} + \frac{N-1}{N-j}\binom{N-j}{j-1}\right\}(-a)^j x^{N+1-2j}.$$

It is straightforward to check that the expression in braces equals

$$\frac{N+1}{N+1-j}\binom{N+1-j}{j}, \tag{13.1.3}$$

and so the result follows by induction. ∎

For the expression in (13.1.3) with N replaced by $N-1$, we have the identity

$$\frac{N}{N-j}\binom{N-j}{j} = 2^{2j+1-N}\sum_{s=j}^{[N/2]}\binom{N}{2s}\binom{s}{j}, \quad N \geq 1;$$

see Gould's collection of combinatorial identities [1, formula (3.120)]. One could use this in conjunction with Proposition 13.1.2 and the binomial theorem to give an alternative proof of Proposition 13.1.3.

A recursive property of Dickson polynomials is given in the next proposition.

Proposition 13.1.4 *For each pair of natural numbers k and n,*

$$V_{kn}(x) = V_n(V_k(x)).$$

Proof For $x = y + y^{-1}$, $y \neq 0$, the assertion follows from Proposition 13.1.1. Hence the result holds for all x. ∎

By (13.1.2), $V_n(x, a^2)$ is a homogeneous polynomial in x and a. We restate this as Proposition 13.1.5.

Proposition 13.1.5 *For each pair of complex numbers a, b,*

$$V_n(x\sqrt{a}, ab) = a^{n/2}V_n(x,b).$$

It is convenient for later use to introduce the polynomial $F_n(x,a)$ defined by

(13.1.4) $$F_n(x,a) = V_{2n}(\sqrt{x}, a),$$

and put

(13.1.5) $$F_n(x,1) = F_n(x).$$

Two simple properties of $F_n(x)$ are given in the next proposition.

Proposition 13.1.6 *For each natural number n and each complex number a,*

(a) $F_n(xa,a) = a^n F_n(x)$,
(b) $F_n(x+2) = V_n(x)$.

Proof (a) By Proposition 13.1.5,

$$F_n(xa,a) = V_{2n}(\sqrt{xa},\, a) = a^n V_{2n}(\sqrt{x}) = a^n F_n(x).$$

(b) By Proposition 13.1.4 and (13.0.1)(ii),

$$F_n(x+2) = V_{2n}(\sqrt{x+2}\,) = V_n(V_2(\sqrt{x+2}\,)) = V_n(x).$$ ■

13.2 Formulas for the ordinary Brewer sums Λ_n

Throughout the rest of this chapter, p denotes an odd prime, a is an integer not divisible by p, and n is a natural number.

We fix a generator γ of the cyclic group $F_{p^2}^*$ and set $\tau = \gamma^{(p+1)/2}$ and $g = \tau^2 = \gamma^{p+1}$. Observe that g is a primitive root (mod p) and that $\tau^p = -\tau$. Since $\tau \notin F_p$,

$$F_{p^2} = \{a + b\tau : a, b \in F_p\}.$$

Now set $\theta = \gamma^{p-1}$. Then $H = <\theta>$ is the cyclic subgroup of $F_{p^2}^*$ of order $p+1$. For $0 \le m \le p-1$, $(1+\tau m)^{p-1} = (1-\tau m)/(1+\tau m)$, and so the p elements $(1+\tau m)^{p-1}$ are distinct elements ($\neq -1$) of H. Hence

$$H = \{(1-\tau m)/(1+\tau m) : 0 \le m \le p-1\} \cup \{-1\}.$$

If $y \in H$, then $y + y^{-1} = y + y^p = \operatorname{tr}(y) \in F_p$. Thus we can define, for each natural number n,

(13.2.1) $$\Theta_n = \sum_{y \in H} \left(\frac{y^n + y^{-n}}{p}\right).$$

Let $D = \gcd(n, p+1)$. As y runs through the elements of H, y^n runs through the same elements as y^D does with the same multiplicity. Thus

(13.2.2) $$\Theta_n = \Theta_D = \sum_{y \in H} \left(\frac{y^D + y^{-D}}{p}\right).$$

Further, for $y \in H$,

$$(\gamma y)^p = g(\gamma y)^{-1},$$

and so

$$(\gamma y)^n + g^n(\gamma y)^{-n} = (\gamma y)^n + (\gamma y)^{np} = \operatorname{tr}((\gamma y)^n) \in F_p.$$

Thus we may define

$$\Theta_n(g) = \sum_{y \in H} \left(\frac{(\gamma y)^n + (\gamma y)^{np}}{p} \right) = \sum_{y \in H} \left(\frac{(\gamma y)^n + g^n(\gamma y)^{-n}}{p} \right). \tag{13.2.3}$$

We also define

$$\Omega_n(a) = \sum_{x=1}^{p-1} \left(\frac{x^n + a^n x^{-n}}{p} \right), \tag{13.2.4}$$

and put

$$\Omega_n = \Omega_n(1) = \sum_{x=1}^{p-1} \left(\frac{x^n + x^{-n}}{p} \right). \tag{13.2.5}$$

Let $d = \gcd(n, p-1)$. As x runs through the elements of F_p^*, x^n runs through the same elements as x^d does with the same multiplicity. Hence

$$\Omega_n(a) = \Omega_d(a) = \sum_{x=1}^{p-1} \left(\frac{x^d + a^n x^{-d}}{p} \right). \tag{13.2.6}$$

The relationship between the Brewer sum Λ_n (defined in (13.0.4)) and the sums Ω_n and Θ_n is given in the next theorem.

Theorem 13.2.1 *For $n \geq 1$,*

$$2\Lambda_n = \Omega_n + \Theta_n.$$

Proof We define sets S and T by

$$S = \{m \in F_p : \left(\frac{m^2 - 4}{p} \right) = 1\} \text{ and } T = \{m \in F_p : \left(\frac{m^2 - 4}{p} \right) = -1\}.$$

Clearly $F_p = S \cup T \cup \{-2, 2\}$. By Theorem 2.1.2,

$$\begin{aligned} |S| &= \frac{1}{2} \sum_{\substack{m=0 \\ m^2 \not\equiv 4 \pmod{p}}}^{p-1} \left(1 + \left(\frac{m^2 - 4}{p} \right) \right) = \frac{1}{2}(p-2) + \frac{1}{2} \sum_{m=0}^{p-1} \left(\frac{m^2 - 4}{p} \right) \\ &= \frac{1}{2}(p-2) + \frac{1}{2}(-1) = \frac{1}{2}(p-3). \end{aligned}$$

Then

$$|T| = p - 2 - |S| = p - 2 - \frac{1}{2}(p-3) = \frac{1}{2}(p-1).$$

Let $y \in H - \{-1,1\}$. Set $y = a + b\tau$, where $a, b \in F_p$. Then

$$(y + y^{-1})^2 - 4 = (y - y^{-1})^2 = (y - y^p)^2 = \{(a + b\tau) - (a - b\tau)\}^2 = 4b^2 g,$$

which shows that $\left(\frac{(y + y^{-1})^2 - 4}{p}\right) = -1$, and thus $y + y^{-1} \in T$. By Proposition 13.1.1 it follows that

$$\Theta_n = \sum_{y \in H} \left(\frac{V_n(y + y^{-1})}{p}\right) = 2\sum_{m \in T}\left(\frac{V_n(m)}{p}\right) + \sum_{m \in \{-2,2\}}\left(\frac{V_n(m)}{p}\right).$$

Since $(x + x^{-1})^2 - 4 = (x - x^{-1})^2$, we also find that

$$\Omega_n = \sum_{x=1}^{p-1} \left(\frac{V_n(x + x^{-1})}{p}\right) = 2\sum_{m \in S}\left(\frac{V_n(m)}{p}\right) + \sum_{m \in \{-2,2\}}\left(\frac{V_n(m)}{p}\right).$$

Addition of these two equalities gives

$$\Theta_n + \Omega_n = 2\sum_{m=0}^{p-1}\left(\frac{V_n(m)}{p}\right) = 2\Lambda_n. \qquad \blacksquare$$

The next theorem shows how the Brewer sum Λ_n can be expressed in terms of the Jacobsthal sums in (6.1.1) and (6.1.2) and the sum $E(\chi)$ defined in (12.2.1). We recall that for a character χ on F_{p^2}, the sum $E(\chi)$ is given by

$$E(\chi) = \sum_{b=0}^{p-1} \chi(1 + b\tau). \tag{13.2.7}$$

Theorem 13.2.2 *Let $n \geq 1$, $d = \gcd(n, p-1)$, and $D = \gcd(n, p+1)$. If n is odd,*

$$2\Lambda_n = \begin{cases} \phi_{2d}(1) + \left(\frac{2}{p}\right) \sum_{j=0}^{2D-1} E(\psi^{2j+1}), & \text{if } p \equiv 1 \pmod 4, \\ 0, & \text{if } p \equiv 3 \pmod 4, \end{cases}$$

where ψ is any character on F_{p^2} of order $4D$.

If n is even,

$$2\Lambda_n = \begin{cases} \psi_{2d}(1) + \left(\frac{2}{p}\right) \sum_{j=0}^{D-1} E(\chi^{2j+1}), & \text{if } (p+1)/D \text{ is odd}, \\ \psi_{2d}(1), & \text{if } (p+1)/D \text{ is even}, \end{cases}$$

where χ is any character on F_{p^2} of order $2D$.

Proof First we examine the case when n is odd and $p \equiv 3 \pmod 4$. Replacing x by $-x$ in (13.2.5) and replacing y by $-y$ in (13.2.1), we find that $\Omega_n = 0$ and $\Theta_n = 0$, so that $\Lambda_n = 0$ by Theorem 13.2.1.

Secondly we suppose that n is odd and that $p \equiv 1 \pmod 4$. By (13.2.6),

$$\begin{aligned}\Omega_n = \Omega_d &= \sum_{x=1}^{p-1}\left(\frac{x^d + x^{-d}}{p}\right) = \sum_{x=1}^{p-1}\left(\frac{x^{-d}}{p}\right)\left(\frac{x^{2d}+1}{p}\right)\\ &= \sum_{x=1}^{p-1}\left(\frac{x}{p}\right)\left(\frac{x^{2d}+1}{p}\right) = \phi_{2d}(1).\end{aligned}$$

Next we evaluate Θ_n. Let λ be a character on F_{p^2} of order $2D(p-1)$, and let $\psi = \lambda^{(p-1)/2}$. Then ψ is a character on F_{p^2} of order $4D$. By Proposition 11.4.1, the restriction ψ^* of the character ψ to F_p has order 2. Thus

$$\Theta_n = \Theta_D = \sum_{y \in H} \psi(y^D + y^{-D}) = \sum_{y \in H} \overline{\psi}^D(y)\psi(y^{2D}+1).$$

Now $\lambda(\gamma)$ is a primitive $2D(p-1)$th root of unity. Since $\theta = \gamma^{p-1}$, $\lambda(\theta)$ is a primitive $2D$th root of unity. Hence

$$\psi^D(\theta) = \lambda^{D(p-1)/2}(\theta) = \left(\lambda^{2D}(\theta)\right)^{(p-1)/4} = 1,$$

and so $\psi^D(y) = 1$ for all $y \in H$. Thus

$$\begin{aligned}\Theta_n &= \sum_{y\in H}\psi(y^{2D}+1) = \sum_{y \in H}\psi(y+1)\sum_{j=1}^{2D}\lambda^j(y)\\ &= \sum_{m=0}^{p-1}\psi\left(\frac{2}{1+m\tau}\right)\sum_{j=1}^{2D}\lambda^j((1+m\tau)^{p-1})\\ &= \sum_{j=1}^{2D}\sum_{m=0}^{p-1}\psi(2)\psi^{2j-1}(1+m\tau) = \left(\frac{2}{p}\right)\sum_{j=1}^{2D}E(\psi^{2j-1}).\end{aligned}$$

Then, by Theorem 13.2.1,

$$2\Lambda_n = \Omega_n + \Theta_n = \phi_{2d}(1) + \left(\frac{2}{p}\right)\sum_{j=0}^{2D-1}E(\psi^{2j+1}).$$

Finally we suppose that n is even. In this case both d and D are even and

$$\Omega_n = \Omega_d = \sum_{x=1}^{p-1}\left(\frac{x^d+x^{-d}}{p}\right) = \sum_{x=1}^{p-1}\left(\frac{x^{-d}}{p}\right)\left(\frac{x^{2d}+1}{p}\right) = \sum_{x=1}^{p-1}\left(\frac{x^{2d}+1}{p}\right) = \psi_{2d}(1).$$

We first assume that $(p+1)/D$ is even. Then $p \equiv 3 \pmod 4$. Replacing y by $y\theta^{(p+1)/(2D)}$ in (13.2.2), we see that $\Theta_n = \Theta_D = 0$, and the asserted result follows from Theorem

13.2.1. Now assume that $(p+1)/D$ is odd. Let μ be a character on F_{p^2} of order $D(p-1)$, and let $\chi = \mu^{(p-1)/2}$. Thus χ has order $2D$, and the restriction χ^* of the character χ to F_p has order 2. Hence

$$\Theta_n = \Theta_D = \sum_{y \in H} \chi(y^D + y^{-D}) = \sum_{y \in H} \overline{\chi}^D(y)\chi(y^{2D}+1).$$

Since $\theta = \gamma^{p-1}$, $\mu(\theta)$ is a primitive Dth root of unity. Thus $\chi^D(\theta) = 1$, and so $\chi^D(y) = 1$ for all y in H. Furthermore, since $\gcd(2D, p+1) = D$, the sequence $\{y^{2D} : y \in H\}$ is just a permutation of the sequence $\{y^D : y \in H\}$. Hence

$$\begin{aligned}\Theta_n &= \sum_{y \in H} \chi(y^{2D}+1) = \sum_{y \in H} \chi(y^D + 1) = \sum_{y \in H} \chi(y+1) \sum_{j=1}^{D} \mu^j(y) \\ &= \sum_{m=0}^{p-1} \chi\left(\frac{2}{1+m\tau}\right) \sum_{j=1}^{D} \mu^j((1+m\tau)^{p-1}) \\ &= \sum_{j=1}^{D} \sum_{m=0}^{p-1} \chi(2)\chi^{2j-1}(1+m\tau) = \left(\frac{2}{p}\right) \sum_{j=0}^{D-1} E(\chi^{2j+1}).\end{aligned}$$

The asserted result now follows from Theorem 13.2.1. ∎

The next theorem gives the Brewer sum Λ_{2n} in terms of Λ_n.

Theorem 13.2.3 *Let* $d = \gcd(n, p-1)$ *and* $D = \gcd(n, p+1)$.
(a) *If n is odd and $p \equiv 1 \pmod 4$, then*

$$\Lambda_{2n} = \Lambda_n + \phi_d(1).$$

(b) *If n is odd and $p \equiv 3 \pmod 4$, then*

$$\Lambda_n = 0, \quad \Lambda_{2n} = \phi_d(1).$$

(c) *If n is even and $p \equiv 1 \pmod 4$, then*

$$\Lambda_{2n} = \begin{cases} \Lambda_n + \frac{1}{2}\phi_{2d}(1), & \text{if } (p-1)/d \text{ is even}, \\ \Lambda_n, & \text{if } (p-1)/d \text{ is odd}. \end{cases}$$

(d) *If n is even and $p \equiv 3 \pmod 4$, then*

$$\Lambda_{2n} = \begin{cases} \Lambda_n, & \text{if } (p+1)/D \not\equiv 2 \pmod 4, \\ \Lambda_n + \frac{1}{2}\left(\frac{2}{p}\right) \sum_{j=0}^{2D-1} E(\psi^{2j+1}), & \text{if } (p+1)/D \equiv 2 \pmod 4, \end{cases}$$

where ψ is any character of order $4D$ on F_{p^2}.

Proof (a) Suppose that n is odd and $p \equiv 1 \pmod 4$. By Theorem 13.2.2,

$$2\Lambda_n = \phi_{2d}(1) + \left(\frac{2}{p}\right) \sum_{j=0}^{2D-1} E(\psi^{2j+1}),$$

where ψ is any character on F_{p^2} of order $4D$. As $\gcd(2n, p+1) = 2D$, $\gcd(2n, p-1) = 2d$, and $(p+1)/2D$ is odd, Theorem 13.2.2 yields

$$2\Lambda_{2n} = \psi_{4d}(1) + \left(\frac{2}{p}\right) \sum_{j=0}^{2D-1} E(\psi^{2j+1}).$$

Hence

$$2(\Lambda_{2n} - \Lambda_n) = \psi_{4d}(1) - \phi_{2d}(1) = \psi_{2d}(1),$$

by Theorem 6.1.13. As d is odd, $\left(\frac{m}{p}\right) = \left(\frac{m^{-d}}{p}\right)$ for each $m \in F_p^*$, and so

$$\begin{aligned}\phi_d(1) &= \sum_{m=1}^{p-1} \left(\frac{m}{p}\right)\left(\frac{m^d+1}{p}\right) = \sum_{m=1}^{p-1} \left(\frac{m^{-d}}{p}\right)\left(\frac{m^d+1}{p}\right) \\ &= \sum_{m=1}^{p-1} \left(\frac{1+m^{-d}}{p}\right) = \sum_{m=1}^{p-1} \left(\frac{1+m^d}{p}\right) = \psi_d(1).\end{aligned}$$

Since $\psi_{2d}(1) = \phi_d(1) + \psi_d(1)$, by Theorem 6.1.13, it follows that $\psi_{2d}(1) = 2\phi_d(1)$, and thus $\Lambda_{2n} = \Lambda_n + \phi_d(1)$.

(b) Suppose that n is odd and $p \equiv 3 \pmod 4$. By Theorem 13.2.2, $\Lambda_n = 0$. As $\gcd(2n, p-1) = 2d$, $\gcd(2n, p+1) = 2D$, and $(p+1)/2D$ is even, we have, by Theorems 13.2.2 and 6.1.13, $2\Lambda_{2n} = \psi_{4d}(1) = \psi_{2d}(1) + \phi_{2d}(1)$. Now as $(p-1)/\gcd(2d, p-1) = (p-1)/2d$ is odd, Proposition 6.1.10 gives $\phi_{2d}(1) = 0$. Thus $2\Lambda_{2n} = \psi_{2d}(1) = \psi_d(1) + \phi_d(1) = 2\phi_d(1)$, as in part (a).

(c) Suppose that n is even and $p \equiv 1 \pmod 4$. Here d is even, $D \equiv 2 \pmod 4$, and $(p+1)/D$ is odd. We first examine the case when $(p-1)/d$ is even. We have $\gcd(2n, p-1) = d\gcd(2n/d, (p-1)/d) = d\gcd(2, (p-1)/d) = 2d$, and $\gcd(2n, p+1) = D\gcd(2n/D, (p+1)/D) = D\gcd(2, (p+1)/D) = D$. Hence, by Theorem 13.2.2,

$$2\Lambda_n = \psi_{2d}(1) + \left(\frac{2}{p}\right) \sum_{j=0}^{D-1} E(\chi^{2j+1})$$

and

$$2\Lambda_{2n} = \psi_{4d}(1) + \left(\frac{2}{p}\right) \sum_{j=0}^{D-1} E(\chi^{2j+1}),$$

where χ is any character on F_{p^2} of order $2D$, so that, by Theorem 6.1.13,

$$2(\Lambda_{2n} - \Lambda_n) = \psi_{4d}(1) - \psi_{2d}(1) = \phi_{2d}(1).$$

If $(p-1)/d$ is odd, then $\gcd(2n, p-1) = d\gcd(2n/d, (p-1)/d) = d$, and $\gcd(2n, p+1) = D\gcd(2n/D, (p+1)/D) = D$. Hence, by Theorem 13.2.2,

$$2\Lambda_n = \psi_{2d}(1) + \left(\frac{2}{p}\right)\sum_{j=0}^{D-1} E(\chi^{2j+1}) = 2\Lambda_{2n},$$

where χ is any character on F_{p^2} of order $2D$.

(d) Finally, we suppose that n is even and $p \equiv 3 \pmod 4$. Here $d \equiv 2 \pmod 4$, D is even, and $(p-1)/d$ is odd. Also $\gcd(2n, p-1) = d\gcd(2n/d, (p-1)/d) = d$. Furthermore

$$\begin{aligned}\gcd(2n, p+1) &= D\gcd(2n/D, (p+1)/D) = D\gcd(2, (p+1)/D)\\ &= \begin{cases} 2D, & \text{if } (p+1)/D \text{ is even},\\ D, & \text{if } (p+1)/D \text{ is odd}.\end{cases}\end{aligned}$$

If $(p+1)/D \equiv 0 \pmod 4$, Theorem 13.2.2 gives

$$2\Lambda_n = \psi_{2d}(1),\ 2\Lambda_{2n} = \psi_{2d}(1),$$

so that $\Lambda_{2n} = \Lambda_n$.

If $(p+1)/D \equiv 2 \pmod 4$, Theorem 13.2.2 gives

$$2\Lambda_n = \psi_{2d}(1)$$

and

$$2\Lambda_{2n} = \psi_{2d}(1) + \left(\frac{2}{p}\right)\sum_{j=0}^{2D-1} E(\psi^{2j+1}),$$

where ψ is any character on F_{p^2} of order $4D$, so that

$$\Lambda_{2n} = \Lambda_n + \frac{1}{2}\left(\frac{2}{p}\right)\sum_{j=0}^{2D-1} E(\psi^{2j+1}).$$

If $(p+1)/D$ is odd, Theorem 13.2.2 gives

$$2\Lambda_n = \psi_{2d}(1) + \left(\frac{2}{p}\right)\sum_{j=0}^{D-1} E(\chi^{2j+1})$$

and

$$2\Lambda_{2n} = \psi_{2d}(1) + \left(\frac{2}{p}\right)\sum_{j=0}^{D-1} E(\chi^{2j+1}),$$

where χ is any character on F_{p^2} of order $2D$, so that $\Lambda_{2n} = \Lambda_n$. ■

Define the character sum

$$B_n = \sum_{x=0}^{p-1}\left(\frac{x+2}{p}\right)\left(\frac{V_n(x)}{p}\right). \tag{13.2.8}$$

We conclude this section by evaluating B_n in terms of Brewer sums.

Theorem 13.2.4 *For $n \geq 1$,*

$$B_n = \Lambda_{2n} - \Lambda_n.$$

Proof By Proposition 13.1.4 with $k = 2$,

$$V_{2n}(x) = V_n(V_2(x)) = V_n(x^2 - 2).$$

Hence,

$$\begin{aligned}
\Lambda_{2n} &= \sum_{x=0}^{p-1} \left(\frac{V_{2n}(x)}{p} \right) = \sum_{x=0}^{p-1} \left(\frac{V_n(x^2-2)}{p} \right) \\
&= \sum_{x=0}^{p-1} \left(\frac{V_n(x-2)}{p} \right) \left\{ 1 + \left(\frac{x}{p} \right) \right\} = \sum_{x=0}^{p-1} \left(\frac{V_n(x)}{p} \right) \left\{ 1 + \left(\frac{x+2}{p} \right) \right\} \\
&= \sum_{x=0}^{p-1} \left(\frac{V_n(x)}{p} \right) + \sum_{x=0}^{p-1} \left(\frac{x+2}{p} \right) \left(\frac{V_n(x)}{p} \right) = \Lambda_n + B_n,
\end{aligned}$$

so that $\Lambda_{2n} - \Lambda_n = B_n$. ■

13.3 Evaluation of the Brewer sums $\Lambda_1, \Lambda_2, \Lambda_3, \Lambda_4$, and Λ_6

In this section we use the results of the previous two sections, as well as the values of certain Jacobsthal sums and Eisenstein sums, to determine $\Lambda_1, \Lambda_2, \Lambda_3, \Lambda_4$, and Λ_6.

Theorem 13.3.1 $\Lambda_1 = 0$.

Proof We have

$$\Lambda_1 = \sum_{x=0}^{p-1} \left(\frac{V_1(x)}{p} \right) = \sum_{x=0}^{p-1} \left(\frac{x}{p} \right) = 0.$$ ■

Theorem 13.3.2 $\Lambda_2 = -1$.

Proof By Theorem 2.1.2,

$$\Lambda_2 = \sum_{x=0}^{p-1} \left(\frac{V_2(x)}{p} \right) = \sum_{x=0}^{p-1} \left(\frac{x^2-2}{p} \right) = -1.$$ ■

Theorem 13.3.3 *If $p \equiv 1 \pmod 4$, then*

$$\Lambda_3 = \begin{cases} 2(-1)^{(\mathrm{ind}_g(-3))/2+(p-1)/4} a_4, & \text{if } \mathrm{ind}_g(-3) \equiv 0 \pmod 2, \\ 2(-1)^{(\mathrm{ind}_g(-3)-1)/2+(p-1)/4} b_4, & \text{if } \mathrm{ind}_g(-3) \equiv 1 \pmod 2. \end{cases}$$

Alternatively, if $p \equiv 1 \pmod 4$*, then it follows that*

$$\Lambda_3 = \begin{cases} 2(-1)^{(p-1)/4}a_4, & \text{if } b_4 \equiv 0 \pmod 3, \\ 2(-1)^{(p-1)/4}b_4, & \text{if } a_4 \equiv -b_4 \pmod 3, \\ 2(-1)^{(p+3)/4}a_4, & \text{if } a_4 \equiv 0 \pmod 3, \\ 2(-1)^{(p+3)/4}b_4, & \text{if } a_4 \equiv b_4 \pmod 3. \end{cases}$$

If $p \equiv 3 \pmod 4$*, then*

$$\Lambda_3 = 0.$$

Proof We have

$$\Lambda_3 = \sum_{x=0}^{p-1}\left(\frac{V_3(x)}{p}\right) = \sum_{x=0}^{p-1}\left(\frac{x^3-3x}{p}\right) = \sum_{x=0}^{p-1}\left(\frac{x(x^2-3)}{p}\right) = \phi_2(-3).$$

Suppose first that $p \equiv 1 \pmod 4$. By Theorem 6.2.1 and Corollary 6.1.8,

$$\phi_2(-3) = \begin{cases} 2(-1)^{(p-1)/4}a_4, & \text{if } \mathrm{ind}_g(-3) \equiv 0 \pmod 4, \\ 2(-1)^{(p-1)/4}b_4, & \text{if } \mathrm{ind}_g(-3) \equiv 1 \pmod 4, \\ 2(-1)^{(p+3)/4}a_4, & \text{if } \mathrm{ind}_g(-3) \equiv 2 \pmod 4, \\ 2(-1)^{(p+3)/4}b_4, & \text{if } \mathrm{ind}_g(-3) \equiv 3 \pmod 4. \end{cases}$$

This completes the proof of the first part of the theorem.

Now let χ be the quartic character (mod p) given by $\chi(g) = i$. By Lemma 7.0.1,

$$G^3(\chi) \equiv \overline{\chi}^3(3)G(\chi^3) = \chi(3)G(\chi^3) \pmod 3,$$

so that, by Theorem 1.1.4(a),

$$\chi(-3) = \chi(-1)\chi(3) \equiv \frac{G^4(\chi)}{\chi(-1)G(\chi)G(\chi^3)} = G^4(\chi)/p \pmod 3.$$

Now, from the proof of Theorem 7.2.1, $G^4(\chi) = pK^2(\chi)$, so that, by Theorem 3.2.1,

$$i^{\mathrm{ind}_g(-3)} = \chi(-3) \equiv K^2(\chi) = (a_4 + ib_4)^2 \pmod 3.$$

If $b_4 \equiv 0 \pmod 3$, then $a_4 \not\equiv 0 \pmod 3$. Thus $i^{\mathrm{ind}_g(-3)} \equiv a_4^2 \equiv 1 \pmod 3$, that is, $\mathrm{ind}_g(-3) \equiv 0 \pmod 4$. If $a_4 \equiv 0 \pmod 3$, then $b_4 \not\equiv 0 \pmod 3$, so that $i^{\mathrm{ind}_g(-3)} \equiv -b_4^2 \equiv -1 \pmod 3$, that is, $\mathrm{ind}_g(-3) \equiv 2 \pmod 4$. If $a_4b_4 \not\equiv 0 \pmod 3$, then we have $a_4b_4 \equiv \pm 1 \pmod 3$, and so $a_4 \equiv \pm b_4 \pmod 3$. If $a_4 \equiv -b_4 \pmod 3$, then $i^{\mathrm{ind}_g(-3)} \equiv a_4^2(1-i)^2 \equiv -2i \equiv i \pmod 3$, so that $\mathrm{ind}_g(-3) \equiv 1 \pmod 4$. If $a_4 \equiv b_4 \pmod 3$, then $i^{\mathrm{ind}_g(-3)} \equiv a_4^2(1+i)^2 \equiv 2i \equiv -i \pmod 3$, so that $\mathrm{ind}_g(-3) \equiv 3 \pmod 4$. In summary,

$$\text{(13.3.1)} \qquad \mathrm{ind}_g(-3) = \begin{cases} 0 \pmod 4, & \text{if } b_4 \equiv 0 \pmod 3, \\ 1 \pmod 4, & \text{if } a_4 \equiv -b_4 \pmod 3, \\ 2 \pmod 4, & \text{if } a_4 \equiv 0 \pmod 3, \\ 3 \pmod 4, & \text{if } a_4 \equiv b_4 \pmod 3. \end{cases}$$

This completes the proof when $p \equiv 1 \pmod 4$.

If $p \equiv 3 \pmod 4$, Proposition 6.1.2 yields $\phi_2(-3) = 0$, so that $\Lambda_3 = 0$. ■

Theorem 13.3.4

$$\Lambda_4 = \begin{cases} -1 + 2(-1)^{(p-1)/8} a_8, & \text{if } p \equiv 1 \pmod 8, \\ -1 - 2a_8, & \text{if } p \equiv 3 \pmod 8, \\ -1, & \text{if } p \equiv 5 \text{ or } 7 \pmod 8). \end{cases}$$

Proof We make use of Theorem 13.2.3 with $n = 2$, so that $d = \gcd(n, p-1) = \gcd(2, p-1) = 2$ and $D = \gcd(n, p+1) = \gcd(2, p+1) = 2$. Moreover,

$$(p-1)/d \text{ is even } \iff p \equiv 1 \pmod 4$$

and

$$(p+1)/D \equiv 2 \pmod 4 \iff p \equiv 3 \pmod 8.$$

If $p \equiv 1 \pmod 8$, then, by Theorems 13.2.3(c), 13.3.2, and 6.2.3,

$$\Lambda_4 = \Lambda_2 + \frac{1}{2}\phi_4(1) = -1 + 2(-1)^{(p-1)/8} a_8.$$

If $p \equiv 3 \pmod 8$, then, by Theorem 13.2.3(d) and Theorem 13.3.2,

$$\Lambda_4 = -1 - \frac{1}{2}\{E(\psi) + E(\psi^3) + E(\psi^5) + E(\psi^7)\},$$

where ψ is any character of order 8 on F_{p^2}.

Now let γ be a generator of the cyclic group $F_{p^2}^*$. Set $g = \gamma^{p+1}$ and $Z = \text{ind}_g 2$. As $p \equiv 3 \pmod 8$, Z is odd. Choose ψ to be the character of order 8 such that $\psi(\gamma) = \beta$, where $\beta = e^{2\pi i/8}$. Then

$$\psi(2) = \psi(g^Z) = \psi(\gamma^{(p+1)Z}) = \psi(\gamma)^{(p+1)Z} = \beta^{(p+1)Z} = \beta^{4Z} = (-1)^Z = -1.$$

By Theorems 12.2.1 and 12.7.1(b),

$$E(\psi) = \psi(2)E_2(\psi) = -E_2(\psi) = a_8 + ib_8\sqrt{2},$$

so that by Theorem 12.2.2,

$$E(\psi) + E(\psi^3) + E(\psi^5) + E(\psi^7) = 4a_8,$$

and thus $\Lambda_4 = -1 - 2a_8$.

If $p \equiv 5 \pmod 8$, then, by Theorem 13.2.3(c), Theorem 13.3.2, and Proposition 6.1.2,

$$\Lambda_4 = \Lambda_2 + \frac{1}{2}\phi_4(1) = -1.$$

If $p \equiv 7 \pmod 8$, then, by Theorems 13.2.3(d) and 13.3.2,

$$\Lambda_4 = \Lambda_2 = -1. \qquad \blacksquare$$

The evaluation of Λ_4 given in Theorem 13.3.4 was used in 1986 by Greene and Stanton [2, eq. (4.17)] to prove the character sum evaluation

$$(13.3.2)\ \sum_{x,y\in F_p} \left(\frac{2xy(1+x)(1+y)(x+y)}{p}\right) = \begin{cases} -p, & \text{if } p\equiv 5 \text{ or } 7 \pmod 8, \\ 4a_8^2-p, & \text{if } p\equiv 1 \text{ or } 3 \pmod 8). \end{cases}$$

The formula (13.3.2) had been conjectured five years earlier by Evans, Pulham, and Sheehan [1].

Theorem 13.3.5

$$\Lambda_6 = \begin{cases} -1+2a_3+2(-1)^{(\text{ind}_g(-3))/2+(p-1)/4}a_4, & \textit{if } p\equiv 1 \pmod{12}, \\ -1+2(-1)^{(\text{ind}_g(-3)-1)/2+(p-1)/4}b_4, & \textit{if } p\equiv 5 \pmod{12}, \\ -1+2a_3, & \textit{if } p\equiv 7 \pmod{12}, \\ -1, & \textit{if } p\equiv 11 \pmod{12}. \end{cases}$$

Proof We make use of Theorem 13.2.3 with $n=3$. Here

$$d=\gcd(n,p-1)=\gcd(3,p-1)=\begin{cases} 3, & \text{if } p\equiv 1 \pmod 3, \\ 1, & \text{if } p\equiv 2 \pmod 3, \end{cases}$$

so that

$$\Lambda_6 = \begin{cases} \Lambda_3+\phi_3(1), & \text{if } p\equiv 1 \pmod{12}, \\ \Lambda_3+\phi_1(1), & \text{if } p\equiv 5 \pmod{12}, \\ \phi_3(1), & \text{if } p\equiv 7 \pmod{12}, \\ \phi_1(1), & \text{if } p\equiv 11 \pmod{12}. \end{cases}$$

If $p\equiv 1 \pmod{12}$, then $\text{ind}_g(-3)\equiv 0 \pmod 2$, so that the result follows from Theorems 13.3.3 and 6.2.2.

If $p\equiv 5 \pmod{12}$, then $\text{ind}_g(-3)\equiv 1 \pmod 2$, so that, since $\phi_1(1)=-1$, the result follows from Theorem 13.3.3.

If $p\equiv 7 \pmod{12}$, then, by Theorem 6.2.2, $\phi_3(1)=-1+2a_3$, so that $\Lambda_6=-1+2a_3$. If $p\equiv 11 \pmod{12}$, then $\Lambda_6=\phi_1(1)=-1$. ■

Appealing to Theorems 13.2.4 and 13.3.1–13.3.5, we obtain the values of B_1, B_2, and B_3, defined in (13.2.8).

Theorem 13.3.6

$$B_1 = -1,$$

$$B_2 = \begin{cases} 2(-1)^{(p-1)/8}a_8, & \textit{if } p\equiv 1 \pmod 8, \\ -2a_8, & \textit{if } p\equiv 3 \pmod 8, \\ 0, & \textit{if } p\equiv 5 \textit{ or } 7 \pmod 8, \end{cases}$$

$$B_3 = \begin{cases} -1+2a_3, & \textit{if } p\equiv 1 \pmod 3, \\ -1, & \textit{if } p\equiv 2 \pmod 3. \end{cases}$$

13.4 Evaluation of the Brewer sums Λ_5 and Λ_{10}

We first evaluate Λ_5.

Theorem 13.4.1 *Let p be a prime with $p \neq 2$, 5. If $p \equiv 1 \pmod{20}$ and $5 \nmid a_4$, then*

$$\Lambda_5 = 4(-1)^{(p-1)/4} a_{20}/c_{20},$$

where a_{20} and c_{20} are defined in Theorem 3.8.2.

If $p \equiv 9 \pmod{20}$ and $5 \nmid a_4$, then

$$\Lambda_5 = 4(-1)^{(p-1)/4} a_{20},$$

where a_{20} is defined in Theorem 12.11.1.

In all other cases

$$\Lambda_5 = 0.$$

Proof We take $n = 5$ in Theorem 13.2.2, so that

$$d = \gcd(n, p-1) = \gcd(5, p-1) = \begin{cases} 5, & \text{if } p \equiv 1 \pmod{5}, \\ 1, & \text{if } p \not\equiv 1 \pmod{5}, \end{cases}$$

and

$$D = \gcd(n, p+1) = \gcd(5, p+1) = \begin{cases} 1, & \text{if } p \not\equiv 4 \pmod{5}, \\ 5, & \text{if } p \equiv 4 \pmod{5}. \end{cases}$$

We now separate cases according to the value of p modulo 20. If $p \equiv 3 \pmod 4$, then $\Lambda_5 = 0$ by Theorem 13.2.2. Hence we need only consider $p \equiv 1, 9, 13, 17 \pmod{20}$.

If $p \equiv 1 \pmod{20}$, then, by Theorem 13.2.2,

$$2\Lambda_5 = \phi_{10}(1) + \left(\frac{2}{p}\right)\{E(\psi) + E(\psi^3)\},$$

where ψ is any character on F_{p^2} of order 4. By Theorem 6.2.7,

$$\phi_{10}(1) = \begin{cases} 2(-1)^{(p-1)/4} a_4, & \text{if } 5 \mid a_4, \\ (-1)^{(p-1)/4}(2a_4 + 8a_{20}/c_{20}), & \text{if } 5 \nmid a_4. \end{cases}$$

Now let γ be a generator of $F_{p^2}^*$, and set $g = \gamma^{p+1}$. Choose the character ψ so that $\psi(\gamma) = i$. By Proposition 11.4.1 the restriction character ψ^* is of order 2, so that $\psi(2) = \psi^*(2) = \left(\frac{2}{p}\right)$. By Theorems 12.2.1 and 12.4.1,

$$E(\psi) = \psi(2)E_2(\psi) = \left(\frac{2}{p}\right)(-1)^{(p+3)/4}(a_4 + ib_4),$$

so that

$$\left(\frac{2}{p}\right)\{E(\psi)+E(\psi^3)\} = 2(-1)^{(p+3)/4}a_4.$$

Hence

$$\Lambda_5 = \begin{cases} 0, & \text{if } 5 \mid a_4, \\ 4(-1)^{(p-1)/4}a_{20}/c_{20}, & \text{if } 5 \nmid a_4. \end{cases}$$

If $p \equiv 9 \pmod{20}$, then, by Theorem 13.2.2,

$$2\Lambda_5 = \phi_2(1) + (-1)^{(p-1)/4}\sum_{j=0}^{9} E(\psi^{2j+1}),$$

where ψ is a character on F_{p^2} of order 20 as in Theorem 12.11.1. By Theorem 6.2.1, $\phi_2(1) = 2(-1)^{(p-1)/4}a_4$. From Theorem 12.11.1 and (12.11.7), we see that, if $5 \nmid a_4$,

$$E(\psi) = E(\psi^3) = E(\psi^7) = E(\psi^9) = a_{20} + ib_{20}\sqrt{5},$$
$$E(\psi^{11}) = E(\psi^{13}) = E(\psi^{17}) = E(\psi^{19}) = a_{20} - ib_{20}\sqrt{5},$$

and, if $5 \mid a_4$,

$$E(\psi) = -E(\psi^3) = -E(\psi^7) = E(\psi^9) = -b_{20}\sqrt{5} + ia_{20},$$
$$E(\psi^{11}) = -E(\psi^{13}) = -E(\psi^{17}) = E(\psi^{19}) = -b_{20}\sqrt{5} - ia_{20}.$$

Since $\psi(2) = \left(\frac{2}{p}\right)$, Theorems 12.2.1 and 12.4.1 yield

$$E(\psi^5) = -(a_4 + ib_4),\ E(\psi^{15}) = -a_4 + ib_4,$$

so that

$$\sum_{j=0}^{9} E(\psi^{2j+1}) = \begin{cases} 8a_{20} - 2a_4, & \text{if } 5 \nmid a_4, \\ -2a_4, & \text{if } 5 \mid a_4, \end{cases}$$

and hence

$$\Lambda_5 = \begin{cases} 4(-1)^{(p-1)/4}a_{20}, & \text{if } 5 \nmid a_4, \\ 0, & \text{if } 5 \mid a_4. \end{cases}$$

If $p \equiv 13$ or $17 \pmod{20}$, Theorem 13.2.2 yields

$$2\Lambda_5 = \phi_2(1) + \left(\frac{2}{p}\right)\{E(\psi)+E(\psi^3)\},$$

where ψ is any character on F_{p^2} of order 4. By Theorem 6.2.1,

$$\phi_2(1) = 2(-1)^{(p-1)/4}a_4,$$

and, exactly as in the case $p \equiv 1 \pmod{20}$, we obtain

$$\left(\frac{2}{p}\right)\{E(\psi) + E(\psi^3)\} = 2(-1)^{(p+3)/4} a_4,$$

so that $\Lambda_5 = 0$. ■

From Theorems 13.2.3 and 13.4.1, we can now determine the value of Λ_{10}.

Theorem 13.4.2 *Let p be a prime with $p \neq 2, 5$. Then*

$$\Lambda_{10} = \begin{cases} -1, & \text{if } p \equiv 3, 7, 13, 17 \text{ or } 19 \pmod{20}, \text{ or} \\ & \quad p \equiv 9 \pmod{20} \text{ and } 5 \mid a_4, \\ -1 + 4a_5, & \text{if } p \equiv 11 \pmod{20}, \text{ or} \\ & \quad p \equiv 1 \pmod{20} \text{ and } 5 \mid a_4, \\ -1 + 4(-1)^{(p-1)/4} a_{20}, & \text{if } p \equiv 9 \pmod{20} \text{ and } 5 \nmid a_4, \\ -1 + 4a_5 + 4(-1)^{(p-1)/4} a_{20}/c_{20}, & \text{if } p \equiv 1 \pmod{20} \text{ and } 5 \nmid a_4. \end{cases}$$

Proof By Theorem 13.2.3(a)(b) with $n = 5$ and $d = \gcd(5, p-1)$,

$$\Lambda_{10} = \Lambda_5 + \phi_d(1). \tag{13.4.1}$$

If $p \equiv 3, 7, 13, 17$ or $19 \pmod{20}$, or $p \equiv 9 \pmod{20}$ and $5 \mid a_4$,

$$\Lambda_{10} = 0 + \phi_1(1) = -1.$$

If $p \equiv 11 \pmod{20}$, or $p \equiv 1 \pmod{20}$ and $5 \mid a_4$, then Theorem 6.2.4 yields

$$\Lambda_{10} = 0 + \phi_5(1) = -1 + 4a_5.$$

If $p \equiv 9 \pmod{20}$ and $5 \nmid a_4$,

$$\Lambda_{10} = 4(-1)^{(p-1)/4} a_{20} + \phi_1(1) = -1 + 4(-1)^{(p-1)/4} a_{20}.$$

If $p \equiv 1 \pmod{20}$ and $5 \nmid a_4$,

$$\Lambda_{10} = 4(-1)^{(p-1)/4} a_{20}/c_{20} + \phi_5(1) = -1 + 4a_5 + 4(-1)^{(p-1)/4} a_{20}/c_{20}.$$ ■

From (13.4.1) and the equality $B_5 = \Lambda_{10} - \Lambda_5$, we obtain the evaluation of B_5 stated in the next theorem.

Theorem 13.4.3 *Let p be a prime with $p \neq 2, 5$. Then*

$$B_5 = \begin{cases} -1 + 4a_5, & \text{if } p \equiv 1 \pmod{5}, \\ -1, & \text{if } p \equiv 2, 3, 4 \pmod{5}. \end{cases}$$

13.5 Evaluation of the Brewer sum Λ_8

Theorem 13.5.1 *Let p be an odd prime. Then*

$$\Lambda_8 = \begin{cases} -1 + 2a_8 + 4(-1)^{(p-1)/16} a_{16}, & \text{if } p \equiv 1 \pmod{16}, \\ -1 - 2a_8, & \text{if } p \equiv 3, 9 \text{ or } 11 \pmod{16}, \\ -1, & \text{if } p \equiv 5, 13 \text{ or } 15 \pmod{16}, \\ -1 + 4a_{16}, & \text{if } p \equiv 7 \pmod{16}. \end{cases}$$

Proof We take $n = 4$ in Theorem 13.2.3, so that

$$d = \gcd(n, p-1) = \gcd(4, p-1) = \begin{cases} 4, & \text{if } p \equiv 1 \pmod 4, \\ 2, & \text{if } p \equiv 3 \pmod 4, \end{cases}$$

and

$$D = \gcd(n, p+1) = \gcd(4, p+1) = \begin{cases} 2, & \text{if } p \equiv 1 \pmod 4, \\ 4, & \text{if } p \equiv 3 \pmod 4. \end{cases}$$

If $p \equiv 1 \pmod 4$, then

$$(p-1)/d \equiv 0 \pmod 2 \iff p \equiv 1 \pmod 8,$$

and, if $p \equiv 3 \pmod 4$, then

$$(p+1)/D \equiv 2 \pmod 4 \iff p \equiv 7 \pmod{16}.$$

Hence Theorem 13.2.3 gives

$$\Lambda_8 = \begin{cases} \Lambda_4, & \text{if } p \equiv 3, 5, 11, 13, \text{ or } 15 \pmod{16}, \\ \Lambda_4 + \dfrac{1}{2}\phi_8(1), & \text{if } p \equiv 1 \text{ or } 9 \pmod{16}, \\ \Lambda_4 + \dfrac{1}{2}\displaystyle\sum_{j=0}^{7} E(\psi^{2j+1}), & \text{if } p \equiv 7 \pmod{16}, \end{cases}$$

where ψ is any character of order 16 on F_{p^2}. Appealing to Theorem 13.3.4, we obtain

$$\Lambda_8 = \begin{cases} -1, & \text{if } p \equiv 5, 13 \text{ or } 15 \pmod{16}, \\ -1 - 2a_8, & \text{if } p \equiv 3 \text{ or } 11 \pmod{16}. \end{cases}$$

Appealing to Theorems 13.3.4 and 6.2.6, we obtain

$$\Lambda_8 = -1 + 2a_8 + 4(-1)^{(p-1)/16} a_{16}, \quad \text{if } p \equiv 1 \pmod{16}.$$

Appealing to Theorem 13.3.4 and Proposition 6.1.2, we have

$$\Lambda_8 = -1 - 2a_8, \quad \text{if } p \equiv 9 \pmod{16}.$$

It remains to examine the case $p \equiv 7 \pmod{16}$. By Theorem 13.3.4, $\Lambda_4 = -1$, and, from (12.12.3)–(12.12.6),

$$\sum_{j=0}^{7} E(\psi^{2j+1}) = 8a_{16}.$$

Hence $\Lambda_8 = -1 + 4a_{16}$. ■

From Theorems 13.3.4 and 13.5.1, we obtain the value of $B_4 = \Lambda_8 - \Lambda_4$.

Theorem 13.5.2 *Let p be an odd prime. Then*

$$B_4 = \begin{cases} 4(-1)^{(p-1)/16} a_{16}, & \textit{if } p \equiv 1 \pmod{16}, \\ 4a_{16}, & \textit{if } p \equiv 7 \pmod{16}, \\ 0, & \textit{otherwise}. \end{cases}$$

13.6 Formulas for the generalized Brewer sums $\Lambda_n(a)$

We first show that every generalized Brewer sum $\Lambda_{2n}(a)$ can be evaluated in terms of ordinary Brewer sums.

Theorem 13.6.1 *For $n \geq 1$ and each integer a not divisible by the odd prime p,*

$$\Lambda_{2n}(a) = \left(\frac{a}{p}\right)^{n+1} \Lambda_{2n} + \left(\frac{a}{p}\right)^{n} \left(1 - \left(\frac{a}{p}\right)\right) \Lambda_n.$$

Proof We have

$$\begin{aligned}
\Lambda_{2n}(a) &= \sum_{x=0}^{p-1} \left(\frac{V_{2n}(x,a)}{p}\right) = \sum_{x=0}^{p-1} \left(\frac{F_n(x^2,a)}{p}\right) && \text{(by (13.1.4))} \\
&= \sum_{x=0}^{p-1} \left(\frac{F_n(x,a)}{p}\right) \left\{1 + \left(\frac{x}{p}\right)\right\} = \sum_{x=0}^{p-1} \left(\frac{F_n(xa,a)}{p}\right) \left\{1 + \left(\frac{xa}{p}\right)\right\} \\
&= \sum_{x=0}^{p-1} \left(\frac{a^n F_n(x)}{p}\right) \left\{1 + \left(\frac{xa}{p}\right)\right\} && \text{(by Proposition 13.1.6(a))} \\
&= \sum_{x=0}^{p-1} \left(\frac{a^n F_n(x+2)}{p}\right) \left\{1 + \left(\frac{a}{p}\right)\left(\frac{x+2}{p}\right)\right\} \\
&= \left(\frac{a}{p}\right)^{n} \sum_{x=0}^{p-1} \left(\frac{V_n(x)}{p}\right) + \left(\frac{a}{p}\right)^{n+1} \sum_{x=0}^{p-1} \left(\frac{V_n(x)}{p}\right)\left(\frac{x+2}{p}\right) && \text{(by Proposition 13.1.6(b))} \\
&= \left(\frac{a}{p}\right)^{n} \Lambda_n + \left(\frac{a}{p}\right)^{n+1} B_n && \text{(by (13.2.8))}
\end{aligned}$$

$$= \left(\frac{a}{p}\right)^{n+1} (\Lambda_{2n} - \Lambda_n) + \left(\frac{a}{p}\right)^n \Lambda_n \qquad \text{(by Theorem 13.2.4)}$$
$$= \left(\frac{a}{p}\right)^{n+1} \Lambda_{2n} + \left(\frac{a}{p}\right)^n \left(1 - \left(\frac{a}{p}\right)\right) \Lambda_n. \qquad \blacksquare$$

This theorem shows that once the ordinary Brewer sums are known, we need only examine the Brewer sum $\Lambda_n(a)$ for odd positive integers n. Moreover, if g is a primitive root $\pmod p$, we see from Proposition 13.1.5 that

$$V_n(g^k x, g^{2k} b) = g^{kn} V_n(x, b)$$

for every positive integer k. Hence

$$\Lambda_n(g^{2k} b) = (-1)^{kn} \Lambda_n(b).$$

Thus it remains to evaluate $\Lambda_n(g)$ for a primitive root $g \pmod p$ and an odd positive integer n. This is done in Theorem 13.6.2. Before stating and proving Theorem 13.6.2, we note the following simple extension of Theorem 13.2.1 (see Problem 4):

$$(13.6.1) \qquad 2\Lambda_n(g) = \Omega_n(g) + \Theta_n(g), \quad n \geq 1.$$

Theorem 13.6.2 *Let n be an odd positive integer, and let g be a primitive root modulo the odd prime p. Set $d = \gcd(n, p-1)$ and $D = \gcd(n, p+1)$. Let ψ be a character on F_{p^2} of order $4D$ chosen such that $\psi^D(\gamma) = i$, where γ is a generator of $F_{p^2}^*$ such that $g = \gamma^{p+1}$. If $p \equiv 1 \pmod 4$, then*

$$\Lambda_n(g) = \frac{1}{2}\left\{(-1)^{(n-d)/2} \phi_{2d}(g^d) - i(-1)^{(n-D)/2}\left(\frac{2}{p}\right) \sum_{j=0}^{2D-1} (-1)^j E(\psi^{2j+1})\right\},$$

and if $p \equiv 3 \pmod 4$, then

$$\Lambda_n(g) = 0.$$

Proof By (13.2.3), (13.2.6), and (13.6.1),

$$(13.6.2) \qquad 2\Lambda_n(g) = \sum_{j=1}^{p-1} \left(\frac{j^d + g^n j^{-d}}{p}\right) + \sum_{y \in H} \left(\frac{(\gamma y)^n + (\gamma y)^{np}}{p}\right).$$

Suppose first that $p \equiv 3 \pmod 4$. Replacing j by $-j$ and y by $-y$ in (13.6.2), we obtain $2\Lambda_n(g) = -2\Lambda_n(g)$, so that $\Lambda_n(g) = 0$. Now suppose that $p \equiv 1 \pmod 4$. Replacing j by $g^{(n-d)/2d} j$ in the first sum on the right side of (13.6.2), we find that it becomes

$$(-1)^{(n-d)/2} \sum_{j=1}^{p-1} \left(\frac{j^d + g^d j^{-d}}{p}\right) = (-1)^{(n-d)/2} \phi_{2d}(g^d).$$

It thus remains to determine the second sum on the right side of (13.6.2).

Let λ be a character on F_{p^2} of order $2D(p-1)$ chosen so that $\lambda^{D(p-1)/2}(\gamma) = i$. Then $\psi = \lambda^{(p-1)/2}$. By Proposition 11.4.1, ψ^* has order 2. Clearly $\psi^D(y) = 1$ for all $y \in H$, as $\psi^D(\theta) = \psi^D(\gamma^{p-1}) = i^{p-1} = 1$. This shows that the second sum on the right side of (13.6.2) is

$$\sum_{y \in H} \psi^n(\gamma y)\psi(1 + (\gamma y)^{n(p-1)}) = \psi(\gamma^n) \sum_{y \in H} \psi(1 + \theta^n y^{n(p-1)}).$$

Since $\gcd(n(p-1), p+1) = 2D$, the sequence $\{y^{n(p-1)} : y \in H\}$ is a permutation of the sequence $\{y^{2D} : y \in H\}$. Hence, setting $\tau = \gamma^{p+1}$,

$$\begin{aligned}
\sum_{y \in H} \psi(1 + \theta^n y^{n(p-1)}) &= \sum_{y \in H} \psi(1 + \theta^n y^{2D}) = \sum_{y \in H} \psi(1 + \theta^n y) \sum_{j=1}^{2D} \lambda^j(y) \\
&= \sum_{y \in H} \psi(1 + y) \sum_{j=1}^{2D} \lambda^j(\theta^{-n} y) = \sum_{y \in H} \psi(1 + y) \sum_{j=1}^{2D} \lambda^{-jn}(\theta)\lambda^j(y) \\
&= \sum_{m=0}^{p-1} \psi\left(\frac{2}{1 + m\tau}\right) \sum_{j=1}^{2D} \lambda^{-jn(p-1)}(\gamma)\lambda^{j(p-1)}(1 + m\tau) \\
&= \psi(2) \sum_{j=1}^{2D} \psi^{-2j}(\gamma^n) \sum_{m=0}^{p-1} \psi^{2j-1}(1 + m\tau) \\
&= \psi(2) \sum_{j=1}^{2D} \overline{\psi}^{2j}(\gamma^n) E(\psi^{2j-1}) = \left(\frac{2}{p}\right) \sum_{j=0}^{2D-1} \overline{\psi}^{2j+2}(\gamma^n) E(\psi^{2j+1}),
\end{aligned}$$

and so

$$\psi(\gamma^n) \sum_{y \in H} \psi(1 + \theta^n y^{n(p-1)}) = \left(\frac{2}{p}\right) \sum_{j=0}^{2D-1} \overline{\psi}^{2j+1}(\gamma^n) E(\psi^{2j+1}).$$

As $D-1$ and $(n/D)-1$ are both even, $(n/D) - 1 \equiv D((n/D) - 1) \equiv n - D \pmod 4$. Hence, as $\psi(\gamma^D) = i$, $\psi(\gamma^n) = i^{n/D} = i^{n-D+1} = i(-1)^{(n-D)/2}$. Thus $\overline{\psi}^{2j+1}(\gamma^n) = -i(-1)^j(-1)^{(n-D)/2}$, and the result follows. ■

13.7 Evaluation of the Brewer sums $\Lambda_1(a), \Lambda_2(a), \Lambda_3(a), \Lambda_4(a)$, and $\Lambda_6(a)$

Theorem 13.7.1 $\Lambda_1(a) = 0$.

Proof We have

$$\Lambda_1(a) = \sum_{x=0}^{p-1} \left(\frac{V_1(x,a)}{p}\right) = \sum_{x=0}^{p-1} \left(\frac{x}{p}\right) = 0.$$

■

Theorem 13.7.2 $\Lambda_2(a) = -1$.

Proof By Theorems 13.3.1, 13.3.2, and 13.6.1,

$$\Lambda_2(a) = \Lambda_2 + \left(\frac{a}{p}\right)\left(1 - \left(\frac{a}{p}\right)\right)\Lambda_1 = -1. \qquad \blacksquare$$

Theorem 13.7.3 *Let p be an odd prime, let a be an integer not divisible by p, and let g be a primitive root of p. If $p \equiv 1 \pmod 4$, define the integers a_4 and b_4 uniquely as in Theorems 3.2.1 and 3.2.2, that is,*

$$p = a_4^2 + b_4^2,\ a_4 \equiv -\left(\frac{2}{p}\right) \pmod 4,\ b_4 \equiv a_4 g^{(p-1)/4} \pmod p.$$

Then

$$\Lambda_3(a) = \begin{cases} 2(-1)^{(p-1)/4}a_4, & \text{if } \operatorname{ind}_g(-3a) \equiv 0 \pmod 4, \\ 2(-1)^{(p-1)/4}b_4, & \text{if } \operatorname{ind}_g(-3a) \equiv 1 \pmod 4, \\ 2(-1)^{(p+3)/4}a_4, & \text{if } \operatorname{ind}_g(-3a) \equiv 2 \pmod 4, \\ 2(-1)^{(p+3)/4}b_4, & \text{if } \operatorname{ind}_g(-3a) \equiv 3 \pmod 4. \end{cases}$$

If $p \equiv 3 \pmod 4$, then

$$\Lambda_3(a) = 0.$$

Proof First we examine the case $p \equiv 1 \pmod 4$. Since $\left(\frac{-3}{p}\right) = \left(\frac{p}{3}\right)$,

$$\operatorname{ind}_g(-3) \equiv \begin{cases} 0 \pmod 2, & \text{if } p \equiv 1 \pmod 3, \\ 1 \pmod 2, & \text{if } p \equiv 2 \pmod 3. \end{cases} \tag{13.7.1}$$

Thus, by Theorem 13.3.3,

$$\Lambda_3(1) = \begin{cases} 2(-1)^{(\operatorname{ind}_g(-3))/2+(p-1)/4}a_4, & \text{if } p \equiv 1 \pmod 3, \\ 2(-1)^{(\operatorname{ind}_g(-3)-1)/2+(p-1)/4}b_4, & \text{if } p \equiv 2 \pmod 3. \end{cases} \tag{13.7.2}$$

Define γ as in Theorem 13.6.2. Since

$$d = \gcd(3, p-1) = \begin{cases} 3, & \text{if } p \equiv 1 \pmod 3, \\ 1, & \text{if } p \equiv 2 \pmod 3, \end{cases}$$

and

$$D = \gcd(3, p+1) = \begin{cases} 1, & \text{if } p \equiv 1 \pmod 3, \\ 3, & \text{if } p \equiv 2 \pmod 3, \end{cases}$$

Theorem 13.6.2 yields

$$\Lambda_3(g) = \begin{cases} \dfrac{1}{2}\left(\phi_6(g^3) + i\left(\dfrac{2}{p}\right)(E(\psi_1) - E(\psi_1^3))\right), & \text{if } p \equiv 1 \pmod 3, \\ \dfrac{1}{2}\left(-\phi_2(g) - i\left(\dfrac{2}{p}\right)\displaystyle\sum_{j=0}^{5}(-1)^j E(\psi_2^{2j+1})\right), & \text{if } p \equiv 2 \pmod 3, \end{cases}$$

where, for $p \equiv 1 \pmod 3$, ψ_1 is a character on F_{p^2} of order 4 such that $\psi_1(\gamma) = i$, and, for $p \equiv 2 \pmod 3$, ψ_2 is a character on F_{p^2} of order 12 such that $\psi_2(\gamma) = e^{2\pi i/12}$.

Let $p \equiv 1 \pmod 3$. Since $\psi_1(2) = \left(\frac{2}{p}\right)$ by Proposition 11.4.1, it follows from Theorems 12.2.1 and 12.4.1 that

$$E(\psi_1) = \psi_1(2)E_2(\psi_1) = -a_4 - ib_4.$$

Thus

$$i(E(\psi_1) - E(\psi_1^3)) = 2b_4.$$

Furthermore, by Theorem 6.2.5 and (13.3.1),

$$\phi_6(g^3) = \begin{cases} 2(-1)^{(p-1)/4}b_4, & \text{if } \operatorname{ind}_g(-3) \equiv 0 \pmod 4, \\ -6(-1)^{(p-1)/4}b_4, & \text{if } \operatorname{ind}_g(-3) \equiv 2 \pmod 4. \end{cases}$$

Putting these results together, we find that, if $p \equiv 1 \pmod 3$,

$$\Lambda_3(g) = 2(-1)^{(\operatorname{ind}_g(-3))/2+(p-1)/4}b_4. \tag{13.7.3}$$

Suppose now that $p \equiv 2 \pmod 3$, so that $p \equiv 5 \pmod{12}$. By Theorem 12.13.1, (12.13.1), and (12.13.3),

$$\begin{aligned} E(\psi_2) &= E(\psi_2^5) = -\delta b_4 + i\delta a_4, \\ E(\psi_2^7) &= E(\psi_2^{11}) = -\delta b_4 - i\delta a_4, \\ E(\psi_2^3) &= -a_4 - ib_4, \ E(\psi_2^9) = -a_4 + ib_4, \end{aligned}$$

so that

$$-i\sum_{j=0}^{5}(-1)^j E(\psi_2^{2j+1}) = 4\delta a_4 + 2b_4,$$

where $\delta = \pm 1$, $\delta \equiv a_4 b_4 \pmod 3$. By Theorem 6.2.1,

$$\phi_2(g) = 2(-1)^{(p-1)/4}b_4,$$

so that, if $p \equiv 2 \pmod 3$,

$$\Lambda_3(g) = 2\delta\left(\frac{2}{p}\right)a_4.$$

By (13.3.1),

$$\delta = (-1)^{(1+\text{ind}_g(-3))/2},$$

so that, if $p \equiv 2 \pmod 3$

$$\Lambda_3(g) = 2(-1)^{(\text{ind}_g(-3)+1)/2+(p-1)/4} a_4. \tag{13.7.4}$$

Next, recalling that $\Lambda_n(g^{2k}b) = (-1)^{kn}\Lambda_n(b)$, we deduce from (13.7.1)–(13.7.4) that

$$\begin{aligned} \Lambda_3(a) &= \Lambda_3(g^{\text{ind}_g a}) \\ &= \begin{cases} (-1)^{(\text{ind}_g a)/2}\Lambda_3(1), & \text{if } \text{ind}_g\, a \equiv 0 \pmod 2, \\ (-1)^{(\text{ind}_g a - 1)/2}\Lambda_3(g), & \text{if } \text{ind}_g\, a \equiv 1 \pmod 2, \end{cases} \\ &= \begin{cases} 2(-1)^{(\text{ind}_g(-3a))/2+(p-1)/4} a_4, & \text{if } \text{ind}_g(-3a) \equiv 0 \pmod 2, \\ 2(-1)^{(\text{ind}_g(-3a)-1)/2+(p-1)/4} b_4, & \text{if } \text{ind}_g(-3a) \equiv 1 \pmod 2. \end{cases} \end{aligned}$$

Finally we consider the case $p \equiv 3 \pmod 4$. By Theorems 13.2.2 and 13.6.2 with $n = 3$, we have $\Lambda_3(1) = \Lambda_3(g) = 0$, so that $\Lambda_3(a) = 0$. ■

Theorem 13.7.4

$$\Lambda_4(a) = \begin{cases} -1 + 2\left(\dfrac{a}{p}\right)(-1)^{(p-1)/8} a_8, & \textit{if } p \equiv 1 \pmod 8, \\ -1 - 2\left(\dfrac{a}{p}\right) a_8, & \textit{if } p \equiv 3 \pmod 8, \\ -1, & \textit{if } p \equiv 5 \textit{ or } 7 \pmod 8. \end{cases}$$

Proof By Theorem 13.6.1,

$$\Lambda_4(a) = \left(\frac{a}{p}\right)\Lambda_4 + \left(1 - \left(\frac{a}{p}\right)\right)\Lambda_2 = \Lambda_2 - \left(\frac{a}{p}\right)B_2.$$

The result now follows from Theorems 13.3.2 and 13.3.6. ■

Theorem 13.7.5

$$\Lambda_6(a) = \begin{cases} -1 + 2a_3 + 2\left(\dfrac{a}{p}\right)(-1)^{(\text{ind}_g(-3))/2+(p-1)/4} a_4, & \textit{if } p \equiv 1 \pmod{12}, \\ -1 + 2\left(\dfrac{a}{p}\right)(-1)^{(\text{ind}_g(-3)-1)/2+(p-1)/4} b_4, & \textit{if } p \equiv 5 \pmod{12}, \\ -1 + 2a_3, & \textit{if } p \equiv 7 \pmod{12}, \\ -1, & \textit{if } p \equiv 11 \pmod{12}. \end{cases}$$

Proof By Theorem 13.6.1,

$$\Lambda_6(a) = \Lambda_6 + \left(\left(\frac{a}{p}\right) - 1\right)\Lambda_3 = B_3 + \left(\frac{a}{p}\right)\Lambda_3.$$

The result now follows from Theorems 13.3.3 and 13.3.6. ■

We have now completed our evaluation of $\Lambda_n(a)$ for $n = 1, 2, 3, 4$, and 6. Provided the values of the relevant Jacobsthal and Eisenstein sums are known, the methods of this chapter can be used to determine $\Lambda_n(a)$ for other small values of n. We leave it as a challenge to the reader to evaluate $\Lambda_5(a)$ (Problem 8), $\Lambda_8(a)$ (Problem 10), $\Lambda_{10}(a)$ (Problem 11), and $\Lambda_{12}(a)$ (Problem 12).

Exercises 13

1. Using

$$\log(1 - yt) = -\sum_{n=1}^{\infty} y^n \frac{t^n}{n}, \quad |t| < 1/|y|,$$

expand both sides of the identity

$$\log(1 - x_1 t) + \log(1 - x_2 t) = \log(1 - (x_1 + x_2 - t x_1 x_2)t)$$

in powers of t. Equate coefficients of t^n to obtain Waring's formula

$$x_1^n + x_2^n = \sum_{j=0}^{[n/2]} \frac{n}{n-j} \binom{n-j}{j} (-1)^j (x_1 + x_2)^{n-2j} (x_1 x_2)^j.$$

By making appropriate choices for x_1 and x_2 in Waring's formula, deduce the formula of Proposition 13.1.3.

2. Let p be an odd prime. Let a, b, and c be integers with a not divisible by p. Set $d = b^2 - 4ac$. Prove that

$$\begin{aligned} &\operatorname{card}\left\{ m \in F_p : \left(\frac{am^2 + bm + c}{p}\right) = 1 \right\} \\ &= \frac{1}{2}\left\{ \left(1 + \left(\frac{a}{p}\right) - \left(\frac{ad^2}{p}\right)\right) p - \left(\frac{a}{p}\right) - \left(\frac{d}{p}\right) - 1 \right\}, \end{aligned}$$

and

$$\operatorname{card}\left\{ m \in F_p : \left(\frac{am^2 + bm + c}{p}\right) = 0 \right\} = 1 + \left(\frac{d}{p}\right).$$

Deduce the formulas for $|S|$ and $|T|$ used in the proof of Theorem 13.2.1.

3. Let p be an odd prime. Prove that the polynomial $x^{-1} V_p(x)$ is irreducible over Q.

4. Prove (13.6.1).

5. Let p be an odd prime with $p \equiv 1$ or $9 \pmod{20}$. Berndt and Evans [2, pp. 416, 417] have defined the integers A_4 and B_4 by

$$A_4 = -\left(\frac{2}{p}\right)a_4, \quad B_4 = -\left(\frac{2}{p}\right)b_4,$$

and the integers A_{20} and B_{20} by

$$p = A_{20}^2 + 5B_{20}^2, \quad A_{20} \equiv \begin{cases} A_4 \pmod 5, & \text{if } 5 \nmid A_4, \\ B_4 \pmod 5, & \text{if } 5 \mid A_4. \end{cases}$$

If $p \equiv 1 \pmod{20}$, prove that

$$A_{20} = \begin{cases} (-1)^{(p+3)/4} a_{20}/c_{20}, & \text{if } 5 \nmid a_4, \\ i(-1)^{(p-1)/4} a_{20}/c_{20}, & \text{if } 5 \mid a_4, \end{cases}$$

where a_{20} and c_{20} are defined in Theorem 3.8.2.

If $p \equiv 9 \pmod{20}$, prove that $A_{20} = (-1)^{(p-1)/4} a_{20}$, where a_{20} is defined in Theorem 12.11.1.

6. Berndt and Evans [2, Theorem 5.20] prove that, in the notation of Problem 5,

$$\Lambda_5 = \begin{cases} -4A_{20}, & \text{if } p \equiv 1 \pmod{20},\ 5 \nmid A_4, \\ 4A_{20}, & \text{if } p \equiv 9 \pmod{20},\ 5 \nmid A_4. \end{cases}$$

Show that this is in agreement with Theorem 13.4.1.

7. Let p be a prime with $p \neq 2, 5$, and let g be a primitive root of p. Prove that

$$\Lambda_5(g) = \begin{cases} i4(-1)^{(p+3)/4} a_{20}/c_{20}, & \text{if } p \equiv 1 \pmod{20} \text{ and } 5 \mid a_4, \\ 4(-1)^{(p-1)/4} a_{20}, & \text{if } p \equiv 9 \pmod{20} \text{ and } 5 \mid a_4, \\ 0, & \text{otherwise.} \end{cases}$$

8. Use Theorem 13.4.1 and Problem 7 to determine $\Lambda_5(a)$.

9. Use Theorems 13.2.3 and 13.3.5 with $n = 6$ to determine Λ_{12} and $B_6 = \Lambda_{12} - \Lambda_6$. (One needs the evaluation of the sum $E(\chi)$ when $p \equiv 11 \pmod{24}$ and χ is a character of order 24 on F_{p^2}; see Problem 16 of Exercises 12.)

10. Use Theorems 13.3.4, 13.5.1, and 13.6.1 to evaluate $\Lambda_8(a)$.

11. Use Theorems 13.4.1, 13.4.1, and 13.6.1 to evaluate $\Lambda_{10}(a)$.

12. Use Theorem 13.3.5, Theorem 13.6.1, and Problem 9 to evaluate $\Lambda_{12}(a)$.

Notes on Chapter 13

The character sums known as Brewer sums Λ_n were first studied by Brewer [2] in 1961. In this paper Brewer obtained the fundamental result $2\Lambda_n = \Omega_n + \Theta_n$ (Theorem 13.2.1) and evaluated Λ_3, Λ_4, and Λ_5. His method required congruences for binomial coefficients of the type proved in Chapter 9 and Section 12.9. In 1966 Brewer [3] introduced the sum $\Lambda_n(a)$, and evaluated it for $n = 5$. See Problem 8. Leprévost and Morain [1, p. 179] used the evaluation of this Brewer sum to count the number of points on a certain elliptic curve with complex multiplication by $Z[\sqrt{-5}]$. In 1963 Whiteman [12] determined the value of Λ_4 and in 1965 [13], [14] the value of Λ_5 without using the congruences for binomial coefficients modulo p used by Brewer. In 1968, S. F. Robinson [1] gave a formula for $\Lambda_{2n}(a)$ when n is odd, and used it to evaluate $\Lambda_6(a)$ and $\Lambda_{10}(a)$. In 1972 Giudici, Muskat, and Robinson [1] evaluated $\Lambda_7(a), \Lambda_8(a), \Lambda_9(a), \Lambda_{10}(a), \Lambda_{12}(a), \Lambda_{14}(a)$, and $\Lambda_{18}(a)$. Leonard and Williams [5] have given another proof of Brewer's evaluation of B_2, and Williams [28] has given a very simple proof that $B_2 = 0$ when $p \equiv 5$ or $7 \pmod 8$ (Theorem 13.3.6). The same proof of this latter result has been given by Radoux [1], who also proved a special case of the evaluation of B_2 given in Theorem 13.3.6. See also P. Chowla [1].

The approach in this chapter to the theory of Brewer sums follows that of Berndt and Evans [2]. This theory unifies the existing results in the literature, which had been obtained by a variety of methods. No use is made of the theory of cyclotomic numbers. Ordinary Brewer sums and generalized Brewer sums are expressed in terms of Jacobsthal sums over F_p and Eisenstein sums over F_{p^2}. If the values of the relevant Jacobsthal sums and Eisenstein sums are known, then it is an easy matter to determine the value of the Brewer sum under consideration.

Propositions 13.1.1–13.1.6 are taken from Berndt and Evans [2, §5.1]. Theorem 13.2.1, as well as its extension (13.6.1), are due to Brewer [2, Lemma 2], [3, Lemma 2]. Theorem 13.2.2 is a result of Berndt and Evans [2, Theorem 5.9] with a misprint corrected (in [2, p. 408], the words "even" and "odd" should be interchanged). Theorem 13.2.3 generalizes a result of S. F. Robinson [1, Theorem 5.6] and is due to Berndt and Evans [2, Corollary 5.10]. Theorem 13.2.4 can be found in the work of Berndt and Evans [2, Theorem 5.11]. Theorem 13.3.3 is a more precise form of a theorem of Brewer [2, Theorem 1]. Theorem 13.3.4 is due to Brewer [2, Theorem 3]. Proofs have also been given by Berndt and Evans [2, Theorems 5.15 and 5.17], Leonard and Williams [5], and Whiteman [12]. Theorem 13.3.5 is due to S. F. Robinson [1, eq. (6.11)]; see also Berndt and Evans [2, Theorem 5.16].

The evaluation of Λ_5 given in Theorem 13.4.1 is due to Brewer [2]. Our proof follows that of Berndt and Evans [2, Theorem 5.20]. Theorem 13.4.2 is due essentially to S. F.

Robinson [1, §7] and Giudici, Muskat, and Robinson [1, §9]; see also Berndt and Evans [2, Theorem 5.22].

The evaluation of Λ_8 given in Theorem 13.5.1 is a variant of that of Giudici, Muskat, and Robinson [1, §6]; see also Berndt and Evans [2, Theorem 5.19].

Theorems 13.6.1 and 13.6.2 are due to Berndt and Evans [2, Theorems 5.12 and 5.14].

Theorem 13.7.3 is a more precise form of a result originally due to Brewer [2]; see also Berndt and Evans [2, §5.31]. Theorems 13.7.4 and 13.7.5 are due to Giudici, Muskat, and Robinson [1, p. 336].

Problem 3 is taken from Berndt and Evans [2, Proposition 5.3]. For Problem 4, see, for example, Berndt and Evans [2, Lemma 5.13]. For Problem 9, see Berndt and Evans [2, Theorem 5.18]. Regarding Problems 10–12, see Giudici, Muskat, and Robinson [1, §6 and §9].

We now describe generalizations of the double sum evaluation in (13.3.2). For a prime $p > 3$ and an integer $t \not\equiv 0 \pmod p$, define the double sum of Legendre symbols

$$I(t) = \sum_{x,y \in F_p} \left(\frac{xy(1+x)(1+y)(x+ty)}{p} \right).$$

This is a character sum analogue over F_p of the integral formula for a special hypergeometric ${}_3F_2$ series. Clearly $I(t^{-1}) = \left(\frac{t}{p}\right) I(t)$. Formula (13.3.2) yields an evaluation of $I(1)$. Evans [7, §6] evaluated a generalization of $I(-1)$. That evaluation has been used by Koike [1] to prove congruences for Apéry numbers, and his work has been extended by K. Ono [1, §5]. Koike [1, p. 465] posed the problem of evaluating $I(-4)$ (or equivalently, $I(-1/4)$). This was solved by K. Ono [1], who also evaluated $I(t)$ for several other values of t. The results are:

$$\begin{aligned} I(-1) &= 4a_4^2 - p\left(1 + (-1)^{(p-1)/2}\right), \\ I(1) &= \left(\frac{2}{p}\right)\left(4a_8^2 - p\right), \\ I(-1/4) &= \left(\frac{3}{p}\right)\left(4a_3^2 - p\right), \\ I(8) &= 4a_4^2 - p, \end{aligned}$$

and

$$I(-1/64) = \left(\frac{7}{p}\right)\left(4a_7^2 - p\right).$$

Here $p = a_4^2 + b_4^2$ with a_4 odd when $p \equiv 1 \pmod 4$, while a_4 is defined to be 0 if $p \equiv 3 \pmod 4$. Similarly, if $p \equiv 1$ or $3 \pmod 8$, then $p = a_8^2 + 2b_8^2$, while $a_8 = 0$ otherwise; if $p \equiv 1 \pmod 3$, then $p = a_3^2 + 3b_3^2$, while $a_3 = 0$ otherwise; and if $p \equiv 1$, 2, or 4 (mod 7), then $p = a_7^2 + 7b_7^2$, while $a_7 = 0$ otherwise.

Chapter 14

A General Eisenstein Reciprocity Law

The aim of this chapter is to use the properties of Gauss and Jacobi sums that we have developed to give a proof of a general Eisenstein reciprocity law for the kth–power residue symbol in cyclotomic fields, where k is a prime power. This law was first proved by Eisenstein [12] for an odd prime k in 1850. That special case is the beautiful Eisenstein reciprocity law (Corollary 14.5.10) mentioned by Weil in his introduction to Kummer's *Werke* [CW, vol. 1, p. 9]. It and related reciprocity laws due to Kummer [4], [5], [7] were used extensively in the nineteenth century in work on Fermat's last theorem. The reader will find the precise statement of our extension of Eisenstein's law for prime powers k in Theorem 14.3.1, following the definition of a primary element of the ring of integers in the cyclotomic field $Q(\beta)$, where $\beta = \exp(2\pi i/k)$. We will refer to Theorem 14.3.1 simply as Eisenstein's reciprocity law.

14.1 A quotient of Gauss sums

In this section, we discuss a quotient of Gauss sums that will be needed for the definition of primary elements in Section 14.2. Let $k = l^n$ ($\neq 2, 4$), where l is a prime and n is a positive integer. Set $\beta = \exp(2\pi i/k)$, $\lambda = 1-\beta$, and $K = Q(\beta)$. Let P be a prime ideal of O_K not dividing k, so that the absolute norm of P is $N(P) = |O_K/P| = p^r$, where p is rational prime whose order (mod k) is r. Let χ_P denote the kth–power residue character (mod P) on O_K defined in (11.2.1). For a rational integer a, define

$$\Phi_a(P) = \begin{cases} G_r^a(\chi_P)/G_r(\chi_P^a), & \text{if } k \nmid a, \\ G_r^a(\chi_P), & \text{if } k \mid a, \end{cases} \tag{14.1.1}$$

where as usual $G_r(\chi_P)$ is the Gauss sum over F_{p^r}. The expression on the right side of (14.1.1) was already studied by Jacobi in 1827 [1, p. 399], in the case $r = 1$. Define $\Phi_a(O_K) = 1$. We extend the domain of Φ_a to all nonzero ideals of O_K by multiplicativity. For $0 \neq \alpha \in O_K$, we define $\Phi_a(\alpha) = \Phi_a(\alpha O_K)$. Thus $\Phi_a(\delta) = 1$ for every unit $\delta \in O_K$. Observe that $\Phi_a(P) \in K$, because

$$\Phi_a(P) = \prod_{j=1}^{a-1} J_r(\chi_P, \chi_P^j), \quad \text{if } 1 \le a \le k-1,$$

and

$$\Phi_k(P) = p^r \chi_P(-1) \prod_{j=1}^{k-2} J_r(\chi_P, \chi_P^j)$$

(see Problem 1 of Exercises 2). Note furthermore that $\Phi_a(P) \in O_K$ if a is nonnegative.

Let $R(k)$ denote a reduced residue system (mod k). For real a, $L(a)$ denotes the least nonnegative residue of a (mod k), so that

$$[a/k] = (a - L(a))/k. \tag{14.1.2}$$

Throughout this chapter ν denotes an element of O_K with ν and k relatively prime, and u is a rational prime with $\gcd(u, k) = 1$. Let s be the order of u (mod k). For $t \in R(k)$ define $\sigma_t \in \mathrm{Gal}(K/Q)$ by $\sigma_t(\beta) = \beta^t$.

The next theorem gives the prime ideal factorization of $\Phi_a(P)$ in K.

Theorem 14.1.1 *For any integer a,*

$$\Phi_a(P)O_K = \prod_{t\in R(k)} \sigma_{-t}^{-1}(P^{[aL(t)/k]}). \tag{14.1.3}$$

Proof If $k \mid a$, then (14.1.3) follows from Theorem 11.2.8. Now suppose that $k \nmid a$, and write $a = kb + c$, where b and c are integers such that $0 < c < k$. Then

$$\Phi_a(P) = \Phi_{kb}(P)\Phi_c(P),$$

and so it suffices to prove (14.1.3) for $0 < a < k$. Again by Theorem 11.2.8,

$$\Phi_a(P) = \frac{G_r^a(\chi_P)}{G_r(\chi_P^a)} = \prod_{t\in R(k)} \sigma_{-t}^{-1}(P^{(aL(t)-L(ta))/k}),$$

and so the result follows from (14.1.2). ∎

Theorem 14.1.2 *Let $\nu \in O_K$ be coprime with k, and let a be any integer. Then*

$$\prod_{t\in R(k)} \sigma_{-t}^{-1}(\nu)^{[aL(t)/k]} = \Phi_a(\nu)\eta_a(\nu)\epsilon_a(\nu), \tag{14.1.4}$$

where $\epsilon_a(\nu)$ is a power of β, and

$$\eta_a(\nu) = \begin{cases} 1, & \text{if } l = 2, \\ \pm 1, & \text{if } l > 2. \end{cases} \tag{14.1.5}$$

Proof By the multiplicativity of $\Phi_a(P)$ and Theorem 14.1.1,

$$\Phi_a(\nu)O_K = \prod_{t\in R(k)} \sigma_{-t}^{-1}(\nu)^{[aL(t)/k]}O_K.$$

Thus

$$\Phi_a(\nu)\delta = \prod_{t\in R(k)} \sigma_{-t}^{-1}(\nu)^{[aL(t)/k]} \tag{14.1.6}$$

for some unit $\delta \in O_K$. By Theorem 2.1.13, it remains to prove that $|\delta|^2 = 1$. By (14.1.1) (see Problem 2),

$$|\Phi_a(\nu)|^2 = \begin{cases} N(\nu O_K)^a, & \text{if } k \mid a, \\ N(\nu O_K)^{a-1}, & \text{if } k \nmid a. \end{cases} \tag{14.1.7}$$

Since

$$[aL(t)/k] + [aL(-t)/k] = \begin{cases} a, & \text{if } k \mid a, \\ a-1, & \text{if } k \nmid a, \end{cases}$$

for $t \in R(k)$ (see Problem 1),

$$|\prod_{t \in R(k)} \sigma_{-t}^{-1}(\nu)^{[aL(t)/k]}|^2 = \begin{cases} N(\nu O_K)^a, & \text{if } k \mid a, \\ N(\nu O_K)^{a-1}, & \text{if } k \nmid a. \end{cases} \tag{14.1.8}$$

Thus $|\delta|^2 = 1$ by (14.1.6)–(14.1.8). ∎

Theorem 14.1.3 *Let l be an odd prime. Let $\nu_1, \nu_2 \in O_K$ be coprime with k, and let a be any integer. Then, in the notation of Theorem 14.1.2,*

$$\eta_a(\nu_1\nu_2) = \eta_a(\nu_1)\eta_a(\nu_2), \quad \textit{and} \quad \epsilon_a(\nu_1\nu_2) = \epsilon_a(\nu_1)\epsilon_a(\nu_2).$$

Proof This is a simple exercise. See Problem 3. ∎

14.2 Primary integers of cyclotomic fields

If $l = 2$, let $c = 3$; if $l > 2$, let c be a primitive root $\pmod{l^m}$ for every integer $m \geq 1$. For example, when $l > 2$, we could take for c any primitive root of l for which $c^{l-1} \not\equiv 1 \pmod{l^2}$, in which case $l^e \parallel c^{\phi(l^e)} - 1$ for each positive integer e; see the book of Nagell [2, Theorem 65, p. 107].

Given $\nu \in O_K$ coprime with k, define the integer M by

$$M = (N(\nu O_K) - 1)/k. \tag{14.2.1}$$

Recall that $\epsilon_a(\nu)$ is the power of β defined by (14.1.4).

Definition *An element $\nu \in O_K$ is said to be primary if ν is coprime with k and $\epsilon_c(\nu) = (-1)^M$.*

The condition $\epsilon_c(\nu) = (-1)^M$ is equivalent to

$$\begin{cases} \epsilon_c(\nu) = 1, & \text{if } l > 2, \\ \epsilon_3(\nu) = (-1)^M, & \text{if } l = 2. \end{cases} \tag{14.2.2}$$

When $l > 2$, this definition appears to depend on the choice of the primitive root c (mod l^m). However, we will show in Theorem 14.5.8 that the definition is actually independent of the choice of the primitive root c.

As examples, we determine the primary integers in K for $k = 8$ and $k = 3$ in the next two theorems.

Theorem 14.2.1 *Let $K = Q(\beta)$, where $\beta = e^{2\pi i/8}$, so that $k = 8$, $l = 2$, and $n = 3$. Let $\nu \in O_K$ be coprime with 8. Then*

$$\nu \text{ is primary} \Longleftrightarrow \nu \equiv 1 \text{ or } 1 + \beta + \beta^3 \pmod 2.$$

Proof For $t \in R(8)$, set $\nu_t = \sigma_t(\nu)$. By Theorem 14.1.2,

$$\Phi_2(\nu) = \overline{\epsilon_2(\nu)\nu\nu_3} \tag{14.2.3}$$

and

$$\Phi_3(\nu) = \overline{\epsilon_3(\nu)}\nu^2\nu_3\nu_5 = \overline{\epsilon_3(\nu)}\nu^2\nu_3\overline{\nu}_3. \tag{14.2.4}$$

Let $\lambda = 1 - \beta$, so that $2 = \epsilon\lambda^4$ for the unit $\epsilon = -i(\sqrt{2}+1)^2$ in O_K.

Let P denote a prime ideal dividing νO_K. Then, by Theorem 2.1.8, for the octic character χ_P,

$$K_r(\chi_P) \equiv K_r(\chi_P^2) \equiv K_r(\overline{\chi}_P) \equiv -1 \pmod{\lambda^5},$$

and thus

$$J_r(\chi_P, \chi_P) = \overline{\chi}_P(4)K_r(\chi_P) \equiv -\overline{\chi}_P(4) \pmod{\lambda^5}, \tag{14.2.5}$$
$$J_r(\chi_P, \chi_P^2) = \chi_P(-4)K_r(\chi_P^2) \equiv -\chi_P(-4) \pmod{\lambda^5}, \tag{14.2.6}$$

and

$$\overline{J_r(\chi_P, \chi_P)} = \chi_P(4)K_r(\overline{\chi}_P) \equiv -\chi_P(4) \pmod{\lambda^5}, \tag{14.2.7}$$

where in (14.2.6) we have used Theorem 3.3.3 and the fact that $\chi^4(2) = 1$. Now

$$\Phi_2(P) = G_r^2(\chi_P)/G_r(\chi_P^2) = J_r(\chi_P, \chi_P),$$

so that, by (14.2.5) and (14.2.7),

$$|\Phi_2(P)|^2 = J_r(\chi_P, \chi_P)\overline{J_r(\chi_P, \chi_P)} \equiv 1 \pmod{\lambda^5}.$$

Taking the product over all P dividing νO_K counting multiplicity, we find that

$$|\Phi_2(\nu)|^2 \equiv 1 \pmod{\lambda^5}. \tag{14.2.8}$$

Further, by (14.2.5) and (14.2.6),

$$\Phi_3(P) = G_r^3(\chi_P)/G_r(\chi_P^3) = J_r(\chi_P, \chi_P)J_r(\chi_P, \chi_P^2) \equiv \chi_P(-1) \pmod{\lambda^5}.$$

Taking the product of these congruences over all P dividing νO_K, we see from (1.1.3) that

(14.2.9) $$\Phi_3(\nu) \equiv (-1)^{(N(\nu O_K)-1)/k} = (-1)^M \pmod{\lambda^5}.$$

Then, from (14.2.3), (14.2.4), (14.2.8), and (14.2.9),

$$\nu\nu_3\overline{\nu}\,\overline{\nu}_3 \equiv 1 \pmod{\lambda^5}$$

and

$$\overline{\epsilon_3(\nu)}\nu^2\nu_3\overline{\nu}_3 \equiv (-1)^M \pmod{\lambda^5}.$$

Thus,

$$\begin{aligned}\nu \text{ is primary} \quad &\Longleftrightarrow \epsilon_3(\nu) = (-1)^M \Longleftrightarrow \nu^2\nu_3\overline{\nu}_3 \equiv 1 \pmod{\lambda^5}\\ &\Longleftrightarrow \nu^2\nu_3\overline{\nu}_3 \equiv \nu\nu_3\overline{\nu}\,\overline{\nu}_3 \pmod{\lambda^5} \Longleftrightarrow \nu \equiv \overline{\nu} \pmod{\lambda^5}.\end{aligned}$$

As $\nu \in O_K$ and $\lambda \nmid \nu$,

$$\nu = \sum_{i=0}^{(k/2)-1} a_i\lambda^i$$

for some integers a_i with a_0 odd. Thus

$$\nu \equiv \epsilon_0 + \epsilon_1\lambda + \epsilon_2\lambda^2 + \epsilon_3\lambda^3 \pmod{\lambda^5},$$

where $\epsilon_0 = \pm 1$ and, for $i = 1, 2, 3$, each $\epsilon_i = 0$ or 1. Now $\overline{\lambda} = \beta^3\lambda$, so that

$$\overline{\nu} = \sum_{i=0}^{(k/2)-1} a_i\beta^{3i}\lambda^i,$$

and thus,

$$\begin{aligned}\overline{\nu} &\equiv \epsilon_0 + \epsilon_1\beta^3\lambda + \epsilon_2\beta^6\lambda^2 + \epsilon_3\beta\lambda^3\\ &= \epsilon_0 + \epsilon_1(1-\lambda)^3\lambda + \epsilon_2(1-\lambda)^6\lambda^2 + \epsilon_3(1-\lambda)\lambda^3\\ &\equiv \epsilon_0 + \epsilon_1(\lambda + \lambda^2 + \lambda^3 + \lambda^4) + \epsilon_2(\lambda^2 + \lambda^4) + \epsilon_3(\lambda^3 + \lambda^4) \pmod{\lambda^5}.\end{aligned}$$

Hence

$$\begin{aligned}\nu \equiv \overline{\nu} \pmod{\lambda^5} &\Longleftrightarrow \epsilon_1(\lambda^2 + \lambda^3 + \lambda^4) + \epsilon_2\lambda^4 + \epsilon_3\lambda^4 \equiv 0 \pmod{\lambda^5}\\ &\Longleftrightarrow \epsilon_1\lambda^2 + \epsilon_1\lambda^3 + (\epsilon_1 + \epsilon_2 + \epsilon_3)\lambda^4 \equiv 0 \pmod{\lambda^5}\\ &\Longleftrightarrow \epsilon_1 = 0 \text{ and } \epsilon_2 + \epsilon_3 = 0 \text{ or } 2.\end{aligned}$$

This proves that ν is primary if and only if $\nu \equiv \pm 1$ or $\pm 1 + \lambda^2 + \lambda^3 \pmod{\lambda^5}$. Since $2 \equiv \lambda^4 \pmod{\lambda^5}$, ν is primary if and only if

$$\nu \equiv 1 \text{ or } 1 + \lambda^2 + \lambda^3 \pmod{\lambda^4}.$$

Since

$$\begin{aligned} 1+\lambda^2+\lambda^3 &= 1+(1-\beta)^2+(1-\beta)^3 \\ &= 3-5\beta+4\beta^2-\beta^3 \equiv 1+\beta+\beta^3 \pmod{\lambda^4}, \end{aligned}$$

the result follows. ■

Theorem 14.2.2 *Let $K = Q(\beta)$, where $\beta = e^{2\pi i/3}$, so that $k = l = 3$ and $n = 1$. Let $\lambda = 1 - \beta$ and choose the primitive root $c = 2$ in (14.2.2). Let $\nu \in O_K$ be coprime with 3, i.e., $\lambda \nmid \nu$. Then ν is primary if and only if $\nu \equiv \pm 1 \pmod 3$. (Thus the definitions of primary in Sections 14.2 and 8.1 agree for $k = 3$.)*

Proof By Theorem 14.1.2,

$$\nu = \eta_2(\nu)\epsilon_2(\nu)\Phi_2(\nu).$$

Let P denote a prime ideal dividing νO_K. By Theorems 2.1.3(a) and 2.1.7, for the cubic character χ_P,

$$\Phi_2(P) = G_r^2(\chi_P)/G_r(\chi_P^2) = J_r(\chi_P, \chi_P) \equiv -p^r \equiv -1 \pmod 3.$$

Then, by the multiplicativity of Φ_2,

$$\Phi_2(\nu) \equiv \pm 1 \pmod 3.$$

First suppose that ν is primary, so that by (14.2.2), $\epsilon_2(\nu) = 1$. Then

$$\nu = \eta_2(\nu)\Phi_2(\nu) \equiv \pm 1 \pmod 3,$$

as desired. Conversely, suppose that $\nu \equiv \pm 1 \pmod 3$. It remains to show that $\epsilon_2(\nu) = 1$. If ν is a unit, then $\nu = \pm 1$, so that the assertion $\epsilon_2(\nu) = 1$ follows easily from (14.1.4). In view of Theorem 8.1.5, we may thus suppose that ν is a product of Eisenstein primes $\equiv \pm 1 \pmod 3$. By Theorem 14.1.3, it suffices to prove that $\epsilon_2(\nu) = 1$ when $\nu \in O_K$ is a prime $\equiv -1 \pmod 3$. Either ν is a rational prime $p \equiv 2 \pmod 3$ of norm p^2, or ν is a prime $\pi \equiv -1 \pmod 3$ of norm $p \equiv 1 \pmod 3$. First suppose that ν is a rational prime $p \equiv 2 \pmod 3$. Then, writing $P = pO_K$, we have

$$\Phi_2(\nu) = J_2(\chi_P, \chi_P) = p = \nu,$$

by Theorem 12.2.4 for $p > 2$, and by direct calculation for $p = 2$. Thus $\eta_2(\nu)\epsilon_2(\nu) = 1$, so that $\epsilon_2(\nu) = \pm 1$, and hence $\epsilon_2(\nu) = 1$, since $\epsilon_2(\nu)$ is a power of β. Next suppose that $\nu = \pi \equiv -1 \pmod 3$, with $p = \pi\bar{\pi}$. Then, writing $P = \pi O_K$, we have

$$\Phi_2(\nu) = J(\chi_P, \chi_P) = \pi = \nu,$$

by Lemma 8.1.6, and so $\epsilon_2(\nu) = 1$ as in the previous case. ■

14.3 Statement of Eisenstein's reciprocity law

Let $K = Q(\beta)$ with $\beta = e^{2\pi i/k}$. For a prime ideal P of O_K not dividing k, the kth–power residue character χ_P defined in (11.2.1) is often denoted by the symbol

$$\chi_P(\alpha) = \left(\frac{\alpha}{P}\right)_k, \quad \alpha \in O_K.$$

We proceed to define the kth–power residue symbol $\left(\frac{\alpha}{\delta}\right)_k$, where $\delta \in O_K$ is coprime with k. If δ is a unit of O_K, define

$$\left(\frac{\alpha}{\delta}\right)_k = \begin{cases} 1, & \text{if } 0 \neq \alpha \in O_K, \\ 0, & \text{if } \alpha = 0. \end{cases}$$

Otherwise $\delta O_K = \prod_{i=1}^n P_i$ for (not necessarily distinct) prime ideals P_i of O_K with $P_i \not| k$, and we define

$$\left(\frac{\alpha}{\delta}\right)_k = \prod_{i=1}^n \left(\frac{\alpha}{P_i}\right)_k.$$

Now that we have defined the kth–power residue symbol on O_K and primary elements of O_K, we can state Eisenstein's reciprocity law.

Theorem 14.3.1 (*Eisenstein's reciprocity law.*) *Let $k = l^n$ ($\neq 2, 4$), where l is a prime and n is a positive integer. Let u be a rational prime with* $\gcd(u,k) = 1$, *and let ν be a primary integer of $K = Q(e^{2\pi i/k})$. Then*

(a) $$\left(\frac{\nu}{u}\right)_k = \left(\frac{u}{\nu}\right)_k, \quad \text{if } l > 2,$$

(b) $$\left(\frac{\nu}{u}\right)_k = \left(\frac{(-1)^{(u-1)/2}u}{\nu}\right)_k, \quad \text{if } l = 2.$$

This theorem will be deduced with the aid of a general reciprocity relation (Theorem 14.4.1), which we establish in the next section. The deduction of assertion (a) from Theorem 14.4.1 is given in Section 14.5 and that of (b) in Section 14.6.

14.4 A general reciprocity relation

In the following reciprocity relation, no reference is made to primary integers (cf. Theorem 14.3.1).

Theorem 14.4.1 *Let $k = l^n$ ($\neq 2, 4$), where l is a prime and n is a positive integer. Let u be a rational prime coprime with k. Set $K = Q(\beta)$, where $\beta = \exp(2\pi i/k)$, and* *let ν be an integer of K coprime with k. Then*

$$\epsilon_u(\nu)^{\overline{u}}\left(\frac{\nu}{u}\right)_k = \left(\frac{u}{\nu}\right)_k,$$

where $\overline{u}$ denotes the inverse of u (mod k).

Proof We may assume that ν and u are coprime, since otherwise both sides vanish. Let s denote the order of u (mod k). Let H be a set of $\phi(k)/s$ integers such that

$$R(k) = \bigcup_{t\in H}\bigcup_{j=0}^{s-1}\{t\overline{u}^j\},$$

where $R(k)$ is a reduced residue system (mod k). By (14.1.4) with $a = u$,

$$\eta_u(\nu)\epsilon_u(\nu)\Phi_u(\nu) = \prod_{t\in H}\prod_{j=0}^{s-1}\sigma_{-t\overline{u}^j}^{-1}(\nu)^{[uL(t\overline{u}^j)/k]} = \prod_{t\in H}\sigma_{-t}^{-1}\left(\prod_{j=0}^{s-1}\sigma_{u^j}(\nu)^{[uL(t\overline{u}^j)/k]}\right).$$

Since $\nu \in O_K = Z[\beta]$, we have $\sigma_{u^j}(\nu) \equiv \nu^{u^j}$ (mod u). Hence

$$\eta_u(\nu)\epsilon_u(\nu)\Phi_u(\nu) \equiv \prod_{t\in H}\sigma_{-t}^{-1}(\nu)^{\sum_{j=0}^{s-1}u^j[uL(t\overline{u}^j)/k]} \pmod u. \tag{14.4.1}$$

By (14.1.2) the summation in (14.4.1) equals

$$\begin{aligned}(14.4.2)\quad k^{-1}\sum_{j=0}^{s-1}u^j(uL(t\overline{u}^j) - L(uL(t\overline{u}^j))) &= k^{-1}\sum_{j=0}^{s-1}(u^{j+1}L(t\overline{u}^j) - u^jL(t\overline{u}^{j-1}))\\ &= (u^s-1)L(tu)/k.\end{aligned}$$

Thus (14.4.1) becomes

$$\eta_u(\nu)\epsilon_u(\nu)\Phi_u(\nu) \equiv \prod_{t\in H}\sigma_{-t}^{-1}(\nu)^{(u^s-1)L(tu)/k} \pmod u. \tag{14.4.3}$$

For the remainder of the proof we let U denote a prime ideal factor of uO_K. Since, for $x \in O_K$,

$$\left(\frac{x}{U}\right)_k \equiv x^{(u^s-1)/k} \pmod U,$$

it follows from (14.4.3) that

$$\eta_u(\nu)\epsilon_u(\nu)\Phi_u(\nu) \equiv \prod_{t\in H}\left(\frac{\sigma_{-t}^{-1}(\nu)}{U}\right)_k^{L(tu)} = \prod_{t\in H}\left(\frac{\sigma_{-t}^{-1}(\nu)}{U}\right)_k^{tu} \pmod U.$$

Now, whenever $t \in R(k)$,

$$\left(\frac{x}{U}\right)_k^t = \sigma_t\left(\frac{x}{U}\right)_k = \left(\frac{\sigma_t(x)}{\sigma_t(U)}\right)_k, \tag{14.4.4}$$

so that

$$\eta_u(\nu)\epsilon_u(\nu)\Phi_u(\nu) \equiv \prod_{t\in H}\left(\frac{\overline{\nu}}{\sigma_t(U)}\right)_k^u \pmod{U}.$$

By (11.1.3),

$$uO_K = \prod_{t\in H}\sigma_t(U).$$

Thus

$$\eta_u(\nu)\epsilon_u(\nu)\Phi_u(\nu) \equiv \left(\frac{\overline{\nu}}{u}\right)_k^u \equiv \left(\frac{\nu}{u}\right)_k^{-u} \pmod{U}. \tag{14.4.5}$$

Now, by Theorem 1.1.3, for any prime ideal P in O_K dividing ν,

$$G_r^u(\chi_P) \equiv \overline{\chi}_P^u(u)G_r(\chi_P^u) \pmod{u}.$$

Thus

$$\left(\frac{u}{P}\right)_k^{-u} = \overline{\chi}_P^u(u) \equiv G_r^u(\chi_P)/G_r(\chi_P^u) \pmod{u},$$

so that

$$\left(\frac{u}{\nu}\right)_k^{-u} \equiv \Phi_u(\nu) \pmod{u}.$$

Then, from (14.4.5),

$$\eta_u(\nu)\epsilon_u(\nu)\left(\frac{u}{\nu}\right)_k^{-u} \equiv \left(\frac{\nu}{u}\right)_k^{-u} \pmod{U}. \tag{14.4.6}$$

Since both sides of (14.4.6) are $2k$th roots of unity, they must, in fact, be equal by Theorem 2.1.9, that is,

$$\eta_u(\nu)\epsilon_u(\nu)\left(\frac{u}{\nu}\right)_k^{-u} = \left(\frac{\nu}{u}\right)_k^{-u}. \tag{14.4.7}$$

Raising both sides of (14.4.7) to the kth power, we obtain $\eta_u(\nu) = 1$, so that (14.4.7) becomes

$$\epsilon_u(\nu)\left(\frac{u}{\nu}\right)_k^{-u} = \left(\frac{\nu}{u}\right)_k^{-u}. \tag{14.4.8}$$

Then, raising both sides of (14.4.8) to the $\overline{u}$th power, we deduce the assertion of the theorem. ■

14.5 Proof of Eisenstein's reciprocity law for $l > 2$

Let $K = Q(\beta)$, with $\beta = e^{2\pi i/k}$ and $k = l^n$, where l is an odd prime. In this section, we prove a series of results (culminating with Lemma 14.5.7) that lead up to the proof of Eisenstein's reciprocity law (Theorem 14.3.1) in the case $l > 2$. We also prove some corollaries of Theorem 14.3.1. Recall from Corollary 2.1.10 that $lO_K = (\lambda O_K)^{\phi(k)}$, where $\lambda = 1 - \beta$, so that $\lambda^2 \mid l$.

Definition *When $l > 2$, the integer ν of K is said to be semiprimary if ν is coprime with k and $\nu \equiv x \pmod{\lambda^2}$ for some rational integer x (with $l \nmid x$.)*

Example When $k = 3$, that is, $l = 3$, $n = 1$, and $\beta = e^{2\pi i/3}$, the concepts of primary and semiprimary coincide, as

$$\begin{aligned}\nu \text{ is semiprimary} &\iff \nu \equiv x \pmod{\lambda^2} \text{ for some rational integer } x \not\equiv 0 \pmod 3\\ &\iff \nu \equiv \pm 1 \pmod{\lambda^2} \iff \nu \text{ is primary,}\end{aligned}$$

by Theorem 14.2.2. In fact, the concepts of primary and semiprimary coincide for *any* odd prime k, as will follow from Lemma 14.5.2, (14.2.2), and Theorem 14.5.4.

Note that the product of semiprimary integers is again semiprimary. As $\beta^l \equiv 1 \pmod{\lambda^2}$, β^l is semiprimary, and hence μ and $-\mu$ are semiprimary for any l^{n-1}th root of unity μ.

Theorem 14.5.1 *Let $l > 2$. Let $\nu \in O_K$ be coprime with k. Then there is a unique integer m with $0 \leq m < l$ such that $\beta^m \nu$ is semiprimary.*

Proof Since $\nu \in O_K = Z[\beta] = Z[\lambda]$, we have $\nu \equiv x + y\lambda \pmod{\lambda^2}$ for some $x, y \in Z$. Note that $\gcd(x, l) = 1$. Let m denote the least nonnegative integer satisfying $m \equiv yx^{-1} \pmod{l}$. Since $\beta^m = (1-\lambda)^m \equiv 1 - m\lambda \pmod{\lambda^2}$,

$$\beta^m \nu \equiv (1 - m\lambda)(x + y\lambda) \equiv x + \lambda(y - mx) \equiv x \pmod{\lambda^2}.$$

Thus $\beta^m \nu$ is semiprimary.

Suppose that $\beta^{m_1}\nu$ and $\beta^{m_2}\nu$ are both semiprimary for integers m_1 and m_2 with $0 \leq m_1 < m_2 < l$. Then $\beta^{m_1}\nu \equiv x_1 \pmod{\lambda^2}$ and $\beta^{m_2}\nu \equiv x_2 \pmod{\lambda^2}$ for some integers x_1 and x_2. Hence $\beta^t \equiv b \pmod{\lambda^2}$ for $t = m_2 - m_1$ and $b \equiv x_1^{-1}x_2 \pmod{l}$. Note that $0 < t < l$. Applying the automorphism σ_t^{-1}, we obtain $\beta \equiv b \pmod{\lambda^2}$.

Taking this congruence modulo λ $(= 1 - \beta)$, we see that $b \equiv 1 \pmod{l}$. As $\lambda^2 \mid l$, we see that $b \equiv 1 \pmod{\lambda^2}$, and so $\beta \equiv 1 \pmod{\lambda^2}$, which is a contradiction. ■

Lemma 14.5.2 *Let $l > 2$. Let $\nu \in O_K$ be coprime with k. If ν is semiprimary, then $\epsilon_a(\nu)$ is a power of β^l for each integer a.*

Proof As ν is semiprimary, there is a rational integer x such that $\nu \equiv x \pmod{\lambda^2}$. Hence

$$\prod_{t \in R(k)} \sigma_{-t}^{-1}(\nu)^{[aL(t)/k]} \equiv \prod_{t \in R(k)} x^{[aL(t)/k]} \pmod{\lambda^2}.$$

By (14.1.8) with $\nu = x$, the right side above equals

$$\begin{cases} x^{a\phi(k)/2}, & \text{if } k \mid a, \\ x^{(a-1)\phi(k)/2}, & \text{if } k \nmid a. \end{cases}$$

Since $x^{\phi(k)/2} \equiv \pm 1 \pmod{k}$, it follows that

(14.5.1) $$\prod_{t \in R(k)} \sigma_{-t}^{-1}(\nu)^{[aL(t)/k]} \equiv \pm 1 \pmod{\lambda^2}.$$

Using the two formulas following (14.1.1), we can express $\Phi_a(P)$ in terms of Jacobi sums of the form $J(\chi_P, \chi_P^j)$, $1 \le j \le k-2$, and each such Jacobi sum is congruent to $-1 \pmod{\lambda^2}$ by Theorem 2.1.11. Thus

(14.5.2) $$\Phi_a(\nu) \equiv \pm 1 \pmod{\lambda^2}.$$

Appealing to (14.1.4), (14.1.5), (14.5.1), and (14.5.2), we deduce that

(14.5.3) $$\epsilon_a(\nu) \equiv \pm 1 \pmod{\lambda^2}.$$

As $\beta^h \not\equiv -1 \pmod{(1-\beta)^2}$ for any integer h, the plus sign must hold in (14.5.3). If $\epsilon_a(\nu)$ were a primitive kth–root of unity, then an application of an appropriate automorphism to (14.5.3) would yield the contradiction $\beta \equiv 1 \pmod{\lambda^2}$. Thus $\epsilon_a(\nu)^{k/l} = 1$. ∎

Theorem 14.5.3 *Let $l > 2$. Let $\nu \in O_K$ be coprime with k. Then there is a unique kth root of unity μ such that $\mu\nu$ is primary.*

Proof By Theorem 14.1.3,

$$\epsilon_c(\nu\beta^m) = \epsilon_c(\nu)\epsilon_c(\beta^m)$$

for every integer m. Thus it suffices to show that there is a unique integer $m \pmod{k}$ for which

$$\epsilon_c(\beta^m) = \overline{\epsilon_c(\nu)}.$$

Since β^m is a unit in O_K, $\Phi_c(\beta^m) = 1$. It follows easily from Theorem 14.1.2 that

$$\eta_c(\beta^m) = 1 \text{ and } \epsilon_c(\beta^m) = \beta^{-me_c},$$

where

$$e_c \equiv \sum_{b \in R(k)} b^{-1}[cL(b)/k] \equiv \sum_{j=0}^{\phi(k)-1} c^j[cL(c^{-j})/k] \pmod{k}.$$

By (14.4.2) with u, s, and t replaced by c, $\phi(k)$, and 1, respectively, we see that

$$e_c \equiv (c^{\phi(k)} - 1)L(c)/k \pmod{k}.$$

Thus

$$\epsilon_c(\beta^m) = \beta^{-m\{c(c^{\phi(k)}-1)/k\}}.$$

The expression in braces is an integer coprime with k, by the definition of c in Section 14.2. Therefore there is a unique integer $m \pmod{k}$ for which $\epsilon_c(\beta^m)$ equals the kth root of unity $\overline{\epsilon_c(\nu)}$. ■

Theorem 14.5.4 *Let $l > 2$. If ν is primary, then ν is semiprimary.*

Proof Assume that ν is primary. By Theorem 14.5.1 there is a unique integer m with $0 \le m < l$ such that $\nu\beta^m$ is semiprimary. We wish to show that $m = 0$.

Since ν is primary, $\epsilon_c(\nu) = 1$. Further, as $\nu\beta^m$ is semiprimary, $\epsilon_c(\nu\beta^m)$ is a power of β^l by Lemma 14.5.2.

By Theorem 14.1.3,

$$\epsilon_c(\nu\beta^m) = \epsilon_c(\nu)\epsilon_c(\beta^m) = \epsilon_c(\beta^m),$$

so that $\epsilon_c(\beta^m)$ is a power of β^l. But from the proof of Theorem 14.5.3,

$$\epsilon_c(\beta^m) = \beta^{-me_c},$$

where e_c is an integer coprime with k. Thus $l \mid m$. Since $0 \le m < l$, it follows that $m = 0$. ■

Lemma 14.5.5 *Let $l > 2$. If ν is semiprimary, then $\epsilon_k(\nu) = 1$.*

Proof We prove the assertion of the lemma by induction on n. In the case $n = 1$, we have $k = l$, and the result follows by Lemma 14.5.2.

Suppose now that $n > 1$, and assume that the result is true when n is replaced by $n - 1$. Since

$$G_r^k(\chi_P) = G_r^{k/l}(\chi_P^l)\left(G_r^l(\chi_P)/G_r(\chi_P^l)\right)^{k/l},$$

it can be deduced, using Theorem 14.1.2, that (see Problem 4)

$$\epsilon_k(\nu) = \epsilon^*_{k/l}(\nu^*)\epsilon_l(\nu)^{k/l}, \tag{14.5.4}$$

where $\epsilon^*_{k/l}(\nu^*)$ corresponds to $\epsilon_k(\nu)$ in Theorem 14.1.2 in the case that β is replaced by β^l and ν is replaced by the element

$$\nu^* = \prod_{j=0}^{l-1} \sigma_{1+(jk/l)}(\nu) \in Z[\beta^l]. \tag{14.5.5}$$

It can be seen that, if ν is semiprimary,

$$\nu^* \equiv x \pmod{(1-\beta^l)^2}$$

for some rational integer x. Thus ν^* is semiprimary. Hence, by the inductive hypothesis, $\epsilon^*_{k/l}(\nu^*) = 1$. Also $\epsilon_l(\nu)^{k/l} = 1$ by Lemma 14.5.2. Thus $\epsilon_k(\nu) = 1$ by (14.5.4). ■

Lemma 14.5.6 *Let $l > 2$. Suppose that ν is semiprimary. If x and y are integers with $x \equiv y \pmod{k}$, then $\epsilon_x(\nu) = \epsilon_y(\nu)$.*

Proof Let x and y be rational integers with $x \equiv y \pmod{k}$. Then a simple calculation shows that (see Problem 5)

$$\Phi_y(\nu) = \Phi_x(\nu)\Phi_{y-x}(\nu) = \Phi_x(\nu)\Phi_k(\nu)^{(y-x)/k}. \tag{14.5.6}$$

Hence, from (14.1.4), we deduce that (see Problem 6)

$$\epsilon_y(\nu) = \epsilon_x(\nu)\epsilon_k(\nu)^{(y-x)/k}. \tag{14.5.7}$$

As ν is semiprimary, it follows from Lemma 14.5.5 that $\epsilon_k(\nu) = 1$, so that $\epsilon_y(\nu) = \epsilon_x(\nu)$. ■

Lemma 14.5.7 *Let $l > 2$. Suppose that ν is primary and y is an integer coprime with k. Then $\epsilon_y(\nu) = 1$.*

Proof For each positive integer a (see Problem 7),

$$\Phi_c(\nu)^{c^a}\sigma_c(\Phi_{c^a}(\nu)) = \Phi_{c^{a+1}}(\nu), \tag{14.5.8}$$

so that (see Problem 8)

$$\epsilon_{c^a}(\nu) = \epsilon_c(\nu)^{ac^{a-1}}. \tag{14.5.9}$$

Since ν is primary, the right side of (14.5.9) equals 1, and so $\epsilon_{c^a}(\nu) = 1$. Further, as $\gcd(y,k) = 1$, we may choose the integer a so that $y \equiv c^a \pmod{k}$. Then, by Theorem 14.5.4 and Lemma 14.5.6, we conclude that $\epsilon_y(\nu) = 1$, as asserted. ■

We are now in a position to prove Eisenstein's reciprocity law when $l > 2$.

Proof of Theorem 14.3.1(a) As ν is primary, Lemma 14.5.7 gives $\epsilon_u(\nu) = 1$. Then, from the reciprocity relation Theorem 14.4.1, we conclude that $\left(\frac{\nu}{u}\right)_k = \left(\frac{u}{\nu}\right)_k$, which is the assertion of Theorem 14.3.1(a). ■

Although it is only a simple consequence of Lemma 14.5.7, we designate the next result as a theorem to stress its importance.

Theorem 14.5.8 *Let $l > 2$. Let c and c' be primitive roots* (mod l^m) *for every positive integer m. Let ν be an integer of K. Then ν is primary with respect to c if and only if ν is primary with respect to c'.*

Proof Suppose ν is primary with respect to c. As $\gcd(c', k) = 1$, Lemma 14.5.7 gives $\epsilon_{c'}(\nu) = 1$, so that ν is primary with respect to c'. ■

The next two results are both corollaries of Eisenstein's reciprocity law for $l > 2$ (Theorem 14.3.1(a)).

Corollary 14.5.9 *Let $l > 2$. Let ν be an integer of K. Let u be a rational prime. If $u^{l-1} \equiv 1 \pmod{k}$ and ν is semiprimary, then*

$$\left(\frac{\nu}{u}\right)_k = \left(\frac{u}{\nu}\right)_k.$$

Proof By Theorem 14.5.3 there is an integer m such that $\beta^m \nu$ is primary. By Theorem 14.5.4, $\beta^m \nu$ is semiprimary. By hypothesis, ν is semiprimary. Thus, by Theorem 14.5.1 and the remarks preceding Theorem 14.5.1, we have $l \mid m$. Further, as $\beta^m \nu$ is primary, Theorem 14.3.1(a) gives

$$\left(\frac{u}{\nu}\right)_k = \left(\frac{u}{\beta^m \nu}\right)_k = \left(\frac{\beta^m \nu}{u}\right)_k = \left(\frac{\beta^l}{u}\right)_k^{m/l} \left(\frac{\nu}{u}\right)_k.$$

Thus it remains to show that $\left(\frac{\beta^l}{u}\right)_k = 1$. As in Section 14.4, we let U denote a prime ideal factor of uO_K, and let s be the order of $u \pmod{k}$. Then, by (11.1.3),

$$uO_K = \prod_{t \in H} \sigma_t(U), \tag{14.5.10}$$

where H is a set of $\phi(k)/s$ integers such that

$$\bigcup_{t \in H} \bigcup_{j=0}^{s-1} \{t\overline{u}^j\}$$

is a reduced residue system (mod k). Then, by (14.5.10),

$$\left(\frac{\beta^l}{u}\right)_k = \prod_{t \in H} \left(\frac{\beta^l}{\sigma_t(U)}\right)_k = \prod_{t \in H} \beta^{l(u^s-1)/k} = \beta^{l(u^s-1)\phi(k)/(ks)}.$$

As $s \mid l-1$,

$$l\phi(k) = l\phi(l^n) = l(l^n - l^{n-1}) = l^n(l-1) = k(l-1) \equiv 0 \pmod{ks},$$

so that $l(u^s-1)\phi(k)/ks$ is a multiple of $u^s - 1$, and thus a multiple of k. This shows that $\beta^{l(u^s-1)\phi(k)/ks} = 1$, as required. ■

Corollary 14.5.10 *Let $l > 2$, and let u be a rational prime with $l \nmid u$. If ν is semiprimary and $k = l$, then*

$$\left(\frac{\nu}{u}\right)_k = \left(\frac{u}{\nu}\right)_k .$$

Proof Clearly, $u^{l-1} \equiv 1 \pmod{k}$, and so the result follows from Corollary 14.5.9. ∎

The next theorem gives a simple characterization of primary integers for those primes l for which $2^{l-1} \not\equiv 1 \pmod{l^2}$. It is known that the only primes less than $4 \cdot 10^{12}$ which do not satify this condition are $l = 1093$ and $l = 3511$; see Crandall, Dilcher, and Pomerance [1]. The congruence

$$2^{1092} \equiv 1 \pmod{1093^2}$$

was discovered by Meissner [1] in 1913, and the congruence

$$2^{3510} \equiv 1 \pmod{3511^2}$$

by Beeger [1] in 1922. A short proof of each of these congruences has been given by Guy [1].

Theorem 14.5.11 *Let $l > 2$. Suppose that $2^{l-1} \not\equiv 1 \pmod{l^2}$. Then ν is primary if and only if $\epsilon_2(\nu) = 1$.*

Proof If ν is primary then $\epsilon_2(\nu) = 1$ by Lemma 14.5.7.

Conversely suppose that $\epsilon_2(\nu) = 1$. By Theorem 14.5.3 there is a unique integer m with $0 \le m < k$ such that $\nu\beta^m$ is primary. To complete the proof of the theorem, we must show that $m = 0$. By Theorem 14.1.3,

$$\epsilon_2(\nu\beta^m) = \epsilon_2(\nu)\epsilon_2(\beta^m).$$

Since $\epsilon_2(\nu) = 1$ and $\epsilon_2(\nu\beta^m) = 1$ by Lemma 14.5.7, $\epsilon_2(\beta^m) = 1$. But the proof of Theorem 14.5.3 shows that

$$1 = \epsilon_2(\beta^m) = \beta^{-me_2},$$

where

$$e_2 \equiv \sum_{b \in R(k)} b^{-1}[2L(b)/k] \pmod{k}.$$

To show that $m = 0$, it remains to prove that $\gcd(e_2, k) = 1$.

Let a be the least positive integer such that $2 \equiv c^a \pmod{k}$. Then, by (14.1.2),

$$\begin{aligned}
e_2 &\equiv \sum_{j=1}^{\phi(k)} c^j [c^a L(\bar{c}^j)/k] \equiv k^{-1} \sum_{j=1}^{\phi(k)} \left\{c^{j+a} L(\bar{c}^j) - c^j L(\bar{c}^j c^a)\right\} \\
&\equiv k^{-1} \sum_{j=1}^{a} \left\{c^{\phi(k)+j} L(\bar{c}^j c^a) - c^j L(\bar{c}^j c^a)\right\} \\
&\equiv \frac{(c^{\phi(k)} - 1)}{k} \sum_{j=1}^{a} c^a = ac^a(c^{\phi(k)} - 1)/k \pmod{k},
\end{aligned}$$

where $\bar{c}$ denotes the inverse of $c \pmod{k}$. By the definition of c given in Section 14.2, $(c^{\phi(k)} - 1)/k$ is coprime with k. It remains to show that $\gcd(a, l) = 1$. Suppose for the purpose of contradiction that $a = lb$ for some integer b. If $k = l$, then

$$c^b \equiv c^{lb} = c^a \equiv 2 \pmod{l},$$

which contradicts the minimality of a. If $k = l^n$ with $n \geq 2$, then

$$2^{l-1} \equiv c^{a(l-1)} = c^{b\phi(l^2)} \equiv 1 \pmod{l^2},$$

which contradicts the hypothesis of the theorem. ∎

14.6 Proof of Eisenstein's reciprocity law for $l = 2$

In this section, we prove a series of results (culminating with Lemma 14.6.8) leading to a proof of Eisenstein's reciprocity law (Theorem 14.3.1) in the case $l = 2$. We also prove some corollaries of Theorem 14.3.1.

Throughout this section, $l = 2$ and $k = 2^n$ with $n \geq 3$, so that the cases $k = 2, 4$ are excluded. Also, $R(k)$ denotes a reduced residue system $\pmod{k}$, $K = Q(\beta)$ with $\beta = e^{2\pi i/2^n}$, $O_K = Z[\beta]$, $\lambda = 1 - \beta$, and ν is an element of O_K coprime with 2. Define $\sigma_t \in \mathrm{Gal}(K/Q)$ by $\sigma_t(\beta) = \beta^t$ for each $t \in R(k)$. By Corollary 2.1.10, $2 = \lambda^d \delta$, where $d = \phi(2^n) = 2^{n-1}$ and δ is a unit of O_K. Thus $\lambda^4 \mid 2$.

Lemma 14.6.1 *Let $l = 2$ and let $\nu \in O_K$ be coprime with 2. Then $\epsilon_a(\nu)$ is a power of β^2 for each integer a.*

Proof Since $d = 2^{n-1} = \phi(2^n) = [K : Q]$ and $Z[\beta] = Z[\lambda]$, there are integers $b_0, b_1, \ldots, b_{d-1}$ such that

$$\nu = \sum_{j=0}^{d-1} b_j \lambda^j.$$

Hence $\nu \equiv b_0 + b_1\lambda \pmod{\lambda^2}$, and, replacing λ by $1 - \beta$, we see that there are integers g and h with

$$\nu \equiv g + h\beta \pmod{\lambda^2}. \tag{14.6.1}$$

Since $\lambda^4 \mid 2$, it follows from (14.6.1) that

$$\nu \equiv 0, 1, \beta, \text{ or } 1 + \beta \pmod{\lambda^2}.$$

As ν and 2 are coprime, so are ν and λ. Thus $\nu \not\equiv 0 \pmod{\lambda^2}$. Also $\nu \not\equiv 1+\beta \pmod{\lambda^2}$, for otherwise $\nu \equiv 1 + \beta \equiv 0 \pmod{\lambda}$. Thus $\nu \equiv \beta^m \pmod{\lambda^2}$, where $m = 0$ or 1. As

$\beta^2 \equiv 1 \pmod{\lambda^2}$, we have $\sigma_t(\nu) \equiv \beta^m \pmod{\lambda^2}$ for each $t \in R(k)$. It follows that, for each integer a,

$$\prod_{t \in R(k)} \sigma_{-t}^{-1}(\nu)^{[aL(t)/k]} \equiv \beta^{m \sum_{t \in R(k)} [aL(t)/k]} \pmod{\lambda^2}.$$

By Problem 1,

$$2 \sum_{t \in R(k)} [aL(t)/k] = \sum_{t \in R(k)} [aL(t)/k] + \sum_{t \in R(k)} [aL(-t)/k] \equiv 0 \pmod 4,$$

so that

$$\sum_{t \in R(k)} [aL(t)/k] \equiv 0 \pmod 2.$$

Hence, by Theorem 14.1.2,

$$\epsilon_a(\nu)\Phi_a(\nu) = \prod_{t \in R(k)} \sigma_{-t}^{-1}(\nu)^{[aL(t)/k]} \equiv 1 \pmod{\lambda^2}.$$

But, as in (14.5.2),

$$\Phi_a(\nu) \equiv 1 \pmod{\lambda^2},$$

so that $\epsilon_a(\nu) \equiv 1 \pmod{\lambda^2}$. If $\epsilon_a(\nu)$ were an odd power of β, then we would have $\epsilon_a(\nu) \equiv \beta \pmod{\lambda^2}$, which is impossible, since $\beta \not\equiv 1 \pmod{\lambda^2}$. Thus $\epsilon_a(\nu)$ is a power of β^2, as asserted. ■

Theorem 14.6.2 *Let $l = 2$ and let $\nu \in O_K$ be coprime with 2. There are exactly two kth roots of unity μ for which $\mu\nu$ is primary. If μ is one of these two kth roots of unity, then the other is $-\mu$.*

Proof Let m be any integer with $0 \leq m < k$. Exactly as in the proof of Theorem 14.5.3,

$$\epsilon_3(\nu\beta^m) = \epsilon_3(\nu)\beta^{-m \sum_{t \in R(k)} t^{-1}[3L(t)/k]}.$$

By Problem 1, $[3L(t)/k] + [3L(-t)/k] = 2$ for each $t \in R(k)$, and since $\bigcup_{b=0}^{1} \bigcup_{a=1}^{k/4} \{3^a(-1)^b\}$ is a reduced residue system $\pmod k$, we have, as in the proof of Theorem 14.5.3,

$$\begin{aligned}\sum_{t \in R(k)} t^{-1}[3L(t)/k] &\equiv 2\sum_{a=1}^{k/4} 3^a[3L(3^{-a})/k] = 2\sum_{a=1}^{k/4} \left(3^{a+1}L(3^{-a}) - 3^a L(3^{1-a})\right)/k \\ &= 2 \cdot 3(3^{k/4} - 1)/k \pmod k,\end{aligned}$$

where 3^{-a} denotes the inverse of $3^a \pmod k$. As $h := 3(3^{k/4} - 1)/k$ is an odd integer,

$$\beta^{-m \sum_{t \in R(k)} t^{-1}[3L(t)/k]} = \beta^{-2hm}.$$

Further, by Lemma 14.6.1, $\epsilon_3(\nu) = \beta^{2g}$ for some integer g, so that

$$\epsilon_3(\nu\beta^m) = \beta^{2(g-hm)}.$$

As h is odd, there is an odd integer w such that $hw \equiv 1 \pmod{k/2}$, and there are exactly two values of m in the range $0 \le m < k$ for which $\nu\beta^m$ is primary, equivalently, $\epsilon_3(\nu\beta^m) = (-1)^M$, namely,

$$m_1 = L(gw + Mk/4) \text{ and } m_2 = L(gw + Mk/4 + k/2). \qquad \blacksquare$$

Definition *When $l = 2$, an integer ν of K with* $\gcd(\nu, k) = 1$ *is said to be semiprimary if* $\nu \equiv 1 \pmod{\lambda^2}$.

Theorem 14.6.3 *Let $l = 2$. Suppose that ν is an integer of K coprime with 2. Then there is a unique integer $m \in \{0, 1\}$ such that $\nu\beta^m$ is semiprimary.*

Proof As in the proof of Lemma 14.6.1, $\nu \equiv 1$ or $\beta \pmod{\lambda^2}$. Since $\beta^2 \equiv 1 \pmod{\lambda^2}$, $\nu\beta^m$ is semiprimary for some $m \in \{0, 1\}$. Clearly ν and $\beta\nu$ cannot both be semiprimary, for otherwise

$$\lambda\nu = (1-\beta)\nu = \nu - \beta\nu \equiv 1 - 1 \equiv 0 \pmod{\lambda^2},$$

contradicting the fact that ν is coprime with 2. $\blacksquare$

Example Let $l = 2$, $n = 3$, $k = 8$, $\beta = e^{2\pi i/8}$, $\lambda = 1 - \beta$, $K = Q(\beta)$, and let $\nu \in O_K$ be coprime with 2. Then, since $\lambda^4 \mid\mid 2$,

$$\begin{aligned} \nu \text{ is semiprimary} &\iff \nu \equiv 1 \pmod{\lambda^2} \\ &\iff \nu \equiv 1, 1+\lambda^2, 1+\lambda^3, \text{ or } 1+\lambda^2+\lambda^3 \pmod 2 \\ &\iff \nu \equiv 1, \beta^2, \beta+\beta^2+\beta^3, \text{ or } 1+\beta+\beta^3 \pmod 2. \end{aligned}$$

If $\nu \equiv \beta^2$ or $\beta + \beta^2 + \beta^3 \pmod 2$, then we see from Theorem 14.2.1 that ν is semiprimary but not primary. However, if $\nu \equiv \beta^2$ or $\beta+\beta^2+\beta^3 \pmod 2$, then $\beta^2\nu$ is primary.

Lemma 14.6.4 *Let $l = 2$. If ν is semiprimary, then $\epsilon_3(\nu)$ is a power of β^4.*

Proof Assume that ν is semiprimary, so that $\nu \equiv 1 \pmod{\lambda^2}$. Then $\nu \equiv 1$ or $1 + \lambda^2 \pmod{\lambda^3}$. The second possibility is equivalent to $\nu \equiv \beta^2 \pmod{\lambda^3}$. Hence $\nu \equiv \beta^{2m} \pmod{\lambda^3}$ for some integer $m \in \{0, 1\}$. Thus, for $t \in R(k)$,

$$\sigma_t(\nu) \equiv \beta^{2mt} \pmod{\lambda^3},$$

so that

$$\prod_{t\in R(k)} \sigma_{-t}^{-1}(\nu)^{[3L(t)/k]} \equiv \beta^{-2m\sum_{t\in R(k)} t^{-1}[3L(t)/k]} = \beta^{-4mh} \pmod{\lambda^3},$$

where h is defined in the proof of Theorem 14.6.2. Since

$$\beta^4 = (1-\lambda)^4 = 1 - 4\lambda + 6\lambda^2 - 4\lambda^3 + \lambda^4 \equiv 1 \pmod{\lambda^3},$$

we have

(14.6.2) $$\prod_{t\in R(k)} \sigma_{-t}^{-1}(\nu)^{[3L(t)/k]} \equiv 1 \pmod{\lambda^3}.$$

We next show that

(14.6.3) $$\Phi_3(\nu) \equiv 1 \pmod{\lambda^3}.$$

To show (14.6.3), in view of the formula for Φ_3 in terms of Jacobi sums given after (14.1.1), it suffices to prove that

(14.6.4) $$J_r(\chi_P, \chi_P) J_r(\chi_P, \chi_P^2) \equiv 1 \pmod{\lambda^3},$$

where P is a prime ideal factor of νO_K with $N(P) = p^r$. Let ρ denote the quadratic character on F_{p^r}. Then the restriction of ρ to F_p is the trivial character or the Legendre symbol according to whether r is even or odd. In the latter case $r = 1$, since $r \mid \phi(k) = 2^{n-1}$, and hence $p \equiv 1 \pmod{k}$. It follows that, whatever r is, $\rho(2) = 1$, that is, 2 is a square in F_{p^r}. Thus $\chi_P(4)$ is a power of β^4, and so $\chi_P(4) \equiv 1 \pmod{\lambda^3}$. Since $K_r(\chi_P) \equiv 1 \pmod{2}$ by Theorem 2.1.8,

(14.6.5) $$J_r(\chi_P, \chi_P) = \overline{\chi}_P(4) K_r(\chi_P) \equiv 1 \pmod{\lambda^3}.$$

Now

$$\begin{aligned} J_r(\chi_P, \chi_P^2) &= \sum_{\alpha\in F_{p^r}} \chi_P(\alpha)\chi_P^2(1-\alpha) \\ &= \sum_{\alpha\in F_{p^r}} (\chi_P(\alpha)-1)(\chi_P^2(1-\alpha)-1) - p^r \equiv -p^r \equiv 1 \pmod{\lambda^3}, \end{aligned}$$

that is,

(14.6.6) $$J_r(\chi_P, \chi_P^2) \equiv 1 \pmod{\lambda^3}.$$

From (14.6.5) and (14.6.6), we deduce (14.6.4), and thus (14.6.3).

From (14.1.4), (14.6.2), and (14.6.3), we see that $\epsilon_3(\nu) \equiv 1 \pmod{\lambda^3}$, so that $\epsilon_3(\nu)$ is a power of β^4 as asserted. ■

Theorem 14.6.5 *Let $l = 2$. If ν is primary, then ν is semiprimary.*

Proof Assume that ν is primary. By Theorem 14.6.3, $\beta^m\nu$ is semiprimary for a unique integer $m \in \{0,1\}$. We wish to show that $m = 0$. Since ν is primary, $\epsilon_3(\nu) = (-1)^M$ by (14.2.2). Now, as $8 \mid k$, we have $-1 = (\beta^4)^{k/8}$, so that $\epsilon_3(\nu)$ is a power of β^4. Also $\epsilon_3(\nu\beta^m)$ is a power of β^4 by Lemma 14.6.4. From the proof of Theorem 14.6.2,

$$\epsilon_3(\nu\beta^m) = \epsilon_3(\nu)\beta^{-m\sum_{t\in R(k)} t^{-1}[3L(t)/k]} = \epsilon_3(\nu)\beta^{-2hm},$$

where h is odd. Hence $m \equiv 0 \pmod 2$, and so $m = 0$. ∎

Lemma 14.6.6 *Let $l = 2$. If ν is semiprimary, then $\epsilon_k(\nu) = 1$.*

Proof The proof will be accomplished by induction on n. To begin the induction we take $n = 3$ and show that $\epsilon_8(\nu) = 1$. In this case $k = 8$, $\beta = e^{2\pi i/8}$, and $K = Q(\beta)$. From Corollary 2.1.10, $2 = \epsilon(1-\beta)^4 = \epsilon\lambda^4$, where ϵ is a unit of O_K, so that $2(1-\beta) = \epsilon\lambda^5$ and $2 \equiv \lambda^4 \pmod{\lambda^5}$. Let P be a prime ideal factor of νO_K with $N(P) = p^r$. From Theorem 2.1.8,

$$K_r(\chi_P) \equiv K_r(\chi_P^2) \equiv -1 \pmod{\lambda^5}.$$

Hence

$$p^r K_r^4(\chi_P) K_r^2(\chi_P^2) \equiv 1 \pmod{\lambda^5}.$$

But, by Theorem 2.1.3(a), the left side is just $G_r^8(\chi_P)$, and so, by (14.1.1),

$$\Phi_8(P) = G_r^8(\chi_P) \equiv 1 \pmod{\lambda^5}.$$

Hence, by the multiplicativity of Φ_8,

(14.6.7) $$\Phi_8(\nu) \equiv 1 \pmod{\lambda^5}.$$

Since ν is semiprimary, $\nu \equiv 1 \pmod{\lambda^2}$, and so there is an integer α of O_K such that $\nu = 1 + \lambda^2\alpha$. Hence

$$\nu^2 = (1+\lambda^2\alpha)^2 = 1 + 2\lambda^2\alpha + \lambda^4\alpha^2 \equiv \mu \pmod{\lambda^5},$$

where

$$\mu = \begin{cases} 1, & \text{if } \alpha \equiv 0 \pmod{\lambda}, \\ -1, & \text{if } \alpha \equiv 1 \pmod{\lambda}. \end{cases}$$

Thus

$$\begin{aligned}\prod_{t\in R(8)} \sigma_{-t}^{-1}(\nu)^{L(t)} &= \sigma_7(\nu)\sigma_5(\nu)^3\sigma_3(\nu)^5\sigma_1(\nu)^7 \\ &\equiv \sigma_7(\nu)\mu\sigma_5(\nu)\sigma_3(\nu)\mu\sigma_1(\nu) = N(\nu O_K) \pmod{\lambda^5}.\end{aligned}$$

Now $N(\nu O_K)$ is a product of factors of the form p^r with $p^r \equiv 1 \pmod 8$, so that $N(\nu O_K) \equiv 1 \pmod 8$. This shows that

(14.6.8) $$\prod_{t\in R(8)} \sigma_{-t}^{-1}(\nu)^{L(t)} \equiv 1 \pmod{\lambda^5}.$$

In view of (14.1.4), (14.6.7), and (14.6.8), we conclude that $\epsilon_8(\nu) \equiv 1 \pmod{\lambda^5}$, so that $\epsilon_8(\nu) = 1$.

We now assume $n \geq 4$ and complete the induction as in the proof of Lemma 14.5.5, except that at the end, Lemma 14.6.1 is invoked in place of Lemma 14.5.2. ∎

Lemma 14.6.7 *Let $l = 2$. Suppose that ν is semiprimary. Let x and y be rational integers such that $x \equiv y \pmod{k}$. Then*

$$\epsilon_x(\nu) = \epsilon_y(\nu).$$

Proof One argues as in the proof of Lemma 14.5.6, except that at the end, Lemma 14.6.6 is invoked instead of Lemma 14.5.5. ∎

Lemma 14.6.8 *Let $l = 2$. Suppose that ν is primary. Let y be an odd integer. Then*

$$\epsilon_y(\nu) = \begin{cases} 1, & \text{if } y \equiv 1 \pmod{4}, \\ (-1)^M, & \text{if } y \equiv -1 \pmod{4}, \end{cases}$$

where M is defined in (14.2.1).

Proof Arguing as for (14.5.9), we have, for each positive integer a,

$$\epsilon_{3^a}(\nu) = \epsilon_3(\nu)^{a3^{a-1}}. \tag{14.6.9}$$

Since ν is primary, $\epsilon_3(\nu) = (-1)^M$, and (14.6.9) becomes

$$\epsilon_{3^a}(\nu) = (-1)^{Ma} = \begin{cases} 1, & \text{if } a \text{ is even}, \\ (-1)^M, & \text{if } a \text{ is odd}, \end{cases} = \begin{cases} 1, & \text{if } 3^a \equiv 1 \pmod{4}, \\ (-1)^M, & \text{if } 3^a \equiv -1 \pmod{4}. \end{cases}$$

As y is an odd integer, and

$$\bigcup_{b=0}^{1} \bigcup_{a=1}^{k/4} \{3^a(-1)^b\} = R(k),$$

we can choose $\epsilon = \pm 1$ and a positive integer a so that

$$y \equiv \epsilon\, 3^a \pmod{k}.$$

First we assume that $\epsilon = 1$. Since ν is semiprimary by Theorem 14.6.5, we may apply Lemma 14.6.7 to deduce that

$$\epsilon_y(\nu) = \epsilon_{3^a}(\nu) = \begin{cases} 1, & \text{if } y \equiv 1 \pmod{4}, \\ (-1)^M, & \text{if } y \equiv -1 \pmod{4}, \end{cases}$$

as required.

Next we assume that $\epsilon = -1$. Since

$$\chi_P(-1) = \left(\frac{-1}{P}\right)_k = (-1)^{(N(P)-1)/k},$$

it follows by multiplicativity that

$$\left(\frac{-1}{\nu}\right)_k = (-1)^M. \tag{14.6.10}$$

Now, by Theorem 14.1.2,

$$\prod_{t\in R(k)} \sigma_{-t}^{-1}(\nu)^{[yL(t)/k]+[-yL(t)/k]} = \epsilon_y(\nu)\epsilon_{-y}(\nu)\Phi_y(\nu)\Phi_{-y}(\nu). \tag{14.6.11}$$

A simple calculation using Theorem 1.1.4(a), (14.1.1), and (14.6.10) shows that

$$\Phi_y(\nu)\Phi_{-y}(\nu) = (-1)^M/N(\nu O_K). \tag{14.6.12}$$

Also, by Problem 1,

$$\prod_{t\in R(k)} \sigma_{-t}^{-1}(\nu)^{[yL(t)/k]+[-yL(t)/k]} = \prod_{t\in R(k)} \sigma_{-t}^{-1}(\nu)^{-1} = N(\nu O_K)^{-1}. \tag{14.6.13}$$

From (14.6.11)–(14.6.13), we deduce that

$$\epsilon_y(\nu)\epsilon_{-y}(\nu) = (-1)^M. \tag{14.6.14}$$

Then, from (14.6.14),

$$\begin{aligned}
\epsilon_y(\nu) &= (-1)^M\epsilon_{-y}(\nu) = (-1)^M\epsilon_{3^a}(\nu) \\
&= \begin{cases} (-1)^M, & \text{if } 3^a \equiv 1 \pmod 4, \\ 1, & \text{if } 3^a \equiv -1 \pmod 4, \end{cases} \\
&= \begin{cases} (-1)^M, & \text{if } y \equiv -1 \pmod 4, \\ 1, & \text{if } y \equiv 1 \pmod 4. \end{cases}
\end{aligned}$$

This completes the proof of the asserted equality for all odd integers y. ■

We are now ready to prove Eisenstein's reciprocity law when $l = 2$.

Proof of Theorem 14.3.1(b) By Lemma 14.6.8,

$$\epsilon_u(\nu) = \begin{cases} 1, & \text{if } u \equiv 1 \pmod 4, \\ (-1)^M, & \text{if } u \equiv -1 \pmod 4. \end{cases}$$

Then, appealing to (14.6.10), we obtain

$$\epsilon_u(\nu) = \left(\frac{-1}{\nu}\right)_k^{(u-1)/2} = \left(\frac{(-1)^{(u-1)/2}}{\nu}\right)_k.$$

Then, by Theorem 14.4.1,

$$\left(\frac{(-1)^{(u-1)/2}}{\nu}\right)_k^{\overline{u}} \left(\frac{\nu}{u}\right)_k = \left(\frac{u}{\nu}\right)_k,$$

so that, as $\overline{u}$ is odd,

$$\left(\frac{\nu}{u}\right)_k = \left(\frac{(-1)^{(u-1)/2}u}{\nu}\right)_k$$

as asserted. ■

The following corollary is analogous to Corollary 14.5.9.

Corollary 14.6.9 *Let $l = 2$ and let u be a rational prime. If $u \equiv 1 \pmod{k}$ and ν is semiprimary, then*

$$\left(\frac{\nu}{u}\right)_k = \left(\frac{(-1)^{(u-1)/2}u}{\nu}\right)_k.$$

Proof The proof proceeds exactly as for Corollary 14.5.9, except that Theorems 14.6.2, 14.6.5, 14.6.3, and 14.3.1(b) are used in place of Theorems 14.5.3, 14.5.4, 14.5.1, and 14.3.1(a), respectively. ■

Corollary 14.6.10 (*Eisenstein's law of octic reciprocity.*) *Let $\beta = e^{2\pi i/8}$, and let ν be an integer of $Q(\beta)$ satisfying $\nu \equiv 1$ or $1+\beta+\beta^3 \pmod{2}$. Let u be an odd prime. Then*

$$\left(\frac{\nu}{u}\right)_8 = \left(\frac{(-1)^{(u-1)/2}u}{\nu}\right)_8. \tag{14.6.15}$$

Proof This follows from Theorems 14.2.1 and 14.3.1. ■

14.7 Application of Eisenstein's reciprocity law to Wieferich's theorem

We now use Eisenstein's reciprocity law to prove the following famous result of Wieferich [1] on the Fermat equation.

Theorem 14.7.1 *Let l be an odd prime. Suppose that x, y, and z are nonzero integers such that $x^l + y^l + z^l = 0$ and $l \nmid xyz$. Then $2^{l-1} \equiv 1 \pmod{l^2}$.*

Proof Without loss of generality we may assume that x, y, and z are coprime in pairs. Since $x^l + y^l + z^l = 0$, not all of x, y, and z are odd. By symmetry, we can suppose that y is even. Thus x and z are odd. Also, by Fermat's little theorem, we have $0 = x^l + y^l + z^l \equiv x + y + z \pmod{l}$, so that $x + y \equiv -z \not\equiv 0 \pmod{l}$, and thus $(x+y)^{l-1} \equiv 1 \pmod{l}$.

Let $\beta = e^{2\pi i/l}$, $\lambda = 1 - \beta$, and $K = Q(\beta)$. Since $x^l + y^l = -z^l$,

$$(x+y)(x+\beta y)\cdots(x+\beta^{l-1}y) = (-z)^l.$$

For $j = 0, 1, \ldots, l-1$ let A_j denote the principal ideal $(x+\beta^j y)O_K$, and let $B = zO_K$, so that

$$A_0 A_1 \cdots A_{l-1} = B^l. \tag{14.7.1}$$

For $0 \le r < s \le l-1$ we show that A_r and A_s are coprime. Suppose not. Then there exists a prime ideal P such that $P \mid A_r$ and $P \mid A_s$. Thus $x + \beta^r y \in A_r \subseteq P$ and $x + \beta^s y \in A_s \subseteq P$. Hence $(\beta^s - \beta^r)x \in P$ and $(\beta^s - \beta^r)y \in P$. Since x and y are relatively prime, it follows that $\beta^s - \beta^r \in P$, so that $\lambda O_K \subseteq P$. But λO_K and P are both prime ideals (since λO_K has prime norm by Theorem 2.1.9), and so we must have $P = \lambda O_K$. Then $\lambda O_K \mid B$, so that $l \mid z$, which is a contradiction. Hence A_r and A_s are coprime for $0 \le r < s \le l-1$, and in view of (14.7.1) this shows that each A_j $(j = 0, 1, \ldots, l-1)$ is a perfect lth power, say $A_j = B_j^l$. Now set

$$\alpha = (x+y)^{l-2}(x+\beta y) \in O_K, \quad t = (x+y)^{l-2}y \in Z.$$

We note that $l \nmid t$. The principal ideal αO_K is a perfect lth power, as

$$\alpha O_K = A_0^{l-2}A_1 = (B_0^{l-2}B_1)^l.$$

Since αO_K is an lth power, $\left(\frac{2}{\alpha}\right)_l = 1$. Further, as x is odd and y is even, $\alpha \equiv 1 \pmod 2$, so that $\left(\frac{\alpha}{2}\right)_l = \left(\frac{1}{2}\right)_l = 1$. By (11.1.3), the prime ideal factorization of $2O_K$ is of the form

$$2O_K = U_1 U_2 \cdots U_{(l-1)/s},$$

where s is the least positive integer such that $2^s \equiv 1 \pmod l$ and each prime ideal U_j has norm 2^s. Hence

$$\left(\frac{\beta}{2}\right)_l = \prod_{j=1}^{(l-1)/s} \left(\frac{\beta}{U_j}\right)_l = \prod_{j=1}^{(l-1)/s} \beta^{(2^s-1)/l} = \beta^{(l-1)(2^s-1)/(sl)}.$$

Furthermore,

$$\begin{aligned}\alpha &= (x+y)^{l-2}(x+y-\lambda y) = (x+y)^{l-1} - (x+y)^{l-2}y\lambda \\ &= (x+y)^{l-1} - t\lambda \equiv 1 - t\lambda \pmod l,\end{aligned}$$

so that, as λ^{l-1} is an associate of l by Corollary 2.1.10, we have $\alpha \equiv 1 - t\lambda \pmod{\lambda^2}$. Hence $\beta^{-t}\alpha = (1-\lambda)^{-t}\alpha \equiv (1+t\lambda)(1-t\lambda) \equiv 1 \pmod{\lambda^2}$, so that $\beta^{-t}\alpha$ is semiprimary. By Eisenstein's reciprocity theorem in the form given by Corollary 14.5.10,

$$\left(\frac{2}{\alpha}\right)_l = \left(\frac{2}{\beta^{-t}\alpha}\right)_l = \left(\frac{\beta^{-t}\alpha}{2}\right)_l = \left(\frac{\beta}{2}\right)_l^{-t}\left(\frac{\alpha}{2}\right)_l.$$

Thus $\left(\frac{\beta}{2}\right)_l^t = 1$, so that

$$t \cdot \frac{(l-1)}{s} \cdot \frac{(2^s - 1)}{l} \equiv 0 \pmod{l}.$$

As $l \nmid t$ and $l \nmid (l-1)/s$, we must have $(2^s - 1)/l \equiv 0 \pmod{l}$, and thus $2^s - 1 \equiv 0 \pmod{l^2}$. But $s \mid l - 1$, so that $2^{l-1} - 1 \equiv 0 \pmod{l^2}$ as asserted. ■

As we mentioned prior to Theorem 14.5.11, the only primes $l < 4 \cdot 10^{12}$ satisfying $2^{l-1} \equiv 1 \pmod{l^2}$ are $l = 1093$ and $l = 3511$. Thus Wieferich's result shows that the first case of Fermat's last theorem holds for every odd prime exponent $l < 4 \cdot 10^{12}$ except possibly for $l = 1093$ and $l = 3511$. Recently, using much deeper methods, Wiles [1] has succeeded in proving Fermat's last theorem for every exponent $l \geq 3$.

Exercises 14

1. Let a and k be integers with $k > 1$. Define $L(t)$ as in (14.1.2). For an integer t coprime with k, prove that

$$[aL(t)/k] + [aL(-t)/k] = \begin{cases} a, & \text{if } k \mid a, \\ a - 1, & \text{if } k \nmid a. \end{cases}$$

This equality was used in Sections 14.1 and 14.6.

2. With the notation of Section 14.1, prove that

$$|\Phi_a(\nu)| = \begin{cases} N(\nu O_K)^{(a-1)/2}, & \text{if } k \nmid a, \\ N(\nu O_K)^{a/2}, & \text{if } k \mid a. \end{cases}$$

3. Using the multiplicativity of $\Phi_a(\nu)$ and Theorem 14.1.2, prove Theorem 14.1.3.

4. Prove the assertion (14.5.4) in the proof of Lemma 14.5.5.

Hint: If $P \mid \nu O_K$, let $P^* = P \cap Q(\beta^l)$. If $N(P) = p^r$, then $N(P^*) = p^r$ or $N(P^*) = p^{r/l}$, the latter case occurring when p has order $r/l \pmod{k/l}$. In the latter case, show that $P = P^* O_K$ and use the Davenport–Hasse theorem on lifted Gauss sums (Theorem 11.5.2).

5. Prove the assertion (14.5.6) in the proof of Lemma 14.5.6.

6. Deduce (14.5.7) from (14.5.6) and (14.2.7).

7. Prove the assertion (14.5.8) in the proof of Lemma 14.5.7.

8. Deduce (14.5.9) from (14.5.8) by using induction on a.

9. Determine the semiprimary integers of $Q(e^{2\pi i/9})$ as in the Example following Theorem 14.6.3.

10. Determine the primary integers of $Q(e^{2\pi i/9})$ as in Theorem 14.2.1.

11. Give an example of an integer of $Q(e^{2\pi i/9})$ which is semiprimary but not primary.

12. Determine both the semiprimary and primary integers of $Q(e^{2\pi i/27})$, $Q(e^{2\pi i/25})$, and $Q(e^{2\pi i/16})$.

13. Formulate Eisenstein's reciprocity law in the two excluded cases $k = 2$ and $k = 4$.

14. Let l be an odd prime. Let a be an integer not divisible by l. Suppose that the congruence $x^l \equiv a \pmod{p}$ is solvable for all but finitely many primes p. Use Eisenstein's reciprocity law to show that $a = b^l$ for some integer b.

15. Let $k > 0$ and a be integers. Suppose that the congruence $x^k \equiv a \pmod{p}$ is solvable for all but finitely many primes p. Prove that for some integer b, $a = b^k$, if k is not divisible by 8, and $a = b^k$ or $2^{k/2}b^k$, if k is divisible by 8. (Note that 16 is an 8th power modulo each odd prime p, but 16 is not an 8th power.)

16. Prove the evaluations

$$\left(\frac{i}{1+i\sqrt{2}}\right)_8 = i, \quad \left(\frac{1-\sqrt{2}}{3}\right)_8 = 1, \quad \left(\frac{\sqrt{2}}{1+i\sqrt{2}}\right)_8 = i,$$

$$\left(\frac{1-\sqrt{2}}{1+i\sqrt{2}}\right)_8 = e^{2\pi i/8}, \quad \left(\frac{\sqrt{2}}{3}\right)_8 = 1, \text{ and } \left(\frac{i}{3}\right)_8 = -1.$$

17. Use Corollary 14.6.10 (Eisenstein's law of octic reciprocity) to prove the necessary and sufficient condition for 3 to be an octic residue (mod p) given in Theorem 7.3.1.

18. Let $K = Q(e^{2\pi i/k})$ for an integer $k > 1$. Let P and P' denote prime ideals of O_K with norms q and q', respectively, where q, q', and k are pairwise relatively prime. Let χ_P and $\chi_{P'}$ be the corresponding power residue symbols of order k. Prove the reciprocity law

$$\chi_{P'}(G_r^k(\chi_P^{-a})) = \chi_P^a(q'),$$

where a is not divisible by k and $q = p^r$. (This is called the Eisenstein reciprocity law for Gauss sums.)

Hint: Generalize Corollary 7.0.2.

Notes on Chapter 14

This chapter provides a proof of Eisenstein's reciprocity law for the kth–power residue symbol when k is a power of a prime. Eisenstein [12] established his law in the case of a prime k in 1850. His proof used Kummer's work [3] on the prime ideal decomposition of the "reciprocal" factors of a prime $p \equiv 1 \pmod{k}$ in the cyclotomic field $Q(e^{2\pi i/k})$. Eisenstein gave his proof in a note addressed to Jacobi, who communicated it to the Berlin Academy in the *Monatsberichte* for May 30, 1850, p. 189. Readers will find proofs of Eisenstein's reciprocity law for primes k (Corollary 14.5.10) in Cooke [1], Hasse [3, Part II, pp. 77–78], Hilbert [1, §115], Ireland and Rosen [1, pp. 215–220], Landau [3, vol. 3, pp. 277–311], C. J. Moreno [1], Skolem [1], Spearman and Williams [1], and Weil [6]. Elementary proofs for the cases $k = 5$ and $k = 7$ have been given by Spearman [1]. The extension of Eisenstein's law to prime powers k was first given by Western [1], [2] in 1907 and 1908. Western [2] applied the general theory to the cases $k = 8$, 9, and 16 in considerable detail. The proof of Theorem 14.3.1 that we have given in this chapter follows an unpublished manuscript of Evans based on Western's work.

Let $\{\nu, \mu\}$ denote the quotient $\left(\frac{\nu}{\mu}\right)_k / \left(\frac{\mu}{\nu}\right)_k$. Thus in Corollary 14.5.10, $\{\nu, u\} = 1$. Amazingly, Eisenstein [13] showed how to compute $\{\nu, \mu\}$ for arbitrary ν, μ in O_K relatively prime to each other and to the prime k, although he was unable to provide a proof. Eisenstein's formula in [13] specializes to his reciprocity law (Corollary 14.5.10) in the case that μ is a rational integer and ν is semiprimary. Kummer used his own reciprocity law (proved in Kummer [7] and described in H. J. S. Smith [1, p. 120]) to prove and simplify Eisenstein's formula for $\{\nu, \mu\}$ when the prime k is regular; see Kummer [5, pp. 144–145]. As Weil remarked in his introduction to Kummer's *Werke* [CW, vol. 1, pp. 10–11], Kummer in essence showed how to evaluate what are now called Hilbert normresidue symbols. For an exposition of these symbols, see Hasse [3, Part II, pp. 53–64]. Cox [1, pp. 167–168] presents the statement of a general reciprocity law for an arbitrary positive integer k which relates $\{\nu, \mu\}$ to a product of kth–power Hilbert symbols (defined using local class field theory), where ν and μ are integers of an arbitrary algebraic number field containing K. See also Cassels and Fröhlich [1, Exercise 2, p. 351]. Brückner [1] and Vostokov [1] have discussed the explicit computation of the Hilbert symbol. Some of the many reciprocity laws generalizing those of Eisenstein and Kummer have been given by Artin, Fürtwangler, Hasse, Hilbert, and Takagi; see, for example, Artin [1], Artin and Hasse [1], [2], Fürtwangler [1]–[6], and Teiji Takagi [1], [2]. See especially Hasse [3] and the references therein. For expositions on Hilbert's reciprocity law, see the books of O'Meara [1, Chapter 7] and Scharlau [1, pp. 191, 220].

Tate discusses Artin's reciprocity law in Cassels and Fröhlich [1, pp. 162–203]. For Weil's reciprocity law, see the books of Scharlau [1, p. 196] and Milnor and Husemoller [1, p. 132]. For Shimura's reciprocity law, see Shimura [1, p. 157] and the paper of Stark [2, Section 3]. Histories of reciprocity laws may be found in the papers of Frei [1] and Schwermer [1]; the latter paper includes a discussion of Artin's reciprocity law. See also Hasse's article in Cassels and Fröhlich [1, pp. 266–279] and the books of Cox [1] and Lemmermeyer [3].

Wójcik [1] showed that Eisenstein reciprocity in the form presented in Corollary 14.5.9 holds even when k is not a prime power, provided that u is a kth power modulo k and ν satisfies a complicated notion of "semiprimary with respect to k".

Hayes [1] constructed Hecke characters for ray class extensions of global function fields, analogous to the Gauss sum Hecke characters of Weil [7]. He showed that these characters satisfy an analogue of the Eisenstein reciprocity law for Gauss sums (see Problem 18). Hayes [2] employed the Eisenstein reciprocity law for Gauss sums to determine the conductor of Hecke characters defined in terms of Brumer–Stark elements for an Abelian extension of Q.

For an integer $k > 2$, write $\beta = \exp(2\pi i/k), K = Q(\beta), O_K = Z[\beta]$. For a prime ideal P in O_K not dividing k, let $\chi = \chi_P$ denote the kth-power residue symbol $(\ /P)$ on O_K/P. For integers a, b with a, b, and $a+b$ coprime with k, consider the Jacobi sum $J = J(a,b) = J(\chi^a, \chi^b)$. If P is replaced by a different prime ideal P' of O_K not dividing k, write J' in place of J. Using the Eisenstein reciprocity law for Gauss sums as well as the prime factorization of Jacobi sums (Theorem 11.2.3), Helou [5] proved that for any integer N which is the norm of an ideal of O_K coprime with kP, one has the reciprocity law

$$\left(\frac{J}{N}\right) = \left(\frac{N}{J}\right) = \left(\frac{N}{P}\right)^n,$$

where $n = f(a) + f(b) - f(a+b)$ with $f(x) = \left(x^{1+\phi(k)} - x\right)/k$, for the Euler totient function ϕ. As a corollary, one obtains, for odd k and integers c, d with c, d, and $c+d$ coprime with k, the following reciprocity law for Jacobi sums:

$$\left(\frac{J(a,b)}{J'(c,d)}\right) = \left(\frac{J'(c,d)}{J(a,b)}\right).$$

These results generalize previous work of Helou [4].

Theorem 14.7.1 and Problem 14 are applications of Eisenstein's reciprocity law proved in the book of Ireland and Rosen [1, Chapter 14, Section 6]. Problems 15 and 18 are taken from Ireland and Rosen [1, pp. 220, 217].

Research Problems

We present a short list of problems for interested readers to investigate.

1. Find an analogue of Theorem 1.2.2 in the case that $ac + b$ is odd.

2. Find a criterion for determining in which of the three intervals $(-2\sqrt{p}, -\sqrt{p})$, $(-\sqrt{p}, \sqrt{p})$, $(\sqrt{p}, 2\sqrt{p})$ the Gauss sum $g(3)$ belongs.

3. Determine the values of the Jacobi sums for characters (mod p) of order 32.

4. Determine formulas for the cyclotomic numbers of order 32.

5. Give an elementary proof of Theorem 4.2.4, which evaluates $g(4)$.

6. Find a criterion for determining the ambiguous sign in the evaluations of $g(8)$ (Theorem 4.3.1) and $g(12)$ (Theorem 4.4.1).

7. Find a formula for $g(32)$ up to certain sign ambiguities.

8. In Theorem 1.2.2, a reciprocity theorem for generalized quadratic Gauss sums was established. Does there exist an analogous formula for generalized quartic Gauss sums, that is, sums of the type

$$\sum_{n=0}^{c-1} e^{\pi i a n^4/c + \pi i b n^2/c},$$

where $a, b,$ and c are integers with $a, c > 0$? A result in this direction was found by Berndt [1, Corollary 4].

9. D. H. Lehmer [2] has examined partial sums of the quadratic Gauss sum, namely,

$$G_n(m) = \sum_{j=0}^{m-1} e^{2\pi i j^2/n},$$

where m and n are integers such that $1 \leq m \leq n$. Very sharp estimates for $|G_n(m)|$ were obtained. The work was motivated by graphical evidence that showed that, for $m < n/2$, almost all values of $G_n(m)$ lie in the vicinity of $(1 + i)\sqrt{n}/4$ in the complex plane. Do there exist phenomena similar to those observed and established by D. H. Lehmer for the partial sums of cubic or quartic Gauss sums? See Loxton [3].

10. Can the formula of Sullivan and Zannier [1, p. 55] for the incomplete quadratic Gauss sum be extended to incomplete cubic and quartic Gauss sums?

11. If 5 is not a biquadratic residue (mod p), show that H_{20} and M_{20} are not difference sets. (See the Notes in Chapter 5.)

12. Determine if H_{24} and M_{24} ever form difference sets. The formula for $g(24)$ given in Berndt and Evans [1, Theorem 3.32] may be useful.

13. Determine if H_k and M_k are ever difference sets for any $k \geq 20$.

14. Find new classes of sums (6.0.1) that can be explicitly evaluated. The papers of Hadano [1], Parnami, Agrawal, and Rajwade [3], and Williams [30] might be consulted for some possibilities.

15. Establish analogues of Theorems 6.3.1 and 6.3.2 for kth–power residues when $k > 2$.

16. Simplify the definition of "primary" given above (14.2.2).

17. Determine an explicit criterion for 3 to be a 16th power (mod p).

18. Find a proof of the law of biquadratic reciprocity (Theorem 8.2.3) analogous to the simple proof of the law of quadratic reciprocity given at the end of Section 1.2.

19. Reconcile Yamamoto's evaluation [1, Theorem 3] of $G(\chi)$ with that given in Theorem 4.2.3.

20. Evaluate the Jacobsthal sums of orders 12, 16, 24 over an arbitrary finite field.

21. Huard, Spearman, and Williams [1] have given an elementary proof of the Davenport–Hasse product formula (Theorem 11.3.5) for Gauss sums over F_p. Find a purely arithmetic proof of their Theorem 4.

22. Prove the Davenport–Hasse product formula (Theorem 11.3.5) in an elementary manner.

23. From Theorems 13.3.6 and 13.4.3 we see that

$$B_3 = \begin{cases} -1 + 2a_3, & \text{if } p \equiv 1 \pmod 3, \\ -1, & \text{if } p \equiv 2 \pmod 3, \end{cases}$$

$$B_5 = \begin{cases} -1 + 4a_5, & \text{if } p \equiv 1 \pmod 5, \\ -1, & \text{if } p \equiv 2, 3, 4 \pmod 5. \end{cases}$$

Is there a formula for B_k, where k is an odd prime, that extends these two results?

24. Is there a congruence modulo k^r for $\text{ind}_g k$, where r is a positive integer and k is a prime with $k \geq 2$, that reduces to that of Evans [5, eq. (14)] for $k = 2$? See the paper of Burde [6].

25. Find a general congruence that includes the three congruences of Eisenstein in Theorems 12.9.7–12.9.9 as special cases.

26. Determine $\binom{4f+1}{f}$, $\binom{3f}{f}$, and $\binom{3f+1}{f}$ modulo p^2 for primes $p = 8f+3$, $p = 7f+2$, and $p = 7f+4$, respectively; see Theorems 12.9.7–12.9.9.

27. For large primes $p \equiv 1 \pmod 4$, find a fast, efficient algorithm for computing the factor $\left(\frac{p-1}{2}\right)! \pmod p$ in (4.2.12).

28. Prove that the arguments of the Gauss sums $G(\chi)$ of order 8 over F_p are uniformly distributed in $(0, 2\pi)$ as p runs through the primes congruent to (a) 1 (mod 8), (b) 1 (mod 16), (c) 9 (mod 16). Prove similar results for Gauss sums of other orders.

Bibliography

Note. Collected works have been listed as [CW]. For example, see the references for E. Artin.

ABE, M.

[1] On the conductor of algebraic Hecke character generated by Gauss sums, *Mem. Fac. Sci. Kyushu Univ. Ser. A* **46** (1992), 45–56.

ACHARYA, V. V.

[1] *Cyclotomic Numbers of Order* 2ℓ, ℓ *an Odd Prime,* Ph.D. thesis, University of Pune, 1997.

ACHARYA, V. V. and KATRE, S. A.

[1] Cyclotomic numbers of order $2l$, l an odd prime, *Acta Arith.* **69** (1995), 51–74.

ADLEMAN, L., POMERANCE, C., and RUMELY, R.

[1] On distinguishing prime numbers from composite numbers, *Ann. Math.* **117** (1983), 173–206.

ADOLPHSON, A. and SPERBER, S.

[1] Character sums in finite fields, *Compositio Math.* **52** (1984), 325–354.

AIGNER, A.

[1] Kriterion zum 8. und 16. Potenzcharacter der Reste 2 und –2, *Deutsche Math.* **4** (1939), 44–52.

[2] Einige handliche Regeln für biquadratische Reste, *Math. Nachr.* **17** (1959), 219–223.

[3] Quadratische und kubische Restkriterien für das Auftreten einer Fibonacci-Primitivwurzel, *J. Reine Angew. Math.* **274/275** (1975), 139–140.

AKIYAMA, S.

[1] On the pure Jacobi sums, *Acta Arith.* **75** (1996), 97–104.

ALADOV, N.

[1] Distribution of quadratic residues and nonresidues of a prime P in the sequence $1, 2, \ldots, P-1$, *Mat. Sb.* **18** (1896), 61–75. (Russian)

ALDERSON, H. P.

[1] On the quintic character of 2, *Mathematika* **11** (1964), 125–130.

[2] On the septimic character of 2 and 3, *Proc. Cambridge Philos. Soc.* **74** (1973), 421–433.

ANDREWS, G. E.

[1] *The Theory of Partitions*, Addison–Wesley, Reading, Mass., 1976.

[2] Partitions and the Gaussian sum, in *The Mathematical Heritage of C. F. Gauss*, World Scientific Publ. Co., Singapore, 1991, pp. 35–42.

[3] *Number Theory*, Dover, Mineola, New York, 1994.

ANDREWS, G., ASKEY, R., and ROY, R.

[1] *Special Functions,* Cambridge University Press, Cambridge (to appear).

ANKENY, N. C.

[1] Criterion for rth power residuacity, *Pacific J. Math.* **10** (1960), 1115–1124.

AOKI, N.

[1] Abelian fields generated by a Jacobi sum, *Comment. Math. Univ. St. Pauli* **45** (1996), 1–21.

APOSTOL, T. M.

[1] *Introduction to Analytic Number Theory,* Springer–Verlag, New York, 1976.

ARTIN, E.

[1] Beweis des allgemeinen Reziprozitätsgesetzes, *Abh. Math. Sem. Univ. Hamburg* **5** (1927), 353–363. [CW, pp. 131–141]

[CW] *The Collected Papers of Emil Artin*, Addison–Wesley, Reading, Mass., 1965.

ARTIN, E. and HASSE, H.

[1] Über den zweiten Ergänzungssätze zum Reziprozitätsgesetz der l-ten Potenzreste im Körper k_ζ der l-ten Einheitswurzeln und in Oberkörpern von k_ζ, *J. Reine Angew. Math.* **154** (1925), 143–148. [CW (Artin), pp. 125–130]

[2] Die beiden Ergänzungssätze zum Reziprozitätsgesetz der l^n-ten Potenzreste im Körper der l^n-ten Einheitswurzeln, *Abh. Math. Sem. Univ. Hamburg* **6** (1928), 146–162. [CW (Artin), pp. 142–158]

AUSLANDER, L., TOLIMIERI, R., and WINOGRAD, S.

[1] Hecke's theorem in quadratic reciprocity, finite nilpotent groups and the Cooley–Tukey algorithm, *Adv. Math.* **43** (1982), 122–172.

AUTUORE, J.

[1] *Character Sum Analogues of Real Integral Formulas*, Ph.D. thesis, University of California at San Diego, 1990.

AYOUB, R.

[1] *An Introduction to the Analytic Theory of Numbers*, American Mathematical Society, Providence, 1963.

BABAEV, G. and ISMOILOV, D.

[1] Distribution of arguments of Gaussian sums, *Dokl. Akad. Nauk Tadzhik. SSR* **33** (1990), 427–430.

BACH, E. and SHALLIT, J. O.

[1] Factoring with cyclotomic polynomials, *Math. Comp.* **52** (1989), 201–219.

BACHMANN, P.

[1] *Die Lehre von der Kreistheilung und ihre Beziehungen zur Zahlentheorie*, Teubner, Leipzig, 1872.

BAILEY, D.

[1] Two p^3 variations of Lucas' Theorem, *J. Number Theory* **35** (1990), 208–215.

BAMBAH, R. P. and CHOWLA, S.

[1] On the sign of the Gaussian sum, *Proc. Natl. Acad. Sci.* (India) *Sect. A* **13** (1947), 175–176.

BARKAN, P.

[1] Démonstration d'une conjecture de K. Yamamoto sur les sommes de Gauss biquadratique, *C. R. Acad. Sci. Paris Ser. I Math.* **310** (1990), 69–72.

BARRUCAND, P.

[1] Sommes de Gauss et séries singulières de Hardy pour les cubes, *C. R. Acad. Sci. Paris* **250** (1960), 4249–4251.

BARRUCAND, P. and COHN, H.

[1] Note on primes of type $x^2 + 32y^2$, class number, and residuacity, *J. Reine Angew. Math.* **238** (1969), 67–70.

BAUMERT, L. D.

[1] *Cyclic Difference Sets*, Lecture Notes in Mathematics, No. 182, Springer–Verlag, Berlin, 1971.

BAUMERT, L. D. and FREDRICKSEN, H.

[1] The cyclotomic numbers of order eighteen with applications to difference sets, *Math. Comp.* **21** (1967), 204–219.

BAUMERT, L. D. and McELIECE, R. J.

[1] Weights of irreducible cyclic codes, *Information and Control* **20** (1972), 158–175.

BAUMERT, L. D., MILLS, W. H., and WARD, R. L.

[1] Uniform cyclotomy, *J. Number Theory* **14** (1982), 67–82.

BEEGER, N. G. W. H.

[1] On a new case of the congruence $2^{p-1} \equiv 1 \ (p^2)$, *Mess. Math.* **51** (1922), 149–150.

[2] A problem of the theory of numbers and its history, *Nieuw Arch. Wisk.* **22** (1947), 306–309.

BELLMAN, R.

[1] *A Brief Introduction to Theta Functions*, Holt, Rinehart, and Winston, New York, 1961.

BENKÖ, I. and RĂDUICĂ, M.

[1] Les caractères multiplicatifs et la diagonalization de la transformation Fourier sur $L^2(Z_p)$, *Analele Stü. Univ. Al. I. Cuza* IASI **40** (1994), 85–88.

BERNDT, B. C.

[1] On Gaussian sums and other exponential sums with periodic coefficients, *Duke Math. J.* **40** (1973), 145–156.

[2] Periodic Bernoulli numbers, summation formulas and applications, in *Theory and Application of Special Functions* (R. A. Askey, ed.), Academic Press, New York, 1975, pp. 143–189.

[3] Classical theorems on quadratic residues, *Enseignement Math.* **22** (1976), 261–304.

BERNDT, B. C. and CHOWLA, S.

[1] The reckoning of certain quartic and octic Gauss sums, *Glasgow Math. J.* **18** (1977), 153–155.

BERNDT, B. C. and EVANS, R. J.

[1] Sums of Gauss, Jacobi, and Jacobsthal, *J. Number Theory* **11** (1979), 349–398.

[2] Sums of Gauss, Eisenstein, Jacobi, Jacobsthal, and Brewer, *Illinois J. Math.* **23** (1979), 374–437.

[3] Half Gauss sums, *Math. Ann.* **249** (1980), 115–125.

[4] The determination of Gauss sums, *Bull. Am. Math. Soc.* (N.S.) **5** (1981), 107–129; Corrigendum, ibid. **7**(1982), 441.

BERNDT, B. C. and SCHOENFELD, L.

[1] Periodic analogues of the Euler–Maclaurin and Poisson summation formulas with applications to number theory, *Acta Arith.* **28** (1975), 23–68.

BERRY, M. V. and GOLDBERG, J.

[1] Renormalisation of curlicues, *Nonlinearity* **1**(1988), 1–26.

BERRY, M. V. and KLEIN, S.

[1] Integer, fractional, and fractal Talbot effects, *J. Modern Optics* **43** (1996), 2139–2164.

BEYER, G.

[1] Über eine Klasseneinteilung aller kubischen Restcharaktere, *Abh. Math. Sem. Univ. Hamburg* **19** (1954), 115–116.

BICKMORE, C. E.

[1] On the numerical factors of $a^n - 1$, *Mess. Math.* **25** (1896), 1–44.

[2] On the numerical factors of $a^n - 1$ (second notice), *Mess. Math.* **26** (1897), 1–38.

BOCHNER, S.

[1] Remarks on Gaussian sums and Tauberian theorems, *J. Indian Math. Soc.* **15** (1951), 97–104.

BOREVICH, Z. and SHAFAREVICH, I.

[1] *Number Theory*, translated by Newcomb Greenleaf, Academic Press, New York, 1966.

BOYARSKY, M.

[1] p-adic gamma functions and Dwork cohomology, *Trans. Am. Math. Soc.* **257** (1980), 359–369.

BRAUER, A.

[1] Über Sequenzen von Potenzresten, *Sitzungs. d. Preuss. Akad. Wiss., Math.-Phys. Kl.* (1928), 9–16.

BREMSER, P. S.

[1] A generalization of Gaussian sums to vector spaces over finite fields, *Linear Algebra and Appl.* **81** (1986), 35–45.

BRESSOUD, D. M.

[1] On the value of Gaussian sums, *J. Number Theory* **13** (1981), 88–94.

BREWER, B. W.

[1] On the quadratic reciprocity law, *Am. Math. Monthly* **58** (1951), 177–179.

[2] On certain character sums, *Trans. Am. Math. Soc.* **99** (1961), 241–245.

[3] On primes of the form $u^2 + 5v^2$, *Proc. Am. Math. Soc.* **17** (1966), 502–509.

BRILLHART, J.

[1] Concerning the numbers $2^{2p} + 1$, p prime, *Math. Comp.* **16** (1962), 424–430.

BRINKHUIS, J.

[1] Gauss sums and their prime factorization, *Enseignement Math.* **36** (1990), 39–51.

[2] On a comparison of Gauss sums with products of Lagrange resolvents, *Compositio Math.* **93** (1994), 155–170.

BROWN, E.

[1] A theorem on biquadratic reciprocity, *Proc. Am. Math. Soc.* **30** (1971), 220–222.

[2] Quadratic forms and biquadratic reciprocity, *J. Reine Angew. Math.* **253** (1972), 214–220.

[3] Biquadratic reciprocity laws, *Proc. Am. Math. Soc.* **37** (1973), 374–376.

BRÜCKNER, H.

[1] *Explizites Reziprozitätsgesetz und Anwendungen*, Lecture Notes, Universität Essen, (1979), 83 pp.

BRÜDERN, J.

[1] On pairs of diagonal cubic forms, *Proc. London Math. Soc.* **61** (1990), 273–343.

BUCK, N., SMITH, L., SPEARMAN, B. K., and WILLIAMS, K. S.

[1] The cyclotomic numbers of order fifteen, *Math. Comp.* **48** (1987), 67–83.

BUCK, N. and WILLIAMS, K. S.

[1] Sequel to Muskat's evaluation of the cyclotomic numbers of order fourteen, *Carleton Mathematical Series* No. 216, Carleton University, Ottawa, November 1985, 24 pp.

BUELL, D. A. and HUDSON, R. H.

[1] On runs of consecutive quadratic residues and quadratic nonresidues, *BIT* **24** (1984), 243–247.

[2] Sequences in power residue classes, *Internatl. J. Math. Math. Sci.* **9** (1986), 261–266.

BUELL, D. A., HUDSON, R. H., and WILLIAMS, K. S.

[1] Extension of a theorem of Cauchy and Jacobi, *J. Number Theory* **19** (1985), 309–340.

BURDE, K.

[1] Ein rationales biquadratisches Reziprozitätsgesetz, *J. Reine Angew. Math.* **235** (1969), 175–184.

[2] Über allgemeine Sequenzen der Länge 3 von Legendresymbolen, *J. Reine Angew. Math.* **272** (1974), 203–216.

[3] Zur Herleitung von Reziprozitätsgesetzen unter Benutzung von endlichen Körpern, *J. Reine Angew. Math.* **293/294** (1977), 418–427.

[4] Eine Verteilungseigenschaft der Legendre symbole, *J. Number Theory* **12** (1980), 273–277.

[5] Pythagoräische Tripel und Reziprozität in Galoisfeldern, *J. Number Theory* **12** (1980), 278–282.

[6] Ergänzungsgesetze, *Math. Nachr.* **163** (1993), 15–26.

BURGESS, D. A.

[1] Partial Gauss sums. III, *Glasgow Math. J.* **34** (1992), 253–261.

BURNSIDE, W.

[1] On cyclotomic quinquisection, *Proc. London Math. Soc.* **14** (1915), 251–259.

BUSHNELL, C. and FRÖHLICH, A.

[1] *Gauss Sums and p-adic Division Algebras*, Lecture Notes in Mathematics, No. 987, Springer-Verlag, Berlin, 1983.

CARAGIU, M.

[1] Counting the maximal sequences of consecutive quadratic residues modulo p, *Rev. Roumaine Math. Pures Appl.* **38** (1993), 745–749.

CAREY, F. S.

[1] Notes on the division of the circle, *Quart. J. Pure Appl. Math.* **26** (1893), 322–371.

CARLITZ, L.

[1] A theorem on higher congruences, *Bull. Am. Math. Soc.* **41** (1935), 844–846.

[2] On certain higher congruences, *Bull. Am. Math. Soc.* **41** (1935), 907–914.

[3] Criteria for certain higher congruences, *Am. J. Math.* **59** (1937), 618–628.

[4] The number of solutions of certain equations in a finite field, *Proc. Natl. Acad. Sci.* (USA) **38** (1952), 515–519; Errata, ibid. **38** (1952), 618.

[5] Some special equations in a finite field, *Pacific J. Math.* **3** (1953), 13–24.

[6] Certain special equations in a finite field, *Monatsh. Math.* **58** (1954), 5–12.

[7] The number of solutions of some equations in a finite field, *Portug. Math.* **13** (1954), 25–31.

[8] The number of solutions of some special equations in a finite field, *Pacific J. Math.* **4** (1954), 207–217.

[9] The number of solutions of a special quadratic congruence, *Portug. Math.* **14** (1955), 9–14.

[10] The number of solutions of certain types of equations in a finite field, *Pacific J. Math.* **5** (1955), 177–181.

[11] The number of solutions of some equations in a finite field, *J. Math. Soc. Jpn.* **7** (1955), 209–223.

[12] A note on Gauss' "Serierum singularium", *Portug. Math.* **15** (1956), 79–81.

[13] Solvability of certain equations in a finite field, *Quart. J. Math. Oxford* (2) **7** (1956), 3–4.

[14] The number of solutions of a particular equation in a finite field, *Publ. Math. Debrecen* **4** (1956), 379–383.

[15] A note on Gauss' sum, *Proc. Am. Math. Soc.* **7** (1956), 910–911.

[16] The number of points on certain cubic surfaces over a finite field, *Boll. Un. Mat. Ital.* (3) **12** (1957), 19–21.

[17] A note on exponential sums, *Acta Sci. Math.* (Szeged) **21** (1960), 135–143.

[18] A note on multiple Kloosterman sums, *J. Indian Math. Soc.* **29** (1965), 197–200.

[19] A note on multiple exponential sums, *Pacific J. Math.* **15** (1965), 757–765.

[20] A conjecture concerning a certain system of equations in a finite field, *Rev. Roumaine Math. Pures Appl.* **11** (1966), 277–282.

[21] A note on Gauss's sum, *Matematiche* (Catania) **23** (1968), 147–150.

[22] Some formulas related to Gauss's sum, *Rend. Sem. Mat. Univ. Padova* **41** (1968), 222–226.

[23] A note on exponential sums, *Pacific J. Math.* **30** (1969), 35–37.

[24] Gauss sums over finite fields of order 2^n, *Acta Arith.* **15** (1969), 247–265.

[25] Kloosterman sums and finite field extensions, *Acta Arith.* **16** (1969), 179–193.

[26] Reduction formulas for certain multiple exponential sums, *Czechoslovak Math. J.* **20** (1970), 616–627.

[27] A theorem on linear exponential sums, *Univ. Beograd. Publ. Elektrotehn. Fak. Ser. Mat. Fiz.* (1977), 55–56.

[28] A note on exponential sums, *Math. Scand.* **42** (1978), 39–48.

[29] Explicit evaluation of certain exponential sums, *Math. Scand.* **44** (1979), 5–16.

[30] Evaluation of some exponential sums over a finite field, *Math. Nachr.* **96** (1980), 319–330.

CASSELS, J. W. S.

[1] On the determination of generalized Gauss sums, *Arch. Math.* (Brno) **5** (1969), 79–84.

[2] On Kummer sums, *Proc. London Math. Soc.* (3) **21** (1970), 19–27.

[3] Review of E. D. Tabakova, "A numerical investigation of Kummer cubic sums", *Math. Comp.* **29** (1975), 665–666.

CASSELS, J. W. S. and FRÖHLICH, A.

[1] *Algebraic Number Theory*, Academic Press, London, 1967.

CAUCHY, A.

[1] Mémoire sur la théorie des nombres, *Bull. Sci. Math. M. Férussac* **12** (1829), 205–221. [CW: Sér. 1, vol. 2, pp. 88–107]

[2] Mémoire sur la théorie des nombres, *Mém. Acad. Sci. Inst. de France* **17** (1840), 249–768. [CW: Sér. 1, vol. 3, pp. 5–450]

[3] Méthode simple et nouvelle pour la détermination complète des sommes alternées, formées avec les racines primitives des équations binomes, *C. R. Acad. Sci. Paris* **10** (1840), 560–572. [CW: Sér. 1, vol. 5, pp. 152–166]

[CW] *Oeuvres*, Gauthiers Villars, Paris, Sér.1, vol. 1 (1882), vol. 2 (1908), vol. 3 (1911), vol. 4 (1884), vol. 5 (1885), vol. 6 (1888), vol. 7 (1892), vol. 8 (1890), vol. 9 (1896), vol. 10 (1897), vol. 11 (1899), vol. 12 (1900); Sér.2, vol. 1 (1905), vol. 2 (1958), vol. 3 (1897), vol. 4 (1899), vol. 5 (1903), vol. 6 (1887), vol. 7 (1889), vol. 8 (1890), vol. 9 (1891), vol. 10 (1895), vol. 11 (1913), vol. 12 (1916), vol. 13 (1932), vol. 14 (1938), vol. 15 (1974).

CAYLEY, A.

[1] On the binomial equation $x^p - 1 = 0$; trisection and quartisection, *Proc. London Math. Soc.* **11** (1880), 4–17. [CW: vol. 11, pp. 84–96]

[2] On the binomial equation $x^p - 1 = 0$; quinquisection, *Proc. London Math. Soc.* **12** (1881), 15–16. [CW: vol. 11, pp. 314–316]

[3] The binomial equation $x^p - 1 = 0$; quinquisection, second part, *Proc. London Math. Soc.* **16** (1885), 61–63. [CW: vol. 12, pp. 72–73]

[CW] *The Collected Mathematical Papers of Arthur Cayley*, University Press, Cambridge, vol. 1 (1889), vol. 2 (1889), vol. 3 (1890), vol. 4 (1891), vol. 5 (1892), vol. 6 (1893), vol. 7 (1894), vol. 8 (1895), vol. 9 (1896), vol. 10 (1896), vol. 11 (1896), vol. 12 (1897), vol. 13 (1897).

CHANDRASEKHARAN, K.

[1] *Introduction to Analytic Number Theory*, Springer–Verlag, New York, 1968.

CHOWLA, I.

[1] The number of solutions of a congruence in two variables, *Proc. Natl. Acad. Sci.* (India) *Sect. A* **4** (1936), 654–655.

[2] On the number of solutions of some congruences in two variables, *Proc. Natl. Acad. Sci.* (India) *Sect. A* **5** (1937), 40–44.

[3] Generalization of a theorem of Dickson, *Proc. Natl. Acad. Sci.* (India) *Sect. A* **8** (1938), 223–226.

[4] On Waring's problem (mod p), *Proc. Natl. Acad. Sci.* (India) *Sect. A* **13** (1943), 195–220.

CHOWLA, P.

[1] A new proof and generalization of some theorems of Brewer, *Norske Vid. Selsk. Forh.* (Trondheim) **41** (1968), 1–3.

CHOWLA, S.

[1] Some formulae of the Gauss sum type, *Tôhoku Math. J.* **30** (1928), 226–234; Corrigenda, ibid. **32** (1930), 109–110.

[2] Some formulae of the Gauss sum type (II), *Tôhoku Math. J.* **32** (1930), 352–353.

[3] A property of biquadratic residues, *Proc. Natl. Acad. Sci.* (India) *Sect. A* **14** (1944), 45–46.

[4] The cubic character of 2 (mod p), *Proc. Lahore Philos. Soc.* **6** (1944), 12.

[5] A formula similar to Jacobsthal's for the explicit value of x in $p = x^2 + y^2$ where p is a prime of the form $4k + 1$, *Proc. Lahore Philos. Soc.* **7** (1945), 2 pp.

[6] The cubic character of 2 (mod p), *Proc. Lahore Philos. Soc.* **7** (1945), 1 p.

[7] The last entry in Gauss' diary, *Proc. Natl. Acad. Sci.* (USA) **35** (1949), 244–246.

[8] On the congruence $\sum_{i=0}^{s} a_i x_i^k \equiv 0 \pmod{p}$, *J. Indian Math. Soc.* **25** (1961), 47–48.

[9] On a formula of Jacobsthal, *Norske Vid. Selsk. Forh.* (Trondheim) **34** (1961), 105–106.

[10] On Gaussian sums, *Norske Vid. Selsk. Forh.* (Trondheim) **35** (1962), 66–67.

[11] On Gaussian sums, *Proc. Natl. Acad. Sci.* (USA) **48** (1962), 1127–1128.

[12] *The Riemann Hypothesis and Hilbert's Tenth Problem*, Gordon and Breach, New York, 1965.

[13] An algebraic proof of the law of quadratic reciprocity, *Norske Vid. Selsk. Forh.* (Trondheim) **39** (1966), 59.

[14] On the class–number of the function field $y^2 = f(x)$ over $GF(p)$, I, II, *Norske Vid. Selsk. Forh.* (Trondheim) **39** (1966), 86–88; ibid. **40** (1967), 7–10.

[15] Observation on a theorem of Stark, *Norske Vid. Selsk. Forh.* (Trondheim) **40** (1967), 34–36.

[16] On some character sums, *Norske Vid. Selsk. Forh.* (Trondheim) **40** (1967), 62–66.

[17] On Kloosterman's sum, *Norske Vid. Selsk. Forh.* (Trondheim) **40** (1967), 70–72.

CHOWLA, S., COWLES, J., and COWLES, M.

[1] On the number of zeros of diagonal cubic forms, *J. Number Theory* **9** (1977), 502–506.

[2] Congruence properties of the number of solutions of some equations, *J. Reine Angew. Math.* **298** (1978), 101–103.

[3] The number of zeros of $x^3 + y^3 + z^3$ in certain finite fields, *J. Reine Angew. Math.* **299/300** (1978), 406–410.

[4] On the difference of cubes (mod p), *Acta Arith.* **37** (1980), 61–65.

CHOWLA, S., DWORK, B., and EVANS, R. J.

[1] On the mod p^2 determination of $\binom{(p-1)/2}{(p-1)/4}$, *J. Number Theory* **24** (1986), 188–196.

CHUNG, F. R. K.

[1] Several generalizations of Weil sums, *J. Number Theory* **49** (1994), 95–106.

COATES, J. H.

[1] p-adic L-functions and Iwasawa's theory, in *Algebraic Number Fields: L-functions and Galois Properties* (A. Fröhlich, ed.), Academic Press, London, 1977, pp. 269–353.

COHEN, E.

[1] Representations by cubic congruences, *Proc. Natl. Acad. Sci.* (USA) **39** (1953), 119–121.

[2] The number of solutions of certain cubic congruences, *Pacific J. Math.* **5** (1955), 877–886.

COHEN, H.

[1] *A Course in Computational Algebraic Number Theory,* Springer–Verlag, Berlin, 1993.

COHEN, H. and LENSTRA, H. W.

[1] Primality testing and Jacobi sums, *Math. Comp.* **42** (1984), 297–330.

COHEN, S. D.

[1] Quadratic residues and character sums over fields of square order, *J. Number Theory* **18** (1984), 391–395.

[2] Generators in cyclic difference sets, *J. Combin. Theory Ser. A* **51** (1989), 227–236.

[3] Primitive elements and polynomials: existence results, in *Finite Fields, Coding Theory, and Advances in Communications and Computing* (Las Vegas, 1991), Lecture Notes in Pure and Applied Mathematics, No. 141 (G. Mullen and P. Shiue, eds.), Marcel Dekker, New York, 1993, pp. 43–55.

COLEMAN, R. F.

[1] The Gross–Koblitz formula, in *Galois Representations and Arithmetic Algebraic Geometry,* Y. Ihara, ed., Advanced Studies in Pure Mathematics, vol. 12, North–Holland, Amsterdam, 1987, pp. 21–52.

[2] (Corrigenda to the paper above), p-adic Analysis, Lecture Notes in Mathematics, No. 1454, Springer–Verlag, Berlin, 1990, p. 193.

COLLISON, M. J.

[1] The origins of the cubic and biquadratic reciprocity laws, *Arch. Hist. Exact Sci.* **17** (1977), 63–69.

CONRAD, K.

[1] Jacobi sums and Stickelberger's congruence, *Enseignement Math.* **41** (1995), 141–153.

COOKE, G.

[1] *Lectures on the Power Reciprocity Laws of Algebraic Number Theory,* Part 1, mimeographed lecture notes, Cornell University, Ithaca, New York, 1974.

CORSON, H. H.

[1] On some special systems of equations, *Pacific J. Math.* **6** (1956), 449–452.

COSTER, M. J.

[1] Generalization of a congruence of Gauss, *J. Number Theory* **29** (1988), 300–310.

[2] *Supercongruences*, Ph.D thesis, Leiden, 1988.

COX, D. A.

[1] *Primes of the Form $x^2 + ny^2$: Fermat, Class Field Theory, and Complex Multiplication*, Wiley, New York, 1989.

CRANDALL, R., DILCHER, K., and POMERANCE, C.

[1] A search for Wieferich and Wilson primes, *Math. Comp.* **66** (1997), 433–449.

CUNNINGHAM, A.

[1] On 2 as a 16-ic residue, *Proc. London Math. Soc.* **27** (1895), 85–122.

CUNNINGHAM, A. and GOSSET, T.

[1] 4-tic and 3-tic residuacity-tables, *Mess. Math.* **50** (1920), 1–30.

CURTIS, C. W. and SHINODA, K.

[1] Unitary Kloosterman sums and Gelfand–Graev representation of GL_2, preprint, 1997.

DABROWSKI, R.

[1] Generalized Kloosterman sums, *Proc. Sympos. Pure Math.*, vol. 56, American Mathematical Society, Providence, 1994, pp. 219–231.

DANTSCHER, V.

[1] Bemerkungen zum analytischen Beweise des cubischen Reciprocitätsgesetzes, *Math. Ann.* **12** (1877), 241–253.

DARMON, H.

[1] Note on a polynomial of Emma Lehmer, *Math. Comp.* **194** (1991), 795–800.

DAVENPORT, H.

[1] On the distribution of quadratic residues (mod p), *J. London Math. Soc.* **6** (1931), 49–54. [CW: vol. 4, pp. 1451–1456]

[2] On the distribution of quadratic residues (mod p) (second paper), *J. London Math. Soc.* **8** (1933), 46–52. [CW: vol. 4, pp. 1481–1487]

[3] *Multiplicative Number Theory*, 2nd ed., Springer–Verlag, New York, 1980.

[4] *The Higher Arithmetic*, 6th ed. Cambridge University Press, Cambridge, 1992.

[CW] *The Collected Works of Harold Davenport*, vols. 1–4, Academic Press, London, 1977.

DAVENPORT, H. and HASSE, H.

[1] Die Nullstellen der Kongruenzzetafunktionen in gewissen zyklischen Fällen, *J. Reine Angew. Math.* **172** (1934), 151–182. [CW(Davenport): vol. 4, pp. 1488–1519]

DAVIS, K. and WEBB, W.

[1] A binomial coefficient congruence modulo prime powers, *J. Number Theory* **43** (1993), 20-23.

DESARMENIEN, J.

[1] Un analogue des congruences de Kummer pour les q-nombres d'Euler, *Eur. J. Combin.* **3** (1982), 19–28.

DEURING, M.

[1] Die Typen der Multiplikatorenringe elliptischer Funktionenkörper, *Abh. Math. Sem. Univ. Hamburg* **14** (1941), 197–272.

DIAMOND, H. G., GERTH, F. III, and VAALER, J. D.

[1] Gauss sums and Fourier analysis on multiplicative subgroups of Z_q, *Trans. Am. Math. Soc.* **227** (1983), 711–726.

DICKSON, L. E.

[1] Determination of the structure of all linear homogeneous groups in a Galois field which are defined by a quadratic invariant, *Am. J. Math.* **15** (1899), 193–256.[CW: vol. 1, pp. 1–64]

[2] On the last theorem of Fermat, *Quart. J. Pure Appl. Math.* **40** (1908), 27–45. [CW: vol. 5, pp. 453–471]

[3] On the congruence $x^n + y^n + z^n \equiv 0 \pmod{p}$, *J. Reine Angew. Math.* **135** (1909), 134–141. [CW: vol. 1, pp. 420–427]

[4] Lower limit for the number of sets of solutions of $x^e + y^e + z^e \equiv 0 \pmod{p}$, *J. Reine Angew. Math.* **135** (1909), 181–188. [CW: vol. 1, pp. 412–419]

[5] Ternary quadratic forms and congruences, *Ann. Math.* (2) **28** (1927), 333–341. [CW: vol. 1, pp. 641–649]

[6] Cyclotomy and trinomial congruences, *Trans. Am. Math. Soc.* **37** (1935), 363–380. [CW: vol. 2, pp. 555–572]

[7] Cyclotomy when e is composite, *Trans. Am. Math. Soc.* **38** (1935), 187–200. [CW: vol. 2, pp. 541–554]

[8] Cyclotomy, higher congruences, and Waring's problem, *Am. J. Math.* **57** (1935), 391–424. [CW: vol. 3, pp. 227–260]

[9] Cyclotomy, higher congruences, and Waring's problem II, *Am. J. Math.* **57** (1935), 463–474. [CW: vol. 3, pp. 261–272]

[10] Congruences involving only e-th powers, *Acta Arith.* **1** (1936), 161–167. [CW: vol. 3, pp. 461–467]

[11] *History of the Theory of Numbers*, vols. 1–3, Chelsea, New York, 1966.

[CW] *Mathematical Papers*, Chelsea, New York, vols. 1–5 (1975), vol. 6 (1983).

DICKSON, L. E., MITCHELL, H. H., VANDIVER, H. S., and WAHLIN, G. E.

[1] *Algebraic Numbers*, Chelsea, New York, 1967.

DIEUDONNÉ, J.

[1] Carl Friedrich Gauss, A bicentenary, *Southeast Asian Bull. Math.* **2** (1978), 61–70.

DIMITRAKOUDIS, D.

[1] Gauss'sche Summen mit Grössencharakter, *Math. Balkanica* **8** (1978), 75–82.

DIRICHLET, G. L.

[1] Recherches sur les diviseurs premiers d'une classe de formules du quatrième degré, *J. Reine Angew. Math.* **3** (1828), 35–69. [CW: vol. 1, pp. 63–98]

[2] Ueber eine neue Anwendung bestimmter Integrale auf die Summation endlicher oder unendlicher Reihen, *Abh. K. Preuss. Akad. Wiss.* (1835), 391–407. [CW: vol. 1, pp. 237–256]

[3] Sur l'usage des intégrales définies dans la sommation des séries finies ou infinies, *J. Reine Angew. Math.* **17** (1837), 57–67. [CW: vol. 1, pp. 257–270]

[4] Recherches sur diverses applications de l'analyse infinitésimale à la théorie des nombres, *J. Reine Angew. Math.* **21** (1840), 134–155. [CW: vol. 1, pp. 473–496]

[5] Über den biquadratischen Charakter der Zahl "Zwei", *J. Reine Angew. Math.* **57** (1860), 187–188. [CW: vol. 2, pp. 259–262]

[6] *Vorlesungen über Zahlentheorie,* 4th ed., edited and supplemented by R. Dedekind, Chelsea, New York, 1968.

[CW] *Werke*, vols. 1–2 (bound as one), Chelsea, New York, 1969.

DÖRGE, K.

[1] Zur Verteilung der quadratischen Reste, *Jber. Deutsch. Math.-Verein.* **38** (1929), 41–49.

DRMOTA, M. and SKALBA, M.

[1] Sign changes of the Thue–Morse fractal function and Dirichlet L-series, *Manuscripta Math.* **86** (1995), 519–541.

DU, D. Z. and HWANG, F. K.

[1] Existence of symmetric skew balanced starters for odd prime powers, *Proc. Am. Math. Soc.* **104** (1988), 660–667.

DUKE, W.

[1] On multiple Salié sums, *Proc. Am. Math. Soc.* **114** (1992), 623–625.

DUKE, W. and IWANIEC, H.

[1] A relation between cubic exponential and Kloosterman sums, *Contemp. Math.* **143** (1993), 255–258.

DUNTON, M.

[1] Non-trivial solutions of $ax^3 + by^3 \equiv c \pmod{p}$, *Norske Vid. Selsk. Forh.* (Trondheim) **33** (1960), 45–46.

EICHLER, M.

[1] The quadratic reciprocity law and the elementary theta function, *Glasgow Math. J.* **27** (1985), 19–30.

EISENSTEIN, G.

[1] Beiträge zur Kreistheilung, *J. Reine Angew. Math.* **27** (1844), 269–278. [CW: vol. 1, pp. 45–54]

[2] Beweis des Reciprocitätssatzes für die cubischen Reste in der Theorie der aus dritten Wurzeln der Einheit zusammengesetzten complexen Zahlen, *J. Reine Angew. Math.* **27** (1844), 289–310. [CW: vol. 1, pp. 59–80]

[3] Nachtrag zum cubischen Reciprocitätssatze für die aus dritten Wurzeln der Einheit zusammengesetzten complexen Zahlen. Criterien des cubischen Characters der Zahl 3 und ihrer Theiler, *J. Reine Angew. Math.* **28** (1844), 28–35. [CW: vol. 1, pp. 81–88]

[4] La loi de réciprocité tirée des formules de M. Gauss, sans avoir déterminé préalablement le signe du radical, *J. Reine Angew. Math.* **28** (1844), 41–43. [CW: vol. 1, pp. 114–116]

[5] Lois de réciprocité, *J. Reine Angew. Math.* **28** (1844), 53–67. [CW: vol. 1, pp. 126–140]

[6] Einfacher Beweis und Verallgemeinerung des Fundamentaltheorems für die biquadratischen Reste, *J. Reine Angew. Math.* **28** (1844), 223–245. [CW: vol. 1, pp. 141–163]

[7] Applications de l'algèbre à l'arithmétique transcendante, *J. Reine Angew. Math.* **29** (1845), 177–184. [CW: vol. 1, pp. 291–298]

[8] Beiträge zur Theorie der elliptischen Functionen. I, *J. Reine Angew. Math.* **30** (1846), 185–210. [CW: vol. 1, pp. 299–324]

[9] Beiträge zur Theorie der elliptischen Functionen. VI, *J. Reine Angew. Math.* **35** (1847), 153–274. [CW: vol. 1, pp. 357–478]

[10] Zur Theorie der quadratischen Zerfällung der Primzahlen $8n+3$, $7n+2$ und $7n+4$, *J. Reine Angew. Math.* **37** (1848), 97–126. [CW: vol. 2, pp. 506–535]

[11] Über die Irreductibilität und einige andere Eigenschaften der Gleichung, von welcher die Theilung der ganzen Lemniscate abhängt, *J. Reine Angew. Math.* **39** (1850), 160–179. [CW: vol. 2, pp. 536–555]

[12] Beweis der allgemeinsten Reciprocitätsgesetze zwischen reelen und complexen Zahlen, *Monatsb. Deutsch. Akad. Wiss. Berlin* (1850), 189–198. [CW: vol. 2, pp. 712–721]

[13] Über ein einfaches Mittel zur Auffindung der höheren Reciprocitätsgesetze und der mit ihnen zu verbindenden Ergänzungssätze, *J. Reine Angew. Math.* **39** (1850), 351–364. [CW: vol. 2, pp. 623–636]

[CW] *Mathematische Werke*, vols. 1–2, Chelsea, New York, 1989.

ESTERMANN, T.

[1] On the sign of the Gaussian sum, *J. London Math. Soc.* **20** (1945), 66–67.

[2] On Kloosterman's sum, *Mathematika* **8** (1961), 83–86.

EULER, L.

[CW] *Opera Omnia*, Ser. 1, vols. 1–29, Teubner, Leipzig, 1911–56.

EVANS, R. J.

[1] Generalizations of a theorem of Chowla on Gaussian sums, *Houston J. Math.* **3** (1977), 343–349.

[2] Resolution of sign ambiguities in Jacobi and Jacobsthal sums, *Pacific J. Math.* **81** (1979), 71–80.

[3] Unambiguous evaluations of bidecic Jacobi and Jacobsthal sums, *J. Austral. Math. Soc.* **28** (1979), 235–240.

[4] Bioctic Gauss sums and sixteenth power residue difference sets, *Acta Arith.* **38** (1980), 37–46.

[5] The 2^r-th power character of 2, *J. Reine Angew. Math.* **315** (1980), 174–189.

[6] Tables of cyclotomic numbers of order twenty-four, *Math. Comp.* **35** (1980), 1036–1038.

[7] Identities for products of Gauss sums over finite fields, *Enseignement Math.* **27** (1981), 197–209.

[8] Pure Gauss sums over finite fields, *Mathematika* **28** (1981), 239–248.

[9] Rational reciprocity laws, *Acta Arith.* **39** (1981), 281–294.

[10] Period polynomials for generalized cyclotomic periods, *Manuscripta Math.* **40** (1982), 217–243.

[11] Twenty-fourth power residue difference sets, *Math. Comp.* **40** (1983), 677–683.

[12] The octic period polynomial, *Proc. Am. Math. Soc.* **87** (1983), 389–393.

[13] Determinations of Jacobsthal sums, *Pacific J. Math.* **110** (1984), 49–58.

[14] The octic and bioctic character of certain quadratic units, *Utilitas Math.* **25** (1984), 153–157.

[15] Congruences for minimal polynomials of Gauss sums, *Bull. Instit. Math. Acad. Sinica* **14** (1986), 169–185.

[16] Hermite character sums, *Pacific J. Math.* **122** (1986), 357–390.

[17] Residuacity of primes, *Rocky Mountain J. Math.* **19** (1989), 1069–1081.

[18] Character sums over finite fields, in *Finite Fields, Coding Theory, and Advances in Communications and Computing* (Las Vegas, 1991), Lecture Notes in Pure and Applied Mathematics, No. 141, Marcel Dekker, New York, 1993, pp. 57–73.

[19] Character sums as orthogonal eigenfunctions of adjacency operators for Cayley graphs, *Contemp. Math.* **168** (1994), 33–50.

[20] Selberg–Jack character sums of dimension 2, *J. Number Theory* **54** (1995), 1–11.

[21] Congruences for Jacobi sums, *J. Number Theory*, to appear.

EVANS, R. J. and HILL, J. R.

[1] The cyclotomic numbers of order sixteen, *Math. Comp.* **33** (1979), 827–835.

EVANS, R. J., PULHAM, J., and SHEEHAN, J.

[1] On the number of complete subgraphs contained in certain graphs, *J. Combin. Theory Ser. B* **30** (1981), 364–371.

FABRYKOWSKI, J.

[1] On quadratic residues and nonresidues in difference sets modulo m, *Proc. Am. Math. Soc.* **122** (1994), 325–331.

FARSI, C. and WATLING, N.

[1] Quartic algebras, *Can. J. Math.* **44** (1992), 1167–1191.

[2] Cubic algebras, *J. Operator Theory* **30** (1993), 243–266.

FISHER, B.

[1] Distinctness of Kloosterman sums, *Contemp. Math.* **133** (1992), 81–102.

[2] Kloosterman sums as algebraic integers, *Math. Ann.* **301** (1995), 485–505.

FRAME, J. S.

[1] A short proof of quadratic reciprocity, *Am. Math. Monthly* **85** (1978), 818–819.

FREI, G.

[1] The reciprocity law from Euler to Eisenstein, in *The Intersection of History and Mathematics,* Sci. Networks Hist. Stud. 15, Birkhäuser, Basel, 1994, pp. 67–90.

FRICKE, R.

[1] *Lehrbuch der Algebra,* vol. 1 (1924), vol. 2 (1926), vol. 3 (1928), Friedr. Vieweg und Sohn, Braunschweig.

FRIESEN, C., MUSKAT, J. B., SPEARMAN, B. K., and WILLIAMS, K. S.

[1] Cyclotomy of order 15 over $GF(p^2)$, $p \equiv 4, 11 \pmod{15}$, *Internatl. J. Math. Math. Sci.* **9** (1986), 665–704.

FRIESEN, C., SPEARMAN, B. K., and WILLIAMS, K. S.

[1] Another proof of Eisenstein's law of cubic reciprocity and its supplement, *Rocky Mountain J. Math.* **16** (1986), 395–402.

FRÖBERG, C.–E.

[1] New results on the Kummer conjecture, *BIT* **14** (1974), 117–119.

FRÖHLICH, A.

[1] *Galois Module Structure of Algebraic Integers*, Springer-Verlag, New York, 1983.

FULTON, J.

[1] Gauss sums and solutions to simultaneous equations over $\mathrm{GF}(2^y)$, *Acta Arith.* **35** (1979), 17–24.

FUNAKURA, T.

[1] A generalization of the Chowla-Mordell theorem on Gaussian sums, *Bull. London Math. Soc.* **24** (1992), 424–430.

FURTADO GOMIDE, E.

[1] On the theorem of Artin-Weil, *Bol. Soc. Math. São Paulo* **4** (1949), 1–18. (Portuguese)

FURTWÄNGLER, P.

[1] Über das Reziprozitätsgesetz zwischen l-ten Potenzreste in algebraschen Zahlkörpern, wenn l eine ungerade Primzahl bedeutet, *Math. Ann.* **58** (1904), 1–50.

[2] Die Reziprozitätsgesetze für Potenzreste mit Primzahlexponenten in algebraischen Zahlkörpern. I, *Math. Ann.* **67** (1909), 1–31.

[3] Die Reziprozitätsgesetze für Potenzreste mit Primzahlexponenten in algebraischen Zahlkörpern. II, *Math. Ann.* **72** (1912), 346–386.

[4] Die Reziprozitätsgesetze für Potenzreste mit Primzahlexponenten in algebraischen Zahlkörpern. III, *Math. Ann.* **74** (1913), 413–429.

[5] Über die Reziprozitätsgesetze für Primzahlpotenzexponenten, *J. Reine Angew. Math.* **157** (1927), 15–25.

[6] Über die Reziprozitätsgesetze für ungerade Primzahlexponenten, *Math. Ann.* **98** (1928), 539–543.

FURUTA, Y.

[1] A reciprocity law of the power residue symbol, *J. Math. Soc. Jpn.* **10** (1958), 46–54.

GAO, S., GATHEN, J. von zur, and PANARIO, D.

[1] Gauss periods and fast exponentiation in finite fields, in *Proceedings LATIN '95*, Lecture Notes in Computer Science, No. 911, Springer–Verlag, Berlin, 1995, pp. 311–322.

GATHEN, J. von zur and SHPARLINSKI, I. E.

[1] Orders of Gauss periods in finite fields, in *Algorithms and Computations* (Cairns, 1995), Lecture Notes in Computer Science, No. 1004, Springer, Berlin, 1995, pp. 208–215.

GAUSS, C. F.

[1] *Mathematisches Tagebuch* 1796–1814, *Ostwalds Klassiker der exakten Wissenschaften*, Bd. 256, Akad. Verlagsgesell., Geest und Portig K.-G., Leipzig, 1981. [CW: vol. 10, pp. 483–574]

[2] *Disquisitiones Arithmeticae*, Fleischer, Leipzig, 1801. [CW: vol. 1, pp. 1–478] *Disquisitiones Arithmeticae*, translated from Latin into English by A. A. Clarke, Yale University Press, 1966.

Untersuchungen über Höhere Arithmetik, translated from Latin into German by H. Maser, Chelsea, New York, 1965.

Recherches Arithmétiques, translated from Latin into French by A.–C.–M. Poullet-Delisle, Blanchard, Paris, 1979.

[3] Summatio quarumdam serierum singularium, *Comment. Soc. Reg. Sci. Gottingensis* **1** (1811). [CW: vol. 2, pp. 9–45, 155–158]

[4] Theorematis fundamentalis in doctrina de residuis quadraticis, *Comment. Soc. Reg. Sci. Gottingensis* **4** (1818), 18 pp. [CW: vol. 2, pp. 47–64]

[5] Theoria residuorum biquadraticorum, Commentatio prima, *Comment. Soc. Reg. Sci. Gottingensis* **6** (1828), 28 pp. [CW: vol. 2, pp. 65–92]

[6] Theoria residuorum biquadraticorum, Commentatio secunda, *Comment. Soc. Reg. Sci. Gottingensis* **7** (1832), 56 pp. [CW: vol. 2, pp. 93–148]

[7] Disquisitionum circa aequationes puras ulterior evolutio. [CW: vol. 2, pp. 243–265]

[CW] *Werke*, Königlichen Gesellschaft der Wissenschaften, Göttingen. vol. 1 (1863), vol. 2 (1876), vol. 3 (1876), vol. 4 (1873), vol. 5 (1877), vol. 6 (1874), vol. 7 (1877), vol. 8 (1900), vol. 9 (1903), vol. 10 (1917), vol. 11 (1927), vol. 12 (1929). Reprinted by Georg Olms Verlag, Hildesheim, 1975.

GENOCCHI, A.

[1] Sulla formula sommatoria di Eulero, e sulla teorica de' residui quadratici, *Ann. Sci. Mat. Fis.* (Roma) **3** (1852), 406–436.

[2] Note sur la theorie des residus quadratiques, *Acad. Roy. Sci. Lettres Belgique* (Brussels) **25** (1854), 56 pp.

GÉRARDIN, P. and LI, W.–C.

[1] Identities on quadratic Gauss sums, *Trans. Am. Math. Soc.* **321** (1990), 159–182.

GIUDICI, R. E., MUSKAT, J. B., and ROBINSON, S. F.

[1] On the evaluation of Brewer's character sums, *Trans. Am. Math. Soc.* **171** (1972), 317–347.

GLASHAN, J. C.

[1] Quinquisection of the cyclotomic equation, *Am. J. Math.* **21** (1899), 270–275.

GOLDSCHEIDER, F.

[1] Das Reziprozitätsgesetz der 8-ten Potenzreste, *Wis. Beitr. z. Progr. d. Luisenstädischen Realgymn, Berlin*, 1899, 29 pp.

GORDON, D.

[1] The prime power conjecture is true for $n < 2,000,000$, *Electron. J. Combin.* **1** (1994), research paper 6, approximately 7 pp.

GOSSET, T.

[1] On the law of quartic reciprocity, *Mess. Math.* **41** (1911), 65–90.

GOULD, H. W.

[1] *Combinatorial Identities*, Morgantown Printing and Binding Co., W. Va., 1972.

GOUVÊA, F. Q. and YUI, N.

[1] *Arithmetic of Diagonal Hypersurfaces over Finite Fields,* Cambridge University Press, Cambridge, 1995.

GRANT, D.

[1] A proof of quintic reciprocity using the arithmetic of $y^2 = x^5 + 1/4$, *Acta Arith.* **75** (1996), 321–337.

GRANVILLE, A.

[1] Arithmetic properties of binomial coefficients (electronic paper on the Internet), http://www.math.uga.edu/~andrew/Binomial/.

GRANVILLE, A., LI, S., and SUN, Q.

[1] On the number of solutions of the equation $\sum_{i=1}^{n} x_i/d_i \equiv 0 \pmod 1$ and of diagonal equations in finite fields, *Sichuan Daxue Xuebao* **32** (1995), 243–248.

GRAS, G.

[1] Sommes de Gauss sur les corps finis, *Publ. Math. Fac. Sci. Bescançon* 1977–1978, No. 1, 71 pp.

GRAY, J.

[1] A commentary on Gauss's mathematical diary, 1796–1814, with an English translation, *Expositiones Math.* **2** (1984), 97–130.

GREENE, J.

[1] *Character Sum Analogues for Hypergeometric and Generalized Hypergeometric Functions over Finite Fields*, Ph.D. thesis, University of Minnesota, 1984.

[2] Hypergeometric functions over finite fields, *Trans. Am. Math. Soc.* **301** (1987), 77–101.

GREENE, J. and STANTON, D.

[1] The triplication formula for Gauss sums, *Aequationes Math.* **30** (1986), 134–141.

[2] A character sum evaluation and Gaussian hypergeometric series, *J. Number Theory* **23** (1986), 136–148.

GROSS, B. and KOBLITZ, N.

[1] Gauss sums and the p-adic Γ-function, *Ann. Math.* (2) **109** (1979), 569–581.

GROSSWALD, E.

[1] *Representations of Integers as Sums of Squares*, Springer–Verlag, Berlin, 1985.

GUINAND, A. P.

[1] Gauss sums and primitive characters, *Quart. J. Math.* (Oxford) **16** (1945), 59–63.

GURAK, S. J.

[1] Relative norms of Gauss sums, *J. Number Theory* **25** (1987), 353–359.

[2] Factors of period polynomials for finite fields, I, *Contemp. Math.* **166** (1994), 309–333.

[3] Factors of period polynomials for finite fields, II, *Contemp. Math.* **168** (1994), 127–138.

[4] On the last factor of the period polynomial for finite fields, *Acta Arith.* **71** (1995), 391–400.

[5] On the minimal polynomials for certain Gauss periods over finite fields, in *Finite Fields and Applications* (Glasgow, 1995), London Math. Soc. Lecture Note Series, No. 233, S. Cohen and H. Niederreiter, eds., Cambridge University Press, Cambridge, 1996, pp. 85–96.

[6] On the middle factor of the period polynomial for finite fields, preprint, 1996.

GUY, R. K.

[1] The primes 1093 and 3511, *Math. Student* **35** (1967), 204–206.

GYOJA, A. and KAWANAKA, N.

[1] Gauss sums of prehomogeneous vector spaces, *Proc. Jpn. Acad. Ser. A* **61** (1985), 19–22.

GYOJA, A. and ONO, T.

[1] A note on Jacobi sums. II, *Proc. Jpn. Acad. Ser. A* **69** (1993), 91–93.

HABICHT, W.

[1] Ein elementarer Beweis des kubischen Reziprozitätsgesetzes, *Math. Ann.* **139** (1960), 343–365.

HADANO, T.

[1] Conductor of elliptic curves with complex multiplication and elliptic curves of prime conductor, *Proc. Jpn. Acad. Ser. A* **51** (1975), 92–95.

HALBERSTADT, E.

[1] Sommes de Gauss, symbole de Jacobi, *Revue de Mathématiques Spéciales* **103** (1992), 137–143.

HALL, M., JR.

[1] *Combinatorial Theory,* 2nd ed., Wiley, New York, 1986.

HALTER-KOCH, F.

[1] Kriterien zum 8. Potenzcharakter der Reste 3, 5 und 7, *Math. Nachr.* **44** (1970), 129–144.

HAN, W. B.

[1] On counting absolute trace of powers in $GF(p^m)$, *Acta Math. Sinica (N.S.)* **4** (1988), 266–269.

HANNAY, J. H. and BERRY, M. V.

[1] Quantization of linear maps on a torus–Fresnel diffraction by a periodic grating, *Physica* D 1 (1980), 267–290.

HARDY, K. and WILLIAMS, K. S.

[1] Table of primitive roots of primes less than 50,000, *Carleton Mathematical Series*, No. 207, Carleton University, Ottawa, June 1984, 86 pp.

[2] Table of solutions of the diophantine system

$72p = 2x_1^2 + 42(x_2^2 + x_3^2 + x_4^2) + 343(x_5^2 + 3x_6^2)$,

$12x_2^2 - 12x_4^2 + 147x_5^2 - 441x_6^2 + 56x_1x_6 + 24x_2x_3 - 24x_2x_4 + 48x_3x_4 + 98x_5x_6 = 0$,

$12x_3^2 - 12x_4^2 + 49x_5^2 - 147x_6^2 + 28x_1x_5 + 28x_1x_6 + 48x_2x_3 + 24x_2x_4 + 24x_3x_4 + 490x_5x_6 = 0$, for primes $p \equiv 1 \pmod 7$, $p \leq 49,939$, *Carleton Mathematical Series*, No. 208, Carleton University, Ottawa, June 1984, 47 pp.

[3] Evaluation of the infinite series $\sum_{\substack{n=1 \\ (\frac{n}{p})=1}}^{\infty} \left(\frac{n}{p}\right)_4 n^{-1}$, *Proc. Am. Math. Soc.* **109** (1990), 597–603.

HASSE, H.

[1] Der 2^n–te Potenzcharakter von 2 im Körper der 2^n–ten Einheitswurzeln, *Rend. Circ. Mat. Palermo* (2) **7** (1958), 185–244.

[2] *Vorlesungen über Zahlentheorie*, zweite Aufl., Springer–Verlag, Berlin, 1964.

[3] *Bericht über neuere Untersuchungen und Probleme aus der Theorie der algebraischen Zahlkörper*, dritte Aufl., Physica–Verlag, Würzburg, 1970.

HAYASHI, Harry S.

[1] The number of solutions of certain quintic congruences, *Duke Math. J.* **33** (1966), 747–756.

HAYASHI, Heima

[1] On a simple proof of Eisenstein's reciprocity law, *Mem. Fac. Sci. Kyushu Univ. Ser. A.* **28** (1974), 93–99.

HAYES, D. R.

[1] Hecke characters and Eisenstein reciprocity in function fields, *J. Number Theory* **43** (1993), 251–292.

[2] The conductors of Eisenstein characters in cyclotomic number fields, *Finite Fields Appl.* **1** (1995), 278–296.

HEATH–BROWN, D. R. and PATTERSON, S. J.

[1] The distribution of Kummer sums at prime arguments, *J. Reine Angew. Math.* **310** (1979), 111–130.

HECKE, E.

[1] Eine neue Art von Zetafunktionen und ihre Beziehungen zur Verteilung der Primzahlen II, *Math. Z.* **6** (1920), 11–51.

[2] *Lectures on the Theory of Algebraic Numbers,* translated by G. Brauer, J. Goldman, and R. Kotzen, Springer–Verlag, New York, 1981.

[CW] *Mathematische Werke*, 3rd ed., Vandenhoeck & Ruprecht, Göttingen, 1983.

HELOU, C.

[1] An explicit 2^nth reciprocity law, *J. Reine Angew. Math.* **389** (1988), 64–89.

[2] On rational reciprocity, *Proc. Am. Math. Soc.* **108** (1990), 861–866.

[3] Quintic reciprocity, *Proc. Am. Math. Soc.* **117** (1993), 877–884.

[4] Reciprocity for Jacobi sums, *J. Number Theory* **44** (1993), 214–221.

[5] Power residue character of Jacobi sums, *J. Number Theory* **49** (1994), 107–117.

HELVERSEN–PASOTTO, A.

[1] L'identité de Barnes pour les corps finis, *C. R. Acad. Sci. Paris* **286** (1978), 297–300.

[2] Über Identitäten von Charaktersummen, *Séminaire Lotharingien de Combinatoire* (Saint–Nabor, 1992), Publ. Inst. Rech. Math. Av., No. 498, Univ. Louis Pasteur, Strasbourg, 1992, pp. 65–71.

[3] Character sum identities in analogy with special functions identities, *Advances Math.* **128** (1997), 186-189.

HILBERT, D.

[1] Die Theorie der algebraischen Zahlkörper, *Jber. Deutsch. Math.-Verein.* **4** (1897), 175–546. [CW: vol. 1, pp. 63–363]

[CW] *Gesammelte Abhandlungen*, vols. 1–3, Chelsea, New York, 1981.

HIRAMATSU, T. and ISHII, N.

[1] Quartic residuacity and cusp forms of weight one, *Comment. Math. Univ. St. Pauli* **34** (1985), 91–103.

HOPF, H.

[1] Über die Verteilung quadratischer Reste, *Math. Z.* **32** (1930), 222–231.

HUA, L.-K.

[1] On exponential sums, *Sci. Record* (new series) **1** (1957), 1–4. [CW: pp. 277–280]

[2] *Introduction to Number Theory,* Springer–Verlag, New York, 1982.

[CW] *Selected Papers*, Springer–Verlag, New York, 1983.

HUA, L.-K. and VANDIVER, H. S.

[1] On the existence of solutions of certain equations in a finite field, *Proc. Natl. Acad. Sci.* (USA) **34** (1948), 258–263. [CW (Hua): pp. 454–495]

[2] Characters over certain types of rings with application to the theory of equations in a finite field, *Proc. Natl. Acad. Sci.* (USA) **35** (1949), 94–99. [CW (Hua): pp. 460–465]

[3] On the number of solutions of some trinomial equations in a finite field, *Proc. Natl. Acad. Sci.* (USA) **35** (1949), 477–481. [CW (Hua): pp. 471–475]

[4] On the nature of the solutions of certain equations in a finite field, *Proc. Natl. Acad. Sci.* (USA) **35** (1949), 481–487. [CW (Hua): pp. 477–483]

HUARD, J. G., SPEARMAN, B. K., and WILLIAMS, K. S.

[1] An arithmetic approach to the Davenport–Hasse relation over $GF(p)$, *Rocky Mountain J. Math.* **25** (1995), 1341–1350.

[2] Integral bases for quartic fields with quadratic subfields, *J. Number Theory* **51** (1995), 87–102.

HUDSON, R. H.

[1] Diophantine determinations of $3^{(p-1)/8}$ and $5^{(p-1)/8}$, *Pacific J. Math.* **111** (1984), 49–55.

[2] Products and sums of powers of binomial coefficients $(\text{mod } p)$ and solutions of certain quaternary diophantine systems, *Math. Comp.* **43** (1984), 603–613.

HUDSON, R. H. and WILLIAMS, K. S.

[1] Some new residuacity criteria, *Pacific J. Math.* **91** (1980), 135–143.

[2] A new criterion for 7 to be a fourth power (mod p), *Israel J. Math.* **38** (1981), 221–230.

[3] Resolution of ambiguities in the evaluation of cubic and quartic Jacobsthal sums, *Pacific J. Math.* **99** (1982), 379–386.

[4] An application of a formula of Western to the evaluation of certain Jacobsthal sums, *Acta Arith.* **51** (1982), 261–276.

[5] Cauchy-type congruences for binomial coefficients, *Proc. Am. Math. Soc.* **85** (1982), 169–174.

[6] A new formulation of the law of octic reciprocity for primes $\equiv \pm 3 \pmod 8$ and its consequences, *Internatl. J. Math. Math. Sci.* **5** (1982), 565–584.

[7] Congruences for representations of primes by binary quadratic forms, *Acta Arith.* **41** (1982), 311–322.

[8] Extensions of theorems of Cunningham–Aigner and Hasse–Evans, *Pacific J. Math.* **104** (1983), 111–132.

[9] Binomial coefficients and Jacobi sums, *Trans. Am. Math. Soc.* **281** (1984), 431–505.

HULL, R.

[1] The numbers of solutions of congruences involving only kth powers, *Trans. Am. Math. Soc.* **34** (1932), 908–937.

HURT, N.

[1] Kloosterman sums and their applications: a review, *Results Math.* **29** (1996), 16–41.

HURWITZ, A.

[1] Über die Kongruenz $ax^e + by^e + cz^e \equiv 0 \pmod p$, *J. Reine Angew. Math.* **136** (1909), 272–292. [CW: vol. 2, pp. 430–445]

[CW] *Mathematische Werke*, Birkhäuser, Basel, vol. 1 (1962), vol. 2 (1963).

IHARA, Y.

[1] Profinite braid groups, Galois representations, and complex multiplications, *Ann. Math.* **123** (1986), 43–106.

IRELAND, K. and ROSEN, M. I.

[1] *A Classical Introduction to Modern Number Theory*, 2nd ed., Springer–Verlag, New York, 1990.

ISHIBASHI, H.

[1] The Terwilliger algebras of certain association schemes over the Galois rings of characteristic 4, *Graphs and Combin.* **12** (1996), 39–54.

ISHIBASHI, M., SATO, H., and SHIRATANI, K.

[1] On the Hasse invariants of elliptic curves, *Kyushu J. Math.* **48** (1994), 307–321.

ISHII, N.

[1] Cusp forms of weight one, quartic reciprocity and elliptic curves, *Nagoya Math. J.* **98** (1985), 117–137.

[2] On the quartic residue symbol of totally positive quadratic units, *Tokyo J. Math.* **9** (1986), 53–65.

[3] Quadratic units and congruences between Hilbert modular forms, in *Investigations in Number Theory*, T. Kubota, ed., Advanced Studies in Pure Mathematics, vol. 13, Academic Press, Boston, 1988, pp. 261–275.

[4] On the eighth power residue of totally positive quadratic units, in *Investigations in Number Theory*, T. Kubota, ed., Advanced Studies in Pure Mathematics, vol. 13, Academic Press, Boston, 1988, pp. 413–431.

ISHIKAWA, T.

[1] On the number of zeros of diagonal forms, *Math. Comp.* **64** (1995), 841–854.

[2] Some congruences for Fourier coefficients of eta products, *Math. Japonica* **42** (1995), 297–300.

ITO, H.

[1] On a problem of Yamamoto concerning biquadratic Gauss sums, *Proc. Jpn. Acad. Ser. A* **63** (1987), 35–38.

[2] On a problem of Yamamoto concerning biquadratic Gauss sums II, *J. Math. Soc. Jpn.* **40** (1988), 647–660.

IWASAWA, K.

[1] A note on Jacobi sums, in *Istituto Nazionale di Acta Mathematica,* Symposia Mathematica, vol. 15, Academic Press, London, 1975, pp. 447–459.

JACOBI, C. G. J.

[1] Brief an Gauss vom 8. Februar 1827. [CW: vol. 7, pp. 393–400]

[2] De residuis cubicis commentatio numerosa, *J. Reine Angew. Math.* **2** (1827), 66–69. [CW: vol. 6, pp. 233–237]

[3] Über die Kreistheilung und ihre Anwendung auf die Zahlentheorie, *Monatsber. Königl. Akad. Wiss. Berlin* (1837), 127–136. (Same paper as *J. Reine Angew. Math.* **30** (1846), 166–182.) [CW: vol. 6, pp. 254–274]

[CW] *Gesammelte Werke*, vols. 1–8, Chelsea, New York, 1969.

JACOBSTHAL, E.

[1] *Anwendungen einer Formel aus der Theorie der quadratischen Reste*, dissertation, Berlin, 1906.

[2] Über die Darstellung der Primzahlen der Form $4n+1$ als Summe zweier Quadrate, *J. Reine Angew. Math.* **132** (1907), 238–245.

JAGER, H.

[1] The average order of Gaussian sums, in *Studies in Pure Mathematics: To the Memory of Paul Turán*, P. Erdös, ed., Birkhäuser, Basel and Boston, 1983, pp. 381–384.

JAKUBEC, S.

[1] The congruence for Gauss period, *J. Number Theory* **48** (1994), 36–45.

[2] Note on the Jacobi sum $J(\chi, \chi)$, *J. Théorie Nombres Bordeaux* **7** (1995), 461–477.

[3] Criterion for 3 to be eleventh power, *Acta Mathematica et Informatica Universitatis Ostraviensis* **3** (1995), 37–43.

[4] An elementary proof of the Davenport–Hasse relation, preprint.

JOLY, J.–R.

[1] Équations et variétés algébriques sur un corps fini, *Enseignement Math.* **19** (1973), 1–117.

JORDAN, C.

[1] Sur les congruences du second degré, *C. R. Acad. Sci. Paris* **62** (1866), 687–690. [CW: vol. 3, pp. 363–365]

[CW] *Oeuvres*, Gauthier-Villars, Paris, vol. 1 (1961), vol. 2 (1961), vol. 3 (1962), vol. 4 (1964).

JORIS, H.

[1] On the evaluation of Gaussian sums for non-primitive Dirichlet characters, *Enseignement Math.* **23** (1977), 13–18.

JOUX, A. and MORAIN, F.

[1] Sur les sommes de caractères liées aux courbes elliptiques à multiplication complexe, *J. Number Theory* **55** (1995), 108–128.

JUNGNICKEL, D.

[1] Difference sets, in *Contemporary Design Theory, a Collection of Surveys*, J. Dinitz and D. Stinson, eds., Wiley, New York, 1992, pp. 241–324.

[2] *Finite Fields*, Wissenschaftsverlag, Mannheim, 1993.

KAPLAN, P.

[1] Démonstration des lois de réciprocité quadratique et biquadratique, *J. Fac. Sci. Tokyo Sect. I* **16** (1969), 115–145.

KARATSUBA, A.

[1] Distribution of pairs of residues and nonresidues of special form, *Izv. Akad. Nauk AAAR Ser. Mat.* **51** (1987), 994–1009, 1117–1118; translation in *Math. USSR-Izv.* **31** (1988), 307–323.

KATRE, S. A.

[1] Cyclotomic numbers and a conjecture of Snapper, *Indian J. Pure Appl. Math.* **20** (1989), 99–103.

KATRE, S. A. and RAJWADE, A. R.

[1] On the Jacobsthal sum $\Phi_9(a)$ and the related sum $\Psi_9(a)$, *Math. Scand.* **53** (1983), 193–202.

[2] Euler's criterion for quintic nonresidues, *Can. J. Math.* **37** (1985), 1008–1024.

[3] Unique determination of cyclotomic numbers of order five, *Manuscripta Math.* **53** (1985), 65–75.

[4] Complete solution of the cyclotomic problem in F_q for any prime modulus l, $q = p^\alpha$, $p \equiv 1 \pmod{l}$, *Acta Arith.* **45** (1985), 183–199.

[5] On the Jacobsthal sum $\Phi_4(a)$ and the related sum $\Psi_8(a)$, *Ann. Univ. Sci. Budapest Eötvös Sect. Math.* **29** (1986), 3–7.

[6] Jacobsthal sums of prime order, *Indian J. Pure Appl. Math.* **17** (1986), 1345–1362.

[7] Resolution of the sign ambiguity in the determination of the cyclotomic numbers of order 4 and the corresponding Jacobsthal sum, *Math. Scand.* **60** (1987), 52–62.

KATZ, N. M.

[1] *Gauss sums, Kloosterman Sums, and Monodromy Groups,* Annals of Math. Studies, No. 116, Princeton University Press, Princeton, 1988.

KATZ, N. M. and ZHENG, Z.

[1] On the uniform distribution of Gauss sums and Jacobi sums, in *Analytic Number Theory,* vol. 2, B. C. Berndt, H. G. Diamond, and A. J. Hildebrand, eds., Birkhäuser, Boston, 1996, pp. 537–558.

KESAVA MENON, P.

[1] On Gauss's sums, *J. Indian Math. Soc.* **16** (1952), 31–36.

KIDA, M. and ONO, T.

[1] A note on Jacobi sums, *Proc. Jpn. Acad. Ser. A.* **69** (1993), 32–34.

KIM, D. S.

[1] Gauss sums for classical groups over a finite field, in *Number Theory, Geometry and Related Topics* (Iksan City, 1995), Pyungsan Inst. Math. Sci., Seoul, 1996, pp. 97–101.

KLOOSTERMAN, H. D.

[1] On the representation of numbers in the form $ax^2 + by^2 + cz^2 + dt^2$, *Acta Math.* **49** (1926), 407–464.

KNUTH, D. E.

[1] *The Art of Computer Programming,* vols. 1–3, Addison-Wesley, Reading, Mass., 1973, 1981, 1989.

KOBLITZ, N.

[1] *p-adic Numbers, p-adic Analysis, and Zeta-Functions,* Springer–Verlag, New York, 1977.

[2] The number of points on certain families of hypersurfaces over finite fields, *Compositio Math.* **48** (1983), 3–23.

[3] Jacobi sums, irreducible zeta-polynomials, and cryptography, *Can. Math. Bull.* **34** (1991), 229–235.

KOIKE, M.

[1] Hypergeometric series over finite fields and Apéry numbers, *Hiroshima Math. J.* **22** (1992), 461–467.

KONDO, T.

[1] On Gaussian sums attached to the general linear groups over finite fields, *J. Math. Soc. Jpn.* **15** (1963), 244–255.

KONYAGIN, S. V.

[1] Estimates for Gaussian sums and Waring's problem modulo a prime, *Trudy Mat. Inst. Steklov* **198** (1992), 111–124; translation in *Proc. Steklov Inst. Math.* 1994, 105–117.

KOROBOV, N. M.

[1] *Exponential Sums and Their Applications,* translated by Yu. Shakhov, Mathematics and its Applications (Soviet series) 80, Kluwer Academic Publishers, Dordrecht, 1992.

KOSCHMIEDER, L.

[1] Beweis des kubischen Reziprozitätsgesetzes mit Hilfe der elliptischen Funktionen, *Math. Ann.* **83** (1921), 280–285.

KRÄTZEL, E.

[1] Kubische und biquadratische Gausssche Summen, *J. Reine Angew. Math.* **228** (1967), 159–165.

KRAZER, A.

[1] *Lehrbuch der Thetafunktionen*, Chelsea, New York, 1970.

KRONECKER, L.

[1] Sur une formule de Gauss, *J. Math. Pures Appl.* **21** (1856), 392–395. [CW: vol. 4, pp. 171–175]

[2] Über den vierten Gauss'schen Beweis des Reciprocitätsgesetzes für die quadratischen Reste, *Monatsber. K. Preuss. Akad. Wiss. Berlin* (1880), 686–698, 854–860. [CW: vol. 4, pp. 275–294]

[3] Summirung der Gauss'schen Reihen $\sum_{h=0}^{h=n-1} e^{2h^2\pi i/n}$, *J. Reine Angew. Math.* **105** (1889), 267–268. [CW: vol. 4, pp. 295–300]

[4] Über die Dirichletsche Methode der Wertbestimmung der Gauss'schen Reihen, *Mitt. Math. Gesellsch. Hamburg* **2** (1890), 32–36. [CW: vol. 4, pp. 301–307]

[CW] *Werke*, vols. 1–5, Chelsea, New York, 1968.

KUANG, J. H.

[1] A note on Weil representation of $SL_2(F_q)$ and Soto–Andrade sums, *Contemp. Math.* **168** (1994), 207–212.

KUBERT, D. S.

[1] Jacobi sums and Hecke characters, *Am. J. Math.* **107** (1985), 253–280.

KUBERT, D. S. and LICHTENBAUM, S.

[1] Jacobi sum Hecke characters and Gauss-sum identities, *Compositio Math.* **48** (1983), 55–87.

KUBOTA, T.

[1] On a special kind of Dirichlet series, *J. Math. Soc. Jpn.* **20** (1968), 193–207.

[2] Some results concerning reciprocity and functional analysis, *Actes, Congrès Intern. Math.* 1970, vol. I, Gauthier–Villars, Paris, 1971, pp. 395–399.

[3] Some results concerning reciprocity law and real analytic automorphic functions, Proc. Sympos. Pure Math., vol. 20, American Mathematical Society, Providence, 1971, pp. 382–395.

KUMMER, E. E.

[1] Eine Aufgabe, betreffend die Theorie der cubischen Reste, *J. Reine Angew. Math.* **23** (1842), 285–286. [CW: vol. 1, pp. 143–144]

[2] De residuis cubicis disquisitiones nonnullae analyticae, *J. Reine Angew. Math.* **32** (1846), 341–359. [CW: vol. 1, pp. 145–163]

[3] Ueber die Divisoren gewisser Formen der Zahlen, welche aus der Theorie der Kreisteilung entstehen, *J. Reine Angew. Math.* **30** (1846), 107–116. [CW: vol. 1, pp. 193–202]

[4] Allgemeine Reciprocitätsgesetze für beliebig hohe Potenzreste, *Monatsber. Königl. Akad. Wiss. Berlin* (1850), 154–165. [CW: vol. 1, pp. 345–357]

[5] Über die Ergänzungssätze zu den allgemeinen Reciprocitätsgesetzen, *J. Reine Angew. Math.* **44** (1852), 93–146. [CW: vol. 1, pp. 485–538]

[6] Theorie der idealen Primfaktoren der complexen Zahlen, welche aus den Wurzeln der Gleichung $w^n = 1$ gebildet sind, wenn n eine zusammengesetzte Zahl ist, *Math. Abh. Akad. Wiss. Berlin* (1856), 1–47. [CW: vol. 1, pp. 583–629]

[7] Über die allgemeinen Reciprocitätsgesetze unter den Resten und Nichtresten der Potenzen, *Math. Abh. Akad. Wiss. Berlin* (1859), 19–159. [CW: vol. 1, pp. 699–839]

[CW] *Collected Papers*, vols. 1–2, Springer–Verlag, New York, 1975.

LACHAUD, G.

[1] Sommes d'Eisenstein et nombre de points de certaines courbes algébriques sur les corps finis, *C. R. Acad. Sci. Paris Sér. I Math.* **305** (1987), 729–732.

LAHTONEN, J.

[1] Examples of small hybrid sums, in *Finite Fields and Applications* (Glasgow, 1995), London Math. Soc. Lecture Note Series, No. 233, S. Cohen and H. Niederreiter, eds., Cambridge University Press, Cambridge, 1996, pp. 155–161.

LAKKIS, K.

[1] Special product relations between local Gauss sums, in *Number Theory*, vol. 2 (Budapest, 1987), Colloq. Math. Soc. Janós Bolyai , vol. 51, North–Holland, Amsterdam, 1990, pp. 787–793.

LAM, T. Y. and LEUNG, K. H.

[1] Vanishing sums of mth roots of unity in finite fields, *Finite Fields Appl.* **2** (1996), 422–438.

LAMPRECHT, E.

[1] Calculation of general Gauss sums and quadratic Gauss sums in finite rings, in *Théorie des Nombres* (Quebec, 1987), de Gruyter, Berlin, 1989, pp. 561–573.

LANDAU, E.

[1] Über das Vorzeichen der Gaussschen Summe, *Nachr. Gesell. Wiss. Göttingen Math.-Phys. Kl.* (1928), 19–20. [CW: vol. 9, pp. 39–40]

[2] *Elementary Number Theory*, 2nd ed., Chelsea, New York, 1966.

[3] *Vorlesungen über Zahlentheorie*, vols. 1–3 (bound as one), Chelsea, New York, 1969.

[CW] *Collected Works*, vols. 1–9, Thales Verlag, Essen, 1985.

LANDSBERG, G.

[1] Zur Theorie der Gaussschen Summen und der linearen Transformation der Thetafunctionen, *J. Reine Angew. Math.* **111** (1893), 234–253.

LANG, S.

[1] *Cyclotomic Fields I and II*, Springer–Verlag, New York, 1990.

LANG, S. and WEIL, A.

[1] Number of points of varieties in finite fields, *Am. J. Math.* **76** (1954), 819–827. [CW (Weil): vol. 2, pp. 165–173]

LANGEVIN, P.

[1] Some sequences with good autocorrelation properties, *Contemp. Math.* **168** (1994), 213–216.

[2] A new class of two weight codes, in *Finite Fields and Applications* (Glasgow, 1995), London Math. Soc. Lecture Note Series, No. 233, S. Cohen and H. Niederreiter, eds., Cambridge University Press, Cambridge, 1996, pp. 181–187.

[3] Calculs de certaines sommes de Gauss, *J. Number Theory* **63** (1997), 59–64.

LANGEVIN, P. and ZANOTTI, J.-P.

[1] Linear codes with balanced weight distribution, *Appl. Algebra Engrg. Comm. Comput.* **6** (1995), 299–307.

LAZARUS, A. J.

[1] Gaussian periods and units in certain cyclic fields, *Proc. Am. Math. Soc.* **115** (1992), 961–968.

[2] Cyclotomy and delta units, *Math. Comp.* **61** (1993), 295–305.

[3] The sextic period polynomial, *Bull. Austral. Math. Soc.* **49** (1994), 293–304.

LEBESGUE, V.–A.

[1] Recherches sur les nombres I, II, III, *J. Math. Pures Appl.* **2** (1837), 253–292; ibid. **3** (1838), 113–131, 132–144.

[2] Démonstration de quelque théorèmes relatifs aux résidus et aux non-résidus quadratiques, *J. Math. Pures Appl.* **7** (1842), 137–159.

[3] Suite du mémoire sur les applications du symbole $\left(\frac{a}{b}\right)$, *J. Math. Pures Appl.* **15** (1850), 215–237.

[4] Démonstration de quelques formules d'un mémoire de M. Jacobi, *J. Math. Pures Appl.* **19** (1854), 289–300.

[5] Démonstration de l'irreductibilité de l'équation aux racines primitives de l'unité, *J. Math. Pures Appl.* **24** (1859), 105–110.

[6] Note sur les congruences, *C. R. Acad. Sci. Paris* **51** (1860), 9–13.

[7] Sur une congruence du deuxième degré à plusieurs inconnues, *C. R. Acad. Sci. Paris* **62** (1866), 868–872.

LEGENDRE, A.-M.

[1] *Théorie des Nombres*, vols. 1–2, Blanchard, Paris, 1955.

LEHMER, D. H.

[1] Review of "The cyclotomic numbers of order ten" by A. L. Whiteman, *Math. Reviews* **22** (1961), 780.

[2] Incomplete Gauss sums, *Mathematika* **23** (1976), 125–135. [CW: vol. 2, pp. 772–782]

[CW] *Selected Papers*, vols. 1–3, Charles Babbage Research Centre, Winnipeg, 1981.

LEHMER, D. H. and LEHMER, E.

[1] On the cubes of Kloosterman sums, *Acta Arith.* **6** (1960), 15–22. [CW (D. H. Lehmer): vol. 2, pp. 705–712]

[2] Cyclotomy for nonsquarefree moduli, *Analytic Number Theory,* M. Knopp, ed., Lecture Notes in Mathematics, No. 899, Springer–Verlag, Berlin, 1981, pp. 276–300.

[3] Cyclotomy with short periods, *Math. Comp.* **41** (1983), 743–758.

[4] Multiple sums of cyclotomic numbers, *Utilitas Math.* **23** (1983), 223–239.

[5] The sextic period polynomial, *Pacific J. Math.* **111** (1984), 341–355.

[6] Cyclotomic resultants, *Math. Comp.* **48** (1987), 211–216.

[7] The Lehmer project, *Math. Comp.* **61** (1993), 313–317.

LEHMER, E.

[1] On the quintic character of 2, *Bull. Am. Math. Soc.* **55** (1949), 62–63.

[2] The quintic character of 2 and 3, *Duke Math. J.* **18** (1951), 11–18.

[3] On residue difference sets, *Can. J. Math.* **5** (1953), 425–432.

[4] On the number of solutions of $u^k + D \equiv w^2 \pmod p$, *Pacific J. Math.* **5** (1955), 103–118.

[5] Problem 4636 with solution by L. Carlitz, *Am. Math. Monthly* **63** (1956), 584–587.

[6] On the location of Gauss sums, *Math. Tables Aids Comput.* **10** (1956), 194–202.

[7] Criteria for cubic and quartic residuacity, *Mathematika* **5** (1958), 20–29.

[8] On Euler's criterion, *J. Austral. Math. Soc.* **1** (1959), 64–70.

[9] On Jacobi functions, *Pacific J. Math.* **10** (1960), 887–893.

[10] On the quadratic character of the Fibonacci root, *Fibonacci Quart.* **4** (1966), 135–138.

[11] Artiads characterized, *J. Math. Anal. Appl.* **15** (1966), 118–131.

[12] On the divisors of the discriminant of the period equation, *Am. J. Math.* **90** (1968), 375–379.

[13] On some special quartic reciprocity laws, *Acta Arith.* **21** (1972), 367–377.

[14] On the cubic character of quadratic units, *J. Number Theory* **5** (1973), 385–389.

[15] On the quartic character of quadratic units, *J. Reine Angew. Math.* **268/269** (1974), 294–301.

[16] Rational reciprocity laws, *Am. Math. Monthly* **85** (1978), 467–472.

[17] Cyclotomy of order twice a prime, *Rocky Mountain J. Math.* **15** (1985), 481–489.

[18] On special primes, *Pacific J. Math.* **118** (1985), 471–478.

[19] Connection between Gaussian periods and cyclic units, *Math. Comp.* **50** (1988), 535–541.

[20] An indeterminate in number theory, *J. Austral. Math. Soc.* **46** (1989), 469–472.

LEICHTLE, C., AVERBUKH, I. SH., and SCHLEICH, W. P.

[1] Multilevel quantum beats: an analytical approach, *Physical Review A* **54** (1996), 5299–5312.

LEMMERMEYER, F.

[1] Rational quartic reciprocity, *Acta Arith.* **67** (1994), 387–390.

[2] Rational quartic reciprocity II, *Acta Arith.* **80** (1997), 273–276.

[3] *Reciprocity laws: their evolution from Euler to Artin*, Springer-Verlag, to appear.

LENSTRA, A. K.

[1] Primality testing, in *Cryptology and Computational Number Theory*, C. Pomerance, ed., Proc. Sympos. Appl. Math., vol. 42, American Mathematical Society, Providence, 1990, pp. 13–25.

LEONARD, P. A., MORTIMER, B. C., and WILLIAMS, K. S.

[1] The eleventh power character of 2, *J. Reine Angew. Math.* **286/287** (1976), 213–222.

LEONARD, P. A. and WILLIAMS, K. S.

[1] The septic character of 2, 3, 5 and 7, *Pacific J. Math.* **52** (1974), 143–147.

[2] A Diophantine system of Dickson, *Rend. Accad. Naz. Lincei* **56** (1974), 145–150.

[3] The cyclotomic numbers of order seven, *Proc. Am. Math. Soc.* **51** (1975), 295–300.

[4] The cyclotomic numbers of order eleven, *Acta Arith.* **26** (1975), 367–383.

[5] Jacobi sums and a theorem of Brewer, *Rocky Mountain J. Math.* **5** (1975), 301–308; Erratum, ibid. **6** (1976), 509.

[6] A rational sixteenth power reciprocity law, *Acta Arith.* **33** (1977), 365–377.

[7] Evaluation of certain Jacobsthal sums, *Boll. Un. Mat. Ital.* **15** (1978), 717–723.

LEPRÉVOST, F. and MORAIN, F.

[1] Revêtements de courbes elliptiques à multiplication complexe par des courbes hyperelliptiques et sommes de caractères, *J. Number Theory* **64** (1997), 165–182.

LERCH, M.

[1] Zur Theorie der Gausschen Summen, *Math. Ann.* **57** (1903), 554–567.

LI, B.–H. and LI, T.–J.

[1] Generalized Gaussian sums & Chern–Simons–Witten–Jones invariants of lens-spaces, *J. Knot Theory Ramifications* **5** (1996), 183–224.

LI, W.–C.

[1] Various aspects of the Barnes' identity, in *Séminaire de Théorie des Nombres* (Paris, 1984–85), Progress in Mathematics, vol. 63, Birkhäuser, Boston, 1986.

[2] Estimates of character sums arising from finite upper half planes, in *Finite Fields and Applications* (Glasgow, 1995), London Math. Soc. Lecture Note Series, No. 233, S. Cohen and H. Niederreiter, eds., Cambridge University Press, Cambridge, 1996, pp. 219–228.

LIBRI, G.

[1] Mémoire sur la théorie des nombres, *J. Reine Angew. Math.* **9** (1832), 169–188.

[2] Mémoire sur la théorie des nombres, *J. Reine Angew. Math.* **9** (1832), 261–276.

LIDL, R., MULLEN, G. L., and TURNWALD, G.

[1] *Dickson Polynomials*, Pitman Monograph 65, Longman Sci. Tech., Harlow, UK, 1993.

LIDL, R. and NIEDERREITER, H.

[1] *Finite Fields*, Addison-Wesley, Reading, Mass., 1983 (2nd edition, Cambridge University Press, Cambridge, 1997).

LIENEN, H. von

[1] *Regeln für kubische und höhere Potenzreste*, Habilitationsschrift, Bochum, 1970.

[2] Primzahlen als achte Potenzreste, *J. Reine Angew. Math.* **266** (1974), 107–117.

LIN, C. Y.

[1] On the norm of double Gaussian sums, *Chin. J. Math.* **10** (1982), 53–60.

LINDELÖF, E.

[1] *Le Calcul des Résidus*, Chelsea, New York, 1947.

LIVERANCE, E.

[1] A formula for the root number of a family of elliptic curves, *J. Number Theory* **51** (1995), 288–305.

LLORENTE, P.

[1] On kth power residuacity, *Rev. Un. Mat. Argentina* **29** (1976/77), 60–67.

LOESER, F.

[1] Arrangements d'hyperplans et sommes de Gauss, *Ann. Sci. École Norm. Sup.* (4) **24** (1991), 379–400.

LOMONT, J. S.

[1] A generalization and refinement of Madivanane's theorem on general involutional transformations, *J. Math. Phys.* **26** (1985), 2087–2091.

LOXTON, J. H.

[1] On the determination of Gauss sums, *Sém. Delange–Pisot–Poitou* 1976/77, 12 pp.

[2] Some conjectures concerning Gauss sums, *J. Reine Angew. Math.* **297** (1978), 153–158.

[3] The graphs of exponential sums, *Mathematika* **30** (1983), 153–163.

LUCAS, E.

[1] *Théorie des Nombres*, 2nd ed., Gauthier–Villars, Paris, 1961.

MANN, H. B.

[1] *Addition Theorems*, Wiley, New York, 1965.

MASLEY, J. M. and MONTGOMERY, H. L.

[1] Cyclotomic fields with unique factorization, *J. Reine Angew. Math.* **286/287** (1976), 248–256.

MATHEWS, G. B.

[1] *Theory of Numbers*, 2nd ed., Chelsea, New York, 1961.

MATTHEWS, C. R.

[1] *Gauss Sums and Elliptic Functions*, Ph.D. thesis, University of Cambridge, 1978.

[2] Gauss sums and elliptic functions. I, The Kummer sum, *Invent. Math.* **52** (1979), 163–185.

[3] Gauss sums and elliptic functions. II, The quartic sum, *Invent. Math.* **54** (1979), 23–52.

MAUCLAIRE, J.–L.

[1] Sommes de Gauss modulo p^α I, *Proc. Jpn. Acad. Ser. A* **59** (1983), 109–112.

[2] Sommes de Gauss modulo p^α II, *Proc. Jpn. Acad. Ser. A* **59** (1983), 161–163.

McELIECE, R. J.

[1] Irreducible cyclic codes and Gauss sums, in *Combinatorics*, M. Hall Jr. and J. H. van Lint, eds., Reidel Publishing, Dordrecht–Holland, 1975, pp. 185–202.

McELIECE, R. J. and RUMSEY, H.

[1] Euler products, cyclotomy, and coding, *J. Number Theory* **4** (1972), 302–311.

McFARLAND, R. L.

[1] Difference sets in abelian groups of order $4p^2$, *Mitt. Math. Sem. Giessen* (1989), 1–70.

McGETTRICK, A. D.

[1] A result in the theory of Weierstrass elliptic functions, *Proc. London Math. Soc.* **25** (1972), 41–54.

[2] On the biquadratic Gauss sum, *Proc. Cambridge Philos. Soc.* **71** (1972), 79–83.

MEISSNER, W.

[1] Über die Teilbarkeit von 2^p-2 durch das Quadrat der Primzahl $p = 1093$, *Sitzungsberichte Akad. d. Wiss. Berlin* (1913), 663–667.

MERRILL, K. D. and WALLING, L. H.

[1] On quadratic reciprocity over function fields, *Pacific J. Math* **173** (1996), 147–150.

MERTENS, F.

[1] Über die Darstellung der Legendre'schen Symbole der biquadratischen, kubischen und bikubischen Reste durch Thetareihen, *Sitzungs. d. k. Akad. d. Wiss. Wien Abt.* II a **105** (1906), 1339–1360.

MIKI, H.

[1] On the l-adic expansion of certain Gauss sums and its applications, in *Galois Representations and Arithmetic Algebraic Geometry*, Y. Ihara, ed., Advanced Studies in Pure Mathematics, vol. 12, North–Holland, Amsterdam, 1987, pp. 87–118.

[2] On the conductor of the Jacobi sum Hecke character, *Compositio Math.* **92** (1994), 23–41.

[3] On Shioda's problem about Jacobi sums. II, *Acta Arith.* **73** (1995), 373–377.

[4] Jacobi sums and the Hilbert symbol for a power of two, *J. Math. Soc. Jpn.* **48** (1996), 367–408.

MILLER, J. C. P.

[1] Alfred Edward Western, *J. London Math. Soc.* **38** (1963), 278–281.

MILNOR, J. W. and HUSEMOLLER, D.

[1] *Symmetric Bilinear Forms*, Spring–Verlag, New York, 1973.

MITCHELL, H. H.

[1] On the generalized Jacobi–Kummer cyclotomic function, *Trans. Am. Math. Soc.* **17** (1916), 165–177.

MONTGOMERY, H. L., VAUGHAN, R. C., and WOOLEY, T. D.

[1] Some remarks on Gauss sums associated with kth powers, *Math. Proc. Cambridge Philos. Soc.* **118** (1995), 21–33.

MONZINGO, M. G.

[1] On the distribution of quadratic residues, in *A Collection of Manuscripts Related to the Fibonacci Sequence*, Fibonacci Assoc., Santa Clara, Calif., 1980, pp. 94–97.

[2] An elementary evaluation of the Jacobsthal sum, *J. Number Theory* **22** (1986), 21–25.

MOORE, R. A.

[1] *Proofs of the Law of Quadratic Reciprocity*, M.A. thesis, Wake Forest University, Winston–Salem, N. C., 1977.

MORDELL, L. J.

[1] On a simple summation of the series $\sum_{s=0}^{n-1} e^{2s^2\pi i/n}$, *Mess. Math.* **48** (1918), 54–56.

[2] The definite integral $\int_{-\infty}^{\infty} e^{ax^2+bx}/(e^{cx}+d)dx$ and the analytic theory of numbers, *Acta Math.* **61** (1933), 323–360.

[3] The number of solutions of some congruence in two variables, *Math. Z.* **37** (1933), 193–209.

[4] On Lehmer's congruence associated with cubes of Kloosterman's sums, *J. London Math. Soc.* **36** (1961), 335–339.

[5] On a cyclotomic resolvent, *Archiv Math.* **13** (1962), 486–487.

[6] The sign of the Gaussian sum, *Illinois J. Math.* **6** (1962), 177–180.

[7] Solvability of the equation $ax^2 + by^2 = p$, *J. London Math. Soc.* **41** (1966), 517–522.

MORENO, C. J.

[1] *The Eisenstein reciprocity law according to Landau*, (mimeographed notes), 1974, 38 pp.

[2] Sur le problème de Kummer, *Enseignement Math.* **20** (1974), 45–51.

[3] *Algebraic Curves over Finite Fields*, Cambridge University Press, Cambridge, 1991.

MORENO, O.

[1] Counting traces of powers over $GF(2^m)$, *Congr. Numer.* **29** (1980), 673–679.

MORITA, Y.

[1] A p-adic analogue of the Γ-function, *J. Fac. Sci. Univ. Tokyo Sect. I A Math.* **22** (1975), 255–266.

MORLAYE, B.

[1] Démonstration élémentaire d'un théorème de Davenport et Hasse, *Enseignement Math.* **18** (1972), 269–276.

MORLEY, F.

[1] Note on the congruence $2^{4n} = (-)^n(2n)!/(n!)^2$, where $2n+1$ is prime, *Ann. Math.* **9** (1895), 168–170.

MORTON, P.

[1] On the eigenvectors of Schur's matrix, *J. Number Theory* **12** (1980), 122–127.

MOSER, L.

[1] *An Introduction to the Theory of Numbers* (mimeographed lecture notes), Canadian Math. Congress, University of Alberta, 1957.

MULLEN, G. L. and SHPARLINSKI, I. E.

[1] Open problems and conjectures in finite fields, in *Finite Fields and Applications* (Glasgow, 1995), London Math. Soc. Lecture Note Series, No. 233, S. Cohen and H. Niederreiter, eds., Cambridge University Press, Cambridge, 1996, pp. 243–268.

MUSKAT, J. B.

[1] On certain prime power congruences, *Abh. Math. Sem. Univ. Hamburg* **26** (1963), 102–110.

[2] On the solvability of $x^e \equiv e \pmod{p}$, *Pacific J. Math.* **14** (1964), 257–260.

[3] Criteria for solvability of certain congruences, *Can. J. Math.* **16** (1964), 343–352.

[4] The cyclotomic numbers of order fourteen, *Acta Arith.* **11** (1966), 263–279.

[5] Reciprocity and Jacobi sums, *Pacific J. Math.* **20** (1967), 275–280.

[6] On Jacobi sums of certain composite orders, *Trans. Am. Math. Soc.* **134** (1968), 483–502.

[7] Use of computers in cyclotomy, in *Computers in Number Theory*, A. O. L. Atkin and B. J. Birch, eds., Academic Press, London, 1971, pp. 141–147.

[8] On simultaneous representations of primes by binary quadratic forms, *J. Number Theory* **19** (1984), 263–282.

MUSKAT, J. B., SPEARMAN, B. K., and WILLIAMS, K. S.

[1] Predictive criteria for the representation of primes by binary quadratic forms, *Acta Arith.* **70** (1995), 215–278.

MUSKAT, J. B. and WHITEMAN, A. L.

[1] The cyclotomic numbers of order twenty, *Acta Arith.* **17** (1970), 185–216.

MUSKAT, J. B. and ZEE, Y.–C.

[1] Sign ambiguities of Jacobi sums, *Duke Math. J.* **40** (1973), 313–334.

[2] On the uniqueness of solutions of certain Diophantine equations, *Proc. Am. Math. Soc.* **49** (1975), 13–19.

MYERSON, G.

[1] On the number of zeros of diagonal cubic forms, *J. Number Theory* **11** (1979), 95–99.

[2] Period polynomials and Gauss sums for finite fields, *Acta Arith.* **39** (1981), 251–264.

NAGELL, T.

[1] Sur quelques problèmes dans la théorie des restes quadratiques et cubiques, *Arkiv Math.* **3** (1956), 211–222.

[2] *Introduction to Number Theory*, 2nd ed., Chelsea, New York, 1964.

NAKAHARA, T.

[1] A simple proof for non-monogenesis of the rings of integers in some cyclic fields, in *Advances in Number Theory* (Kingston, 1991), Oxford University Press, New York, 1993, pp. 167–173.

NARKIEWICZ, W.

[1] The work of C. F. Gauss in algebra and number theory, *Festakt und Tagung aus Anlass des 200. Geburtstages von Carl Friedrich Gauss* (Berlin, 1977), *Abh. Akad. Wiss. DDR, Abt. Math. Naturwiss. Tech.*, 1978, **3**, Akademie–Verlag, Berlin, 1978, pp. 75–82.

[2] *Elementary and Analytic Theory of Algebraic Numbers*, Springer–Verlag, Berlin, and PWN-Polish Scientific Publishers, Warsaw, 1990.

NASHIER, B. S., and RAJWADE, A. R.

[1] Determination of a unique solution of the quadratic partition for primes $p \equiv 1 \pmod 7$, *Pacific J. Math.* **72** (1977), 513–521.

NEMCHENOK, J.

[1] Imprimitive Gaussian sums and theta functions over a number field, *Trans. Am. Math. Soc.* **338** (1993), 465–478.

NEUMANN, O.

[1] Zur Genesis der algebraischen Zahlentheorie. I., *NTM Schr. Geschichte Natur. Tech. Medizin* **16(2)** (1979), 22–39.

[2] Zur Genesis der algebraischen Zahlentheorie. II., *NTM Schr. Geschichte Natur. Tech. Medizin* **17(1)** (1980), 32–48.

[3] Zur Genesis der algebraischen Zahlentheorie. III., *NTM Schr. Geschichte Natur. Tech. Medizin* **17(2)** (1980), 38–58.

[4] Two proofs of the Kronecker–Weber theorem "according to Kronecker, and Weber", *J. Reine Angew. Math.* **323** (1981), 105–126.

NEUMANN, J. von and GOLDSTINE, H. H.

[1] A numerical study of a conjecture of Kummer, *Math. Tables and Others Aids to Computation* **7** (1953), 133–134.

NIEDERREITER, H.

[1] Exponential sums over finite fields, *Math. J. Okayama Univ.* **27** (1985), 147–158.

ODONI, R.

[1] On Gauss sums (mod p^n), $n \geq 2$, *Bull. London Math. Soc.* **5** (1973), 325–327.

O'MEARA, O. T.

[1] *Introduction to Quadratic Forms*, Springer–Verlag, New York, 1971.

ONO, K.

[1] Values of Gaussian hypergeometric series, *Trans. Am. Math. Soc.* **350** (1998), 1205–1223.

[2] Tate–Shafarevich groups of the congruent number elliptic curve, *Acta Arith.* **81** (1997), 247–252.

ONO, T.

[1] On a generalization of Jacobi sums, *J. Fac. Sci. Univ. Tokyo* **28** (1982), 823–828.

[2] A note on Jacobi sums. III, *Proc. Jpn. Acad. Ser. A* **69** (1993), 272–274.

OSKOLKOV, K.

[1] On functional properties of incomplete Gauss sums, *Can. J. Math.* **43** (1991), 182–212.

OZEKI, M.

[1] On certain generalized Gauss sums, *Proc. Jpn. Acad. Ser. A.* **58** (1982), 223–226.

PADMA, R. and VENKATARAMAN, S.

[1] Elliptic curves with complex multiplication, *J. Number Theory* **61** (1996), 274–282.

PALEY, R. E. A. C.

[1] On orthogonal matrices, *J. Math. Phys.* **12** (1933), 311–320.

PARNAMI, J. C., AGRAWAL, M. K., PALL, S., and RAJWADE, A. R.

[1] A new proof of the Leonard and Williams criterion for 3 to be a 7th power, *J. Indian Math. Soc.* **45** (1981), 129–134.

[2] Criterion for 2 to be an lth power, *Acta Arith.* **43** (1984), 361–365.

PARNAMI, J. C., AGRAWAL, M. K., and RAJWADE, A. R.

[1] A congruence relation between the coefficients of the Jacobi sum, *Indian J. Pure Appl. Math.* **12** (1981), 804–806.

[2] Jacobi sums and cyclotomic numbers for a finite field, *Acta Arith.* **41** (1982), 1–13.

[3] Some identities involving character sums and their applications, *J. Indian Math. Soc.* **54** (1989), 125–132.

[4] Some linear relations between certain binomial coefficients modulo a prime, *Bull. Soc. Math. Belg.-Tijdschr. Belg. Wisk. Gen.* **41** (1989), 339–345.

PATTERSON, S. J.

[1] A cubic analogue of the theta series, *J. Reine Angew. Math.* **296** (1977), 125–161.

[2] A cubic analogue of the theta series II, *J. Reine Angew. Math.* **296** (1977), 217–220.

[3] On Dirichlet series associated with cubic Gauss sums, *J. Reine Angew. Math.* **303/304** (1978), 102–125.

[4] On the distribution of Kummer sums, *J. Reine Angew. Math.* **303/304** (1978), 126–143.

[5] The distribution of general Gauss sums at prime arguments, in *Recent Progress in Analytic Number Theory*, vol. 2, H. Halberstam and C. Hooley, eds., Academic Press, New York, 1981, pp. 171–182.

[6] The distribution of certain special values of the cubic Legendre symbol, *Glasgow Math. J.* **27** (1985), 165–184.

[7] Tori in metaplectic covers of GL_2 and applications to a formula of Loxton–Matthews, *Math. Proc. Cambridge Philos. Soc.* **100** (1986), 249–263.

[8] Metaplectic forms and Gauss sums I, *Compositio Math.* **62** (1987), 343–366.

[9] The distribution of general Gauss sums and similar arithmetic functions at prime arguments, *Proc. London Math. Soc.* **54** (1987), 193–215.

PÉPIN, T.

[1] Mémoire sur les lois de réciprocité relatives aux résidus de puissances, *Pontif Accad. Sci.* **31** (1877), 40–148.

PERALTA, R.

[1] On the distribution of quadratic residues and nonresidues modulo a prime number, *Math. Comp.* **58** (1992), 433–440.

PEREL'MUTER, G. I.

[1] Incomplete Gauss sums in finite fields, *Diskret. Mat.* **3** (1991), 68–73.

[2] On the best polynomial nonresidue and incomplete Gaussian sums in finite fields, *Mat. Zametki* **53** (1993), 83–88, 155.

PEROTT, J.

[1] Sur l'équation $t^2 - Du^2 = -1$, *J. Reine Angew. Math.* **102** (1888), 185–223.

PERRON, O.

[1] Bemerkungen über die Verteilung der quadratischen Reste, *Math. Z.* **56** (1952), 122–130.

POLLARD, H. and DIAMOND, H. G.

[1] *The Theory of Algebraic Numbers*, 2nd ed., Mathematical Association of America, Washington, D.C., 1975.

POULAKIS, D.

[1] Évaluation d'une somme cubique de caracteres, *J. Number Theory* **27** (1987), 41–45.

RABIN, M. O. and SHALLIT, J. O.

[1] Randomized algorithms in number theory, *Commun. Pure Appl. Math.* **39** (1986), S239–S256.

RADEMACHER, H.

[1] *Lectures on Elementary Number Theory*, Blaisdell, New York, 1964.

RADOUX, C.

[1] Annulation et grandes valeurs des sommes de Brewer, *Ann. Soc. Sci. Bruxelles* **100** (1986), 43–51.

RAJWADE, A. R.

[1] Arithmetic on curves with complex multiplication by $\sqrt{-2}$, *Proc. Cambridge Philos. Soc.* **64** (1968), 659–672.

[2] Arithmetic on curves with complex multiplication by the Eisenstein integers, *Proc. Cambridge Philos. Soc.* **65** (1969), 59–73.

[3] On rational primes p congruent to 1 (mod 3 or 5), *Proc. Cambridge Philos. Soc.* **66** (1969), 61–70.

[4] A note on the number of solutions N_p of the congruence $y^2 \equiv x^3 - Dx \pmod p$, *Proc. Cambridge Philos. Soc.* **67** (1970), 603–606.

[5] The period equation for primes p congruent to 1 (mod 5), *Proc. Cambridge Philos. Soc.* **69** (1971), 153–155.

[6] On the congruence $y^2 \equiv x^5 - a \pmod p$, *Proc. Cambridge Philos. Soc.* **74** (1973), 473–475.

[7] The number of solutions of the congruence $y^2 \equiv x^6 - a \pmod p$, *Indian J. Pure Appl. Math.* **4** (1973), 325–332.

[8] Certain classical congruences via elliptic curves, *J. London Math. Soc.* **8** (1974), 60–62.

[9] Notes on the congruence $y^2 \equiv x^5 - a (\mathrm{mod}\ p)$, *Enseignement Math.* **21** (1975), 49–56.

[10] Some congruences in algebraic integers and rational integers, *Indian J. Pure Appl. Math.* **7** (1976), 431–435.

[11] The diophantine equation $y^2 = x(x^2+21Dx+112D^2)$ and the conjectures of Birch and Swinnerton-Dyer, *J. Austral. Math. Soc.* **24** (1977), 286–295.

[12] Cyclotomy–a survey article, *Math. Student* **48** (1980), 70–111.

[13] On a conjecture of Williams, *Bull. Soc. Math. Belg.-Tijdschr. Belg. Wisk. Gen.* **36** (1984), 1–4.

RAJWADE, A. R. and PARNAMI, J. C.

[1] A new cubic character sum, *Acta Arith.* **40** (1982), 347–356.

RANKIN, R. A.

[1] Generalized Jacobsthal sums and sums of squares, *Acta Arith.* **49** (1987), 5–14.

RETI, Z.

[1] Multiple Gauss sums, to appear.

REUSCHLE, C. G.

[1] Mathematische Abhandlung, enthaltend neue Zahlentheoretische Tabellen, *Programm zum Schlusse des Schuljahrs 1855–56 am Königlichen Gymnasium zu Stuttgart,* 1856, 61 pp.

RICHMAN, D.

[1] Some remarks on the number of solutions to the equation $f(x_1) + \cdots + f(x_n) = 0$, *Stud. Appl. Math.* **71** (1984), 263–266.

RISHI, D. B., PARNAMI, J. C., and RAJWADE, A. R.

[1] Evaluation of a cubic character sum using the $\sqrt{-19}$ division points of the curve $y^2 = x^3 - 2^3 \cdot 19x + 2 \cdot 19^2$, *J. Number Theory* **19** (1984), 184–194.

ROBINSON, R. M.

[1] Numbers having m small mth roots $\pmod p$, *Math. Comp.* **61** (1993), 393–413.

ROBINSON, S. F.

[1] Theorems on Brewer sums, *Pacific J. Math.* **25** (1968), 587–596.

ROUSSEAU, G.

[1] On the quadratic reciprocity law, *J. Austral. Math. Soc. Ser. A* **51** (1991), 423–425.

[2] On the Jacobi symbol, *J. Number Theory* **48** (1994), 109–111.

ROWE, D. E.

[1] Gauss, Dirichlet, and the law of quadratic reciprocity, *Math. Intell.* **10**(2) (1988), 13–25.

RUBIN, K.

[1] Kolyvagin's system of Gauss sums, in *Arithmetic Algebraic Geometry* (Texel, 1989), Progress in Mathematics, vol. 89, Birkhäuser, Boston, 1991, pp. 309–324.

RYSER, H. J.

[1] *Combinatorial Mathematics*, Carus monographs, no. 14, Mathematical Association of America, Wiley, New York, 1963.

SAITO, H.

[1] A generalization of Gauss sum and its applications to Siegel modular forms and L-functions associated with the vector space of quadratic forms, *J. Reine Angew. Math.* **416** (1991), 91–142.

SALIÉ, H.

[1] Über die Kloostermanschen Summen $S(u, v; q)$, *Math. Z.* **34** (1932), 91–109.

[2] Zur Abschätzung der Fourierkoeffizienten ganzer Modulformen, *Math. Z.* **36** (1933), 263–278.

[3] Über die Verteilung der quadratischen Reste, *Math. Z.* **37** (1933), 594–602.

SANDS, J.

[1] Abelian fields and the Brumer–Stark conjecture, *Compositio Math.* **53** (1984), 337–346.

SARNAK, P.

[1] *Some Applications of Modular Forms*, Cambridge University Press, Cambridge, 1990.

SCHAAR, L.

[1] Mémoires sur les intégrales Eulériennes et sur la convergence d'une certaine classe de séries, *Mémoires couronnés et mémoires des savants étrangers, l'Académie Royale des Sciences, des Lettres et des Beaux-Arts de Belgique* **22** (1848), 1–16.

SCHARLAU, W.

[1] *Quadratic and Hermitian Forms,* Springer–Verlag, New York, 1985.

SCHMID, H. L.

[1] Relationen zwischen verallgemeinerten Gaussschen Summen, *J. Reine Angew. Math.* **176** (1937), 189–191.

SCHMIDT, C.-G.

[1] Gauss sums and the classical gamma function, *Bull. London Math. Soc.* **12** (1980), 344–346.

SCHMIDT, W. M.

[1] *Equations over Finite Fields: An Elementary Approach,* Lecture Notes in Mathematics, No. 536, Springer–Verlag, Berlin, 1976,

SCHOOF, R.

[1] Elliptic curves over finite fields and the computation of square roots mod p, *Math. Comp.* **44** (1985), 483–494.

SCHOOF, R. and WASHINGTON, L.

[1] Quintic polynomials and real cyclotomic fields with large class numbers, *Math. Comp.* **50** (1988), 543–556.

SCHROEDER, M. R.

[1] *Number Theory in Science and Communication*, 2nd ed., Springer–Verlag, Berlin, 1986.

SCHRUTKA, L. von

[1] Eine Beweis für die Zerlegbarkeit der Primzahlen von der Form $6n + 1$ in ein einfaches und ein dreifaches Quadrat, *J. Reine Angew. Math.* **140** (1911), 252–265.

SCHWERMER, J.

[1] Über Reziprozitätsgesetze in der Zahlentheorie, in *Arithmetik und Geometrie,* Math. Miniaturen No. 3, Birkhäuser, Basel–Boston, 1986, pp. 29–69.

SCOTT, C. A.

[1] The binomial equation $x^p - 1 = 0$, *Am. J. Math.* **8** (1886), 216–264.

SEPANSKI, M. R.

[1] Konstant's conjecture and the geometry of Cartan subalgebras in the rank two and exceptional cases, *J. Algebra* **182** (1996), 347–382.

SHANKS, D.

[1] Two theorems of Gauss, *Pacific J. Math.* **8** (1958), 609–612.

[2] Review of C.–E. Fröberg, "Kummer's Förmodan"(Lund University, 1973), *Math. Comp.* **29** (1975), 331–333.

SHIMURA, G.

[1] *Introduction to the Arithmetic Theory of Automorphic Functions,* Princeton University Press, Princeton, N.J., 1971.

SHIODA, T.

[1] Some observations on Jacobi sums, in *Galois Representations and Arithmetic Algebraic Geometry,* Y. Ihara, ed., Advanced Studies in Pure Mathematics, vol. 12, North–Holland, Amsterdam, 1987, pp. 119–135.

SHIRATANI, K.

[1] Die Diskriminante der Weierstrassschen elliptischen Funktionen und das Reziprozitätsgesetz in besonderen imaginär-quadratischen Zahlkörpern, *Abh. Math. Sem. Univ. Hamburg* **31** (1967), 51–61.

SHIRATANI, K. and YAMADA, M.

[1] On rationality of Jacobi sums, *Colloq. Math.* **73** (1997), 251–260.

SHPARLINSKI, I. E.

[1] On Gaussian sums for finite fields and elliptic curves, in *Algebraic Coding* (Paris, 1991), Lecture Notes in Computer Science, No. 573, Springer–Verlag, Berlin, 1992, pp. 5–15.

SIEGEL, C. L.

[1] Über die analytische Theorie der quadratischen Formen, *Ann. Math.* **36** (1935), 527—606. [CW: vol. 1, pp. 326–405]

[2] Über die analytische Theorie der quadratischen Formen III, *Ann. Math.* **38** (1937), 212–291. [CW: vol. I, pp. 469–548]

[3] Über das quadratische Reziprozitätsgesetz in algebraischen Zahlkörpern, *Nachr. Akad. Wiss. Göttingen Math.-Phys. Kl. II* (1960), 1–16. [CW: vol. 3, pp. 334–349]

[4] Moduln abelschen Funktionen, *Nachr. Akad. Wiss. Göttingen Math.-Phys. Kl. II* (1963), 365–427. [CW: vol. 3, pp. 373–435]

[CW] *Gesammelte Abhandlungen*, vols. 1–3 (1966), vol. 4 (1979), Springer–Verlag, Berlin.

SINGH, S. and RAJWADE, A. R.

[1] The number of solutions of the congruence $y^2 \equiv x^4 - a \pmod p$, *Enseignement Math.* **20** (1974), 265–273.

SKINNER, C. M.

[1] Solvability of $\wp$-adic diagonal equations, *Acta Arith.* **75** (1996), 251–258.

SKOLEM, T. A.

[1] Remarks on proofs by cyclotomic formulas of reciprocity laws for power residues, *Math. Scand.* **9** (1961), 229–242.

SMALL, C.

[1] *Arithmetic of Finite Fields*, Marcel Dekker, New York, 1991.

SMITH, H. J. S.

[1] *Report on the Theory of Numbers*, Chelsea, New York, 1965. [CW: vol. 1, pp. 38–364]

[CW] *Collected Mathematical Papers*, vols. 1–2, Chelsea, New York, 1979.

SMITH, R. A.

[1] On n-dimensional Kloosterman sums, *J. Number Theory* **11** (1979), 324–343.

[2] Estimates for exponential sums, *Proc. Am. Math. Soc.* **79** (1980), 365–368.

SNAPPER, E.

[1] Finite fields, integral matrices, and Diophantine equations, *J. Algebra* **97** (1985), 267–277.

SPACKMAN, K. W.

[1] Simultaneous solutions to diagonal equations over finite fields, *J. Number Theory* **11** (1979), 100–115.

[2] On the number and distribution of simultaneous solutions to diagonal congruences, *Can. J. Math.* **33** (1981), 421–436.

SPEARMAN, B. K.

[1] *Some Elementary Proofs of Certain Reciprocity Laws*, M.Sc thesis, Carleton University, Ottawa, Canada, 1976.

SPEARMAN, B. K. and WILLIAMS, K. S.

[1] A simple proof of Eisenstein's reciprocity law from Stickelberger's theorem, *Indian J. Pure Appl. Math.* **17** (1986), 169–174.

[2] Two short papers in number theory, *Carleton Mathematical Series*, No. 7, 1991, 7 pp.

SPIRIDONOV, V. and ZHEDANOV, A. S.

[1] Zeros and orthogonality of the Askey–Wilson polynomials for q a root of unity, to appear.

STARK, H. M.

[1] Dirichlet's class number formula revisited, *Contemp. Math.* **143** (1993), 571–577.

[2] Counting points on CM elliptic curves, *Rocky Mountain J. Math.* **26** (1996), 1115–1138.

STEPANOV, S. A.

[1] Proof of the Davenport–Hasse relations, *Mat. Zametki* **27** (1980), 3–4; *Math. Notes* (1980), 3–4. (Russian)

[2] Character sums and coding theory, in *Finite Fields and Applications* (Glasgow, 1995), London Math. Soc. Lecture Note Series, No. 233, S. Cohen and H. Niederreiter, eds., Cambridge University Press, Cambridge, 1996, pp. 355–378.

STERN, M.

[1] Eine Bemerkung zur Zahlentheorie, *J. Reine Angew. Math.* **32** (1846), 89–90.

STERNECK, R. von

[1] On the distribution of quadratic residues and nonresidues of a prime, *Mat. Sb.* **20** (1898), 269–284. (Russian).

STICKELBERGER, L.

[1] Ueber eine Verallgemeinerung der Kreisteilung, *Math. Ann.* **37** (1890), 321–367.

STORER, T.

[1] *Cyclotomy and Difference Sets*, Markham, Chicago, 1967.

STYER, R.

[1] Evaluating symplectic Gauss sums and Jacobi symbols, *Nagoya J. Math.* **95** (1984), 1–22.

SUBBARAO, M. V.

[1] The algebra of biquadratic residues, *J. Madras Univ. Sect. B* **30** (1960), 123–131.

SULLIVAN, F. and ZANNIER, U.

[1] A note on incomplete quadratic Gauss sums, *Boll. Un. Mat. Ital.* (7) **6-A** (1992), 49–57.

SUN, Q.

[1] On diagonal equations over finite fields, *Finite Fields Appl.* **3** (1997), 175–179.

SUN, Q. and WAN, D. Q.

[1] On the solvability of the equation $\sum_{i=1}^{n} x_i/d_i \equiv 0 \pmod 1$ and its applications *Proc. Am. Math. Soc.* **100** (1987), 220–224.

[2] On the Diophantine equation $\sum_{i=1}^{n} x_i/d_i \equiv 0 \pmod 1$, *Proc. Am. Math. Soc.* **112** (1991), 25–29.

SUN, Q. and YUAN, P.-Z.

[1] On the number of solutions of diagonal equations over a finite field, *Finite Fields Appl.* **2** (1996), 35–41.

SUN, Z.-H.

[1] Rational cubic reciprocity law and cubic congruences, *Acta Arith.*, to appear.

SUN, Z.–W.

[1] Exact m–covers and the linear form $\sum_{s=1}^{k} x_s/n_s$, *Acta Arith.* **81**(1997), 175–198.

SYLVESTER, J. J.

[1] Sur les équations à 3 et à 4 périodes des racines de l'unité, *Compte Rendu de la Association Française*, Reims, 1880, pp. 96–98. [CW: vol. 3, pp. 438–439]

[CW] *Collected Mathematical Papers of James Joseph Sylvester*, vols. 1–4, Chelsea, New York, 1973.

TABAKOVA, E. D.

[1] A numerical investigation of Kummer cubic sums, *Inst. Appl. Math. USSR Acad. Sci.*, Moscow, preprint no. 98, 1973, 22 pp. (Russian)

TAKAGI, TEIJI

[1] Ueber das Reciprocitätsgesetz in einem beliebigen algebraischen Zahlkörper, *J. College Sci. Tokyo* **44** (1922), 1–50. [CW: pp. 179–216]

[2] On the law of reciprocity in the cyclotomic corpus, *Proc. Phys.-Math. Soc. Jpn.* **4** (1922), 173–182. [CW: pp. 217–225]

[CW] *Collected Papers of Teiji Takagi*, Iwanami Shoten Publishers, Tokyo, 1973.

TAKAGI, TSUYOSHI

[1] Product formula of the cubic Gauss sum modulo the product of the primes, *J. Number Theory* **62** (1997), 298–306.

TANNER, H. W. L.

[1] On the binomial equation $x^p - 1 = 0$: quinquisection, *Proc. London Math. Soc.* **18** (1886/1887), 214–234.

[2] On complex primes formed with the fifth roots of unity, *Proc. London Math. Soc.* **24** (1893), 223–272.

THAINE, F.

[1] On the p-part of the ideal class group of $Q(\zeta_p + \zeta_p^{-1})$ and Vandiver's conjecture, *Michigan Math. J.* **42** (1995), 311–344.

[2] Properties that characterize Gaussian periods and cyclotomic numbers, *Proc. Am. Math. Soc.* **124** (1996), 35–45.

[3] Properties that characterize Jacobi sums, preprint, 1996.

THAKUR, D. S.

[1] Behavior of function field Gauss sums at infinity, *Bull. London Math. Soc.* **25** (1993), 417–426.

TIETÄVÄINEN, A.

[1] On pairs of additive equations, *Ann. Univ. Turku Ser. AI* **112** (1967), 7 pp.

TIHANYI, M.

[1] Kobös maradékok ismertetö jele, *Mat. Fiz. Lapok* **27** (1918), 76–79. (Hungarian)

TOLIMIERI, R.

[1] The construction of orthonormal bases diagonalizing the discrete Fourier transform, *Adv. Appl. Math.* **5** (1984), 56–86.

UEHARA, T.

[1] On a congruence relation between Jacobi sums and cyclotomic units, *J. Reine Angew. Math.* **382** (1987), 199–214.

VAIDYANATHASWAMY, R.

[1] The algebra of cubic residues, *J. Indian Math. Soc.* **21** (1957), 57–66.

VAN DER CORPUT, J. G.

[1] Zahlentheoretische Abschätzungen, *Math. Ann.* **84** (1921), 53–79.

VAN DER VLUGT, M.

[1] Hasse–Davenport curves, Gauss sums, and weight distributions of irreducible cyclic codes, *J. Number Theory* **55** (1995), 145–159.

VANDIVER, H. S.

[1] On the number of solutions of certain non-homogenous trinomial equations in a finite field, *Proc. Natl. Acad. Sci.* (USA) **31** (1945), 170–175.

[2] On the number of solutions of some general types of equations in a finite field, *Proc. Natl. Acad. Sci.*(USA) **32** (1946), 47–52.

[3] Cyclotomy and trinomial equations in a finite field, *Proc. Natl. Acad. Sci.* (USA) **32** (1946), 317–319.

[4] On some special trinomial equations in a finite field, *Proc. Natl. Acad. Sci.* (USA) **32** (1946), 320–326.

[5] Limits for the number of solutions of certain types of equations in a finite field, *Proc. Natl. Acad. Sci.* (USA) **33** (1947), 236–242.

[6] Applications of cyclotomy to the theory of nonhomogenous equations in a finite field, *Proc. Natl. Acad. Sci.* (USA) **34** (1948), 62–66.

[7] Cyclotomic power characters and trinomial equations in a finite field, *Proc. Natl. Acad. Sci.* (USA) **34** (1948), 196–203.

[8] Quadratic relations involving the numbers of solutions of certain types of equations in a finite field, *Proc. Natl. Acad. Sci.* (USA) **35** (1949), 681–685.

[9] On a generalization of a Jacobi exponential sum associated with cyclotomy, *Proc. Natl. Acad. Sci.* (USA) **36** (1950), 144–151.

[10] On cyclotomy and extensions of Gaussian type quadratic relations involving numbers of solutions of conditional equations in finite fields, *Proc. Natl. Acad. Sci.* (USA) **38** (1952), 981–991.

[11] On trinomial equations in a finite field, *Proc. Natl. Acad. Sci.* (USA) **40** (1954), 1008–1010.

[12] On the properties of certain trinomial equations in a finite field, *Proc. Natl. Acad. Sci.* (USA) **41** (1955), 651–653.

[13] On cyclotomic relations and trinomial equations in a finite field, *Proc. Natl. Acad. Sci.* (USA) **41** (1955), 775–780.

[14] On distribution problems involving the number of solutions of certain trinomial congruences, *Proc. Natl. Acad. Sci.* (USA) **45** (1959), 1635–1641.

VAUGHAN, R. C.

[1] Sommes trigonométriques sur les nombres premiers, *C. R. Acad. Sci. Paris* **285** (1977), 981–983.

VINOGRADOV, A. I.

[1] On the cubic Gaussian sum, *Izv. Akad. Nauk SSS Ser. Mat.* **31** (1967), 123–148; **33** (1969), 455. (Russian)

VOSTOKOV, S. V.

[1] An explicit form of the reciprocity law, *Izv. Akad. Nauk SSSR Ser. Mat.* **42** (1978), 1288–1321, 1439. (Russian)

WALUM, H.

[1] On the distribution of quadratic residues modulo a prime, *J. Number Theory* **15** (1982), 248–251.

WAMELEN, P. B. van

[1] Proof of the Evans–Root conjectures for Selberg character sums, *J. London Math. Soc.* **48** (1993), 415–426.

WAN, D. Q.

[1] Some arithmetic properties of the minimal polynomials of Gauss sums, *Proc. Am. Math. Soc.* **100** (1987), 225–228.

[2] Zeros of diagonal equations over finite fields, *Proc. Am. Math. Soc.* **103** (1988), 1049–1052.

[3] A Chevalley–Warning approach to p-adic estimates of character sums, *Proc. Am. Math. Soc.* **123** (1995), 45–54.

[4] Minimal polynomials and distinctness of Kloosterman sums, *Finite Fields Appl.* **1** (1995), 189–203.

WANG, J.

[1] On the Jacobi sums modulo p^n, *J. Number Theory* **39** (1991), 50–64.

[2] On the Poincaré series for diagonal forms, *Proc. Am. Math. Soc.* **116** (1992), 607–611.

[3] The reducibility of the generalized cyclotomic equations modulo an odd integer, *Chin. Sci. Bull.* **37** (1992), 281–284.

[4] On the Jacobi sums for finite commutative rings with identity, *J. Number Theory* **48** (1994), 284–290.

WASHINGTON, L. C.

[1] *Introduction to Cyclotomic Fields*, 2nd ed., Springer–Verlag, New York, 1997.

WASHIO, T., SHIMAURA, A., and SHIRATANI, K.

[1] On certain congruences for Gauss sums, *Sci. Bull. Fac. Ed. Nagasaki Univ.* **55** (1996), 1–8.

WATABE, M.

[1] An arithmetical application of elliptic functions to the theory of cubic residues, *Proc. Jpn. Acad. Ser. A* **53** (1977), 178–181.

[2] An arithmetical application of elliptic functions to the theory of biquadratic residues, *Abh. Math. Sem. Univ. Hamburg* **49** (1979), 118–125.

WATERHOUSE, W. C.

[1] The sign of the Gaussian sum, *J. Number Theory* **2** (1970), 363.

WEBER, H.

[1] Über Abel's Summation endlicher Differenzenreihen, *Acta Math.* **27** (1903), 225–233.

[2] *Lehrbuch der Algebra*, 3rd ed., vols. 1–3, Chelsea, New York, 1961.

WEIL, A.

[1] On the Riemann hypothesis in function-fields, *Proc. Natl. Acad. Sci.* (USA) **27** (1941), 345–347. [CW: vol. 1, pp. 277–279]

[2] On some exponential sums, *Proc. Natl. Acad. Sci.* (USA) **34** (1948), 204–207. [CW: vol. 1, pp. 386–389]

[3] Sur les courbes algébriques et les variétés qui s'en deduisent, *Actualités Sci. Ind.* No. 1041, *Publ. Inst. Math. Univ. Strasbourg*, vol. 7 (1945), Hermann, Paris, 1948, 85 pp.

[4] Numbers of solutions of equations in finite fields, *Bull. Am. Math. Soc.* **55** (1949), 497–508. [CW: vol. 1, pp. 399–410]

[5] Jacobi sums as "Grössencharaktere", *Trans. Am. Math. Soc.* **73** (1952), 487–495. [CW: vol. 2, pp. 63–71]

[6] La cyclotomie jadis et naguère, *Enseignement Math.* **20** (1974), 247–263. [CW: vol. 3, pp. 311–327]

[7] Sommes de Jacobi et caractères de Hecke, *Nachr. Akad. Wiss. Göttingen Math.-Phys. Kl. II* (1974), 1–14. [CW: vol. 3, pp. 329–342]

[CW] *Collected Papers*, vols. 1–3, Springer–Verlag, New York, 1979.

WEISS, E.

[1] *Algebraic Number Theory*, McGraw–Hill, New York, 1963.

WESTERN, A. E.

[1] An extension of Eisenstein's law of reciprocity, *Proc. London Math. Soc.* (2) **6** (1907), 16–28.

[2] An extension of Eisenstein's law of reciprocity II, *Proc. London Math. Soc.* (2) **7** (1908), 265–297.

[3] Some criteria for the residues of eighth and other powers, *Proc. London Math. Soc.* (2) **9** (1911), 244–272.

WHITEMAN, A. L.

[1] A note on Kloosterman sums, *Bull. Am. Math. Soc.* **51** (1945), 373–377.

[2] Theorems analogous to Jacobsthal's theorem, *Duke Math. J.* **16** (1949), 619–626.

[3] Theorems on quadratic partitions, *Proc. Natl. Acad. Sci.* (USA) **36** (1950), 60–66.

[4] Finite Fourier series and cyclotomy, *Proc. Natl. Acad. Sci.* (USA) **37** (1951), 373–378.

[5] Cyclotomy and Jacobsthal sums, *Am. J. Math.* **74** (1952), 89–99.

[6] Finite Fourier series and equations in finite fields, *Trans. Am. Math. Soc.* **74** (1953), 78–98.

[7] The sixteenth power residue character of 2, *Can. J. Math.* **6** (1954), 364–373.

[8] The cyclotomic numbers of order sixteen, *Trans. Am. Math. Soc.* **86** (1957), 401–413.

[9] The cyclotomic numbers of order twelve, *Acta Arith.* **6** (1960), 53–76.

[10] The cyclotomic numbers of order ten, in *Proc. Sympos. Appl. Math.*, vol. 10, American Mathematical Society, Providence, 1960, pp. 95–111.

[11] A family of difference sets, *Illinois J. Math.* **6** (1962), 107–121.

[12] A theorem of Brewer on character sums, *Duke Math. J.* **30** (1963), 545–552.

[13] Theorems on Brewer and Jacobsthal sums. I, in *Proc. Sympos. Pure Math.*, vol. 8, American Mathematical Society, Providence, 1965, pp. 44–55.

[14] Theorems on Brewer and Jacobsthal sums. II, *Michigan Math. J.* **12** (1965), 65–80.

WIEFERICH, A.

[1] Zum letzten Fermat'schen Theorem, *J. Reine Angew. Math.* **136** (1909), 293–302.

WILES, A.

[1] Modular elliptic curves and Fermat's last theorem, *Ann. Math.* **141** (1995), 443–551.

WILLIAMS, K. S.

[1] On the number of solutions of a congruence, *Am. Math. Monthly* **73** (1966), 44–49.

[2] On a result of Libri and Lebesgue, *Am. Math. Monthly* **77** (1970), 610–613.

[3] Finite transformation formulae involving the Legendre symbol, *Pacific J. Math.* **34** (1970), 559–568.

[4] Note on the Kloosterman sum, *Proc. Am. Math. Soc.* **30** (1971), 61–62.

[5] A class of character sums, *J. London Math. Soc.* **3** (1971), 67–72.

[6] On Salié's sum, *J. Number Theory* **3** (1971), 316–317.

[7] Note on Salié's sum, *Proc. Am. Math. Soc.* **30** (1971), 393–394.

[8] Exponential sums over $GF(2^n)$, *Pacific J. Math.* **40** (1972), 511–519.

[9] The Kloosterman sum revisited, *Can. Math. Bull.* **16** (1973), 363–365.

[10] Elementary treatment of a quadratic partition of primes $p \equiv 1 \pmod 7$, *Illinois J. Math.* **18** (1974), 608–621.

[11] 3 as a ninth power, *Math. Scand.* **35** (1974), 309–317.

[12] A quadratic partition of primes $\equiv 1 \pmod 7$, *Math. Comp.* **28** (1974), 1133–1136.

[13] Note on the supplement to the law of cubic reciprocity, *Proc. Am. Math. Soc.* **47** (1975), 333–334.

[14] Note on a cubic character sum, *Aequationes Math.* **12** (1975), 229–231.

[15] On Euler's criterion for cubic nonresidues, *Proc. Am. Math. Soc.* **49** (1975), 277–283.

[16] 2 as a ninth power $\pmod p$, *J. Indian Math. Soc.* **39** (1975), 167–172.

[17] On Euler's criterion for quintic nonresidues, *Pacific J. Math.* **61** (1975), 543–550.

[18] The quadratic character of 2 $\pmod p$, *Math. Mag.* **49** (1976), 89–90.

[19] Note on a result of Kaplan, *Proc. Am. Math. Soc.* **56** (1976), 34–36.

[20] Cubic nonresidues $\pmod p$, *Delta* **6** (1976), 23–28.

[21] Note on a result of Barrucand and Cohn, *J. Reine Angew. Math.* **285** (1976), 216–220.

[22] Explicit forms of Kummer's complementary theorems to his law of quintic reciprocity, *J. Reine Angew. Math.* **288** (1976), 207–210.

[23] A rational octic reciprocity law, *Pacific J. Math.* **63** (1976), 563–570.

[24] Explicit criteria for quintic residuacity, *Math. Comp.* **30** (1976), 847–853.

[25] On the supplement to the law of biquadratic reciprocity, *Proc. Am. Math. Soc.* **59** (1976), 19–22.

[26] Note on Burde's rational biquadratic reciprocity theorem, *Can. Math. Bull.* **20** (1977), 145–146.

[27] On Eisenstein's supplement to the law of cubic reciprocity, *Bull. Calcutta Math. Soc.* **69** (1977), 311–314.

[28] Note on Brewer's character sum, *Proc. Am. Math. Soc.* **71** (1978), 153–154.

[29] Remark on an assertion of Chowla, *Norske Vid. Selsk. Skr.* (Trondheim), No. 1 (1979), 3–4.

[30] Evaluation of character sums connected with elliptic curves, *Proc. Am. Math. Soc.* **73** (1979), 291–299.

[31] On the quadratic residues (mod p) in the interval $(0, p/4)$, *Can. Math. Bull.* **26** (1983), 123–124.

WILLIAMS, K. S., FRIESEN, C., and HOWE, L. J.

[1] Criteria for biquadratic residuacity modulo a prime p involving quaternary representations of p, *Can. J. Math.* **37** (1985), 337–370.

WILLIAMS, K. S., HARDY, K. and FRIESEN, C.

[1] On the evaluation of the Legendre symbol $\left(\frac{A + B\sqrt{m}}{p}\right)$, *Acta Arith.* **45** (1985), 255–272.

WILLIAMS, K. S., HARDY, K., and SPEARMAN, B. K.

[1] Explicit evaluation of certain Eisenstein sums, in *Number Theory*, R. A. Mollin, ed., de Gruyter, Berlin, 1990, pp. 553–626.

WINGEN, H.

[1] On the Wedderburn structure of some integral Frobenius group rings, *J. Algebra* **171** (1995), 294–327.

WÓJCIK, J.

[1] An extension of the Eisenstein reciprocity law, *Comment. Math. Prace Mat.* **35** (1995), 277–285.

WOLFMANN, J.

[1] Nombre de points rationnels de courbes algébriques sur des corps finis associées à des codes cycliques, *C. R. Acad. Sci. Paris Sér. I Math.* **305** (1987), 345–348.

[2] The number of solutions of certain diagonal equations over finite fields, *J. Number Theory* **42** (1992), 247–257.

[3] New results on diagonal equations over finite fields from cyclic codes, *Contemp. Math.* **168** (1994), 387–395.

WRIGHT, D. J.

[1] Cubic character sums of cubic polynomials, *Proc. Am. Math. Soc.* **100** (1987), 409–413.

WU, P.

[1] *A Rational Reciprocity Law*, Ph.D. thesis, University of Southern California, Los Angeles, 1975.

YAMADA, M.

[1] Hadamard matrices generated by an adaptation of generalized quaternion type array, *Graphs Combin.* **2** (1986), 179–187.

[2] Distance-regular digraphs of girth 4 over an extension ring of $Z/4Z$, *Graphs Combin.* **6** (1990), 381–394.

[3] Supplementary difference sets and Jacobi sums, *Discrete Math.* **103** (1992), 75–90.

YAMAMOTO, K.

[1] On Gaussian sums with biquadratic residue characters, *J. Reine Angew. Math.* **219** (1965), 200–213.

[2] On a conjecture of Hasse concerning multiplicative relations of Gaussian sums, *J. Combin. Theory Ser. A* **1** (1966), 476–489.

[3] On Jacobi sums and difference sets, *J. Combin. Theory Ser. A* **3** (1967), 146–181.

[4] The gap group of multiplicative relationships of Gaussian sums, in *Istituto Nazionale di Acta Mathematica*, Symposia Mathematica, vol. 15, Academic Press, London, 1975, pp. 427–440.

[5] On congruences arising from relative Gauss sums, in *Number Theory and Combinatorics*, J. Akiyama, ed., World Scientific Publishing Co., Singapore, 1985, pp. 423–446.

YAMAMOTO, K. and YAMADA, M.

[1] Williamson Hadamard matrices and Gauss sums, *J. Math. Soc. Jpn.* **37** (1985), 703–717.

YAMAUCHI, M.

[1] Some identities on the character sum containing $x(x-1)(x-\lambda)$, *Nagoya Math. J.* **42** (1971), 109–113.

YEUNG, K. M.

[1] On congruences for binomial coefficients, *J. Number Theory* **33** (1989), 1–17.

YOKOYAMA, A.

[1] On the Gaussian sum and the Jacobi sum with its application, *Tôhoku Math. J.* (2) **16** (1964), 142–153.

YOUNG, P. T.

[1] On the Gross–Koblitz formula, *Contemp. Math.* **168** (1994), 379–386.

[2] On Jacobi sums, multinomial coefficients, and p-adic hypergeometric functions, *J. Number Theory* **52** (1995), 125–144.

YUI, N.

[1] On the norms of algebraic numbers associated to Jacobi sums, *J. Number Theory* **47** (1994), 106–129.

ZEE, Y.–C.

[1] The Jacobi sums of orders thirteen and sixty and related quadratic decompositions, *Math. Z.* **115** (1970), 259–272.

[2] The Jacobi sums of order twenty-two, *Proc. Am. Math. Soc.* **28** (1971), 25–31.

ZHEDANOV, A. S.

[1] Gauss sums and orthogonal polynomials, *Internatl. J. Modern Phys.* (*A*) **12** (1997), 289–294.

Notation

Note that certain symbols have different meanings in different contexts.

p	prime
$q = p^r$	power of the prime p, r a positive integer
F_q	finite field with q elements and characteristic p, $(q = p^r)$
F_q^*	$F_q - \{0\}$
γ	generator of F_q^* $(=g$ when $r = 1)$
k	a positive integer dividing $(q-1)$ (or $p-1$ when $r = 1$)
f	$(q-1)/k$ $(=(p-1)/k$ when $r = 1)$
$\vert S\vert = \text{card } S$	number of elements in the finite set S
Z	domain of rational integers
Q	field of rational numbers
Q_p	field of p-adic rationals
Z_p	ring of p-adic integers in Q_p
Z_p^*	group of units in Z_p
$\gcd(a, b)$	greatest common divisor of the integers a and b (not both zero)
$\text{lcm}(a, b)$	least common multiple of the integers a and b (not both zero)
k^*	$\dfrac{k}{\gcd(k, (q-1)/(p-1))}$ (Theorem 12.10.1)
g	primitive root modulo p
$\text{ind}_g m$	index of the integer $m \not\equiv 0 \pmod p$ with respect to g (see p. 68)
Z	$\text{ind}_g 2$
T	$\text{ind}_g 3$
R	$\text{ind}_g 5$
$\phi(m)$	Euler's phi function
$L(a)$	least nonnegative integer congruent to a modulo k
$\text{sgn}\, x$	sign of the nonzero real number x
$\left(\dfrac{u}{v}\right)$	quadratic residue symbol (Section 1.2)
$\left(\dfrac{a}{p}\right)_{2^r}$	rational residue symbol of order 2^r (Section 8.3)
$R,\ R_k,\ R(k)$	group of reduced residues $(\text{mod } k)$

H_k	set of kth–power residues (mod p)
M_k	$H_k \cup \{0\}$
$\binom{m}{n}$	binomial coefficient
$(s,t)_k$	cyclotomic number of order k (Section 2.2)
$(s_1,\ldots,s_l)_{l,k}$	generalized cyclotomic number of length l and order k (see (2.7.1))
$\overline{z}$	complex conjugate of the complex number z
$\mathrm{Re}\, z = \mathrm{Re}(z)$	real part of the complex number z
$\mathrm{Im}\, z = \mathrm{Im}(z)$	imaginary part of the complex number z
$\beta,\ \beta_k$	$\exp(2\pi i/k)$
λ	$1-\beta$ (Sections 14.1 and 14.6)
K	algebraic number field, usually $Q(\beta)$
O_K	ring of integers of the algebraic number field K
$P,\ P_j$	prime ideals of K (see (2.1.8))
M	$Q(\beta_k,\ \beta_p)$ (Section 11.1)
$\wp$	prime ideal of M (Section 11.1)
π	$\beta_p - 1$ (Section 11.1)
$N(\alpha)$	norm of the element $\alpha \in O_K$
$\begin{bmatrix} n \\ m \end{bmatrix}$	Gaussian polynomial (Section 1.3)
$f(x,n)$	$\sum_{j=0}^{n}(-1)^j \begin{bmatrix} n \\ j \end{bmatrix}$ (Section 1.3)
$\theta(z)$	$\sum_{n=-\infty}^{\infty} e^{-\pi n^2 z}$ $(\mathrm{Re}\, z > 0)$ (see page 44)
$\wp(z)$	Weierstrass $\wp$-function (see (4.1.3) and page 162)
χ	a multiplicative character on F_q (Section 1.1)
ρ	quadratic character on F_q
χ^*	restriction of character χ (Section 11.4)
χ'	lift of character χ (Section 11.4)
$\mathrm{tr}(\alpha)$	trace of the element α from F_q to F_p (Section 1.1)
$\mathrm{tr}_{E/F_q}(\alpha)$	trace of the element α of E from E to F_q (Section 11.4)
$N_{E/F_q}(\alpha)$	norm of the element α of E from E to F_q (Section 11.4)
$e(\alpha)$	$e^{2\pi i\, \mathrm{tr}(\alpha)/p} = \exp(2\pi i\ \mathrm{tr}(\alpha)/p)$ (Section 1.1)

$G_r(\beta,\chi)$	$\sum_{\alpha\in F_q} \chi(\alpha)e(\alpha\beta)$ $(\beta\in F_q, q=p^r)$ (Section 1.1)
$G_r(\chi)=G_r(1,\chi)$	$\sum_{\alpha\in F_q} \chi(\alpha)e(\alpha)$ $(q=p^r)$ (Section 1.1)
$G(\beta,\chi)=G_1(\beta,\chi)$	$\sum_{\alpha\in F_p} \chi(\alpha)e^{2\pi i\alpha\beta/p}$ $(\beta\in F_p)$ (Section 1.1)
$G(\chi)=G(1,\chi)$	$\sum_{\alpha\in F_p} \chi(\alpha)e^{2\pi i\alpha/p}$ (Section 1.1)
R_j	$G(\chi^j)+G(\overline{\chi}^j)$ (see (4.0.5))
$g_r(\beta,k)$	$\sum_{\alpha\in F_q} e^{2\pi i\ \mathrm{tr}(\beta\alpha^k)/p}$ $(\beta\in F_q,\ q=p^r,\ k\in Z,\ k\geq 1)$ (Section 1.1)
$g_r(k)=g_r(1,k)$	$\sum_{\alpha\in F_q} e^{2\pi i\ \mathrm{tr}(\alpha^k)/p}$ $(q=p^r,\ k\in Z,\ k\geq 1)$ (Section 1.1)
$g(m,k)=g_1(m,k)$	$\sum_{n=0}^{p-1} e^{2\pi imn^k/p}$ (Section 1.1)
$g(k)=g(1,k)$	$\sum_{n=0}^{p-1} e^{2\pi in^k/p}$ (Section 1.1)
$g(m;k)$	$\sum_{n=0}^{k-1} e^{2\pi imn^2/k}$ $(m\in Z,\ k\in Z,\ k\geq 1)$ (Section 1.1 and page 25)
$g(1;k)$	$\sum_{n=0}^{k-1} e^{2\pi in^2/k}$ (Section 1.1)
$S(a,b,c)$	$\sum_{n=0}^{\lvert c\rvert-1} e^{\pi i(an^2+bn)/c}$ $(a,b,c\in Z,\ ac\neq 0)$ (Section 1.2)
$J_r(\chi,\psi)$	$\sum_{\alpha\in F_q} \chi(\alpha)\psi(1-\alpha)$ (see page 57)
$J_r^*(\chi,\psi)$	$\sum_{\substack{\alpha\in F_q\\ \alpha\neq 0,1}} \chi(\alpha)\psi(1-\alpha)$ (Section 2.5)
$J(\chi,\psi)=J_1(\chi,\psi)$	$\sum_{\alpha\in F_p} \chi(\alpha)\psi(1-\alpha)$ (see page 57)
$K_r(\chi)$	$\chi(4)J_r(\chi,\chi)$ (see page 57)
$K(\chi)=K_1(\chi)$	$\chi(4)J(\chi,\chi)$ (see page 57)
$J_r(\chi_1,\ldots,\chi_k)$	generalized Jacobi sum (see (10.1.1))
$J_0(\chi_1,\ldots,\chi_k)$	(10.1.2)
$\tau_k(m,\chi)$	$\sum_{n=1}^{k-1} \chi(n)e^{2\pi imn/k}$ (see (1.6.1))

$\tau_k(\chi) = \tau_k(1,\chi)$	$\sum_{n=1}^{k-1} \chi(n) e^{2\pi i n/k}$ (see (1.6.8))
$\tau(\chi)$	$\sum_{\alpha \in F_q^*} \chi(\alpha) e(\alpha)$ (see (11.3.1))
$\tau(\chi)$	$\tau_k(\chi)$ (see (1.6.8))
$H(\chi)$	(4.1.4)
$h(\chi)$	(4.2.4)
$S(k)$	Section 5.1
$\phi_k(a)$	Jacobsthal sum $\sum_{m=1}^{p-1} \left(\frac{m}{p}\right)\left(\frac{m^k+a}{p}\right)$ (see (6.1.1))
$\psi_k(a)$	$\sum_{m=1}^{p-1} \left(\frac{m^k+a}{p}\right)$ (see (6.1.2))
$\xi_k(a)$	double Jacobsthal sum $\sum_{m,n=1}^{p-1} \left(\frac{m^k+an}{p}\right)\left(\frac{m^k+an^2}{p}\right)$ (Section 6.5)
η_s	f-nomial period over F_p (Section 10.10)
$P_k(x)$	period polynomial over F_p (Section 10.10)
η_s^*	reduced f-nomial period over F_p(Section 10.10)
$P_k^*(x)$	reduced period polynomial over F_p (Section 10.10)
$\eta_{s,r}$	f-nomial period over F_q (see (12.10.4))
$P_{k.r}(x)$	period polynomial over F_q (see (12.10.5))
$\eta_{s,r}^*$	reduced f-nomial period over F_q (see (12.10.6))
$P_{k.r}^*(x)$	reduced period polynomial over F_q (see (12.10.7))
$E_r(\chi)$	Eisenstein sum (see pages 389 and 400)
$E(\chi)$	(12.2.1)
$V_n(x,a)$	Dickson polynomial (see (13.0.1))
$V_n(x)$	$V_n(x,1)$
$\Lambda_n(a)$	generalized Brewer sum (see (13.0.3))
$\Lambda_n = \Lambda_n(1)$	ordinary Brewer sum (see (13.0.4))
$F_n(x,a)$	$V_{2n}(\sqrt{x},a)$ (see (13.1.4))
$F_n(x)$	$F_n(x,1)$ (see (13.1.5))
Θ_n	(13.2.1)
$\Theta_n(g)$	(13.2.3)
Ω_n	(13.2.5)
$\Omega_n(a)$	(13.2.4)

B_n	(13.2.8)
$\Phi_a(P)$	(14.1.1)
σ_t	automorphism of $Q(\beta)$ $(\beta = e^{2\pi i/k})$ which maps β to β^t $(t \in R(k))$
$I(t)$	double sum of Legendre symbols (see page 467)
$\Phi_k(x)$	the monic irreducible polynomial of β over Z (see page 62)
$\epsilon_a(\nu)$	(14.1.4)
$\eta_a(\nu)$	(14.1.5)
r_3, s_3	Theorem 3.1.3
a_3, b_3	Theorems 3.1.1 and 3.1.2
u_3, v_3	Theorem 3.1.4
a_4, b_4	Theorems 3.2.1 and 3.2.2
a_8, b_8	Theorems 3.3.1, 3.3.2, and 12.7.1
$a_{16}, b_{16}, c_{16}, d_{16}$	Theorems 3.4.1, 3.4.2, and 12.12.1
a_{12}, b_{12}	Theorem 3.5.2
c_{12}	Theorem 3.5.3
a_{24}, b_{24}	Theorems 3.6.1 and 3.6.3, Problem 18 of Exercises 12
d_{24}	(3.6.3)
a_5, b_5, c_5, d_5	Theorems 3.7.2 and 3.7.3
$x_1, x_2, x_3, x_4, x_5, x_6$	Theorem 3.9.3
a_7, b_7	Theorems 3.9.4, 3.9.5 and 12.8.1
x, w, v, u	(3.7.35) and Theorem 3.7.8
a_{20}, b_{20}, c_{20}	Theorems 3.8.2, 3.8.3 and 12.11.1
e_{24}, f_{24}	Problems 16 and 17 of Exercises 12
V	$b_{24}/2$ (see page 120)
λ	(11.2.10)
$N(l, \chi)$	(11.3.3)
$C_r(a, b)$	cyclotomic number over F_q (see page 365)
M	(14.2.1)
$s(a)$	(11.2.4)
$t(a)$	(11.2.4)
$\chi_P(\alpha)$, $\left(\frac{\alpha}{P}\right)_k$	kth–power residue symbol mod P (page 66 and (11.2.1))
$\chi_\pi(\alpha)$, $\left(\frac{\alpha}{\pi}\right)_k$	kth–power residue symbol mod π (Section 14.3)

K_P	P-adic completion of the field K (see page 277)
$t(d_1, \ldots, d_n)$	(10.8.1)
T	$R/\{1, p, p^2, \ldots, p^{r-1}\}$ (Section 11.1)
ρ_3, ρ_4	Theorem 9.2.10
A_s	(9.3.12)
$\Gamma_p(z)$	p-adic gamma function (see (9.3.3))
c	see page 470
$\text{Gal}(K/Q)$	Galois group of K over Q

Author Index

Subject Index

D

K

L

M

N

P

Q

R

S

T

U

V

W

Y

CANADIAN MATHEMATICAL SOCIETY SERIES OF MONOGRAPHS AND ADVANCED TEXTS

Monographies et Études de la Société Mathématique du Canada

EDITORS/RÉDACTEURS: Jonathan M. Borwein and Peter B. Borwein

Author	Title	Vol. No.
Frank H. Clarke	***Optimization and Nonsmooth Analysis**	1
Erwin Klein and Anthony C. Thompson	***Theory of Correspondences: Including Applications to Mathematical Economics**	2
I. Gohberg, P. Lancaster, and L. Rodman	**Invariant Subspaces of Matrices with Applications**	3
Jonathan M. Borwein and Peter B. Borwein	**Pi and the AGM—A Study in Analytic Number Theory and Computational Complexity**	4
Subhashis Nag	**The Complex Analytic Theory of Teichmüller Spaces**	5
Manfred Kracht and Erwin Kreyszig	***Methods of Complex Analysis in Partial Differential Equations with Applications**	6
Ernest J. Kani and Robert A. Smith	***The Collected Papers of Hans Arnold Heilbronn**	7
Victor P. Snaith	***Topological Methods in Galois Representation Theory**	8
Kalathoor Varadarajan	**The Finiteness Obstruction of CTC Wall**	9
John H. Berglund, Hugo D. Jünghenn, and Paul Milne	***Analysis of Semigroups: Function Spaces Compactifications Representation**	10
Robert V. Moody and Arturo Pianzola	**Lie Algebras with Triangular Decompositions**	11
F. A. Sherk, P. McMullen, A. Thompson, and A. Weiss	**Kaleidoscopes: Selected Writings of H. S. M. Coxeter**	12
Alf van der Poorten	**Notes on Fermat's Last Theorem**	13
Peter A. Fillmore	**A User's Guide to Operator Algebras**	14
Peter Lancaster and Kęstutis Šalkauskas	**Transform Methods in Applied Mathematics: An Introduction**	15
Ole A. Nielsen	**An Introduction to Integration and Measure Theory**	16

*Indicates an out-of-print title. *Indique un titre épuisé.*

Canadian Mathematical Society Series *(Continued)*
Collection d'ouvrages de la Société Mathématique du Canada (suite)

Author	Title	Vol. No.
Wieslaw Krawcewicz and Jianhong Wu	**Theory of Degrees with Application to Bifurcations and Differential Equations**	17
Marián J. Fabian	**Gâteaux Differentiability of Convex Functions and Topology: Weak Asplund Spaces**	18
Glen D. Anderson, Mavina K. Vamanamurthy, and Matti K. Vuorinen	**Conformal Invariants, Inequalities, and Quasiconformal Maps**	19
Ivan Singer	**Abstract Convex Analysis**	20
Bruce C. Berndt, Ronald J. Evans, and Kenneth S. Williams	**Gauss and Jacobi Sums**	21

Printed and bound in the UK by
CPI Antony Rowe, Eastbourne

Printed and bound by CPI Group (UK) Ltd, Croydon, CR0 4YY
18/09/2026
14965015-0002